HANDBUCH DER ASTROPHYSIK

HERAUSGEGEBEN VON

G. EBERHARD · A. KOHLSCHÜTTER
H. LUDENDORFF

BAND VI

DAS STERNSYSTEM

ZWEITER TEIL

Springer-Verlag Berlin Heidelberg GmbH
1928

DAS STERNSYSTEM

ZWEITER TEIL

BEARBEITET VON

F. C. HENROTEAU · H. LUDENDORFF
K. G. MALMQUIST · F. J. M. STRATTON

MIT 123 ABBILDUNGEN

Springer-Verlag Berlin Heidelberg GmbH
1928

URSPRÜNGLICH ERSCHIENEN BEI JULIUS SPRINGER IN BERLIN 1928
SOFTCOVER REPRINT OF THE HARDCOVER 1ST EDITION 1928

ISBN 978-3-662-37399-6
ISBN 978-3-662-38148-9 (eBook)
DOI 10.1007/978-3-662-38148-9

Inhaltsverzeichnis.

Chapter 1.

The Radial Velocities of the Stars.

By Dr. K. G. Malmquist, Lund.

(With 3 illustrations.)

Kapitel 2.

Die veränderlichen Sterne.

Von Professor Dr. H. Ludendorff, Potsdam.

(Mit 36 Abbildungen.)

Chapter 3.

Novae.

By Professor F. J. M. Stratton, Cambridge.

(With 15 illustrations.)

Chapter 4.

Double and Multiple Stars.

By Dr. F. C. HENROTEAU, Ottawa.

(With 69 illustrations.)

Chapter 1.

The Radial Velocities of the Stars.

By

K. G. Malmquist-Lund.

With 3 illustrations.

a) Introduction.

1. The two Different Components of Motion. The investigation of the motions of the stars is a very important branch of modern astronomy. A thorough analysis of the motions will discover the mechanical conditions of the stellar system, which conditions, combined with the results concerning the arrangement of the stars in space deduced along other lines of research, will enable the astronomers to find out the structure of the sidereal universe.

The motion of a star naturally resolves itself into two components: one at right angles to the line drawn from the observer to the star, and the other along this line (the "line of sight"). The former component is the star's proper motion, the latter is its motion in the line of sight or its radial velocity. The radial velocity is considered positive if the star is receding from the observer, and negative if it is approaching him.

These two components are of entirely different natures. The proper motion is the star's apparent angular motion on the celestial sphere and is usually defined in seconds of arc per year. To get this component of the star's motion expressed in absolute units, as, for example, in kilometres per second, we must also know the distance of the star. The radial velocity is measured by means of the spectroscope. According to Doppler's principle, the lines in the spectrum of a star are displaced towards the red if the star is receding from the earth, or towards the violet if approaching the earth. The radial velocity is found directly in kilometres per second, so that the actual linear speed, independent of the more doubtful element of distance, is known. In this respect, therefore, the radial velocities have a very great advantage over the proper motions. Also in the determination of the radial velocities there is the other great advantage that, while the radial velocities can be measured accurately at once, the determination of the proper motions, which are all in general very minute owing to the large distances of the stars, requires a long interval of time to elapse between two accurate position observations.

The most serious disadvantage is that we are only able to determine the radial velocities of the brighter stars, as the light is spread over the large area of the spectrum, whereas the point images of very faint stars are available for the determination of proper motions. By more powerful telescopes that may be constructed, the radial velocities will be determined for fainter stars. At present, with the most powerful telescope, the upper limit is about the 12th magnitude.

The greatest value of the radial velocities lies, however, not in their advantage over the proper motions, but—to cite CAMPBELL—in aiding the proper motions to come into their own strategic worth. A very important result of a combination of the two components is the determination of the mean parallaxes for groups of stars. The knowledge of this quantity is of fundamental weight with regard to the distances and the absolute magnitudes of the stars.

For a complete solution of the problem of stellar motion we must know not only the proper motions and the radial velocities, but also the individual distances and masses of the stars. The two latter elements are as yet, however, not very accurately known for any large number of stars and we must therefore for the present rely largely on the results obtained from the radial velocities alone.

2. Historical Notes concerning the Determination of Radial Velocities. The existence of the proper motions was first proved by HALLEY in 1718, who pointed out that some of the apparently bright stars were occupying positions appreciably different from those given in PTOLEMY's Almagest. A century and a half later the first attempts to measure the displacement of the spectral lines by the visual method were made by HUGGINS, in London, 1868, and by VOGEL at Bothkamp, 1871. More extensive observations of this kind were performed by MAUNDER at Greenwich, including forty-eight of the brightest stars. The probable error in the Greenwich observations was 22 km for an evening. These observations have thus demonstrated that the visual observations with medium sized instruments cannot furnish results the certainty of which is of a higher order than the average velocities of the stars themselves.

In the year 1890, using the 36-inch Lick refractor combined with an efficient spectroscope, KEELER[1] gave us the first reliable visually determined radial velocities. For α Bootis he found, from nine measures during 1890 and 1891, a radial velocity of -7 km; for α Tauri, on three evenings, $+55$ km, and for α Orionis, on two evenings, $+16$ km per sec. The probable error of a single evening's observations was 1,8 km. The velocities of these three stars only were measured by KEELER, a fact which shows the difficulty of obtaining accurate results.

Fortunately, at this point of the progress, an incomparably better method was discovered and developed, namely the photographic method. The first attempt to photograph stellar spectra, using wet plates, was made by HUGGINS in 1863, but his photographs did not show any lines. The spectral lines were first recorded on the stellar spectrograms obtained by DRAPER in 1872. The introduction of dry-plates (about 1875) signified an enormous improvement. In 1887 VOGEL at Potsdam, with the assistance of SCHEINER, made the first attempt to record photographically the displacements of the lines in stellar spectra, and to measure them as accurately as possible on the spectrograms. The attempt was successful, and since a new spectrograph, exclusively constructed for this definite purpose, was attached to the 30-cm telescope, the radial velocities of fifty-one of the brighter stars were determined at Potsdam in 1888—1891[2]. The average probable error, for single observations, was 2,6 km. Thus the Potsdam observers were able, with their medium sized instrument, to obtain photographically radial velocities for second- and third-magnitude stars of the same accuracy as those obtained with the largest existing telescope by the visual method

[1] Lick Publ 3, p. 195—196 (1894).

[2] Publ Astroph Obs Potsdam 7 (1) (1892).

for the first-magnitude stars. Not until the photographic method was adapted to the largest telescopes was it possible to see its actual efficiency. In 1896 a spectrograph, planned by CAMPBELL, and named after the donor the Mills spectrograph, was attached to the Lick telescope and the accuracy attained surpassed all expectations. For the brighter stars containing good lines the probable errors were reduced to 0,5 km per sec, and to 0,3 km for bright stars containing the best quality of lines. A still better spectrograph — the new Mills spectrograph — was constructed in 1902 and attached to the same instrument. To obtain radial velocities of the stars in the Southern Hemisphere, an expedition — the D. O. MILLS Expedition — was sent from the Lick Observatory to Cerro San Cristobal, Santiago, Chile, where radial velocity determinations were carried out with a 37-inch reflecting telescope and spectrographs. Since that time the Lick Observatory has furnished a great portion of the radial velocities known. On account of the pressing want of radial velocities most of the largest telescopes were engaged in the problem. In America: the Allegheny, Columbus, Detroit, Lowell, Mount Wilson, Ottawa, Victoria and Yerkes observatories; in Europe: the Bonn, Cambridge, Paris, Potsdam, Pulkowa and Wien observatories; and in South Africa: the Cape of Good Hope observatory have been contributing to the labour of determining radial velocities. Among these the Victoria and the Mount Wilson observatories, with their great reflectors, are for the present the leading contributors. At present the radial velocities of more than three thousand stars are published or prepared for publication.

3. Monographs dealing with the Subject of the Present Chapter. In his Stellar Motions[1], CAMPBELL has given an excellent account of the development of the spectrographic method and of the results obtained concerning the stellar motions up to the beginning of the year 1910. The mathematical treatment of the stellar velocities, especially from the standpoint of the two-stream hypothesis, is clearly outlined in EDDINGTON's Stellar Movements[2]. In his lectures delivered at the University of California, 1924, CHARLIER[3] gives, in a condensed form, the results of his important researches on the motion of the stars. A complete account of the ellipsoidal hypothesis is to be found in this memoir.

b) Lists and Catalogues of Radial Velocities.

4. Compilation of Lists and Catalogues of Radial Velocities. On account of the many observatories engaged in the determination of radial velocities, the published results are to be found in many different publications. In the following pages a compilation of lists and catalogues of radial velocities is given. This compilation comprises those stars which are assigned to be of constant radial velocity; the stars which are found to be of variable velocity are only included if the velocity of the centre of mass of the system is computed or estimated. For details concerning the stars with variable radial velocities see the chapters treating of the double and the variable stars. It is endeavoured to make the compilation as complete as possible with respect to the modern determinations, as far as the works bearing on the topic were accessible.

Allegheny.

FRANK C. JORDAN, The radial velocities of twenty-six stars. Publ Allegheny Obs 2, p. 121–124. 1911.

[1] Yale University Press (1913).
[2] MACMILLAN and Co., London (1914).
[3] The motion and the distribution of the stars. Mem Univ California, 7 (1926).

The catalogue contains radial velocities for twenty-six stars of the B and A type fainter than magnitude 4,5. The spectrograms are obtained with the Mellon spectrograph in connection with the Cassegrain form of the 30-inch Keeler Memorial reflector.

Bonn.

F. KÜSTNER, Radial velocities for 99 stars of the second and third spectral classes observed at Bonn. Ap J 27, p. 301—324. 1908.

The catalogue contains radial velocities for nearly all stars of the second and third spectral classes down to the fourth visual, or fifth photographic, magnitude, which could be observed in Bonn with the photographic 30-cm refractor by REPSOLD and STEINHEIL. Observations on the moon showed that the radial velocities obtained require a small negative correction amounting to about —1,0 km. The probable error of a plate is found to be 0,64 km.

F. KÜSTNER, Radialgeschwindigkeiten von 227 Sternen des Spektraltypus F bis M, beobachtet 1908 bis 1913 am Bonner 30-cm-Refraktor. A N 198, p. 409—448. 1914.

The catalogue is an extension of the preceding catalogue down to the sixth photographic magnitude. On account of the smaller dispersion used the probable error of a plate is greater than before, it is found to be 1,33 km. The systematic correction as derived from observations on the moon is found to be —1,0 km, agreeing with the former determination. The radial velocities are further compared with CAMPBELLS determinations (Lick Bull 229) and the differences Lick—Bonn are given.

Cambridge.

H. F. NEWALL, On some spectroscopic determinations of velocity in the line of sight made at the Cambridge Observatory. M N 57, p. 567—577. 1897.

Radial velocities of seven stars determined with the 25-inch telescope.

H. F. NEWALL, Velocity in the line of sight. Selected stars. Cambridge Observatory, I. M N 63, p. 296—301. 1903; II. M N 65, p. 651—655. 1905.

Radial velocities of ten standard stars.

Cape.

S. S. HOUGH, Determination of radial velocities and wavelenghts. Cape Ann 10, Part I. 1911.

Table XX contains the detailed results of radial velocity determinations for 33 bright stars, and table XXI a summary of results. The spectrograms are obtained with a 4-prism spectrograph attached to the 24-inch refractor.

J. HALM, A spectrographic determination of the constant of aberration and the solar parallax. Ibid. 10, Part III. 1909.

Detailed radial velocities for six stars.

J. LUNT, The radial velocities of 60 southern stars. Ap J 47, p. 201—205. 1918.

Table I contains the provisional radial velocities for 60 stars of spectral types F, G, K, and M south of the equator and brighter than the visual magnitude 5,50. From a comparison of the results obtained at Lick and Mt. Wilson the following systematic differences are found: Cape—Lick —0,5 km (10 stars) and Cape—Mt. Wilson +2,0 km (9 stars).

J. LUNT, The radial velocities of 119 stars observed at the Cape. Ap J 48, p. 261—278. 1918.

The radial velocities of 119 stars down to magnitude 3,7 are given. The stars are divided into two classes: 76, which appear to have constant velocities, and 43, which are either known or suspected to be spectroscopic binaries. Eighteen stars in the first class were very frequently observed for a continuation of the solar parallax work and the results for these stars are given separately in table I. The spectrograms are obtained with the 24-inch refractor. The probable error of the velocity from a single plate is about 0,75 km. The systematic difference Lick—Cape is +0,08 km (76 stars), and Chile—Cape —0,3 km (33 stars).

J. LUNT, The radial velocities of 185 stars observed at the Cape. Ap J 50, p. 161—173. 1919.

In continuation of the preceding paper the radial velocities are given here for 185 stars of magnitudes 3,7 to 4,6. As before, the stars are divided into two classes, the first containing 122, and the second 63 stars. The systematic difference Lick—Cape is found to be +0,7 km (120 stars), and Chile—Cape +0,5 km (41 stars).

J. LUNT, The radial velocities of 120 southern stars, observed at the Cape. Cape Ann 10, Part V. 1921.

The radial velocities here given are a continuation of those in the list published in Ap J 47. Systematic difference: Cape—Lick +0,5 km (13 stars), Cape—Mt. Wilson +1,7 km (3 stars).

Columbus.

H. C. LORD, Some observations on stellar motions in the line of sight made at the Emerson McMillin Observatory. Ap J 8, p. 65—69. 1898.

Radial velocities for five stars determined with a 12,5-inch equatorial.

H. C. LORD, Observations of the radial velocities of thirty-one stars made at the Emerson McMillin Observatory. Ap J 21, p. 297—322. 1905.

Detroit.

PAUL W. MERRILL, Spectroscopic observations of stars of class Md. Publ Obs Univ Michigan, 2, p. 45—70. 1916.

In table II the radial velocities from bright lines of 43 Md stars are given, and in table VII the radial velocities from both bright and dark lines for 13 stars. The spectrograms are obtained with a one-prism spectrograph in connection with the 37,5-inch Cassegrain reflector.

W. CARL RUFUS, An investigation of the spectra of stars belonging to class R of the Draper classification. Ibid. 2, p. 103—143. 1916.

In table V the radial velocities for ten stars of class R are given.

F. HENROTEAU, Radial velocities of the antapex group of stars. Ibid. 3, p. 53—60. 1923.

The radial velocities for twelve stars are determined.

LEWIS L. MELLOR, Spectrographic study of the early Potsdam velocity stars, not known to be binaries. Ibid. 3, p. 61—105. 1923.

The radial velocities for twenty-six of the fifty-one early Potsdam velocity stars are redetermined.

Lick.

W. W. CAMPBELL, The Mills spectrograph of the Lick Observatory. Ap J 8, p. 123—156. 1898.

On pages 148—151 the first results of the radial velocity determinations with the MILLS spectrograph are published. A list of radial velocities for eleven stars is given.

W. W. CAMPBELL, On the motion of the brighter class B stars, Lick Bull 195, p. 108—117. 1911.

The catalogue contains the radial velocities for 225 class B stars most of which have been observed by the Lick Observatory, on Mount Hamilton and on Cerro San Cristobal. Some of the results published by the Yerkes, Allegheny, Ottawa, Potsdam and Pulkowa observatories are included. Twenty-two stars have visual magnitudes fainter than 5,0, but they do not render the material non-homogeneous, for the selection of faint stars was not made on account of any peculiarities attaching to these stars. Two hundred and three are equal to or brighter than 5,0. The corresponding number of class B stars in the Revised Harvard Photometry is 350, and consequently there are 145 class B stars not given in the catalogue, chiefly because there are many spectroscopic binary systems uninvestigated, and partly because the spectral lines in many of them are too ill-defined for satisfactory measurement.

W. W. CAMPBELL, Preliminary radial velocities of 212 brighter class A stars. Lick Bull 211, p. 20—29. 1912.

The catalogue contains radial velocities for 212 class A stars, most of which have been observed by the Lick Observatory, on Mount Hamilton and on Cerro San Cristobal; for a few of the stars the radial velocities have been taken from the published results of other observatories.

H. C. WILSON, On the real motion of 100 stars of large proper motion, whose radial velocities have been determined. Lick Bull 214, p. 51—62. 1912.

The catalogue contains the radial velocities of 100 stars of large proper motion, distributed over the entire sky, including a few results, obtained and published by other observatories in the northern hemisphere.

W. H. WRIGHT, Radial velocities of 150 stars south of declination —20°. Lick Publ 9, Part IV. 1911.

This is a detailed catalogue of the radial velocities of 150 stars determined by the D. O. MILLS Expedition on Cerro San Cristobal. The individual results for each star are first given, and at the end a general catalogue. For a star whose spectrum contains good and fairly numerous lines, the probable error of a single observation is 0,5 km. Where the spectral lines are broad and diffuse and few in number the corresponding quantity is naturally much larger, but it is not supposed to exceed 3 km for a good observation of any star in the catalogue. Since there are, on the average, four or more observations of each star, the probable error of the catalogue value will be about half that of a single observation.

W. W. CAMPBELL, The radial velocities of 915 stars. Lick Bull 229, p. 114—128. 1913.

In the present catalogue all available results obtained up to that time at Mount Hamilton or at Cerro San Cristobal for the brighter stars of classes F, G, K and M have been included, even if observations of the same stars had been published in the earlier catalogues mentioned above. Moreover, the radial velocities for some stars of classes B and A, additional to or different from those published before, are given. The results for a few stars obtained at other observatories are included. Those stars whose velocities are suspected to be variable and those whose spectra do not contain lines measurable with reasonable accuracy with the spectrographs employed are not included in the publication.

Except these two classes of stars, this and the preceding lists will include all stars brighter than the visual magnitude 5,0 and many fainter stars.

J. H. MOORE, Radial velocities of twenty-five stars of SECCHI's fourth type. Lick Bull 342, p. 160—168. 1922.

The catalogue contains the radial velocities of twenty-five class N stars. The spectrograms of twenty of these stars were obtained at Mount Hamilton and the five others at Cerro San Cristobal.

LEAH B. ALLEN, The radial velocities of twenty southern variable stars of class Me. Lick Bull 369, p. 71—75. 1925.

A catalogue containing the radial velocities of twenty stars of class Me is given. They are determined from spectrograms made with a one-prism spectrograph attached to the 37,5-inch reflecting telescope at Cerro San Cristobal in 1922 and 1923, the faintest star being of the 7,5 magnitude at maximum.

P. VAN DE KAMP, A determination of the sun's velocity with respect to stars of magnitude 9 to 10. Lick Bull 374, p. 95—98. 1926.

The radial velocities of 105 stars between visual magnitudes 9 and 10 have been determined and are given in the catalogue; 53 of these are within 50° of the antapex, 52 within 50° of the apex of solar motion. In the determination the light one-prism spectrograph, with a 6-inch camera, was used in combination with the 36-inch telescope at Mount Hamilton. The average probable error of the radial velocity derived from a good spectrum of an A and F type star amounts to 12 km. For the faint G type stars the same error is 7,2 km, and for the faint K type stars 5,2 km.

At the meeting of the Royal Astronomical Society, July 24, 1925, CAMPBELL announced the approaching publication of a new great catalogue of radial velocities. The velocities of nearly three thousand stars have hitherto been observed, and within the past year, the systematic corrections to the radial velocities as obtained with different instruments have been determined and applied to the individual observations in MS. form prior to publication[1].

Lowell.

V. M. SLIPHER, Observations of standard velocity stars with the Lowell spectrograph. Ap J 22, p. 318—340. 1905.

Radial velocities of ten standard velocity stars. The spectrograms are obtained by means of a spectrograph attached to the 61-cm Clark refractor.

Mount Wilson.

WALTER S. ADAMS and ARNOLD KOHLSCHÜTTER, The radial velocities of one hundred stars with measured parallaxes. Mt Wilson Contr 79; Ap J 39, p. 341—349. 1914.

The catalogue contains the radial velocities of one hundred stars fainter than magnitude 5,5 on the visual scale for which observations of parallax are available. The instrument employed is the 60-inch reflector combined with the Cassegrain spectrograph adapted for use with one prism. Two cameras have been used, giving different dispersions, one for stars between magnitudes 5,5 and 6,5, the other for stars fainter than 6,5.

WALTER S. ADAMS, The radial velocities of five hundred stars, Mt Wilson Contr 105; Ap J 42, p. 172—194. 1915.

In this catalogue the radial velocities of the following classes of stars are given:

[1] Obs 48, p. 274 (1925).

1. A and B type stars, mainly between magnitudes 5 and 6,5.

2. A, F, G, K and M type stars of magnitudes 5,5 to 6,5 which have very small proper motions.

3. Stars with measured parallaxes, most of which have very large proper motions. The magnitudes of these stars are chiefly between 5,5 and 8,5. (For the radial velocities of 100 of these stars see the preceding catalogue.)

In addition a number of brighter stars have been observed for which determinations of radial velocity have been published from other observatories. The instrument employed is the 60-inch reflector combined with the Cassegrain spectrograph. Different optical combinations were used. With a few exceptions, three or more observations have been secured for all stars in the catalogue. The average probable error of the radial velocities is 1,0 km. A comparison with the Lick Observatory results indicates a small systematic difference, about half a km for the B, A, K and M stars, and about one km for the F and G stars, in the direction of larger negative or smaller positive values for Mount Wilson.

WALTER S. ADAMS and ALFRED H. JOY, The radial velocities of 1013 stars. Mt Wilson Contr 258; Ap J 57, p. 149—176. 1923.

The material in this catalogue is made up almost wholly of stars of types F, G, K and M, which have been observed not only for radial velocities, but also for determination of absolute magnitudes. Therefore the material is not homogeneous as regards proper motion, apparent magnitude or spectral type. The spectrograms used have all been obtained at the Cassegrain focus of the 60-inch and 100-inch reflectors, the larger instrument having been employed almost wholly for stars fainter than the eighth visual magnitude, and for a small number of stars south of —30° decl. The probable errors of the mean radial velocities vary from 0,2 km to over 3 km, with an average of about 1,35 km. A comparison with measurements from the Lick Observatory for 109 stars of types F to M shows a systematic difference M. W.—L. O. of —0,12 km, and for 83 stars measured at the Dominion Astrophysical Observatory the systematic difference M. W.—D. A. O. is +0,8 km. In the case of 16 B and A stars the mean difference M. W. minus others is +1,6 km.

PAUL W. MERRILL, The radial velocities of long-period variable stars. Mt Wilson Contr 264; Ap J 58, p. 215—262. 1923.

In table III of this investigation are given the measurements of the emission-line velocities for 112 long period variable stars and of the absorption-line velocities for 43 of them. The observations have been secured at Mount Wilson since 1919 in continuation of the program begun at Detroit in 1913. The instruments employed were one-prism spectrographs to the 60-inch and 100-inch telescopes. Collected radial velocity data for 133 stars are given in table V together with periods and new estimates of spectral types. Forty-seven stars have radial velocities found from both bright and dark lines.

R. F. SANFORD, Radial velocities of stars of spectral class R. Mt Wilson Contr 276; Ap J 59, p. 339—355. 1924.

Table I of this investigation gives the radial velocities of twenty-nine stars of spectral class R, one of spectral class S, and one variable. The spectrograms are obtained with the 100-inch reflector. In table III the radial velocities of 8 class N stars are given.

W. S. ADAMS and A. H. JOY, A list of stars with radial velocities exceeding 50 km/sec. Publ A S P 38, p. 121—124. 1926.

A list of 69 stars with radial velocities exceeding 50 km/sec is given. No previous results have been published for any of these stars with the few exceptions listed in the accompanying notes.

Ottawa.

J. B. CANNON, Measures of radial velocity of Boss 4826, 7 Virginis, Boss 4721, 59 Herculis and μ Virginis. Ottawa Publ 4, No. 2. 1916.

J. B. CANNON, Measures of the radial velocities of fourteen stars. Ibid 4, No. 17. 1918.

W. E. HARPER, Measures of radial velocity of 23 Comae Berenices, δ Serpentis, and χ Serpentis. Ibid. 4, No. 20. 1919.

W. E. HARPER, Measures of radial velocity of $\varkappa$ Cassiopeiae, g Persei, 69 Tauri and ε Cygni. Ibid. 4, No. 21. 1919.

W. E. HARPER, Radial velocities of 30 stars. Ibid. 4, No. 22. 1919.

F. HENROTEAU and J. P. HENDERSON, A spectrographic study of early class B stars. Ibid. 5, No. 1. 1920.

Radial velocity determinations for ten B stars.

F. HENROTEAU, A spectrographic study of early class B stars (second paper). Ibid. 5, No. 3. 1921.

Radial velocity determinations for 25 B stars.

F. HENROTEAU, A spectrographic study of early class B stars (third paper). Ibid. 5, No. 8. 1922.

Radial velocity determinations for 31 stars.

F. HENROTEAU, A spectrographic study of stars of classes A and F. Ibid. 8, No. 5. 1923.

Radial velocity determinations for 11 stars.

Potsdam.

H. C. VOGEL, Untersuchung über die Eigenbewegung der Sterne im Visionsradius auf spektrographischem Wege. Publ Astroph Obs Potsdam 7 (1). 1892.

Radial velocities of fifty-one stars (see Introduction).

Pulkowa.

A. BELOPOLSKY (On the astrophysical work at Pulkowo in the year 1897). R A G 4, p. 429. (Russian.)

Radial velocities of nine stars in the Pleiades, one double star and six variable stars.

A. BELOPOLSKY, Spectrographic observations of standard velocity stars at Pulkowa. Ap J 19, p. 85—104. 1904.

Radial velocities for eight standard velocity stars.

Victoria.

J. S. PLASKETT, W. E. HARPER, R. K. YOUNG, H. H. PLASKETT, The radial velocities of 594 stars. Victoria Publ 2, No. 1. 1921.

The Boss stars north of the equator whose velocities had not previously been determined or for which observations for velocity had not been obtained, were divided between Mt. Wilson and Victoria, the general plan being for Victoria to take the even minutes and for Mt. Wilson the odd minutes of right ascension. Table I of the present publication contains the mean radial velocities of 537 stars between the fifth and eighth magnitudes assigned to be of constant velocity. In table II is given the velocity of the system of 22 spectroscopic binaries, whose orbits have been determined at Victoria, and table III contains the radial velocities for 35 suspected binaries, obtained from the individual velocities, generally by taking the arithmetic mean. The spectrograms are obtained with the 72-inch reflector with spectrograph at the Cassegrain focus. For the stars of spectral types F to M the probable errors of the mean velocity range between 0,1 km and 1,0 km, for a single plate between 0,2 km and 2,5 km. For the B and A stars with fairly sharp metallic lines the probable errors range between 0,5 km and 1,5 km for the mean velocity, and between 1,2 km and 3,5 km for a single plate. For the stars of earlier types with very diffuse and broad lines the corresponding values are 1,0 km and 3,8 km for the mean, 2,5 km and 10,0 km for a single plate. The average difference Lick—D. A. O. is found to be +0,80 km.

J. W. CAMPBELL, The spectroscopic orbit of H. R. 6532 and the radial velocities of ten stars. Ibid. 2, No. 5. 1922.

The radial velocities of ten stars are given.

W. E. HARPER, The radial velocities of 125 stars. Ibid. 2, No. 10. 1923.

The velocities of 125 stars mostly of spectral types F, G, K and M whose parallaxes have been determined trigonometrically, are given.

J. S. PLASKETT, The O-type stars. Ibid. 2, No. 16. 1924.

In table VI the individual velocities from 528 spectra obtained from 80 O and B stars are given. The probable error of a single plate is found to be about 4 km. In tables VII and VIII a summary of the velocities of the O and B stars is tabulated.

W. H. CHRISTIE, The radial velocities of fourteen stars. J. Can. R A S 18, p. 165—168. 1924.

STANLEY SMITH, Orbit of the spectroscopic binary Boss 6070 and the radial velocities of fifteen stars. Victoria Publ 3, No. 5. 1925.

The radial velocities of fifteen stars are given.

Wien.

ADOLF HNATEK, Beobachtungen von Radialgeschwindigkeiten am Coudé-Spektrographen der Wiener Sternwarte. A N 195, p. 171—174. 1913.

Radial velocities for five stars.

ADOLF HNATEK, Beobachtungen von Radialgeschwindigkeiten. A N 196, p. 389—392. 1913.

Radial velocities for ten stars. Systematic difference W—Lick = —0,6 km (8 stars).

ADOLF HNATEK, Beobachtungen von Radialgeschwindigkeiten. A N 197, p. 185—188. 1914.

Radial velocities for 24 stars.

ADOLF HNATEK, Beobachtungen von Radialgeschwindigkeiten am Wiener Coudé-Spektrographen. A N 210, p. 245—248. 1919.

Radial velocities for 15 stars.

ADOLF HNATEK, Radial velocities of eighteen stars. Ap J 52, p. 198—200. 1920.

Radial velocities of ten binaries and of eight other stars.

ADOLF HNATEK, Radialgeschwindigkeiten von 26 Sternen. A N 214, p. 277—280. 1921.

Yerkes.

WALTER S. ADAMS, Some miscellaneous radial velocity determinations with the Bruce spectrograph. Ap J 18, p. 67—69. 1903.

Radial velocities of five stars.

EDWIN B. FROST and WALTER S. ADAMS, Spectrographic observations of standard velocity stars (1902—1903). Ap J 18, p. 237—277. 1903.

Radial velocities of 13 stars.

EDWIN B. FROST and WALTER S. ADAMS, Radial velocities of twenty stars having spectra of the Orion type. Yerkes Publ 2, p. 145—250. 1903.

Detailed radial velocity determinations for twenty stars. A summary of results is given on page 247.

GEORG E. HALE, FERDINAND ELLERMAN, and J. A. PARKHURST, The spectra of stars of SECCHI's fourth type. Ibid. p. 253—385. 1903.

On page 341 the mean radial velocities of eight stars are given.

WALTER S. ADAMS, The radial velocities of the brighter stars in the Pleiades. Ap J 19, p. 338—343. 1904.

Radial velocities for six stars.

J. C. KAPTEYN and EDWIN B. FROST, On the velocity of the sun's motion through space as derived from the radial velocity of Orion stars. Ap J 32, p. 83—90. 1910.

Radial velocities for 61 O and B stars are collected from published data and from unpublished observations at Yerkes.

EDWIN B. FROST, Preliminary note on the sun's velocity with respect to the stars of spectral type A. Mem. Spettr. Italiani, Ser. 2^a, Vol. 1, p. 26—28. 1912.

Radial velocities for 33 stars.

C. A. MANEY, On the system of the brighter A stars. A J 29, p. 53—61. 1915.

Radial velocities for 284 A stars are given. Besides the 212 stars used by CAMPBELL (Lick Bull 211) 72 more stars have been added, most of them from measures at Yerkes made by the writer and by others.

EDWIN B. FROST, STORRS B. BARRETT, and OTTO STRUVE, Radial velocities of 368 Helium stars. Ap J 64, p. 1—77. 1926.

A detailed catalogue of 368 B stars.

The spectrograms are obtained with the Bruce spectrograph attached to the 40-inch refractor. The probable error for the average radial velocity is 3,5 km/sec; for a single spectrogram 9,0 km/sec.

5. Comprehending Catalogue. A catalogue containing all stars for which radial velocities have been determined is of course highly desirable. In 1920 VOÛTE published his First catalogue of radial velocities[1] which embraces all radial velocities accessible in the publications present in the library of the Royal Observatory at the Cape of Good Hope up to that time. The catalogue contains the radial velocities of 2071 objects. Corrigenda to this catalogue have been published by MARGARETTA PALMER[2]. Owing to the rapid progress of the radial velocity determinations a new complete catalogue would fill an actual want.

6. Two Special Lists of Radial Velocities. In table I the radial velocities (ϱ) for the stars brighter than the visual magnitude 2^m,00 are given. Moreover the Boss number, name, coordinates, visual magnitude, spectrum and total proper motion are tabulated. The radial velocities are taken from VOÛTE's catalogue with the addition of some more recent determinations.

The stars with radial velocities $\geqq 100$ km/sec have been brought together from different catalogues and given in table II. In the first column the letter B before the numbers refers to the Boss Catalogue, C to no. 18 of the Cincinnati Publications and L to LUYTEN's catalogue of stars of large proper motion given in Lick Bull 344.

[1] Natuurkundig Tijdschrift voor Ned.-Indie Deel 80, 2, p. 95—177. Visser and Co., Weltevreden (1920).

[2] A J 35, p. 99—100. (1923).

Table I. Radial Velocities of the Apparently Brightest Stars.

Boss No.	Name	α_{1900}	δ_{1900}	m_v	Sp	μ	ϱ
363	α Eri	1^h 34^m,0	−57° 45′	0^m,60	B5	0″,088	—
772	α Per	3 17 ,2	+49 30	1 ,90	F5	0 ,039	− 2,4km
1077	α Tau	4 30 ,2	+16 19	1 ,06	K5	0 ,203	+54,5
1246	α Aur	5 9 ,3	+45 54	0 ,21	G0	0 ,437	+30,2
1250	β Ori	5 9 ,7	− 8 19	0 ,34	B8p	0 ,001	+22,6
1303	γ Ori	5 19 ,8	+ 6 16	1 ,70	B2	0 ,020	+19
1304	β Tau	5 20 ,0	+28 31	1 ,78	B8	0 ,180	+11
1370	ε Ori	5 31 ,1	− 1 16	1 ,75	B0	0 ,002	+26,3
1468	α Ori	5 49 ,8	+ 7 23	0 ,92	Ma	0 ,029	+21,3
1609	β C Ma	6 18 ,3	−17 54	1 ,99	B1	0 ,007	+33
1622	α Car	6 21 ,7	−52 38	−0 ,86	F0	0 ,018	+20,2
1690	γ Gem	6 31 ,9	+16 29	1 ,93	A0	0 ,065	−12,3
1732	α C Ma	6 40 ,7	−16 35	−1 ,58	A0	1 ,316	− 8
1804	ε C Ma	6 54 ,7	−28 50	1 ,63	B1	0 ,001	+28,2
1839	δ C Ma	7 4 ,3	−26 14	1 ,98	F8p	0 ,005	+34
1979	α Gem	7 28 ,2	+32 6	1 ,99	A0	0 ,203	+ 2,6
2008	α C Mi	7 34 ,1	+ 5 29	0 ,48	F5	1 ,242	− 4,3
2031	β Gem	7 39 ,2	+28 16	1 ,21	K0	0 ,625	+ 3,6
2233	ε Car	8 20 ,5	−59 11	1 ,74	K0	0 ,032	+11,7
2493	β Car	9 12 ,1	−69 18	1 ,80	A0	0 ,193	−16
2698	α Leo	10 3 ,0	+12 27	1 ,34	B8	0 ,248	+ 3,1
2933	α U Ma	10 57 ,6	+62 17	1 ,95	K0	0 ,139	− 8
3237	α¹ Cru	12 21 ,0	−62 33	1 ,58	B1	0 ,047	+ 7
3263	γ Cru	12 25 ,6	−56 33	1 ,61	Mb	0 ,273	+21,5
3328	β Cru	12 41 ,9	−59 9	1 ,50	B1	0 ,056	+13
3363	ε U Ma	12 49 ,6	+56 30	1 ,68	A0p	0 ,115	−11,9
3476	α Vir	13 19 ,9	−10 38	1 ,21	B2	0 ,055	+ 1,6
3566	η U Ma	13 43 ,6	+49 49	1 ,91	B3	0 ,119	− 6
3615	β Cen	13 56 ,8	−59 53	0 ,86	B1	0 ,041	+12
3662	α Boo	14 11 ,1	+19 42	0 ,24	K0	2 ,282	− 5,0
3735	{α¹ Cen	14 32 ,8	−60 25	0 ,33	G0	} 3 ,680	−22,2
	{α² Cen	14 32 ,8	−60 25	1 ,70	K5		
4193	α Sco	16 23 ,3	−26 13	1 ,22	Ma	0 ,034	− 3,1
4250	α Tr A	16 38 ,1	−68 51	1 ,88	K2	0 ,032	− 3,7
4439	λ Sco	17 26 ,8	−37 2	1 ,71	B2	0 ,036	+ 3
4645	ε Sgr	18 17 ,5	−34 26	1 ,95	A0	0 ,139	−12
4722	α Lyr	18 33 ,6	+38 41	0 ,14	A0	0 ,346	−13,8
5062	α Aql	19 45 ,9	+ 8 36	0 ,89	A5	0 ,655	−33
5320	α Cyg	20 38 ,0	+44 55	1 ,33	A2p	0 ,001	− 4
5916	α Ps A	22 52 ,1	−30 9	1 ,29	A3	0 ,365	+ 6,7

Table II. Stars with Radial Velocity $\geqq$ 100 km/sec.

Designation	α_{1900}	δ_{1900}	m_v	Sp	μ	ϱ
v. Maanen No. 2	0^h 43^m,9	+ 4° 54′	12^m,3	dF3	3″,01	+238km
B D +23° 123	0 48 ,9	+23 32	8 ,8	R3	0 ,15	−234
C 149	1 3 ,3	+61 1	7 ,8	F5	0 ,64	−325
RR Cet	1 27 ,0	+ 0 50	var.	F0	0 ,07	−102
R Ari	2 10 ,4	+24 35	var.	M3e	0 ,04	+114
C 340	2 32 ,6	+30 24	7 ,2	G0	0 ,62	−100
C 407	3 2 ,5	+25 58	8 ,0	A3p	0 ,86	−144
C 489	3 35 ,3	− 3 32	6 ,7	F5	0 ,75	+114
C 560	4 8 ,6	+22 6	8 ,9	A8	0 ,54	+338
C 602	4 34 ,5	+41 56	7 ,3	G0	0 ,69	+105
R Pic	4 43 ,5	−49 26	var.	Md	0 ,07	+208
C 675	5 7 ,7	−44 59	9 ,2	K2	8 ,75	+242
C 756	5 57 ,3	+19 23	9 ,0	F9	0 ,93	−191
B 1511	5 59 ,2	−26 17	5 ,2	K2	0 ,10	+181
B 1517	6 0 ,6	−32 10	5 ,6	B4	0 ,12	+100
X Mon	6 52 ,4	− 8 56	var.	M4e	0 ,01	+157
B 2053	7 41 ,9	−33 59	5 ,4	F9	1 ,71	+100

Table II. (Continuation.)

Designation	α_{1900}	δ_{1900}	m_v	Sp	μ	ϱ
C 935	$7^h 47^m,2$	+30 55	$8^m,2$	F7	1″,96	−242 km
A. G. Wash. 3498	8 36 ,1	−15 59	9 ,4	A7	0 ,56	+200
B 2412	8 54 ,0	−15 45	5 ,9	F5	0 ,32	+122
S Car	10 6 ,2	−61 4	var.	Md	0 ,11	+289
SU Dra	11 32 ,2	+67 53	var.	A2	0 ,13	−173
C 1640	12 47 ,9	−17 57	8 ,3	F3	0 ,88	+144
C 1666	12 56 ,1	−26 50	8 ,2	F4	0 ,54	+226
RV UMa	13 29 ,4	+54 30	var.	F0	—	−177
V UMi	13 37 ,0	+74 49	var.	gM5	—	−158
BD +34° 2476	13 54 ,8	+34 23	9 ,3	A3sp	0 ,54	−164
BD +6° 2932	14 38 ,5	+ 6 15	9 ,5	dG1	0 ,93	−139
C 1988	14 54 ,2	−21 36	8 ,5	F4	0 ,78	+158
C 2018	15 4 ,7	−15 59	9 ,9	G9	3 ,68	+306
C 2019	15 4 ,7	−15 54	9 ,2	G0	3 ,68	+290
BD −3° 3746	15 8 ,9	− 3 26	9 ,2	dM0	0 ,69	−106
S Lib	15 15 ,6	−20 2	var.	M2e	0 ,20	+295
C 2101	15 37 ,7	−10 36	7 ,3	A2p	1 ,18	−170
BD +32° 2697	16 11 ,4	+32 25	8 ,5	K4	0 ,04	−164
VX Her	16 26 ,2	+18 36	var.	A3	—	−380
R Dra	16 32 ,4	+66 58	var.	M5e	0 ,04	−138
RW Dra	16 33 ,7	+58 3	var.	A5	—	−108
C 2341	17 29 ,9	+ 6 4	8 ,5	F1	0 ,62	−148
C 2348	17 33 ,9	+18 37	9 ,1	F0	0 ,28	−240
C 2374	17 47 ,6	− 7 53	7 ,6	G0	0 ,25	−125
Barnard's Star	17 52 ,9	+ 4 25	9 ,7	Mb	10 ,30	−117
C 2392	18 0 ,7	+ 4 39	6 ,8	G1	0 ,30	−124
T Her	18 5 ,3	+31 0	var.	M3e	0 ,01	−125
W Lyr	18 11 ,5	+36 38	var.	M5e	—	−174
RZ Lyr	18 39 ,9	+32 42	var.	A2	—	−220
B 4789	18 49 ,7	−60 20	5 ,1	K0	0 ,14	+177
BD +41° 3306	19 15 ,7	+41 28	7 ,6	G8	0 ,66	−125
BD +35° 3659	19 27 ,6	+35 57	9 ,5	F0s	0 ,56	−172
C 2558	19 29 ,7	+32 59	6 ,6	G2	0 ,52	−162
XZ Cyg	19 30 ,4	+56 10	var.	A	0 ,11	−195
RT Cyg	19 40 ,8	+48 32	var.	M3e	0 ,03	−115
Z Cyg	19 58 ,6	+49 46	var.	M5e	—	−165
B 5166	20 4 ,6	−36 21	5 ,3	K3	1 ,63	−131
C 2654	20 17 ,7	−21 40	8 ,1	G0p	1 ,18	−179
β G C 10404 Br	20 35 ,1	+21 22	8 ,5	gK4	—	−106
C 2750	21 7 ,5	+23 45	8 ,0	dF9	0 ,46	−103
RR Aqr	21 9 ,8	− 3 19	var.	M3e	—	−182
L 673	21 41 ,0	+43 51	11 ,3	A	0 ,64	−354
L 680	21 50 ,8	+32 10	10 ,8	dG0	0 ,73	−178
BD +20° 5071	21 59 ,7	+20 34	8 ,8	R3	0 ,02	−383
C 3093	23 30 ,4	+30 27	6 ,7	G1	0 ,59	−103
BD −3° 5751	23 57 ,0	− 3 23	9 ,9	R0	—	−136

c) The Solar Motion derived from the Radial Velocities of the Stars.

7. Theory of the Determination of the Solar Motion. The radial velocity of a star as given in the catalogues is the star's radial velocity with reference to the sun. This is also the case with the proper motion. But now the sun itself is not at rest, and therefore a part of the observed displacement must be attributed to the solar motion. This part of a star's motion may be termed the group motion, and the residuum is the star's peculiar motion. In order to obtain the peculiar velocities of the stars, we must accordingly determine the solar motion.

What then is the motion of the sun? The sun and the stars are moving in a space entirely devoid of fixed marks of reference and therefore it is, in reality, a matter of convention how to divide the observed motion between the sun and the stars. We are now going to show how the standard of rest, to which the peculiar motions are to be referred, has been defined in practice.

We refer the velocity of a star to two different systems of rectangular coordinates which, according to the nomenclature of CHARLIER, we denote as K_2 and K_3.

The system K_2 is the common astronomical equator system of coordinates. It has the positive Z-axis directed towards the north pole, the positive X-axis towards $\alpha = 0°$, and the positive Y-axis towards $\alpha = +90°$.

		K_2		
		X''	Y''	Z''
	X	γ_{11},	γ_{21},	γ_{31}
K_3	Y	γ_{12},	γ_{22},	γ_{32}
	Z	γ_{13},	γ_{23},	γ_{33}

The system K_3 has the positive Z-axis directed towards the star, the positive X and Y-axes are directed towards increasing right ascension and declination.

The direction cosines relating the axes of these two systems of coordinates are given through the scheme to the left,

where

$$\gamma_{11} = -\sin\alpha\,, \qquad \gamma_{21} = +\cos\alpha\,, \qquad \gamma_{31} = 0\,,$$
$$\gamma_{12} = -\sin\delta\cos\alpha\,, \qquad \gamma_{22} = -\sin\delta\sin\alpha\,, \qquad \gamma_{32} = +\cos\delta\,,$$
$$\gamma_{13} = +\cos\delta\cos\alpha\,, \qquad \gamma_{23} = +\cos\delta\sin\alpha\,, \qquad \gamma_{33} = +\sin\delta\,.$$

α and δ are the right ascension and declination of the star.

Let U, V, W be the star's linear velocity components with reference to the sun along the axes of the system K_3, and U'', V'', W'' the corresponding components in the system K_2, and we have, according to the above scheme, the relations

$$\left.\begin{aligned} U'' &= \gamma_{11} U + \gamma_{12} V + \gamma_{13} W\,,\\ V'' &= \gamma_{21} U + \gamma_{22} V + \gamma_{23} W\,,\\ W'' &= \gamma_{31} U + \gamma_{32} V + \gamma_{33} W\,;\end{aligned}\right\}(1) \quad \text{and} \quad \left.\begin{aligned} U &= \gamma_{11} U'' + \gamma_{21} V'' + \gamma_{31} W''\,,\\ V &= \gamma_{12} U'' + \gamma_{22} V'' + \gamma_{32} W''\,,\\ W &= \gamma_{13} U'' + \gamma_{23} V'' + \gamma_{33} W''\,.\end{aligned}\right\}(2)$$

If r is the distance of the star, μ_α and μ_δ its proper motions in right ascension and declination and ϱ its observed radial velocity, we have

$$U = k\,r\,\mu_\alpha\cos\delta\,, \qquad V = k\,r\,\mu_\delta\,, \qquad W = \varrho\,, \tag{3}$$

where k is a constant dependent on the units employed. Consequently if the distance, proper motion and radial velocity of a star are known, the three linear velocity components in the system K_3 are computed from (3), and then the corresponding components in the system K_2 are obtained from (1).

But

$$\left.\begin{aligned} U'' &= U''_1 + U''_0\,,\\ V'' &= V''_1 + V''_0\,,\\ W'' &= W''_1 + W''_0\,,\end{aligned}\right\} \tag{4}$$

where U''_1, V''_1, W''_1 are the components in the system K_2 of the peculiar motion and U''_0, V''_0, W''_0 the corresponding components of the group motion. As the group motion is the motion of the standard of rest relative to the sun, the components of this motion in the system K_2 are of course the same for all stars considered, and with reversed signs equal to the components of the solar motion with respect to the chosen standard of rest.

BRAVAIS[1] considered the problem of determining the sun's motion relative to the centre of gravity of an arbitrary group of stars, the sun itself included. Let m be the mass of a star and we have from (4)

$$\left.\begin{aligned} m\,U'' &= m\,U_1'' + m\,U_0'' , \\ m\,V'' &= m\,V_1'' + m\,V_0'' , \\ m\,W'' &= m\,W_1'' + m\,W_0'' , \end{aligned}\right\} \tag{5}$$

and summing for all stars of the group

$$\sum m\,U'' = \sum m\,U_1'' + U_0'' \sum m, \text{ etc.}$$

Adding the corresponding equations for the sun, which take the forms

$$0 = -U_0'' + U_0'', \text{ etc. (assuming } m_\odot = 1) ,$$

we get

$$\sum m\,U'' = \sum m\,U_1'' - U_0'' + U_0'' (1 + \sum m) , \text{ etc.}$$

If we refer the peculiar motion to the centre of gravity of the group, the sun itself included, we shall have

$$\sum m\,U_1'' - U_0'' = 0, \quad \sum m\,V_1'' - V_0'' = 0 , \quad \sum m\,W_1'' - W_0'' = 0 , \tag{6}$$

and consequently the equations for the determination of this components of the group motion in this case are

$$\left.\begin{aligned} U_0'' (1 + \sum m) &= \sum m\,U'' , \\ V_0'' (1 + \sum m) &= \sum m\,V'' , \\ W_0'' (1 + \sum m) &= \sum m\,W'' . \end{aligned}\right\} \tag{7}$$

This is the most simple form of BRAVAIS' equations, which determine, free from any hypothesis, the solar motion relative to the adopted standard of rest. The equations (7), however, contain some quantities which are known only to a small extent. The most doubtful element is the mass, and therefore we cannot in general take into account this element. Assuming the masses to be equal and excluding the sun from the group of stars under consideration, the equations (7) transform into

$$\left.\begin{aligned} N \cdot U_0'' &= \sum U'' , \\ N \cdot V_0'' &= \sum V'' , \\ N \cdot W_0'' &= \sum W'' , \end{aligned}\right\} \tag{8}$$

where N is the number of stars.

The standard of rest to which the motions are referred in this case, is defined through the equations

$$\sum U_1'' = 0 , \qquad \sum V_1'' = 0 , \qquad \sum W_1'' = 0 . \tag{9}$$

This standard of rest is called by WEERSMA[2] the geometrical centre of gravity. We will term it the centroid of the stars.

The distances of the stars are not known either with sufficient accuracy for any very large number of stars over the whole sky, and therefore the most reliable determinations of the solar motion at present are based on the radial velocities of the stars. The common method starts from the third of the equations (2)

$$\gamma_{13}\,U'' + \gamma_{23}\,V'' + \gamma_{33}\,W'' + K = \varrho , \tag{10}$$

[1] J. Mathém. Liouville 8. 1843.

[2] Groningen Publ 21, p. 5. 1908.

where $\varrho = W$ is the observed radial velocity. We have added a constant error term K, which will be discussed further on. Using the relations (4) the equation (10) is easily transformed into

$$\gamma_{13} U_0'' + \gamma_{23} V_0'' + \gamma_{33} W_0'' + K = \varrho - \varrho_1, \tag{11}$$

where

$$\varrho_1 = \gamma_{13} U_1'' + \gamma_{23} V_1'' + \gamma_{33} W_1'' \tag{12}$$

is the peculiar radial velocity of the star. Forming the usual normal equations from (11) we get

$$\left.\begin{aligned} d_1 U_0'' + e_1 V_0'' + f_1 W_0'' + g_1 K &= \sum \gamma_{13} \varrho - \sum \gamma_{13} \varrho_1, \\ d_2 U_0'' + e_2 V_0'' + f_2 W_0'' + g_2 K &= \sum \gamma_{23} \varrho - \sum \gamma_{23} \varrho_1, \\ d_3 U_0'' + e_3 V_0'' + f_3 W_0'' + g_3 K &= \sum \gamma_{33} \varrho - \sum \gamma_{33} \varrho_1, \\ d_4 U_0'' + e_4 V_0'' + f_4 W_0'' + g_4 K &= \sum \varrho - \sum \varrho_1, \end{aligned}\right\} \tag{13}$$

where

$$\left.\begin{aligned} d_1 &= \sum \gamma_{13} \gamma_{13}, & e_1 &= \sum \gamma_{13} \gamma_{23}, & f_1 &= \sum \gamma_{13} \gamma_{33}, & g_1 &= \sum \gamma_{13}, \\ d_2 &= \sum \gamma_{23} \gamma_{13}, & e_2 &= \sum \gamma_{23} \gamma_{23}, & f_2 &= \sum \gamma_{23} \gamma_{33}, & g_2 &= \sum \gamma_{23}, \\ d_3 &= \sum \gamma_{33} \gamma_{13}, & e_3 &= \sum \gamma_{33} \gamma_{23}, & f_3 &= \sum \gamma_{33} \gamma_{33}, & g_3 &= \sum \gamma_{33}, \\ d_4 &= \sum \gamma_{13}, & e_4 &= \sum \gamma_{23}, & f_4 &= \sum \gamma_{33}, & g_4 &= N. \end{aligned}\right\} \tag{14}$$

We now assume the standard of rest to be defined through the equations

$$\sum \gamma_{13} \varrho_1 = 0, \qquad \sum \gamma_{23} \varrho_1 = 0, \qquad \sum \gamma_{33} \varrho_1 = 0, \tag{15}$$

and add the condition

$$\sum \varrho_1 = 0, \tag{15*}$$

which determines the K-term. The equations (15) signify that the sums of the projections of the peculiar radial velocities along the three axes of the system of coordinates K_2 are all equal to zero.

With this assumption the equations (13) get the forms

$$\left.\begin{aligned} d_1 U_0'' + e_1 V_0'' + f_1 W_0'' + g_1 K &= k_1, \\ d_2 U_0'' + e_2 V_0'' + f_2 W_0'' + g_2 K &= k_2, \\ d_3 U_0'' + e_3 V_0'' + f_3 W_0'' + g_3 K &= k_3, \\ d_4 U_0'' + e_4 V_0'' + f_4 W_0'' + g_4 K &= k_4, \end{aligned}\right\} \tag{16}$$

where

$$k_1 = \sum \gamma_{13} \varrho, \quad k_2 = \sum \gamma_{23} \varrho, \quad k_3 = \sum \gamma_{33} \varrho, \quad k_4 = \sum \varrho. \tag{17}$$

From the equations (16) the unknown quantities U_0'', V_0'', W_0'', and K are computed.

The only assumption made is expressed through the equations (15), viz. that the sums of the projections of the peculiar radial velocities of the stars on the three axes of coordinates are zero. Let S be the velocity of the sun directed towards a point with right ascension A and declination D, and we have, according to the definition of the group motion,

$$\left.\begin{aligned} S \cos D \cos A &= -U_0'', \\ S \cos D \sin A &= -V_0'', \\ S \sin D &= -W_0''. \end{aligned}\right\} \tag{18}$$

A and D are the coordinates of the apex. From these equations we get the elements of the solar motion:

$$S = \sqrt{U_0''^2 + V_0''^2 + W_0''^2}, \tag{19}$$

$$\operatorname{tg} A = \frac{V_0''}{U_0''}, \tag{20}$$

$$\sin D = -\frac{W_0''}{S}. \tag{21}$$

Thus the computation is performed in the following way:

1. The three direction cosines γ_{13}, γ_{23} and γ_{33} are computed for each star.
2. The sums (14) and (17) are formed.
3. The equations (16) are solved.
4. The velocity of the sun and the coordinates of the apex are computed from (19), (20), and (21).

This method of computation will be referred to as the individual method (i).

If the number of stars is great, it is convenient to combine neighbouring stars into groups. Instead of using individual values of the radial velocities and the direction cosines, we use the mean radial velocity and the direction cosines for the centre of each group. In this way we get an equation of the form (10) for each group, which equations are then solved in the same manner as before. We will refer to this method as the group method (g). If the groups are chosen in a suitable manner, the resulting normal equations take a very simple form.

CHARLIER, for instance, divides the sky into 48 „squares", all of the same area, in the following way: The zone between the equator and the parallel $\delta = +30°$ is divided into 12 squares C_1 to C_{12}, C_1 containing those stars whose right ascension (α) lies between $0°$ and $30°$, C_2 those with α between $30°$ and $60°$, and so on. The zone between $\delta = +30°$ and $\delta = +66° 26',6$ is divided into 10 squares $B_1 - B_{10}$ and the calotte around the north pole into two, A_1 and A_2. The southern hemisphere is divided in the same manner, 12 squares D_1 to D_{12} between $\delta = 0°$ and $\delta = -30°$, 10 squares $E_1 - E_{10}$ between $\delta = -30°$ and $= -66° 26',6$ and two F_1 and F_2, around the south pole. The direction cosines for the different squares have been tabulated by CHARLIER[1] and by GYLLENBERG[2]. Owing to the symmetrical arrangement of these squares, the normal equations take a very simple form. Using the group method, we get

$$\left.\begin{aligned} 16{,}2344\, U_0'' &= \sum \gamma_{13}\,\bar{\varrho}\,, \\ 16{,}3552\, V_0'' &= \sum \gamma_{23}\,\bar{\varrho}\,. \\ 15{,}4132\, W &= \sum \gamma_{33}\,\bar{\varrho}\,, \\ 48\, K &= \sum \bar{\varrho}\,, \end{aligned}\right\} \tag{22}$$

where $\bar{\varrho}$ is the mean value of the radial velocities in a square, and the summation is to be extended to all the 48 squares.

In the group method each group is given the same weight. We may as well, in forming the normal equations, give to each group a weight proportional to the number of stars in the group. This modification may be termed the weighted group method (wg).

[1] Mem. Univ. California 7, p. 56.

[2] Lund Medd Ser. II, No. 13, p. 14.

Sometimes the velocity of the sun is computed from the radial velocity on the assumption of a given apex. In this case the equations of condition are

$$S\cos d + K = \varrho - \varrho_1, \tag{23}$$

where d is the angular distance of the star from the assumed antapex of the solar motion. In this case the normal equations have the following form

$$S\sum\cos^2 d + K\sum\cos d = \sum\varrho\cos d\,, \tag{24}$$

$$S\sum\cos d + K\cdot N = \sum\varrho\,, \tag{24*}$$

where N is the number of stars. The assumptions are:

$$\sum\varrho_1\cos d = 0\,,\qquad \sum\varrho_1 = 0\,. \tag{25}$$

In forming the normal equations for the solution of the elements of the solar motion, it is often necessary to reject some stars which would unduly influence the results. It is mainly the members of moving clusters and the stars with very large velocities which must be considered. The question of rejection is a very delicate one, and it is shown in the ninth paragraph of this section how different investigators have proceeded. In general, all stars with peculiar radial velocities exceeding 60 km per sec are excluded. This limit is chosen from statistical considerations, but in the light of modern results it seems to have a real signification.

It may be mentioned in this connection that HERTZSPRUNG[1] has proposed a method of treating the equations in order to neutralize the abnormally great radial velocities.

8. The Solar Motion derived from Visual Observations. The first attempts to derive the solar motion from the radial velocities were based on visual observations. The results are given in the following table III.

Table III. The Solar Motion from Visual Observations.

Investigator	A	D	S km/sec
Kövesligethy	(261°)	(+35°)	64
Homann	320	+41	39,3 ± 4,2
,,	310	+70	48,5 ± 23,1
,,	279	+14	24,5 ± 15,8

KÖVESLIGETHY'S[2] solution is based on about 70 stars, visually observed at Greenwich before 1881. The position of the apex was assumed from previous proper-motion solutions.

HOMANN'S first solution[3] is founded upon the Greenwich observations, the two others[4] on HUGGINS' and SEABROKE'S observations resp. Owing to the great uncertainty in the visual determinations of radial velocities, mentioned in the introduction, these first attempts have only an historical interest. The failure of the visual method is evident from the large probable errors.

9. The Solar Motion derived from Photographic Radial Velocity Observations when no Regard is taken to the Spectral Type. The first determinations of the solar motion from photographic observations were based on the fifty-one Potsdam radial velocities referred to in the introduction. KEMPF[5], using this

[1] B A N I, 16 (1922).

[2] A N 114, p. 327 (1886).

[3] Beiträge zur Untersuchung der Sternbewegungen und der Lichtbewegung durch Spektralmessungen. Inaug.-Diss. Berlin (1885).

[4] A N 114, p. 25–26 (1886).

[5] A N 132, p. 81–82 (1893).

material, performed three different solutions of the solar motion. In the first solution, including all of the fifty-one stars, he got

$$A = 206° \pm 12°, \quad D = +46° \pm 9°, \qquad S = 18{,}6 \pm 3{,}0 \text{ km per sec (p. e.)}.$$

In the second solution the radial velocities of five Orion stars were combined into one resulting velocity, owing to the fact that these stars seemed to have essentially equal velocities; and similarly for five Ursa Major stars and for three Leo stars. The result was

$$A = 160° \pm 20°, \quad D = +50° \pm 14°, \qquad S = 13{,}0 \pm 3{,}3 \text{ km per sec (p. e.)}.$$

Assuming the coordinates of the apex of the solar motion as $A = 267°$, $D = +31°$ (derived by L. STRUVE from the proper motions), he obtained $S = 12{,}3 \pm 3{,}0$ km per sec. RISTEEN[1], using the same material, rejected Sirius and Procyon on account of irregularities of motion and three other stars with uncertain velocities. Moreover the five Ursa Major stars were combined into one velocity. He got

$$A = 218°, \quad D + 45°, \qquad S = 17{,}5 \text{ km per sec.}$$

The coordinates of the apex obtained from the Potsdam observations differ considerably from those obtained from proper motions and from later radial velocity determinations. Therefore the agreement of the solar velocity results with recent determinations, based on a much greater material, must be considered as merely accidental. The first more reliable determination of the solar motion is that of CAMPBELL[2]. His determination was based on 280 stars whose radial velocities had been measured at Mount Hamilton up to the end of the year 1900 and found to be apparently constant. (A great many of these have since been found to be variable.) The stars were divided into 80 groups, by combining neighbouring stars; for each group the mean of the individual velocities was taken as the velocity of the group. An equation of the form (10) (without the K-term) was then formed for each group, and then the 80 equations were combined into the three usual normal equations. The following elements of the solar motion were obtained:

$$A = 277°{,}5 \pm 4°{,}8, \; D = +20°{,}0 \pm 5°{,}9, \; S = 19{,}89 \pm 1{,}52 \text{ km per sec (p. e.)}.$$

No stars were rejected, on account of very high speed or for other reasons, in the solution.

CAMPBELL's second solution[3] was based on the radial velocities which were known up to the beginning of the year 1910. Before going to treat the new data at hand, it was necessary to decide whether any of the velocities should be rejected. This is a very delicate question; it is, of course, desirable to reject as few stars as possible, but on the other hand, we cannot but reject some stars which would unduly influence the result. It is mainly two kinds of stars which must be considered in this connection, namely the members of moving clusters and the stars with very large velocities. For a moving cluster the question is this: how many members of the group is the computer to include in the solution? This question has been solved by CAMPBELL in the following way: The five Ursa Major stars are represented in the solution by two stars, the Pleiades by two stars and the fourteen bright members of the Hyades by three stars.

[1] A J 13, p. 74–75 (1893).
[2] Ap J 13, p. 80–89 (1901).
[3] Lick Bull 196, p. 125 (1911). Stellar Motions, p. 173–189 (1913).

In the case of double stars, for which the radial velocities of both components have been measured, each pair of stars is, of course, used as one star whose velocity is the velocity of the centre of mass of the system.

Concerning the stars with very large radial velocities, CAMPBELL applied the following method to decide which were to be rejected. He assumed that the solar motion is 19,9 km per sec towards $A = 275°$, $D = +30°$. Let ϱ be the star's observed radial velocity with reference to the solar system, and ϱ_1 the peculiar radial velocity of the same star, then this peculiar velocity is computed from the observed velocity by the aid of the relation (23) (without the K-term)

$$\varrho_1 = \varrho - 19{,}9 \cos d\,,$$

where d is the angular distance of the star from the assumed antapex. From his first solution the numerical average peculiar velocity for the 280 stars was found to be 17,1 km per sec[1], and now he decided to reject those stars for which the peculiar radial velocity was equal to or exceeded three and a half times the average velocity 17,1 km per sec; that is to say all those greater than 60 km per sec.

After the rejections there remained 1047 stars, mainly brighter than the fifth visual magnitude, as a basis for the solution. The stars are not uniformly, but quite satisfactorily, distributed over the entire sphere. As before, the neighbouring stars are combined into groups, 172 in number, each group containing on the average about six stars. Combining the 172 equations into three normal equations and solving them, he obtained:

$$A = 272°{,}0 \pm 2°{,}5\,,\quad D = +27°{,}4 \pm 3°{,}0\,,\quad S = 17{,}77 \pm 0{,}62 \text{ km per sec (p. e.)}.$$

Another solution was made with an equation of the form (10) (without the K-term) for each star with the following result:

$$A = 273°{,}5, \quad D = +28°{,}0, \quad S = 17{,}73 \text{ km per sec.}$$

The two solutions are remarkably accordant.

In 1911 the elements of the solar motion were redetermined from the radial velocities of 1193 objects[2]. No observed velocities were rejected on account of their magnitude, and no members of moving clusters were excluded. Neighbouring stars were combined into groups, 80 in number, and the following results were obtained:

$$A = 268°{,}5, \quad D = +25°{,}1, \quad S = 19{,}5 \text{ km per sec}, \quad K = +1{,}91 \text{ km per sec.}$$

The constant term K is of considerable size, $+1{,}91$ km, by which amount each observed radial velocity appears to be systematically too great. Another solution from the same data, omitting the constant term K, led to the following results:

$$A = 268°{,}5, \quad D = +25°{,}3, \quad S = 19{,}5 \text{ km per sec.}$$

The elements of the solar motion are apparently unaffected by the inclusion or omission of the term K.

CAMPBELL's latest treatment of the problem is of a recent date. The radial velocities for nearly three thousand stars have been determined at Lick and at Cerro San Cristobal, but in order to get a homogeneous material, several hundred results were rejected. Two or three hundred stars were rejected because of extremely poor lines in their spectra, and about two hundred because their variable velocities had not yet been investigated. As in his second solution,

[1] Ap J 13, p. 84. (1901).
[2] Lick Bull 196, p. 127. (1911).

only a few members of each of the moving clusters were used, and moreover all stars with peculiar radial velocities exceeding 70 km per sec were rejected. Thus there remained 2034 stars for the solution of the solar motion. The sky was divided into 94 equal areas, and the average radial velocity for each area was formed. The number of stars in the areas varied from 9 to 55. Two solutions were made: 1) giving equal weight to the 94 mean velocities, and 2) giving weights to the 94 mean velocities proportional to the numbers of stars involved. The elements of the solar motion obtained from these solutions were[1]:

1) $A = 268°,8$, $D = +27°,8$, $S = 19,2$ km per sec, $K = 1,28$ km per sec.

2) $A = 269°,0$, $D = +26°,6$, $S = 18,8$ km per sec, $K = 1,43$ km per sec.

HOUGH and HALM[2] have investigated the solar motion from the radial velocities of 166 stars observed at the Cape. In accordance with CAMPBELL's arrangement, neighbouring groups of stars were combined and from the results given on page 91 of their investigation we derive the following elements

$$A = 257°,2, \quad D = +28°,7, \quad S = 22,5 \text{ km per sec.}$$

Adding 45 stars whose radial velocities were known from other observations, we get from 211 stars comprised within the parallel +30° and the south pole

$$A = 258°,5, \quad D = +30°,0, \quad S = 23,8 \text{ km per sec.}$$

Combining these 211 stars with the 280 stars used by CAMPBELL in his first solution they get (page 100)

$$A = 271°,2 \pm 3°,3, \quad D = +25°,6 \pm 3°,7, \quad S = 20,85 \pm 0,95 \text{ km per sec (p. e.).}$$

From the radial velocities published by CAMPBELL in the Lick Bull 195, p. 211 and 229 B. BOSS[3] derives from 1321 stars of all spectral types

$$A = 268°,9, \quad D = +28°,7, \quad S = 21,6 \text{ km per sec.}$$

The increase of the solar motion as compared with CAMPBELL's values is, according to BOSS, due to the fact that more large velocities were used by him.

In 1915 an extensive investigation of the stellar velocity distribution as derived from the radial velocities was made by GYLLENBERG[4]. This investigation was based on all the radial velocity observations up to that time, in all 1640 stars. The stars are chiefly of bright magnitudes and the material may be considered complete down to the visual magnitude 5,0. Those stars (in all 44) whose velocities exceeded 66,3 km per sec were rejected. In order to obtain the elements of the solar motion, the sky was divided into 48 squares of equal areas, and for each area the mean value of the observed radial velocities was computed. The number of stars in the areas varied from 12 to 69. An equation of condition was formed for each square, and then the 48 equations were combined into the three usual normal equations, giving equal weight to all squares. In this case the normal equations take the simple form (22). The following elements of the solar motion, with their m. e., for all 1596 stars were obtained:

$$A = 270°,5 \pm 2°,5, \quad D = +28°,6 \pm 2°,2, \quad S = 19,84 \pm 0,76 \text{ km per sec,}$$
$$K = +2,4 \text{ km per sec.}$$

---

[1] Obs 48, p. 274 (1925).
[2] M N 70, p. 85 (1909).
[3] A J 28, p. 165 (1914).
[4] Lund Medd Ser. II, No. 13. Kungl. Fysiogr. Säll Handl. Bd. 26, No. 10 (1915.)

In the investigation the velocities are expressed in siriometres per stellar year. To transform them into kilometres per second we have to multiply by 4,7375.

The material was further divided into two groups with regard to magnitude, the one embracing all stars brighter than 5,0, the other those fainter than this limit; and into two groups with regard to galactic latitude, the one embracing all stars within galactic latitudes $\pm 30°$, the other the rest. The resulting elements of the solar motion for these different groups of stars are found in tables V and VI.

A new solution, based on all radial velocity observations available up to the beginning of the year 1924, was performed by GYLLENBERG and MALMQUIST[1]. Only stars of the spectral types B, A, F, G, K and M were considered and moreover only stars brighter than the sixth visual magnitude. The limiting of the stars examined to this magnitude will prevent a selection that otherwise would considerably affect the results. Only a short glance at the stars of fainter magnitudes in the radial velocity catalogues will make clear that stars of large proper motions, dwarf stars, are very much favoured.

Concerning the exclusion of observations, the principle of rejecting as few stars as possible was followed. Amongst the spectral types B and M no star was rejected, 3 stars of type A and 11 stars of type F were excluded on account of large proper motions and radial velocities. For types G and K it was tried to separate between the giant and the dwarf stars, following a method given by MALMQUIST[2]. The stars considered as dwarfs were excluded. The sky was, as before, divided into 48 squares of equal areas, but now the galactic, instead of the equatorial plane, was used as the plane of symmetry. In forming the normal equations, each square is given a weight equal to the number of stars in the square. Before the entering of the radial velocities into the solution they were corrected for the K-effect, using the values found by GYLLENBERG[3]. For B, K and M stars the values $+4{,}3$, $+3{,}6$ and $+5{,}3$ km resp. were used, for A, F and G stars no corrections were applied, as the values found are considerably smaller than their mean errors. Yet a systematic term ΔK was introduced in order to find whether the corrections applied are valid for the larger material used here or not.

From the radial velocities of the 2189 stars used, they got

$$A = 270°{,}6 \pm 2°{,}3, \quad D = +29°{,}7 \pm 2°{,}0, \quad S = 19{,}95 \pm 0{,}70 \text{ km per sec},$$
$$\Delta K = -0{,}06 \pm 0{,}41 \text{ km per sec (m. e.)}.$$

The value found for ΔK shows that the correction for the K-effect is satisfactory.

The elements obtained are remarkably accordant with those obtained by GYLLENBERG, 1915.

The material was further divided into two groups, one embracing the squares between galactic latitudes $+30°$ and $-30°$, the other the rest. The results for these two groups are given in table VI.

In an investigation on the stars of high velocity OORT[4], starting from the elements of the solar motion derived by GYLLENBERG and MALMQUIST, has investigated the influence on the sun's velocity of different modes of rejection. Without any exclusion he got from 2289 stars

$$S = 20{,}8 \text{ km per sec.}$$

---

[1] Lund Medd Ser I, No. 108. Ark Mat Astr Fys 19A, No. 11 (1925).

[2] Lund Medd Ser. II, No. 32. Kungl. Svenska Vet. Akad. Handl. III (1), No. 2 (1924). pp. 40 foll.

[3] Loc. cit. p. 24. See table VIII. [4] Groningen Publ 40, pp. 67 foll (1926).

Excluding all stars fainter than 6,0 and all stars brighter than 6,0 with proper motions exceeding 0″,299 or peculiar radial velocities larger than 62 km per sec, he obtained from 2091 stars

$$S = 19{,}6 \text{ km per sec.}$$

FORBES'[1] determination was based on the radial velocities in VOÛTE's catalogue. Omitting the 138 nebulae, 10 clusters and one star (BARNARD's) there remain 1922 radial velocities. In deducing the elements of the solar motion, the method used by CAMPBELL in his Stellar Motions was followed with the exception that no rejection of any stars was made. He got

$$A = 270°, \quad D = +27°, \quad S = 22{,}00 \text{ km per sec.}$$

From the list of the radial velocities of 537 stars given in Victoria Publ., 2, No. 1, PARASKEVOPOULOS[2] derives

$$A = 271°{,}4, \quad D = +31°{,}6, \quad S = 20{,}7 \text{ km per sec,}$$

and from 743 stars south of the equator, whose radial velocities are given in VOÛTE's catalogue,

$$A = 272°{,}2, \quad D = +29°{,}6, \quad S = 25{,}4 \text{ km per sec.}$$

Combining both materials he got

$$A = 271°{,}6 \pm 3°{,}0, \; D = +30°{,}3 \pm 3°{,}0, \; S = 23{,}33 \pm 1{,}03 \text{ km per sec (p. e.).}$$

In his lectures, mentioned in the introduction, CHARLIER[3] has determined the motion of the sun for all stars brighter than the visual magnitude 6,0 for which the radial velocities were known up to the autumn of 1922. The sky was divided into 48 squares of equal areas, and for each area the mean value of the observed radial velocities was computed. All stars, having a velocity component deviating more than three times the dispersion from the mean value of the component in each square, were excluded. In this manner 191 stars were eliminated, so that 1986 stars remain for the solution. Equal weight was given to all squares, so that the normal equations take the simple form (22). The elements obtained were

$$A = 267°{,}3, \quad D = +30°{,}2, \quad S = 18{,}99 \text{ km per sec}, \quad K = +0{,}91 \text{ km per sec.}$$

FESSENKOFF[4] has made an attempt to introduce into the calculation the masses of the stars considered as the weights of the equations of condition. The masses are uniquely determined from the peculiar radial velocities on special assumptions. The catalogues used are VOÛTE's First catalogue of radial velocities and ADAMS' The radial velocities of 1013 stars, and the peculiar velocities have been computed with the following assumed elements of the solar motion

$$A = 270°, \quad D = +30°, \quad S = 20 \text{ km per sec.}$$

From the radial velocities of 2666 stars the following elements are obtained:

$$A = 271°{,}3, \quad D = +30°{,}5, \quad S = 19{,}42 \text{ km per sec.}$$

10. Collation of the Results from the Preceding Paragraph. In table IV a collation of the results for the elements of the solar motion obtained by the different investigators is given. The third column indicates the method used, the abbreviations *i*, *g*, and *wg* are defined in § 7, *s* indicates that a special method is used. The fourth and fifth columns give the numbers of stars and groups

[1] MN 82, p. 174—177 (1922). [2] AJ 34, p. 181—182 (1922). [3] Mem. Univ. California, 7 (1926). [4] R A J II, Part 2 (1925).

Table IV. The Elements of the Solar Motion derived from Radial Velocities without Regard to Spectral Type.

Investigator	Year	Method	Stars	Groups	A	D	L	B	S	K
KEMPF	1893	i	51	—	206°	+46°	59°	+68°	18,6	—
KEMPF	1893	i	41	—	160	50	128	58	13,0	—
KEMPF	1893	i	41	—	(267)	(35)	(28)	(26)	12,3	—
RISTEEN	1893	i	42	—	218	45	45	62	17,5	—
CAMPBELL	1901	g	280	80	277,5	20,0	17	12	19,9	—
HOUGH, HALM.	1909	wg	166	60	257,2	28,7	18	32	22,5	—
HOUGH, HALM.	1909	wg	211	83	258,5	30,0	20	31	23,8	—
HOUGH, HALM.	1909	wg	491	163	271,2	25,6	20	19	20,8	—
CAMPBELL	1910	g	1042	172	272,0	27,4	21	19	17,8	—
CAMPBELL	1910	i	1042	—	273,5	28,0	23	18	17,7	—
CAMPBELL	1911	g	1193	80	268,5	25,1	18	21	19,5	+ 1,9
CAMPBELL	1911	g	1193	80	268,5	25,3	18	21	19,5	—
B. BOSS.	1914	i	1321	—	268,9	28,7	22	22	21,6	—
GYLLENBERG	1915	g	1596	48	270,5	28,6	23	21	19,8	+ 2,4
FORBES	1922	g	1922	72	270	27	21	20	22,0	—
PARASKEVOPOULOS	1922	wg	537	163	271,4	31,6	26	21	20,7	—
PARASKEVOPOULOS	1922	wg	743	216	272,2	29,6	24	20	25,4	—
PARASKEVOPOULOS	1922	wg	1280	379	271,6	30,3	24	20	23,3	—
GYLLENBERG, MALMQUIST.	1924	wg	2189	48	270,6	29,7	23	21	20,0	corr.
CAMPBELL.	1925	g	2034	94	268,8	27,8	21	22	19,2	+ 1,3
CAMPBELL.	1925	wg	2034	94	269,0	26,6	20	21	18,8	+ 1,4
FESSENKOFF.	1925	s	2666	—	271,3	30,5	24	21	19,4	—
CHARLIER	1926	g	1986	48	267,3	30,2	23	24	19,0	+ 0,9
OORT	1926	s	2289	—	(270)	(31)	(25)	(22)	20,8	—
OORT	1926	s	2091	—	(270)	(31)	(25)	(22)	19,6	—

used in the solution, the sixth and seventh columns contain the right ascension and declination of the solar apex. In the eighth and ninth columns the galactic longitude (L) and latitude (B) of the apex are given. The galactic system of coordinates is that used by PICKERING and CHARLIER and defined through the following position of the galactic pole:

$$\alpha = 190°, \quad \delta = +28°.$$

For the transformation formulas, see CHARLIER[1]. The tenth column gives the velocity of the sun in km per sec, and the eleventh the K-term expressed in the same unit.

Leaving out of consideration the earlier determinations, the agreement between the results is very satisfactory. Combining the results from the most recent determinations, we get for the elements of the solar motion as derived from the radial velocities of the stars the values

$$A = 270°, \qquad D = +30°,$$

or

$$L = 23°, \qquad B = +22°,$$

and

$$S = 19{,}5 \text{ km per sec.}$$

GYLLENBERG has, as already mentioned, divided his material into two groups with respect to magnitude. The results obtained, with their m. e., are given in table V. Within the limits of error the results for the two groups are accordant.

Table V. The Solar Motion as derived from Different Magnitude-Groups.

m	No. Stars	A	D	S km/sec	K km/sec
$\leq 4{,}9$	1114	271°,5 ± 2°,9	+26°,3 ± 2°,5	19,4 ± 0,9	+2,5
$\geq 5{,}0$	482	269 ,4 ± 5 ,1	+30 ,6 ± 4 ,4	21,1 ± 1,5	+0,9

[1] Lund Medd Ser. II, No. 19, p. 11.

The investigation of GYLLENBERG and that of GYLLENBERG and MALMQUIST both indicate that the velocity of the sun is larger for stars in the galactic plane than for stars surrounding the pole of the galaxy. If we divide the stars into two groups with respect to galactic latitude, one embracing all stars the galactic latitudes of which are numerically less than 30° ("galactic stars") and the other the rest ("non-galactic stars"), the following values are obtained:

Table VI. Solar Motion in Terms of Galactic Latitude.

Group	Investigator	No. Stars	A	D	S km/sec
Galactic Stars	G G and M	941 1285	265° 262 ± 3	+28° $+33 \pm 3$	21,4 $21{,}9 \pm 0{,}9$
Non-galactic Stars	G G and M	655 904	271 270 ± 3	+31 $+30 \pm 3$	19,3 $17{,}7 \pm 1{,}1$

SEARES[1] has found from a comparison of the secular parallaxes of stars of different apparent magnitude with their annual parallaxes as derived from the luminosity and density functions that the velocity of the solar system varies with brightness (both absolute and apparent). The direct determinations seem however not to confirme these results.

In an investigation of OORT and DOORN[2] a normal solar motion for the absolutely brightest stars of types F to M is obtained. (Compare also § 11, Table XII.)

VAN DE KAMP[3] has derived the sun's velocity from the radial velocities of 104 stars of mean visual magnitude 9,2 within 50° from either the antapex or apex of solar motion. He found

$$S = 18 \pm 2{,}2 \text{ km/sec.}$$

HERTZSPRUNG[4] has determined the velocity of the sun from the radial velocities of 316 stars with small proper motions ($\mu < 0'',0205$; all B-stars rejected). He got

$$S = 18{,}5 \pm 1{,}0 \text{ km/sec.}$$

Excluding the very great and the very small values of $\varrho/\cos\lambda$, where ϱ is the radial velocity and λ the distance from apex, a somewhat larger value of S was obtained, namely

$$S = 19{,}2 \pm 1{,}0 \text{ km/sec.}$$

11. The Solar Motion as derived from the Different Spectral Types separately. In order to get a homogeneous material for the investigation of the motion of the stars a separate treatment of the different spectral types is of importance. The first extensive investigation of the solar motion as derived from the different spectral types is due to CAMPBELL[5]. His results are collected in table VII.

Table VII. Solar Motion and Spectral Type (CAMPBELL).

Type	No. Stars	S km/sec	K km/sec
B–B9	222	19,9	+3,9
B–B5	177	20,7	+4,7
B–B9	225	20,2	+4,1
A, B8–B9	222	16,8	+1,6
A	177	16,8	+1,0
F	185	15,8	+0,1
G	128	16,0	−0,2
G	123	13,9	+0,2
K	382	21,2	+2,8
K	369	18,9	+1,9
M	73	22,6	+3,9
M	70	20,2	+4,6

The first three solutions are those published in Lick Bull 195, p. 101 to 124. In the first solu-

[1] Mt Wilson Contr 281. Ap J 60, p. 50 (1924). [2] B A N 3, p. 71 (1925).
[3] Lick Bull 374, p. 98–99 (1926). [4] A N 208, p. 183 (1919).
[5] Lick Bull 196, p. 125–135 (1911).

tion the coordinates of the apex were assumed to be $\alpha = 270°,5$, $\delta = +34°,3$, as determined by L. BOSS[1] from the proper motions. As, however, the proper motions of types B8 and B9 resemble more closely those of types A than those of early B types and as further their distribution over the sphere points in the same direction, the 45 B8 and B9 stars were excluded and a new solution for the stars of types B—B5 was made. The third solution is performed with an assumed apex midway between the value of BOSS and that obtained by CAMPBELL from the radial velocities of all spectral types taken together. The solution is based on all the 225 stars of types B—B9 (including three stars accidentally omitted from solutions 1 and 2).

Concerning the other groups in the table, the fourth solution is based on the stars of types B8, B9 and A, assuming BOSS' apex. The other solutions are based upon equations for the individual stars, assuming the apex to be $\alpha = 270°$, $\delta = +30°$. For the types G, K and M two solutions were made. In the first solution all stars of each type are included, in the second those stars are excluded for which the peculiar radial velocity was in excess of 50 km/sec. We find that the exclusion of the abnormally high velocities has in every case reduced the velocity of the sun.

The lowest value of sun's velocity came out for the stars of type G. The velocity appears to increase as we pass towards the earlier spectral types in one direction and towards the later spectral types in the other direction.

GYLLENBERG, in the cited investigation, obtained the following elements of the solar motion for the different spectral types (with m. e.):

Table VIII. Solar Motion and Spectral Type (GYLLENBERG).

Type	No. Stars	*A*	*D*	*S* km/sec	*K* km/sec
B	284	286°,2 ± 2°,9	+25°,5 ± 2°,4	22,1 ± 0,9	+4,3
A	291	261 ,5 ± 4 ,8	14 ,5 ± 3 ,4	19,7 ± 1,4	+0,1
F	237	262 ,5 ± 6 ,6	17 ,3 ± 5 ,7	19,5 ± 2,0	+0,2
G	208	279 ,5 ± 7 ,4	23 ,8 ± 6 ,5	18,8 ± 2,2	−0,8
K	486	273 ,3 ± 5 ,2	31 ,1 ± 4 ,5	19,5 ± 1,6	+3,6
M	85	270 ,0 ± 12 ,4	28 ,4 ± 10 ,8	21,0 ± 3,7	+5,3

The solutions were performed by the aid of the weighted group method; not only the velocity of the sun but also the coordinates of the apex are obtained. The same correlation between spectral type and sun's velocity as found by CAMPBELL reappears here, the amplitude of the variation, however, is smaller.

In the determinations of GYLLENBERG and MALMQUIST the variation of sun's velocity with spectral type is less pronounced. Their results (with m. e.) are collected in the following table:

Table IX. Solar Motion and Spectral Type (GYLLENBERG, MALMQUIST).

Type	No. Stars	*A*	*D*	*S* km/sec	*K* km/sec
B	240	285° ± 3°	+36° ± 3°	23,4 ± 1,1	+4,6
A	588	262 ± 3	+21 ± 3	17,4 ± 1,2	0,0
F	314	280 ± 6	+25 ± 6	18,7 ± 2,0	+1,9
G	235	267 ± 8	+21 ± 8	17,3 ± 2,6	−0,4
K	664	273 ± 5	+33 ± 5	23,6 ± 1,6	+3,0
M	148	260 ± 10	+44 ± 10	19,9 ± 3,5	+1,0

FESSENKOFF[2] has also treated the different spectral types separately and the following elements of the solar motion (with m. e.) are obtained:

[1] A J 26, Nos. 12, 14 (1910). [2] Loc. cit.

Table X. Solar Motion and Spectral Type (FESSENKOFF).

Type	No. Stars	A	D	S km/sec
B	236	268°,7 ± 8°,1	+36°,9 ± 3°,8	22,7 ± 1,3
A	355	266 ,4 ± 6 ,2	+29 ,3 ± 0 ,5	14,1 ± 2,2
F	454	269 ,5 ± 1 ,3	+28 ,4 ± 3 ,1	19,8 ± 1,4
G	601	277 ,1 ± 4 ,1	+26 ,5 ± 1 ,4	19,9 ± 1,3
K	795	271 ,0 ± 3 ,6	+31 ,2 ± 3 ,0	19,1 ± 1,0
M	225	272 ,7 ± 5 ,0	+34 ,6 ± 5 ,8	22,5 ± 2,0

The small value of S obtained for the A-stars is noticeable. The comparatively large velocity for the G-stars may be due to the inclusion of the dwarf stars, which according to the method of rejection are practically eliminated in the first three investigations mentioned. From other investigations it is known that the dwarf stars give a larger value of the velocity of the sun than the giant stars of the same spectral type. (Compare STRÖMBERG, Ap J 56, p. 265. 1922.)

As a result we may establish that there seems to be such a correlation between the spectral type and the velocity of the sun that the velocity is largest for the B and M stars with a minimum for the intermediate types, but with regard to the mean errors of the determinations the observed correlation cannot be considered as significant.

In Lund Medd Ser. I, No. 108 a further division with respect to galactic latitude for the stars of types A, F, G and K was made. The resulting elements and their m. e. are as follows:

Table XI. Solar Motion in Terms of Spectral Type and Galactic Latitude.

Type		No. Stars	A	D	S km/sec
A	*g.*	328	251° ± 5°	+27° ± 5°	21,4 ± 1,6
	n.-g.	260	254 ± 5	+19 ± 5	13,1 ± 1,8
F	*g.*	182	297 ± 8	+13 ± 8	17,5 ± 2,7
	n.-g.	132	255 ± 9	+20 ± 9	21,2 ± 3,1
G	*g.*	126	268 ± 10	+21 ± 10	19,0 ± 3,5
	n.-g.	109	254 ± 11	+21 ± 11	11,5 ± 3,7
K	*g.*	372	259 ± 6	+35 ± 6	27,1 ± 2,0
	n.-g.	292	276 ± 7	+37 ± 7	21,7 ± 2,3

There is in general a good agreement in the apex positions for the different groups. The types A and G, however, show an abnormally small value of the velocity of the sun for the non-galactic regions. The large difference between the results from the galactic and non-galactic stars is not reflected in the results from the proper motions of the same stars, as shown in the cited investigation. The explanation of the observed divergence may be the existence of a stream so situated that the sun's motion, if determined from the radial velocities, is particularly influenced, but if determined from the proper motions practically unaffected.

STRÖMBERG[1] has treated the radial velocities of 1405 stars of spectral types F to M (the M-dwarfs being excluded). For the elements of the solar motion he got

$$A = 270°{,}9 \pm 3°{,}3, \quad D = +\,29°{,}2 \pm 3°{,}4, \quad S = 21{,}5 \pm 1{,}0 \text{ km/sec},$$
$$K = +\,0{,}36 \pm 0{,}60 \text{ km/sec}.$$

The material was further grouped according to absolute magnitude with the results given (with m. e.) in table XII. Here $\overline{M}$ is the mean absolute magnitude

[1] Mt Wilson Contr 144. Ap J 47, p. 7—37 (1918).

(distance unit 10 parsecs) in the group, $\overline{m}$ the mean apparent magnitude and θ the average peculiar radial velocity.

In an extensive investigation on the motion of the stars as derived from all available radial velocities STRÖMBERG[1] has made a still more minute division

Table XII. Solar Motion in Terms of Spectral Type and Absolute Magnitude.

Type	No. Stars	$\overline{M}$	$\overline{m}$	A	D	S km/sec	K km/sec	θ km/sec
F and G	211	0,31	4,68	251°,6 ± 6°,6	+22°,7 ± 6°,1	19,4 ± 1,8	+0,2 ± 1,8	11,4
	177	1,44	5,42	267 ,5 ± 11 ,0	+36 ,3 ± 12 ,4	16,6 ± 2,4	+0,0 ± 1,5	14,6
	167	2,76	5,27	271 ,6 ± 9 ,0	+34 ,6 ± 10 ,0	22,0 ± 2,8	−2,1 ± 1,7	16,2
	170	5,29	6,41	279 ,7 ± 9 ,5	(+ 7 ,4 ± 9 ,9)	(26,5 ± 4,1)	−1,9 ± 2,7	23,9
	725	2,32	5,40	268 ,1 ± 4 ,1	25 ,2 ± 4 ,9	19,94 ± 1,38	−0,69 ± 0,85	17,15
K	122	0,54	4,22	282 ,4 ± 9 ,2	+32 ,8 ± 8 ,8	23,6 ± 2,6	+ 2,0 ± 1,6	13,6
	245	1,41	4,86	268 ,6 ± 8 ,3	+36 ,5 ± 9 ,2	20,5 ± 2,4	+ 1,1 ± 1,4	16,6
	99	2,58	5,12	282 ,1 ± 9 ,9	+16 ,1 ± 9 ,4	26,4 ± 4,6	+ 5,8 ± 2,4	18,1
	79	7,07	7,41	(285 ,9 ± 28 ,9)	(−13 ,0 ± 34 ,4)	(15,3 ± 6,6)	(−11,1 ± 4,9)	26,5
	545	2,25	5,13	277 ,5 ± 5 ,5	+32 ,2 ± 5 ,8	22,23 ± 1,73	+1,05 ± 1,00	18,46
M	135	1,5	4,98	264 ,8 ± 7 ,4	+26 ,3 ± 8 ,2	27,0 ± 3,3	+1,25 ± 1,86	16,9

of the material. To obtain a grouping in classes of different velocity-dispersion, the apparent magnitude and the proper motion were used in combination with the spectral type. It was found that the velocity of the sun is not a constant vector, but shows a continous change with the velocity-dispersions of the groups. A more detailed report of this investigation will be given in the next section in connection with the results concerning the velocity distribution.

12. The *K*-term. In the year 1910 KAPTEYN and FROST[2] made a determination of the velocity of the sun from the radial velocities of the B-type stars which brought out a peculiar fact. It was found that the value of the velocity derived from the stars near the apex was very different from the value derived from the stars near the antapex. They found:

near apex . . $S = 18{,}4 \pm 1{,}4$ km/sec (p. e.) from 32 stars,
near antapex $S = 28{,}4 \pm 1{,}4$ km/sec (p. e.) from 29 stars.

One of the possible explanations was to assume the radial velocities of the B-stars to be affected with a systematical error amounting to about +5 km/sec. The existence of such an "error" was proved by CAMPBELL in his investigations and termed by him the K-term. In table VII of § 11 are given the values of this K-term for the different spectral types as derived by him. For the early B-stars the K-term is nearly equal to +5 km/sec in conformity with the results of KAPTEYN and FROST. When we pass through the spectral series, the effect decreases to a minimum in type G and afterwards increases again to a maximum for type M. The same systematic variation with spectral type was found by GYLLENBERG (§ 11, table VIII). The investigation of GYLLENBERG and MALMQUIST gave values consistent with those obtained before (§ 11, table IX) except for the M-stars for which the constant K is reduced to +1,0 km/sec. FREUNDLICH and V. D. PAHLEN[3] also find a small value, +1,9 km/sec and GERASIMOVIČ[4] finds a negative, but small value of the K-term for the M-stars.

The K-term may be due to different causes: to the use of inaccurate normal wave-lengths in the reduction of the spectrograms, to causes within the stars, to the EINSTEIN effect or to a real motion of the stars.

[1] Mt Wilson Contr. 293. Ap J 61, p. 363—388 (1925). [2] Ap J 32, p. 83—90 (1910).
[3] A N 218, p. 369—400 (1923). [4] A N 221, p. 163—168 (1924).

ALBRECHT[1] has discussed the normal wave-lengths of the lines upon which the B-type velocities are based. On the basis of new wave-lengths for oxygen, nitrogen, silicon and helium the total reduction in the K-term would amount to about 2 km for B0—B2, 1 km for B3, and 0,3 km for B5 to B8.

Many of the solar lines have been found to have a slight displacement towards the red, which was first attributed to pressure in the reversing layer. Through the investigation of EVERSHED[2] it was shown that a pressure-effect could not account for the observed displacements, but that they might very well be explained by rapid convection currents. The maximum observed displacements were of the order of one km per sec.

CAMPBELL[3] has pointed out that it would not be surprising to find that this effect in stars of type B amounts to 4,5 km per sec corresponding to the amount of the K-term for this type. The recent results of ST. JOHN and ADAMS[4] lend support to CAMPBELL's hypothesis that downward convection currents may account — at least in part — for the K-term in the B-stars.

Another explanation of the K-term is founded on the theory of EINSTEIN. According to this theory there must be a displacement of the spectral lines towards the red for all stars depending on the mass and density of the stars by the formula[5]

$$K = 0{,}634\, M^{\frac{2}{3}}\, \delta^{\frac{1}{3}}\ \text{km/sec}, \tag{26}$$

where the mass M and the density δ are expressed in units of the sun.

Assuming with DE SITTER $M = 10$, $\delta = \frac{1}{10}$ for the B-stars, we get $K = +1{,}4$ km/sec, which is only one third of the observed value. Thus the EINSTEIN effect alone does not account for the great value of the K-term for the B-stars.

There seems to be a connection between the K-term and the absolute magnitude of the B-stars. LUDENDORFF[6] has shown that for stars of type B the K-term is a function of the star's place in the classification of LOCKYER, a fact confirmed by GERASIMOVIČ[7]. The absolutely brightest B-stars show the strongest K-effect.

The K-term may also be interpreted as a real motion of the stars. A positive value of K would then signify that the stars on the average are moving out from our sun. GYLLENBERG[8] found for the stars of types B and M a distinct relation between the constant K and the angular distance from the vertices, K was found to be larger in the vertex directions. A similar result was obtained by FREUNDLICH and v. D. PAHLEN[9] for the B- and M-stars and also for those stars of types A, F, G and K for which the proper motions were smaller than 0″,04.

The various possible explanations of the K-term have been briefly alluded to here. We must, however, for the present consider this question as one of the unsolved astronomical problems.

d) The Distribution of Stellar Velocities as derived from the Radial Motions.

13. Introduction. Suppose the three linear velocity components of a star to be known and, besides, the elements of the solar motion. We are then able to compute the linear peculiar velocity of the star. [Compare sec. c, § 7 for-

[1] Ap J 55, p. 361 (1922); 57, p. 57 (1923). [2] Kodaikanal Obs Bull 36 (1913).
[3] Lick Bull 257, p. 82 (1914).
[4] Mt Wilson Contr 279. Ap J 60, p. 43—49 (1924).
[5] DE SITTER, M N 76, p. 719 (1916). [6] A N 190, p. 197 (1912).
[7] Bull. de l'Académie Impériale des Sciences, St. Petersbourg (1916).
[8] Lund Medd Ser. II, No. 13 (1915). [9] A N 218, p. 369 (1923).

mula (4) and (18).] The distribution of the linear peculiar velocities for a group of stars we will term the velocity-distribution of the group.

The earlier investigations concerning the velocity-distribution were based on the proper motions alone, as no reliable radial velocities and distances were available at that time. It was necessary to make an assumption concerning the form of the velocity-distribution and to try to represent the observed proper motions through this assumption.

The most simple assumption was that the peculiar velocities are distributed at random or "haphazard". Through the investigations of KOBOLD[1], however, it became evident that the observed proper motions could not be represented by this simple assumption. The analysis decidedly indicated a preferential motion of the stars along a line parallel with the Milky Way.

From the proper motions of the AUWERS-BRADLEY Catalogue KAPTEYN[2] deduced distinct evidence of such a systematic motion, and formulated his well-known hypothesis of the two star-streams. According to him the systematic arrangement of the stars' motions is accounted for by the assumption that the stars form two great streams moving in opposite directions, one stream moving towards the vertex (vertex I), the other towards the anti-vertex (vertex II). KAPTEYN's conclusions were supported by subsequent investigations, especially those of EDDINGTON and DYSON.

But certain peculiarities in the observed motions remained unaccounted for by the two-stream hypothesis. In order to smooth out the still remaining discrepancies HALM[3] introduced a third stream designated as Drift O. It was found that this stream had no systematic motion in relation to the centroid of the stars, and that it included all the stars of the spectral type B, but that it was by no means exclusively confined to this type alone.

The observed systematic motion of the stars may, however, be accounted for without assuming two or more star streams, as was shown by SCHWARZSCHILD[4]. Instead of the old hypothesis of a random distribution he assumed that on the average the components of the peculiar motion are greater along a certain line than along any other. This line coincides with the line joining the two vertices of KAPTEYN. This hypothesis is termed the ellipsoidal hypothesis on account of the form of the frequency-function which is chosen to represent the peculiar velocities.

A generalization of this hypothesis was given by CHARLIER[5], who, according to the theory of mathematical statistics as developed by him, derived the general form of the frequency-function of the velocities of the stars. CHARLIER's theory may be termed the generalized ellipsoidal theory. The ellipsoidal hypothesis is a special case of the generalized theory, and so is the two star stream hypothesis too. In the cited investigation the generalized theory was applied to the proper motions. It was shown that there was no positive evidence for accepting the hypothesis of two star-streams as an explanation of the distribution of the observed proper motions of the stars.

14. The Mathematical Expressions for the Different Hypotheses. To get a better survey of the different hypotheses, we will put them into mathematical forms. The best mathematical equivalent of a random distribution of the velocities of the stars is a distribution according to MAXWELL's law.

---

[1] A N 125, p. 65. (1890); Halle Nova Acta 64, No. 5 (1895).

[2] Brit. Ass. Rep. 1905, p. 257.

[3] M N 71, p. 610 (1911).

[4] Göttinger Nachrichten p. 191 (1908).

[5] Lund Medd Ser. II, No. 9 (1913).

Let U, V, W be the components in a system of rectangular coordinates of the peculiar velocity of a star and denote by $F(U, V, W)$ the frequency-function of the velocities. Thus

$$F(U, V, W)\, dU\, dV\, dW$$

is the number of stars with components of velocity between $U \pm \frac{1}{2} dU$, $V \pm \frac{1}{2} dV$, and $W \pm \frac{1}{2} dW$.

For a random distribution we get

$$F(U, V, W) = \frac{N}{\sigma^3 \sqrt{(2\pi)^3}} e^{-\frac{1}{2\sigma^2}(U^2+V^2+W^2)}, \tag{27}$$

where N is the total number of stars and σ the dispersion in the velocity-components. Instead of the dispersion, σ, the modulus of precision, h, is often used. Between these two quantities we have the relation

$$h = \frac{1}{\sigma\sqrt{2}}. \tag{28}$$

The dispersion, however, is to be preferred on account of its signification. We have

$$\sigma^2 = \overline{U^2} = \overline{V^2} = \overline{W^2}. \tag{29}$$

The square of the dispersion is simply equal to the mean value of the squares of the velocity-components.

The mean velocity in space, Ω, is related to the dispersion by the equation

$$\Omega = \sigma \sqrt{\frac{8}{\pi}}, \tag{30}$$

and the average velocity, Θ, in a given direction by

$$\Theta = \sigma \sqrt{\frac{2}{\pi}}. \tag{31}$$

Consequently we have

$$\Omega = 2\Theta. \tag{32}$$

According to the two star-stream hypothesis the stars are divided into two great streams. Suppose for the one stream the total number of stars to be N_1 and the velocity components of the stream to be U_1, V_1, W_1 and for the other N_2, U_2, V_2, W_2 resp. Subtracting the velocity components of the stream from the components of the peculiar velocities, we get the components of the internal motions. These internal motions are supposed to have a random distribution, and thus we get

$$F(U, V, W) = F_1(U, V, W) + F_2(U, V, W), \tag{33}$$

where

$$F_1 = \frac{N_1}{\sigma_1^3 \sqrt{(2\pi)^3}} e^{-\frac{1}{2\sigma_1^2}[(U-U_1)^2+(V-V_1)^2+(W-W_1)^2]} \tag{34}$$

and

$$F_2 = \frac{N_2}{\sigma_2^3 \sqrt{(2\pi)^3}} e^{-\frac{1}{2\sigma_2^2}[(U-U_2)^2+(V-V_2)^2+(W-W_2)^2]}. \tag{35}$$

In the three stream hypothesis another stream

$$F_0 = \frac{N_0}{\sigma_0^3 \sqrt{(2\pi)^3}} e^{-\frac{1}{2\sigma_0^2}[U^2+V^2+W^2]} \tag{36}$$

is added, so that

$$F(U, V, W) = F_0 + F_1 + F_2. \tag{37}$$

The ellipsoidal distribution is a modification of the MAXWELLIAN distribution. In a special system of coordinates — the vertex system — where one of the axes (we assume the U-axis) is directed towards the vertex, the frequency function has the form

$$F(U, V, W) = \frac{N}{\sigma_3 \sigma_2 \sigma_1 \sqrt{(2\pi)^3}} e^{-\frac{U^2}{2\sigma_3^2} - \frac{V^2}{2\sigma_2^2} - \frac{W^2}{2\sigma_1^2}}. \tag{38}$$

Here the dispersions along the three principal axes are different. To get the surfaces of equal frequency, we have to put the exponent in (38) equal to a constant. We then get

$$\frac{U^2}{\sigma_3^2} + \frac{V^2}{\sigma_2^2} + \frac{W^2}{\sigma_1^2} = C, \tag{39}$$

which is the equation of an ellipsoid. Thus the surfaces of equal frequency are ellipsoids, from which the hypothesis has taken its name. For $C = 1$ (39) is the equation of the velocity ellipsoid.

SCHWARZSCHILD assumed $\sigma_1 = \sigma_2$, viz. that the ellipsoid was a prolate spheroid, and the same assumption was made by CHARLIER in his first treatment of the proper motions. The solution of the problem on the assumption of three unequal axes was first performed for the radial velocities by GYLLENBERG[1] and for the proper motions by WICKSELL[2].

In the generalized hypothesis of CHARLIER the frequency function of the peculiar velocities has the form

$$F(U, V, W) = \varphi(U, V, W) + \sum A_{ijk} \frac{\partial^{i+j+k} \varphi}{\partial U^i \partial V^j \partial W^k}, \quad (i + j + k \geqq 3) \tag{40}$$

where — in the vertex system — $\varphi(U, V, W)$ has the form (38). The constants of the function $\varphi(U, V, W)$ and the A_{ijk} are termed the characteristics of the frequency function. These characteristics are determined from the moments N_{ijk} which are defined through

$$N_{ijk} = \frac{1}{N} \iiint_{-\infty}^{+\infty} U^i V^j W^k F(U, V, W)\, dU\, dV\, dW \tag{41}$$

or — in practice —

$$N_{ijk} = \text{Mean of } (U^i V^j W^k) = \overline{U^i V^j W^k}. \tag{42}$$

For $A_{ijk} = 0$ we get the ellipsoidal hypothesis.

15. The Correlation between Velocity and Spectral Type. One of the first remarkable results from the radial velocities was the definitive statement of a distinct correlation between linear velocity and spectral type. As early as 1892 it was pointed out by MONK[3] that the proper motions of the stars of later spectral types are on the average larger than those of the earlier types. The explanation of this phenomenon is not necessarily the existence of a correlation between spectral type and linear velocity, it may be wholly accounted for by greater parallaxes on the average for the stars of later spectral types. In 1910, however, CAMPBELL[4] and KAPTEYN[5] pointed out independently that the average radial velocity, and consequently also the average linear speed, increases continually as we pass through the spectral series in the order B to M.

---

[1] Lund Medd Ser. I, No. 59. (1914); Ser. II, No. 13 (1915).
[2] Lund Medd Ser. I, No. 60. (1914); Ser. II, No. 12 (1915).
[3] Astronomy and Astrophysics 11, p. 874 (1892); 12, pp. 8, 513 (1893).
[4] Lick Bull 196 (1911). [5] Mt Wilson Contr 45. Ap J 31, p. 258 (1910).

Table XIII contains the average radial velocities for the different spectral types as deduced by CAMPBELL[1] and by GYLLENBERG[2].

Table XIII. Average Radial Velocity and Spectral Type.

Spetral type	CAMPBELL		GYLLENBERG	
	No. of Stars	Average Rad. Veloc. km/sec	No. of Stars	Average Rad. Veloc. km/sec
B	225	6,5	247	7,0 ± 0,3
A	177	10,9	263	11,7 ± 0,6
F	185	14,4	237	14,4 ± 0,7
G	128	15,0	208	15,8 ± 0,8
K	382	16,8	486	15,9 ± 0,5
M	73	17,1	85	17,2 ± 1,3

16. The Radial Velocities and the Star-Stream Hypotheses. As early as 1909 HOUGH and HALM[3] made an attempt to test the hypothesis of the two star-streams by the aid of the radial velocities. The material consisted of 166 stars observed at the Cape with 45 additional stars from other sources, in all 211 stars comprised within the parallel +30° and the south pole. Besides these, CAMPBELL's 280 stars were used. (Compare § 9.) They concluded that the discordances indicated in the radial velocity investigation can be completely reconciled with the hypothesis of the two star-streams by the additional hypothesis of a variability in the density of mixture of the two streams. The conclusions, however, are based on a scanty material and cannot therefore be regarded as very trustworthy.

In the cited investigation of CAMPBELL he also examined the peculiar radial velocities in order to see how they would bear upon the question of the two star-streams. The position of the vertices, as determined from the proper motions, are[4]

$$\text{Stream I,}\ \alpha = 18^h\,12^m,\ \delta = -12°\,;$$

$$\text{Stream II,}\ \alpha = 6^h\,12^m,\ \delta = +12°\,.$$

1193 peculiar radial velocities were tabulated in terms of the angular distances of the stars from the vertex of Stream II, and the results are given in table XIV.

The average radial velocity of the stars midway between the two vertices is found to be less than the average velocity at the vertices, thus indicating a preferential motion towards and away from the vertices. But quantitatively the preference is very much smaller than the one deduced from the proper motions. In a special investigation of the B-stars[5] it was found that these stars do not show any preferential motion with reference to the assumed vertices.

Table XIV. Average Radial Velocity in Terms of Angular Distance from Vertex.

Distance from Vertex	No. of Stars	Average Rad. Veloc. km/sec
0°– 30°	94	13,7
30 – 60	188	14,6
60 – 90	269	12,8
90 –120	296	11,9
120 –150	251	14,5
150 –180	99	14,6

[1] Loc. cit. [2] Lund Medd Ser. II, No. 13.
[3] M N 70, p. 85 (1909).
[4] Mt Wilson Comm 1, p. 2 (1915).
[5] Lick Bull 195 (1911).

The peculiar radial velocities were also classified in terms of galactic latitude:

Table XV. Average Radial Velocity in Terms of Galactic Latitude.

Gal. Lat.	±90° – ±60°		±60° – ±30°		±30° – 0°	
Type	No.	km/sec	No.	km/sec	No.	km/sec
B	7	5,4	27	5,6	191	7,1
A	18	5,6	61	9,2	98	13,0
F	23	12,6	56	12,4	107	15,3
G	11	11,3	46	15,2	71	15,3
K	44	13,8	109	17,4	229	17,1
M	12	17,7	26	19,2	35	15,9

The radial velocities for the A-stars near the Milky Way are appreciably greater than the radial velocities of the stars in high galactic latitudes. The results for the types F, G and K point in the same direction, though less markedly, but the M-stars do not show any dependence of radial velocities upon galactic latitude. The results for the B-stars point in the same direction as those for the A-stars, but the number of B-stars in high galactic latitudes is too small to give any reliable results.

KAPTEYN and ADAMS[1] grouped the peculiar radial velocities of one thousand stars of spectral types F to M both according to distance (λ) from the nearest vertex and according to amount of proper motion. The results of this grouping are given in table XVI. $\bar{\varrho}_1$ and $\bar{\varrho}_2$ are the average values of the peculiar radial velocity.

Table XVI. Average Values of the Peculiar Radial Velocity ϱ in km.

Type	Proper Motion	λ 60° to 90°		λ 0° to 49°		$\bar{\varrho}_2/\bar{\varrho}_1$	
		No.	$\bar{\varrho}_1$	No.	$\bar{\varrho}_2$	Obs.	Comp.
F	0″,000 to 0″,029	25	8,9	22	13,6	1,53	1,63
	0 ,030 „ 0 ,069	9	8,5	11	12,5	1,47	1,56
	0 ,070 „ 0 ,149	17	12,3	17	20,5	1,67	1,58
	0 ,150 „ 0 ,249	22	11,5	12	18,9	1,64	1,55
	0 ,250 „ 0 ,499	24	16,7	8	21,1	1,26	1,54
	≧0″,500	30	18,0	18	34,9	1,93	1,39
G	0″,000 to 0″,026	37	6,9	32	12,8	1,86	1,55
	0 ,027 „ 0 ,049	10	9,6	13	12,8	1,33	1,53
	0 ,050 „ 0 ,099	18	9,5	7	12,3	1,30	1,63
	0 ,100 „ 0 ,499	13	15,0	11	23,6	1,58	1,41
	≧0″,500	51	22,9	31	45,6	1,99	1,51
K	0″,000 to 0″,025	47	10,9	39	12,3	1,13	1,52
	0 ,026 „ 0 ,039	37	12,2	24	11,9	0,98	1,47
	0 ,040 „ 0 ,059	21	14,9	21	21,3	1,43	1,49
	0 ,060 „ 0 ,079	24	10,7	16	15,1	1,41	1,50
	0 ,080 „ 0 ,099	26	15,7	10	14,0	0,89	1,52
	0 ,100 „ 0 ,119	12	20,7	15	22,6	1,09	1,58
	0 ,120 „ 0 ,149	24	16,5	10	23,8	1,44	1,43
	0 ,150 „ 0 ,199	24	15,6	10	24,5	1,57	1,45
	0 ,200 „ 0 ,299	18	16,5	13	24,3	1,47	1,56
	0 ,300 „ 0 ,599	12	33,3	7	47,0	1,41	1,52
	≧0″,600	25	23,5	28	23,0	0,98	1,52
M	0″,000 to 0″,029	18	12,9	15	18,1	1,40	1,52
	0 ,030 „ 0 ,089	19	15,0	11	21,0	1,40	1,52
	0 ,090 „ 0 ,499	14	17,9	7	28,9	1,61	1,56
	≧0″,500	4	53,2				

[1] Mt Wilson Comm 1; Wash Nat Ac Proc 1, p. 14 (1915).

We find that the average velocity of the stars near the vertices is considerably greater than that of the other stars, as is seen from the values of the ratio $\bar{\varrho}_2/\bar{\varrho}_1$ in the seventh column. This same ratio can be calculated from the elements of the two star-streams as derived from the proper motions. Using the theory developed by KAPTEYN[1], the corresponding values of the ratio were calculated and given in the last column of the table. With the exception of some groups of type K the agreement of the observed and the theoretical values is satisfactory. There is no systematic change in the values of $\bar{\varrho}_2/\bar{\varrho}_1$ with the amount of proper motion, which proves that the preferential motion is not limited to the nearest stars, but extends to very great distances.

From the table it is seen that the average value of ϱ increases with the proper motions. This may be explained in any of the following ways:

1. The peculiar velocity of the stars decreases with the distance.
2. The absolutely bright stars move more slowly than the fainter ones.
3. The distribution of the velocities of the stars is not in accordance with MAXWELL's law, the large velocities being in excess.

The third explanation was tested for the K-stars in the following way. The stars between $\lambda = 60°$ and $\lambda = 90°$ were arranged according to amount of radial velocity. It was found that the distribution of these velocities did not agree with MAXWELL's law, but could be expressed as the sum of two MAXWELLIAN distributions with different dispersions. Such a distribution explains the change of velocity with proper motion in a satisfactory manner.

They also tried to determine the absolute magnitude effect and found an indication of a change of radial velocity per unit of absolute magnitude of 1,1 km in the direction that the bright stars move more slowly.

HALM[2] has investigated the radial velocities of the 1888 stars in VOÛTE's catalogue. The peculiar radial velocities were arranged into groups with reference to the vertex. The argument of the grouping was cos λ, where λ denotes the angle between star and vertex. In each of these groups the numbers of stars were counted whose velocities lie between 0—5, 5—10, etc. km/sec and curves were plotted with the velocities as abscissae and the percentage numbers as ordinates. An analysis of these curves showed that the random or ellipsoidal hypotheses could not be accepted even as a rough approximation. The agreement between the observed distribution and that computed from the two-stream hypothesis was found to be very unsatisfactory too. The introduction of a third stream, Drift O, led to a satisfactory representation of the observed distributions, if allowance was made for systematic differences for the large velocities, where the observed numbers as a rule exceeded the computed values, so that about 5 per cent of the stars were not accounted for.

Of the remaining stars 35 per cent were found to belong to stream I and 35 per cent to stream II. Each of these streams was found to have a stream-velocity equal to 16,4 km/sec and an internal velocity dispersion amounting to 15,4 km/sec. Drift O included 25 per cent of the stars. The stars of this drift appeared to possess a common motion of recession amounting to about 4 km/sec, which motion is identical with the K-term. The velocity dispersion was found to be 7,7 km/sec. The results also supported the conclusion already drawn from the proper motions that the B-stars are closely associated with this third stream, but that the other types too are represented.

17. The Radial Velocities and the Ellipsoidal Hypothesis. In the cited investigations of CAMPBELL and of KAPTEYN and ADAMS it was shown that the average radial velocity was greater near the vertices than at right angles to

[1] M N 72, p. 743 (1912). [2] M N 80, p. 682—692 (1920).

the line joining the two vertices. This result was interpreted as a confirmation of the two-stream hypothesis, but it may as well be interpreted as a confirmation of the ellipsoidal hypothesis. This hypothesis is the most convenient for the mathematical discussion of the radial velocities. The problem is to determine, from the observed radial motions, the magnitudes and the directions of the three principal axes of the velocity-ellipsoid.

This problem was, as already mentioned in § 14, first solved by GYLLENBERG but later on the method of solution has been simplified by CHARLIER. For the details of this method we refer to CHARLIER's California Lectures. Here only a short synopsis of the principle will be given.

We assume that the observed radial velocities have been corrected for the solar motion and for the K-term (supposed to represent a systematic error in the measured radial velocities). For each star we thus get an equation of the form (12)

$$\varrho_1 = \gamma_{13} U_1'' + \gamma_{23} V_1'' + \gamma_{33} W_1'',$$

where ϱ_1 is the peculiar radial velocity. Squaring this equation we get

$$\left.\begin{aligned} \varrho_1^2 = \gamma_{13}\gamma_{13} U_1''^2 + \gamma_{23}\gamma_{23} V_1''^2 + \gamma_{33}\gamma_{33} W_1''^2 + 2\gamma_{13}\gamma_{23} U_1'' V_1'' \\ + 2\gamma_{23}\gamma_{33} V_1'' W_1'' + 2\gamma_{33}\gamma_{13} W_1'' U_1'' \end{aligned}\right\} \tag{43}$$

for each star. The equations (43) are then solved through the method of least squares, and thus the mean values of $U_1''^2$, $V_1''^2$, $W_1''^2$, $U_1'' V_1''$, $V_1'' W_1''$, and $W_1'' U_1''$ are obtained. But according to the definition (42) of the moments we have

$$\begin{aligned} N_{200} &= \overline{U_1''^2}, & N_{110} &= \overline{U_1'' V_1''}, \\ N_{020} &= \overline{V_1''^2}, & N_{011} &= \overline{V_1'' W_1''}, \\ N_{002} &= \overline{W_1''^2}, & N_{101} &= \overline{W_1'' U_1''}; \end{aligned}$$

where the N_{ijk} are the second order moments $(i + j + k = 2)$ in the system K_3.

With these moments we form the equation of JACOBI

$$J(t) = \begin{vmatrix} N_{200} - t, & N_{110}, & N_{101} \\ N_{110}, & N_{020} - t, & N_{011} \\ N_{101}, & N_{011}, & N_{002} - t \end{vmatrix} = 0. \tag{44}$$

The three roots of this equation are equal to the square of the three principal axes of the velocity ellipsoid, σ_1^2, σ_2^2, σ_3^2.

The direction cosines (l, m, n) of the axes are given by the equations:

$$\frac{l}{J_{11}} = \frac{m}{J_{12}} = \frac{n}{J_{13}} = \frac{1}{\sqrt{J_{11}^2 + J_{12}^2 + J_{13}^2}}, \tag{45}$$

where J_{ij} denotes the subdeterminant of $J(t)$.

The results obtained by GYLLENBERG[1] are given in table XVII.

Table XVII. The Axes of the Velocity Ellipsoids and their Directions.

Type	No. Stars	σ_3 km/sec	α_3	δ_3	σ_2 km/sec	α_2	δ_2	σ_1 km/sec	α_1	δ_1
B and A . .	510	14,7	280°,1	−19°,5	8,2	304°,2	+68°,8	12,9	192°,9	+ 8°
F and G . .	445	23,5	253 ,7	+ 0 ,5	14,3	344 ,1	+53 ,0	17,4	163 ,3	+37,9
K and M . .	571	21,4	248 ,9	−14 ,1	19,0	275 ,2	+76 ,0	20,2	158 ,5	+ 2,2
All	1526	20,1	264 ,1	− 5 ,2	14,5	336 ,5	+59 ,5	17,9	177 ,1	+29,9

σ_3, α_3, δ_3 give the magnitude and direction of the largest axis. The direction coincides approximately with the vertex as deduced from the two-stream hypo-

[1] Lund Medd Ser. II, No. 13.

thesis. The next three columns give the length and the direction of the shortest axis. It is remarkable that the shortest, as well as the longest, axis coincides approximately with the plane of the Milky Way. The last three columns contain the intermediate axis, which is directed towards the pole of the Milky Way.

A similar result is obtained by CHARLIER in his California Lectures[1]. He got, from 1986 stars with known radial velocities:

$$\begin{aligned} \sigma_3 &= 19{,}5 \text{ km/sec}, & l_3 &= 341°{,}2, & b_3 &= -\ 5°{,}7, \\ \sigma_2 &= 13{,}4 \text{ km/sec}, & l_2 &= \ 69°{,}6, & b_2 &= -16°{,}2, \\ \sigma_1 &= 15{,}6 \text{ km/sec}, & l_1 &= 270°{,}0, & b_1 &= +72°{,}8. \end{aligned}$$

As the velocity distribution is found to be closely related to the plane of the Milky Way, it is convenient directly to determine the moments in a galactic system of coordinates. This was done by GYLLENBERG, who used a galactic system defined in the following way: The positive Z-axis is directed towards the pole of the Milky Way ($\alpha = 191°{,}2$, $\delta = +28°{,}2$), the positive X-axis is directed towards a point $\alpha = 274°{,}6$, $\delta = -12°{,}0$ situated in the Milky Way and approximately coinciding with the principal vertex. The dispersions of the peculiar velocities in the three directions, which will be practically identical with the axes of the velocity ellipsoid, were computed from the radial velocities and found to be:

Table XVIII. The Dispersions of the Velocities in the three Main Directions in the Galactic System.

Type	No. Stars	σ_x km/sec	σ_y km/sec	σ_z km/sec	Ratio
B	247	8,8 ± 2,6	9,5 ± 1,0	6,0 ± 0,8	(1,52)
A	263	18,6 ± 1,2	11,1 ± 3,6	12,7 ± 2,5	0,64
F	237	21,5 ± 1,6	16,0 ± 1,4	15,7 ± 3,3	0,74
G	208	23,3 ± 1,8	12,5 ± 5,0	21,5 ± 4,2	0,73
K	486	21,2 ± 0,9	18,9 ± 0,8	19,7 ± 1,8	0,91
M	85	20,7 ± 4,9	22,9 ± 12,5	21,0 ± 9,5	1

Adopting the spheroid as a sufficiently close approximation, we may take the mean of the too nearly equal axes and compute the ratio of this mean to the third axis. This ratio is given in the last column of table XVIII.

With the exception of the stars of type B the velocity dispersions agree well with the solutions of the velocity ellipsoid given in table XVII. The stars of type B are found to have a circular distribution in the plane of the Milky Way, the velocity distribution seems to be an oblate spheroid, flattened in the plane of the Milky Way. For the remaining types it seems as if the dispersion in the direction of the principal vertex were constant and as if the mean velocities in the other directions gradually increased so that for the M-stars the velocity distribution is almost spherical.

In order to examine if there was any marked deviation from the ellipsoidal hypothesis, the moments of the third and fourth orders referred to the galactic system of coordinates were computed and the following values of the skewness (S) and the excess (E) were obtained:

Table XIX. The Values of the Skewness (S) and the Excess (E). All Stars $\leqq 4^m{,}9$.

	X	Y	Z
S	−0,13 ± 0,11	+0,01 ± 0,11	+0,01 ± 0,14
E	+0,00 ± 0,04	+0,11 ± 0,04	+0,12 ± 0,04

[1] Loc. cit. p. 26.

S and E are computed from the relations

$$S_x = -\frac{N_{300}}{2\,(N_{200})^{\frac{3}{2}}}\,, \qquad E = \frac{1}{8}\left[\frac{N_{400}}{(N_{200})^2} - 3\right].$$

We find a negative skewness in the vertex direction and a positive excess along the two other axes. The mean errors, however, are large, and we can only conclude that the deviations — if real — from the ellipsoidal hypothesis are small.

The method of determining the magnitudes and directions of the axes of the velocity-ellipsoid given by EDDINGTON and HARTLEY[1] is — in reality — not very different from the method developed above. The main difference is that instead of the square (ϱ_1^2) of the peculiar radial velocity, the peculiar radial velocity itself — without regard to sign — is used at the right hand side of equation (43). They consider that through this procedure the equations are given more adequate weights. The investigation was based on the radial velocities published by CAMPBELL in Lick Bull 211 and 229. Their results are collected in table XX.

Table XX. The Axes of the Velocity Ellipsoids and their Directions.

Type	No. Stars	σ_3	α_3	δ_3	σ_2	α_2	δ_2	σ_1	α_1	δ_1	Ratio
A	212	16,0	278°,7	−15°,7	9,8	342°,8	+56°,9	6,9	197°,3	+28°,3	0,52
F	199	20,8	272 ,9	−16 ,8	12,4	65 ,9	−71 ,3	9,2	180 ,4	− 8 ,0	0,52
G	152	30,0	248 ,8	+13 ,5	18,6	1 ,2	+57 ,9	11,8	151 ,3	+28 ,5	0,51
K−K2	366	20,8	281 ,3	−19 ,9	16,9	66 ,1	−66 ,1	11,8	186 ,7	−12 ,7	0,69
K5−M	157	24,6	280 ,2	− 7 ,7	18,0	11 ,6	+ 9 ,8	12,7	152 ,6	−77 ,4	0,62

From the table it is seen that the direction of the greatest axis agrees well with the position of the vertex derived from proper motions, except for type G, which shows a large deviation. For type A the least axis points to the galactic pole, but for the other types this axis seems to have no distinct relation to the Milky Way. Adopting a spheroid as a sufficient representation, the mean of the two smaller axes was taken to represent the minor axis of the spheroid. The ratio of this to the major axis is given in the last column of the table. These ratios are throughout smaller than those computed from GYLLENBERG's results. This fact may be due to the different methods of weighting the equations and to the different modes of exclusion of large velocities. EDDINGTON and HARTLEY omitted two stars only (with velocities 229 and 158 km/sec resp.), GYLLENBERG rejected 44 stars whose velocities exceeded 66,3 km/sec.

The stars of the spectral type A were divided into two groups, one including A0 and Ap, the other the rest. These two groups showed quite different velocity distributions. In the former group the two greatest axes were found to be nearly equal (14,1 and 11,6 km/sec), the least axis (5,7 km/sec) was directed to the galactic pole. The other group showed a well defined vertex direction and a very prolated spheroid.

The later spectral types were divided into giant and dwarf stars, and it was shown that the velocities of the dwarfs were much greater than those of the giants, especially in the direction of the vertices.

In an investigation based on the radial velocities of the stars of spectral types F to M STRÖMBERG[2] has divided the stars into three groups with respect to the absolute magnitudes. The average radial velocity is namely more closely

[1] M N 75, p. 521—530 (1915).
[2] Mt Wilson Contr 144; Ap J 47, p. 7—37 (1918).

related to absolute magnitude than to spectral type. In table XXI the data for the three groups are given. Here M and m are the absolute and the apparent magnitude, and θ the average radial velocity.

Table XXI. Groups for Study of Preferential Motion.

Group	No. Stars	$\overline{M}$	$\overline{m}$	θ km
I	509	0,76	4,83	13,11
II	513	2,08	5,09	17,13
III	260	6,05	6,82	25,88

The velocity-distributions of these three groups were studied with the aid of a general expression for a surface representing a continuous variation of the average radial velocity with direction. The results from the terms of even degree are given in table XXII.

The maximum axis, which can be assumed to be identical with the axis of preferential motion in the ellipsoidal and two-stream hypotheses, lies in all cases near the galactic equator, the intermediate axis lies in all cases nearest to the galactic pole. These results are in accordance with those obtained by GYLLENBERG.

Table XXII. Vertices of Motion. Terms of Even Degree.

Group	Maximum axis			Minimum axis			Third direction		
	θ km	α	δ	θ km	α	δ	θ km	α	δ
I	16,1	98°,0	+ 5°,1	10,4	236°,0	+83°,2	12,4	187°,7	− 4°,5
II	21,0	85 ,9	+10 ,3	14,1	348 ,4	+35 ,7	16,8	189 ,4	+52 ,3
III	32,4	99 ,6	+34 ,1	17,3	303 ,7	+53 ,4	25,6	197 ,6	+11 ,7

Adding the assymmetrical terms, it is found that the two axes of maximum radial velocity are not quite in a straight line. There is an indication that the stars in question are mainly moving around the center of the local system as determined by CHARLIER and WALKEY, with a preferential motion in the galactic plane.

A similar investigation of the stars of spectral type A has been performed by SHAJN[1]. The investigation was based on VOÛTE's catalogue, and the material was treated in three separate combinations: (1) all stars of type A (362 stars); (2) the stars of types A1—A9 (203 stars); (3) stars of type A0 (including Ap, 159 stars). Some stars with radial velocities larger than 40 km per sec were omitted as well as stars with common motion. The results concerning the principal axes are given in table XXIII.

Table XXIII. Vertices of Motion. Terms of Even Degree.

Type	Maximum axis.			Minimum axis			Third direction		
	θ km	α	δ	θ km	α	δ	θ km	α	δ
A	14,4	100°,0	+23°,0	6,3	192°,5	+17°,0	10,5	312°,1	+63°,2
A1—A9	16,2	98 ,6	+18 ,5	5,9	200 ,5	+29 ,0	9,6	340 ,6	+55 ,1
A0	15,6	72 ,7	+62 ,3	4,9	198 ,4	+21 ,0	10,8	298 ,1	+20 ,4

The greatest and the mean axes lie in the galactic plane, and the least axis is consequently directed towards its pole. The different behaviour of the early and the late A-stars as found by EDDINGTON and HARTLEY is also indicated here.

[1] M N 83, p. 338—340 (1923).

18. The Asymmetry of Stellar Motion. In the investigations concerning the motion of the stars which we have hitherto been dealing with, the very high velocities — as a rule all stars with velocities greater than about 60 km per sec — have been excluded. These large velocities, however, are of very great interest as has been shown during the last few years. As early as 1914 it was found by ADAMS and KOHLSCHÜTTER[1] that the large peculiar velocities in the northern hemisphere were predominantly negative. In an investigation by B. BOSS[2], based on space-velocities, it was shown that the stars of high velocities had a marked tendency to move towards a region limited by galactic longitudes 140°—340°, a result independently found by ADAMS and JOY[3]. STRÖMBERG[4] found that out of about 100 stars of space-velocity larger than 100 km per sec. not a single one had its apex between galactic longitudes 334° and 143°.

The asymmetry of the large motions as indicated by the radial velocities is illustrated in fig. 1. This figure is reproduced from OORT's[5] investigation on the stars of high velocity. The figure represents the distribution in galactic longitude of the peculiar radial velocities larger than 100 km per sec.

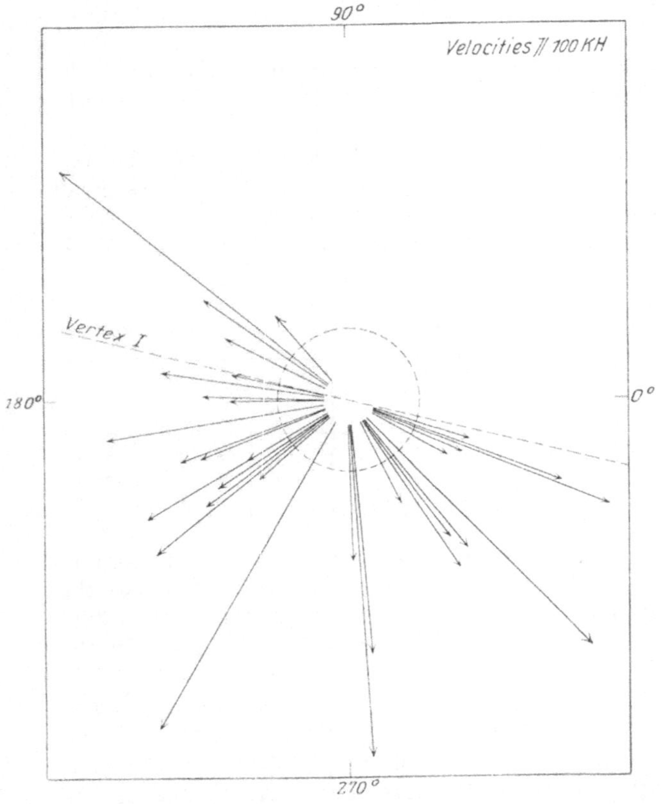

Fig. 1. The distribution in galactic longitude of the peculiar radial velocities larger than 100 km per sec.

The radial velocities have been projected on the galactic plane, but the length of each arrow is proportional to the total radial velocity. The dotted circle has a radius corresponding to 63 km per sec, the dotted line indicates the direction of the true vertices.

An extensive investigation concerning the asymmetry based on the radial velocities has been performed by STRÖMBERG[6]. He has, as already mentioned in section c, § 11, made an analysis of the velocity distribution of the stars based on all available radial velocities. He determined for different groups

[1] Mt Wilson Contr 79; Ap J 39, p. 341 (1914).
[2] Pop Astr 26, p. 686 (1918); Publ. American Astr. Soc. 4, p. 11 (1918).
[3] Mt Wilson Contr 163; Ap J 49, p. 179 (1919).
[4] Mt Wilson Contr 275; Ap J 59, p. 229 (1922).
[5] Groningen Publ 40 (1926).
[6] Mt Wilson Contr 293; Ap J 61, p. 363—388 (1925).

of stars the group-motion relative to the sun, and the velocity distributions relative to the centroids of the groups, assuming ellipsoidal frequency-functions for the internal motions. To obtain a grouping into classes of different velocity-dispersions within each spectral type, the apparent magnitudes and proper motions were used. The stars between certain limits of spectral subdivision were taken together, and each such group was subdivided according to values of the quantity $H = m + 5 \log \mu$, where m is the apparent magnitude, and μ the proper motion.

For each group the components of the group-motion are determined from the radial velocities by the aid of equations of the form (10) and then the elements of the solar motion from (18). In the determination of the axes of the velocity ellipsoids the directions of the principal axes have been assumed to be the same for all the groups, namely:

	α	δ	l	b
a	271°,5	−17°,1	341°,1	− 0°,7
b	327 ,6	+61 ,2	71 ,0	+ 5 ,6
c	188 ,9	+22 ,5	259 ,5	+84 ,4

The letters a, b and c are identical with the letters σ_3, σ_2 and σ_1 in (38) and represent the velocity dispersions along the three axes, and, at the same time, the axes of the velocity ellipsoid. These quantities were determined from the radial velocities in the following way. On a map circles were drawn around the six points where the three axes intersect the sphere. These circles had in general a radius of 45°. The average peculiar radial velocity was formed for the stars within the circles, opposite areas being combined. From the three average velocities thus found for each group, the dispersions were derived on the assumption of an ellipsoidal distribution for the velocities. The average peculiar velocity was used in the computations instead of the mean square velocity in order to lessen the influence of isolated large velocities.

A very detailed table of the results for 50 groups is given, and we will here give an extract for 43 of the groups in table XXIV. The first 29 groups include the common spectral types. Groups 30 and 31 are Cepheids of long and short periods, and group 32 contains the elements for the spectral type O, derived from PLASKETT's[1] radial velocities. In group 33 are given the results of R. E. WILSON[2] from a study of the space-velocities of the O-stars. Group 34 includes the stars with c-characteristics (sharp spectral lines), groups 35 and 36 contain the spectral type P (bright-line nebulae), whose radial velocities have been determined by CAMPBELL and MOORE[3]. In the latter solution six objects of very high radial velocity have been omitted. Group 37 contains results from the radial velocities of the long-period variables, determined by MERRILL[4]. As, according to his investigation, the Me-stars with periods between 150 and 210 days have in general higher velocities than the other Me-stars, the former were treated separately as group 38. Groups 39 and 40 contain the globular clusters and the non-galactic nebulae (continuous spectra) resp. A list of all measured radial velocities of the objects in these two groups has been given by STRÖMBERG[5]. The great majority of the determinations are made by SLIPHER. Groups 41 and 42 are stars with very large velocities, and group 43 contains stars with stationary calcium lines, which lines are probably due to calcium clouds in space[6]. This system of clouds shows a very small velocity-dispersion.

[1] Victoria Publ 2, No. 16 (1924). [2] A J 36, No. 1 (1924).
[3] Lick Publ 13, pt. 4 (1918).
[4] Mt Wilson Contr 264; Ap J 58, p. 215 (1923).
[5] Mt Wilson Contr 292; Ap J 61, p. 353 (1925).
[6] Cf. PLASKETT, Victoria Publ 2, No. 16, p. 342 (1924).

Table XXIV.

Group No.	Type	H	No.	S	A	D	a	b	c	x'	y'	z'
1	M0—M9	≦ −2,0	75	14,6	269°	+36°	19,4	17,1	16,5	− 7,0	− 12,1	− 4,3
2	M0—M9	− 1,9 to + 3,0	97	24,8	300	+51	39,0	20,4	15,0	− 3,6	− 24,6	− 0,6
3	K4—K9	≦ −2,0 . . .	77	22,8	293	+19	20,4	13,1	19,2	−14,3	− 17,4	+ 3,5
4	K4—K9	− 1,9 to + 3,0	90	34,4	262	+31	29,6	23,9	22,6	−19,4	− 25,5	−12,5
5	K4—M9	≧ +3,1 . . .	79	19,6	301	+12	52,2	11,1	35,1	−12,5	− 13,7	+ 6,5
6	G9—K3	≦ −3,0 . . .	124	20,1	279	+31	16,7	13,5	12,2	−10,8	− 16,7	− 2,6
7	G9—K3	− 2,9 to − 2,0 .	120	14,8	269	+34	19,6	13,6	16,6	− 7,6	− 11,9	− 4,2
8	G9—K3	− 1,9 to − 1,0 .	102	14,9	288	+28	23,2	22,0	16,3	− 7,9	− 12,5	+ 0,2
9	G9—K3	− 0,9 to 0,0 . .	139	23,5	275	+42	26,0	22,7	16,5	− 9,0	− 20,9	− 5,6
10	G9—K3	+ 0,1 to + 1,0 .	109	21,4	264	+43	28,9[1]	20,4	19,2	− 8,2	− 18,2	− 7,9
11	G9—K3	+ 1,1 to + 3,0 .	100	42,8	284	+27	41,9	28,0	30,4	−24,3	− 35,2	− 1,2
12	G9—K3	+ 3,1 to + 12,0	104	26,1	273	+55	33,7	34,3	18,3	− 4,5	− 24,5	− 7,8
13	G0—G8	≦ −2,0 . . .	144	17,0	251	+44	17,6	12,4	13,6	− 5,8	− 13,4	− 8,8
14	G0—G8	−1,9 to +1,0 .	131	20,6	271	+34	25,8[2]	16,5	9,4	−10,7	− 16,9	− 5,2
15	G0—G8	+1,1 to +5,0 .	154	27,8	283	+17	44,0	21,7	12,7	−19,6	− 19,7	+ 0,6
16	G0—G8	+5,1 to +10,0	111	41,6	287	+19	46,2	32,4	32,1	−27,2	− 31,2	+ 3,1
17	F0—F9	≦ −2,0 . . .	86	22,1	254	+31	19,7	9,0	15[3]	−12,0	− 15,4	−10,4
18	F0—F9	−1,9 to 0,0 . .	70	19,3	277	+40	24,0	18,6	13,2	− 7,9	− 17,2	− 3,9
19	F0—F9	+0,1 to +3,0 .	246	18,6	259	+21	28,9	19,3	10,7	−12,7	− 11,7	− 6,9
20	F0—F9[4]	+3,1 to +5,0 .	82	36,1	277	0	48,4	14,8	18,1	−31,9	− 16,8	+ 0,9
21	F0—F9[5]	+5,1 to +14,0	38	108,1	330	+67	122	76	55	+25,5	−105,0	− 1,6
22	B6—A9	≦ −3,0. . . .	188	22,0	275	+52	13,0	12,0	8,7	− 4,7	− 20,6	− 6,0
23	B6—A9	−2,9 to − 2,0 .	138	14,8	250	+26	12,5	14,1	7,2	− 8,9	− 9,0	− 7,6
24	B6—A9	−1,9 to − 1,0 .	133	17,1	259	+10	17,7	17,0	5,7	−13,9	− 8,3	− 5,5
25	B6—A9	−0,9 to 0,0 . .	130	23,1	269	+10	15,4	10,6	10,7	−19,0	− 12,6	− 3,8
26	B6—A9	+0,1 to + 2,7 .	91	26,8	281	+13	12,1	14,1	10,0	−20,1	− 17,7	+ 0,5
27	B6—A9	−2,9 to + 2,7 .	492	18,3	267	+15	16,4	14,6	9,1	−14,1	− 10,9	− 4,0
28	B0—B5	≦ −4,0. . . .	123	21,4	287	+31	8,1	8,4	8,4	−10,9	− 18,5	− 0,2
29	B0—B5	−3,9 to + 0,1 .	126	21,0	285	+41	7,4	13,0	10,6	− 7,7	− 19,4	− 2,3
30	Cepheids[6] Per. > $2^d,0$	—	37	11,5	283	+24	17,3	10,0	(10)	− 7,2	− 9,0	− 0,2
31	Cepheids[7] Per. < $0,^d7$	—	26	109	306	+47	74	74	74	−12	−107	+ 8
32	O5—O9	—	49	35,7	275	+53	19,6	33,9	(19)	− 7,3	− 33,5	− 9,8
33	O5—O9	—	41	30,0	263	+47	30	30	30	− 9,2	− 26,1	−11,5
34	c-stars	—	66	26,1	262	+31	12,9	17,5	(12)	−14,6	− 19,4	− 9,6
35	P	—	107	28,4	268	+49	64,7	23,9	(15)	− 9,1	− 25,6	− 9,4
36	P[8]	—	101	30,2	288	+44	45,8	30,7	(15)	− 9,5	− 28,6	− 2,3
37	M1e—M6e	—	86	65,2	287	+41	52	57	46	−23,7	− 60,5	− 5,6
38	M2e—M5e Per. 150^d — 210^d	—	13	162	301	+57	88	88	88	− 7	−161	−11
39	Globular clusters	—	18	286	306	+62	117	117	117	+14	−285	−22
40	Non-galactic nebulae	—	44	344	305	+56	300	300	300	−14	−344	−8
41	Stars of max. space-velocity	—	22	281	303	+60	207	108	82	+40	−278	+20
42	Stars of rad. veloc. > 100 km	—	49	236	299	+45	—	—	—	−58	−229	+5
43	Calcium clouds	—	64	20,4	276	+37	5	5	—	−9,2	−17,7	−4,2

The first column in the table gives the group-number, the second the types, the third the limits of H, the fourth the number of objects in the group. The

[1] One star with $\varrho = +181$ km/sec omitted.
[2] One star with $\varrho = +183$ km/sec omitted.
[3] From five stars only. [4] Four A stars included.
[5] Four A stars and three G stars included. [6] W Virginis, $\varrho = -60$ km/sec omitted.
[7] VX Herculis, $\varrho = -380$ km/sec omitted.
[8] Six nebulae of velocities higher than 100 km/sec omitted.

elements of the solar motion (S, A, D) and the dispersions (a, b, c) are next given. The last three columns show the component of the group-motion in a special system of coordinates which will be defined later on.

The dispersions increase in general with H, which shows that the velocity-components are not independent of one another. In general S increases with H and with the dispersions, as would be expected if the velocity distribution were asymmetrical. The results are more clearly arranged in fig. 2, given by STRÖMBERG. In this figure are given the projections of the velocity ellipsoids

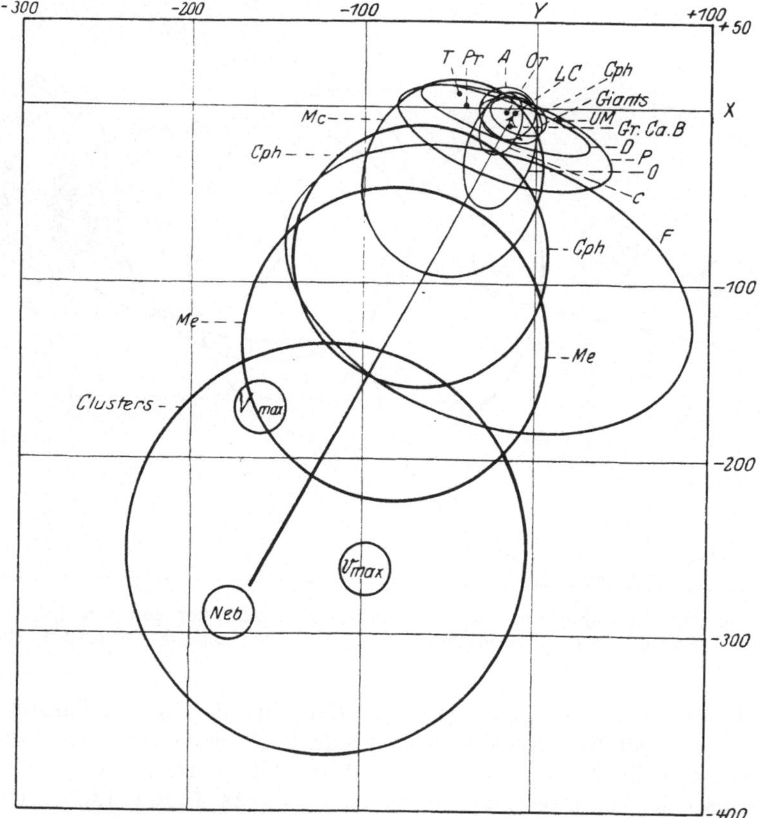

Fig. 2. Projection of a number of velocity-ellipsoids on the galactic plane.

on the galactic plane. The sun is at the origin, and a line from the origin to the centre of an ellipse or circle indicates the group-motion projected on the galactic plane, and the size of the axes of the ellipses indicates the dispersions a and b. v_{max} and V_{max} are the groups 41 and 42 resp.

The group motion increases along a given direction (in the third quadrant) as the ellipses grow larger. The opposite direction ($l = 61°,5$, $b = +9°$) is the direction of motion of objects with small dispersion relative to objects of large dispersion. Projecting the velocity ellipsoids on three new axes (x', y', z'), of which y' has the direction just mentioned and the x'-axis has zero galactic latitude, we get fig. 3, where Y_1 lies in the galactic plane in longitude $61°,5$.

The components of the group motion projected on the three new axes are given in the last three columns of table XXIV (termed x', y', z'). The z'-component varies very little, the x'-component somewhat more, but not systematically.

The y'-component, on the other hand, varies considerably, and this variation is correlated with the dispersion (b) along the same axis. The relation between y' and b can (with the exception of groups 22, 28 and 29) be represented by the parabola

$$y' = -pb^2 + \beta', \tag{46}$$

where

$$p = 0{,}0192 \text{ sec/km},$$
$$\beta' = -10{,}0 \text{ km/sec}.$$

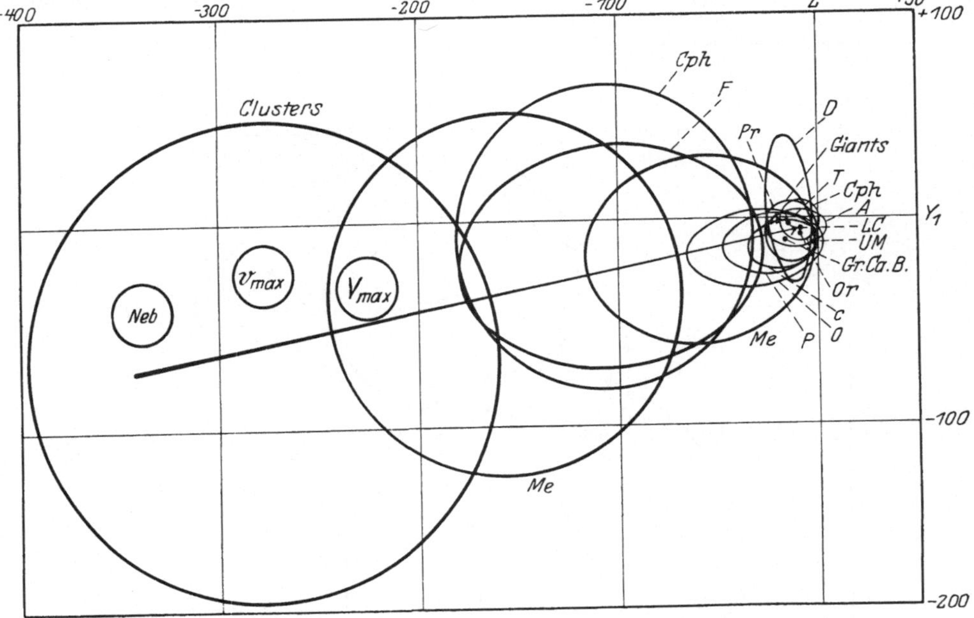

Fig. 3. Projection of a number of velocity ellipsoids on a plane perpendicular to the galactic plane: The Y_1-axis lies in galactic longitude 61°,5 and the Z-axis points towards the north galactic pole.

By the aid of this relation it is shown that the velocity distribution for the objects studied can be expressed as a product of two symmetrical functions S_1 and S_2 with different centres. The centre of S_1 corresponds to a velocity of 14,8 km/sec in the direction $\alpha = 88°,7$, $\delta = -22°,0$, and the centre of S_2 corresponds to a velocity of 300 km/sec in the direction $\alpha = 127°$, $\delta = -56°$. The first distribution is interpreted as the velocity distribution within our local system, and the second as a velocity-restriction in a universal world-frame in which the clusters and spirals are statistically at rest. For further details we refer to the cited investigation and to the analogous treatment of the space-velocities[1].

According to STRÖMBERG the asymmetry is a general property of stellar motion which can be traced also among the stars of small velocity dispersion. In opposition to this, OORT, in the cited investigation, expresses the opinion that the asymmetry is restricted to the stars with high velocities[2]. He finds that there exists a rather sharply defined limit of about 63 km/sec below which the asymmetry disappears, and puts forth that there is good reason to believe that this limit roughly corresponds to the velocity of escape from the system

[1] Mt Wilson Contr 275; Ap J 59, p. 228 (1924).

[2] Compare the discussion in Obs 49, p. 280 and p. 302—304.

of slow moving stars. The high velocity stars may in all probability be regarded as foreigners coming from outside and not as belonging to our local stellar system.

The interpretation of the asymmetry has also been discussed by HAAS[1], KOHLSCHÜTTER[2] and more extensively by LINDBLAD[3]. LINDBLAD assumes that the stellar system has a general motion of rotation around an axis perpendicular to the galactic plane. The asymmetry of stellar velocities of great size is interpreted as due to a general decrease of the speed of rotation with increasing velocity dispersion. A relation between the linear speed of rotation and the dispersion of the velocity distribution is derived. With a presumable supposition concerning our position in the stellar system this relation corresponds very well with the relation (46) empirically found by STRÖMBERG.

e) The Radial Velocities combined with other Attributes of the Stars.

19. The Moving Clusters. In the preceding sections we have dealt with the results derived from the radial velocities alone. But combined with other attributes, proper motion, distance, absolute magnitude and mass we have been able not only to confirm the results already obtained, but also to draw some other conclusions concerning the stars. We will try here to make a short synopsis of the more important results.

The investigation of the proper motion of the stars has revealed a number of groups of stars for each of which the proper motions all appear to converge towards a single point in the sky. This fact indicates that the members of such a group — a moving cluster — have equal and parallel space-velocities. If the convergent point is determined and moreover the radial velocity of one or more members is known, we are able to determine the linear velocity perpendicular to the line of sight; and then the distance of every star of the cluster can be computed with a fair degree of accuracy. Therefore the investigations of the moving clusters have contributed very much to our knowledge of the distances and absolute magnitudes of the stars.

The best known of the moving clusters are the Taurus and the Ursa Major clusters, the Pleiades, the Praesepe, the Perseus and the Scorpio-Centaurus clusters. We refer to the monograph of RASMUSON[4]: A Research on Moving Clusters, and to CHARLIER's California Lectures.

20. The Mean Parallaxes. According to equations (2) and (3) of § 7 we have for each star of known proper motion

$$\gamma_{11}\frac{U''}{r} + \gamma_{21}\frac{V''}{r} + \gamma_{31}\frac{W''}{r} = k\,\mu_\alpha \cos\delta\,, \tag{47}$$

$$\gamma_{12}\frac{U''}{r} + \gamma_{22}\frac{V''}{r} + \gamma_{32}\frac{W''}{r} = k\,\mu_\delta\,. \tag{48}$$

Making use of the relations (4) and observing that

$$\gamma_{11}\,U_1'' + \gamma_{21}\,V_1'' + \gamma_{31}\,W_1'' = U_1\,, \tag{49}$$

$$\gamma_{12}\,U_1'' + \gamma_{22}\,V_1'' + \gamma_{32}\,W_1'' = V_1\,, \tag{50}$$

where U_1 and V_1 are the components of the linear peculiar motion along the axes

---

[1] Veröff. Sternw. Babelsberg 3, Heft 3 (1923). [2] Seeliger-Festschrift p. 120 (1924).
[3] Ap J 62, p. 191 (1925); Ark Mat Astr Fys 19 A, 21, 27, 35; 19 B, 7, (1925 and 1926); Nova Acta Reg. Soc. Scient. Upsaliensis. vol. extraord. ed. (1927); M N 87, p. 553 to 564 (1927). [4] Lund Medd Ser. II, No. 26 (1921).

X and Y of the system of coordinates K_3, we get from (47) and (48)

$$\gamma_{11}\frac{U_0''}{r}+\gamma_{21}\frac{V_0''}{r}+\gamma_{31}\frac{W_0''}{r}=k\,\mu_\alpha\cos\delta-\frac{U_1}{r}, \tag{51}$$

$$\gamma_{12}\frac{U_0''}{r}+\gamma_{22}\frac{V_0''}{r}+\gamma_{32}\frac{W_0''}{r}=k\,\mu_\delta-\frac{V_1}{r}. \tag{52}$$

These equations are analogous to equation (11) and treated in a similar manner. The quantities $\frac{U_1}{r}$ and $\frac{V_1}{r}$ are certainly unknown, but in the normal equations they will vanish if we define the standard of rest so that the sum of the projections of the linear peculiar tangential velocities along the three axes of coordinates in the system K_2 are all equal to zero, provided that the distances are independent of the peculiar velocities.

From the normal equations we are thus able to determine the quantities

$$a=\overline{\left(\frac{1}{r}\right)}U_0'', \qquad b=\overline{\left(\frac{1}{r}\right)}V_0'', \qquad c=\overline{\left(\frac{1}{r}\right)}W_0'',$$

where $\overline{\left(\frac{1}{r}\right)}$ is the mean parallax of the stars considered. From (20) and (21) it follows that

$$\operatorname{tg} A=\frac{b}{a}, \tag{53}$$

$$\sin D=-\frac{c}{\sqrt{a^2+b^2+c^2}} \tag{54}$$

and from (19)

$$\overline{\left(\frac{1}{r}\right)}=\frac{\sqrt{a^2+b^2+c^2}}{S}. \tag{55}$$

Consequently we can determine, from the proper motions alone, the coordinates of the apex from (53) and (54), and moreover, if the value of the sun's velocity is known from the radial velocities, also the mean parallax from (55).

As pointed out by CHARLIER[1], it is more advantageous to introduce, instead of the distance r of a star, another parameter R defined through the relation

$$R=r\cdot 10^{-0,2\,\mathrm{m}}=10^{-0,2\,\mathrm{M}}. \tag{56}$$

This new parameter, which is a function of the absolute magnitude, M, only, offers a great advantage also from a theoretical point of view. From the proper motions and apparent magnitudes of the stars the mean value of $1/R$ — the mean reduced parallax — may be calculated in quite the same way as the mean parallaxes from the proper motions. The relation between the mean or the mean reduced parallax and the characteristics of the absolute magnitude curve has been investigated by MALMQUIST[2].

If, for a group of stars, both the proper motions and the radial velocities are known, the equations (10), (47) and (48) may be treated together, as has been done by GYLLENBERG and MALMQUIST[3].

The mean parallax of a group of stars may also be determined on the assumption that the elements of the solar motion are known.

From the proper motions in right ascension and declination the proper motion components v in the direction of the sun's antapex and τ at right angles

---

[1] A N 201, p. 9 (1915); Lund Medd Ser. II, No. 14 (1916).

[2] Lund Medd Ser. II, No. 22 (1920); Ser. I, No. 100 (1922).

[3] Lund Medd Ser. I, No. 108 (1925).

to this direction are computed. Denoting by d the angular distance of the star from the assumed antapex, we obtain

$$\left(\frac{\overline{1}}{r}\right) = k \frac{\sum v \sin d}{S \sum \sin^2 d}, \tag{57}$$

which equation corresponds to equation (24) for the radial velocities.

The component τ has also been used for the computation of mean parallaxes through the formula

$$\left(\frac{\overline{1}}{r}\right) = \frac{k\bar{\tau}}{\bar{\varrho}_1}, \tag{58}$$

where ϱ_1 is the peculiar radial velocity, and the mean values at the right hand side are taken without regard to signs. This equation postulates however that the peculiar velocities are distributed at random. For other velocity-distributions the formula becomes more complicated.

The mean parallaxes have contributed very much to our knowledge of the distances and absolute magnitudes of the stars. The reduction curves for obtaining spectroscopic absolute magnitudes are — for the giant stars — chiefly based on the mean parallaxes. Extensive determinations of mean parallaxes have been performed in Groningen and in Lund. We refer to the Groningen Publ. and Lund Medd.

21. The Relationship between Radial Velocity and Absolute Magnitude. As already mentioned in § 16, Kapteyn and Adams have found an indication of a change of radial velocity per unit of absolute magnitude of 1,1 km/sec. Using spectroscopic absolute magnitudes and absolute magnitudes calculated from a modified formula connecting mean parallax with proper motion, Adams and Strömberg[1] derived, from 1300 stars of types F to M, an increase in the radial velocity of about 1,5 km/sec for a decrease in brightness of one magnitude. It was shown that this effect cannot be ascribed to distance, to the law of velocity distribution or to the effect of stream motion. Adams, Strömberg and Joy[2] found an increase of about 1,2 km/sec from the radial velocities. For the space-velocities an increase of 3 km/sec per unit absolute magnitude was found. The latter result, however is, affected by a systematic error, as pointed out by Eddington and Douglas[3]. The spectroscopic absolute magnitudes are used in order to convert the proper motions into linear motions. If by accidental error we have got an absolute magnitude too bright, the star is placed too far away and the deduced linear motion is too great; and vice versa. Thus the accidental error in the absolute magnitudes produces a considerable systematic error in the deduced change of velocity. In the cited paper it is shown how the influence of accidental errors in the spectroscopic absolute magnitudes is eliminated.

22. The Space-Velocities. Thanks to the spectroscopic method of determining stellar distances we are now able to compute the space-velocities for nearly two thousand stars. It is — of course — a great advantage to study the velocity-distribution of the stars directly from the space-velocities. But still it remains to determine the systematic errors in the spectroscopic absolute magnitudes and to extend the determinations of absolute magnitudes over the southern hemisphere before we have got a homogeneous material.

Boss, Raymond and Wilson[4] studied the space-velocities of 520 stars. The amount of solar motion for the swiftly moving stars was found to be larger

[1] Mt Wilson Contr No. 131; Ap J 45, p. 293—305 (1917).
[2] Mt Wilson Contr No. 210; Ap J 54, p. 9—27 (1921).
[3] MN 83, p. 112—118 (1923). [4] A J 35, p. 26 (1923).

than for the slower stars. The velocity-distribution is flattened towards the galaxy and elongated in the vertex-direction. The large space-velocities show a strong tendency to move towards the region comprised within galactic longitudes 130°—340°.

The space-velocities of 1300 stars of spectral types F to M have been investigated by STRÖMBERG[1]. The dwarf stars give a much larger solar motion than the giant stars. The velocity-distribution of different groups of stars was studied in different ways. The giant stars of types F to M form a single group with an approximately ellipsoidal distribution with the shortest axis nearly in the direction of the galactic pole. Among the stars of spectral types A7 to F9 there is in addition to the general group of giants another stream, containing about 20 per cent of the stars the motion of which coincides with that of the Taurus group. There is an asymmetry in the distribution of the velocities, especially marked among the stars of very high velocity.

332 stars of spectral types B7 to F2 are studied in the same way[2]. The velocity of the sun is found somewhat smaller than that found for the other spectral types on account of the presence of a large proportion of stars whose motion is nearly the same as that of the Ursa Major group.

The investigation shows the existence of three well-defined groups, the Central group (69 per cent), the Ursa Major group (23 per cent) and the Taurus group (8 per cent). The elements of ellipsoidal distribution-functions for all three groups are given.

CHARLIER[3] has discussed the space-velocities of 1418 stars apparently brighter than the sixth magnitude. A dissection of the material into two components of approximately the same number of stars in both components (two star-stream hypothesis) was found to be decidedly out of question. It is, however, possible to obtain a dissection into two MAXWELLIAN components, one (component I) containing 91 per cent of the stars, the other (component II) 9 per cent.

Relative to the centroid of the stars the component I moves with a velocity of 5,55 km per sec in the direction towards a point in the galactic plane having a galactic longitude of 341°,2 which coincides with the vertex (in Sagittarius). The component II moves with a velocity of 55,93 km per sec in the opposite direction. The Taurus cluster has approximately the same velocity and also nearly the same convergent as component II. The dispersion in the velocities is found to be very different for the two components. In the component I the dispersion is 16,1 km per sec, in the component II it amounts to 29,6 km per sec.

FESSENKOFF and OGORODNIKOFF[4] have determined the solar motion from the space-velocities. The masses of the stars too are taken into account by the aid of the relation between absolute magnitude and mass derived by EDDINGTON[5].

In the Publications de l'Institut Astrophysique de Russie, Vol. III, Part. 2, 1926, a catalogue of the equatorial components of the space-velocities for 1470 stars is given.

[1] Mt Wilson Contr 245; Ap J 56, p. 265—294 (1922).
[2] Mt Wilson Contr 257; Ap J 57, p. 77—85 (1923).
[3] Lund Medd Ser. I, No. 109 (1925).
[4] R A J II, Part 1, p. 37 (1925); III, Part 1, p. 36 (1926). [5] M N 84, p. 308 (1924).

Kapitel 2.

Die veränderlichen Sterne.

Von

H. LUDENDORFF-Potsdam.

Mit 36 Abbildungen.

a) Allgemeines.

1. Definition der veränderlichen Sterne. Schon seit langer Zeit ist es bekannt, daß es Sterne gibt, deren scheinbare Helligkeit Veränderungen unterworfen ist; man nennt diese Objekte „veränderliche Sterne". Diejenigen unter ihnen, bei denen die Helligkeitsänderungen im wesentlichen in einem einmaligen, starken und rasch verlaufenden Aufleuchten bestehen, nennt man „neue Sterne" und behandelt sie, wie es auch in diesem Werke geschieht, getrennt von den eigentlichen Veränderlichen; ganz folgerichtig ist dies nicht, denn die neuen Sterne sind von den veränderlichen keineswegs scharf unterschieden, sondern es gibt Übergangsformen zwischen den beiden Klassen von Himmelskörpern. Zudem unterliegen die neuen Sterne, nachdem sie wieder schwach geworden sind, vielfach kleinen Helligkeitsschwankungen und sind dann demnach veränderliche Sterne; bisweilen hat man auch schon mit Hilfe vorhandener photographischer Aufnahmen des betreffenden Teiles des Himmels an Sternen, die nachher „neue Sterne" wurden, kleine Helligkeitsschwankungen vor dem Aufleuchten feststellen können. Trotzdem ist die Unterscheidung zwischen den neuen und den veränderlichen Sternen so allgemein üblich, daß sie auch hier beibehalten werden soll.

Unter den veränderlichen Sternen gibt es nun eine große Gruppe von Objekten, bei denen die Änderungen der Helligkeit, wie wir sicher wissen, nicht dem betreffenden Sterne inhärent, sondern nur scheinbar sind und durch unsere Stellung im Weltraume zu ihm bedingt werden. Es sind dies die sog. „Bedeckungs"- oder „Verfinsterungs-Veränderlichen" (eclipsing stars), auch „Algol-Sterne" bzw. „β Lyrae-Sterne" genannt. Wie die spektrographischen Beobachtungen im Verein mit den photometrischen beweisen, haben wir in diesen enge Doppelsternsysteme vor uns, deren Bahnebenen so im Raume gelagert sind, daß sie genau oder nahezu durch die Erde gehen. Die beiden Komponenten verdecken sich also bei jedem Umlauf zeitweise ganz oder zum Teil, und infolge dieses Verfinsterungsvorganges ist die Helligkeit des auch im Fernrohr einfach erscheinenden Sternes für einen irdischen Beobachter veränderlich. Diese Bedeckungs-Veränderlichen werden in dem vorliegenden Werke in dem Kapitel über die Doppelsterne behandelt, wohin sie weit mehr gehören, als in das über veränderliche Sterne. Freilich ist auch diese Abtrennung nicht ohne Willkür. Es scheint nämlich, als ob auch bei gewissen anderen Veränderlichen, und zwar denen der R Coronae-Klasse, Bedeckungen eine Rolle spielen, allerdings nicht solche durch eine zweite Komponente, sondern durch kosmische Nebel- oder

Staubmassen. Aber als sicher kann dies nicht angesehen werden, und höchst wahrscheinlich sind es nicht allein Bedeckungen, die den Lichtwechsel der R Coronae-Sterne erklären. So ist es denn geboten, diese Sterne zusammen mit den eigentlichen Veränderlichen zu behandeln, von denen man annimmt, daß sie wirklichen Helligkeitsschwankungen unterliegen, deren wahre Ursache man allerdings noch nicht anzugeben vermag.

2. Geschichtliche Entwicklung der Kenntnis der veränderlichen Sterne. Nur ganz kurz kann hier die geschichtliche Entwicklung unserer Kenntnis von den veränderlichen Sternen gestreift werden. Wer sich näher dafür interessiert, sei z. B. auf die ausführlichen Darlegungen in J. G. HAGEN, Die veränderlichen Sterne[1], verwiesen. Der erste Stern, bei dem eine Veränderlichkeit bemerkt wurde, war *o* Ceti, der eben wegen dieser Veränderlichkeit später den Namen Mira Ceti erhielt. Im August 1596 sah ihn der friesische Pfarrer und Astronom DAVID FABRICIUS als Stern zweiter Größe, konnte ihn aber einige Wochen darauf nicht mehr wahrnehmen. Auch 1609 sah er den Stern wieder hell, aber er ist merkwürdigerweise seiner Entdeckung nicht weiter nachgegangen. So wurde die Veränderlichkeit von *o* Ceti im Jahre 1638 von HOLWARDA unabhängig wieder entdeckt. Nur langsam nahm dann im 17. und 18. Jahrhundert die Zahl der bekannten veränderlichen Sterne zu, und im Jahre 1800 zählte man deren erst elf (nach der Reihenfolge der Entdeckung *o* Ceti, β Persei, χ Cygni, R Hydrae, R Leonis, η Aquilae, β Lyrae, δ Cephei, α Herculis, R Coronae, R Scuti, von denen zwei, β Persei und β Lyrae, Bedeckungs-Veränderliche sind). Merkwürdig ist es, daß man diese wenigen Objekte, die doch als höchst interessant erscheinen mußten, nur recht sporadisch beobachtet hat. Dies blieb so auch noch in den ersten vier Jahrzehnten des 19. Jahrhunderts. Erst um das Jahr 1840 trat eine Wendung ein. Damals bildete ARGELANDER seine Stufenschätzungsmethode aus und begann mit seinen systematischen Beobachtungen der Veränderlichen. Bald nach ihm begannen HEIS und SCHMIDT ihre Beobachtungsreihen, später OUDEMANS, WINNECKE, SCHÖNFELD und KRUEGER (die letzteren vier erst nach 1850). Von weiteren Beobachtern aus der Mitte und zweiten Hälfte des vorigen Jahrhunderts sind HIND, POGSON, KNOTT und BAXENDELL zu nennen, ferner E. C. PICKERING, GORE, GROVER, ŠAFAŘÍK, PLASSMANN, HARTWIG u. a. m.

Auch die Zahl der Neuentdeckungen veränderlicher Sterne nahm in der zweiten Hälfte des vorigen Jahrhunderts rasch zu. Ein 1850 von ARGELANDER aufgestelltes Verzeichnis enthält 24 Veränderliche, ein SCHÖNFELDscher Katalog von 1875 schon 143 und der dritte CHANDLERsche Katalog (1896) sogar 393. Ist auch so in der zweiten Hälfte des 19. Jahrhunderts ein bedeutender Aufschwung in der Erforschung der Veränderlichen zu verzeichnen, so waren doch zu jener Zeit die Beobachtungen auch hellerer Objekte dieser Art immerhin zuweilen noch recht lückenhaft, und es macht sich das bei Spezialuntersuchungen über einzelne Sterne recht schmerzlich fühlbar.

Ungefähr um die Jahrhundertwende entwickelte sich dann ein sehr viel größeres Interesse an den veränderlichen Sternen, und man kann sagen, daß dieses Interesse seitdem immer mehr gewachsen ist. Auf dem Gebiete der Beobachtungen macht sich dies dadurch bemerkbar, daß verschiedene astronomische Gesellschaften sich der Organisation der Beobachtungen angenommen und ihre Mitglieder zu Beobachtungen angeregt haben. Zuerst war es die schon 1890 gegründete Variable Star Section of the British Astronomical Association, die auf den Plan trat und eine Reihe von Veränderlichen außerordentlich sorg-

[1] Erster Band, S. 10ff. (Freiburg i. Br., Herdersche Verlagshandlung, 1913).

fältig beobachtete. 1911 wurde die American Association of Variable Star Observers ins Leben gerufen, die ihre Beobachtungen zahlreicher Veränderlicher (1924 waren es 450 Sterne, von denen 70 Beobachter nahezu 20000 Beobachtungen lieferten) fortlaufend in der Zeitschrift „Popular Astronomy" veröffentlicht. Seit 1921 publiziert die Association française d'observateurs d'étoiles variables ihre Beobachtungen im „Bulletin de l'Observatoire de Lyon" und seit 1922 die Nordisk Astronomisk Selskab die ihrigen in den „Astronomischen Nachrichten", seit 1926 auch in der „Nordisk Astronomisk Tidsskrift". Auch die auf dem Harvard-Observatorium angestellten bzw. gesammelten Beobachtungen verdienen besonders hervorgehoben zu werden. Fast alle die genannten Beobachtungen werden nach Schätzungsmethoden gemacht. Dazu treten noch zahlreiche Beobachtungen von Astronomen und Amateuren, die außerhalb jener Organisationen stehen.

Auch die Bearbeitungen einzelner Veränderlicher, sowie die allgemeinen Untersuchungen und die theoretischen Betrachtungen auf diesem Gebiete haben in den letzten zwei Jahrzehnten an Zahl und Bedeutung ungemein zugenommen, wie aus den weiteren Ausführungen hervorgehen wird.

3. Die Literatur über die veränderlichen Sterne. Die Literatur über die veränderlichen Sterne ist ganz außerordentlich umfangreich. Zahllose Abhandlungen verschiedensten Umfanges sind über dieses Gebiet in den astronomischen Zeitschriften und in den Sternwartenpublikationen oder als selbständige Veröffentlichungen erschienen. Es können hier zur Einführung nur die wichtigsten Publikationen dieser Art aufgezählt werden, wobei in erster Linie die Literatur berücksichtigt wird, die für die Gegenwart noch von Bedeutung ist.

α) Lehrbücher. An erster Stelle ist zu nennen:

J. G. HAGEN, S. J. Die veränderlichen Sterne. Erster Band. Geschichtlich-technischer Teil. Freiburg im Breisgau: Herder & Co., Verlagsbuchhandlung. (Zugleich Bd. 5 der Pubblicazioni della Specola Vaticana.)

Das monumentale Werk ist in vier Lieferungen 1913—1921 erschienen und umfaßt insgesamt 811 Seiten. Die erste Lieferung (1913) ist betitelt: Die Ausrüstung des Beobachters. Dieser Titel kennzeichnet den Inhalt nicht gerade glücklich. Wir finden hier ausführliche geschichtliche Angaben über das Studium der veränderlichen Sterne, Betrachtungen über die Klassifizierung, die Anzahl und die Verteilung, die Verzeichnisse und die Nomenklatur der Veränderlichen, sowie schließlich Ausführungen über die Beobachtungsinstrumente, die Sternkarten und die Ephemeriden für die veränderlichen Sterne. Auch Ratschläge für das Beobachtungsprogramm werden gegeben.

Die zweite Lieferung (Die Beobachtung der veränderlichen Sterne) ist 1914 erschienen und behandelt in allergrößter Ausführlichkeit die Beobachtung durch Stufenschätzung unter Ausschluß anderer Beobachtungsmethoden. Die dritte Lieferung (1920) trägt den Titel „Die Berechnung der Beobachtungen", die vierte (1921) denjenigen „Die Elemente des Lichtwechsels".

Der zweite Band des hier besprochenen Werkes (Die veränderlichen Sterne. Zweiter Band. Mathematisch-physikalischer Teil. Von Dr. J. STEIN S. J.) ist im gleichen Verlage 1924 erschienen, zugleich als Bd. 6 der Pubblicazioni della Specola Vaticana (383 Seiten). Er gibt eine vollständige Darstellung der Theorien, die aufgestellt worden sind, um den Lichtwechsel der verschiedenen Klassen von veränderlichen Sternen zu erklären. Auch die neuen Sterne und die Bedeckungs-Veränderlichen sind in den Kreis der Betrachtungen gezogen.

Das Werk von HAGEN und STEIN ist so umfangreich, daß es nur für den Spezialforscher auf dem behandelten Gebiete in Frage kommen kann. Viel kürzer und für einen größeren Leserkreis bestimmt sind die beiden folgenden Lehrbücher:

CAROLINE E. FURNESS, An Introduction to the Study of Variable Stars. Boston and New York: Houghton Mifflin Company 1915. (327 S.)

K. SCHILLER, Einführung in das Studium der veränderlichen Sterne. Leipzig: Johann Ambrosius Barth 1923. (383 S.)

Alle drei genannten Lehrbücher beschäftigen sich weit mehr mit den Beobachtungen und ihren Hilfsmitteln, mit ihrer Reduktion und mit der Theorie der Veränderlichen als mit ihnen selbst und ihren Eigenschaften. Es ist daher zu hoffen, daß die im folgenden enthaltenen Ausführungen eine empfindliche Lücke in der Literatur ausfüllen werden.

β) Kataloge von Veränderlichen. Die älteren Kataloge der Veränderlichen (z. B. von ARGELANDER, POGSON, SCHÖNFELD, GORE, CHANDLER u. a.) können hier übergangen werden, da sie nur noch historische Bedeutung haben und nicht mehr benutzt werden. Ein ausführliches Verzeichnis derselben findet sich im Vorwort zum ersten Bande des gleich nachher zu besprechenden Werkes „Geschichte und Literatur der veränderlichen Sterne". Gelegentlich noch benutzt wird:

Second Catalogue of Variable Stars by ANNIE J. CANNON. Annals of the Harvard College Observatory 55, Part I. 1907. (Supplement dazu ebenda, Part II, S. 272.)

Das grundlegende Werk über veränderliche Sterne, das niemand entbehren kann, der sich mit diesen Himmelskörpern theoretisch oder praktisch beschäftigt, aber ist das folgende:

Geschichte und Literatur des Lichtwechsels der bis Ende 1915 als sicher veränderlich anerkannten Sterne nebst einem Katalog der Elemente ihres Lichtwechsels. Herausgegeben im Auftrage (Bd. I: und auf Kosten) der Astronomischen Gesellschaft von G. MÜLLER und E. HARTWIG. Leipzig: in Kommission bei Poeschel & Trepte. Erster Band, 1918, XIX + 401 S. Zweiter Band, 1920, VIII + 468 S. Dritter Band, 1922, 137 S.

Dieses große Werk, das infolge seines umständlichen Titels meist abgekürzt als „Geschichte und Literatur der veränderlichen Sterne" bezeichnet wird, und das wir im folgenden kurz mit „G. u. L." zitieren werden, ist einer Anregung G. MÜLLERS zu verdanken, die dieser in einer Vorstandssitzung der Astronomischen Gesellschaft im Jahre 1900 gab. MÜLLER blieb auch bei der Durchführung des Werkes, das von der Astronomischen Gesellschaft unter ihre Auspizien genommen und zum großen Teil auch finanziert wurde, die treibende Kraft.

In der G. u. L. finden sich für jeden bis Ende 1915 als veränderlich anerkannten Stern Angaben über die Entdeckung und über die den Stern betreffenden Untersuchungen, namentlich auch über den Charakter der Lichtkurve. Für sehr viele Sterne werden neue Elemente des Lichtwechsels abgeleitet. Besonders wichtig sind die vollständigen Zusammenstellungen der Literatur über jeden Stern. Auch die vorhandenen genauen Ortsbestimmungen, die etwaige Nummer in der Bonner Durchmusterung, die vorhandenen Karten der Umgebung des Sternes und die Vergleichsternfolgen, sowie die veröffentlichten Lichtkurven werden angeführt; wir erhalten ferner, soweit möglich, Auskunft über die Spektren und Farben der einzelnen Sterne.

In dieser Weise werden im ersten und zweiten Bande der G. u. L. 1687 nach der Rektaszension geordnete Sterne behandelt. Am Schluß des zweiten Bandes werden in Anhang I auf die gleiche Art 32 neue Sterne bearbeitet (von E. ZINNER mit Unterstützung von G. VAN BIESBROECK). Anhang II (von C. HOFFMEISTER) behandelt die veränderlichen Sterne in Sternhaufen.

Der dritte Band der G. u. L. bringt zunächst den eigentlichen Katalog, der in übersichtlicher Form die wichtigsten Daten für die 1687 Veränderlichen zusammenfaßt. Daran schließt sich ein „Zusatzkatalog der Elemente des Lichtwechsels für 320 in den Jahren 1915 bis 1920 neubenannte veränderliche Sterne und für 15 Novae mit ihrer Geschichte und Literatur". Den Schluß des dritten Bandes nehmen Hilfstafeln ein (Lichtzeit für kurzperiodische Veränderliche — Julianische Tage für den Zeitraum von 1600 bis 2000 — Verwandlung von Stunden und Minuten in Dezimalteile des Tages).

Die Sammlung der Literatur über die einzelnen veränderlichen Sterne wird auf der Sternwarte Berlin-Neubabelsberg fortgesetzt, und es ist geplant, von Zeit zu Zeit Ergänzungsbände zu der G. u. L. erscheinen zu lassen, um das Werk stets auf dem Laufenden zu erhalten. Ferner werden in unregelmäßigen Zeitintervallen seitens der Kommission der Astronomischen Gesellschaft für die veränderlichen Sterne Benennungslisten der neuentdeckten Veränderlichen mit Literaturangaben veröffentlicht, und zwar in den Astronomischen Nachrichten. Bisher liegen seit Erscheinen des dritten Bandes der G. u. L. folgende derartige Listen vor: A N 215, S. 185 (96 Sterne); 217, S. 369 (87 Sterne); 223, S. 41 (203 Sterne); 224, S. 129 (238 Sterne); 227, S. 161 (173 Sterne); 228, S. 353 (61 Sterne). Als eine unter Umständen sehr brauchbare Ergänzung zu der G. u. L. können die folgenden Publikationen dienen:

Maxima and Minima of Variable Stars of Long Period. By Annie J. Cannon. Harv Ann 55, Part II. 1909.

(Es sind in dieser Arbeit alle beobachteten Maxima und Minima der langperiodischen Veränderlichen in sehr übersichtlicher Form zusammengestellt.)

Ferner: Maxima and Minima of Two Hundred and Seventy-two Long Period Variable Stars during the Years 1900—1920. By L. Campbell. Harv Ann 79, Part 2. 1926.

Will man die nach der Veröffentlichung der G. u. L. erschienene Literatur über einen Veränderlichen aufsuchen, so benutzt man dazu mit Vorteil den „Astronomischen Jahresbericht, bearbeitet im Astronomischen Rechen-Institut zu Berlin", dessen einzelne Bände die astronomische Literatur für jedes Kalenderjahr vollständig anführen und speziell auch für die alphabetisch nach Sternbildern geordneten Veränderlichen die Literaturnachweise geben.

γ) Ephemeriden der veränderlichen Sterne wurden seit 1870 bis 1926 für jedes Kalenderjahr in der „Vierteljahrsschrift der Astronomischen Gesellschaft" veröffentlicht. Die für die Jahre 1891 bis 1923 rühren sämtlich von E. Hartwig her; die späteren sind von P. Guthnick, R. Prager und E. Heise bearbeitet. Die eigentlichen Ephemeriden bezogen sich zuletzt nur auf die langperiodischen Sterne; es wurde aber gleichzeitig ein vollständiges Verzeichnis der jeweils bekannten veränderlichen Sterne mit deren Elementen gegeben, abgesehen von den in Sternhaufen zusammengedrängten; die Ephemeriden für 1926 enthalten 2671 Veränderliche. Seit 1927 erscheinen diese Ephemeriden in den „Kleineren Veröffentlichungen der Universitätssternwarte zu Berlin-Babelsberg". Weitere Ephemeriden für Veränderliche langer Periode finden sich in den „Circulars" des Harvard College Observatory, Angaben über die voraussichtliche jeweilige Helligkeit solcher Sterne in den „Bulletins" desselben Observatoriums (Bimonthly Compilation of Current Data on Variable Stars of Long Period). Für die kurzperiodischen Veränderlichen findet man Vorausberechnungen der Maxima und Minima laufend in der Zeitschrift „Popular Astronomy", speziell für Bedeckungs-Veränderliche in den Zirkularen der Krakauer Sternwarte. Für besonders interessante oder helle Veränderliche werden Ephemeriden auch in verschiedenen populären astronomischen Zeitschriften und Jahrbüchern abgedruckt.

δ) Karten veränderlicher Sterne. Spezialkarten der Umgebung veränderlicher Sterne sind namentlich für die schwächeren unter diesen zur Identifizierung für den Beobachter unumgänglich nötig. Eine Sammlung solcher Karten gibt J. G. HAGENS großes Werk „Atlas Stellarum Variabilium" (käuflich durch Herder & Co., Freiburg i. Br.), erschienen 1899 bis 1908. Dieser Atlas enthält in 6 Serien auf 311 Karten die Umgebung von 338 Veränderlichen. Aus dem Katalog im 3. Bande der G. u. L. ist sofort ersichtlich, welche Veränderlichen in diesem Atlas berücksichtigt sind. Dieser gibt auf seinen Textblättern auch Vergleichsternfolgen an. Wichtige Ergänzungen zum Text liefern HAGENS „Aggiunte al Catalogo dell' Atlas Stellarum Variabilium" (Specola Vaticana 11, 1916), solche zu den Karten, namentlich was sehr schwache Vergleichsterne betrifft, HAGENS „Aggiunte alle Carte dell' Atlas Stellarum Variabilium" (Specola Vaticana 12, 1922, mit 37 Tafeln).

Auch an anderen Stellen sind Karten für Veränderliche einzeln oder in kleineren Sammlungen veröffentlicht. Die G. u. L. gibt, wie schon erwähnt, für jeden Stern die nötigen Hinweise. Die großen Himmelsatlanten, z. B. der der Bonner Durchmusterung, sind gleichfalls unter Umständen für die Identifizierung der veränderlichen Sterne von Nutzen.

ε) Vergleichsternfolgen. Helligkeitssequenzen für die Vergleichsterne der einzelnen Veränderlichen werden in der G. u. L. nachgewiesen. Größere Zusammenstellungen von solchen Sequenzen finden sich außer in HAGENS Atlas vor allem noch an folgenden Stellen:

FLEMING, WILLIAMINA P., A Photographic Study of Variable Stars. Harv Ann 47, Part I. 1907.

CAMPBELL, L., Comparison Stars for 252 Variables of Long Period. Harv Ann 57, Part II. 1908.

CAMPBELL, L., Comparison Stars for 279 Variables. Harv Ann 63, Part II. 1913.

(Harvard College Observatory), SCHÖNFELDS Comparison Stars for Variables. Harv Ann 64, No. III. 1912.

GRAFF, K., Photometrische Helligkeiten und Farben schwacher Sterne in der Umgebung von 55 Veränderlichen. A N 213, S. 33. 1921.

Ferner findet man Helligkeitsfolgen von Vergleichsternen in verschiedenen der unter ζ) aufgezählten Sammlungen von Beobachtungen.

ζ) Sammlungen von Beobachtungen veränderlicher Sterne. Die wichtigsten derartigen Sammlungen sind im folgenden zusammengestellt; solche, die in Zeitschriften veröffentlicht sind, werden hier nur ausnahmsweise angeführt. Die unzähligen, meist in Zeitschriften veröffentlichten Abhandlungen, die Beobachtungen eines bestimmten Veränderlichen oder einer geringen Zahl solcher Sterne enthalten, können hier nicht berücksichtigt werden, sondern es muß für diese auf die G. u. L. und den Astronomischen Jahresbericht verwiesen werden. Viele von diesen Arbeiten geben auch nicht die einzelnen Beobachtungen, sondern nur deren Resultate wieder.

ARGELANDER, F. W. A., Beobachtungen und Rechnungen über veränderliche Sterne. Bonner Beobachtungen Bd. 7. 1869.

ARGELANDER, F. W. A., Nachgelassene Beobachtungen veränderlicher Sterne. Bonn 1898.

(ARGELANDER, F. W. A.), Observations of Variable Stars by ARGELANDER. Harv Ann 33, No. IV. 1900.

(Enthält eine Bearbeitung des größten Teiles der ARGELANDERschen Beobachtungen durch E. C. PICKERING.)

(BAXENDELL, J.), Baxendell's Observations of Variable Stars. Edited by H. H. Turner and Mary A. Blagg. M N 73 (1912) bis 78 (1918).

VAN BIESBROECK, G., et CASTEELS, L., Études sur les étoiles variables. Annales de l'Obs. R. de Belgique 13, fasc. 2. 1913.

CAMPBELL, L., Observations of Seventy-five Variable Stars of Long Period during the Years 1902—1905. Harv Ann 57, Part I. 1907.

CAMPBELL, L., Observations of Three Hundred and Twenty-eight Variable Stars during the Years 1906—1910. Harv Ann 63, Part I. 1912.

CAMPBELL, L., Observations of Three Hundred and Twenty-three Variable Stars during the Years 1911—1916. Harv Ann 79, Part I. 1918.

(Die in diesen drei Publikationen zusammengestellten Beobachtungen rühren von verschiedenen Beobachtern her.)

ENEBO, S., Beobachtungen veränderlicher Sterne, angestellt auf Dombaas (Norwegen). Teil I (1906) bis IX (1917). (Sonderabdruck aus Archiv for Mathematik og Naturvidenskab.)

FLEMING, WILLIAMINA P., Photographic Observations of Variable Stars during the Years 1886 to 1905. Harv Ann 47, Part II. 1912.

FURNESS, CAROLINE E., and WHITNEY, MARY W., Observations of Variable Stars Made during the Years 1901—1912. Publ. of the Vassar College Obs. No. 3. 1913.

GRAFF, K., Beiträge zur Untersuchung des Lichtwechsels veränderlicher Sterne. Mitt. der Hamburger Sternwarte Nr. 8. 1905.

HAGEN, J. G., Observations of Variable Stars Made in the Years 1884—1890. Georgetown College Observatory. 1901.

HARTWIG, E., Beobachtungen veränderlicher Sterne. Veröffentl. der Sternwarte zu Bamberg. Bd. I, Heft I u. II. 1910, 1913.

(Harvard College Observatory), Photographic Observations of Seven Circumpolar Variables. Harv Ann 84, No. 1.

(Harvard College Observatory), Photographic Observations of Six Circumpolar Variables. Harv Ann 84, No. 3.

(Harvard College Observatory), Discussion of Thirteen Circumpolar Variables. Harv Ann 84, No. 4.

HASSENSTEIN, W., Beobachtungen von veränderlichen Sternen in den Jahren 1920—1923. Publikationen des Astrophysikalischen Observatoriums zu Potsdam Nr. 81. 1925.

(HEIS, E., und KRUEGER, A.), Beobachtungen veränderlicher Sterne von EDUARD HEIS aus den Jahren 1840—1877 und von ADALBERT KRUEGER aus den Jahren 1853—1892, herausgegeben von J. G. HAGEN S. J. Berlin 1903.

INNES, R. T. A., Results of Observations of Variable Stars. Annals of the Cape Observatory 9, Part II. 1903.

JOST, E., Helligkeitsmessungen von langperiodischen Veränderlichen nach Beobachtungen von E. JOST, P. MOSCHICK und G. VAN BIESBROECK. Mitt. der Großh. Sternwarte zu Heidelberg (Astronomisches Institut) Nr. 17. 1909.

(KNOTT, G.), Observations of Twenty-three Variable Stars by the Late GEORGE KNOTT. Edited by H. H. TURNER. Mem R A S 52. 1899.

LACCHINI, G. B., Osservazioni di variabili a lungo periodo. R. Osservatorio di Capodimonte. Contributi Nr. 10. 1914.

LAZZARINO, O., Osservazioni fotometriche durante gli anni 1912—1913. R. Osservatorio di Capodimonte. Contributi Nr. 7. 1914.

LUYTEN, W. J., Observations of Variable Stars. Annalen van de Sterrewacht te Leiden. Deel XIII, Tweede Stuk. 1922.

NIJLAND, A. A., Beobachtungen von Cepheiden. Recherches Astronomiques de l'Observatoire d'Utrecht VIII. 1923.

OUDEMANS, J. A. C., Zweijährige Beobachtungen der meisten jetzt bekannten veränderlichen Sterne. Aus den Abhandlungen der mathematisch-physischen

Classe der Königlich Niederländischen Akademie der Wissenschaften. Amsterdam 1856.

PARKHURST, H. M., Observations of Variable Stars. Harv Ann 29, No. IV. 1893.

PARKHURST, J. A., Researches in Stellar Photometry during the Years 1894 to 1906. Washington: Carnegie Institution 1906.

(PEEK, C. E.), Observations of Variable Stars Made at the Rousdon Observatory under the Direction of the Late Sir C. E. PEEK. Edited by H. H. TURNER. Mem R A S 55. 1904.

PICKERING, E. C., Observations of Variable Stars Made with the Meridian Photometer during the Years 1892—1898. Harv Ann 46, Part II. 1904.

PLASSMANN, J., Beobachtungen veränderlicher Sterne. I. Münster (1888), II.—III. Köln 1890—1891, IV. Warendorf 1895, V.—X. Münster 1900—1912.

(POGSON, N. R.), Observations of Thirty-one Variable Stars by the Late N. R. POGSON. Edited by C. L. BROOK. Mem R A S 58. 1908.

PRAČKA, L., Beiträge zur Untersuchung des Lichtwechsels veränderlicher Sterne. Heft 1—3. Prag 1909—1912.

(ŠAFAŘÍK, V.), L. PRAČKA: Untersuchungen über den Lichtwechsel älterer veränderlicher Sterne. Nach den Beobachtungen von Prof. Dr. V. ŠAFAŘÍK in Prag. Bd. I (Prag 1910), Bd. II (Prag 1916).

(SCHMIDT, J.), Observations of Variable Stars by SCHMIDT. Harv Ann 33, No. VI. 1900.

(Enthält die Bearbeitung eines Teiles der Beobachtungen von J. SCHMIDT, deren Manuskripte auf der Sternwarte in Bonn und in nicht ganz vollständiger Abschrift auf dem Observatorium in Potsdam aufbewahrt werden.)

SCHÖNFELD, E., Beobachtungen von veränderlichen Sternen. Sitzungsberichte der Kais. Akademie der Wissenschaften in Wien, Math.-naturw. Klasse. I. Abteilung. Bd. 42. 1860. II. Abteilung. Bd. 44. 1861.

(SCHÖNFELD, E.), Observations of Variable Stars. Harv Ann 33, No. V. 1900.

(Enthält die Bearbeitung eines Teiles der obigen Beobachtungen SCHÖNFELDS.)

SCHÖNFELD, E., Beobachtungen veränderlicher Sterne, herausgegeben von W. VALENTINER. Veröffentl. der Großh. Sternwarte zu Heidelberg (Astrometrisches Institut), Bd. I. 1900.

(SCHWERD, FR. M.), SCHWERDS Beobachtungen veränderlicher Sterne in den Jahren 1823—1833 und 1849—1859. Bearbeitet von R. MÜLLER. Publikationen des Astrophysikalischen Observatoriums zu Potsdam Nr. 82. 1925.

STRATONOW, W., Observations d'étoiles variables. Publications de l'Observatoire de Tachkent Nr. 5. 1901.

TOWNLEY, S. D., Observations of Telescopic Variable Stars of Long Period. Publications of the Washburn Observatory VI, Part 3. 1892.

WENDELL, O. C., Observations of Circumpolar Variable Stars during the Years 1889—1899. Harv Ann 37, Part I. 1900.

WENDELL, O. C., Observations of Fifty-eight Variable Stars of Long Period during the Years 1890—1901. Harv Ann 37, Part II. 1902.

(Die in diesen beiden Publikationen zusammengestellten Beobachtungen rühren von verschiedenen Beobachtern her.)

WENDELL, O. C., Photometric Observations Made with the Fifteen Inch East Equatorial during the Years 1892 to 1902. Harv Ann 69, Part I. 1909.

WENDELL, O. C., Photometric Observations Made with the Fifteen Inch East Equatorial during the Years 1903 to 1912. Harv Ann 69, Part II. 1913.

WILSING, J., Beobachtungen veränderlicher Sterne in den Jahren 1881—1885. Publikationen des Astrophysikalischen Observatoriums zu Potsdam Nr. 37. 1897.

Wilson, H. C., Observations of Variable Stars of Long Period. Publications of the Goodsell Observatory No. 8. 1920.

Beobachtungen der British Astronomical Association: Memoirs of the British Astronomical Association 1 (1893), 3 (1895), 5 (1897), 11 (1903), 15 (1906), 18 (1912), 22 (1918), 25 (1924).

(Die daraus abgeleiteten Lichtkurven sind in „Appendix to Vols 15 and 18" und „Appendix to Vols 22 and 25" wiedergegeben.)

Beobachtungen der American Association of Variable Star Observers: Monthly Report of the American Association of Variable Star Observers (in Popular Astronomy von Jahrgang 1911 an, auch alljährlich gesammelt besonders abgedruckt).

Beobachtungen der Association française d'observateurs d'étoiles variables: Im Bulletin de l'Observatoire de Lyon von Jahrgang 1921 an.

Beobachtungen der Nordisk Astronomisk Selskab: In den Astronomischen Nachrichten von Bd. 215 (1922) an und in der Nordisk Astronomisk Tidsskrift von 1926 an.

Die im vorstehenden angeführten Sammlungen von Beobachtungen enthalten teils die Helligkeitsschätzungen (um solche, nicht um eigentliche photometrische Messungen handelt es sich zumeist) in ihrer Originalform, meist mit den daraus abgeleiteten Größen, teils auch nur letztere.

η) Weitere Literatur. Die weitere Literatur über die veränderlichen Sterne braucht hier nicht besonders angeführt zu werden, da sie im Verlaufe der folgenden Ausführungen zitiert werden wird.

4. Die Nomenklatur der veränderlichen Sterne. Schon um die Mitte des vorigen Jahrhunderts, als die Zahl der bekannten Veränderlichen allmählich immer mehr anwuchs, machte sich das Bedürfnis geltend, für sie besondere Bezeichnungen einzuführen. Argelander machte daher den Vorschlag, die Veränderlichen innerhalb eines jeden Sternbildes nach der Reihenfolge ihrer Entdeckung mit den großen lateinischen Buchstaben R bis Z zu benennen. Nur für die wenigen veränderlichen Sterne, die schon andere, allgemein übliche Bezeichnungen haben, z. B. Algol oder β Persei, o Ceti, χ Cygni, wollte er diese beibehalten wissen. Diese Vorschläge wurden allgemein angenommen; so heißt der erste im Sternbilde Virgo entdeckte Veränderliche R Virginis, der zweite S Virginis usf. Sobald nun aber in einem Sternbilde mehr als neun Veränderliche bekannt wurden, entstanden Schwierigkeiten. Zur Behebung derselben wurden verschiedene Vorschläge gemacht; wer sich näher für diese Entwicklung interessiert, sei auf Hagen, Die veränderlichen Sterne Bd. I, S. 78ff., verwiesen. Schließlich wurde einer von Hartwig auf der Versammlung der Astronomischen Gesellschaft in Straßburg 1881 gegebenen Anregung Folge geleistet: Der zehnte Veränderliche eines Sternbildes wird hiernach mit RR, der elfte mit RS, ... der neunzehnte mit SS usf. bezeichnet. So war für im ganzen 54 Veränderliche in jedem Sternbilde gesorgt. Aber auch dies reichte angesichts der großen Zahl von Neuentdeckungen bald nicht mehr aus. Wollte man nun nicht zu Abweichungen von der alphabetischen Reihenfolge der Buchstaben innerhalb einer Bezeichnung oder aber zu Kombinationen von drei Buchstaben greifen, so blieb nichts anderes übrig, als auch die R vorangehenden Buchstaben des Alphabets heranzuziehen. Man läßt also[1] auf ZZ der Reihe nach folgen AA, AB, ... bis AZ, dann BB bis BZ, ... QQ bis QZ. So stehen für jedes Sternbild 9 einfache Buchstaben und 325 Kombinationen von je zweien, im ganzen also 334 Bezeichnungen zur Verfügung. Die Veränderlichen in Sternhaufen,

[1] A N 176, S. 181 (1907).

in den MAGELLANschen Wolken und ähnlichen Gebilden werden nicht in diese Bezeichnungsweise einbezogen, sondern innerhalb jedes dieser Gebilde mit Nummern versehen; die neuen Sterne werden innerhalb jedes Sternbildes nach der Reihenfolge ihrer Entdeckung numeriert und z. B. mit N Sagittarii 1, N Sagittarii 2 usw. bezeichnet; früher sind jedoch die neuen Sterne einfach ebenso wie die Veränderlichen bezeichnet worden (z. B. T Scorpii, T Aurigae), und man hat ihnen diese Bezeichnung gelassen, um nicht Verwirrung anzurichten; T Scorpii ist also z. B. identisch mit Nova Scorpii 1. Neuerdings bezeichnet man die Novae durch Sternbild und Jahr des Aufleuchtens.

Gegen die jetzt übliche Benennung der Veränderlichen jedes Sternbildes mit einem bzw. mit der Kombination von zwei Buchstaben ist nun einzuwenden, daß auch sie schließlich nicht ausreichen wird, sobald die Zahl der Veränderlichen 334 in einem Sternbilde übersteigt; in der Tat war man im Sternbilde Carina Ende 1926 bereits bis FR gelangt. Von verschiedenen Seiten ist daher diese Bezeichnungsweise beanstandet worden. Schon 1865 schlug CHAMBERS vor, die Veränderlichen eines jeden Sternbildes einfach der Reihenfolge der Entdeckung nach zu numerieren (z. B. 1 var Virginis, 2 var Virginis usw.). Denselben Vorschlag wiederholte ANDRÉ 1899 in seinem Traité d'Astronomie stellaire; NIJLAND kam 1913 darauf zurück[1], ebenso kurze Zeit darauf TOWNLEY[2]. Auf der Versammlung der International Astronomical Union in Cambridge (England) 1925 hat man daher den Beschluß gefaßt, diese Numerierung vom 335sten Veränderlichen jeden Sternbildes an zu beginnen. Auf QZ Carinae würden also folgen V. 335 Carinae, V. 336 Carinae usw. Wahrscheinlich dürfte dieser Beschluß allgemein anerkannt werden.

Die definitive Namengebung der Veränderlichen erfolgt durch die Veränderlichen-Kommission der Astronomischen Gesellschaft in ihren schon früher (Ziff. 3 unter β) erwähnten Benennungslisten. Die neuentdeckten Veränderlichen werden nach einem Vorschlage von H. KREUTZ[3] provisorisch nach der Reihenfolge ihrer Entdeckung innerhalb jedes Kalenderjahres numeriert, also z. B. mit 1.1924 Cygni usw. bezeichnet.

Das Harvard-Observatorium numeriert die dort entdeckten Veränderlichen ohne Rücksicht auf die Jahre durch.

Außer der eigentlichen Benennung der Veränderlichen ist hier noch eine häufig angewendete Numerierung derselben zu erwähnen, die von den Sternbildern unabhängig ist, und die E. C. PICKERING eingeführt hat. Diese Numerierung wird auf folgende Art erhalten: Man rundet zunächst die Position des Veränderlichen für 1900 auf Zehntel Zeitminuten in AR und ganze Minuten in Deklination ab. Die Nummer des Veränderlichen besteht dann aus sechs Ziffern; die ersten beiden geben die Stunde der AR, die dritte und vierte die volle Minute der AR, die fünfte und sechste den vollen Grad der Deklination. Ist letztere negativ, so werden alle sechs Ziffern kursiv gedruckt. So erhält z. B. der Veränderliche X Tauri [AR (1900) $3^h 47^m 50^s$, Dekl. (1900) $+ 7° 28',6$] die Nummer 034707. Diese Art der Numerierung ist zwar insofern zweckmäßig, als sie eine ungefähre Ortsangabe ersetzt, aber sie wird nicht dauernd beibehalten werden können, da es in fernerer Zukunft keinen Sinn haben wird, die Numerierung von dem Ort des Sternes im Jahre 1900 abhängig zu machen.

5. Die Lichtkurven der veränderlichen Sterne. Die Lichtkurve eines veränderlichen Sternes erhält man, wenn man die Zeiten (meist in julianischen Tagen ausgedrückt) als Abszissen, die zugehörigen Helligkeiten (in Größenklassen ausgedrückt) als Ordinaten in ein rechtwinkliges Koordinatensystem einträgt und

[1] A N 199, S. 215 (1914). [2] Publ A S P 27, S. 209 (1915).
[3] A N 154, S. 77 (1900.)

die erhaltenen Punkte durch eine Kurve verbindet. Aus der Lichtkurve kann man die Zeiten der Helligkeitsmaxima und -minima direkt ablesen oder durch exaktere Methoden ermitteln. Näheres über die Konstruktion der Lichtkurven, ihre Auswertung, rechnerische Darstellung usw. findet sich in HAGENS zitiertem Werk, und wir verweisen ferner auf eine Abhandlung von W. HASSENSTEIN „Untersuchungen über einige allgemeine Eigenschaften einförmiger Lichtkurven"[1]. Für das folgende werden die an diesen Stellen zu findenden Ausführungen nicht gebraucht. Von der Beschaffenheit der Lichtkurven der verschiedenen Klassen von veränderlichen Sternen wird in unseren weiteren Betrachtungen ausführlich die Rede sein.

6. Die Klassifikation der veränderlichen Sterne. Die Klassifikation der Veränderlichen bietet insofern große Schwierigkeiten, als wir die Ursachen der Lichtschwankungen außer bei den Bedeckungs-Veränderlichen nicht kennen, also auch nicht als maßgebend für die Klassifizierung verwenden können. Immerhin wird man das wenige, was man über die physische Beschaffenheit der Veränderlichen weiß, hauptsächlich auch die spektralanalytischen Ergebnisse, bei der Klassifizierung mitsprechen lassen; wir werden also z. B. einen Stern wie W Ursae majoris, von dem wir wissen, daß er ein Bedeckungs-Veränderlicher ist, nicht in dieselbe Klasse wie ζ Geminorum einreihen, obwohl die Lichtkurven beider Sterne große Ähnlichkeit haben, denn wir wissen, daß der Lichtwechsel von ζ Geminorum durch ganz andere Ursachen hervorgerufen sein muß. Aus diesem Grunde ist auch eine Klassifikation, die sich einzig und allein auf die Beschaffenheit der Lichtkurven stützt, heutzutage nicht mehr brauchbar.

Angesichts der großen Wichtigkeit, die die Frage der Klassifikation bietet, muß hier ausführlich auf die verschiedenen, bisher vorgeschlagenen Systeme eingegangen werden. Es sei betont, daß manche der hier folgenden Erörterungen nur für den mit den Haupteigenschaften der verschiedenen Klassen schon einigermaßen vertrauten Leser verständlich sein werden. Diese Eigenschaften können aber erst weiterhin bei der eingehenden Besprechung der einzelnen Klassen näher erörtert werden.

Wir übergehen hier die ältesten Klassifikationen, die sich auf ein noch ganz unzureichendes Material stützten, nämlich die von PIGOTT[2], OLBERS[3] und H. J. KLEIN[4]. E. C. PICKERING hat dann 1881 eine Einteilung aufgestellt[5], die lange Zeit im Gebrauch war, ja manchmal sogar jetzt noch angewandt wird. Auch im Second Catalogue of Variable Stars[6] wird sie gebraucht. PICKERING stellt folgende Klassen auf:

I. Neue Sterne (z. B. Brahes Nova von 1572 und die Nova Coronae von 1866 = T Coronae);

II. Sterne, die große Lichtschwankungen in Perioden von mehreren Monaten oder Jahren erleiden (z. B. o Ceti und χ Cygni);

III. Sterne, die kleine Lichtschwankungen nach bisher unbekannten Gesetzen erleiden (z. B. α Orionis und α Cassiopeiae);

IV. Sterne, deren Helligkeit kontinuierlich veränderlich ist; die Änderungen vollziehen sich sehr regelmäßig in Perioden, die einige Tage nicht überschreiten (z. B. β Lyrae und δ Cephei);

V. Sterne, welche in Intervallen von wenigen Tagen und in regelmäßiger Wiederkehr für kurze Zeit eine Verminderung ihrer Helligkeit erleiden (z. B. β Persei und S Cancri).

---

[1] A N 219, S. 373. (1923). [2] Phil Trans 76, S. 214 (1786).
[3] Lindenaus und Bohnenbergers Zeitschr. 2, S. 181 (1816).
[4] Sirius 6, S. 278 (1874). [5] Proc Amer Acad, New Series, 8, S. 17 u. 257 (1881).
[6] Harv Ann 55, Part I (1907).

Schon bei Erscheinen des Second Catalogue war diese Einteilung eigentlich völlig unbrauchbar. Klasse IV umfaßt z. B. so verschiedene Sterne wie β Lyrae, von dem schon damals bekannt war, daß er ein Bedeckungs-Veränderlicher ist, und δ Cephei, den typischen Vertreter einer Klasse, bei der der Lichtwechsel sicher nicht durch Bedeckungen zustande kommt. Auch ist in der Einteilung kein Platz für unregelmäßige Veränderliche mit großen Lichtschwankungen, wie R Coronae und U Geminorum. In der Tat ist es ganz verfehlt, wenn, wie es im Second Catalogue geschieht, diese Sterne als zur Klasse II gehörig bezeichnet werden. (In den Anmerkungen wird übrigens auf das Bedenkliche dieser Einreihung hingewiesen.)

Eine wesentliche Verbesserung seiner Klassifikation nahm PICKERING 1911 vor[1]. Er behält hier die obigen 5 Klassen bei, führt aber folgende Unterabteilungen ein:

Ia. Normale neue Sterne (z. B. Nova T Aurigae 1, Nova Persei 2);

Ib. Neue Sterne in Nebelflecken (z. B. Nova S Andromedae 1, Nova Z Centauri 1);

IIa. Langperiodische Veränderliche des normalen Typus (z. B. o Ceti, χ Cygni);

IIb. Sterne, die gewöhnlich schwach sind, aber in unregelmäßigen Intervallen plötzlich hell werden und dann allmählich wieder abnehmen (z. B. U Geminorum, SS Cygni, SS Aurigae);

IIc. Sterne, die gewöhnlich hell sind, aber in unregelmäßigen Intervallen schwach werden und dann ihre normale Helligkeit wieder erreichen (z. B. R Coronae, RY Sagittarii, SU Tauri);

III. Sterne mit unregelmäßigen Änderungen (z. B. α Orionis und R Scuti);

IV. Sterne mit kurzen, sehr nahe regelmäßigen Perioden (z. B. δ Cephei und ζ Geminorum);

V. Definition wie früher (z. B. β Persei und δ Librae), doch werden jetzt zu dieser Klasse auch Sterne wie β Lyrae gerechnet, die abwechselnd schwache und weniger schwache Minima haben.

Diese Klassifikation ist der früheren weit vorzuziehen, leidet aber auch noch an erheblichen Mängeln. So ist es nicht angängig, die Mira-Sterne (IIa), U Geminorum-Sterne (IIb) und R Coronae-Sterne (IIc) in einer Hauptklasse (II) zu vereinigen. Auch ist zwischen Sternen wie α Orionis und R Scuti (beide in III) sicherlich ein Unterschied zu machen. Die Teilung der Klasse I in Ia und Ib dürfte wohl überflüssig sein.

STANLEY WILLIAMS stellte 1912 eine Klassifikation auf[2], die nicht zu empfehlen ist. Sie kann hier übergangen werden, zumal die Bedenken dagegen von S. D. TOWNLEY[3] ausführlich dargelegt worden sind. TOWNLEY selbst schlägt folgende Einteilung vor:

I. Neue Sterne;

II. Langperiodische Veränderliche;

III. Unregelmäßige Veränderliche;

IV. Kurzperiodische Veränderliche [a) δ Cephei-Typus, b) ζ Geminorum-Typus, c) „Cluster“- oder Antalgol-Typus];

V. Bedeckungs-Veränderliche [a) Algol-Typus, b) β Lyrae-Typus];

VI. Veränderliche in den MAGELLANschen Wolken.

Es ist zunächst nicht einzusehen, zu welchem Zwecke die Klasse VI eingeführt ist, da die Veränderlichen in den MAGELLANschen Wolken sich nicht von anderen zu unterscheiden scheinen. Dann aber ist die Klasse III viel zu

[1] Harv Circ 166 (1911). [2] J B A A 23, S. 133 (1912/13).
[3] Publ A S P 25, S. 239 (1913).

umfassend, da sie Objekte der verschiedensten Art (z. B. α Orionis, R Coronae, R Scuti) enthalten müßte. Gerade die Definition der verschiedenen Arten von unregelmäßigen Veränderlichen bietet Schwierigkeiten und darf bei keiner Klassifikation mit Stillschweigen übergangen werden.

Im gleichen Jahre, 1913, wie TOWNLEY veröffentlichte auch A. A. NIJLAND seine Vorschläge hinsichtlich der Einteilung der veränderlichen Sterne[1]; er hat sie kurz darauf in den A N[2] rekapituliert. Nach letzterer Quelle ist die Klassifikation NIJLANDS folgende:

I. Regelmäßige Veränderliche:
a) Sterne wie Algol,
b) Sterne wie β Lyrae,
c) Sterne wie ζ Geminorum,
d) Sterne wie δ Cephei,
e) Cumuliden (Antalgol-Sterne),
f) Sterne wie S Sagittae;

II. halbregelmäßige Veränderliche:
a) Sterne wie Mira Ceti,
b) Sterne wie U Geminorum,
c) andere halbregelmäßige Sterne (SS Cygni, RV Tauri, η und W Geminorum);

III. unregelmäßige Veränderliche:
a) neue Sterne,
b) andere unregelmäßige Sterne (R Coronae, RX Andromedae usw.).

Auch diese Einteilung genügt den heutigen Ansprüchen nicht mehr. Die Einreihung der Bedeckungs-Veränderlichen (Ia, Ib) und der δ Cephei-Sterne in weiterem Sinne (Ic—If) in eine Klasse I ist nicht angängig. Die Klasse IIc enthält Objekte der verschiedensten Art, auch gehört SS Cygni in die Klasse IIb. Ebenso umfaßt auch IIIb ganz verschiedenartige Objekte, wie R Coronae und die roten unregelmäßigen Veränderlichen mit kleinen Lichtschwankungen.

Ebenfalls im Jahre 1913 veröffentlichte J. G. HAGEN in seinem schon öfters zitierten Werke[3] eine Einteilung, die sich in noch höherem Grade als die von NIJLAND allein auf äußere Merkmale der Lichtkurven stützt. Von diesem Standpunkte aus ist sie einwandfrei, aber, wie schon erwähnt, kommt eine solche Klassifikation für die Gegenwart nicht mehr in Betracht, da wir die physikalische Beschaffenheit der Veränderlichen nicht außer acht lassen dürfen, wenn wir auch die Ursache des Lichtwechsels noch nicht kennen.

Im Katalog der Veränderlichen, der im dritten Bande der G. u. L. abgedruckt ist, kommt nur eine ziemlich rohe Einteilung in 6 Klassen zur Verwendung, die wohl kaum beansprucht, als wirkliche Klassifikation angesehen zu werden.

Einen Fortschritt bedeuten dann die Klassifikationen von P. GUTHNICK (1921) und von K. GRAFF (1922). GUTHNICK[4] unterscheidet folgende Klassen von Veränderlichen nach ihren typischen Vertretern:

I. Neue Sterne,
II. R Coronae,
III. U Geminorum,
IV. μ Cephei, η Geminorum, R Scuti,
V. Mira Ceti,
VI. R Sagittae,
VII. δ Cephei, ζ Geminorum, RR Lyrae (δ Cephei-, ζ Geminorum-, Antalgol-Typus),
VIII. Bedeckungs-Veränderliche (Algol- und β Lyrae-Typus).

---

[1] Hemel en Dampkring 10, S. 184 (1913).
[2] A N 199, S. 209 (1914). [3] I. Band, S. 56.
[4] Die Kultur der Gegenwart. III. Band: Astronomie S. 435. Unter Redaktion von J. HARTMANN. Leipzig u. Berlin: Teubner 1921.

Ein Mangel dieser Einteilung ist der, daß Klasse IV so verschiedene Objekte wie μ Cephei, η Geminorum und R Scuti umfaßt. R Scuti gehört offenbar in die Klasse VI, und η Geminorum ist, wie wir später sehen werden, in die Mira Ceti-Klasse einzureihen, wenn er auch in dieser Klasse eine extreme Form bildet. Keinen Platz in dieser Klassifikation haben die Nova-ähnlichen Veränderlichen, wie T Pyxidis.

GRAFFS[1] Klassifikation unterscheidet 4 Hauptklassen (unter Fortlassung der neuen Sterne):

I. Die roten, meist periodischen Veränderlichen:
  a) langperiodische Veränderliche vom Mira-Typus,
  b) unregelmäßige Veränderliche vom Typus μ Cephei;

II. die gelblichweißen, nichtperiodischen Veränderlichen:
  a) Sterne mit Andeutungen von Gesetzmäßigkeiten (Typus U Geminorum),
  b) unregelmäßige Veränderliche (Typus R Coronae);

III. die Cepheiden:
  a) Sterne mit mehrtägigen Perioden (Typus δ Cephei),
  b) kurzperiodische Sterne (Typus RR Lyrae);

IV. die Bedeckungs-Veränderlichen:
  a) Sterne mit lichtschwachen Begleitern (Algol-Typus),
  b) Sterne mit hellen Komponenten (Typus β Lyrae).

Das größte Bedenken, das sich gegen diese Klassifikation richtet, besteht darin, daß sie keinen Platz für die Sterne vom Typus RV Tauri, R Scuti, R Sagittae bietet, falls man diese nicht etwa auch in die Klasse IIa einordnen will, wo sie dann mit den ganz von ihnen verschiedenen U Geminorum-Sternen vereint wären. Auch ist zu bemerken, daß die R Coronae-Sterne keineswegs alle gelblichweiß sind; S Apodis besitzt z. B. ein Spektrum der Klasse R.

Den folgenden Ausführungen wird eine Klassifikation zugrunde gelegt, die H. LUDENDORFF in seiner Abhandlung „Über die Beziehungen der verschiedenen Klassen der veränderlichen Sterne"[2] angewandt hat. Sie dürfte dem heutigen Stande der Kenntnis am besten entsprechen, ohne daß sie deswegen den Anspruch erheben kann, etwas Endgültiges zu bieten. Nur ganz vereinzelte veränderliche Sterne lassen sich in dieser Klassifikation nicht unterbringen und diese vielleicht zum Teil auch nur, weil ihr Lichtwechsel noch nicht genau genug bekannt ist. Auch muß von vornherein hervorgehoben werden, daß die einzelnen Klassen der Veränderlichen nicht durchweg scharf gegeneinander abgegrenzt sind; so gibt es z. B. Sterne, die Übergangsformen zwischen dem Mira Ceti-Typus und dem δ Cephei-Typus zu bilden scheinen, so daß man nach dem heutigen Stande der Kenntnis nicht entscheiden kann, zu welchem dieser beiden Typen sie gerechnet werden müssen. Auf die Aufzählung der Klassen folgt eine kurze Definition derselben; ausführlicher werden die Eigenschaften der einzelnen Klassen dann in den folgenden Unterabteilungen dargelegt. Es liegt in der Natur der Sache, daß die Definitionen nicht völlig scharf sein können. Auch lassen sich die Klassen nicht, wie es bei den Spektralklassen der Fixsterne der Fall ist, in eine zwangsläufige Reihenfolge bringen, sondern es läßt sich in der Reihenfolge eine gewisse Willkür nicht vermeiden. Die Beziehungen der Klassen zueinander werden später behandelt werden.

Die Klassen sind folgende:

I. Neue Sterne (Novae), z. B. N Persei 2 (1901), N Aquilae 3 (1918);

---

[1] SCHEINER-GRAFF, Astrophysik, S. 386. Leipzig u. Berlin: Teubner 1922.

[2] In: Probleme der Astronomie. Festschrift für HUGO VON SEELIGER. Berlin: Julius Springer 1924.

II. Nova-ähnliche Sterne, z. B. T Pyxidis, η Carinae;
III. R Coronae-Sterne, z. B. R Coronae, RY Sagittarii, SU Tauri;
IV. U Geminorum-Sterne, z. B. U Geminorum, SS Aurigae, SS Cygni;
V. Mira-Sterne, z. B. o Ceti, χ Cygni;
VI. μ Cephei-Sterne, z. B. μ Cephei, ϱ Persei;
VII. RV Tauri-Sterne, z. B. RV Tauri, R Scuti, R Sagittae;
VIII. langperiodische δ Cephei-Sterne, z. B. δ Cephei, η Aquilae;
IX. kurzperiodische δ Cephei-Sterne, z. B. RR Lyrae und die überwiegende Mehrzahl der Veränderlichen im Sternhaufen ω Centauri;
X. Bedeckungs-Veränderliche, z. B. β Persei, β Lyrae, W Ursae majoris.

Die Klassen II, III, IV, VI und VII könnte man unter der Bezeichnung „unregelmäßige Veränderliche" zusammenfassen; es handelt sich aber in diesen Klassen um Objekte so verschiedener Art, daß eine solche Zusammenfassung höchstens aus Gründen der Bequemlichkeit statthaft erscheint. Die genannten Klassen folgen daher in der obigen Übersicht auch nicht alle unmittelbar aufeinander.

Die Definitionen der einzelnen Klassen sind kurz folgende:

I. Neue Sterne. Sterne mit einem meist nur ziemlich kurze Zeit währenden Helligkeitsmaximum, zu dem die Helligkeit sehr rasch ansteigt, und das etwa vorhandene andere Maxima bei weitem an Bedeutung übertrifft, so daß letztere höchstens als sekundär bezeichnet werden können. Die Abnahme der Helligkeit nach dem Hauptmaximum erfolgt weit langsamer als die Zunahme.

II. Nova-ähnliche Sterne. Sterne, die manche Merkmale der neuen Sterne zeigen, aber sich doch auch wieder von diesen in anderer Hinsicht unterscheiden. Die wenigen Objekte dieser Art werden weiterhin näher besprochen werden; eine strengere Definition läßt sich nicht aufstellen.

III. R Coronae-Sterne. Diese sind dadurch charakterisiert, daß sie manchmal lange Zeit bei konstanter Helligkeit im Maximum bleiben. Diese Zeiten normaler Helligkeit werden in unregelmäßigen Intervallen unterbrochen durch meist scharf einsetzende Minima, in denen die Sterne oft ebenfalls lange Zeit unter bedeutenden Lichtschwankungen verharren.

IV. U Geminorum-Sterne. Die Helligkeit dieser Sterne wächst aus einem Minimum heraus sehr rasch und in unregelmäßigen Intervallen bedeutend an und nimmt dann wieder ungefähr bis zu der früheren Minimalgröße ab; die Abnahme erfolgt langsamer als die Zunahme. Bei vielen dieser Sterne ist die Helligkeit im Minimum längere Zeit konstant oder nahezu konstant.

V. Mira-Sterne. Die Lichtschwankungen umfassen im allgemeinen mehrere Größenklassen und gehen in meist nicht sehr stark veränderlichen Perioden von etwa 90^d bis zu etwa 600^d vor sich.

VI. μ Cephei-Sterne. Es sind dies rötliche oder rote Sterne, die unregelmäßigen, in den meisten Fällen ziemlich kleinen Lichtschwankungen unterliegen.

VII. RV Tauri-Sterne. Veränderliche, bei denen zwischen zwei Hauptminima in der Regel ein sekundäres Minimum eintritt; die sekundären Minima bleiben aber zuweilen aus, auch vertauschen sie sich zuweilen mit den Hauptminima, und die Lichtkurven, sowie manchmal auch die Abstände der Hauptminima, sind stark veränderlich.

VIII. Langperiodische δ Cephei-Sterne. Der nicht sehr umfangreiche Lichtwechsel geht sehr regelmäßig in Perioden von etwa 1^d bis zu etwa 45^d vor sich. Der Lichtwechsel kann nicht durch Verfinsterungserscheinungen erklärt werden.

IX. Kurzperiodische δ Cephei-Sterne. Ebenso wie VIII, doch ist die Periode kürzer als etwa 1^d. Eine besondere Abart dieser Klasse sind die sog. β Cephei-Sterne, auf die wir später näher eingehen werden.

X. Bedeckungs-Veränderliche. Die Lichtschwankungen rühren von der Bedeckung durch einen Begleiter her.

Die Mannigfaltigkeit der Erscheinungen, die die veränderlichen Sterne bieten, ist so groß, daß man manche Objekte in eine der aufgezählten Klassen einordnen wird, auch wenn die Definitionen, wie sie oben gegeben sind, nicht völlig passen. Z. B. wird man einen periodischen Veränderlichen wie S Vulpeculae, der eine Periode von etwa 68^d hat, entweder in die Klasse der Mira-Sterne (Perioden nach Definition 90^d bis 600^d) oder in die der langperiodischen δ Cephei-Sterne (Perioden 1^d bis 45^d) einzuordnen haben. In welche Klasse er wirklich gehört, oder ob er etwa in der Tat eine Übergangsform zwischen beiden Klassen bildet, wird sich erst entscheiden lassen, wenn seine spektralen Eigentümlichkeiten näher bekannt sind. Solche Beispiele lassen sich noch mehr anführen. Erwähnt sei noch von vornherein, daß die hauptsächlich von P. GUTHNICK und R. PRAGER photoelektrisch gemessenen Veränderlichen mit sehr kleinen Helligkeitsamplituden hier einstweilen außer acht gelassen worden sind. Es ist sehr wohl möglich, daß für sie noch neue Klassen aufgestellt werden müssen, doch wissen wir noch nicht genug über sie. Veränderliche, die sich absolut nicht in die obige Klasseneinteilung einfügen lassen, gibt es nur in sehr geringer Zahl; sie sind besonders zu besprechen.

7. Allgemeine Untersuchungen über die veränderlichen Sterne. Da die verschiedenen Klassen der Veränderlichen z. T. sehr verschiedene Eigenschaften haben, so werden sie jetzt bei statistischen Untersuchungen, die die Auffindung von Gesetzmäßigkeiten bezwecken, voneinander getrennt behandelt. Früher dagegen faßte man manchmal die Gesamtheit der Veränderlichen als Einheit auf und suchte aus dem Gesamtmaterial allgemeine Gesetze abzuleiten. Diese Untersuchungen haben meist nur noch historische Bedeutung, und es können nur wenige von ihnen hier erwähnt werden, aus denen einige besonders interessante Eigenschaften der Veränderlichen hervorgehen. Schon HARDING versuchte in seinen „Astronomischen Ephemeriden" für 1831 gewisse Regeln über die Veränderlichen aufzustellen, die aber im ganzen naturgemäß noch ziemlich vage sind. SCHÖNFELD erneuerte diesen Versuch in einem Vortrage, der im „29. Jahresbericht des Mannheimer Vereins für Naturkunde" (1863) abgedruckt ist. Es ist SCHÖNFELD schon damals aufgefallen, daß unter den Perioden die ganz kurzen von wenigen Tagen und die von etwa einem Jahre vorwiegen, daß bei den meisten Veränderlichen die Helligkeit rascher zu- als abnimmt, und daß auffallend viele Veränderliche rötlich gefärbt sind. Diese ganz allgemeinen Gesetze sind später durch ein weit größeres Beobachtungsmaterial durchaus bestätigt worden. Um das Gesetz über die Häufigkeit der Periodenlängen klar hervortreten zu lassen, genügt es, hier eine kleine Tabelle abzudrucken, die P. KEMPF in der fünften Auflage von NEWCOMB-ENGELMANNS „Populärer Astronomie" 1914 gegeben hat (siehe die nebenstehende Tabelle).

Periodenlänge	Zahl der Veränderlichen
$<10^d$	183
10^d bis 100^d	64
100 „ 200	64
200 „ 300	194
300 „ 400	182
400 „ 500	54
$>500^d$	12

Über die Farbe der Veränderlichen fand schon J. SCHMIDT[1] das merkwürdige Gesetz, daß diese Objekte um so stärker rot gefärbt sind, je länger die Periode des Lichtwechsels ist. Auch dieses Gesetz hat sich in der Folge bestätigt. Die ausführlichste allgemeine Untersuchung darüber rührt von BELIAWSKY[2] her; sie beruht auf dem zweiten Harvard-Katalog der veränderlichen

[1] A N 80, S. 12 (1873). [2] A N 177, S. 209 (1908).

Sterne. BELIAWSKY drückt die Farben in der OSTHOFFschen Skala aus (0 = weiß, 10 = hellrot, 12 = dunkelrot). In der folgenden von BELIAWSKY aufgestellten Tabelle enthält die erste Kolumne die Klassen der Veränderlichen nach E. C. PICKERING, die zweite die betreffenden Periodenintervalle, die dritte die mittleren Perioden, die vierte die entsprechenden mittleren Farben und die letzte die Zahl der benutzten Sterne:

Klasse	Periode	Mittlere Periode	Mittlere Farbe	Anzahl der Sterne
V	Algol-Typus	—	0,8	13
IV	kurzperiod.	10^d	2,4	20
II	$<100^d$	80	3,4	4
II	100^d bis 200^d	163	5,1	24
II	200 „ 250	226	4,1	34
II	250 „ 300	274	5,5	35
II	300 „ 350	325	5,8	49
II	350 „ 400	374	6,7	36
II	400 „ 450	418	7,4	24
II	450 „ 500	474	7,9	7
III	unregelm.	—	7,3	44

Die allgemeine Verteilung der Veränderlichen nach der galaktischen Breite ist von E. ZINNER[1] untersucht worden. Er stützt sich dabei auf den Katalog der Veränderlichen in HARTWIGS Ephemeriden für 1912. Die Anhäufungen von Veränderlichen in Sternhaufen und Nebeln, sowie in einzelnen besonders auf Veränderliche untersuchten Teilen des Himmels wurden ausgeschlossen. So gelangte ZINNER zu der folgenden Tabelle [n bedeutet die Zahl der Veränderlichen in jeder Zone galaktischer Breite; in der vorletzten Kolumne steht die „Dichte“ der Veränderlichen in jeder Zone, bezogen auf diejenige in der Zone +10° bis —10° als Einheit, in der letzten die entsprechende „Dichte“ für die Gesamtheit der Sterne (nach SEELIGERS Sternabzählungen)]:

Galaktische Breite	n	Dichte: Veränderliche Sterne	Dichte: Alle Sterne
+90° bis +70°	10	0,14	0,35
+70 „ +50	62	0,29	0,37
+50 „ +30	101	0,31	0,45
+30 „ +10	248	0,63	0,68
+10 „ −10	422	1,00	1,00
−10 „ −30	238	0,60	0,77
−30 „ −50	93	0,29	0,47
−50 „ −70	43	0,20	0,41
−70 „ −90	17	0,23	0,38

Es ergibt sich aus dieser Tabelle, daß die Veränderlichen stärker nach der Milchstraße hin konzentriert sind als die Gesamtheit der Sterne. Dieses Phänomen dürfte hauptsächlich von der starken Konzentration der langperiodischen δ Cephei-Sterne und der Bedeckungs-Veränderlichen in der Nähe der Milchstraße herrühren.

8. Die Anzahl der veränderlichen Sterne. Über die Zahl der bisher bekannten veränderlichen Sterne haben wir bereits früher (Ziff. 3 γ) eine Angabe gemacht. Die Ephemeriden für 1926 in der Vierteljahrsschrift der Astronomischen Gesellschaft zählen 2671 Veränderliche auf, wobei aber die in Sternhaufen zusammengedrängten ausgeschlossen sind. In der Großen MAGELLANschen Wolke sind allein über 800, in der Kleinen fast 1000 veränderliche Sterne auf dem Harvard-Observatorium aufgefunden worden Überhaupt nimmt die Zahl der Veränderlichen naturgemäß stark zu, sobald man auch die sehr schwachen Sterne in den Kreis der Beobachtungen zieht. So fand M. WOLF[2] zahlreiche schwache Veränderliche in der Milchstraße in den Gegenden von δ und γ Aquilae, β Cygni, γ Lyrae und γ Sagittae, deren Lichtwechsel noch nicht näher untersucht ist. Auch die Milchstraßenwolken im Sagittarius und im Scutum enthalten zahlreiche Veränderliche[3], und solcher Fälle ließen sich noch mehr anführen. Die Gesamt-

[1] A N 190, S. 17 (1911).
[2] A N 164 bis 172 an verschiedenen Stellen.
[3] Harv Bull 804, 809; Harv Circ 265 (1924).

zahl der Veränderlichen abzuschätzen, ist daher nur möglich, wenn man eine untere Helligkeitsgrenze, die die Sterne in ihrem Helligkeitsmaximum mindestens erreichen, festsetzt. So glaubte E. C. PICKERING[1] 1907 auf Grund der speziellen Untersuchung einiger Regionen der photographischen Harvard Map of the Sky auf das Vorhandensein neuer Veränderlicher hin aussprechen zu dürfen, daß damals am nördlichen Himmel etwa ein Drittel, am südlichen etwa die Hälfte derjenigen Veränderlichen unbekannt waren, die im Maximum die Größe 10,5 erreichen.

H. SHAPLEY[2] findet, daß 3% der Sterne bis zur Größe 6,0 veränderlich sind.

Wir gehen nach diesen allgemeinen Erörterungen nunmehr zur Betrachtung der einzelnen Klassen von veränderlichen Sternen über. Wie schon erwähnt, werden die neuen Sterne und die Bedeckungs-Veränderlichen in anderen Kapiteln dieses Werkes behandelt.

b) Die Nova-ähnlichen Veränderlichen.

9. Definition der Nova-ähnlichen Veränderlichen. Eine scharfe Definition der Nova-ähnlichen Veränderlichen läßt sich nicht geben, und eine scharfe Grenze zwischen ihnen und den eigentlichen neuen Sternen zu ziehen ist unmöglich. Es handelt sich hier um einige Objekte, die in vieler Beziehung mit den neuen Sternen große Ähnlichkeit haben, in anderer sich aber doch wieder von ihnen unterscheiden. Besonders H. SHAPLEY hat in einer Arbeit[3] „Novae and Variable Stars" auf diese Sterne aufmerksam gemacht.

10. Die einzelnen Nova-ähnlichen Veränderlichen. Diese Sterne sind so eigentümlich und interessant, daß sie einzeln besprochen werden müssen, zumal ihre Zahl nur gering ist.

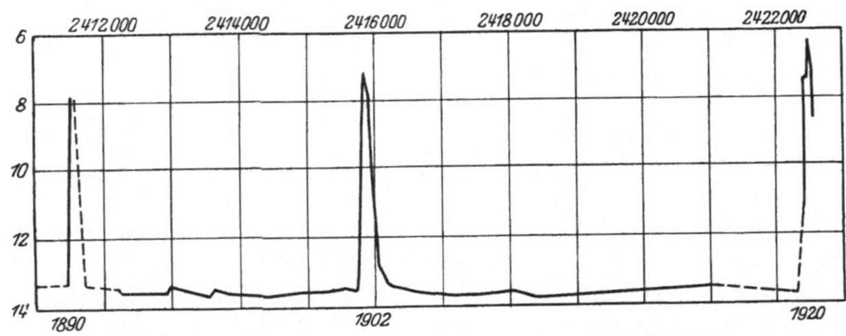

Abb. 1. Photographische Lichtkurve von T Pyxidis.

1. T Pyxidis ist wohl das schönste Beispiel für einen Nova-ähnlichen Veränderlichen. Die Helligkeit dieses Sternes liegt in der Regel zwischen 13^m und 14^m (phot.) und ist während dieser Zeit konstant oder höchstens ganz geringen Schwankungen unterworfen. Wie man auf photographischen Aufnahmen festgestellt hat, ist der Stern aber sowohl 1890 wie 1902 plötzlich bis etwa zur Größe 7,5 (phot.) aufgeflammt, und die Helligkeit hat nachher wieder ziemlich rasch bis zu dem ursprünglichen geringen Betrage abgenommen. Ein drittes Aufleuchten erfolgte[4] wiederum sehr rasch 1920, und der Stern wurde diesmal noch etwas heller als 1890 und 1902. Die Lichtkurve ist in Abb. 1 nach der in SHAPLEYS zitierter Abhandlung enthaltenen Abbildung wiedergegeben.

[1] Harv Circ 130 (1907). [2] Ap J 41, S. 305 (1915) = Mt Wilson Contr 99.
[3] Publ A S P 33, S. 185 (1921). [4] Harv Bull 716 (1920).

Während des Maximums von 1920 ist auch das Spektrum des Sternes untersucht worden. W. S. ADAMS und A. H. JOY[1] stellten mit Hilfe von Aufnahmen, die mit dem 100zölligen Reflektor des Mt. Wilson-Observatoriums gewonnen waren, fest, daß das Spektrum von T Pyxidis damals die charakteristischen Eigentümlichkeiten des Spektrums der neuen Sterne besaß. Die Wasserstoffemission trat in Form von breiten Banden auf, auf deren brechbarerer Seite sich sehr starke, in mehrere Komponenten geteilte Absorptionsbanden desselben Elementes befanden. Außerdem waren schwächere Emissionsbanden sichtbar, die sich meist mit Funkenlinien des Fe und anderer Elemente identifizieren ließen, sowie verwaschene Absorptionslinien des O und N. Das Spektrum hatte große Ähnlichkeit mit dem der Nova Aquilae von 1918 zehn Tage nach ihrer Entdeckung. Die Absorptionslinien waren proportional der Wellenlänge um starke Beträge verschoben. Eine weitere Notiz über das Spektrum hat M. L. HUMASON[2] gegeben.

Trotz des ganz den neuen Sternen entsprechenden spektralen Verhaltens von T Pyxidis kann man diesen Stern doch nicht als eigentliche Nova betrachten, denn er ist in 30 Jahren dreimal zu etwa gleich hellen Maxima aufgeflammt. Allenfalls könnte man ihn zu den U Geminorum-Sternen rechnen.

2. η Carinae. Dieser auch als η Argus bezeichnete Stern verhält sich ganz anders als T Pyxidis, hat aber entschieden auch Ähnlichkeit mit den neuen Sternen. Die Helligkeiten in älterer Zeit waren:

Ende des 17. Jahrhunderts	3^m bis 4^m
Mitte des 18. Jahrhunderts	2^m „ 3^m
1811—1815	etwa 4^m
1822—23	wahrscheinlich 2^m
1827	1^m

1829 war η Carinae dann wieder etwas schwächer, 1834—1837 1^m—2^m. Ende 1837 wurde der Stern sehr hell und der weitere Lichtwechsel bis 1902 ist

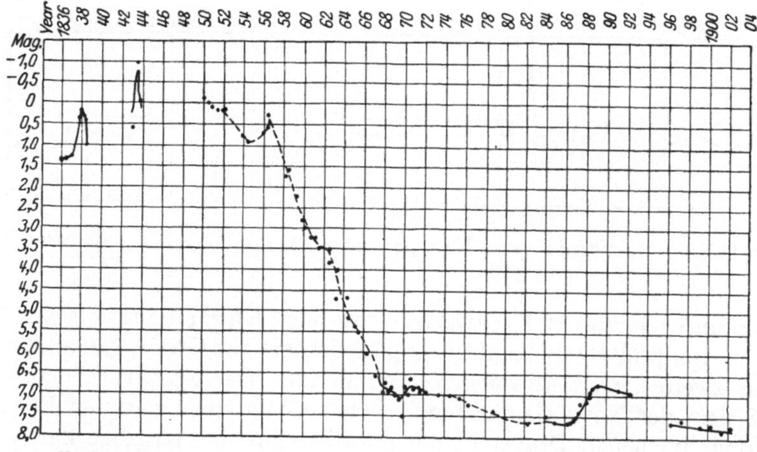

Abb. 2. Visuelle Lichtkurve von η Carinae nach INNES. (Annals of the Cape Obs, Vol. IX.)

aus der Abb. 2 zu ersehen. Am hellsten war der Stern wohl Anfang 1843; seitdem ist er schwach geworden, und nach Miss CANNON[3] ist die Helligkeit von Mai 1895 bis 1920 ziemlich unverändert geblieben (8^m,3 phot.).

[1] Pop Astr 28, S. 514 (1920). [2] Publ A S P 32, S. 200 (1920).
[3] Pop Astr 28, S. 524 (1920).

Die Helligkeitsänderungen um das Maximum herum und auch die darauf folgende Abnahme sind also sehr viel langsamer verlaufen, als es bei neuen Sternen sonst der Fall ist. Man muß daher Bedenken tragen, η Carinae zu den neuen Sternen zu rechnen, obwohl dies von mancher Seite geschieht. Die G. u. L. zählt η Carinae zu den unregelmäßigen Veränderlichen.

Das Spektrum dieses Sterns ist in den letzten Jahrzehnten mehrfach untersucht worden, zuerst wohl von D. GILL[1] auf Grund von Aufnahmen aus dem Jahre 1899. Er hob hervor, daß das Spektrum Ähnlichkeit mit dem der Nova T Aurigae von 1891 habe. Auch Miss A. J. CANNON fand[2] darin zahlreiche helle Banden, ähnlich wie bei der eben genannten Nova, außerdem aber ein Absorptionsspektrum der Klasse F5; sie konstatierte, daß das Spektrum veränderlich sei. Eine eingehende Untersuchung rührt von J. H. MOORE and R. F. SANFORD[3] her; sie fanden, daß das Spektrum 1912 bis 1914 im wesentlichen aus hellen Linien bestand, von denen eine Anzahl mit Funkenlinien des Fe, Ti und Cr identifiziert werden konnten. Die hellen Wasserstofflinien waren vielleicht doppelt; Absorptionslinien ließen sich nicht feststellen. Das Spektrum erinnerte an das der neuen Sterne „in the early period of their history", und das ist sehr auffällig, da das Maximum von η Carinae so weit zurückliegt. Änderungen der Radialgeschwindigkeit waren nicht zu erkennen. Miss CANNON[4] hat dann später noch bemerkt, daß die dunkeln Linien nur auf Platten aus den Jahren 1892 bis 1893 sichtbar waren, auf Platten aus dem Jahre 1895 und späterer Zeit nicht mehr. Das Spektrum scheint sich während der Helligkeitsabnahme des Sternes 1894 bis 1895 geändert zu haben.

Weitere Untersuchungen des Spektrums rühren von F. E. BAXANDALL[5], J. LUNT[6], W. M. WORSSELL[7], C. D. PERRINE[8] und P. DAVIDOVICH[9] her.

Es ist bekannt, daß η Carinae in einem großen, unregelmäßigen Nebel liegt.

3. RS Ophiuchi. Wieder anders verhält sich der Veränderliche RS Ophiuchi, dessen photographische Lichtkurve Abb. 3 zeigt. Im allgemeinen besitzt er

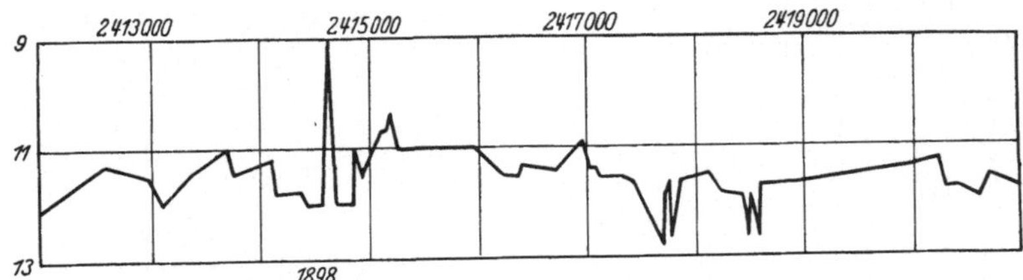

Abb. 3. Lichtkurve von RS Ophiuchi nach Publ A S P 33, S. 191.

einen ganz unregelmäßigen Lichtwechsel zwischen den photographischen Größen 11 und 13. Im Juni 1898 leuchtete er plötzlich bis zur Größe 8,9 auf, und 1900 folgte ein zweites, weit schwächeres Maximum. Das Spektrum[10] ähnelte im Juli 1898 dem der Nova Geminorum 1 von 1903 in einer gewissen Phase ihrer Erscheinung; im neuen Draper-Katalog wird es für die damalige Zeit als Ocp bezeichnet. PICKERING bezeichnet den Stern als Nova Ophiuchi 3, die G. u. L. aber als unregelmäßigen Veränderlichen. Ein sehr ähnliches Spektrum besitzen

[1] M N 61, App. S. 66 (1901).
[2] Harv Circ 59 (1901); Harv Ann 28, S. 175 (1901); 76, S. 36 (1916).
[3] Lick Bull 8, S. 55 u. 134 (1914—15). [4] Pop Astr 28, S. 524 (1920).
[5] M N 79, S. 619 (1919). [6] M N 79, S. 621 (1919). [7] Union Circ 46 (1919).
[8] Publ A S P 38, S. 117 (1926). [9] Harv Bull 837 (1926).
[10] Harv Circ 99 (1905); Berichtigung Harv Ann 76, S. 30 (1916).

CI Cygni und Y Coronae australis, über deren Lichtwechsel wir noch nichts Näheres wissen.

4. P Cygni = Nova Cygni 1 von 1600. Dieser Stern wird in der Regel, und zwar auch in der G. u. L., zu den neuen Sternen gerechnet; er ist aber in seinem Lichtwechsel doch so von den normalen neuen Sternen verschieden, daß er wohl besser der hier besprochenen Klasse von Veränderlichen beizuzählen ist. Die G. u. L.[1] beschreibt seinen Lichtwechsel wie folgt: „Zusammenfassend läßt sich sagen, daß der Stern etwa von 1600 bis 1606 von der 3. Größe war, dann abnehmend im Jahre 1626 anscheinend unsichtbar für das bloße Auge wurde. Im Jahre 1655 wieder sichtbar, war er 1657 bis 1659 wieder von der 3. Größe ($3^m,5$ P.D.), nahm darauf rasch ab und war von 1662 bis 1682 bald als sehr kleiner Stern sichtbar, bald unsichtbar. Dann wieder zunehmend, wurde er von 1715 an als $5^m,2$ P.D. beobachtet, welche Helligkeit er bis zur Gegenwart beibehielt mit Ausnahme der Jahre 1781 bis 1786, wo er um $^1/_3{}^m$ schwächer gewesen zu sein scheint. Jedenfalls gleicht sein Lichtwechsel wenig dem der gewöhnlichen neuen Sterne.“ Das Spektrum, das im neuen Draper-Katalog als B1p bezeichnet wird, ist charakterisiert durch helle und dunkle Linien des H und He, und zwar liegen die dunkeln Linien auf der brechbareren Seite der hellen, wie es auch bei den neuen Sternen der Fall ist. Das Spektrum ist hauptsächlich auf dem Harvard-Observatorium[2] und (von P. W. Merrill) auf dem Lick-Observatorium[3] untersucht worden; auf die Einzelheiten dieser Untersuchungen kann hier nicht eingegangen werden.

Andere Veränderliche, die ein Spektrum besitzen, das dem von P Cygni ähnlich ist, sind S Doradus[4], AG Pegasi[5] und AG Carinae[6]; über den Lichtwechsel derselben wissen wir indessen nur sehr wenig. Es gibt ferner eine Reihe von Sternen, die ebenfalls ein solches Spektrum besitzen, von denen aber noch nicht bekannt ist, ob sie überhaupt veränderlich sind[7].

Die vier besprochenen Sterne (T Pyxidis, η Carinae, RS Ophiuchi, P Cygni) sind wohl die ausgesprochensten Vertreter der hier behandelten Klasse von Veränderlichen. Man kann natürlich kaum viel dagegen einwenden, wenn sie von anderer Seite zu den eigentlichen neuen Sternen gerechnet werden. Es gibt nun noch einige weitere Fälle, die vielleicht hierher gehören. In erster Linie ist dies

5. Z Andromedae. Die Einordnung dieses Veränderlichen in eine andere Klasse ist kaum möglich, doch paßt er auch nicht allzu gut in die der Nova-ähnlichen. Eine photographische Lichtkurve von Z Andromedae für 1887 bis 1910 ist in Harv Circ 168 gegeben; danach hatte der Stern 1901 ein Maximum ($9^m,1$), dem Lichtschwankungen vorangingen und folgten, während derer die Helligkeit bedeutend kleiner war als in diesem Maximum. Abb. 4 gibt die Lichtkurve für 1911 bis 1923 wieder. Es zeigen sich hier drei fast gleich hohe Maxima, die heller sind als das von 1901. Nach Miss Cannon[8] ähnelt das Spektrum dem der neuen Sterne; nach H. H. Plaskett[9] hat der Stern ein Spektrum der Klasse A mit hellen Linien, in welchem auch die Nebellinien sichtbar sind, die nach Plasketts Ansicht von einer den Stern umgebenden Nebelhülle herrühren. Ein ähnliches Spektrum wie Z Andromedae hat nach dem neuen Draper-Katalog SY Muscae, über dessen Lichtwechsel wir noch nichts Näheres wissen, und nach

---

[1] Zweiter Band, S. 445. [2] Harv Ann 28, S. 101, Remark 165 (1897).
[3] Lick Bull 6, S. 156 (1911); 8, S. 24 (1913). [4] Harv Bull 814 (1925).
[5] Lundmark, A N 213, S. 93 (1921); vgl. ferner A N 224, S. 146 (1925); Harv Bull 762 (1922).
[6] Harv Ann 56, S. 183 (1912).
[7] Harv Ann 76, S. 31 (1916); Harv Bull 801 (1924).
[8] Harv Circ 168 (1911); Harv Ann 76, S. 27 (1916). [9] Pop Astr 31, S. 658 (1923).

Harv Bull 826 (1925) CM Aquilae, der 1914 und 1925 um etwa 1^m bzw. 2^m heller war als gewöhnlich.

Vielleicht könnte auch Z Geminorum zur Gruppe der Nova-ähnlichen Veränderlichen gehören, doch wissen wir über die Helligkeitsänderungen dieses Sternes äußerst wenig. RX Puppis, ein Stern, den SHAPLEY[1] ebenfalls hierher rechnet, dürfte wohl eher ein R Coronae-Stern sein und wird später behandelt werden.

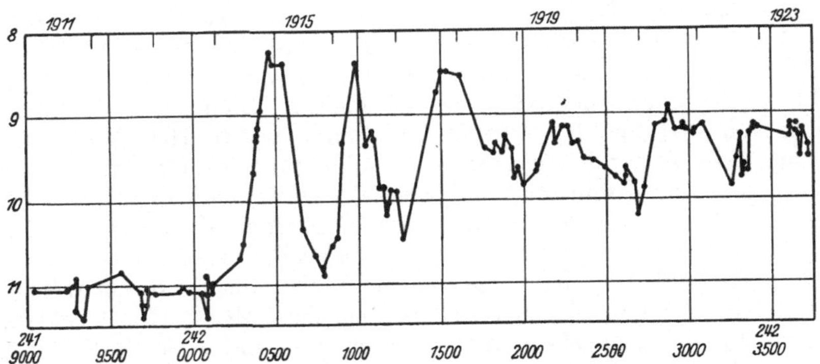

Abb. 4. Lichtkurve von Z Andromedae nach Harv Bull 797.

Auch RY Scuti ist ein Veränderlicher, der vielleicht hier in Betracht kommt. Nach P. W. MERRILL[2] hat dieser Stern ein Spektrum der Klasse Bp, in dem die Linien des H und He, sowie die Nebellinie λ4658 hell sind. Über den geringen Lichtwechsel dieses Sternes wissen wir nichts Näheres.

Endlich sei an dieser Stelle auch noch der höchst merkwürdige Stern DH Carinae erwähnt, an dem E. HERTZSPRUNG[3] ein einmaliges ganz kurzes Aufleuchten um $1^m,8$ beobachtete, das in wenig mehr als einer halben Stunde erfolgt sein muß. $1^h,3$ später hatte der Stern schon wieder eine Größenklasse an Helligkeit verloren. Ob der Stern zur Klasse der Nova-ähnlichen Veränderlichen gehört, ist durchaus zweifelhaft, er könnte auch eine richtige Nova oder aber ein U Geminorum-Stern sein. Ähnlich liegen die Dinge bei dem gleichfalls von HERTZSPRUNG[4] beobachteten Veränderlichen EP Carinae, bei dem das Aufleuchten freilich nicht ganz so plötzlich geschah wie bei DH Carinae.

11. Weitere Bemerkungen über die Nova-ähnlichen Veränderlichen. Wir geben zunächst eine Zusammenstellung der fünf Sterne, die wir mit mehr oder weniger Berechtigung zu dieser Klasse zählen können, mit ihren Örtern für 1900 und dem Abstande g vom galaktischen Äquator [die oben erwähnten zweifelhaften Fälle sind fortgelassen (s. nebenstehende Tabelle)].

	AR	Dekl	g
T Pyxidis	9^h $0^m,5$	$-31°$ $59'$	11°
η Carinae	10 41 ,2	-59 10	0
RS Ophiuchi . . .	17 44 ,8	$-$ 6 41	10
P Cygni.	20 14 ,1	+37 43	1
Z Andromedae . .	23 28 ,9	+48 16	12

Wie die neuen Sterne selbst, so bevorzugen also auch die Nova-ähnlichen Veränderlichen die Nähe der Milchstraße.

Es sei hier noch erwähnt, daß es auch unter den allgemein zu den eigentlichen neuen Sternen gerechneten Objekten einige gibt, die von dem normalen Ver-

[1] Publ A S P 33, S. 192 (1921). [2] Publ A S P 34, S. 134 u. 295 (1922).
[3] B A N 2, S. 87 (1924). [4] B A N 2, S. 209 (1925).

halten der neuen Sterne ziemlich stark abweichen und so den Unterschied zwischen diesen und den hier betrachteten Sternen noch geringer erscheinen lassen. Lediglich als Beispiele führen wir folgende Tatsachen an: Die Nova T Aurigae von 1891, die Nova Persei 1 von 1887 und die Nova Aquilae 4 von 1919 hatten ziemlich langandauernde Maxima. Die Nova Ophiuchi 5 hatte ein dreifaches Maximum, die Nova Ophiuchi 4 ein doppeltes (Lichtkurven dieser beiden Novae finden sich in der zitierten Arbeit von SHAPLEY). Die Nova Sagittarii 5 ist gewöhnlich ein Stern 14. Größe und leuchtete 1919 bis zur 7. Größe auf; aber schon 1901 hatte sie ein schwächeres Maximum 11. Größe.

Die Ursache der Helligkeitsschwankungen der Nova-ähnlichen Veränderlichen muß natürlich sehr ähnlich derjenigen der Lichtschwankungen der neuen Sterne sein. Über letztere wird in einem andern Kapitel dieses Werkes berichtet. Es sei hier nur erwähnt, daß es eine völlig befriedigende Erklärung des Verhaltens der neuen Sterne noch nicht gibt.

c) Die Veränderlichen der R Coronae-Klasse.

12. Definition der R Coronae-Sterne. Das charakteristische Kennzeichen dieser Klasse von Veränderlichen besteht darin, daß die Helligkeit manchmal lange Zeit konstant ist. Die Zeiten dieser „normalen" Helligkeit werden regellos unterbrochen durch oft sehr plötzlich einsetzende Minima, die zuweilen viele Jahre andauern, und während derer der Stern regellosen Helligkeitsschwankungen unterliegt. Nach einem solchen Minimum erreicht der Stern wieder seine normale Helligkeit. Ein Anwachsen über diese hinaus findet niemals oder nur höchst selten statt.

Die R Coronae-Sterne sind Gegenstand einer Untersuchung von H. LUDENDORFF[1] gewesen; eine Ergänzung zu der in dieser Arbeit enthaltenen Liste von R Coronae-Sternen hat er später[2] gegeben.

Auch bei dieser Klasse von Veränderlichen, die in mehr als einer Hinsicht ganz besonderes Interesse verdient, erweist sich eine Besprechung der einzelnen Objekte als nötig, und zwar werden zunächst diejenigen behandelt, die nach dem heutigen Stande der Kenntnis sicher oder höchst wahrscheinlich zur R Coronae-Klasse gehören und die charakteristischen Merkmale derselben besitzen; darauf folgen dann die, bei denen die Zugehörigkeit zu dieser Klasse wegen zu geringen Umfanges des Beobachtungsmateriales noch zweifelhaft ist, oder die in ihrem Verhalten gewisse Abweichungen von der gegebenen Definition zeigen.

13. Die sicher oder höchst wahrscheinlich zur R Coronae-Klasse gehörigen Veränderlichen.

1. R Coronae borealis. Wir beginnen mit diesem Stern, der der Klasse seinen Namen gegeben hat, und dessen Lichtwechsel am besten bekannt ist. Eine Spezialuntersuchung über ihn ist von H. LUDENDORFF[3] ausgeführt worden, die auch eine Lichtkurve für die Zeit von 1843 bis 1905 gibt. Eine Ergänzung dieser Lichtkurve für die Zeit von 1825 bis 1832 findet sich bei R. MÜLLER, „Schwerds Beobachtungen veränderlicher Sterne"[4], und eine Lichtkurve für die Zeit bis 1923 im Harv Circ 247. Nach letzterer gibt Abb. 5 eine Darstellung der Lichtschwankungen für 1892 bis 1923; es ist hierzu zu bemerken, daß die in dieser Kurve angedeuteten sehr kleinen Helligkeitsschwankungen um die Normalgröße (6,0) herum höchst wahrscheinlich nicht reell sind.

---

[1] A N 209, S. 273 (1919).
[2] Seeliger-Festschrift S. 91 (1924).
[3] Publ Astrophys Obs Potsdam 19, Nr. 57 (1908). Nachtrag dazu A N 201, S. 439 (1915).
[4] Publ Astrophys Obs Potsdam 25, Nr. 82 (1925).

Die Hauptresultate der Untersuchung von LUDENDORFF, die durch den weiteren Verlauf der Helligkeitsschwankungen bisher bestätigt worden sind, lassen sich kurz in folgende Sätze zusammenfassen:

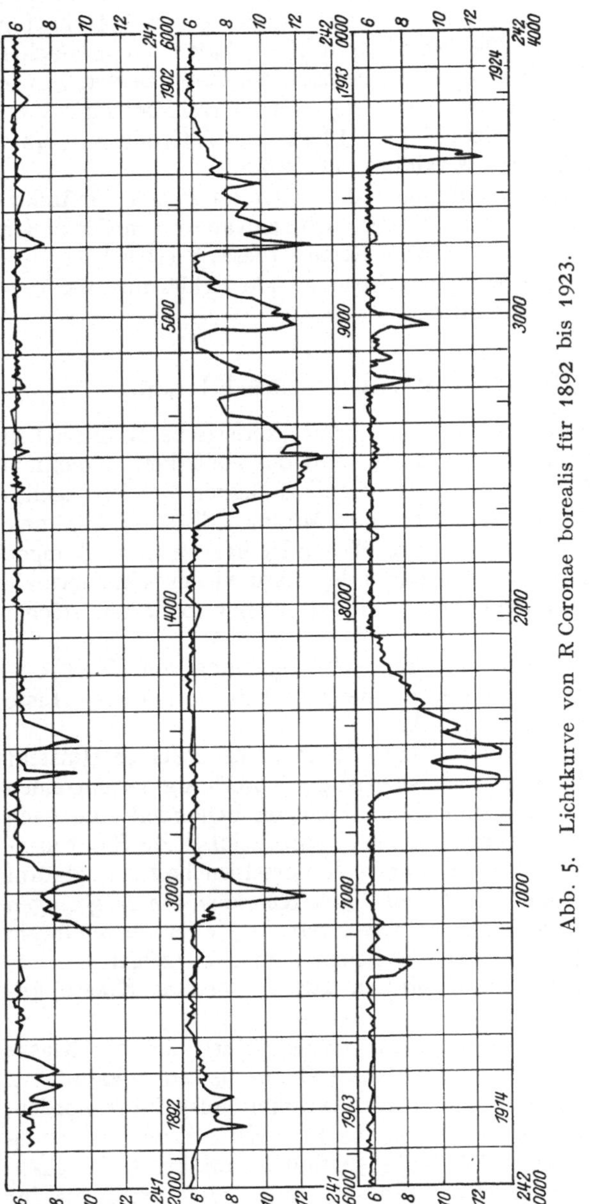

Abb. 5. Lichtkurve von R Coronae borealis für 1892 bis 1923.

α) R Coronae ist oft lange Zeit hindurch von konstanter Helligkeit, und zwar besitzt er dann stets sehr nahe die Größe 6,0. Kleine säkulare, insgesamt 0,2 bis 0,3 Größenklassen betragende Änderungen dieser normalen Größe sind angedeutet.

β) Ein Anwachsen der Helligkeit von R Coronae über die normale Größe hinaus hat bisher höchst wahrscheinlich noch niemals stattgefunden, solange dieser Stern beobachtet worden ist.

γ) Die Zeiten normaler Helligkeit werden anscheinend regellos unterbrochen durch zahlreiche Minima von verschiedenster Dauer und verschiedenster Minimalhelligkeit. Die Helligkeitsabnahme bei Beginn eines Minimums geschieht zumeist sehr schnell, die Wiederzunahme am Ende eines Minimums meist weit langsamer. Dazwischen liegen Lichtschwankungen von größter Unregelmäßigkeit. Die kleinste beobachtete Helligkeit ist etwa $15^{m},0$. Das längste bisher beobachtete Minimum dauerte fast elf Jahre. In einigen Fällen haben die Minima, was die Form der Lichtkurve anbetrifft, Ähnlichkeit mit denen von Bedeckungs-Veränderlichen.

Das Spektrum von R Coronae ist sehr merkwürdig. Wir übergehen hier die älteren Beobachtungen von ESPIN und auf dem Harvard-Observatorium. 1903 und 1905 gelangen auf dem Yerkes-Observatorium zwei Aufnahmen des Spektrums mit weitem Spalt; nach E. B. FROST[1] war es ähnlich dem von α Persei

[1] Ap J 22, S. 215 (1905).

(cF5), und die Radialgeschwindigkeit ergab sich zu +14 km. Auf dem Potsdamer Observatorium wurden von 1902 bis 1906 sieben, 1913 bis 1915 sechs Aufnahmen des Spektrums gewonnen, alle zur Zeit normaler Helligkeit des Sternes. Diese Platten sind von H. LUDENDORFF[1] bearbeitet worden. Er fand eine konstante Radialgeschwindigkeit von rund +25 km; das Spektrum zeigte ebenfalls große Ähnlichkeit mit dem von α Persei, doch war merkwürdigerweise die Wasserstofflinie $H\gamma$ weder als Absorptions- noch als Emissionslinie sichtbar (die übrigen Linien des H waren nicht in dem aufgenommenen Spektralbezirk enthalten). Auf den erwähnten Yerkes-Platten fehlten ebenfalls $H\gamma$ und $H\delta$, wie FROST nachträglich feststellte. Auf zwei Potsdamer Spektralaufnahmen geringer Dispersion aus dem Jahre 1906 fehlten $H\gamma$, $H\delta$ und $H\zeta$ ($H\varepsilon$ wird durch H überdeckt).

Im neuen Draper-Katalog ist das Spektrum als G0p bezeichnet, und es ist dazu bemerkt: „The spectrum is peculiar and on photographs of small dispersion appears to contain bright lines, but it is not certain whether these are lines or spaces. The *G* band is absent as a distinct absorption band.“ Eine ausführliche Untersuchung des Spektrums rührt von A. H. JOY und M. L. HUMASON her[2]. Vier Platten (1922 Februar bis Juli), die während der normalen Helligkeit des Sternes aufgenommen sind, ergaben ein cG0-Spektrum. Drei Platten, die während des Minimums der Helligkeit im Mai 1923 aufgenommen waren (Helligkeiten des Veränderlichen Mai 17: $12^m,3$, Mai 23: $11^m,4$, Mai 24: $11^m,3$), ergaben, daß diejenigen Funkenlinien des Ti, die während des Helligkeitsmaximums sehr starke Absorptionslinien waren, nunmehr als scharfe Emissionslinien erschienen. Außerdem waren noch mehrere andere helle Linien sichtbar; *H* und *K* waren Mai 17 scharfe Emissionslinien, Mai 24 breite Emissionsbanden. Das Absorptionsspektrum war im Minimum ziemlich unverändert, doch waren die Linien nicht so scharf wie im Maximum. Beim Anwachsen der Helligkeit (1923 Juni 10: $10^m,2$, Juni 11: $10^m,1$) waren die Emissionslinien bereits verschwunden, und kurze Zeit darauf (Juni 24: $8^m,0$) hatte das Spektrum dasselbe Aussehen wie im Maximum.

Änderungen der Radialgeschwindigkeit ließen sich nicht feststellen. Die hellen Linien, die im Minimum auftraten, waren gegen die Absorptionslinien nach Violett um einen Betrag verschoben, der einer Differenz der Radialgeschwindigkeiten von 20 km entspricht. Dies ist ungemein bemerkenswert, denn die hellen Linien im Spektrum der Mira-Sterne zeigen ein ganz ähnliches Verhalten. —

Es folgt nun die Besprechung der übrigen sicher oder höchst wahrscheinlich zur R Coronae-Klasse gehörigen Sterne in der Reihenfolge ihrer Rektaszension.

2. X Persei hat bisher nur kleine Helligkeitsschwankungen im Betrage von $0^m,7$ erkennen lassen. Man würde kaum in Versuchung kommen, diesen Stern mit R Coronae zu vergleichen, wenn er nicht nach den zahlreichen photometrischen Messungen von G. MÜLLER und P. KEMPF von Anfang 1888 bis Anfang 1891 völlig konstante Helligkeit ($6^m,3$) besessen hätte. Dann nahm die Helligkeit bis 1898 stetig um $0^m,6$ ab. Bis 1907 hat der Stern unregelmäßige Schwankungen zwischen den Grenzen $6^m,3$ und $7^m,0$ vollführt.

Eine Lichtkurve für die Zeit von 1888 bis 1907 haben MÜLLER und KEMPF[3] gegeben. Die Helligkeitsänderungen verlaufen viel langsamer als bei R Coronae. Für die spätere Zeit (bis 1925) liegen hauptsächlich Beobachtungen von A. N BROWN vor[4]. Danach sind die Änderungen jedenfalls nur gering gewesen,

[1] Publ Astrophys Obs Potsdam 19, Nr. 57 (1908); A N 201, S. 439 (1915).
[2] Publ A S P 35, S. 325 (1923). [3] A N 175, S. 161 (1907).
[4] M N 86, S. 101 (1926).

und die Periode von einem Jahr, die BROWN angedeutet findet, dürfte vielleicht auf systematische Fehler der Helligkeitsschätzungen zurückzuführen sein.

Nach Aufnahmen des Harvard-Observatoriums[1] ist das Spektrum von X Persei Fp; die Linien $H\alpha$, $H\beta$, $H\gamma$ und die He-Linien D_3 und λ 4471 sind hell. Der neue Draper-Katalog bezeichnet das Spektrum mit B0p und bemerkt, daß $H\beta$ und mehrere andere Linien hell, sowie daß die dunkeln Linien sehr schwach sind. Beobachtungen auf dem Mt. Wilson-Observatorium[2] bestätigen die Einordnung in die Klasse B; auch $H\alpha$ ist danach hell.

3. T Tauri. Einen ganz besonders interessanten Fall R Coronae-ähnlicher Veränderlichkeit bietet der durch die unmittelbare Nachbarschaft veränderlicher Nebel ausgezeichnete Stern T Tauri. Es ist bekannt, daß seine Helligkeitsschwankungen mit denen dieser Nebel nicht in Zusammenhang stehen; die Veränderungen der letzteren sollen daher hier nicht weiter diskutiert werden[3]. Eine Skizze der T Tauri umgebenden Nebel gibt F. G. PEASE[4]. Der Veränderliche selbst besitzt einen kleinen, veränderlichen Nebelschweif.

Während aus der Zeit vor 1863 nur wenige vereinzelte Beobachtungen von T Tauri vorhanden sind, ist dieser Stern in den Jahren 1863 bis 1887 von verschiedenen Beobachtern ziemlich eingehend beobachtet worden, und LUDENDORFF hat auf Grund dieser Beobachtungen eine Lichtkurve des Sternes für das angegebene Zeitintervall entwerfen können[5]. Auffallend ist an der Kurve zunächst die auch für R Coronae so charakteristische jahrelange Konstanz im Maximum ($9^m,8$), die wir hier für die Jahre 1876 bis Anfang 1881, vielleicht sogar bis 1882 feststellen können. Es scheint allerdings, als ob T Tauri im Gegensatz zu R Coronae mitunter auch in anderen Helligkeiten als der maximalen längere Zeit hindurch konstant ist, z. B. in dem tiefen Minimum von 1866. Zuweilen erfolgen aber auch, wie bei R Coronae, rasche Schwankungen im Minimum. Von 1863 bis zur Jahreswende 1875/76 hat T Tauri niemals die Maximalhelligkeit erreicht; so lange Minima kommen aber auch bei R Coronae vor. Übrigens scheint sich das Minimum von T Tauri noch weiter rückwärts erstreckt zu haben. Nach der Kurve scheinen die Lichtschwankungen von T Tauri im allgemeinen langsamer vor sich zu gehen als bei R Coronae, auch sind sie kleiner ($9^m,8$ bis $13^m,6$) als bei letzterem Stern. — Von 1887 bis 1904 sind nur ganz sporadische Beobachtungen von T Tauri vorhanden; sie lehren uns nichts Neues und, soweit sie reichen, hat ein wesentliches Überschreiten der früheren Maximalhelligkeit in diesem Intervall nicht stattgefunden. Ende 1904 setzen dann die leider nicht sonderlich genauen und auch nicht sehr zahlreichen Beobachtungen von L. CAMPBELL und Miss A. J. CANNON ein, die bis 1910 reichen. Sie geben zeitweise wiederum konstante Maximalhelligkeit (etwa $10^m,0$), sowie Minima bis herab zur 12. Größe. Anfang 1916 scheint der Stern in einem flachen Minimum (etwa 11^m) gewesen zu sein. Von 1917 an bis 1923 war die Helligkeit nach den wenig zahlreichen Beobachtungen der American Association of Variable Star Observers vermutlich meist die normale.

Nach dem Gesagten scheint T Tauri, wie R Coronae, eine normale Maximalhelligkeit (etwa $9^m,7$ bis $10^m,0$) zu besitzen, die er oft lange beibehält; natürlich ist es möglich, daß während der Zeiten normaler Helligkeit kleine Schwankungen innerhalb der Grenzen der Beobachtungsfehler vorhanden sind. Merkliche Überschreitungen dieser Maximalhelligkeit sind bisher noch nicht mit Sicherheit festgestellt.

---

[1] Harv Ann 56, S. 182 u. 184 (1912).
[2] Ap J 61, S. 404 u. 409 (1925) = Mt Wilson Contr 294.
[3] Vgl. BARNARD in M N 55, S. 442 (1895); 59, S. 372 (1899).
[4] Ap J 45, S. 89 (1917) = Mt Wilson Contr 127. [5] A N 209, S. 277 (1919).

Wie R Coronae und X Persei, so besitzt auch T Tauri ein merkwürdiges Spektrum. Es ist zuerst von W. S. ADAMS und F. G. PEASE[1] untersucht worden. Das Absorptionsspektrum ist danach ein solches der Klasse F 5. Die Linien des Wasserstoffs sind aber durch Emissionslinien verdeckt, und in diesem Punkte erinnert es an das von R Coronae, in welchem, wie wir sahen, nach den Potsdamer und Yerkes-Aufnahmen die Wasserstofflinien weder als Absorptions- noch als Emissionslinien sichtbar sind, ein Zustand, den man als den Beginn des Auftretens von Wasserstoffemissionslinien auffassen kann. Außer den Wasserstofflinien sind bei T Tauri auch noch die Ca-Linien H und K, ferner Linien des He und einige andere nicht identifizierte Linien hell. Das Emissionsspektrum von T Tauri ähnelt sehr dem des Wolf-Rayet-Sternes BD + 36° 3639. Die hellen Linien ragen über das kontinuierliche Spektrum hervor, was auf eine ausgedehnte Atmosphäre hindeutet. — Im Gegensatz zu dem eben geschilderten Befunde sagt F. G. PEASE[2]: „Mr. ADAMS finds that the spectrum of T Tauri is of type Md with additional bright lines.“ Wie der Widerspruch zu erklären ist, wird nicht mitgeteilt.

Eine weitere Untersuchung des Spektrums von T Tauri rührt von R. F. SANFORD her[3]. Die Absorptionslinien ergaben eine Radialgeschwindigkeit von +29 km. An hellen Linien fand er $H\beta$, $H\gamma$, $H\delta$, $H\zeta$, H, K, ferner Funkenlinien des Fe, aber keine Linien des He; die hellen Linien ergeben ungefähr dieselbe Radialgeschwindigkeit wie die dunkeln. E. HUBBLE[4] bezeichnet das Spektrum als Gpe.

4. AB Aurigae. Über den Lichtwechsel dieses Sternes ist erst sehr wenig bekannt, nach einer Notiz von H. SHAPLEY[5] kann aber kaum ein Zweifel bestehen, daß er zur R Coronae-Klasse gehört. In der Regel besitzt der Stern die Größe 7,2 (phot.); auf einigen von 800 Platten des Harvard-Observatoriums, die von 1898 bis 1923 aufgenommen waren, war er aber merklich schwächer (geringste Helligkeit $8^m,4$). Um einen Bedeckungs-Veränderlichen kann es sich nach SHAPLEY nicht handeln; dieser macht noch folgende Bemerkung über den Stern: „Moreover, photographs of long exposure indicate that AB Aurigae is situated at the edge of an obscured region. Probably, therefore, its abnormal variability is to be attributed to occultations by cosmic dust clouds.“ Das Spektrum ist A 0.

5. T Orionis. Dieser Veränderliche ist durch seine Lage bemerkenswert; er liegt nämlich an der Begrenzung des großen gekrümmten Nebelbogens, der, von der Mitte des Orionnebels ausgehend, sich nach Südosten hinzieht, und ist auf photographischen Aufnahmen des Orionnebels sehr leicht aufzufinden. Die Helligkeitsschätzungen von BOND und SAFFORD 1863 bis 1865 ergeben Schwankungen zwischen $9^m,3$ und $10^m,5$. Sie sind vereinbar mit der Annahme, daß Zeiten konstanter oder nahezu konstanter Helligkeit unterbrochen werden durch Zeiten geringerer Helligkeit, erstrecken sich aber nur über kurze Intervalle. Nach J. SCHMIDTs Beobachtungen scheint es, als ob T Orionis von 1868 bis 1877, soweit die Beobachtungen reichen, nur geringe Schwankungen um die Größe $9^m,7$ ausgeführt hat, nur zu Anfang 1877 scheint ein flaches Minimum stattgefunden zu haben. SCHMIDT gibt für die genannten Jahre zwar verschiedene zweifelhafte und unsichere Maxima und Minima an, aber es ist bekannt, daß er zuweilen mehr aus seinen Beobachtungen herauslas, als sie wirklich gaben.

[1] Publ A S P 27, S. 132 (1915).
[2] Ap J 45, S. 91 (1917) = Mt Wilson Contr 127.
[3] Publ A S P 32, S. 59 (1920).
[4] Ap J 56, S. 182 (1922) = Mt Wilson Contr 241.
[5] Harv Bull 798 (1924).

Wahrscheinlich dürfte sich die Helligkeit von T Orionis in dem erwähnten Intervall, von dem Minimum 1877 abgesehen, überhaupt nicht geändert haben. 1878 fand dann ein Minimum 13^m statt mit sehr raschem Helligkeitsanstieg. 1879, 1880, 1881 traten weitere Minima ein, die mit Zeiten normaler Helligkeit abwechselten. Die Beobachtungen von H. M. PARKHURST und J. H. EADIE (1885 bis 1888, 1892) geben, graphisch ausgeglichen, Helligkeiten zwischen $10^m,0$ und $13^m,0$, die also alle unterhalb der normalen Helligkeit liegen. Entschieden für den R Coronae-Charakter des Sternes sprechen die Schätzungen der Harvard-Beobachter aus den Jahren 1904 bis 1910, die längere Zeiten der Konstanz im normalen Licht (etwa $9^m,7$) und Minima bis zur Größe 12 ,8 ergeben. Endlich ergeben die Beobachtungen der American Association of Variable Star Observers für 1914 bis 1925 meist konstante Helligkeit, mehrfach unterbrochen durch nicht sehr tiefe Minima (etwa bis $11^m,5$). Die konstante Helligkeit scheint in den letzten Jahren etwas kleiner gewesen zu sein als früher. Zeitweise zeigen übrigens die Schätzungen der Mitglieder der genannten Gesellschaft sehr erheb liche Abweichungen voneinander.

Überblickt man das gesamte für T Orionis vorhandene Beobachtungsmaterial, so muß man auf eine ziemlich enge Verwandtschaft dieses Sternes mit R Coronae schließen. Über das Spektrum von T Orionis scheint nichts bekannt zu sein.

6. SU Tauri ist ein typischer Vertreter der R Coronae-Klasse. In den Jahren 1886 bis 1908[1] war die Helligkeit im allgemeinen völlig oder nahezu konstant, doch treten vier tiefe Minima auf, und zwar 1891, 1899, 1904 bis 1905 und 1908. Die Jahresmittel der Helligkeit für die übrigen Jahre des genannten Intervalls liegen alle zwischen $10^m,15$ und $10^m,35$, nur 1900 ist das Mittel $10^m,59$ und 1901 $10^m,39$; wahrscheinlich hatte sich der Stern in diesen beiden Jahren noch nicht völlig von dem vorangegangenen Minimum erholt. Als mittlere photographische „normale Größe" wird man $10^m,3$ annehmen können. Für 1909 bis 1925 liegen visuelle Schätzungen durch Beobachter des Harvard-Observatoriums und der American Association of Variable Star Observers vor. Danach hat sich das Minimum, welches 1908 begonnen hatte, bis 1911 erstreckt; dann traten neue Minima 1912, 1913 bis 1914, Ende 1916 und 1924 ein. Besonders interessant ist das Minimum (zirka 13^m) von 1916, welches ganz symmetrisch ist und wie das eines Algol-Sternes aussieht; man kann sich kaum des Eindrucks erwehren, daß es durch eine Bedeckungserscheinung zustande gekommen sein muß. Dieses Minimum von SU Tauri erinnert ganz außerordentlich an das von R Coronae im Jahre 1861 und noch mehr an das desselben Sternes von 1895.

SU Tauri sinkt in manchen Minima bis unter die 14. Größe (visuell) herab. Die visuelle Normalgröße 1911 bis 1925 war rund $9^m,5$, und wenn man annimmt, daß die Normalgröße vor 1908 dieselbe war wie nach 1911, so erhält man, da die photographische Normalgröße vor 1908 etwa $10^m,3$ war, für diesen Stern den Farbenindex $+0^m,8$. Über das Spektrum von SU Tauri ist wenig bekannt. Mrs. W. P. FLEMING[2] bemerkt darüber: „It appears to be of class G, and may resemble that of R Coronae borealis." Nach einer Aufnahme vom 29. Dezember 1921[3] ergab sich ebenfalls, daß es dem der Sonne ähnelt, aber die Dispersion war nicht genügend, um festzustellen, ob es die Eigentümlichkeiten der Spektren von R Coronae und RY Sagittarii hat.

7. Z Canis majoris. Nach Beobachtungen auf dem Harvard-Observatorium[4] ist die Helligkeit während langer Zeiten völlig oder nahezu konstant. Seit 1899

[1] Harv Circ 151 (1909). [2] Harv Ann 56, S. 209 (1912).
[3] Harv Bull 762 (1922). [4] Harv Circ 225 (1921).

können aber zwei tiefe Minima festgestellt werden (1908 Februar bis 1909 April, Ende 1920 bis 1921 April). Im Spektrum sind $H\beta$ und $H\gamma$ hell; außerdem sind zahlreiche dunkle Linien vorhanden, darunter solche des He. Im Minimum bleibt $H\gamma$ hell.

8. RX Puppis ist auf dem Harvard-Observatorium photographisch beobachtet worden[1]. Anfang 1890 war der Stern 12. Größe, 1894 bis 1899 dagegen nahezu konstant, ungefähr $11^m,4$. Bis 1904/05 nahm die Helligkeit ab (Minimum $14^m,1$) und dann allmählich wieder zu, so daß Anfang 1912 die Größe $12^m,0$ erreicht war. Von 1913 bis 1924 ist die Helligkeit konstant geblieben[2]. Besonders interessant ist RX Puppis durch sein Spektrum. Nach dem neuen Draper-Katalog sind außer den Wasserstofflinien auch die Nebellinien λ 3869, 4363, 4688 hell, und das Spektrum gleicht dem einer Nova zu der Zeit, wo sie sich in einen Gasnebel verwandelt.

9. UW Centauri. Das Beobachtungsmaterial für diesen Stern ist leider nur klein und nicht ausführlich veröffentlicht. Nach photographischen Aufnahmen des Harvard-Observatoriums[3] lag die Helligkeit von 1901 März 26 bis 1904 Januar 9 stets zwischen $10^m,0$ und $10^m,6$, 1904 Februar 3 hatte sie bis $12^m,2$ abgenommen, April 16 bis $14^m,7$. Ferner lag 1897 Januar 5 bis Mai 29 die Helligkeit stets zwischen $11^m,0$ und $11^m,3$, Juni 9 war dagegen die Größe $12^m,0$, Juli 1 $12^m,9$, Juli 22 $< 16^m$. Während die beiden plötzlichen Abstürze der Helligkeit hiernach durchaus denen von R Coronae ähnlich sind, ist es auffällig, daß die völlig oder nahezu konstante Helligkeit vor den beiden Minima etwas verschieden ist. Das Spektrum gehört zur Klasse K. Abgesehen von der hervorgehobenen Abweichung, die vielleicht durch einen langsamen Beginn des zweiten Minimums erklärt werden könnte, ist wohl an der Zugehörigkeit von UW Centauri zur R Coronae-Gruppe kaum zu zweifeln.

10. S Apodis. In Harv Bull 761 (1921) wird kurz mitgeteilt, daß der Stern zur R Coronae-Klasse gehört. Die visuelle Amplitude erstreckt sich von $10^m,0$ bis $<13^m,4$, die photographische von $10^m,6$ bis $15^m,8$. Nach neueren Beobachtungen der American Association of Variable Star Observers war der Stern von Ende 1922 bis Anfang 1925 nahezu konstant (etwa $10^m,0$), nur Mitte 1924 fand ein sich über etwa 150^d ausdehnendes Minimum statt, in welchem der Stern bis zur Größe $12^m,8$ abnahm, und bei dem die Abnahme der Helligkeit etwas rascher vor sich ging als die Zunahme. Nach dem neuen Draper-Katalog ist das Spektrum R3.

11. U Lupi. Die Lichtschwankungen dieses Sternes sind wenig bekannt, doch gibt R. T. A. INNES[4] seine Lichtkurve für die Zeit 1896 bis 1901. In diesem Zeitintervall sind Minima eingetreten: 1896 ($10^m,3$), 1898 (das erste $11^m,0$, vom zweiten ist nur der Anfang beobachtet), 1899 ($10^m,0$). Im übrigen war der Stern konstant, etwa von der Größe $9^m,2$. Die kleinen Schwankungen im Gesamtbetrage von etwa $0^m,5$, die INNES' Beobachtungen außerhalb der genannten Minima anzudeuten scheinen, brauchen wohl nicht reell zu sein, wie man erkennt, wenn man z. B. INNES' Beobachtungen des Algolsternes RR Velorum betrachtet. Vielleicht ist noch ein Teil der 1901 beobachteten Schwankungen als reell anzusehen. Bemerkenswert ist, daß der Stern sich nie um mehr als einen innerhalb der Grenzen der Beobachtungsfehler liegenden Betrag über die normale Helligkeit erhebt.

Die Farbe von U Lupi ist nach INNES „full yellow", das Spektrum dürfte also zwischen G und K liegen; sonst ist über das Spektrum nichts bekannt.

---

[1] Harv Circ 182 (1914) mit Lichtkurve. [2] Harv Bull 809 (1924).
[3] Harv Circ 170 (1912). [4] Cape Ann 9, S. 114 B (1902).

12. RT Serpentis. Auf zahlreichen, von 1891 bis 1908 auf dem Harvard-Observatorium und in Heidelberg aufgenommenen Platten war der Stern unsichtbar[1]. Im Juli 1909 war seine Größe $13^m,9$ (phot.), dann nahm allmählich die Helligkeit zu und erreichte 1915 den Wert $10^m,5$. Von 1915 bis 1923 war die photographische Größe $10^m,3$ bis $10^m,6$[2]. Nach ziemlich verstreuten Beobachtungen der American Association of Variable Star Observers aus den Jahren 1922 bis 1924 war die visuelle Größe in dieser Zeit ständig etwa $9^m,5$. Es ist wohl sicher anzunehmen, daß es sich hier um einen R Coronae-Stern handelt, der vor 1915 ein außerordentlich lang ausgedehntes Minimum gehabt hat und sich seitdem in der Phase konstanter Helligkeit befindet. H. SHAPLEY hat den Stern zum Gegenstande einer längeren Betrachtung gemacht[3]. Er scheint am meisten zu der Ansicht zu neigen, daß es sich hier um einen Stern handelt, der hinter einer absorbierenden kosmischen Wolke gestanden hat und nunmehr infolge seiner Bewegung hinter derselben hervorgekommen ist.

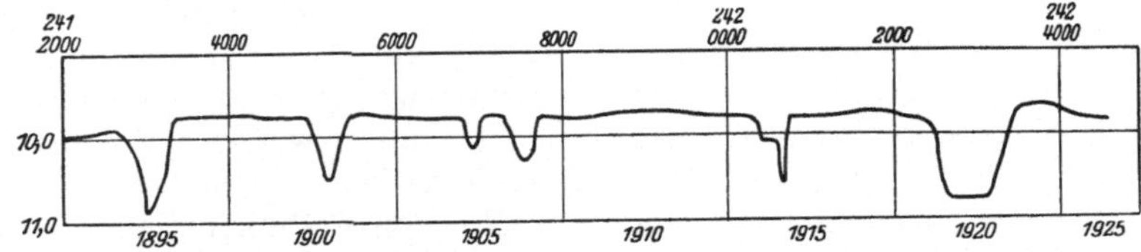

Abb. 6. Lichtkurve von XX Ophiuchi nach Harv Circ 292.

Das Spektrum von RT Serpentis ist auf dem Mt. Wilson-Observatorium untersucht worden. W. S. ADAMS und A. H. JOY klassifizierten es anfangs[4] als F0 mit heller $H\beta$-Linie; sie bemerken, daß die Intensität im brechbareren Teile stärker abzufallen scheint als es gewöhnlich bei Sternen dieses Spektraltypus der Fall ist. (Vielleicht erklärt dies, daß der Stern photographisch beträchtlich schwächer zu sein scheint, als visuell.) Später wurde das Spektrum von RT Serpentis auf dem Mt. Wilson-Observatorium als A8p ($H\beta$ hell)[5] und schließlich[6] als cA8 klassifiziert. Die spektroskopische Parallaxe ergab sich zu $0'',003$, die entsprechende absolute Helligkeit ist $+2^m,9$; für einen *c*-Stern ist diese absolute Helligkeit auffällig gering.

Auch die Radialgeschwindigkeit wurde auf dem Mt. Wilson-Observatorium gemessen; in dem nicht angegebenen Zeitintervall, das die sieben aufgenommenen Platten umfassen, hat sie sich allmählich von +125 km auf +44 km verringert.

13. XX Ophiuchi ist ein typischer R Coronae-Stern, dessen photographische Lichtkurve nach Harvard-Beobachtungen wir in Abb. 6 wiedergeben. Die Lichtschwankungen sind, wie aus dieser Kurve hervorgeht, ziemlich klein. Höchst eigentümlich ist das Spektrum dieses Sternes. Im Henry Draper Catalogue ist es als Bp (mit hellen Linien des H) bezeichnet. P. W. MERRILL[7] fand auf Platten aus den Jahren 1921 bis 1923 70 helle Linien in dem Spektrum, von denen 25 dem Eisen, und zwar ganz vorwiegend dessen Funkenspektrum, angehörten. Im Jahre 1925 waren dagegen nach MERRILL[8] die Absorptionslinien des Ti-Funkenspektrums das Charakteristische in dem Spektrum. Dieses ist also stark veränderlich.

[1] G. u. L., 3. Bd., S. 93. [2] Harv Bull 753 (1921), 789 (1923).
[3] Publ A S P 31, S. 226 (1919). [4] Pop Astr 28, S. 514 (1920).
[5] Ap J 53, S. 75 (1921) = Mt Wilson Contr 199.
[6] Publ A S P 36, S. 139 (1924). [7] Publ A S P 36, S. 225 (1924).
[8] Publ A S P 38, S. 45 (1926).

14. RY Sagittarii gleicht in seinem Lichtwechsel R Coronae außerordentlich; die vorhandenen Beobachtungen sind freilich infolge der stark südlichen Lage des Sternes nicht angenähert so vollständig wie für R Coronae. Die normale Größe ist etwa $7^m,0$, und zeitweise sinkt die Helligkeit unter die 13. Größe hinab. Eine Abweichung von R Coronae besteht darin, daß in einem Falle ein Überschreiten der normalen Helligkeit sichergestellt zu sein scheint: 1896 Oktober 19 erhielt nämlich E. C. PICKERING mit dem Meridianphotometer für den Veränderlichen die Größe $6^m,1$ und am gleichen Tage E. E. MARKWICK durch Schätzung die Größe $5^m,9$ in der Skala der Uranometria Argentina = $5^m,7$ der Harvard-Skala; auch 12 Tage später findet MARKWICK dieselbe Helligkeit.

Das Spektrum ähnelt nach dem neuen Draper-Katalog dem von R Coronae und wird mit G0p bezeichnet[1].

15. SY Cephei, photographisch von C. R. D'ESTERRE[2] entdeckt und beobachtet, zeigt bei konstanter, lange anhaltender Maximalhelligkeit (etwa $11^m,5$) gelegentlich scharfe Minima, die nicht regelmäßig wiederzukehren scheinen, über deren Verlauf aber Näheres noch nicht bekannt ist.

14. Sterne, die vielleicht zur R Coronae-Klasse gehören.

1. R Monocerotis. Der Stern liegt, wie T Tauri, an der Spitze eines kleinen, einem Kometenschweife ähnelnden Nebels (N G C 2261). Der Veränderliche ist bisher noch nicht sehr eingehend beobachtet worden. Er führt unregelmäßige Lichtschwankungen ungefähr zwischen 11^m und 13^m aus und scheint sich nur selten ein wenig über die 11. Größe zu erheben. Beobachtungen des Harvard-Observatoriums und der American Association of Variable Star Observers scheinen darauf hinzudeuten, daß die Helligkeit manchmal lange Zeit konstant ist bei ungefähr 11^m bis $11^m,5$. Der Verdacht, daß hier Zugehörigkeit zur R Coronae-Klasse vorliegt, ist ziemlich stark.

Der von R Monocerotis ausgehende Nebel ist, ebenso wie der bei T Tauri, gleichfalls veränderlich nach den Untersuchungen von E. HUBBLE[3], die durch solche von H. KNOX SHAW und C. C. L. GREGORY[4] sowie von C. O. LAMPLAND[5] bestätigt werden. Das Spektrum von R Monocerotis ebenso wie das des Nebels gleicht nach V. M. SLIPHER[6] dem einer Nova in frühem Stadium (Absorptionslinien auf der brechbareren Seite der Emissionslinien des H).

2. WY Velorum. Nach photographischen Beobachtungen auf dem Harvard-Observatorium[7] (101 Platten, 1890 bis 1922) hat die Helligkeit in dem Intervall von 1890 bis 1901 allmählich von $9^m,8$ bis $9^m,2$ zugenommen und seit 1902 langsam wieder abgenommen. Im Mai 1922 war die Größe $10^m,1$. Das Spektrum ist Ma, enthält aber fünf helle Linien oder Banden, welche mit einigen der stärksten hellen Linien unbekannten Ursprunges im Spektrum von η Carinae identisch zu sein scheinen.

Falls hier nicht Zugehörigkeit zur R Coronae-Klasse vorliegt, was sich noch nicht entscheiden läßt, wäre der Stern wohl in die Klasse der roten unregelmäßigen Veränderlichen einzuordnen.

3. TX Ophiuchi. Auf Grund von 94 nicht veröffentlichten Beobachtungen sagt E. ZINNER[8] über den Lichtwechsel folgendes: „Die Lichtkurve zeigt deutlich, daß der Stern meistens die Helligkeit $10^m,1$ hat, die er gelegentlich verläßt, um bis zu $10^m,6$ und seltener bis $11^m,3$ abzunehmen. Diese Lichtschwächen, die

[1] Siehe auch Harv Ann 56, S. 210, 211, 224 (1912).
[2] A N 194, S. 152 (1913); 199, S. 297 (1914).
[3] Ap J 44, S. 190 (1916); 45, S. 351 (1917).
[4] Helwan Obs Bull 1, S. 178 (1920); S. 235 (1921).
[5] Pop Astr 26, S. 249 (1918); 27, S. 31 (1919); 29, S. 632 (1921); 34, S. 621 (1926).
[6] Lowell Obs Bull 3, S. 63 (1918). [7] Harv Bull 783 (1923).
[8] Astron. Abhandlungen. Ergänzungshefte zu den A N 4, Nr. 3, S. 3 u. 11 (1922).

tiefen und die halbtiefen, folgen nicht, wie bei den Verfinsterungssternen, aufeinander, noch erscheinen sie regelmäßig nach einer bestimmten Zeit wieder, sondern bleiben gelegentlich aus. Die eintreffenden gleichen sich, natürlich mit dem Unterschiede, daß die tiefen Lichtschwächen 50 bis 60 Tage, die halbtiefen nur 40 Tage dauern." Die eingetroffenen Minima liegen um ganzzahlige Vielfache von $67^d,5$ voneinander entfernt. Später hat ZINNER[1] die Ansicht ausgesprochen, daß der Stern zur R Coronae-Klasse gehöre. Da das Beobachtungsmaterial wenig umfangreich und die Lichtschwankungen nur gering sind, wird man dies noch nicht als festgestellt betrachten dürfen. Das Spektrum des Sternes ist unbekannt.

4. R Coronae australis liegt wie T Tauri und R Monocerotis an der Spitze eines kleinen Nebels (N G C 6729). Über den Lichtwechsel ist wenig bekannt. Eine größere Beobachtungsreihe (1861 bis 1879) von SCHMIDT ist unzuverlässig wegen des tiefen Standes des Sternes über dem Horizont von Athen. Beobachtungen von R. T. A. INNES (1899 bis 1901) ergeben unregelmäßige Schwankungen. Spätere Beobachtungen von INNES[2] und von H. KNOX SHAW[3] widersprechen sich teilweise. In Harv Bull 806 (1924) ist über den Lichtwechsel folgendes gesagt: „R Coronae australis is so deeply involved in nebulosity that observations of its brightness are likely to be influenced by the nebulosity. The photographs, however, clearly confirm its variability between about 11,5 and 13,5. It is generally near maximum brightness. Out of more than a hundred observations, it was found as faint as magnitude 13,0 on only three plates." Hiernach könnte man Zugehörigkeit zur R Coronae-Klasse vermuten.

Über die Veränderlichkeit von S und T Coronae australis, die R benachbart sind, wissen wir noch weniger.

Der Nebel bei R Coronae australis ist, ebenso wie die bei T Tauri und bei R Monocerotis, veränderlich. Er ist in letzter Zeit besonders von H. KNOX SHAW photographisch beobachtet worden[4]. Zwischen den Veränderungen des Nebels (die sich nicht nur auf die Helligkeit, sondern auch auf die Form erstrecken) und denen von R Coronae australis besteht kein Zusammenhang; es kann aber sein, daß der Nebel einige Tage, nachdem R am hellsten gewesen ist, an Helligkeit zunimmt. Auch andere Sterne in dem Nebel und in dessen Umgebung sind veränderlich[5].

Das Spektrum von R Coronae australis und des Nebels gleicht nach V. M. SLIPHER[6] dem von R Monocerotis und dem zugehörigen Nebel (Spektrum einer Nova in frühem Stadium). E. HUBBLE[7] hat dagegen festgestellt, daß das Spektrum Gp ist. Er sagt darüber: „R Coronae australis has bright unsymmetrically reversed hydrogen and enhanced iron lines on an absorption spectrum that is approximately a G-type but which has contradictory characteristics. The spectrum resembles that of T Tauri except that it has no bright *H* and *K*".

5. UV Cassiopeiae. Dieser von C. R. D'ESTERRE entdeckte Veränderliche war von 1911 September bis wahrscheinlich Mitte März 1913 konstant von der Größe $12^m,3$ (phot.). Von Anfang Mai bis August 1913 war er dann außerordentlich schwach und hatte im Februar 1914 wieder die 15. Größe erreicht[8].

Nach E. ZINNER[9] sollen auch SU Aquarii und W Scuti zur R Coronae-Klasse gehören. Von ersterem Stern wird man noch mehr Beobachtungen ab-

[1] A N 224, S. 270 (1925). [2] Union Circ 36, S. 283 (1916).
[3] Helwan Obs Bull 1, S. 188 (1920).
[4] M N 76, S. 646 (1916); Helwan Obs Bull 1, S. 141 (1915), S. 182 (1920); 2, S. 71 (1924).
[5] Union Circ 31 (1915); 33, 36 (1916); 37 (1917).
[6] Lowell Obs Bull 3, S. 66 (1918).
[7] Ap J 56, S. 181 (1922) = Mt Wilson Contr 241.
[8] A N 196, S. 301 (1913); 198, S. 271 (1914). [9] A N 224, S. 270 (1925).

warten müssen, ehe man ihn zu dieser Klasse rechnet. Nach den Ephemeriden für 1926 in der Vierteljahrsschrift der Astronomischen Gesellschaft ist er ein Bedeckungs-Veränderlicher. W Scuti ist nach C. HOFFMEISTER[1] gleichfalls ein Bedeckungs-Veränderlicher.

Ferner hat sich TV Cephei, dessen Zugehörigkeit zur R Coronae-Klasse LUDENDORFF für nicht ganz ausgeschlossen, wenn auch für unwahrscheinlich hielt, ebenfalls als Bedeckungs-Veränderlicher erwiesen[2].

VY Pegasi endlich, den LUDENDORFF in der Seeliger-Festschrift[3] als fraglichen R Coronae-Stern bezeichnet, hat sich als langperiodischer Veränderlicher herausgestellt[4].

15. Allgemeines über die R Coronae-Sterne[5]. Wir geben zunächst eine Liste der im vorstehenden behandelten R Coronae-Sterne; über dem Strich stehen diejenigen Sterne, welche sicher oder höchst wahrscheinlich zu dieser Klasse gehören, unter dem Strich diejenigen, bei denen dies zweifelhafter ist. *g* bedeutet den Abstand vom galaktischen Äquator.

	A R (1900)	Dekl (1900)	*g*	Spektrum
X Persei	$3^h 49^m,1$	+30° 45′	16°	B0pe
T Tauri	4 16 ,2	+19 18	20	Gpe
AB Aurigae . . .	4 49 ,4	+30 24	8	A0
T Orionis	5 30 ,9	− 5 32	19	unbekannt
SU Tauri	5 43 ,2	+19 2	3	G
Z Canis majoris . .	6 59 ,0	−11 24	1	Pec (Helle Linien)
RX Puppis . . .	8 10 ,7	−41 24	3	Spektrum einer Nova in späterem Stadium
UW Centauri . . .	12 37 ,6	−53 59	9	K
S Apodis	14 59 ,4	−71 40	12	R3
R Coronae borealis	15 44 ,5	+28 28	50	cG0p
U Lupi	15 54 ,5	−29 38	16	unbekannt. Farbe „full yellow“
RT Serpentis . . .	17 34 ,3	−11 53	9	cA8pe
XX Ophiuchi . .	17 38 ,6	− 6 14	11	Pec
RY Sagittarii . . .	19 10 ,1	−33 42	21	G0p
SY Cephei	22 10 ,3	+62 2	5	unbekannt
R Monocerotis . .	6 33, 7	+ 8 50	3	Spektrum einer Nova in frühem Stadium
WY Velorum . . .	9 18, 7	−52 8	1	Ma mit einigen hellen Linien
TX Ophiuchi . . .	16 59, 1	+ 5 7	25	unbekannt
R Coronae austr. .	18 55, 2	−37 6	20	Gpe
UV Cassiopeiae . .	22 58, 1	+59 4	1	unbekannt

Zunächst ist die starke Zusammendrängung der R Coronae-Sterne nach der Milchstraße auffällig. Nur einer von den 20 Sternen hat eine 25° übersteigende galaktische Breite, nämlich R Coronae borealis selbst mit $g = 50°$. Die R Coronae-Sterne verhalten sich also hinsichtlich ihrer Verteilung in galaktischer Breite ebenso wie die neuen Sterne und die Nova-ähnlichen Veränderlichen. Merkwürdigerweise liegt die am weitesten von der Milchstraße entfernte Nova (Nova T Coronae von 1866, $g = 47°$) nicht weit von R Coronae borealis, dem am weitesten von der Milchstraße entfernten R Coronae-Stern.

Die galaktische Anhäufung ist ja nun eine Eigenschaft, die die neuen Sterne und die R Coronae-Sterne mit mehreren anderen Klassen von Sternen teilen, und man kann daraus nicht auf irgendwelche Beziehungen zwischen den beiden Gruppen von Sternen schließen. Auffälliger aber ist es, daß einige unter den R Coronae-Sternen in ihrem Spektrum Eigentümlichkeiten zeigen, die ebenfalls an die Novae erinnern: RX Puppis und R Monocerotis, nach SLIPHER auch

[1] A N 214, S. 6 (1921). [2] B Z 5, S. 22 (1923). [3] S. 91.
[4] B Z 5, S. 41 u. 49 (1923).
[5] Vgl. hierzu LUDENDORFF, A N 209, S. 273 (1919).

R Coronae australis, haben ein Nova-Spektrum. Die Spektren von R Coronae borealis und RT Serpentis haben c-Charakter, wie die Absorptionsspektren mancher neuen Sterne nahe ihrem Maximum. Das Spektrum von S Apodis gehört der höchst seltenen Spektralklasse R an, wie das der Nova Z Centauri in ihrem gegenwärtigen Stadium. Ferner ist hervorzuheben, daß T Tauri, R Monocerotis, R Coronae australis und T Orionis in Nebeln liegen; einen engen Zusammenhang mit Nebeln nimmt man aber vielfach auch bei den neuen Sternen an. Es scheinen also tatsächlich gewisse Parallelen zwischen den neuen Sternen und den Sternen der R Coronae-Klasse angedeutet zu sein, so unwahrscheinlich das zunächst auch klingen mag. Diese Beziehungen werden aber verständlicher, wenn man sich überlegt, auf welche Weise die merkwürdigen Lichtschwankungen der R Coronae-Sterne zu erklären sein können. Das charakteristische Merkmal der Lichtkurven — die oft sehr lange andauernde Konstanz in einer normalen Helligkeit, unterbrochen durch meist scharf einsetzende Minima — findet eine einfache Deutung, wenn wir annehmen, daß sich vor dem Stern zeitweise absorbierende kosmische Nebel- oder Staubmassen vorüber bewegen, deren Vorhandensein im Weltraum gegenwärtig allgemein zugegeben wird. Ist eine solche Nebelwolke regelmäßig geformt, so können Minima zustandekommen, die denen der Algol-Sterne ähneln, wie wir solche bei R Coronae und SU Tauri festgestellt haben. Sind dagegen die Nebelwolken von unregelmäßiger Struktur und wechselnder Durchsichtigkeit, so kann die Lichtkurve während des Minimums die verschiedensten Formen annehmen.

Nach dieser einfachen Hypothese würden die R Coronae-Sterne als Bedeckungs-Veränderliche anzusehen sein, nur würde bei ihnen die Bedeckung nicht durch eine zweite Komponente, sondern eben durch eine kosmische Wolke geschehen. Einen Übergang von ihnen zu den eigentlichen Bedeckungs-Veränderlichen würde dann der merkwürdige Stern ε Aurigae bilden, bei dem die alle 27 Jahre wiederkehrenden Verfinsterungen nach den Untersuchungen von H. LUDENDORFF[1] vielleicht durch ein umlaufendes wolkenähnliches Gebilde verursacht werden, das schwerlich stabil sein dürfte.

Die obige einfache Hypothese würde zur Erklärung des Lichtwechsels der R Coronae-Sterne genügen, wenn nicht ein so erheblicher Teil dieser Sterne so ausgesprochen merkwürdige Spektren hätte, und wenn nicht bei R Coronae selbst und XX Ophiuchi sich das Spektrum mit der Helligkeit änderte (ob dies nicht auch bei den anderen R Coronae-Sternen der Fall ist, wissen wir noch nicht). Diese Umstände zwingen zu der Annahme, daß sich jene Sterne nicht einfach h i n t e r, sondern vielmehr wenigstens zeitweise i n den Wolken befinden müssen. Die Bewegung durch die Wolke hindurch wird in dem Spektrum jene Eigentümlichkeiten erzeugen, so wie das nach der neuerdings freilich etwas zweifelhaft gewordenen SEELIGERschen Theorie auch bei den neuen Sternen zutrifft; nur sind bei den R Coronae-Sternen die Änderungen augenscheinlich nicht so durchgreifender Natur und daher auch nicht mit so starker Helligkeitszunahme verbunden, die Absorption durch die Wolke überwiegt jedenfalls die letztere. Ein gelegentliches Anwachsen über die normale Helligkeit, wie wir es bei RY Sagittarii in einem Falle festgestellt haben, läßt sich aber hiernach leicht erklären. Die durch die Bewegung innerhalb der Wolke entstandenen hellen Linien können auch in den auf die Minima folgenden Zeiten normaler Helligkeit, während derer sich kein absorbierendes Medium zwischen Beobachter und Stern befindet, bestehen bleiben, zumal es recht wohl möglich ist, daß sich der Stern auf seiner Wanderung durch die Wolke mit einer Nebelatmosphäre umgeben hat.

---

[1] Sitzungsber. d. Preuß. Akad. d. Wiss. 1924, S. 49.

Die engen Beziehungen zwischen Nebelflecken und R Coronae-Sternen, die wir in einigen Fällen konstatieren konnten, sprechen dafür, daß an dem Grundgedanken dieser Hypothese etwas Richtiges sein kann, wenn sich auch noch allerlei Einwände gegen sie erheben lassen. Es sei übrigens erwähnt, daß H. KNOX SHAW in seinen oben zitierten Arbeiten über R Coronae australis und den zugehörigen Nebel zu ähnlichen Ansichten über diesen Stern gekommen ist.

Zum Schlusse dieses Abschnittes möge hier auf eine Gruppe von Veränderlichen hingewiesen werden, bei denen vielleicht der Lichtwechsel sich auch durch kosmische Wolken erklären ließe, und die demnach in Beziehung zu den R Coronae-Sternen stehen könnten. Es sind die schwachen Veränderlichen im Orionnebel, über deren Lichtwechsel noch sehr wenig bekannt ist. C. HOFFMEISTER[1] und H. SHAPLEY[2] haben neuerdings Beobachtungen einiger von ihnen veröffentlicht. Nach SHAPLEY gibt es innerhalb eines Grades vom Zentrum des Nebels 80 solcher Veränderlicher, deren Lichtwechsel ganz regellos zu sein scheint; er macht darauf aufmerksam, daß es sich hier augenscheinlich um Zwergsterne handelt, während sonst fast alle veränderlichen Sterne (außer einigen Bedeckungs-Veränderlichen) Riesensterne sind. Der im vorstehenden behandelte R Coronae-Stern T Orionis ist der bei weitem hellste unter dieser Gruppe von Sternen.

d) Die Veränderlichen der U Geminorum-Klasse.

16. Definition der U Geminorum-Sterne. In unregelmäßigen, aber doch in den meisten Fällen Anzeichen einer gewissen Regelmäßigkeit verratenden Intervallen leuchten diese Sterne meist sehr rasch zu einem Maximum auf und nehmen darauf langsamer wieder ab. Viele von ihnen verharren verhältnismäßig lange im Minimum, und bei einigen ist die Helligkeit im Minimum lange Zeit nahezu konstant, und zwar ist dann die Minimalhelligkeit in verschiedenen Minima immer nahezu dieselbe.

17. Die einzelnen U Geminorum-Sterne. Die wenigen (im ganzen 11) Objekte dieser Art sollen hier in der Reihenfolge ihrer Rektaszension besprochen werden. Über manche von ihnen ist erst sehr wenig bekannt.

1. UV Persei. Dieser Stern ist hauptsächlich von J. VAN DER BILT und A. A. NIJLAND verfolgt und von letzterem in einer zusammenfassenden Arbeit behandelt worden[3]. In der Regel ist der Stern sehr schwach und im zehnzölligen Utrechter Refraktor unsichtbar ($< 14^m$). Die sehr selten eintretenden Maxima zerfallen in zwei Typen: lange und kurze. Bei den langen erreicht der Stern die Größe $12^m,2$, und er bleibt etwa 16^d oberhalb der Größe 14,0. Für den Verlauf eines solchen langen Maximums gibt NIJLAND folgende mittlere Lichtkurve:

Phase	Größe	Phase	Größe
0^d	$<14^m,0$	10^d	$12^m,68$
2	12 ,27	12	13 ,03
4	12 ,20	14	13 ,43
6	12 ,25	16	13 ,90
8	12 ,44	18	$<$ 14 ,0

Die meisten der langen Maxima liegen ziemlich genau um ein ganzzahliges Vielfaches von $131^d,48$ auseinander, aber es tritt keineswegs etwa alle 131^d ein solches Maximum ein. Bei den kurzen Maxima, in denen der Stern etwa die Größe $12^m,5$ erreicht, ist er nur drei Tage oberhalb der Größe 14,0. Ihr Eintreten scheint an keinerlei Regel gebunden zu sein.

---

[1] Mitteilungen der Sternwarte Sonneberg Nr. 3 (1923).

[2] Ap J 49, S. 260 (1919) = Mt Wilson Contr 156; Harv Circ 254 (1924); Harv Bull 803 (1924).

[3] A N 221, S. 243 (1924).

Das kürzeste Zeitintervall, das bisher zwischen zwei Maxima gefunden worden ist, beträgt 142^d (zwei lange Maxima), wenn wir von fraglichen Beobachtungen absehen. Eine obere Grenze für das Intervall läßt sich nicht angeben. Über das Verhalten des Sternes im Minimum wissen wir nichts; die Minimalgröße ist aber sicher $< 16^m$ (phot.).

2. TZ Persei. Dieser Stern ist neuerdings von K. GRAFF näher untersucht worden[1]. Ungefähr alle 21^d leuchtet er rasch (in etwa 6^d bis 7^d) aus einem Minimum ($15^m,3$) zu einem Maximum ($12^m,5$) auf und nimmt dann langsam wieder ab. Das Minimum ist, ebenso wie das Maximum, spitz, wenn auch nicht so spitz wie letzteres; die Helligkeit ist im Minimum nicht längere Zeit konstant. Die Lichtkurve gleicht der eines δ Cephei-Sternes, doch unterscheidet sie sich von dieser durch die Unregelmäßigkeit der Wiederkehr der Maxima. Auch ist die Amplitude der Lichtschwankungen ($2^m,8$) für einen δ Cephei-Stern sehr groß. GRAFF rechnet daher den Stern zu den U Geminorum-Sternen. Die Farbe des Sternes ist weiß.

3. BI Orionis. Über diesen Stern liegt eine Untersuchung von C. HOFFMEISTER[2] vor. Dieser hat 13 Maxima beobachtet (1916 bis 1918), deren Zeiten sich durch die Formel

$$M = 2420954^d + 24^d,64\,E + 6^d \sin(15°,7\,E + 169°)$$

befriedigend darstellen ließen. (Die beobachtete Zwischenzeit zweier Maxima schwankt von $23^d,0$ bis $26^d,3$.) Ältere Bergedorfer Beobachtungen verlangen aber eine kürzere mittlere Periode (etwa 20^d). Die größte Helligkeit ist $13^m,2$, im Minimum scheint die Helligkeit unter 16^m zu sinken; die Amplitude ist also $> 2^m,8$, wäre also für einen δ Cephei-Stern sehr groß. Die Gestalt der Lichtkurve scheint etwas veränderlich zu sein, entspricht aber im ganzen der des δ Cephei-Typus. HOFFMEISTER glaubt nach allem den Stern zur U Geminorum-Klasse rechnen zu dürfen, wenn der Lichtwechsel auch regelmäßiger ist als bei anderen Sternen dieser Klasse. Es scheint sich hier, ebenso wie bei dem oben besprochenen Stern TZ Persei, um eine Übergangsform von der δ Cephei- zur U Geminorum-Klasse zu handeln.

4. SS Aurigae ist ein typischer Vertreter der U Geminorum-Klasse. Über das Verhalten des Sternes im Minimum ist wenig bekannt, doch ist nach Beobachtungen der American Association of Variable Star Observers anzunehmen, daß SS Aurigae während des Minimums lange Zeit hindurch von konstanter Helligkeit ist, und zwar ungefähr von der 14. bis 15. Größe. Diese Minimalhelligkeit scheint in den verschiedenen Minima immer ziemlich dieselbe zu sein. Aus ihr erhebt sich der Stern in unregelmäßigen Intervallen plötzlich zu einem Maximum; der Abstand zweier aufeinander folgender Maxima beträgt nach den bisherigen Beobachtungen etwa 30^d bis 140^d. Man kann lange und kurze Maxima unterscheiden, die sich aber nicht regelmäßig abwechseln, wie es bei U Geminorum der Fall ist. A. A. NIJLAND[3] gibt für den mittleren Helligkeitsverlauf während der langen und der kurzen Maxima folgende Zahlen:

Phase	Langes Maximum	Kurzes Maximum	Phase	Langes Maximum	Kurzes Maximum	Phase	Langes Maximum	Kurzes Maximum
$-0^d,5$	$14^m,0$	$14^m,0$	5^d	$10^m,8$	$12^m,3$	12^d	$12^m,3$	$14^m,25$
0	12 ,0	12 ,0	6	10 ,8	12 ,8	13	12 ,8	14 ,25
+0 ,5	11 ,2	11 ,2	7	10 ,9	13 ,3	14	13 ,3	14 ,25
1	11 ,0	11 ,0	8	11 ,0	13 ,8	15	13 ,8	14 ,25
2	10 ,8	11 ,1	9	11 ,1	14 ,0	16	14 ,0	14 ,25
3	10 ,8	11 ,3	10	11 ,4	14 ,2	17	14 ,2	14 ,25
4	10 ,8	11 ,8	11	11 ,7	14 ,25	18	14 ,25	14 ,25

[1] A N 225, S. 253 (1925). [2] A N 218, S. 316 (1923). [3] A N 227, S. 353 (1926).

Die Amplitude beträgt, wenn man die Minimalgröße zu $15^m,0$ annimmt, $4^m,2$. Die langen Maxima haben große Ähnlichkeit mit denen von U Geminorum. Bisweilen kommen auch abnorme Maxima mit langsamerer Helligkeitszunahme vor. Das Spektrum[1] ist im Maximum der Helligkeit nahezu kontinuierlich mit schwachen, breiten, dunkeln Banden des H und He. Die Farbe des Sternes wird als weiß bezeichnet.

5. U Geminorum ist der Gegenstand einer eingehenden Untersuchung von J. VAN DER BILT[2] gewesen, die fast das gesamte Beobachtungsmaterial von 1856 bis Anfang 1907 umfaßt. Die Resultate dieser Arbeit sind in der G. u. L.[3] wie folgt zusammengefaßt: „Nach VAN DER BILT lassen sich sämtliche hinreichend beobachtete Maxima unter die beiden Formen „lange" und „kurze" einordnen, obwohl innerhalb derselben zweifellos nicht unbeträchtliche Abweichungen von der normalen Form vorkommen, die aber im Vergleich zu den Unregelmäßigkeiten im Lichtwechsel des verwandten Veränderlichen SS Cygni unbedeutend sind. Die langen Maxima verlaufen durchschnittlich folgendermaßen: Normalhelligkeit $13^m,5$ (J. A. PARKHURSTS Skala), plötzlicher Anstieg auf $10^m,25$ in 2^d, in weiteren 4^d bis zum Maximum $9^m,5$, Maximum ziemlich flach (Dauer etwa 4^d); Abnahme bedeutend langsamer als Zunahme, anfangs $0^m,8$ in 6^d, dann schneller, die Normalhelligkeit in 14^d nach dem Maximum wieder erreicht, ganze Dauer der langen Erscheinung $20^1/_2{}^d$. Die langen Erscheinungen sind in ihrem Verlauf größeren Schwankungen unterworfen. Die kurzen Maxima erheben sich ebenso unvermittelt aus der Normalhelligkeit, Anstieg bis zum Maximum $3^1/_2{}^d$, Maximum spitz, Abnahme sofort schnell und 9^d dauernd, ganze Dauer der kurzen Maxima 12^d, Helligkeit $9^m,8$. Der Verlauf der kurzen Maxima schwankt weniger als der der langen. Die Minimalhelligkeit, die den normalen Zustand des Sternes zu bedeuten scheint, ist nicht völlig konstant, sondern ähnlich wie bei SS Cygni, jedoch in weit geringerem Maße, kleinen allgemeinen Schwankungen und vorübergehenden Aufhellungen unterworfen, die vielleicht bis zu einer Größenklasse anwachsen können ... Die Zwischenzeiten zwischen den einzelnen Maxima sind im höchsten Grade ungleich, und in ihren Schwankungen sind bisher keinerlei Gesetzmäßigkeiten zu erkennen gewesen. Nach VAN DER BILT sind die äußersten beobachteten Zwischenzeiten 62^d und 152^d, und jeder der zwischenliegenden Werte scheint gleich häufig auftreten zu können." Die langen und die kurzen Maxima wechseln regelmäßig miteinander ab.

Die Resultate VAN DER BILTS sind durch den weiteren Verlauf der Lichtschwankungen seit 1907 in allen wesentlichen Punkten bestätigt worden. A. A. NIJLAND[4] hat namentlich hervorgehoben, daß das Gesetz der Abwechslung der langen und kurzen Maxima bisher keine Ausnahme erlitten hat; er betont auch noch, daß der Helligkeitsaufstieg stets steil ist, mehr symmetrisch gestaltete Maxima, wie sie manchmal bei SS Aurigae zu finden sind, bei U Geminorum also nicht vorkommen.

Die Farbe von U Geminorum wird von den Beobachtern als weiß oder bläulichweiß angegeben. Nach älteren Aufnahmen des Harvard-Observatoriums[5] ähnelt das Spektrum einem solchen der Klasse F und hat breite Linien H und K, während die Wasserstofflinien schwach sind. Im neuen Draper-Katalog ist dagegen bemerkt, daß das Spektrum auf den besten Aufnahmen ganz kontinuierlich erscheint.

6. SU Ursae majoris wurde von C. HOFFMEISTER als U Geminorum-Stern erkannt. Die Zwischenzeit der Aufhellungen beträgt etwa 16^d; es ist aber mög-

[1] Annual Report of the Director of the Mount Wilson Obs. for the Year 1922, S. 234.
[2] Recherches astronomiques de l'Observatoire d'Utrecht III (1908).
[3] 1. Band, S. 241. [4] A N 224, S. 299 (1925). [5] Harv Ann 56, S. 210 (1912).

lich, daß starke Unregelmäßigkeiten vorkommen oder manche Erscheinungen ganz ausfallen. Der Bereich des Lichtwechsels ist 11^m bis $< 14^m$. HOFFMEISTER vermutet, daß man auch bei diesem Stern lange und kurze Maxima unterscheiden muß[1].

7. Z Camelopardalis gehört nach A. BRUN[2] ebenfalls zur U Geminorum-Klasse. Nach BRUNS Beobachtungsreihe treten die Maxima ($10^m,3$ bis $11^m,5$) in unregelmäßigen Intervallen von 15^d bis 36^d auf. Die Minima ($12^m,7$ bis $13^m,3$) sind meist ziemlich spitz, aber flacher als die Maxima. Das mittlere Intervall zwischen zwei Maxima ist 23^d.

8. X Leonis. Nach K. GRAFF[3] ist dieser Stern zur U Geminorum-Klasse zu rechnen. Das Aufleuchten geschieht ungefähr alle 16^d. Die extremen von GRAFF beobachteten Helligkeiten sind $11^m,8$ und $15^m,6$.

9. TW Virginis. Über den Lichtwechsel dieses Sternes ist erst sehr wenig bekannt, doch dürfte er wohl sicher ein U Geminorum-Stern sein. Beobachtungen sind in den Beobachtungszirkularen der Astronomischen Nachrichten veröffentlicht. Das kürzeste, bisher beobachtete Intervall zwischen zwei Maxima scheint 23^d zu sein, die größte Helligkeit 10^m bis 11^m; im Minimum ist der Stern $< 14^m$.

10. SS Cygni. Eine Bearbeitung des Lichtwechsels dieses sehr viel beobachteten, höchst merkwürdigen Veränderlichen für die Zeit von der Entdeckung (1896) an bis 1908 hat L. CAMPBELL[4] gegeben. Auf Grund dieser Untersuchung und der weiteren Beobachtungen bis 1914 liefert die G. u. L.[5] eine ausführliche Schilderung des Lichtwechsels, der wir folgendes entnehmen: „Der Lichtwechsel zeigte bis 1907 den auch für U Geminorum eigentümlichen Verlauf, jedoch traten auch vor dem genannten Jahr bereits von Zeit zu Zeit Störungen in Gestalt sog. ‚anormaler' Maxima auf. Während des Jahres 1907 trat nun ein vollständiger Wechsel in der Art der Veränderlichkeit ein, indem die sonst bekannten ‚langen' und ‚kurzen' Maxima gänzlich aufhörten und ihre Stelle von ‚anormalen' Maxima eingenommen wurde, die aber nun nicht mehr durch Perioden gleicher Minimalhelligkeit voneinander getrennt waren, sondern unter fortwährenden unregelmäßigen Schwankungen der Helligkeit aus mehr oder weniger scharf ausgeprägten Minima emporstiegen bzw. in solche übergingen. Erst im Laufe des Jahres 1909 hörte die Periode der großen Störung allmählich auf, und seit 1910 ist die alte, etwas unregelmäßige Abwechslung zwischen langen und kurzen Maxima wieder die herrschende, mit dem Unterschiede, daß seit 1910 anscheinend keine anormalen Erscheinungen, die auch bereits vor 1907 zeitweise auftraten, mehr mit Sicherheit beobachtet worden sind. In gewöhnlichen Zeiten ist der Lichtwechsel etwa folgender. Nach einem, meist ungefähr einen Monat dauernden Verharren der Helligkeit im Minimum von nahe der 12. Größe, steigt dieselbe plötzlich innerhalb eines Tages um 2^m bis $2^1/_2{}^m$ an und erreicht in den kurzen Maxima nach 2,6, in den langen nach 5,1 Tagen ihren größten Wert von etwa $8^m,3$, und sinkt dann in 9 bzw. $14^1/_2$ Tagen wieder zum Minimum herab. Die langen und die kurzen Maxima wechseln in der Regel längere Zeit mehr oder weniger regelmäßig miteinander ab — häufig treten Folgen von je zwei gleichartigen Maxima auf —, bis dann plötzlich ein anormales Maximum auftritt, dessen Form gänzlich verschieden von der beschriebenen ist. Die vor der großen Störung beobachteten anormalen Maxima hatten nach CAMPBELL im Durchschnitt sehr nahe symmetrischen Verlauf (An- und Abstieg je 9 Tage) und etwas geringere Maximalhelligkeit ($8^m,8$). Die Minimumhelligkeit war im Mittel um 1896/97 = $11^m,3$, 1898 bis 1902 = $11^m,8$, von 1902

[1] B Z 6, S. 36 u. 68 (1924).
[2] Lyon Bull 7, S. 203 (1925).
[3] A N 211, S. 317 (1920).
[4] Harv Ann 64, No. II (1912).
[5] 2. Band, S. 335.

bis 1905 unregelmäßig, von 1905 bis 1907 = $11^m,6$, jedoch ist die hierin ausgesprochene Änderung nicht ganz gesichert. Für die Zeit der großen Störung 1907 bis 1909 läßt sich eine zusammenfassende Beschreibung nicht geben, da in derselben regellos Maxima, scharf ausgeprägte Minima, Perioden unveränderter Helligkeit und mehr oder weniger großer Schwankungen miteinander abwechselten und die normalen Formen ‚kurz' und ‚lang' gar nicht auftraten." Die Zwischenzeiten zwischen zwei aufeinanderfolgenden Maxima sind höchst ungleich; die kleinste, bis 1914 beobachtete Zwischenzeit ist 24^d, die größte 80^d. Im Mittel ergibt sich für die Zeit von 1896 bis 1913 der Wert $50^d,65$; dieser Mittelwert ist aber für verschiedene Zeitintervalle nicht konstant, und zwar erhält man nach der G. u. L.

für die Zeit	das mittlere Intervall
1896—1899	$55^d,4$
1899—1902	52 ,6
1902—1905	48 ,3
1905—1907	41 ,8
1907—1910	48 ,7
1910—1912	57 ,1

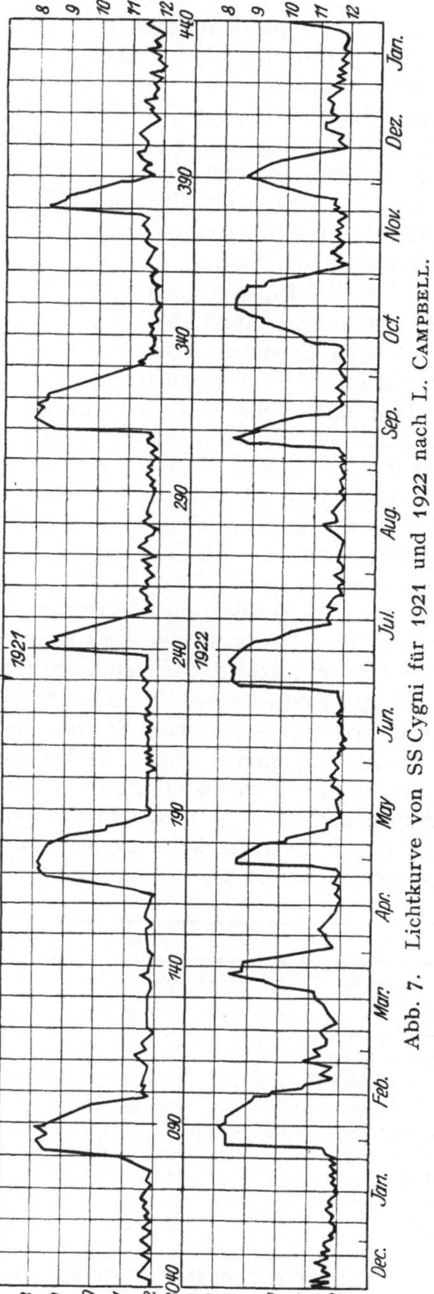

Abb. 7. Lichtkurve von SS Cygni für 1921 und 1922 nach L. CAMPBELL.

Die bisherigen Erörterungen beziehen sich, wie erwähnt, auf den Lichtwechsel bis 1914. Für die Zeit von 1914 bis 1922 hat L. CAMPBELL[1] eine vollständige Lichtkurve gegeben. Das oben über den Lichtwechsel Gesagte behält im allgemeinen auch für dieses Intervall seine Gültigkeit, doch kommen von der zweiten Hälfte 1914 an wieder „anormale" Maxima vor. Die kürzeste Zwischenzeit zwischen zwei Maxima beträgt in diesem Intervall etwa 21^d, doch ist in diesem Falle das zweite Maximum sehr tief (etwa $9^m,6$); das längste Intervall ist etwa 88^d. Eine nähere Diskussion der Lichtkurve steht noch aus; Abb. 7 gibt dieselbe für die Jahre 1921 und 1922 wieder.

Über den mittleren Verlauf der Maxima auf Grund der Beobachtungen in Utrecht von 1905 bis 1916 hat A. A. NIJLAND[2] Angaben gemacht.

E. T. WHITTAKER[3] und D. GIBB[4] haben versucht, den Lichtwechsel von

[1] Pop Astr 31, S. 314 (1923). [2] A N 204, S. 71 (1917).
[3] M N 71, S. 686 (1911). [4] M N 74, S. 678 (1914).

SS Cygni der Periodogrammanalyse zu unterwerfen. Zu irgendwelchen bemerkenswerten Erfolgen haben diese Untersuchungen nicht geführt.

Nach WILLIAMINA P. FLEMING[1] sind im Spektrum von SS Cygni die Wasserstofflinien von veränderlicher Intensität, manchmal scheinen sie auch hell zu sein. Auch *H* und *K* sind auf einigen Aufnahmen als Absorptionslinien sichtbar. Der neue Draper-Katalog bemerkt über das als „peculiar" bezeichnete Spektrum: „On the Harvard photographs, taken at maximum, very narrow dark hydrogen lines are barely seen." W. S. ADAMS und A. H. JOY[2] machen über das Spektrum von SS Cygni folgende Angaben: „Maximum: Spectrum is continuous with faint dark bands of hydrogen and helium 20 angstroms wide. Minimum: Strong bright bands of hydrogen and helium about 20 angstroms wide but not displaced. Possibly a few faint absorption lines. — The spectrum of this star bears considerable resemblance to that of novae." Die Farbe ist nach K. GRAFF[3] 1,3 der OSTHOFFschen Skala.

11. RU Pegasi scheint mit SS Cygni in seinem Lichtwechsel große Ähnlichkeit zu haben. Auch bei diesem Stern hat man lange, kurze und annähernd symmetrische Maxima zu unterscheiden. In dem in der Regel lang andauernden Minimum hat der Stern ungefähr die Größe 12,4. Im Maximum erreicht er ungefähr die Größe 11,1. Die mittlere Periode beträgt nach den Angaben der G. u. L. 70^d, doch kommen Abweichungen bis zu $\pm 40^d$ vor. Nach K. GRAFF ist der Stern gelb. Die Beobachtungen aus den letzten Jahren sind noch nicht zusammenfassend bearbeitet.

Stark der Zugehörigkeit zur U Geminorum-Klasse verdächtig ist RX Andromedae, und eine Diskussion der Beobachtungen dieses Sternes wäre sehr erwünscht. Nach E. ZINNER[4] soll auch RV Cancri ein U Geminorum-Stern sein. Da die Helligkeitsamplitude dieses Sternes aber sehr gering (etwa $0^m,8$) ist und nur wenige Beobachtungen von ihm vorliegen, so wird man gut tun, ihn noch nicht in die Liste der U Geminorum-Sterne aufzunehmen. Die am Schlusse von Ziff. 10 besprochenen Veränderlichen DH und EP Carinae könnten möglicherweise auch U Geminorum-Sterne sein.

18. Allgemeines über die U Geminorum-Sterne. Die folgende Tabelle enthält eine Zusammenstellung der oben näher besprochenen U Geminorum-Sterne. Außer dem Ort für 1900 und der galaktischen Breite *g* sind auch noch gegeben die ungefähre Helligkeitsamplitude und das Intervall zwischen zwei aufeinanderfolgenden Maxima bzw. das mittlere Intervall zwischen zwei Maxima.

	AR (1900)	Dekl (1900)	*g*	*A*	*I*	Mittl. *I*
UV Persei	$2^h\ 3^m,3$	$+56°43'$	4°	$4^m\pm$	142^d u. größer	
TZ „	2 6 ,8	+57 55	3	2 ,8	—	21^d
BI Orionis	5 18 ,7	+ 0 55	18	>2 ,8	19^d — 26^d	
SS Aurigae	6 5 ,8	+47 46	15	4 ,2	25 —103	
U Geminorum . . .	7 49 ,2	+22 16	24	4 ,0	62 —152	
SU Ursae maj. . .	8 3 ,4	+62 54	33	>3	—	16
Z Camelopardalis .	8 14 ,0	+73 26	33	3 ,0	15 — 36	23
X Leonis	9 45 ,6	+12 21	46	2 ,8	—	16
TW Virginis	11 40 ,3	— 3 52	55	>4	23 u. größer	
SS Cygni	21 38 ,8	+43 8	7	3 ,7	21 — 88	51
RU Pegasi	22 9 ,2	+12 12	36	1 ,3	30 —110	70

Es geht zunächst aus der Tabelle hervor, daß einige dieser Objekte ziemlich hohe galaktische Breiten haben. In galaktischer Länge bevorzugen diese Objekte

[1] Harv Ann 56, S. 211 (1912). [2] Pop Astr 30, S. 103 (1922).
[3] A N 197, S. 73 (1913). [4] A N 224, S. 270 (1925).

die Teile des Himmels, die an Me-Sternen, planetarischen Nebeln und neuen Sternen verhältnismäßig arm sind, worauf F. DE ROY aufmerksam gemacht hat[1].

Man kann die U Geminorum-Sterne augenscheinlich in zwei Gruppen einteilen, nämlich in solche mit im allgemeinen lang ausgedehntem Minimum, während dessen der Stern nahezu konstante Helligkeit hat, und solche mit ziemlich spitzem Minimum. Zur ersten Gruppe gehören SS Aurigae, U Geminorum, SS Cygni, RU Pegasi und wohl sicher auch UV Persei, zur zweiten TZ Persei, Z Camelopardalis und augenscheinlich auch BI Orionis, SU Ursae majoris und X Leonis. (Über TW Virginis wissen wir erst sehr wenig.) Die Sterne der ersten Gruppe ähneln in mancher Beziehung gewissen Mira-Sternen, die der zweiten mehr langperiodischen δ Cephei-Sternen. Ob die beiden Gruppen durch Übergangsformen miteinander verbunden sind (es kommen z. B. bei SS Cygni spitze Minima vor), oder ob die U Geminorum-Klasse in zwei verschiedene Klassen zu teilen ist, läßt sich noch nicht entscheiden.

Über die Spektren der U Geminorum-Sterne besitzen wir erst sehr spärliche Kenntnisse, die oben für die einzelnen Sterne dargelegt sind; bemerkenswert ist, daß nach W. S. ADAMS und A. H. JOY das Spektrum von SS Cygni Ähnlichkeit mit dem einer Nova hat. Auch das bei den U Geminorum-Sternen die Regel bildende sehr rasche Aufleuchten erinnert an die neuen Sterne, und den bei den Nova-ähnlichen Sternen besprochenen Veränderlichen T Pyxidis könnte man geradezu einen U Geminorum-Stern nennen, bei welchem freilich die Intervalle zwischen den Maxima sehr lang sind. Vielleicht ist die Ansicht, daß die U Geminorum-Sterne mit den neuen Sternen bzw. den Nova-ähnlichen Veränderlichen eine gewisse Verwandtschaft besitzen, durchaus nicht von der Hand zu weisen.

e) Die Veränderlichen der Mira-Klasse.

19. Definition der Mira-Sterne. Man versteht unter Mira-Sternen (die nach ihrem typischen Vertreter, *o* oder Mira Ceti, so benannt werden) diejenigen Veränderlichen, deren Helligkeit sich in Perioden von etwa 90^d bis zu etwas mehr als 600^d angenähert periodisch ändert; die Helligkeitsschwankungen weisen in verschiedenen Perioden zuweilen erhebliche Verschiedenheiten auf, sowohl was die Amplitude, als auch was die Form die Lichtkurve angeht. Auch die Periode selbst ist mitunter starken Änderungen unterworfen, die sowohl mehr oder weniger plötzlich auftreten als auch säkularer Natur sein können. Die oben angegebenen Grenzen für die Länge der Perioden entsprechen dem heutigen Stande der Kenntnis; die kürzeste Periode, die man bisher bei einem zweifellos der Mira-Klasse angehörigen Veränderlichen (T Centauri) festgestellt hat, ist nämlich 91^d, und die längste, die man bei einem solchen Stern gefunden hat, etwa 610^d (S Cassiopeiae). Es ist natürlich keineswegs ausgeschlossen, daß dieses Periodenintervall durch zukünftige Befunde erweitert wird. In der Tat kennen wir schon jetzt Veränderliche, deren Perioden zwischen 45^d (der oberen Grenze des Periodenwertes für solche Veränderliche, die zweifellos der δ Cephei-Klasse angehören) und 90^d liegen. Wir wissen vorläufig nicht, ob wir diese Sterne mit Perioden von 45^d bis zu 90^d zur δ Cephei-Klasse oder zur Mira-Klasse zu rechnen haben, oder ob sie einen Übergangstypus zwischen diesen beiden Klassen darstellen. Sie sollen mit in dem vorliegenden Abschnitt über die Mira-Sterne behandelt werden. Es ist natürlich auch denkbar, daß sie teils der einen, teils der anderen Klasse angehören.

---

[1] Lyon Bull 7, S. 46 (1925).

20. Vorbemerkungen über die Mira-Sterne. Einige der wichtigsten Eigenschaften der Mira-Sterne sollen hier, ehe wir auf Einzelheiten eingehen, im voraus zusammengestellt werden, da hierdurch das Verständnis des folgenden erleichtert wird.

Die Mira-Sterne sind ohne Ausnahme rötlich oder rot gefärbt und gehören vorwiegend der Spektralklasse M an; doch kommen auch solche der Spektralklassen N, R, S und K vor, sowie einige mit besonderem Spektrum. Die Mira-Sterne der Spektralklasse M zeigen in der Regel außer dem dieser Klasse entsprechenden Absorptionsspektrum noch ein Emissionsspektrum, in dem besonders die Linien der Balmer-Serie des Wasserstoffs hervortreten; das Spektrum dieser Sterne ist also mit Me zu bezeichnen (bis vor kurzem war dafür allgemein die Bezeichnung Md üblich). Es hat sich gezeigt, daß der Lichtwechsel der Mira-Sterne verschiedener Spektralklassen verschiedene Eigentümlichkeiten aufweist, so daß man bei statistischen Untersuchungen nicht, wie es bislang fast allgemein geschah, den Spektralcharakter außer acht lassen darf.

In der weit überwiegenden Mehrzahl der Fälle verlaufen die Lichtkurven der Mira-Sterne glatt, also ohne Buckel oder sekundäre Maxima und Minima. Doch treten bei einer immerhin nicht ganz kleinen Zahl von Mira-Sternen Buckel oder Stufen im aufsteigenden Aste der Lichtkurve auf, die bisweilen den Charakter eines sekundären Maximums annehmen; bei einzelnen Mira-Sternen sind doppelte, ungefähr gleich hohe Maxima mit dazwischen liegendem sekundären Minimum vorhanden, so daß die Lichtkurve der Gestalt nach Ähnlichkeit mit der von β Lyrae hat. Auf dem absteigenden Aste der Lichtkurven kommen bei den Mira-Sternen (im Gegensatze zu den δ Cephei-Sternen) Buckel oder Stufen nur außerordentlich selten und nur in schwacher Ausbildung vor.

Der Anstieg der glatten Lichtkurven ist meist steiler als der Abstieg oder höchstens ist letzterer ebenso steil als ersterer. Nur bei ganz vereinzelten Mira-Sternen mit glatter Lichtkurve ist der Abstieg steiler als der Anstieg, und dann nur in wenig ausgeprägtem Maße.

Die Helligkeitsamplituden der Mira-Sterne sind meist sehr beträchtlich und umfassen eine ganze Anzahl (bis zu etwa 9 oder 10) Größenklassen.

Die Periodenlängen der zweifellos zur Mira-Klasse gehörigen Sterne liegen, wie wir sahen, zwischen 90^d und etwa 610^d. Am häufigsten sind Perioden zwischen 200^d und 400^d (vgl. Ziff. 7). Wie schon erwähnt, sind die Perioden häufig kleineren oder größeren Änderungen unterworfen. Früher suchte man bei der rechnerischen Darstellung der Zeiten der Maxima oder Minima den Änderungen der Periode in der Regel dadurch gerecht zu werden, daß man der Formel Sinusglieder hinzufügte, die sich aber fast in allen Fällen bei der Vorausberechnung des weiteren Verlaufes der Helligkeitsänderungen nicht bewährten. Heutzutage neigt man mehr dazu, sprungweise Änderungen der Periode anzunehmen, also innerhalb verschiedener Zeitintervalle mit verschiedenen, aber konstanten Perioden zu rechnen. Bei einigen wenigen Sternen (R Hydrae, R Aquilae) hat man eine lange Zeit hindurch anhaltende Verkürzung der Periode konstatiert.

Zur vorläufigen Orientierung über die die Mira-Sterne betreffende Literatur sollen hier die Titel einiger Abhandlungen allgemeineren Inhalts über diese Veränderlichen zusammengestellt werden. Die schon in Ziff. 3 angeführte Literatur (insbesondere die Sammlungen von Beobachtungen) ist hier fortgelassen. Zahlreiche weitere Arbeiten werden bei Gelegenheit der folgenden Ausführungen zitiert werden.

CAMPBELL, L., A Tentative Classification of Long-Period Variables. Harv Reprint No. 21. 1925.

GYLLENBERG, W., On the Motion and Distribution of the Long-Period Variable Stars. Lund Medd Série I, Nr. 90. 1918.

HEISKANEN, W., und LUDENDORFF, H., Über die Radialgeschwindigkeiten der Mira Ceti-Sterne. A N 213, S. 297. 1921.

LUDENDORFF, H., Untersuchungen über veränderliche Sterne III. 1. Die veränderlichen Sterne der Spektralklasse N. 2. Die veränderlichen Sterne der Spektralklasse R. A N 217, S. 161. 1922.

LUDENDORFF, H., Untersuchungen über veränderliche Sterne IV. Die veränderlichen Sterne der Spektralklassen K, Ma, Mb, Mc. A N 219, S. 1. 1923.

LUDENDORFF, H., Untersuchungen über veränderliche Sterne V. 1. Statistik der veränderlichen Sterne der Spektralklasse Md. 2. Die Mira-Sterne mit Perioden von mehr als 400^d. A N 220, S. 145. 1924.

LUDENDORFF, H., Untersuchungen über veränderliche Sterne VI. Die unregelmäßig veränderlichen Md-Sterne. A N 220, S. 241. 1924.

LUDENDORFF, H., Untersuchungen über veränderliche Sterne VII. Die Md-Sterne mit Perioden bis zu 200^d. A N 222, S. 17. 1924.

LUDENDORFF, H., Untersuchungen über veränderliche Sterne VIII. 1. Die Beziehungen zwischen Periodenlänge und Form der Lichtkurve bei den Mira-Sternen. 2. Beziehungen zwischen Form der Lichtkurve und Spektraltypus bei den Mira-Sternen. 3. Beziehungen zwischen mittlerer Periodenlänge und Spektraltypus bei den Mira-Sternen. A N 228, S. 369. 1926.

MERRILL, P. W., Spectroscopic Observations of Stars of Class Md. Publ. of the Astr. Obs. of the University of Michigan II, S. 45. 1916.

MERRILL, P. W., Stellar Spectra of Class S. Ap J 56, S. 457. 1922. = Mt Wilson Contr 252.

MERRILL, P. W., The Radial Velocities of Long-Period Varable Stars. Ap J 58, S. 215. 1923. = Mt Wilson Contr 264.

MERRILL, P. W., and STRÖMBERG, G., The Absolute Magnitudes of Long-Period Variable Stars. Ap J 59, S. 97. 1924. = Mt Wilson Contr 267.

MERRILL, P. W., and STRÖMBERG, G., Space-Velocities of Long-Period Variable Stars of Classes Me and Se. Ap J 59, S. 148. 1924. = Mt Wilson Contr 268.

PHILLIPS, T. E. R., Presidential Address. J B A A 27, S. 2. 1916.

THOMAS, H., Eine kritische Darstellung unseres gesamten Wissens und unserer theoretischen Vorstellungen von den Veränderlichen vom Miratypus. Diss. Berlin 1925.

(Diese nur in wenigen Exemplaren vervielfältigte Dissertation war für die Abfassung des vorliegenden Abschnittes von Nutzen.)

TURNER, H. H., On the Classification of Long-Period Variable Stars, and a Possible Physical Interpretation. M N 67, S. 332. 1907.

TURNER, H. H., Note on the Range in Brightness at Maximum of Long-Period Variables. M N 67, S. 489. 1907.

TURNER, H. H., An Example of Professor Karl Pearson's Calculation of Correlation in the Case of the Periodic Inequalities of Long-Period Variables. M N 68, S. 544. 1908.

TURNER, H. H., On the Classification of Long-Period Variable Stars. M N 78, S. 92. 1917.

TURNER, H. H., On the Classification of Long-Period Variable Stars. Third Paper: The Influence of the Period on the Shape of the Curves in Phillips's two Classes. M N 79, S. 371. 1919.

TURNER, H. H., On the Suggested Decrease of Period of Stars in Phillips's Group II. M N 80, S. 273. 1920.

TURNER, H. H., On the Suggested Increase in Period of Variable Stars in Phillips's Group I. M N 80, S. 481. 1920.

TURNER, H. H., On the Changes in Period of o (Mira) Ceti, S Herculis, and R Leonis. M N 80, S. 604. 1920.

WILSON, R. E., The Proper Motions of 315 Red Stars. A J 34, S. 183. 1923.

WILSON, R. E., On the Motions, Parallaxes and Luminosities of the Long-Period Variables and Other Stars of Late Spectral Type. A J 35, S. 125. 1923.

Von Monographien über einzelne Veränderliche der Mira-Klasse seien hier folgende erwähnt:

GUTHNICK, P., Neue Untersuchungen über den veränderlichen Stern o (Mira) Ceti. Nova Acta. Abh. der Kaiserl. Leop.-Carol. Deutschen Akademie der Naturforscher Bd. 79, Nr. 2. Halle 1901.

LUDENDORFF, H., Untersuchungen über den Lichtwechsel von R Hydrae. A N 203, S. 117. 1916.

MÜLLER, R., Untersuchungen über den Veränderlichen R Aquilae. Diss. Berlin 1925.

ROSENBERG, H., Der Veränderliche χ Cygni. Nova Acta. Abh. der Kaiserl. Leop.-Carol. Deutschen Akademie der Naturforscher Bd. 85, Nr. 2. Halle 1906.

21. Klassifikation der Mira-Sterne nach der Gestalt ihrer Lichtkurven. Die Lichtkurven der Mira-Sterne sind von außerordentlich verschiedener Gestalt, und es ergibt sich daher von selbst der Gedanke, diese Veränderlichen nach der Gestalt ihrer Lichtkurven zu klassifizieren und zu versuchen, durch eine Statistik der Lichtkurven Aufschlüsse über die allgemeinen Eigenschaften der Mira-Sterne zu gewinnen. Man kann hierbei nun zwei Wege einschlagen: entweder man kann die Lichtkurven einfach nach ihrem Aussehen klassifizieren, oder man kann es unternehmen, sie (durch eine Fourier-Entwicklung oder durch eine sonstige Formel) zahlenmäßig darzustellen. Letzteres würde verhältnismäßig einfach sein, wenn die Kurven für jeden Stern von Periode zu Periode völlig oder nahezu unveränderlich wären, wie es für die δ Cephei-Sterne in der Regel der Fall ist. Bei den Mira-Sternen treten nun aber sehr oft von Periode zu Periode recht erhebliche Änderungen sowohl in der Form wie in der Amplitude der Kurven auf, wenn auch der Gesamtcharakter der Kurve meist erhalten bleibt. Man wird also, wenn man eine mathematische Darstellung der Kurven im Auge hat, zunächst aus dem Lichtwechsel während einer größeren Anzahl von Perioden eine mittlere Lichtkurve ableiten müssen, was ein ziemlich großes und vollständiges Beobachtungsmaterial erfordert und mit ziemlich erheblicher Arbeit verbunden ist. Auch besteht mitunter die Gefahr, daß bei Ableitung mittlerer Lichtkurven gewisse charakteristische Eigentümlichkeiten des Lichtwechsels stark verwischt werden, z. B. Buckel oder Stufen, wenn sie nicht immer an derselben Stelle der Lichtkurve auftreten. Für solche nicht glatt verlaufenden Lichtkurven kann die mathematische Darstellung natürlich unter Umständen auch schon etwas kompliziert werden. Für die meisten Fälle wird eine Klassifizierung nach dem Augenschein genügen.

Der erste, der in etwas umfassenderer Weise die mathematische Darstellung von Lichtkurven der Mira-Sterne unternahm, war H. H. TURNER[1]. Er hatte aus den auf dem Rousdon Observatory angestellten Beobachtungen für 19 Mira-Sterne mittlere Lichtkurven abgeleitet. Diese stellte er nun numerisch durch eine Fourier-Entwicklung in der Form

$$y = A \sin\vartheta + B \cos\vartheta + C \sin 2\vartheta + D \cos 2\vartheta + E \sin 3\vartheta + F \cos 3\vartheta$$

dar. Für die Koeffizienten $A, B, \ldots$ dient die Amplitude als Einheit, ϑ ist definiert durch $\vartheta = \frac{2\pi T}{P}$, wo T die vom Maximum an gezählte Zeit, P die

[1] Mem R A S 55, S. XCVIII (1904); M N 64, S. 543 (1904).

Periode ist. Bei allen 19 Sternen sind die Koeffizienten C bis F ziemlich klein gegen A und B. Ordnet man die Sterne nach dem Werte von A, so zeigt sich, daß nicht nur die B, sondern auch die anderen Koeffizienten einen Gang mit A zeigen. (Für B ist dies selbstverständlich, da, wenn die übrigen Koeffizienten $= 0$ wären, die als Einheit gewählte Amplitude $= 2\sqrt{A^2 + B^2}$ sein würde.) Später hat TURNER noch die Lichtkurven von 12 weiteren Mira-Sternen in derselben Weise dargestellt und die früheren Schlüsse bestätigt gefunden[1]. Den Grund der Erscheinung, daß alle Koeffizienten einen Gang mit A zeigen, hat J. STEIN[2] dargelegt; er ergibt sich nach ihm aus den allgemeinen Eigenschaften der langperiodischen Lichtkurven und bietet nichts Überraschendes. Auf TURNERS Vergleichung der Lichtkurven der Mira-Sterne mit der Kurve der Sonnenfleckenhäufigkeit werden wir bei Besprechung der Hypothesen über die Mira-Sterne noch zurückkommen.

In seiner zuletzt zitierten Arbeit weist TURNER auf die Wichtigkeit der Größe $\alpha = [2(M - m) - P] : P$, wo $M - m$ der zeitliche Abstand des Maximums von dem vorangehenden Minimum ist, für die Klassifizierung der Lichtkurven hin und glaubt auf Grund des Materials, das der dritte CHANDLERsche Katalog von Veränderlichen bietet, behaupten zu dürfen, daß Sterne mit positivem α nicht nahe den Polen der Milchstraße vorkommen. Diese Erscheinung ist später von H. LUDENDORFF näher untersucht worden, worauf wir noch zurückkommen.

Eine Klassifizierung der Lichtkurven nach dem Augenschein haben dann L. CAMPBELL und E. C. PICKERING vorgenommen[3]. Für 67 Mira-Sterne wurden mittlere Lichtkurven (die z. T. auf zu geringem Material beruhen und daher nicht alle ganz zuverlässig sind) abgeleitet; sie sind auf zwei Tafeln in leider sehr zusammengedrängter Form graphisch dargestellt. Dabei sind folgende 5 Kurventypen unterschieden (die Zahlen n geben an, wie oft der betreffende Typus unter den 67 Kurven vorkommt):

1. Kurven mit breitem Maximum $n = 4$,
2. Kurven mit breitem Minimum $n = 19$,
3. Kurven mit schnellem Aufstieg $n = 18$,
4. Kurven mit schnellem Abstieg $n = 4$,
5. Kurven mit gleichmäßiger Änderung (symmetrische Kurven) $n = 22$.

Diese Einteilung kann nicht als sehr glücklich bezeichnet werden, da z. B. manche der Kurven mit breitem Minimum sehr raschen Aufstieg zeigen, wonach sie also in die dritte Klasse eingeordnet werden könnten. Irgendwelche statistischen Schlüsse werden aus dem Material nicht gezogen.

Im Jahre 1916 veröffentlichte T. E. R. PHILLIPS[4] die Resultate einer mathematischen Analyse von 80 Lichtkurven langperiodischer Veränderlicher. Er schreibt die Fourier-Entwicklung für jede Lichtkurve in folgender Form (die Bezeichnung ist hier gegen die von PHILLIPS etwas abgeändert):

$$M + C_1 \cos(\vartheta - 180°) + C_2 \cos(2\vartheta - \varphi_2) + C_3 \cos(3\vartheta - \varphi_3).$$

Die Kurven wurden alle auf eine einheitliche Amplitude (6^m) reduziert, und es wurden aus ihnen die Konstanten C und die Winkel φ_2 und φ_3 ermittelt. Die Form der Lichtkurve wird besonders durch φ_2 und φ_3 charakterisiert, und PHILLIPS unterscheidet nach den Werten dieser Winkel zwei Gruppen von Sternen. Trägt man nämlich die in Graden ausgedrückten φ_2 als Abszissen, die φ_3 als Ordinaten in ein rechtwinkliges Koordinatensystem ein, so zeigt sich, daß die erhaltenen Punkte in zwei Gruppen zerfallen; innerhalb jeder Gruppe

---

[1] M N 67, S. 332 (1907).

[2] HAGEN u. STEIN, Die Veränderlichen Sterne. Zweiter Band, S. 110.

[3] Harv Ann 57, Part I (1907). [4] J B A A 27, S. 2 (1916).

liegen die Punkte nahe einer geraden Linie. Für Gruppe I (28 Sterne) sind die φ_3 nicht sehr von 200° verschieden, die φ_2 liegen zwischen 233° und 126°; die entsprechende Gerade ist also nahezu parallel der Abszissenachse. Die der Gruppe II (46 Sterne) entsprechende Gerade ist dagegen gegen beide Koordinatenachsen geneigt, und φ_2 liegt hier zwischen 67° und 197°. Sechs Sterne passen nicht in das Schema, oder ihre Gruppeneinordnung ist zweifelhaft. Die Gleichungen der beiden, übrigens nahezu zusammenstoßenden Geraden lauten

$$\text{Gruppe I: } \varphi_3 = 202^\circ{,}1 - 0{,}04\,\varphi_2 \quad (\varphi_2 = 233^\circ \text{ bis } 126^\circ).$$
$$\text{Gruppe II: } \varphi_3 = 1{,}67\,\varphi_2 - 126^\circ{,}8 \quad (\varphi_2 = 67^\circ \text{ bis } 197^\circ).$$

Es ergeben sich noch folgende Unterschiede zwischen den beiden Gruppen:

1. Die Helligkeitsamplitude ist bei den Sternen der Gruppe I durchschnittlich kleiner als für die der Gruppe II.

2. C_2 ist für Gruppe I durchschnittlich kleiner als für Gruppe II.

3. Die Sterne der Gruppe I sind durchschnittlich um fast eine Größenklasse heller als die der Gruppe II.

4. Die Perioden der Sterne der Gruppe I sind durchschnittlich kürzer als die der Sterne der Gruppe II.

Die Betrachtung der Lichtkurven der beiden Gruppen führt zu folgendem Ergebnis: Zur Gruppe I gehören diejenigen Lichtkurven, bei denen die Zeit zwischen einem Minimum und dem folgenden Maximum nicht sehr verschieden ist von der Zeit zwischen dem letzteren und dem nächsten Minimum; bei diesen Kurven ist häufig ein Buckel oder eine Stufe im aufsteigenden Aste vorhanden. Zur Gruppe II gehören diejenigen Lichtkurven, bei denen der Helligkeitsanstieg vom Minimum zum Maximum wesentlich rascher verläuft als die darauf folgende Abnahme vom Maximum zum Minimum. Es ist also in der Regel möglich, schon nach dem bloßen Aussehen der Kurve festzustellen, in welche Gruppe sie gehört, wie es ja auch ganz selbstverständlich ist.

An die Untersuchungen von PHILLIPS knüpft H. H. TURNER, der dieselben für sehr wichtig hält, in mehreren Arbeiten an; in der ersten derselben[1] beschäftigt er sich mit verschiedenen Fragen, die sich aus PHILLIPS' Betrachtungen ergeben, sowie mit seiner eigenen Klassifizierung der Lichtkurven nach der Größe $\alpha = [2(M - m) - P] : P$ (vgl. oben) und den Beziehungen dieser Art der Klassifizierung zu der von PHILLIPS. Zu besonders bemerkenswerten Ergebnissen kommt er dabei nicht.

In seiner zweiten Arbeit[2] untersucht TURNER den Einfluß der Periodenlänge auf die Gestalt der Kurven in PHILLIPS' zwei Gruppen. Er definiert dabei diese Gruppen etwas anders als PHILLIPS. Zu Gruppe I rechnet er alle Sterne, für die $\cos\varphi_2$ positiv, zu Gruppe II alle, für die $\cos\varphi_2$ negativ ist; nur wenige Sterne ändern sich dadurch in ihrer Gruppenzugehörigkeit. Zunächst ergibt sich, daß in beiden Gruppen die Sterne mit sehr langen Perioden die Milchstraße bevorzugen, eine Erscheinung, die durch spätere Untersuchungen auf Grund eines umfangreicheren Materials bestätigt worden ist. Die übrigen Ergebnisse faßt TURNER ungefähr wie folgt zusammen:

1. In Gruppe II zeigen φ_2 und φ_3 „fluctuations", wenn die Periode sich um Vielfache von 45^d ändert, und ähnliche „fluctuations" erscheinen in der Häufigkeit, mit welcher ein bestimmter Wert der Periode auftritt.

2. In Gruppe II ändern sich $k_2' = C_2 : C_1$ und $k_3' = C_3 : C_1$ ständig mit der Periode; auch φ_2 und φ_3 zeigen solche Änderungen.

3. Die Häufigkeit der Periodenwerte für Gruppe I zeigt „fluctuations" mit Änderungen der Periode um Vielfache von 60^d.

[1] MN 78, S. 92 (1917). [2] MN 79, S. 371 (1919).

4. In Gruppe I zeigen k_2', k_3', φ_2 und φ_3 allmähliche Änderungen mit der Periode.

Da das Material, auf dem diese Schlüsse beruhen, nicht sehr groß ist (29 Sterne in Gruppe I, 51 in Gruppe II), so bedürfen sie wohl zum Teil noch der Nachprüfung.

Schließlich weist TURNER noch auf einen interessanten Umstand hin. Wir haben bereits erwähnt, daß die Helligkeitsamplituden für Gruppe I durchschnittlich kleiner sind, als für Gruppe II. Es zeigt sich nun, daß für große Perioden dieser Unterschied zwischen Gruppe I und II in der Tat beträchtlich ist, für kleinere Perioden (etwa um 200^d) aber verschwindet. (Ein größeres Material zeigt, daß letzteres zwar nicht ganz der Fall ist, daß aber der Unterschied jedenfalls für längere Perioden größer ist als für kürzere.) TURNER wirft im Anschluß an seine Wahrnehmung die Frage auf, ob nicht etwa bei den kleinen Periodenwerten die beiden Gruppen ineinander übergehen, d. h. sich Sterne der einen Gruppe in solche der andern verwandeln können.

Gelegentlich einer gemeinsam mit Miss MARY A. BLAGG verfaßten Abhandlung über den Veränderlichen W Cygni ist dann TURNER[1] auf Grund gewisser Erscheinungen, die der komplizierte Lichtwechsel dieses Sternes bietet, zu der Vermutung gekommen, daß die Mira-Sterne ihre Entwicklung als Sterne der Gruppe II mit langen Perioden beginnen, daß die Periode allmählich abnimmt, daß dann die Sterne in solche der Gruppe I übergehen, und daß darauf die Periode wieder zunimmt. Die Vermutung, daß die Perioden der Sterne der Gruppe II abnehmen, die der Sterne der Gruppe I dagegen zunehmen, prüft TURNER in drei Abhandlungen[2]. In der ersten untersucht er die Periodenänderungen der der Gruppe II angehörigen Sterne R Hydrae, S Tauri, U Herculis, R Aquilae, χ Cygni und S Coronae borealis. Daß die Perioden von R Hydrae und R Aquilae stark abgenommen haben, ist eine bekannte Tatsache. Bei den anderen genannten Sternen ist teilweise eine Abnahme der Periode angedeutet, in keinem Falle aber eine Zunahme. In der zweiten Abhandlung prüft TURNER 14 Sterne der Gruppe I und findet bei mehreren von ihnen eine Zunahme der Periode, bei keinem eine Abnahme angedeutet. In der dritten Arbeit endlich beschäftigt er sich noch mit o Ceti, S Herculis und R Leonis. Die Periode von o Ceti (Gruppe II) zeigt keine säkularen Änderungen; bei S Herculis (Gruppe I) ist eine Zunahme nicht ausgeschlossen, bei R Leonis (Gruppe I nach PHILLIPS, II nach TURNER) hat sich die frühere Abnahme der Periode in eine starke Zunahme verwandelt.

TURNER legt seiner Beweisführung kein großes Gewicht bei, und man wird sich dieser Ansicht anschließen müssen. Es ist ja in der Tat kaum zu erwarten, daß sich Änderungen der Periode, die durch eine kosmische Entwicklung bewirkt werden, schon jetzt in den relativ doch sehr kurzen Beobachtungsreihen der Mira-Sterne bemerkbar machen, und in der Tat sind auch gerade bei o Ceti und χ Cygni, den beiden am längsten und eingehendsten beobachteten Sternen dieser Art, solche Änderungen bisher nicht festgestellt worden. Der ausgesprochenste Fall einer säkularen Periodenänderung betrifft R Hydrae. Aber bei diesem Stern scheint die Abnahme der Periode neuerdings aufgehört zu haben, und zudem ist er ein wenig typischer Vertreter der Gruppe II.

Während, wie erwähnt, TURNER der Untersuchung von PHILLIPS großen Wert beilegt, steht J. G. HAGEN ihr gegenüber auf einem wesentlich anderen Standpunkt, den er in drei kurzen Abhandlungen[3] dargelegt hat. Er vergleicht

[1] M N 80, S. 41 (1919).

[2] M N 80, S. 273, 481 u. 604 (1920).

[3] M N 79, S. 572 (1919); A N 209, S. 257 (1919); Ap J 53, S. 179 (1921).

die PHILLIPSsche Klassifizierung der Lichtkurven nach den Winkeln φ_2 und φ_3 mit der schon früher erwähnten graphischen Darstellung von 67 Lichtkurven in Harv Ann 57. Da die PICKERINGschen Kurventypen 1 (breite Maxima) und 4 (schneller Abstieg) nur sehr selten vorkommen, so rechnet HAGEN sie zum Typus 5 (gleichmäßige Änderung). HAGEN unterscheidet schließlich also drei Kurventypen, nämlich:

I. Gleichförmige Kurven (PICKERINGS Klassen 1, 4, 5),

IIa. Kurven mit flachem Minimum (PICKERINGS Klasse 2),

IIb. Kurven mit steilem Aufstieg (PICKERINGS Klasse 3).

Die Vergleichung mit PHILLIPS ergibt nun, daß sich PHILLIPS' Gruppe I im wesentlichen deckt mit HAGENS Klasse I, und daß sich die Sterne von PHILLIPS' Gruppe II auf HAGENS Klassen IIa und IIb verteilen. (Wie bereits erwähnt, hatte auch schon PHILLIPS die geometrische Bedeutung seiner beiden Gruppen erkannt.) HAGEN bezweifelt auf Grund dieses Befundes, daß die periodischen Glieder der Reihenentwicklung von PHILLIPS irgendwelche Wahrheiten enthüllen können, die nicht schon aus dem Studium der Lichtkurven selbst erkennbar sind; im großen ganzen wird man ihm beipflichten müssen.

Nicht sehr glücklich erscheint HAGENS bzw. PICKERINGS Unterscheidung der Klassen IIa und IIb, da flache Minima und steiler Aufstieg zwei Eigenschaften sind, die sich durchaus nicht ausschließen, sondern oft vereint auftreten. Als Beispiel sei der Veränderliche V Camelopardalis genannt, bei dem das breite Minimum reichlich die Hälfte der Periode, der Helligkeitsanstieg nur etwa 0,13 der Periode dauert. Jedenfalls bedürfen HAGENS Klassen IIa und IIb noch einer näheren Definition etwa dadurch, daß bei IIa wohl ein steiler Aufstieg, bei IIb aber kein flaches Minimum auftreten darf.

Wir verlassen nunmehr die Besprechung der an PHILLIPS' Untersuchung anknüpfenden Arbeiten und wenden uns den weiteren Versuchen zu, die Lichtkurven der Mira-Sterne nach ihrer Gestalt zu klassifizieren. Darauf bezügliche Arbeiten liegen vor von W. HASSENSTEIN, H. THOMAS, L. CAMPBELL und H. LUDENDORFF. Die beiden Erstgenannten unternehmen, wie TURNER und PHILLIPS, eine mathematische Darstellung der Lichtkurven, während L. CAMPBELL und H. LUDENDORFF von einer solchen absehen. Die Methode der Kurvendarstellung von HASSENSTEIN[1] erfordert zu ihrem Verständnis ziemlich komplizierte Betrachtungen, und es erscheint daher zweckmäßig, den Leser für ihr Studium auf die Originalabhandlungen zu verweisen. Es sei nur erwähnt, daß HASSENSTEIN die 67 in Harv Ann 57 wiedergegebenen und hier schon mehrfach erwähnten Lichtkurven nach seiner Methode darstellt.

THOMAS gibt in seiner schon in Ziff. 20 angeführten Dissertation (Berlin 1925) eine außerordentlich einfache Art der rechnerischen Darstellung der Lichtkurven an, die allerdings (ebenso wie übrigens die von HASSENSTEIN) der nicht ganz unwesentlichen Beschränkung unterliegt, daß sie nur auf glatt verlaufende Kurven, also Kurven ohne doppelte Maxima oder Minima und ohne Buckel oder Stufen, anwendbar ist. Es sei $y = f(x)$ die Gleichung der Lichtkurve, so wird die mittlere Größe $\bar{y}$ definiert durch

$$\bar{y} = \frac{1}{P}\int_0^P f(x)\, dx\,,$$

wo P die Periode ist. Es sei ferner $y_{\min}$ die Größe im Minimum, $y_{\max}$ die im Maximum, M-m die Zeit zwischen dem Minimum und dem darauffolgenden

[1] A N 219, S. 373 (1923) und besonders A N 221, S. 385 (1924).

Maximum. THOMAS setzt dann

$$S_x = \frac{y_{\min} - \bar{y}}{y_{\min} - y_{\max}}, \qquad S_y = \frac{M - m}{P}$$

und benutzt die beiden Größen S_x und S_y zur Charakterisierung der Form der Lichtkurve. S_x ist offenbar die Differenz zwischen der kleinsten und der mittleren Größe, ausgedrückt in Einheiten der Amplitude $A = y_{\min} - y_{\max}$. Die Größe S_y, die wir hier, wie es allgemein üblich ist, mit ε bezeichnen wollen, ist mit der von TURNER benutzten Größe α verknüpft durch die Gleichung:

$$\alpha = 2S_y - 1 = 2\varepsilon - 1.$$

Es wird angenommen, daß die y in Größenklassen ausgedrückt sind, so daß kleineren Helligkeiten größere Zahlenwerte y entsprechen. S_x und ε liegen beide, wie leicht ersichtlich, zwischen 0 und 1. S_x kann als Maß der Unsymmetrie der Lichtkurve relativ zur x-Achse (Zeitachse) betrachtet werden. Ist das Maximum breiter als das Minimum, so ist $S_x > \frac{1}{2}$, ist das Minimum dagegen breiter als das Maximum, so ist $S_x < \frac{1}{2}$. ε ist das Maß der Unsymmetrie der Kurve relativ zur y-Achse. Ist der Helligkeitsanstieg steiler als der Abstieg, so ist $\varepsilon < \frac{1}{2}$, ist dagegen der Abstieg steiler als der Anstieg, so ist $\varepsilon > \frac{1}{2}$. Die Bestimmung der Größe ε aus der Lichtkurve macht natürlich keine Schwierigkeiten. Um S_x berechnen zu können, muß zunächst $\bar{y}$ bestimmt werden. Dies kann mit hinreichender Annäherung in der Weise geschehen, daß man für n äquidistante Punkte x der Lichtkurve die zugehörigen y abliest und dann

$$\bar{y} = \frac{1}{n} \sum_{i=0}^{i=n-1} y_i$$

bildet. THOMAS hat es als vollkommen ausreichend gefunden, $n = 24$ zu nehmen, und er hat für 116 Lichtkurven von 105 verschiedenen Mira-Sternen (für einige Sterne wurden verschiedene Lichtkurven benutzt) die Größen S_x und ε abgeleitet. Das Material an Kurven, das er dabei benutzt, ist zum größten Teil wiederum Harv Ann 57 sowie J. A. PARKHURSTS „Researches in Stellar Photometry“ (zitiert in Ziff. 3) entnommen; es kommen aber auch zahlreiche Lichtkurven zur Verwendung, die nur auf kürzeren Beobachtungsreihen beruhen und daher nicht als mittlere Lichtkurven angesehen werden können. Wie sehr S_x und ε bei demselben Sterne sich ändern können, wenn man verschiedenen Zeiten entsprechende Lichtkurven verwendet, zeigt das Beispiel von o Ceti, für den THOMAS folgende verschiedene Wertepaare von S_x und ε findet:

$$\begin{array}{ll} S_x = 0{,}44 & \varepsilon = 0{,}36 \\ S_x = 0{,}42 & \varepsilon = 0{,}37 \\ S_x = 0{,}43 & \varepsilon = 0{,}25 \\ S_x = 0{,}39 & \varepsilon = 0{,}28 \\ S_x = 0{,}42 & \varepsilon = 0{,}33 \end{array}$$

Es ist nun die Frage, ob die beiden Größen S_x und ε den Verlauf der Lichtkurven eindeutig definieren. Hierzu äußert sich THOMAS ungefähr wie folgt: Um zu prüfen, ob die Vermutung, daß die beiden Zahlen S_x und ε tatsächlich im wesentlichen den Verlauf der Lichtkurven bestimmen, richtig war, wurde das Material nach Intervallen von ε in Gruppen geteilt. Die erste Gruppe enthält sämtliche Lichtkurven, für die ε zwischen $> 0{,}20$ und $= 0{,}25$, die zweite Gruppe nur Kurven, für die ε zwischen $> 0{,}25$ und $= 0{,}30$ liegt usw., indem jedes Intervall von ε 0,05 Einheiten beträgt. Innerhalb jeder Gruppe wurden nun die Kurven nach der Größe von S_x geordnet. Sollte nun die Vermutung richtig sein, so müßten die so geordneten Kurven innerhalb der einzelnen Gruppen einen stetigen Übergang ineinander zeigen. Dies ist in der Tat der Fall, womit obige Vermutung bestätigt wird. Teilt man die Kurven nach Intervallen von S_x in Gruppen und ordnet innerhalb jeder Gruppe nach ε, so findet ebenfalls ein annähernd stetiger Übergang der einzelnen Kurven ineinander statt. Damit ist, wenigstens für das benutzte Kurvenmaterial, gezeigt, daß der Verlauf der Lichtkurven der Mira-Sterne im wesentlichen durch die Größen S_x und ε, d. h. durch ihre mehr oder weniger große Symmetrie gegen die Koordinatenachsen bestimmt ist. (Streng genommen handelt es sich um den Verlauf der auf die gleiche Amplitude und Periodenlänge reduzierten Lichtkurven, da ja S_x und ε gar nicht vom absoluten Betrag der Periode und Amplitude abhängen. Zur Kenntnis des wahren Verlaufs bedarf es also außer S_x und ε noch der Werte P und A.)

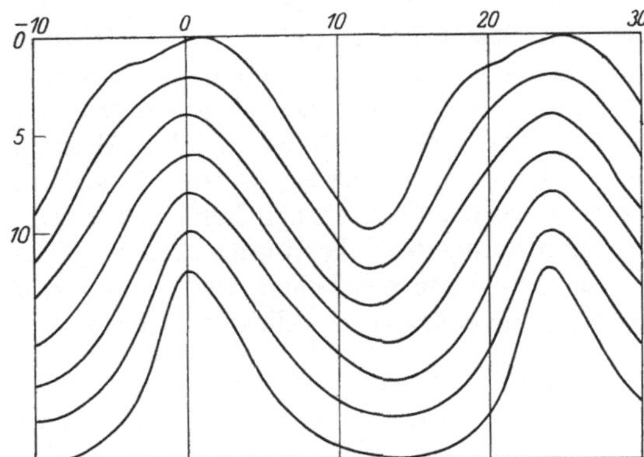

Abb. 8. Die sieben Typen der Lichtkurven der Mira-Sterne nach L. CAMPBELL.

Angesichts der anschaulichen Bedeutung der Größen S_x und ε kann die Definition der Lichtkurven durch diese beiden Größen empfohlen werden. Wie schon erwähnt, ist diese Art der Darstellung aber nur auf glatt verlaufende Kurven anwendbar, und wenn THOMAS auch für eine Reihe von Lichtkurven (z. B. die von T Cassiopeiae, R Aurigae, S Carinae, S Serpentis, RU Herculis, W Lyrae, χ Cygni, T Cephei), die nicht glatt zu verlaufen pflegen, die S_x und ε berechnet, so darf man in diesen Fällen die Bedeutung dieser Größen nicht zu hoch einschätzen. Freilich verschwinden die Ungleichmäßigkeiten häufig mehr oder minder in den mittleren Lichtkurven, wie schon erwähnt wurde, aber die mittlere Lichtkurve hat in solchen Fällen wohl überhaupt nur einen geringen Wert, da sie eine falsche Vorstellung von dem Lichtwechsel, wie er in der Regel erfolgt, vermittelt. Die statistischen Schlüsse, die THOMAS aus dem von ihm bearbeiteten Material an Lichtkurven zieht, werden weiter unten behandelt werden.

Wir wenden uns nun denjenigen Klassifikationen von Lichtkurven zu, die eine rechnerische Darstellung nicht erstreben, nämlich denen von L. CAMPBELL und H. LUDENDORFF. Ersterer hat seine Klassifikation schon in einem vor der American Association of Variable Star Observers im Jahre 1920 gehaltenen Vortrage bekanntgegeben, einem weiteren Kreise ist sie aber erst 1925 zugänglich geworden[1]. L. CAMPBELL hat für 124 Mira-Sterne mittlere Lichtkurven bestimmt und ordnet diese in sieben Typen ein, die er mit I bis VII numeriert. Diese Typen sind in Abb. 8 graphisch dargestellt, und zwar Typus I zu oberst, Typus VII zu unterst. Wie man aus der Abbildung ersieht, gehen die Typen allmählich ineinander über. Alle Kurven sind auf gleiche Periode und gleiche Amplitude reduziert. Sehr auffallend ist, daß der merkwürdige und aus später zu erörternden Gründen besonders interessante Kurventypus, der eine lang andauernde (manchmal die Hälfte der Periodenlänge umfassende) Phase kon-

[1] Harv Reprint 21 (1925).

stanter Helligkeit im Minimum aufweist, unter den CAMPBELLschen Kurven ganz fehlt; der Kurventypus VII entspricht ihm keineswegs.

Die statistischen Folgerungen L. CAMPBELLS werden ebenfalls weiter unten besprochen werden.

H. LUDENDORFF verfährt (s. seine in Ziff. 20 zitierten Arbeiten) bei der Klassifikation der Lichtkurven folgendermaßen: Es werden zunächst für ein längeres Zeitintervall (soweit möglich, für mindestens 10 bis 12 Jahre) die Beobachtungen der einzelnen Sterne graphisch dargestellt. Die so erhaltenen Lichtkurven werden allein nach dem Augenschein in verschiedene Klassen eingeordnet. Auf die Ableitung mittlerer Lichtkurven wird verzichtet, erstens aus dem schon öfter erwähnten Grunde, daß dabei charakteristische Eigenschaften des Lichtwechsels verwischt werden können, und zweitens deshalb, weil in sehr vielen Fällen das Beobachtungsmaterial die Ableitung solcher mittleren Kurven als noch nicht lohnend oder sogar als unmöglich erscheinen läßt, während die Einordnung in die verschiedenen Klassen schon recht wohl angängig ist. Es hat sich schließlich folgende Einteilung und Bezeichnung der Lichtkurven als zweckmäßig erwiesen:

α: Anstieg der Kurve merklich steiler als Abstieg; Minimum, von vereinzelten Ausnahmefällen abgesehen, stets breiter als Maximum.

Unterabteilungen:

α_1: Kurven mit nahezu oder völlig konstanter Phase von beträchtlicher Dauer (etwa $^1/_3$ bis $^1/_2$ der Periodenlänge) im Minimum und meist sehr steilem Helligkeitsanstieg.

α_2: Das Minimum weist keine konstante Phase erheblicher Ausdehnung mehr auf, ist aber noch sehr breit; Anstieg meist sehr steil.

α_3: Minimum nicht mehr so breit wie bei α_2, Anstieg immer noch recht steil.

α_4: Wie α_3, aber Anstieg weniger steil.

β: Anstieg nur noch ganz wenig oder überhaupt nicht mehr steiler als Abstieg; Lichtkurve im wesentlichen symmetrisch.

Unterabteilungen:

β_1: Maximum spitzer als Minimum.

β_2: Maximum ebenso spitz oder flach wie Minimum.

β_3: Maximum flacher als Minimum.

β_4: Das Maximum ist sehr breit und zeigt eine längere Zeit andauernde konstante Phase.

γ: Lichtkurven mit Stufe oder Buckel im aufsteigenden Ast oder mit Doppelmaximum.

Unterabteilungen:

γ_1: Stufe oder Buckel im aufsteigenden Ast.

γ_2: Doppelmaximum.

Gelegentlich werden noch Zwischenstufen, z. B. α_4—γ_1, α_4—β_1 usw., unterschieden. Von diesen abgesehen, umfaßt die Klassifikation 3 Haupttypen mit im ganzen 10 Unterabteilungen. Die wenigen Lichtkurven von Mira-Sternen, die sich nicht in dieses Schema einordnen lassen, werden als „pec" bezeichnet. Es kommt natürlich vor, daß ein Stern während verschiedener Perioden verschiedene Kurvenformen aufweist, z. B. β_1 und β_3, da ja der Lichtwechsel der Mira-Sterne keineswegs regelmäßig verläuft; maßgebend für die Einordnung ist das vorherrschende Verhalten. Meist macht, sofern genügend zahlreiche Beobachtungen vorliegen, die Einordnung keine Schwierigkeiten und geht mit großer Sicherheit vor sich.

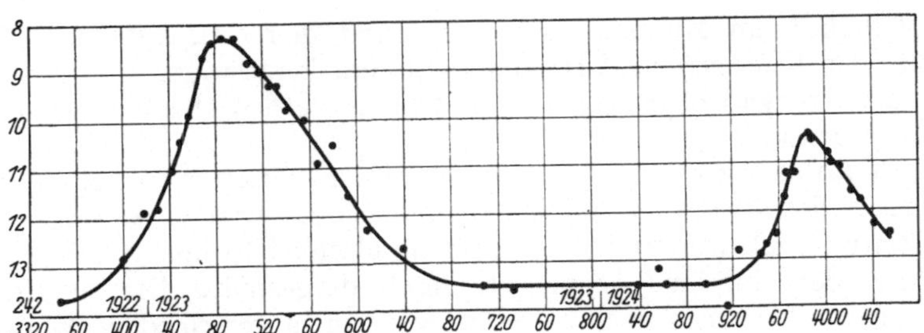

Abb. 9. Lichtkurve der Form α_1. Y Velorum 1922 bis 1924.

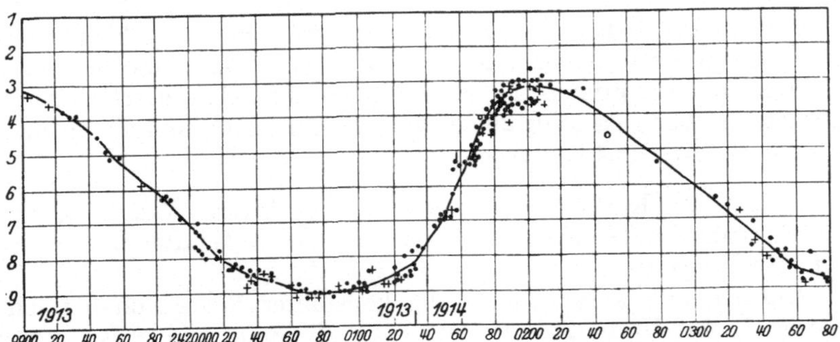

Abb. 10. Lichtkurve der Form α_3. *o* Ceti 1913 bis 1914.

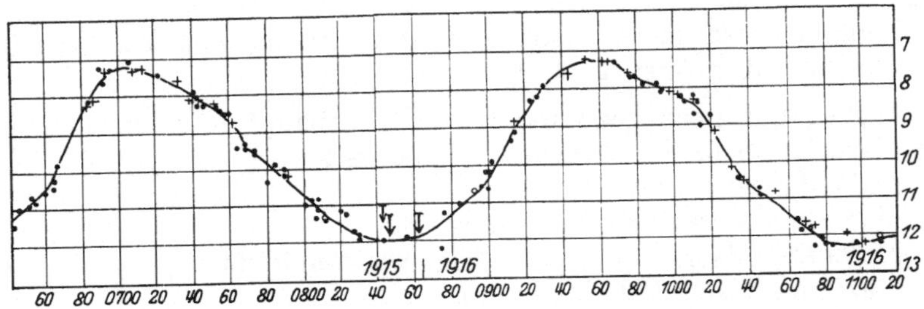

Abb. 11. Lichtkurve der Form α_4. R Draconis 1915 bis 1916.

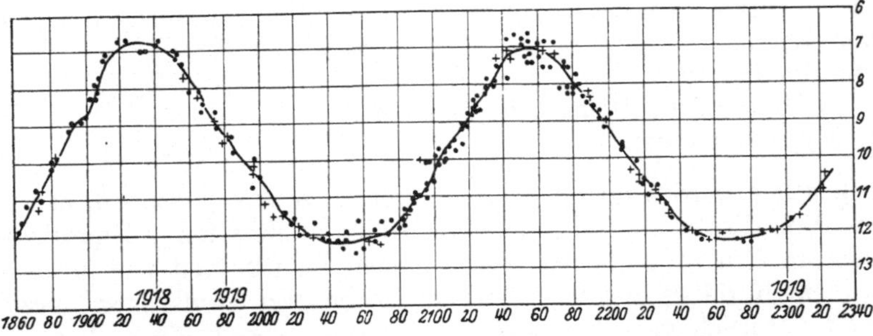

Abb. 12. Lichtkurve der Form β_1. R Bootis 1918 bis 1919.

Ludendorff hat für alle Mira-Sterne mit bekannten Spektren, die unter den 1687 Sternen des Hauptkatalogs der G. u. L. vorkommen, soweit es das vorhandene Beobachtungsmaterial gestattete, die Form der Lichtkurven bestimmt, in der überwiegenden Mehrzahl der Fälle auf Grund neugezeichneter Lichtkurven. Die aus diesem gesamten Material hervorgehenden statistischen Folgerungen werden späterhin näher dargelegt werden. Zunächst geben wir

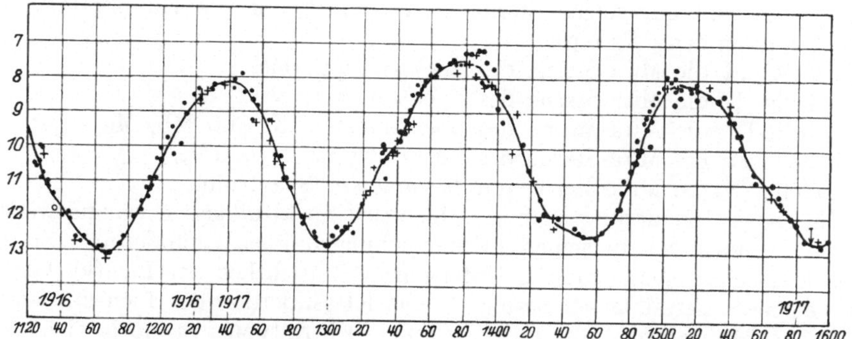

Abb. 13. Lichtkurve der Form β_3. X Camelopardalis 1916 bis 1917.

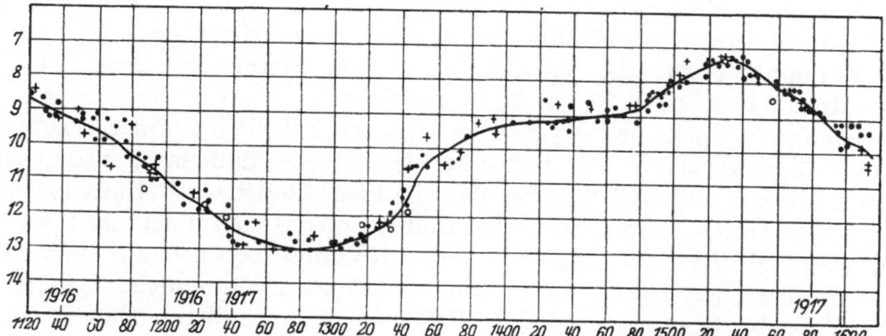

Abb. 14. Lichtkurve der Form γ_1. R Aurigae 1916 bis 1917.

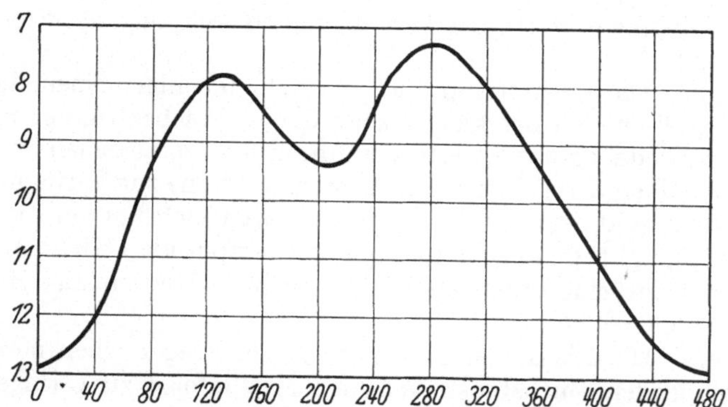

Abb. 15. Lichtkurve der Form γ_2. Mittlere Lichtkurve von R Normae.

noch in den nebenstehenden Abb. 9 bis 15 Beispiele für verschiedene Klassen von Lichtkurven. Abb. 9 beruht auf Beobachtungen der American Association of Variable Star Observers, Abb. 10 bis Abb. 14 sind dem „Appendix to Vols XXII and XXV“ der „Memoirs of the British Astronomical Association“ entnommen und Abb. 15 ist nach Harv Bull Bd. 836 gezeichnet.

22. Tabellen für die Lichtkurven der Mira-Sterne. In den folgenden Tabellen, die zum größten Teil noch nicht anderweitig veröffentlicht sind, finden sich die Resultate der Klassifikation der Lichtkurven der einzelnen Mira-Sterne mit bekannter Periode und bekannten Spektren nach dem soeben dargelegten Schema. Es sind nur solche Sterne berücksichtigt, die im Hauptkatalog der G. u. L. vorkommen. Es enthält

Tabelle I die Mira-Sterne des Spektraltypus Me,

Tabelle II die Mira-Sterne des Spektraltypus Se,

Tabelle III die Mira-Sterne des Spektraltypus K, Ma, Mb, Mc,

Tabelle IV die Mira-Sterne des Spektraltypus N und R,

Tabelle V die Mira-Sterne mit besonderem Spektrum.

Eine solche Sonderung nach verschiedenen Spektralklassen war notwendig, da, wie LUDENDORFF in seinen „Untersuchungen über veränderliche Sterne" dachgewiesen hat, systematische Unterschiede hinsichtlich der Lichtkurven für verschiedene Spektralklassen bestehen. Die Klassifikation der Lichtkurven der zahlreichen Mira-Sterne mit unbekanntem Spektraltypus ist noch nicht ausgeführt worden.

Über die Anordnung der Tabellen ist folgendes zu bemerken:

1. Kolumne: Nummer des Sternes in der G. u. L.; die Angabe soll dazu dienen, dem Benutzer ein etwaiges Nachschlagen zu erleichtern.

2. Kolumne: Name des Veränderlichen. Die Sterne sind innerhalb jeder Tabelle alphabetisch nach Sternbildern geordnet.

3. Kolumne: Spektrum des Sternes. In der Tabelle I ist hier überall der Buchstabe M weggelassen. Es bedeutet hier also be, daß das kontinuierliche Spektrum Mb ist, und daß sich darüber die charakteristischen Emissionslinien lagern; —e bedeutet, daß es noch unbekannt ist, ob das kontinuierliche Spektrum Ma, Mb oder Mc ist. Die Angaben dieser Kolumne beruhen auf dem neuen Draper-Katalog oder, soweit dieser den betreffenden Stern nicht enthält, auf den Angaben der G. u. L., in einigen Fällen auch auf anderen Quellen.

4. Kolumne: Die auf ganze Tage abgerundete Periode P. Die Zahlen beruhen im allgemeinen auf der G. u. L., sind aber häufig verbessert, vor allem nach den Angaben in den Ephemeriden der veränderlichen Sterne für 1925 in der V J S. (Verbesserte Werte der Periode finden sich für viele Sterne in Harv Ann 79, Part 2.)

5. Kolumne: Die Gesamtamplitude A, d. h. im allgemeinen die Differenz zwischen der größten und der kleinsten an dem Stern beobachteten Helligkeit. Es wären hier zweifellos Mittelwerte der Amplitude vorzuziehen, doch standen solche bei Aufstellung der Tabellen nur für wenige Sterne zur Verfügung. Quellen wie unter 4., doch sind oft auch die neugezeichneten Lichtkurven zur Ermittlung von A benutzt; „ph" bedeutet photographische Amplitude. Mittlere Amplituden kann man jetzt für eine erhebliche Zahl von Mira-Sternen aus Harv Ann 79, Part 2, entnehmen.

6. Kolumne: Die Klasse der Lichtkurve. Ein Fragezeichen neben dem die Klasse bezeichnenden Buchstaben bedeutet eine infolge zu geringer Zahl der Beobachtungen bestehende Unsicherheit der Klassifikation; auch wo kein Fragezeichen steht, ist manchmal noch eine allerdings geringe Unsicherheit vorhanden. Steht in dieser Kolumne nur ein Fragezeichen, so bedeutet dies, daß die Klassifikation noch nicht möglich war. Bei der Klasse β kann man oft die Unterabteilung noch nicht angeben. Es steht dann in der 6. Kolumne β ohne Index. Entsprechend kommt auch α ohne Index vor.

7. Kolumne: Bemerkungen, die auf Eigentümlichkeiten des Sternes hinweisen. Längere Bemerkungen finden sich am Fuße der Seiten bzw. Tabellen.

Tabelle I. Me-Sterne mit bekannten Perioden.

Nr.	Stern	Spektrum	P	A	Lichtkurve	Bemerkungen
17	T Andromedae	be	279^d	6^m,0	β_1	
58	U	be	340	4,4	α_1	
42	V	ae	258	6,3	β	
95	W	ce	399	6,8	$\alpha_3-\gamma_1$	Sehr häufig kleine Stufe im Aufstieg.
77	Y	be	221	5,7	α_4	
44	RR	be	330	6,0	β_1	
89	RV	—e?	172	2,7	?	RV Tauri-Stern?
41	RW	ce	430	6,0	α_1?	
1685	SV	ce	314	5,5	α_2	
74	SX	—e?	337	>3,9	?	
1646	TY	ce	140	1,8	β?	
59	UZ	—e	313	6,0	α_1	
1662	R Aquarii	ce	387	4,8	α_1	Nebellinien im Spektrum.
1630	S	be	280	6,5	α_3	Min. meist ziemlich spitz.
1461	T	ae	202	6,6	β_2	
1455	V	be	246	1,9	?	Vielleicht unregelmäßig?
1454	W	—e	383	6,0	α_2?	
1590	X	be	306	6,1	α_2?	
1667	Z	—e	136?	2,2	pec	S. Anm.
1510	RR	—e	181	>4,4	β	
1594	RT	ce	246	4,4	β	
1269	R Aquilae	ce	318	6,2	α_4	Starke Abnahme der Periode.
1357	X	be	348	5,1	α_4	Max. meist sehr spitz.
1408	Z	—e?	130	4,7	$\alpha_4-\beta_2$	
1369	RR	be	398	4,8	α_2	
1373	RS	—e	412	5,4	α_1?	
1335	RT	ae	326	6,1	α_3	
1402	RU	be	274	6,6	α_3	
1338	RV	ae	218	6,2	$\alpha_4-\beta_1$	
1388	SY	ae?	357	5,0	α_4	
1401	TV	be	242	>3,8ph	α_4?	
1080	U Arae	be	229	3,9ph	?	
1085	V	ce	376	>2,6ph	?	
1054	RU	—e	252	>1,8ph	?	
94	R Arietis	ae	186	5,9	β_3	
85	S	—e	292	5,3	α_1?	
141	U	be	372	7,5	α_2	
244	R Aurigae	ce	461	6,8	γ_1	Abstieg manchmal steiler als Aufstieg.
333	U	be	414	5,5	α_1	
368	X	ae	164	5,2	β_3	
370	RR	ae	309	4,0	?	
331	SZ	—e	452	>2,5ph	?	
820	R Bootis	be	223	6,3	β_1	Lk. veränderlich. Ende 1915 Buckel im Aufstieg und im Abstieg. Ende 1916 und Anfang 1917 Lk. = α_3 mit Buckel im Abstieg.
805	S	be	275	5,6	$\alpha_4-\beta_1$	
814	V	be	260	4,9	α pec	Lk. sehr veränderlich. Manchmal Buckel im Aufstieg oder Abstieg. Max. häufig breiter als Min.
794	Z	be	281	6,2	α_3	
207	R Caeli	—e	392	6,4	α_2	
350	V Camelopard.	be	515	6,8	α_1	
204	X	be	142	5,7	β_3	
395	RT	ce	370	>2,8ph	?	
484	R Cancri	be	355	5,8	β_1	Manchmal kleine Einbiegungen der Lk. im Aufstieg? Max. sehr spitz, Min. sehr flach.

Anm. Z Aquarii. Nach photographischen Harvard-Beobachtungen scheinen An- und Abstieg der Lk. sehr steil, die Max. und Min. dagegen meist sehr flach. Manchmal entstehen Algol-ähnliche Lichtkurven. Lk. jedenfalls sehr veränderlich.

Fortsetzung von Tabelle I.

Nr.	Stern	Spektrum	P	A	Lichtkurve	Bemerkungen
496	U Cancri	—e	305ᵈ	>5ᵐ,8	α_1	Maximalhelligkeit sehr veränderlich.
517	W	ae	396	6 ,8	α_2	
780	R Canum ven.	be	325	5 ,2	$\alpha_4-\beta_2$	Lk. stark veränderlich. Gelegentlich Buckel im Aufstieg.
717	U	be	340	>4 ,5ph	?	
451	S Canis min.	ce	334	6 ,0	β_1	Manchmal Stufe im Aufstieg. Zuweilen Abstieg steiler als Aufstieg.
455	T	—e	323	>5 ,2	α_4?	
426	V	ae	367	6 ,2	α_3	
1517	T Capricorni	ae	269	5 ,4	α_4-β_1?	
1497	V	be	276	>4 ,2	α_3?	
1503	Z	—e	181	4 ,2	β?	
1484	RR	ce	238	6 ,7	α_3?	
1433	RU	—e	346	>4 ,9	α_3	
538	R Carinae	be	308	5 ,5	$\alpha_4-\beta_1$	Mittl. Lk. Harv Bull 843.
573	S	ae	149	4 ,0	γ_1	Abstieg etwas steiler als Anstieg. Max. breiter als Min. (Harv Bull 798).
576	Z	—e	384	>4 ,8ph	?	
526	RW	be	317	5 ,5	α_4	
602	RZ	be	273	4 ,6ph	α_3?	
575	SU	ae	231	>5 ,5ph	?	
1678	R Cassiopeiae	ce	427	8 ,4	α_4	Manchmal Stufe im Aufstieg.
19	T	ce	449	5 ,8	$\gamma_1-\gamma_2$	Abstieg steiler als Aufstieg. Max. meist breiter als Min.
1644	V	ce	229	5 ,5	β_2	
1684	Y	be	410	5 ,5	α_2	
1663	Z	be	499	6 ,5	α_1	
45	RV	be	331	7 ,5	α_3	Manchmal kleiner Buckel im Abstieg?
3	SS	be	140	3 ,2	β?	
798	R Centauri	be	561	7 ,7	γ_2	Mittl. Lk. Harv Bull 836.
770	T	ae	91	3 ,4	β_3	Das kontinuierl. Spektrum ist im Max. K9 oder M0 (Harv Circ 253; Lick Bull 12, S. 73).
699	U	ae	221	>6 ,0	β	Mittl. Lk. Harv Bull 844
665	W	be	202	4 ,1ph	β	
662	X	ae	314	5 ,9	α_4	
647	RS	ae	164	4 ,9	$\alpha_4-\beta_3$	Min. wohl etwas spitzer als Max.
779	VX	be	298	2 ,5ph	?	
1507	T Cephei	ce	387	5 ,6	γ_1	Abstieg steiler als Aufstieg.
1500	X	—e	535	8 ,2	α_1?	
34	Y	—e	330	6 ,6	α_3	
98	*o* Ceti	be	331	7 ,6	α_3	Mittl. Lk. A N 220, S. 91.
103	R	be	166	>5 ,9	$\alpha_4-\beta$	Min. und Max. ziemlich gleich spitz.
23	S	ae	322	6 ,3	$\alpha_4-\beta_1$	Lk. scheint stark veränderlich.
111	U	—e	235	6 ,1	α_4	
1676	V	—e	260	5 ,6	α_3?	
1682	W	be	355	8 ,0	α_4	Max. manchmal sehr breit.
53	Z	be	184	4 ,7	β	
491	R Chamaeleontis	be	335	5 ,4	α_4	Mittl. Lk. Harv Bull 842.
347	R Columbae	ae	327	>4 ,0	α_1?	
249	T	be	226	5 ,4	β	Mittl. Lk. Harv Bull 837.
670	R Comae Ber.	—e?	363	7 ,3	α_2	
1184	U Coronae austr.	ae	147	2 ,6	β?	
1167	Z	ce	303	>1 ,9	?	
856	S Coronae bor.	ce	362	7 ,3	α_3 pec	Manchmal kleiner Buckel im Aufstieg oder auch im Abstieg.
930	W	be	236	5 ,7	$\alpha_4-\gamma_1$	Oft kleiner Buckel im Aufstieg.
881	X	be	238	5 ,7	β_2	
888	Z	ae	250	5 ,5	α_4	
684	R Corvi	be	318	6 ,6	β_1	
697	U Crucis	be	351	4 ,1ph	α_3	

Fortsetzung von Tabelle I.

Nr.	Stern	Spektrum	P	A	Lichtkurve	Bemerkungen
1358	χ Cygni	ae	406^d	9^m,0	γ_1	
1391	S	—e	326	7,0	pec	Abstieg steiler als Aufstieg.
1447	V	—e?	418	7,0	$\alpha_4-\beta_2$	
1534	W	—e	259	1,1	γ_2	RV Tauri-Stern. [M N 80, S. 41; 81, S.144]
1379	Z	ce	263	5,8	α_1	
1345	RT	ae	192	5,5	β_3	s. Anm.
1438	ST	—e	336	4,4	$\alpha_4-\beta_1$	
1415	SX	—e	409	6,2	α_2?	
1348	TU	—e	219	5,4	$\alpha_4-\beta$	
1475	UX	ce	560	>5,6	α_1?	
1419	AU	be	440?	>3,6ph	α_2?	
1410	R Delphini	be	286	6,1	α_4	
1452	T	be	331	5,9	β_1	Min. meist recht breit.
1459	V	ae	533	9,4	α_2?	
1472	X	ae	281	5,5	α_2	
978	R Draconis	be	245	6,6	α_4	
1289	U	ce	313	4,8	β_1	Abstieg manchmal rascher als Aufstieg.
1106	V	—e	279	4,8	$\alpha_4-\beta_1$	
1128	W	ae	261	5,8	$\alpha_4-\beta$	
541	Y	be	324	6,0	α_1	
1064	SY	ce	387	>3,0	?	
1508	R Equulei	ae	264	5,5	α_3	
170	T Eridani	be	252	5,5	α_4?	
164	U	be	275	5,1	α_4	
182	W	—e	374	5,0	α_3	
153	RT	ce	380	>4,0	?	
461	S Geminorum	ae	293	6,2	α_2	
444	V	be	276	6,6	α_3	
409	X	be	263	5,4	β_2	
1547	R Gruis	ce	331	7,0	α_4	
1598	S	be	402	6,8	α_4?	
1597	T	—e	137	3,4	β_3	
902	R Herculis	ae	321	6,7	α_3	
1016	S	ce	302	7,2	$\alpha_4-\gamma_1$	Häufig Buckel im Aufstieg und manchmal Abstieg steiler als Aufstieg.
1126	T	ae	165	7,0	β_3	
949	U	ce	410	6,2	α_4	
976	W	ae	278	6,1	β_2	
1051	RS	be	218	5,3	β_3	
1038	RT	ae	293	5,7	α_3	Maximalhelligkeit sehr veränderlich.
916	RU	ce	479	7,2	γ_1	Stufe im Aufstieg.
1104	RY	ce	222	5,7	β_2	
1180	RZ	ce	329	4,9	α_1	
999	UV	ce	343	5,2	?	
131	R Horologii	ce	406	9,0	α_3	
106	S	ae	338	>2,8ph	α_3?	
133	T	be	218	5,5	β_2	
758	R Hydrae	be-ce	404	6,6	β_1	Starke Abnahme der Periode. Vielleicht manchmal kleine Stufe im Aufstieg.
508	S	ae	258	6,0	β	
510	T	be	289	5,7	β	Abstieg manchmal steiler als Aufstieg.
776	W	ce	386	3,0ph	β_1	Abstieg etwas rascher als Aufstieg [Harv Circ 270].
539	X	—e	302	4,5	$\alpha_4-\beta$	
548	RR	ae	340	>5,1	?	
618	RS	—e	331	5,7	β_1	
492	RT	—e	255	2,2	γ_2?	RV Tauri-Stern?

Anm. RT Cygni. Bei dem Anstieg Ende 1917 hatte die Lk. einen Buckel, sonst war sie stets glatt. 1912—23 war die Lk. entschieden nicht so unsymmetrisch, wie in der G. u. L. angegeben.

Fortsetzung von Tabelle I.

Nr.	Stern	Spektrum	P	A	Lichtkurve	Bemerkungen
763	RW Hydrae	ae	370^d	2^m,2ph	?	
1605	R Indi	ae	216	4 ,0	α_4?	
1470	S	be	403	>3 ,9	α_1?	
1615	R Lacertae	be	299	5 ,6	α_3	
1602	S	ce	240	5 ,1	β_2	
549	R Leonis	ce	318	5 ,2	α_4	
565	V	—e	273	4 ,9	α_2	
623	W	be	386	>5 ,3	α_2?	
546	R Leonis min.	be	372	6 ,0	α_4	
559	S	be	293	5 ,2	α_3?	
232	T Leporis	—e	373	5 ,3	β_1	
851	S Librae	—e	192	>5 ,0	β_3	Max. meist breit.
873	U	—e?	227	4 ,5	α_3?	
843	Y	be	275	6 ,2	α_3	
887	RR	be	276	5 ,7	α_4	
857	RS	ce	217	5 ,9	β_2	
839	RT	ae	249	5 ,4	α_3	
862	RU	ce	315	6 ,4	$\alpha_4-\beta_1$	
833	Y Lupi	ae	409	5 ,8	α_1	
406	S Lyncis	be	300	5 ,0	α_1	
1285	S Lyrae	—e	442	5 ,8	α_2?	Lk. sehr veränderlich. Nach J. A. PARKHURST α_3 mit sehr spitzem Min.
1273	V	ce	374	6 ,2	α_2	
1141	W	ae	197	5 ,4	γ_1	Buckel im Aufstieg. Manchmal Abstieg etwas steiler als Aufstieg.
1286	RS	—e	299	>4 ,5	α_4?	
1254	RT	—e	251	5 ,5	α_3	
1205	RW	ce	507	5 ,5	α_1	Maximalhelligkeit sehr stark veränderlich, etwa 9^m,2 bis 13^m,0.
1229	RX	—e	252	5 ,0	α_3	
1292	SS	—e	352	>3 ,8?	α_4-γ_1??	
1148	TU	ce	120	1 ,0	?	
1442	R Microscopii	be	138	4 ,0	β_2	
1524	S	ae	213	6 ,3	β?	
1427	U	ce	333	7 ,0	α_3	
1519	V	be	376	>4 ,2ph	α_3?	
1487	X	ae	390?	>5 ,7ph	?	
386	V Monocerotis	ce	335	6 ,7	β_1	
416	X	be	155	2 ,5	β_3	
415	Y	be	229	4 ,8	β pec	Lk. scheint recht veränderlich. Abstieg manchmal etwas steiler als Aufstieg.
440	RR	—e	392	>6 ,8	α_2?	
874	T Normae	be	242	6 ,3	α_3	
362	R Octantis	ce	405	5 ,0	$\alpha_4-\gamma_1$	Häufig Stufe im Aufstieg.
1058	S	be	259	6 ,0	$\alpha_4-\beta_1$	
1486	T	—e	218	4 ,6	α_4?	
745	U	be	303	5 ,6	$\alpha_4-\beta$	Manchmal Buckel im Aufstieg.
1036	R Ophiuchi	ae	302	7 ,6	α_4	
1182	X	be	339	3 ,0	β_1	Sehr flache Minima (BAN 2, S. 89).
1049	Z	ae	349	6 ,0	α_4	Max. sehr breit, breiter als Min.
1005	RR	ae	297	5 ,0	$\alpha_4-\beta_1$	
1142	RY	ae	151	5 ,8	$\alpha_4-\beta_2$	Lk. veränderlich. Manchmal Buckel im Abstieg oder Anstieg.
1028	SS	ae	179	>4 ,5	β?	
256	S Orionis	ce	421	5 ,6	γ_1	Stufe im Aufstieg.
353	U	ae	377	7 ,0	$\alpha_4-\gamma_1$	Sehr häufig Stufe im Aufstieg.
233	V	—e?	272	>5 ,8	β	
1122	R Pavonis	be	230	4 ,5	$\alpha_4-\beta$?	
1341	T	ae	244	6 ,5	α_3	
1463	U	be	290	>3 ,4	α_3	
1074	W	be	283	3 ,7ph	α_3	

Fortsetzung von Tabelle I.

Nr.	Stern	Spektrum	P	A	Lichtkurve	Bemerkungen
1638	R Pegasi	be	380^d	$6^m,1$	$\alpha_4-\gamma_1$	Häufig Stufe im Aufstieg. Manchmal erst langsamer, dann steiler Aufstieg.
1648	S	ce	318	5 ,8	$\alpha_4-\beta_1$	
1578	T	ce	363	6 ,0	β_1	
1559	V	be	303	6 ,2	α_2	
1647	W	ce	342	5 ,7	β_1	
1580	Y	ce	208	5 ,0	α_3	
1681	Z	ce	320	5 ,3	β_2	Manchmal Stufe im Aufstieg. Zuweilen scheint Abstieg steiler als Aufstieg.
1543	RR	be	265	5 ,3	α_3	
1582	RS	ce	436	5 ,1	α_2	Anstieg manchmal langsam.
1568	RT	—e	215	3 ,5	pec	Lk. sehr veränderlich. Lichtwechsel zeitweise sehr gering. Abstieg scheint oft steiler als Aufstieg.
1600	RV	ce	387	5 ,6	α_2?	
1637	RW	—e	209	4 ,5	β	
1606	SS	ce	423	>4 ,3	?	
1501	TZ	be	122	>3 ,9ph	?	
150	R Persei	ae	210	5 ,9	β	Lk. veränderlich.
101	S	—e	—	5 ,0	pec	Unregelmäßig.
83	U	ce	324	3 ,9	α pec	Lk. veränderlich. Max. meist viel breiter als Min.
105	RR	ae	392	6 ,2	α_1	Aufstieg manchmal nicht sehr steil. Ende 1923 bis Ende 1924 war Lk. = β.
1673	R Phoenicis	ae	269	6 ,2	β	
1679	S	be	157	0 ,8	pec	Lk. sehr veränderlich. Anstieg ziemlich steil, Max. bald spitz, bald flach.
27	T	ce	281	5 ,0	α_2	Mittl. Lk. Harv Bull 840.
1654	V	be	251	4 ,8	?	
211	R Pictoris	ae	333	2 ,8	γ_2	RV Tauri-Stern [A N 225, S. 251].
243	S	be	425	6 ,1	α_4	
247	T	ae	201	5 ,5	β_2	
71	R Piscium	be	394	7 ,0	α_3	
62	S	—e	410	>6 ,5	α_2?	
55	X	ae	354	6 ,0	?	
1588	R Piscis austr.	ae	294	>3 ,0	α_4?	
1563	S	be	272	>2 ,7	α_3?	
437	L_2 Puppis	be	140	3 ,0	γ_1?	Lk. sehr veränderlich.
470	U	be	315	>5 ,5	?	
462	W	ae	121	4 ,2	β_3	Lk. scheint sehr veränderlich.
453	Z	be	515	6 ,5	α_1?	
407	RV	e?	180	>2 ,4ph	?	Das kontinuierliche Spektrum ist vielleicht R.
514	S Pyxidis	ae	207	4 ,8	?	
203	R Reticuli	be	276	6 ,4	α_4	Mittl. Lk. Harv Bull 841.
1307	W Sagittae	ce	279	3 ,2	α_4	
1296	R Sagittarii	be	269	6 ,2	β	
1304	S	ae	232	6 ,8	β?	
1305	Z	ae	443	6 ,0	α_1?	
1364	RR	be	335	7 ,3	α_4	
1411	RT	be	303	5 ,5	$\alpha_4-\beta_1$	
1367	RU	be	242	4 ,4	α_4	
1161	RV	be	320	6 ,0	β	
1282	RX	—e	329	4 ,7	β	
1247	ST	be	395	>5 ,7	α_2?	
1302	SW	—e	291	2 ,6ph	?	
1315	TT	be	334	3 ,2ph	?	
1340	TV	—e	263	4 ,2ph	?	
1277	TW	be	222	3 ,6	?	
1280	TX	ae	242	>2 ,5ph	?	
1297	TY	—e	329	>5 ,3ph	α_1?	

Fortsetzung von Tabelle I.

Nr.	Stern	Spektrum	P	A	Lichtkurve	Bemerkungen
1268	AG Sagittarii	be	326^d	$>4^m$,0ph	?	
1166	AK	ae?	420	>4 ,1ph	?	
1316	AN	be	340	5 ,6ph	?	
899	Z Scorpii	ce	362	4 ,5	β_2	
1025	RR	be	277	6 ,7	β_2	
1019	RS	be	319	6 ,2	α_4	
1032	RT	be	450	>6 ,7	α_1	
1068	RU	—e	357	4 ,4	γ_1	Stufe im Aufstieg. Abstieg manchmal steiler als Aufstieg.
1040	RW	ae	385	>3 ,0	α_1?	
897	RZ	be	161	5 ,0	β_3	Min. manchmal sehr spitz.
1076	SV	—e	253	4 ,4	β	
1053	SW	be	260	4 ,1ph	$\alpha_4-\beta$?	
11	S Sculptoris	ae	358	6 ,7	β_2	Mittl. Lk. Harv Bull 838.
2	V	ae	296	>3 ,2	α_4?	
883	R Serpentis	be	357	7 ,5	$\alpha_4-\gamma_1$	Sehr häufig kleine Stufe im Aufstieg.
854	S	be	365	6 ,2	γ_1	Stufe im Aufstieg.
905	U	be	238	5 ,4	$\alpha_4-\beta_1$	
195	R Tauri	be	323	6 ,4	α_2	
217	V	ae	170	5 ,1	β_2	
205	RX	be	337	4 ,7	α_4	Min. sehr breit.
1400	R Telescopii	ce	462	>5 ,0	α_2?	Manchmal Buckel im Aufstieg?
1156	T	ae	256	>1 ,6ph	?	
1264	U	ce	437	>3 ,2ph	α_1?	
1347	W	be	305	>3 ,0ph	α_3?	
1334	Z	be	230	>4 ,3ph	?	
1265	RU	be	271	5 ,6ph	?	
113	R Trianguli	—e	267	6 ,7	β_2	
1675	R Tucanae	—e	286	>1 ,4	α_4	
20	S	be	241	>3 ,7ph	α_4	
1609	T	ae	251	6 ,0	β	
50	U	be	259	6 ,0	β_2	Mittl. Lk. Harv Bull 840.
608	R Ursae maj.	be	299	7 ,2	α_3	Aufstieg oft zuerst langsam, dann rasch.
702	T	ae	255	7 ,2	α_3	
499	X	ce	249	4 ,8	α_4?	
666	Z	ce	198	2 ,4	γ_2	RV Tauri-Stern [A N 223, S. 281].
756	RR	—e?	232	5 ,3	β?	
708	RS	be	259	6 ,0	α_3	
656	RU	ae	252	>5 ,5	?	
868	S Ursae min.	ce	322	4 ,4	β_2	Manchmal Buckel im Aufstieg; zuweilen Max. spitzer als Min., zuweilen umgekehrt.
767	T	ae	314	5 ,5	β_1	Min. sehr breit.
801	U	ae	330	4 ,8	β_2	
531	Y Velorum	—e?	437	5 ,3	α_1	Maximalhelligkeit sehr veränderlich.
529	RS	ce	421	4 ,7ph	α_4?	
706	R Virginis	ae	146	4 ,9	β_3	
762	S	be	377	6 ,9	α_4	
679	T	be	340	5 ,2	α_2?	
721	U	ae	207	5 ,6	β	
757	V	ae	250	5 ,8	β?	
700	Y	—e?	219	4 ,9	α_4	
808	RS	ce	353	6 ,8	α_3	
672	SU	ae	210	4 ,7	α_4	
436	R Volantis	—e	444	4 ,0	pec	Lk. sehr veränderl., zeitweise γ_1, zeitw. α_1
1490	R Vulpeculae	be	137	5 ,9	β_2	
1443	RU	ae	159	1 ,6	?	Lk. veränderlich.

Gelegentlich finden sich in der 7. Kolumne auch Hinweise auf neuere, noch nicht in der G. u. L. enthaltene Literatur, sofern diese von besonderem Interesse ist.

Anhang zu Tabelle I. Me-Sterne des Hauptkataloges der G. u. L., deren Periode noch unbekannt ist.

Nr.	Stern	A	Nr.	Stern	A	Nr.	Stern	A
962	X Arae	5^m,1ph	87	Y Eridani	2^m,5ph	79	SS Persei	1^m,5
967	Y	>3,7ph	168	U Horologii	>2,5ph	63	W Phoenicis	7,5ph
1014	Z	4,2ph	1505	W Indi	2,4ph	1313	UU Sagittarii	>3,0ph
1055	RR	>2,5ph	1528	X	>3,8ph	1035	TU Scorpii	3,6ph
827	RY Centauri	?	1538	Y	3,4ph	1037	AH	1,7ph
747	TT	3,0ph	1426	T Microscopii	1,0ph	1412	X Telescopii	3,6ph
816	TU	5,0ph	1518	W	>3,2ph	1013	Z Trianguli austr.	2,0ph
785	TW	6,4ph	449	RX Monocerotis	>3,0ph	593	RT Velorum	>1,5ph
1604	ST Cephei	1,2	753	U Muscae	>3,5ph	626	RU	4,6ph
1114	RR Coronae austr.	>2,3ph	1564	V Octantis	3,0ph	525	RW	2,5ph
213	T Doradus	>2,7ph	1377	RR Pavonis	>3,3ph	422	T Volantis	5,0ph
245	U	>2,4ph	1572	RY Pegasi[1]	0,6ph			

[1] RY Pegasi. Die vermutete Periode von 25^d dürfte nicht zutreffen.

Tabelle II. Mira-Sterne mit Spektrum Se.

Nr.	Stern	Spektrum	P	A	Lichtkurve	Bemerkungen
22	R Andromedae	Se	409^d	9^m,2	α_3	Min. manchmal ziemlich spitz. Manchmal kleine Wellen im Auf- und Abstieg.
12	X	Se	343	6,4	α_3	Min. manchmal ziemlich spitz.
812	R Camelopardalis	Se	272	6,5	β_3	Lk. veränderlich, Abstieg oft rascher als Aufstieg.
198	T	Se	372	6,5	γ_1	Manchmal kein Buckel im Aufstieg, sondern breites Max. (Lk. dann β_3).
487	V Cancri	Se	272	5,7	α_4	Min. spitzer als Max.
430	R Canis min.	Se	338	3,8	β_3	
61	S Cassiopeiae	Se	613	8,0	pec	Lk. sehr veränderlich, manchmal α_1, manchmal γ_1 oder andere Formen.
39	U	Se	277	>7,0	$\alpha_4-\beta$	Min. manchmal spitz, manchmal flacher.
47	W	Se?	405	3,6	β_3	Spektrum ähnlich dem von R Andromedae [Harv Circ 221].
1418	SZ Cephei	Se?	327	>4,7	$\alpha_4-\beta$?	Lk. Lyon Bull 8, S. 61.
1336	R Cygni	Se	426	8,2	α_4	Manchmal kleine Wellen im Auf- u. Abstieg.
1407	RS	Se?	413	2,1	$\beta_3-\beta_4$	Vgl. Harv Bull 783.
425	R Geminorum	Se	370	7,4	α_4	Min. manchmal spitz.
463	T	Se	286	5,5	β	Lk. veränderlich, Max. zuweilen ziemlich breit, oft Abstieg steiler als Aufstieg.
855	RW Librae	Se	?	3,1ph	?	
419	R Lyncis	Se	378	7,5	$\beta_2-\gamma_1$	Oft kleiner Buckel im Aufstieg.
225	R Orionis	Se?	377	4,5	β	
1293	T Sagittarii	Se	389	>5,9	β	
713	S Ursae majoris	Se	224	4,4	pec	Lk. sehr veränderlich, α_3, α_4, β, γ_1. Abstieg oft steiler als Aufstieg. Häufig breites Max., spitzes Min.

Anm. Nach dem Draper-Katalog haben die Sterne AA Cygni ($P = 202^d$) und AD Cygni (P unbekannt) N- oder Se-Spektrum, RZ Pegasi ($P = 442^d$) Me- oder Se-Spektrum. Wegen dieser Zweifel sind diese drei Sterne nicht in die Tabelle II aufgenommen worden und ebensowenig in die anderen Tabellen.

Die Tabellen I und V enthalten der Vollständigkeit halber auch einige Sterne, die höchst wahrscheinlich oder sicher zur RV Tauri-Klasse gehören; die Grenze zwischen den Mira-Sternen und den RV Tauri-Sternen ist nämlich in manchen Fällen schwer zu ziehen. Auch der einzige, sicher ganz unregelmäßige Me-Stern, S Persei, ist in die Tabelle I mit aufgenommen. Als Anhang zu Tabelle I

Tabelle III. Mira-Sterne mit Spektrum K, Ma, Mb, Mc.

Nr.	Stern	Spektrum	P	A	Lichtkurve	Bemerkungen
532	UU Carinae	K	202^d	$3^m,0$ph	?	
57	RU Cephei	K 8	110	1 ,0	pec	Lk. Potsd. Publ. Nr. 83; Lk. sehr veränderlich, häufig Nebenminima.
901	RR Herculis	K 5 p	243	1 ,7	?	Lk. scheint veränderlich. Max. breiter als Min.
909	SX	K 2 p	103	1 ,3	α_4	Lick Bull 11, S. 124. Min. spitzer als Max. Linien des H hell um das Max. Im Min. Spektrum Mb.
749	V Canum ven.	Ma	193	2 ,2	α_3?	
1535	AB Cygni		530	1 ,2	β?	s. Anm.
378	η Geminorum		232	0 ,9	β_4	
1516	X Pegasi		202	4 ,7	β_2	Lk. veränderlich. Abstieg manchmal steiler als Aufstieg (M B A A App.).
596	S Sextantis		257	2 ,5	α?	
193	W Tauri		266	4 ,6	β_3	Harv Bull 781.
685	RY Ursae maj.		315	1 ,1	β_3	
773	V Ursae min.		73	1 ,2	pec	Lk. sehr veränderlich.
241	UX Aurigae	Mb	72	0 ,8	α_3?	Lk. A N 223, S. 191. Lk. und P scheinen sehr veränderlich. (Früher $P = 103^d$?)
460	U Canis min.		418	4 ,9	γ_1	Min. spitz u. symmetrisch. P veränderlich.
566	RR Carinae		162:	1 ,3	β?	Lk. scheint veränderlich.
16	T Ceti		161	1 ,5	β?	
144	X		175	4 ,0	β_3	
1326	AF Cygni		88	1 ,6	α	Jap. Journ. of Astronomy 1, S. 207, Lk. A N 225, S. 369. Lk. sehr veränderlich, P ändert sich zwischen 79^d und 94^d.
824	UV Draconis		77	1 ,0	?	
556	Z Leonis		56	1 ,7	α?	
863	R Normae		487	6 ,2	γ_2	Lk. Union Circ 49 und Harv Bull 836.
1279	RW Sagittarii		188	2 ,1	α_4	Lk. M N 84, S. 34. Lk. veränderlich, Max. teils breit, teils spitz.
25	T Sculptoris		200	4 ,2	β_3?	
117	W Trianguli		148	0 ,8	?	Lyon Bull 5, S. 242.
1643	SS Andromedae	Mc	146	1 ,0±	?	
29	TU		315	>4 ,4	$\alpha_4-\beta$?	
1450	Y Aquarii		383:	5 ,5:	α_3?	
124	T Arietis		322	1 ,9	α_1?	
489	Z Cancri		70	0 ,8	?	
518	RS		130	1 ,2	pec	Lk. A N 219, S. 343. Lk. ganz veränderlich, starke Unregelmäßigkeiten im Lichtwechsel.
696	T Canum ven.		287	3 ,9	β_3	
1661	SV Cassiopeiae		272	1 ,7	?	
1539	RU Cygni		462	2 ,3	γ_2 pec	A N 220, S. 243. Nebenmax. entartet zuweilen zu Buckel im Abstieg.
1330	TY		348	6 ,5	β_1	Abstieg häufig steiler als Aufstieg.
1448	S Delphini		276	3 ,5	β	Lk. sehr veränderlich. Max. und Min. manchmal sehr breit.
206	R Doradus		345	2 ,1	γ_1?	
1359	S Pavonis		397	3 ,3	β_3	
127	W Persei		496:	2 ,7	pec	zeitweise β_3. Lk. verläuft in flachen Wellen. Min. meist schärfer als Max. P sehr veränderlich [Harv Bull 784].
972	R Ursae min.		332	1 ,7	$\beta_4-\gamma_2$	A N 218, S. 65.

Anm. AB Cygni. Das Beobachtungsmaterial ist wenig umfangreich und die Periode vielleicht noch nicht ganz sichergestellt, obwohl sie gut zu den Beobachtungen paßt.

geben wir ein Verzeichnis derjenigen Me-Sterne des Hauptkatalogs der G. u. L., deren Periode noch nicht bekannt ist. Es ist möglich, daß sich darunter noch unregelmäßige Me-Sterne befinden, doch ist die Unregelmäßigkeit noch für

Tabelle IV. Mira-Sterne mit Spektrum N oder R.

Nr.	Stern	Spektrum	P	A	Lichtkurve	Bemerkungen
1660	ST Andromedae	Nb	342^d	$2^m,6$	$\beta_3-\beta_4$	
252	S Aurigae	Nb	580	1 ,9	β_4	
384	V	Nb	352	3 ,5	β_2	
251	W	N?	273	6 ,0	α_3	
284	S Camelopardalis	R8	327	3 ,4	$\beta_3-\beta_4$	
511	T Cancri	N	498	2 ,5	β_3	
81	X Cassiopeiae	N?	434	3 ,7	β_2	
766	RV Centauri	Nb	446	3 ,1ph	$\beta_3-\beta_4$	
1537	S Cephei	Nc	489	6	$\beta_3-\beta_4$	Abstieg scheint etwas rascher als Aufstieg.
882	V Coronae borealis	Nb	355	4 ,8	$\alpha_4-\beta_3$	Min. spitzer als Max.
724	V Crucis	Np	377	3 ,2ph	β?	
1422	U Cygni	R8	460	5 ,0	β_3	
1420	WX	Nb	410	3 ,2	$\beta_3-\beta_4$	
1102	T Draconis	N	420	5 ,0	β_1	
619	V Hydrae	N	530	4 ,0	γ_2?	Harv Bull 781. Sekundäres Minimum s. Anm.
228	R Leporis	Nc	419	4 ,4	β_2	
830	S Lupi	Np	345	>3 ,4	?	
1309	U Lyrae	N	458	3 ,5	β_2	
948	V Ophiuchi	Nb	296	3 ,6	β	
148	Y Persei	Nb	259	2 ,2	β_2	
183	SY	N	455?	1 ,7	β_3-β_4?	
66	R Sculptoris	Nb	376	2 ,6	?	
716	RU Virginis	R3p	440	5 ,5	β	Lk. scheint sehr veränderlich.
692	SS	Nc	354	3 ,0	β	

Anm. V Hydrae. Nach L. CAMPBELL (Pop Astr 34, S. 551 [1926]) hat V Hydrae außerdem eine Periode von 17 Jahren mit einer Amplitude von 6^m. Wenn sich dies bestätigt, haben wir hier einen bislang einzig dastehenden Fall vor uns.

Tabelle V. Mira-Sterne mit besonderem Spektrum.

Nr.	Stern	Spektrum	P	A	Lichtkurve	Bemerkungen
1468	RZ Cygni	pec	556^d	$4^m,6$	γ_2	[A N 225, S. 249] RV Tauri-Stern? Anm. 1
1436	Z Delphini	pec	303	6 ,0	α_4	Anm. 2.
1403	RZ Sagittarii	pec	212	1 ,8	β_3?	1903—06 nahezu konstant. Keine neueren Beobachtungen. Anm. 3.

Anm. 1. Nähere Angaben über das Spektrum scheinen nicht vorzuliegen.

Anm. 2. Nach dem Draper-Katalog sind $H\beta$ und $H\gamma$ hell; hellster Teil des Spektrums zwischen $H\beta$ und $H\gamma$.

Anm. 3. Nach dem Draper-Katalog scheint das Spektrum nicht N zu sein; hellster Teil zwischen $H\beta$ und $H\gamma$.

keinen von ihnen erwiesen. Tabelle I mit ihrem Anhang enthält sämtliche Sterne des Hauptkatalogs der G. u. L., von denen bekannt ist, daß sie ein Me-Spektrum besitzen.

In Tabelle III kommen auch einige (im ganzen 6) Sterne vor, die Perioden von weniger als 90^d haben, nämlich solche von 56^d, 70^d, 72^d, 73^d, 77^d, 88^d. Nach unserer Definition der Mira-Sterne gehören diese Objekte eigentlich nicht hierher. Es liegt aber kein Grund vor, sie wegzulassen, denn jene Definition ist ja zunächst willkürlich. Es ist möglich, daß diese Sterne einen Übergang von den Mira- zu den δ Cephei-Sternen bilden (vgl. Ziff. 19).

Von den Mira-Sternen der Spektraltypen K, Ma, Mb, Mc (Tabelle III) und N bzw. R (Tabelle IV) ist eine größere Anzahl nicht in die Tabellen aufgenommen worden, da bei ihnen die Realität der gegenwärtig angenommenen Periodenwerte noch sehr zweifelhaft ist; die Perioden sind bei diesen Sternen z. T. schwer festzustellen, da die Helligkeitsamplituden oft sehr klein sind.

Auf die Angaben der scheinbaren Größen, der Epochen der Maxima usw. verzichten wir hier, da diese Daten bequem in den Ephemeriden nachgeschlagen werden können.

23. Statistische Untersuchungen über die Lichtkurven der Mira-Sterne. Schon in Ziff. 21 haben wir nebenbei einige statistische Ergebnisse über die Mira-Sterne erwähnt, deren wichtigste zunächst hier nochmals kurz zusammengefaßt werden sollen.

T. E. R. PHILLIPS hat gefunden, daß die Helligkeitsamplituden und die Perioden der Sterne seiner Gruppe I durchschnittlich kleiner als bei den Sternen der Gruppe II sind, sowie daß die Sterne der Gruppe I durchschnittlich etwas heller sind als die der Gruppe II.

H. H. TURNER stellte fest, daß Mira-Sterne mit positivem Werte $\alpha = 2\frac{M-m}{P} - 1$ nicht nahe dem Pole der Milchstraße vorkommen, und daß die Mira-Sterne mit sehr langen Perioden die Milchstraße stark bevorzugen. Ferner fand er, daß innerhalb jeder der beiden PHILLIPSschen Gruppen gewisse Konstanten der Kurven sich mit der Periode ändern, und daß für kurze Perioden der Helligkeitsunterschied zwischen Gruppe I und II kleiner ist als für lange.

Weitere statistische Untersuchungen wurden von L. CAMPBELL, H. LUDENDORFF und H. THOMAS im Anschluß an ihre Klassifizierungen der Lichtkurven unternommen. L. CAMPBELL gebührt das Verdienst, in seinem früher zitierten, schon 1920 gehaltenen, aber erst fünf Jahre später allgemein zugänglich gewordenen Vortrage auf den Zusammenhang zwischen der Form der Lichtkurve und der Länge der Periode hingewiesen zu haben. (Allerdings hatte schon PHILLIPS bemerkt, daß die Sterne der Gruppe I durchschnittlich kürzere Perioden besitzen als die der Gruppe II.) CAMPBELL stellt diese Abhängigkeit durch die vorstehende Tabelle dar, in der die römischen Zahlen die Kurventypen (vgl. Abb. 8), die arabischen die Anzahl der den einzelnen Kurventypen angehörigen Sterne bezeichnen. Die letzten drei Periodenintervalle CAMPBELLS sind hier zu einem zusammengefaßt. Einen Unterschied zwischen den verschiedenen Spektraltypen macht CAMPBELL hier nicht, obwohl dies, wie wir sehen werden, geboten ist.

P	I	II	III	IV	V	VI	VII
$100^d - 149^d$	—	3	1	—	—	—	—
150 — 199	—	4	3	2	1	—	—
200 — 249	—	2	3	5	3	1	—
250 — 299	1	4	3	10	2	3	1
300 — 349	2	2	2	3	6	14	1
350 — 399	1	—	2	1	6	8	6
400 — 449	3	—	1	2	4	3	—
450 — 499	1	1	1	—	—	—	—
500 — 649	—	—	—	—	1	1	—
	8	16	16	23	23	30	8

Wenn man von dem Typus I absieht, so kann man sagen, daß die kurzen Perioden unter den Kurventypen mit niedrigen, die langen unter den Kurventypen mit höheren Ordnungszahlen vorwiegen. Dieses Gesetz tritt auch deutlich hervor, wenn man für die jedem einzelnen Kurventypus angehörigen Sterne die Mittelwerte der Perioden bildet. Es ergeben sich dann nach CAMPBELL folgende Zahlen:

Typus	Mittl. Periode	Sterne
I	378^d	8
II	234	16
III	280	16
IV	273	23
V	341	23
VI	348	30
VII	363	8
		124

Ferner ergibt sich, daß die Sterne mit kleiner Amplitude auf die Kurventypen I bis III, die mit den größten Amplituden auf die Typen IV bis VII beschränkt sind. Die Amplituden nehmen mit wachsender Ordnungszahl der Typen zu, die Helligkeiten im Minimum ab, während die Helligkeiten im Maximum ziemlich unverändert bleiben (nur die Typen VI und VII zeigen etwas geringere Maximalhelligkeiten). Das PHILLIPSsche Ergebnis, daß die Sterne seiner Gruppe I merklich heller sind als die seiner Gruppe II, tritt hier also nur schwach hervor (PHILLIPS' Gruppe I entspricht CAMPBELLS Typen I bis III, PHILLIPS' Gruppe II den Typen IV bis VII).

Auf L. CAMPBELLS Untersuchungen folgen dann zeitlich die von ihnen ganz unabhängigen von LUDENDORFF (Untersuchungen über veränderliche Sterne III, IV, V, VII, VIII, genaueres Zitat in Ziff. 20). Die erhaltenen Resultate sollen nunmehr dargelegt werden, wobei zugleich gewisse Ergebnisse von THOMAS (in seiner Dissertation) mitverwertet werden.

Im wesentlichen stützen sich diese Untersuchungen, bei denen die Mira-Sterne nach ihren Spektraltypen gesondert behandelt werden, auf die oben wiedergegebenen Tabellen I bis IV. Es ist aber zu erwähnen, daß die Tabelle I, wie sie oben abgedruckt ist, gegenüber der für die Untersuchungen benutzten entsprechenden Tabelle einige Verbesserungen erfahren hat, die aber gering an Zahl und im allgemeinen ganz unwesentlich sind; sie sind im folgenden nicht berücksichtigt worden.

α) Abhängigkeit der Form der Lichtkurven von der Periodenlänge und Vergleichung der Lichtkurven der verschiedenen Spektraltypen. Die Me-Sterne (Tabelle I) wurden, soweit die Form ihrer Lichtkurven bekannt ist, nach der Periodenlänge geordnet und dann in Gruppen von je 40 zusammengefaßt; dabei wurde R Volantis wegen der starken Veränderlichkeit der Lichtkurve ausgeschlossen. Die folgende Tabelle gibt nun zunächst eine Übersicht darüber, wie oft in jeder dieser Gruppen von je 40 Sternen die Haupttypen der Lichtkurven α (unsymmetrische, glatte Kurven), $\alpha-\beta$ und β (nahezu oder ganz symmetrische, glatte Kurven), γ (Kurven mit Buckel oder Stufe im Aufstieg oder doppelten Maximum und die mit $\alpha-\gamma$ oder $\beta-\gamma$ bezeichneten) sowie die mit pec bezeichneten Kurven (zu diesen wurden auch die mit α pec oder β pec bezeichneten gezählt) vorkommen.

P	α	β / $\alpha-\beta$	γ	pec	Σ
$91^d - 209^d$	1	31	5	3	40
210 − 250	17	20	1	2	40
251 − 279	19	18	2	1	40
279 − 318	29	10	1	0	40
318 − 342	21	17	0	2	40
344 − 396	26	6	7	1	40
$> 398^d$	30	2	8	0	40
Alle	143	104	24	9	280

Das Ergebnis ist folgendes: α-Kurven kommen bei Me-Sternen mit weniger als 210^d Periode so gut wie gar nicht vor, nehmen dann aber mit wachsender Periode rasch an Zahl zu. Die $\alpha-\beta$- und β-Kurven zeigen das umgekehrte Verhalten, und zwar noch deutlicher. Die γ-Kurven bevorzugen die ganz kurzen und die langen

Perioden, die seltenen pec-Kurven scheinen häufiger bei kurzen als bei langen Perioden aufzutreten.

Etwas weiter spezialisiert ist die folgende, ohne nähere Erläuterung verständliche Tabelle:

P	α_1	α_2	α_3	α_4	$\alpha-\beta$	β	γ	pec	Σ
91^d-209^d	0	0	0	1	3	28	5	3	40
210 −250	0	0	4	13	4	16	1	2	40
251 −279	1	1	13	4	5	13	2	1	40
279 −318	4	5	9	11	7	3	1	0	40
318 −342	5	2	7	7	4	13	0	2	40
344 −396	3	10	7	6	0	6	7	1	40
$>398^d$	14	10	1	5	0	2	8	0	40
Alle	27	28	41	47	23	81	24	9	280

Aus dieser Tabelle und auf Grund der früher gegebenen Definition der Kurventypen α_1 bis α_4 folgt, daß bei den Me-Sternen mit α-Kurven bei wachsender Periode das Minimum immer breiter und der Helligkeitsanstieg immer steiler wird.

Die β-Kurven sind in der letzten Tabelle nicht weiter spezialisiert, da bei vielen von ihnen eine Einordnung in die Unterklassen β_1 bis β_4 noch nicht möglich ist. Beschränkt man sich auf die schon genauer klassifizierten β-Kurven, so erhält man für diese die folgende Übersicht:

P	β_1	β_2	β_3	β_4
91^d-209^d	1	6	12	0
210 −250	1	6	1	0
251 −279	1	4	0	0
279 −318	2	0	0	0
318 −342	8	3	0	0
344 −396	4	2	0	0
$>398^d$	1	1	0	0
Alle	18	22	13	0

Hieraus folgt: Bei Me-Sternen mit relativ kurzen Perioden kommen β_1-Kurven (breites Minimum) nur äußerst selten vor, es überwiegen hier vielmehr die β_2-Kurven (Minimum ebenso breit wie Maximum) und die β_3-Kurven (relativ spitzes Minimum). Die β_3-Kurven kommen dagegen bei Me-Sternen mit langen Perioden überhaupt nicht vor. Die β-Kurven verhalten sich also, was die Breite des Minimums angeht, ganz ähnlich wie die α-Kurven. β_4-Kurven (ganz breites Maximum, der sog. η Geminorum-Typus) scheinen bei den Me-Sternen nicht vorzukommen.

Wir haben die Periodenintervalle so gewählt, daß in jedes 40 Sterne fallen. Geht man dagegen von gleichen Periodenintervallen von 50^d aus (nur das erste soll 60^d umfassen und das letzte alle $P > 450^d$), so erhält man folgende Verteilung der Lichtkurven:

	α	β / $\alpha-\beta$	γ	pec	Σ	α	β / $\alpha-\beta$	γ	pec
91^d-150^d	—	11	2	1	14	0%	79%	14%	7%
151 −200	—	15	3	2	20	0	75	15	10
201 −250	18	25	1	2	46	39	56	2	4
251 −300	38	21	2	1	62	61	34	3	2
301 −350	35	24	1	2	62	56	39	2	3
351 −400	23	6	8	1	38	61	16	21	3
401 −450	21	2	4	—	27	78	7	15	0
$>450^d$	8	—	3	—	11	73	0	27	0
	143	104	24	9	280	51	37	9	3

In der rechten Hälfte sind die Häufigkeitszahlen jeder Horizontalreihe in Prozente umgerechnet. Die früher erwähnten Gesetzmäßigkeiten treten auch in dieser Tabelle deutlich hervor.

Bilden wir endlich für jeden Kurventypus den Mittelwert P_m der Perioden, so erhalten wir folgende Zahlen:

α_1	$P_m = 403^d$	$\alpha-\beta$	$P_m = 266^d$
α_2	378	β	244
α_3	301	γ	335
α_4	300		

Die mittlere Periode nimmt also von α_1 bis β ab, die γ-Kurven zeigen dagegen ein besonderes Verhalten. — Sehr starke Unterschiede treten in den P_m für die Unterabteilungen des Kurventypus β auf; es ergibt sich nämlich:

für β_1	$P_m = 334^d$
β_2	249
β_3	159

Die Perioden der Me-Sterne mit β-Kurven sind also durchschnittlich sehr verschieden, je nachdem das Minimum flacher oder spitzer ist als das Maximum; die mit breiterem Minimum haben viel längere Perioden. In der Regel sind dabei die Unterschiede in der Form des Minimums und des Maximums nur ziemlich gering, so daß die Einordnung in die Unterklassen β_1, β_2, β_3 manchmal keineswegs sicher ist.

Hinsichtlich der γ-Lichtkurven sei noch bemerkt, daß unter ihnen die Formen γ_1 und α_4—γ_1 weit überwiegen. Mit γ_2 sind nur drei Kurven bezeichnet, bei deren einer (RT Hydrae) die Klassifikation zweifelhaft ist; die beiden andern γ_2-Kurven gehören den Sternen R Centauri und Z Ursae majoris an. RT Hydrae ist vielleicht, Z Ursae majoris wohl sicher ein RV Tauri-Stern. R Centauri besitzt unter den Mira-Sternen der Spektralklasse Me die längste bekannte Periode (561^d); die kürzeste hat T Centauri (91^d).

Die vorliegenden Untersuchungen über den Zusammenhang zwischen der Form der Lichtkurve und der Periodenlänge bei den Me-Sternen, die auf 280 (mit R Volantis 281) Lichtkurven beruhen, bestätigen die auf 124 Lichtkurven (darunter 92 von Sternen, deren Zugehörigkeit zur Klasse Me bekannt ist) beruhenden von L. Campbell, ergänzen und vervollständigen diese aber zugleich in verschiedener Hinsicht.

Bei den Mira-Sternen der Spektralklasse Se (Tabelle II) läßt sich eine Abhängigkeit der Form der Lichtkurve von der Periode nicht erkennen; bei dem geringen Umfange des Materials an Lichtkurven (nur 18) ist dies aber auch kaum zu erwarten, zumal sehr kurze Perioden ($< 220^d$) nicht vorkommen.

Wesentliche Unterschiede gegenüber den Me-Sternen zeigen sich bei den Se-Sternen in der Form der Lichtkurven. Unter 17 Lichtkurven von Se-Sternen mit Perioden von mehr als 250^d finden sich 5 Kurven des Typus α und 12 der der Typen α—β, β, γ und pec., während die entsprechenden Zahlen bei den Me-Sternen 125 und 75 sind. Insbesondere sind die β-Kurven bei den Se-Sternen relativ häufig, und nicht nur dies, sondern es treten auch Kurven des Typus β_3 (einmal sogar β_3—β_4) auf, die bei Me-Sternen mit langen Perioden überhaupt nicht vorkommen; die Maxima sind also bei diesen Se-Sternen breiter als bei den Me-Sternen gleicher Periode. Auch bei den unter den Se-Sternen vorkommenden α-Kurven sind die Minima zuweilen auffallend spitz, und die Kurventypen α_1 und α_2 kommen trotz der langen Perioden nicht vor, außer gelegentlich bei S Cassiopeiae. Spitze Minima scheinen also besonders charakteristisch für die Se-Sterne zu sein. Es ist ferner auch auffällig, daß unter den wenigen Se-Sternen zwei mit sehr stark veränderlichen Lichtkurven sind, nämlich S Cassiopeiae ($P = 613^d$) und S Ursae majoris ($P = 224^d$), die dem Me-Stern R Volantis ($P = 444^d$) an die Seite treten. Auf einige der in diesem Absatz besprochenen

Eigentümlichkeiten der Se-Sterne hat übrigens schon L. CAMPBELL[1] kurz aufmerksam gemacht.

Wir wenden uns nunmehr dem Studium der Lichtkurven der den Spektralklassen K, Ma, Mb, Mc angehörigen Mira-Sterne zu (Tabelle III). Bei vielen von diesen ist wegen der Kleinheit der Helligkeitsamplituden die Klassifikation der Lichtkurven noch unsicher. Da die Mc-Sterne in Tabelle III durchschnittlich größere Perioden haben als die übrigen, so werden sie von letzteren getrennt behandelt.

Bei den Mira-Sternen der Spektralklassen K, Ma, Mb scheint das Mischungsverhältnis der Kurventypen α, β, γ ungefähr das gleiche zu sein wie bei den Me-Sternen, soweit überhaupt bei der Kleinheit des Materials eine Aussage gestattet ist. Auffällig ist das Vorhandensein zweier infolge starker Veränderlichkeit mit pec bezeichneter Kurven. Im Gegensatz zu den Me-Sternen scheinen die vorkommenden α-Kurven die kleineren Periodenwerte zu bevorzugen. Besonders merkwürdig ist das Vorwiegen der α-Kurven bei den Sternen, deren $P < 90^d$ sind; es ist aber, wie schon mehrfach erwähnt, fraglich, ob diese Sterne noch zu den Mira-Sternen zu rechnen sind.

Bei den Mira-Sternen mit Spektrum Mc sind die α-Kurven relativ seltener (nur 2 unter 12 Lichtkurven) als bei den Me-Sternen; relativ häufig sind die γ-Kurven (3). Im ganzen treten in Tabelle III 5 γ-Kurven unter 32 Lichtkurven auf, während bei den Me-Sternen erst auf etwa 12 eine γ-Kurve kommt. Bemerkenswert ist das Vorkommen einer γ_2-Kurve (R Normae, Spektrum Mb), da wir unter 280 Me-Sternen nur 3 γ_2-Kurven gefunden hatten, und ebenso das Vorkommen einer β_4-Kurve (η Geminorum, Ma), die bei Me-Sternen überhaupt nicht in Erscheinung tritt.

Schließlich ist noch zu erwähnen, daß in Tabelle III β_3-Kurven bei größeren Periodenwerten auftreten als bei den Me-Sternen. Es scheint also, wie bei den Se-Sternen, die Tendenz zum Vorkommen breiter Maxima vorzuliegen. Dem entspricht, daß, von einem fraglichen Fall abgesehen, α_1- und α_2-Kurven in Tabelle III fehlen.

Schließlich sind noch die Lichtkurven der Mira-Sterne mit N- oder R-Spektrum zu diskutieren. Auch hier liegt nur ein kleines Material (22 Lichtkurven) vor, an dem sich ein Zusammenhang zwischen Form der Lichtkurve und Periodenlänge nicht nachweisen läßt. Ganz anders als bei den Me-Sternen ist aber das Mischungsverhältnis der verschiedenen Typen von Lichtkurven; es kommen unter den 22 Lichtkurven vor: 1 α-Kurve, 1 α—β-Kurve, 19 β-Kurven und 1 γ-Kurve, während bei den Me-Sternen mit mehr als 250^d Periode (die kürzeste in Tabelle IV vorkommende Periode ist 259^d) die entsprechenden Zahlen 125, 16, 37, 18 sind. Die β-Kurven überwiegen also bei den N- und R-Sternen ganz enorm. Die einzige reine α-Kurve, die in Tabelle IV auftritt, gehört noch dazu einem Sterne (W Aurigae) an, dessen Zugehörigkeit zur Spektralklasse N zweifelhaft ist. Unter den β-Kurven überwiegen die mit breitem Maximum, die bei den Me-Sternen mit Perioden von mehr als 250^d überhaupt nicht mehr vorkommen.

Die Mira-Sterne mit besonderem Spektrum (Tabelle V) sind zu wenig zahlreich, als daß sich Betrachtungen über ihre Lichtkurven ausführen ließen.

Man kann auf Grund der vorstehenden Ergebnisse sagen, daß die nicht zur Spektralklasse Me gehörigen Mira-Sterne in der Form ihrer Lichtkurven systematische Unterschiede gegen die Me-Sterne aufweisen; ihnen allen gemeinsam ist die Tendenz, relativ zum Minimum breitere Maxima zu zeigen als die

[1] Harv Bull 778 (1922).

Me-Sterne, eine Tendenz, die am stärksten bei den N- und R-Sternen ausgeprägt ist.

Zur Ergänzung der bisherigen Ausführungen seien hier noch einige numerische Angaben gemacht, die freilich Neues nicht ergeben. Wir sahen, daß die Größe $\varepsilon = \frac{M - m}{P}$ als Maß der Unsymmetrie der Lichtkurven relativ zur y-Achse (auf der die Größenklassen abgetragen werden) dienen kann. Je mehr ε von 0,5 verschieden ist, um so unsymmetrischer ist die Lichtkurve. LUDENDORFF fand in Nr. V seiner „Untersuchungen über veränderliche Sterne" für die Me-Sterne folgende Relation zwischen der Periodenlänge und dem Mittelwerte der zu jedem Periodenintervall gehörigen ε:

P	ε	
$\leqq 150^d$	0,48	13 Sterne
$151^d - 250^d$	0,46	60 „
251 − 350	0,45	90 „
351 − 450	0,42	46 „
$>450^d$	0,43	9 „
		218 Sterne

Je länger die Periode ist, um so unsymmetrischer ist also bei den Me-Sternen im Durchschnitt die Lichtkurve. Dieses Resultat stimmt natürlich durchaus überein mit dem Befunde betreffend die Abhängigkeit der Form der Lichtkurven von der Periode. Der Mittelwert von ε für die in obiger Tabelle benutzten 218 Me-Sterne ist 0,444.

Wie schon näher dargelegt wurde, hat THOMAS in seiner zitierten Dissertation gezeigt, daß man die glatten Lichtkurven der Mira-Sterne durch die beiden Größen S_x und ε definieren kann. Er findet nun, daß diese beiden Größen keineswegs unabhängig voneinander sind. Ordnet man die Sterne nach der Größe ε und faßt für verschiedene Intervalle von ε die S_x zu Mitteln $\bar{S}_x$ zusammen, so ergibt sich nach THOMAS folgende Tabelle:

ε	$\bar{S}_x$	
$\leqq 0{,}35$	0,41	6 Sterne
0,35–0,40	0,45	22 „
0,40–0,45	0,47	29 „
0,45–0,50	0,49	39 „
$>0{,}50$	0,51	16 „
		112 Sterne

Schnellem Anstieg zum Maximum entspricht also ein breites Minimum; je langsamer der Anstieg relativ zum Abstieg wird, um so mehr nimmt durchschnittlich die Breite des Maximums relativ zum Minimum zu.

Die Beziehung zwischen Kurvenform und Periodenlänge bei den Me-Sternen stellt THOMAS durch folgende, ohne weiteres verständliche Tabelle dar, in der $\bar{S}_x$ und $\bar{\varepsilon}$ wieder Mittelwerte bedeuten:

P	$\bar{S}_x$	Anzahl	$\bar{\varepsilon}$	Anzahl
$\leqq 150^d$	0,54	5	0,47	30
$150^d - 200^d$	0,53	8		
200 − 250	0,52	13	0,45	88
250 − 300	0,49	15		
300 − 350	0,45	18	0,44	82
350 − 400	0,41	14		
$>400^d$	0,43	15	0,43	22
		88		222

Es ist aus dieser Tabelle sofort zu ersehen, daß die Breite des Maximums relativ zum Minimum mit wachsender Periode abnimmt.

β) Statistik der Periodenlängen der Mira-Sterne der verschiedenen Spektraltypen. Schon A. A. NIJLAND und J. VAN DER BILT haben darauf hingewiesen[1], daß die Häufigkeitsfunktion der Periodenwerte aller Mira-Sterne (ohne Rücksicht auf das Spektrum) wie eine GAUSSsche Fehlerkurve verläuft. LUDENDORFF hat (in Nr. V seiner „Untersuchungen") diese Häufigkeitsfunktion für die Perioden von 314 Me-Sternen abgeleitet. Seine Resultate sind folgende: Der Mittelwert der 314 Perioden ist 302^d, der Zentralwert 303^d. Setzt man $\Delta = (P - 300) : 60$, so ergibt sich als Häufigkeitsfunktion:

$$\varphi(\Delta) = 0{,}4573\, \pi^{-\frac{1}{2}} e^{-(0{,}4573\,\Delta)^2}.$$

[1] Recherches Astronomiques de l'Observatoire d'Utrecht VI, S. 1 (1916).

P	n		Δ	B − R
$90^d - 150^d$	14 =	4,5%	−3	+0,6%
151 −210	28	8,9	−2	−2,3
211 −270	76	24,2	−1	+3,3
271 −330	81	25,8	0	0,0
331 −390	69	22,0	+1	+1,1
391 −450	32	10,2	+2	−1,0
451 −510	8	2,5	+3	−1,4
511 −570	6	1,9	+4	+1,0
	314			

Die nebenstehende Tabelle enthält eine Abzählung der Periodenwerte, und unter B−R sind die Abweichungen der gefundenen Zahlen von den nach der Formel errechneten (in Prozenten) gegeben.

Bei den Mira-Sternen der anderen Spektraltypen ist die Anzahl der bekannten Perioden (vgl. Tabellen II—V) zu klein, als daß man die Häufigkeitsfunktionen ableiten könnte. Wir begnügen uns mit der Angabe der Mittelwerte P_m der Perioden, die wir in folgender Tabelle zusammenstellen:

Spektrum	P_m	
Me	302^d	314 Sterne
Se	361	18 ,,
K, Ma, Mb	207	24 ,,
Mc	305	15 ,,
N, R	404	24 ,,

Die Mira-Sterne der Spektralklassen Se sowie N und R haben also durchschnittlich größere Perioden als die der Klasse Me, die der Klasse Mc dieselben und die der Klassen K, Ma, Mb kleinere. Bei letzteren ist aber zu bemerken, daß unter ihnen 5 Sterne mit $P < 90^d$ vorkommen, deren Zugehörigkeit zur Mira-Klasse zweifelhaft ist (auch unter den Mc-Sternen kommt ein solcher vor). Schließt man diese 5 Sterne aus, so wird für die Klassen K, Ma, Mb die mittlere Periode $P_m = 243^d$, bleibt also immer noch bedeutend kleiner als bei den Me-Sternen. Andererseits nehmen unter den K-, Ma-, Mb-Sternen 3 durch die Länge ihrer Periode eine Sonderstellung ein, nämlich AB Cygni ($P = 530^d$), U Canis minoris ($P = 418^d$) und R Normae ($P = 487^d$). Schließt man auch diese aus, so wird $P_m = 197^d$. Jedenfalls ist man berechtigt zu sagen, daß die Mira-Sterne der Klassen K, Ma, Mb durchschnittlich die kürzesten Perioden haben.

Im folgenden sind noch für jeden Spektraltypus die Sterne mit der kürzesten und der längsten Periode zusammengestellt:

Spektralklasse	Kürzeste Periode		Längste Periode	
Me	91^d	(T Centauri, β_3)	561^d	(R Centauri, γ_2)
Se	224	(S Ursae maj., pec)	613	(S Cassiopeiae, pec)
K, Ma, Mb	56	(Z Leonis, α?)	530	(AB Cygni, β?)
Mc	70	(Z Cancri, ?)	496	(W Persei, pec)
N, R	259	(Y Persei, β_2)	580	(S Aurigae, β_4)

Auffällig ist, daß in dieser Tabelle so viele Sterne vorkommen, deren Lichtkurven mit pec (in diesen Fällen wegen ihrer sehr starken Veränderlichkeit) bezeichnet sind. W Persei könnte man schon beinahe als unregelmäßigen Veränderlichen bezeichnen.

Im neuen Draper-Katalog sind die Me-Sterne, soweit es möglich war, danach unterschieden, ob das kontinuierliche Spektrum Ma, Mb oder Mc ist; die betreffenden Angaben sind in Tabelle I wiedergegeben, wo die Spektra mit Mae, Mbe, Mce bezeichnet sind. Bildet man für die Mira-Sterne, die jeder dieser drei Unterklassen des Spektrums Me angehören, den Mittelwert P_m der Perioden, so findet man:

für die Klasse	Mae	$P_m = 270^d$	77 Sterne
	Mbe	297	120 ,,
	Mce	344	62 ,,

d. h. die Mae-Sterne haben durchschnittlich die kürzesten, die Mce-Sterne die längsten Perioden. Über weitere Zusammenhänge zwischen Eigentümlichkeiten des Spektrums und der Periodenlänge wird später berichtet werden. (Nebenbei sei hier bemerkt, daß ein Zusammenhang zwischen Spektrum und Form der Lichtkurve innerhalb der Klasse Me bei Sternen gleicher Periode nicht erkennbar ist.)

Interessant ist auch das Mischungsverhältnis der Mae-, Mbe-, Mce-Spektren für verschiedene Periodenintervalle, das aus folgender Tabelle ersichtlich ist:

P	Mae	Mbe	Mce
$91^d - 200^d$	16 Sterne	11 Sterne	3 Sterne
201 – 250	17 ,,	20 ,,	8 ,,
251 – 300	17 ,,	34 ,,	3 ,,
301 – 350	12 ,,	30 ,,	18 ,,
351 – 400	10 ,,	13 ,,	14 ,,
$>400^d$	5 ,,	12 ,,	16 ,,
	77 Sterne	120 Sterne	62 Sterne

Für sehr viele Mira-Sterne ist das Spektrum noch unbekannt, und diese sind in den vorangehenden Betrachtungen nicht berücksichtigt worden. Statistiken über die Verteilung der Perioden der Gesamtheit der Mira-Sterne, also ohne jede Rücksicht auf das Spektrum, sind schon mehrfach aufgestellt worden, zuletzt von W. Gyllenberg in seiner in Ziff. 20 zitierten Arbeit und von H. Thomas in seiner Dissertation. Wir geben hier die letztere wieder:

P	Sterne
$50^d - 90^d$	7
90 – 130	22
130 – 170	37
170 – 210	47
210 – 250	89
250 – 290	93
290 – 330	100
330 – 370	82
370 – 410	66
410 – 450	55
450 – 490	14
490 – 530	5
530 – 570	4
570 – 610	2
610 – 650	2
$>650^d$	0
	625

Die Verteilung ist ganz ähnlich wie für die Me-Sterne allein; das kann nicht wundernehmen, denn sicher gehört die überwiegende Mehrzahl auch der Mira-Sterne mit unbekanntem Spektrum der Spektralklasse Me an.

γ) Statistik der Helligkeitsamplituden der Mira-Sterne. Die in den Tabellen I—V angegebenen Helligkeitsamplituden A sind im allgemeinen Maximalamplituden, d. h. sie geben den Helligkeitsunterschied zwischen dem hellsten beobachteten Maximum und dem schwächsten beobachteten Minimum.

Die Häufigkeit der verschiedenen Werte von A für die Mira-Sterne der verschiedenen Spektralklassen ist aus der folgenden Tabelle ersichtlich, in der nur Sterne mit bekannten Perioden und mit bestimmt angegebenen, visuellen Werten von A berücksichtigt sind; für die Me-Sterne sind die Zahlen auch in Prozenten gegeben:

	Me		Se	K Ma, Mb	Mc	N, R
$\leqq 2^m,0$	6 =	2,4%	0	14	6	2
$2^m,1 - 3^m,0$	9	3,7	1	3	3	5
3 ,1 – 4 ,0	12	4,9	2	1	3	7
4 ,1 – 5 ,0	45	18,3	2	3	0	4
5 ,1 – 6 ,0	92	37,4	2	0	1	3
6 ,1 – 7 ,0	63	25,6	3	1	1	0
7 ,1 – 8 ,0	14	5,7	3	0	0	0
$>8^m,0$	5	2,0	2	0	0	0
	246	100,0%	15	22	14	21

Man ersieht aus der Tabelle, daß die Se-Sterne durchschnittlich größere, die übrigen durchschnittlich kleinere Amplituden aufweisen als die Me-Sterne. In der Tat sind die Mittelwerte der A für die verschiedenen Spektralklassen:

Me	$A_m = 5^m,5$	246 Sterne
Se	6 ,0	15 ,,
K, Ma, Mb	2 ,4	22 ,,
Mc	2 ,7	14 ,,
N, R	3 ,7	21 ,,

Es ist aber zu bedenken, daß, wie wir gleich sehen werden, zum mindesten bei den Me-Sternen die Amplitude von der Periode abhängt. Man darf daher, wenn man Sterne verschiedener Spektralklassen in bezug auf ihre Amplituden vergleichen will, diese Vergleichung nur für Sterne gleicher Periode vornehmen; das wird weiter unten geschehen.

Eine Untersuchung über den Zusammenhang zwischen der Periodenlänge und der Amplitude läßt sich nur für die Me-Sterne durchführen, da die Mira-Sterne der übrigen Spektralklassen zu wenig zahlreich sind. Die folgende Tabelle gibt in ihren einzelnen Kolumnen der Reihe nach das Periodenintervall, die zugehörige mittlere Periode P_m, den zugehörigen Mittelwert A_m der A, die Anzahl der Sterne, den graphisch durch eine glatte Kurve ausgeglichenen Wert A_k von A_m und schließlich die Differenz $A_m - A_k$.

P	P_m	A_m	n	A_k	$A_m - A_k$
$\leqq 150^d$	134^d	$3^m,6$	15	$3^m,5$	$+0^m,1$
$151^d - 180^d$	162	3 ,9	11	4 ,0	−0 ,1
181 −210	199	5 ,1	14	4 ,9	+0 ,2
211 −240	226	5 ,3	27	5 ,2	+0 ,1
241 −270	255	5 ,3	34	5 ,5	−0 ,2
271 −300	284	5 ,7	27	5 ,7	0 ,0
301 −330	317	5 ,7	37	5 ,8	−0 ,1
331 −360	343	6 ,0	27	5 ,9	+0 ,1
361 −390	374	6 ,1	18	6 ,1	0 ,0
391 −420	405	6 ,4	17	6 ,1	+0 ,3
421 −460	435	5 ,8	10	6 ,2	−0 ,4
461 −500	480	6 ,8	3	6 ,7	+0 ,1
$> 500^d$	528	7 ,4	6	7 ,4	0 ,0
			246		

Es besteht also eine sehr ausgesprochene Korrelation zwischen P_m und A_m. Die A_k-Kurve steigt erst rasch, dann langsam und dann wieder rasch an; von einer geraden Linie ist sie sehr merklich verschieden. (Daß für die Gesamtheit der Mira-Sterne, ohne Rücksicht auf das Spektrum, ein Zusammenhang zwischen P und A vorhanden ist, ist längst bekannt; dieser Zusammenhang ist z. B. von W. GYLLENBERG untersucht worden.)

Um nun die Mira-Sterne der übrigen Spektralklassen in bezug auf ihre Amplituden mit den Me-Sternen gleicher Periode zu vergleichen, verfahren wir wie folgt: Für jeden Se-Stern entnehmen wir aus der A_k-Kurve die zu seiner Periode gehörige Amplitude A_k und bilden dann die Differenz zwischen der für den betreffenden Se-Stern in Tabelle II gegebenen Amplitude A und diesem Werte A_k. Die Differenzen $A - A_k$ mitteln wir für alle Se-Sterne. Entsprechend verfahren wir für die Mira-Sterne der andern Spektralklassen. Als Mittelwerte der Differenzen $A - A_k$ ergeben sich folgende Werte für die Mira-Sterne

der Spektralklasse	Se	$-0^m,1$
	K, Ma, Mb	−2 ,4
	Mc	−2 ,6
	N, R	−2 ,4,

d. h. in Worten: Die Amplituden der Se-Sterne sind durchschnittlich um $0^m,1$, die der übrigen Klassen um $2^m,4$ bzw. $2^m,6$ bzw. $2^m,4$ kleiner als die Amplituden der Me-Sterne gleicher Periode. Da die Genauigkeit dieser Zahlen nur gering ist, so kann man auch sagen, daß die Se-Sterne durchschnittlich dieselben Amplituden wie Me-Sterne gleicher Periode, die übrigen dagegen um rund $2^m,5$ kleinere besitzen. Schon GYLLENBERG hat übrigens darauf hingewiesen, daß die M-Sterne ohne helle Linien durchschnittlich kleinere Amplituden besitzen als die Me-Sterne.

Es ist nun weiter von Interesse zu entscheiden, ob Me-Sterne von gleicher Periode, aber von verschiedener Form der Lichtkurve dieselben oder aber verschiedene Amplituden haben. Wir fassen einerseits die Sterne mit Lichtkurven der Form α_1 bis α_4, andererseits die mit Kurven der Form β zusammen, und bilden innerhalb gewisser Periodenintervalle die Mittelwerte P_m und A_m. Das Ergebnis dieser Rechnungen ist folgendes:

α-Kurven				β-Kurven			
P	P_m	A_m	n	P	P_m	A_m	n
200^d-249^d	231^d	$5^m,3$	15	91^d-161^d	137^d	$3^m,9$	13
250 −275	261	5 ,6	14	162 −213	191	5 ,5	13
275 −293	283	5 ,7	14	214 −240	227	5 ,6	12
294 −321	310	6 ,1	14	241 −278	262	5 ,8	12
322 −340	331	6 ,0	14	279 −331	316	5 ,5	12
341 −377	362	6 ,4	14	331 −418	360	5 ,8	12
378 −412	400	6 ,1	14				
413 −535	468	6 ,5	14				

Zeichnen wir die A-Kurven für die Sterne, welche α-Kurven, und für diejenigen, welche β-Kurven aufweisen, getrennt, so zeigt sich, daß die beiden A-Kurven, soweit sie sich auf dasselbe Periodenintervall beziehen, sich nur innerhalb der Grenzen der Genauigkeit voneinander unterscheiden. Man kann also einen Unterschied der Amplituden der α- und β-Kurven nicht nachweisen, solange man Sterne gleicher Perioden betrachtet. Insgesamt haben die α-Kurven natürlich größere Amplituden als die β-Kurven, da die zugehörigen Perioden bei ersteren durchschnittlich länger sind als bei letzteren.

Was die relativ seltenen Me-Sterne mit Lichtkurven der Form γ_1 und γ_2 angeht, so haben diejenigen unter ihnen, die Perioden $<300^d$ besitzen, zum Teil auffallend kleine Amplituden; unter diesen Sternen befinden sich sicher einige RV Tauri-Sterne.

C. LUPLAU-JANSSEN hat vor längerer Zeit[1] eine sehr starke Korrelation zwischen A und ε hergeleitet; er benutzt dabei aber nur 40 Mira-Sterne. Eine eingehendere Untersuchung bestätigt sein Ergebnis nicht.

δ) Einige weitere statistische Bemerkungen über die Mira-Sterne der Spektralklasse Me. Für die Häufigkeit des Vorkommens der verschiedenen Werte von ε bei den Me-Sternen hat LUDENDORFF (in Nr. V seiner „Untersuchungen") eine Statistik aufgestellt, die hier in abgekürzter Form wiedergegeben werden möge; sie stützt sich auf die Werte von ε für 218 Me-Sterne:

ε	Anzahl
$\leqq 0{,}30$	6
0,31−0,36	14
0,37−0,39	14
0,40−0,42	32
0,43−0,45	57
0,46−0,48	42
0,49−0,51	32
0,52−0,54	16
0,54−0,60	5

Gerade in den Fällen, in denen ε merklich $>0{,}50$ ist, ist dieser Wert häufig noch unsicher und dürfte wohl durch eine genauere Festlegung verkleinert werden. In obiger Zusammenstellung kommen 34 Me-Sterne vor, deren $\varepsilon<0{,}40$ ist. Merkwürdigerweise sind diese 34 Sterne sehr ungleichmäßig am Himmel verteilt; 11 von ihnen liegen eng zusammengedrängt zwischen RA. $18^h{,}8$ und $20^h{,}8$, Dekl. $-2°$ und $+33°$ in oder sehr nahe der Milchstraße. Ziemlich nahe um dieses Nest herum liegen noch 5 weitere von ihnen. Wenn nun auch dieser Teil des Himmels überhaupt sehr reich an Me-Sternen ist, so ist diese Anhäufung doch sehr merkwürdig. Ein zweites Nest von 9 dieser Sterne liegt von der Milchstraße entfernt zwischen RA. $19^h{,}8$ und $0^h{,}3$, Dekl. $-21°$ und $-62°$.

PHILLIPS hatte gefunden, daß die Sterne seiner Gruppe I (nach unserer Bezeichnung solche mit β- und γ-Lichtkurven) im Durchschnitt heller sind als die der Gruppe II (α-Kurven), und L. CAMPBELL fand, wie wir gesehen haben, für die Minimalhelligkeiten Entsprechendes, während bei dem von ihm benutzten Material diese Erscheinung an den Maximalhelligkeiten nur schwach hervortrat. Bilden wir nach den Angaben der G. u. L. über die Maximalhelligkeiten M und die Minimalhelligkeiten m der Me-Sterne die Mittelwerte dieser Größen für

[1] Ap J 38, S. 200 (1913).

die einzelnen Kurventypen (unter Ausschluß aller photographischen Größen und derjenigen m, für die nur eine obere Grenze gegeben ist), so erhalten wir folgende Tabelle (n = Zahl der benutzten Sterne):

Lichtkurve	M	n	m	n
α_1, α_2	$8^m,8$	49	$13^m,9$	22
α_3, α_4	8 ,3	74	12 ,6	43
$\alpha-\beta$	8 ,1	24	12 ,4	12
β	8 ,0	74	12 ,0	58
γ, $\alpha-\gamma$, $\beta-\gamma$	7 ,0	24	11 ,2	21

Der Gang zeigt sich sowohl in den M wie in den m mit großer Deutlichkeit; er erklärt sich für die α- und β-Kurven höchst wahrscheinlich dadurch, daß, wie wir später sehen werden, die Me-Sterne mit kurzen Perioden, bei denen β-Kurven überwiegen, absolut heller (allerdings auch weiter entfernt) sind als die mit langen Perioden, bei denen die α-Kurven überwiegen.

Sondern wir die Sterne auch noch nach ihrer Periodenlänge, so erhalten wir folgende Mittelwerte für M:

P	α_1, α_2	n	α_3, α_4	n	$\alpha-\beta$	n	β	n	$\alpha-\gamma$, $\beta-\gamma$, γ	n	Alle	n
90^d-200^d	—	—	—	—	} $8^m,6$	8	$8^m,1$	21	} $6^m,9$	6	} $8^m,2$	71
201 —250	—	—	$8^m,5$	16			8 ,2	20				
251 —300	$8^m,8$	6	8 ,7	29	8 ,7	6	7 ,9	12	} 6 ,7	4	} 8 ,3	107
301 —350	8 ,8	11	7 ,9	16	7 ,7	9	8 ,1	14				
351 —400	8 ,7	14	7 ,8	8	—	—	} 7 ,3	7	7 ,2	8		
401 —450	9 ,0	11	7 ,0	5	7 ,0	1			} 7 ,2	6	} 8 ,1	67
$>450^d$	9 ,0	7	—	—	—	—	—	—				

Bei der Gesamtheit der Me-Sterne zeigt sich also kein Gang der scheinbaren Maximalhelligkeit mit der Periode; bei den Sternen mit α_3-, α_4-, $\alpha-\beta$-Kurven könnte man allenfalls eine Zunahme der Maximalhelligkeit mit wachsender Periode vermuten.

Für die m reicht das Material zu einer so weit gehenden Sonderung nicht aus. Wir bilden hier nur die Mittelwerte der m für alle Kurventypen und finden:

P	m	n
90^d-250^d	$10^m,0$	53
251 —350	10 ,3	65
$>350^d$	10 ,9	38

Hier zeigt sich ein Gang, wie er zu erwarten ist, wenn die M von der Periode unabhängig sind; denn die Amplituden nehmen mit wachsender Periode zu.

In teilweisem Widerspruch zu den obigen Ergebnissen steht eine Bemerkung von S. BELJAWSKY[1], daß nämlich die sehr schwachen Mira-Sterne auffällig kleine Perioden haben. Wahrscheinlich klärt sich dieser Widerspruch dadurch auf, daß für unsere oben gegebene Statistik, die auf den Sternen des Hauptkatalogs der G. u. L. beruht, Sterne mit sehr kleinen Maximalhelligkeiten nur in sehr beschränkter Zahl zur Verwendung gekommen sind.

24. Die Änderungen der Perioden der Mira-Sterne. Wie wir schon erwähnt haben, unterliegen die Perioden der Mira-Sterne in vielen Fällen starken Änderungen, und die rechnerische Darstellung der Zeiten der Maxima und Minima wird dadurch erheblich erschwert, die Vorausberechnung unsicher gemacht. Man hat früher meist versucht, den Änderungen der Perioden dadurch Rechnung zu tragen, daß man in die Formel für die Vorausberechnung z. B. der Maxima, die bei konstanter Periode die Form

$$M = T_0 + P \cdot E \quad (E = \text{ganze, positive oder negative Zahl})$$

[1] A N 227, S. 277 (1926).

besitzt, ein periodisches Glied (oder deren mehrere) einführte, also setzte:

$$M = T_0 + P \cdot E + a \sin(\varphi_1 E + \varphi_2).$$

Die Erfahrung hat indessen gelehrt, daß die periodischen Glieder meist nur vorübergehend den Beobachtungen genügen, und man zieht es daher in neuerer Zeit vor, plötzliche Änderungen in der Periode und Sprünge in der Epoche T_0 anzunehmen. Wir werden weiter unten auf diese Dinge ausführlich eingehen.

In einigen wenigen Fällen scheinen die Änderungen der Periode stets im selben Sinne, und zwar in dem einer Abnahme, zu verlaufen. Es handelt sich hier vor allem um die beiden Mira-Veränderlichen R Hydrae und R Aquilae. Den Lichtwechsel von R Hydrae hat H. LUDENDORFF[1] näher studiert, und zwar für die Zeit bis 1914. Gegen Ende des 17. Jahrhunderts war P = etwa 500^d. Die Länge der Periode seit Ende des 18. Jahrhunderts läßt sich durch die beiden folgenden Formeln darstellen ($E = 0$ im Jahre 1848):

$$P = 453^d{,}5 - 0^d{,}107\,E + 0^d{,}0158\,E^2 \quad (1784 \text{ bis } 1848),$$

$$P = 452^d{,}7 - 0^d{,}671\,E - 0^d{,}0038\,E^2 \quad (1848 \text{ bis } 1914).$$

Die Periode hat zwischen 1784 und 1914 sich von 497^d auf 403^d vermindert.

Wie vorsichtig man nun bei solchen Formeln mit der Extrapolation sein muß, beweist der Umstand, daß nach 1914 die Abnahme der Periode plötzlich aufgehört, ja sich vielleicht sogar in eine Zunahme verwandelt hat. Die weiteren Beobachtungen werden lehren, ob es sich dabei nur um eine vorübergehende Erscheinung handelt. Versuche, die Änderungen der Periode von R Hydrae durch ein periodisches Glied darzustellen[2], sind wohl noch verfrüht.

Ganz ähnliche Verhältnisse wie bei R Hydrae liegen bei R Aquilae vor, der von R. MÜLLER[3] näher untersucht worden ist. Für die Periode ergeben sich folgende Formeln:

$$\text{Aus den Maxima} \quad P = 329^d{,}63 - 0^d{,}485\,E,$$

$$\text{aus den Minima} \quad P = 329^d{,}51 - 0^d{,}448\,E.$$

($E = 0$ im Jahre 1890.) Die Periode hat zwischen 1856 und 1924 von 347^d auf 310^d abgenommen.

Auch H. H. TURNER hat sich in einer schon früher besprochenen Arbeit[4] mit der Periodenänderung bei R Hydrae und R Aquilae beschäftigt. Er stellt die Verminderung der Periode von R Hydrae durch die Annahme dar, daß sich diese Periode alle 3415^d um $10^d{,}0$ vermindern kann, dies aber nicht immer tut. Auch bei R Aquilae nimmt er ähnliche Sprünge im Werte der Periode an; R. MÜLLER weist aber, insbesondere betreffs R Aquilae, darauf hin, daß diese TURNERsche Annahme von Sprüngen in den Perioden der beiden Sterne nicht sehr stichhaltig und die Annahme allmählicher Änderungen wohl vorzuziehen sei.

In der zitierten und in zwei weiteren, gleichfalls schon früher erwähnten Arbeiten[5] untersucht TURNER auch noch verschiedene andere Mira-Sterne auf sprunghafte Änderungen der Periode, die stets in gleichem Sinne verlaufen. Seine Ergebnisse sind aber wenig überzeugend, da es sich meist nur um kleine Änderungen bzw. um relativ kurze Beobachtungsreihen handelt. Z. B. ist bei U Herculis, der nach TURNER eine Abnahme der Periode zeigt, letztere nach den neuesten Beobachtungen augenscheinlich wieder länger geworden. Auch

---

[1] A N 203, S. 117 (1916). [2] A N 227, S. 141 (1926).

[3] Dissertation, Berlin 1925. Auszug A N 223, S. 185 (1924).

[4] M N 80, S. 273 (1920). [5] M N 80, S. 481, 604 (1920).

die regelmäßigen Intervalle, die TURNER für die Sprünge der Periode oder auch der Epoche findet, bedürfen jedenfalls noch der Bestätigung. Gerade für die beiden am längsten beobachteten Mira-Sterne, o Ceti und χ Cygni, vermag TURNER säkulare Änderungen der Periode nicht nachzuweisen. Bei χ Cygni findet er eine konstante Periode, aber vier Sprünge der Epoche um je $+28^d$, die um ganzzahlige Vielfache von 1502^d voneinander entfernt sind, bei o Ceti acht bald in einem, bald in anderem Sinne erfolgende Sprünge der Periode, die in Intervallen von ganzzahligen Vielfachen von $20\,P$ (rund 6600^d) erfolgen.

Im Katalog der G. u. L. enthalten auch die Elemente von S Serpentis ein großes säkulares Glied; sie lauten nämlich in etwas anderer Schreibweise:

$$M = 2388796^d + 365^d{,}1\,E + 0^d{,}080\,(E - 36)^2 + 40^d \sin(7^\circ{,}2\,E + 281^\circ).$$

Diese Elemente sind 1908 von TURNER[1] berechnet worden. Während des Intervalles (1829 bis 1904), das TURNERS Diskussion umfaßt, liegen die Werte der Periode ungefähr zwischen 360^d und 375^d. In den letzten 10 Jahren war die Periode nach den neuesten Beobachtungen durchschnittlich 367^d. Das säkulare Glied hat also keine reelle Bedeutung, und die Elemente bedürfen einer Neubestimmung.

Wir wenden uns nun den sog. periodischen Gliedern in den Elementen der Mira-Sterne zu, wie sie im Katalog der G. u. L. für eine große Anzahl von diesen Objekten gegeben sind. Allgemein schreibt man in diesen Fällen die Formel zur Berechnung des Maximums bzw. Minimums in der Form

$$M = T_0 + P_0 E + a \sin(\varphi_1 E + \varphi_2),$$

woraus für P folgt:

$$P = P_0 + 2a \sin\frac{\varphi_1}{2} \cos(\varphi_1 E + \varphi_2).$$

In der Regel ist φ_1 klein, die Periode der Sinusglieder also lang. Als Beispiele führen wir zunächst alle Mira-Sterne an, für die nach dem genannten Katalog $\varphi_1 > 20^\circ$, die Periode des Sinusgliedes also $< 18\,P$ ist:

4	TT Cassiopeiae	$M = T_0 + 400^d E +$	13^d	$\sin(150^\circ E + 340^\circ)$
243	S Pictoris	427	20	$\sin(25\ E + 165)$
361	RS Aurigae	169	11	$\sin(45\ E + 101)$
472	RZ Ursae maj.	129	40	$\sin(36\ E + 270)$
780	R Canum ven.	325	13	$\sin(20\ E + 260)$
882	V Coronae	358	34	$\sin(\ 6\ E + 180) + 12^d \sin(20^\circ E + 320^\circ)$
897	RZ Scorpii	158	20	$\sin(22{,}5\,E + 0^\circ)$
1292	SS Lyrae	352	12	$\sin(45\ E + 0)$

Bei fast allen von diesen Sternen standen zur Ableitung der Elemente nur ziemlich kurze Beobachtungsreihen zur Verfügung, und die Sinusglieder sind daher noch durchaus fraglich. Längere Beobachtungsreihen waren nur für R Canum venaticorum vorhanden (von 22 Jahren Ausdehnung) und für V Coronae (von 1857 an); aber für den ersteren Stern bemerkt die G. u. L., daß das Sinusglied unsicher sei, und bei dem zweiten hat ein Sinusglied nicht ausgereicht, sondern man hat deren zwei einführen müssen, von denen das größere eine sehr lange Periode hat.

Ferner lassen wir ein Verzeichnis derjenigen Mira-Sterne folgen, bei denen nach der G. u. L. das periodische Glied sehr groß, und zwar $> \frac{1}{10}\,P$ werden kann, soweit sie nicht schon in obiger Tabelle mit enthalten sind:

---
[1] M N 68, S. 560 (1908).

17	T Andromedae	$M = T_0 + 281^d E +$	30^d	$\sin(7°,8E + 309°)$
211	R Pictoris	167	20	$\sin(12\ E + 180)$
244	R Aurigae	456	48	$\sin(8,8E + 261)$
256	S Orionis	412	95	$\sin(5,3E + 217)$
484	R Cancri	362	60	$\sin(5,0E + 150)$
489	Z ,,	70	20	$\sin(15\ E + 285)$
508	S Hydrae	256	30	$\sin(4,7E + 242)$
618	RS ,,	338	40	$\sin(12\ E + 0)$
706	R Virginis	145,5	20	$\sin(1,8E + 216) + 4^d,8 \sin(5°,625E + 343°)$
713	S Ursae maj.	227	35	$\sin(5,4E + 194)$
805	S Bootis	270	47	$\sin(4,1E + 357)$
812	R Camelopardalis	270	40	$\sin(4,7E + 200)$
851	S Librae	193	25	$\sin(4,5E + 256)$
899	Z Scorpii	365	38	$\sin(9\ E + 18)$
928	R ,,	223	23	$\sin(5,3E + 190)$
949	U Herculis	406	49	$\sin(7,3E + 327)$
959	T Ophiuchi	365	45	$\sin(6,43E + 219)$
1016	S Herculis	308	47	$\sin(7,5E + 120) + 12^d \sin(15°,0\,E + 60°)$
1025	RR Scorpii	278	32	$\sin(5,8E + 1)$
1293	T Sagittarii	390	80	$\sin(7,3E + 147)$
1490	R Vulpeculae	137	18	$\sin(4,5E + 61)$
1537	S Cephei	486	70	$\sin(12,4E + 311)$
1578	T Pegasi	375	65	$\sin(4,5E + 346)$
1638	R ,,	377	60	$\sin(6,0E + 252)$

Das nähere Studium führt auch hier zu der Überzeugung, daß dem Sinusgliede höchstens in Ausnahmefällen eine andere Bedeutung als die einer reinen Rechnungsgröße zuzuschreiben ist. Bei weitaus den meisten der genannten Sterne umfassen die Beobachtungen, aus denen das Sinusglied abgeleitet ist, weniger oder höchstens nur wenig mehr Zeit, als eine volle Periode des Sinusgliedes beträgt, und man kann daher keineswegs mit Sicherheit erwarten, daß sich die Abweichungen von der konstanten Periode nach demselben Gesetz wiederholen werden. Noch ungünstiger wird die Sachlage in den zahlreichen Fällen sein, in denen der Koeffizient des Sinusgliedes im Verhältnis zu P noch kleiner ist als in den oben angeführten Fällen, oder in denen (wie bei V Coronae, R Virginis, S Herculis in den gegebenen Zusammenstellungen) mehr als ein Sinusglied zur Darstellung der Beobachtungen in Ansatz gebracht ist. Daß die durch die Einführung der Sinusglieder bewirkte Verbesserung der Darstellung der beobachteten Zeiten der Maxima bzw. Minima oft sehr beträchtlich ist, kann natürlich nicht in Abrede gestellt werden; ein der G. u. L.[1] entnommenes Beispiel, das den Stern R Aurigae betrifft, möge dies erhärten. In der folgenden Tabelle enthält die erste Kolumne die Zeiten (in julianischen Tagen) der aus den Beobachtungen abgeleiteten Normalmaxima, die zweite die Differenzen $B-R_1$ zwischen Beobachtung und Rechnung, die man erhält, wenn man die Maxima nach der Formel

$$M = 2401\,508^d + 456^d{,}3\,E,$$

d. h. mit einer möglichst günstig gewählten konstanten Periode berechnet, und die dritte die $B-R_2$, die sich ergeben, wenn man zur Rechnung die Formel

$$M = 2401\,510^d + 456^d{,}3\,E + 48^d \sin(8°{,}8\,E + 261°)$$

verwendet. (Siehe nebenstehende Tabelle.)

Normalmaximum	$B-R_1$	$B-R_2$
240 1916^d	-48^d	$-\ 2^d$
3311	-22	$+19$
5143	-15	$+\ 6$
8380	$+27$	$-\ 1$
241 1136	$+46$	$-\ 4$
2503	$+44$	$-\ 3$
3860	$+32$	$-\ 2$
5228	$+31$	$+17$
6575	$+\ 9$	$+17$
7917	-18	$+10$
9246	-58	-16

Als Gegenstück führen wir nach derselben Quelle[2] ein Beispiel an, bei welchem man für ein langes Zeitintervall (37 Jahre) mit einer konstanten Periode

[1] Band I, S. 141. [2] Bd. I, S. 124.

Normal-maximum	$B-R$
240 4721^d	+ 2^d
5572	+ 2
8633	+ 3
9815	− 6
241 0840	− 1
2207	+ 6
2700	−11
3734	+ 2
5776	+ 4
6964	+ 2
7982	− 1

auskommt. Es handelt sich um V Tauri, dessen Elemente nach der G. u. L. lauten:

$$M = 2404719^d{,}5 + 170^d{,}04\,E.$$

Die $B-R$ sind folgende: (s. nebenstehende Tabelle).

Einen Fall, in welchem das Sinusglied vielleicht doch etwas mehr Bedeutung als die einer Rechnungsgröße hat, bildet der Veränderliche AF Cygni (Spektr. Mb). Die G. u. L. nimmt für ihn noch eine konstante Periode an, S. KANDA[1] hat aber gezeigt, daß für die Maxima folgende Formel anzusetzen ist:

$$M = 2421644^d + 88^d{,}4\,E + 70^d \sin(7^\circ{,}4\,E + 180^\circ),$$

woraus sich für die Periode der Ausdruck

$$P = 88^d{,}4 + 9^d{,}0 \cos(7^\circ{,}4\,E + 180^\circ)$$

ergibt. Während des die Beobachtungen umfassenden Zeitintervalls hat das Sinusglied seine Periode bereits mehr als zweimal durchlaufen. Zu ähnlichen Resultaten ist unabhängig VORONTSOV-VELYAMINOV gelangt[2], der auch die starken Änderungen der Lichtkurve bei diesem Stern näher diskutiert. Auch bei W Hydrae (Me) kommt dem sehr großen periodischen Glied in der die Maxima darstellenden Formel[3]

$$M = 2395345^d + 385^d{,}6\,E + 100^d \sin 7^\circ{,}2\,E$$

vielleicht mehr als rein formale Bedeutung zu.

H. H. TURNER hat die Sinusglieder in den Elementen der Mira-Sterne einer statistischen Untersuchung unterworfen[4], und findet eine Korrelation zwischen P_0 und φ_1 (je größer P_0, desto größer φ_1) und weniger ausgesprochen auch eine solche zwischen P_0 und a (je größer P_0, desto größer a); dagegen ergeben sich a und φ_1 als unabhängig von der Form der Lichtkurve. Später[5] hat TURNER selbst den Glauben an die Bedeutung der Sinusglieder verloren und die Änderungen der Perioden als sprungweise angesehen (vgl. die früheren Ausführungen in dieser Ziffer).

Gegenwärtig neigt man allgemein mehr und mehr dazu, auf das Ansetzen von Sinusgliedern in den Elementen der Mira-Sterne zu verzichten und anstatt dessen, wie TURNER, Sprünge in der Periode und Anfangsepoche anzunehmen. Reiche Erfahrungen auf diesem Gebiete sind auf der Babelsberger Sternwarte gesammelt worden gelegentlich der Berechnung der Ephemeriden der veränderlichen Sterne. P. GUTHNICK berichtet darüber wörtlich, wie folgt[6]: „Wir haben uns entschlossen, die Einführung von Sinusgliedern ganz aufzugeben und statt dessen konstante Elemente für begrenzte Zeitabschnitte einzuführen. Die Länge dieser Zeitabschnitte, für die die jeweilig abgeleiteten Elemente gelten sollen, ist in der Regel ohne besondere Willkür zu bestimmen. Stellt man nämlich die Abweichungen der beobachteten Maxima und Minima von den mit einer konstanten Periode und Ausgangsepoche berechneten bildlich dar, so hat man in nahezu allen Fällen den Eindruck, daß die Periodenänderungen nicht nach einer Sinusformel verlaufen, sondern mehr oder weniger plötzlich eintreten, und daß die Periode zwischen den einzelnen Sprüngen nahezu konstant bleibt. Wir haben es deshalb vorgezogen, zwischen jeder Periodenänderung

[1] Japanese Journal of Astronomy and Geophysics 1, S. 211 (1924).
[2] A N 225, S. 369 (1925); A N 228, S. 135 (1926).
[3] Harv Circ 270 (1924). [4] M N 68, S. 544 (1908). [5] M N 78, S. 538 (1918).
[6] V J S 59, S. 240 (1924).

eine konstante, aus dem betreffenden Zeitabschnitt allein abgeleitete Periode und eine entsprechende Ausgangsepoche zu benutzen. So ergeben sich dann für jeden Stern $n+1$ verschiedene Elementensysteme, wenn die Zahl der Periodenänderungen n beträgt. Die Zahl der notwendig anzunehmenden Periodensprünge übersteigt bisher in keinem Falle drei und sehr selten zwei, wenn man von Sternen wie o Ceti, mit außergewöhnlich langen Beobachtungsreihen, absieht, deren Bearbeitung meistens noch verschoben ist. Vermutlich werden aber mit der Zeit auch Fälle mit häufigeren Periodensprüngen gefunden werden, besonders unter den Grenzfällen zwischen der Miraklasse und den Sternen nach Art von μ Cephei. Bei o Ceti hat die Periode in dem Zeitraum von 1596 bis 1923 anscheinend achtmal eine sprungweise Änderung erlitten. Diese Art der Behandlung scheint uns nach unseren bisherigen Erfahrungen der Wirklichkeit besser zu entsprechen als die frühere, sie hat bisher in keinem Falle versagt und bietet außerdem den Vorteil, daß die Vorausberechnung der Maxima und Minima, solange nicht gerade ein Periodensprung eintritt, mit den instantanen Elementen viel sicherer wird als mit der Benutzung von Fourier-Reihen oder von mittleren Elementen. Es erscheint auch nicht ausgeschlossen, daß auf diese Weise sehr wertvolles statistisches Material gewonnen werden wird, das uns einen tieferen Einblick in die Natur des Lichtwechsels der Mira-Sterne und der ihnen nahestehenden Sterne zu gewähren vermag.

Sehr merkwürdig sind die bisher seltenen Fälle, in denen Epochensprünge oder man kann auch sagen, schnell vorübergehende starke Periodenänderungen auftreten. Unter den bereits bearbeiteten Sternen haben wir zwei oder drei Fälle dieser Art gefunden: V Coronae bor., R Lupi und wahrscheinlich W Aurigae. V Coronae hat eine bisher, nämlich von 1878 bis 1897 und von 1904 bis zur Gegenwart, konstant gebliebene Periode von 355,9 Tagen; $M-m$ ist 158 Tage. Zwischen 1897 und 1904, in welchem Zeitraum Beobachtungen leider nicht vorhanden sind, hat sich die Epoche scheinbar um $+61$ Tage geändert; die Periode ist vorher und nachher merklich die gleiche. Es bleibt die Frage offen, wie schnell der Übergang von der einen Epoche zur anderen vor sich gegangen ist, ob der volle Sprung zwischen zwei aufeinanderfolgenden Maxima oder Minima eingetreten ist, oder ob vorübergehend durch mehrere Lichtwechselperioden hindurch eine starke Änderung der Periodenlänge stattgefunden hat. Ganz ähnlich liegt der Fall R Lupi, leider auch bezüglich des Fehlens der Beobachtungen in der kritischen Zeit. Die Periode beträgt hier 234,5 Tage, $M-m$ ist 117 Tage. Zwischen 1899 und 1917 trat eine Epochenverschiebung von -68 Tagen ein, vorher und nachher ist die Periode die gleiche. Auch hier ist nicht bekannt, wieviel Zeit die Verschiebung in Anspruch genommen hat. Bei W Aurigae, Periode 273,3 Tage, $M-m$ 112 Tage, hat sich die Epoche zwischen 1901 und 1907 um $+54$ Tage verschoben. Vorher und nachher ist die Periode die gleiche. In der kritischen Zeit fehlen auch hier die Beobachtungen. Diese drei Beispiele zeigen besonders eindringlich, wie nötig und wichtig die beständige Überwachung der Mirasterne ist."

Als numerisches Beispiel für die von Guthnick im vorstehenden skizzierte Art der Behandlung sei hier das Ergebnis einer Diskussion der Beobachtungen von R Ursae minoris in dem Intervall von 1883 bis 1922 durch H. Ludendorff angeführt[1]. Dieser Stern (Spektrum Mc) hat eine Lichtkurve von der Form $\beta_4-\gamma_2$. Die Hauptminima lassen sich durch folgende Elemente darstellen:

1883—1889	$H_m = 2409135^d$	$+ 337^d E$
1889—1897	2411720	+ 322
1897—1922	2414246	+ 332

[1] A N 218, S. 65 (1923).

Nach unseren Ausführungen über die periodischen Glieder in den Elementen der Mira-Sterne kann es kaum statthaft erscheinen, an das Vorhandensein dieser Glieder Theorien über die Natur dieser Veränderlichen anzuknüpfen, wie es SHINZO SHINJO und TOSHIMA ARAKI kürzlich getan haben[1].

25. Veränderlichkeit der Lichtkurven und Amplituden der Mira-Sterne. Es ist bereits erwähnt worden, daß bei den Mira-Sternen manchmal sowohl die Gestalt der Lichtkurve als auch die Maximal- und die Minimalhelligkeiten stark veränderlich sind.

Als Beispiele für stark veränderliche Lichtkurven führen wir aus den Tabellen I und II folgende besonders krasse Fälle an:

R Volantis. Spektr. Me. 1899 bis 1900 und 1909 bis 1910 war die Lichtkurve vom Typus γ_1, allerdings ist der Buckel im Anstieg nicht völlig sichergestellt und war bei dem sehr breiten Maximum 1899 wohl sicher nicht vorhanden (vgl. Cape Annals Vol. IX und Transvaal Circ Nr. 5); 1917 bis 1919 war dagegen die Lichtkurve dem Typus α_1 sehr ähnlich, wenn auch das Maximum Ende 1918 einen allmählichen Anstieg zeigte (vgl. Union Circ Nr. 51). Weitere Beobachtungen liegen nicht vor. Der höchst merkwürdige Stern hat also seinen Lichtwechsel vollkommen geändert, doch ist die Periode (444^d) unverändert geblieben.

S Cassiopeiae. Spektr. Se. Hier liegt die Sache ähnlich wie bei R Volantis. Zuweilen ist die Kurve ausgesprochen von der Form α_1, zuweilen von der Form γ_1, zuweilen nähert sie sich auch der symmetrischen Form β.

S Ursae majoris. Spektr. Se. Die Lichtkurve nimmt in anscheinend regellosem Wechsel die Formen α_3, α_4, β, γ_1 an. Der Abstieg ist oft steiler als der Anstieg, das Maximum oft auffallend breit. (Eine ausführliche Lichtkurve ist in den früher zitierten Appendices zu den Memoirs of the British Astronomical Association veröffentlicht.)

Bei den Sternen der Tabelle III (Spektr. K, Ma, Mb, Mc) scheinen auffallend häufig stark veränderliche Lichtkurven vorzukommen. Bei der Kleinheit der Amplituden dieser Sterne ist es aber nicht immer ganz leicht zu entscheiden, inwieweit diese Änderungen reell sind.

Was die Änderungen der Maximal- und der Minimalhelligkeiten der Mira-Sterne angeht, so ist es eine bei der Betrachtung von Lichtkurven solcher Sterne sofort in die Augen springende Tatsache, daß die Maximalhelligkeiten in der Regel weit stärker veränderlich sind als die Minimalhelligkeiten. Genauere Untersuchungen über diese interessante Erscheinung liegen noch nicht vor. Man hat indessen den Eindruck, daß die Maximalhelligkeiten bei den Sternen mit Lichtkurven der Form α_1 (sehr breites Minimum mit konstanter Phase) besonders stark veränderlich sind. Als Beispiele hierfür seien die beiden Veränderlichen Z Tauri und RW Lyrae angeführt. Bei Z Tauri ($P = 502^d$, Spektr. unbekannt) liegt die Minimalhelligkeit, in der der Stern sehr lange verharrt, zwischen 13^m und 14^m, die Maximalhelligkeit steigt manchmal bis fast zu 9^m an, während manchmal (so um die Jahreswende 1913/14) das Maximum ganz auszubleiben scheint oder sehr schwach ist ($12^m,6$ Anfang 1911). Bei RW Lyrae ($P = 507^d$, Spektr. Me) ist die Helligkeit im Minimum etwa $14^m,0$ bis $14^m,5$, die Helligkeit im Maximum liegt zwischen etwa $9^m,0$ und 13^m. Aber auch bei solchen Mira-Sternen, deren Lichtkurve nicht die Form α_1 hat, kommen starke Veränderungen der Maximalhelligkeit vor; z. B. liegt letztere bei RT Herculis ($P = 293^d$, Lichtkurve α_3, Spektr. Me) zwischen $8^m,5$ und etwa $11^m,0$, für *o* Ceti ($P = 331^d$, Lichtkurve α_3, Spektr. Me) gibt die G. u. L. als Grenzwerte der

[1] Japanese Journal of Astronomy and Geophysics 2, Nr. 3 (1924).

Maximalhelligkeit $1^m,7$ und $5^m,2$, als Grenzen der Minimalhelligkeit $8^m,7$ und $10^m,0$. In den Jahren 1902 bis 1919 hat sich bei letzterem Stern die Maximalhelligkeit zwischen $2^m,2$ und $4^m,4$, die Minimalhelligkeit zwischen $8^m,8$ und $9^m,4$ bewegt (nach den Beobachtungen der British Astronomical Association). P. GUTHNICK hat in seiner Monographie über diesen Stern versucht, Gesetzmäßigkeiten für die Änderungen der Maximalhelligkeit zu finden, ist aber nicht zu befriedigenden Ergebnissen gelangt. Auch H. ROSENBERG hat in seiner Monographie über χ Cygni keinerlei Regeln für die (bei den gut bestimmten Maxima) zwischen den Grenzen $3^m,3$ und $6^m,2$ liegenden Maximalhelligkeiten finden können. Dagegen glaubt A. THOM[1] bei verschiedenen Mira-Sternen gewisse Periodizitäten in den Maximal- bzw. Minimalhelligkeiten gefunden zu haben, doch scheinen sich seine Untersuchungen im allgemeinen auf zu kurze Beobachtungsreihen zu stützen.

In einer kleinen Untersuchung, bei der er leider nur 40 Mira-Sterne benutzt, ist H. H. TURNER[2] zu dem Ergebnis gelangt, daß diejenigen Mira-Sterne, bei denen ε groß ist, kleinere Schwankungen der Maximalhelligkeit aufweisen als diejenigen, bei denen ε klein ist, die also steilen Helligkeitsaufstieg haben. Nach den obigen Ausführungen kann dieses Resultat sehr wohl richtig sein; es bedarf jedoch noch weiterer Erhärtung durch Heranziehung einer größeren Zahl von Sternen.

26. Noch einige Bemerkungen über die Lichtkurven der Mira-Sterne. Die Lichtkurven der Mira-Sterne mit unbekanntem Spektrum sind noch nicht in der Weise systematisch untersucht und klassifiziert worden, wie dies für sämtliche im Hauptkatalog der G. u. L. enthaltenen Mira-Sterne mit bekanntem Spektrum in den Tabellen I bis V geschehen ist. In den folgenden Ausführungen ist daher stets nur von diesen letzteren die Rede.

Auf besondere Eigentümlichkeiten der Lichtkurven ist bereits in den genannten Tabellen hingewiesen worden. Interesse verdienen namentlich die Lichtkurven mit Buckel oder Stufe im Aufstieg ($\alpha-\gamma_1$, $\beta-\gamma_1$, γ_1) und diejenigen, die in der Regel oder häufig ein deutlich ausgeprägtes Doppelmaximum besitzen ($\beta-\gamma_2$, $\gamma_1-\gamma_2$, γ_2). Wir zählen hier die letzteren nochmals auf:

Stern	Spektrum	P	A	Lichtkurve	
T Cassiopeiae	Me	449^d	$5^m,8$	$\gamma_1-\gamma_2$	
R Centauri	,,	561	7 ,7	γ_2	
W Cygni	,,	259	1 ,1	γ_2	RV Tauri-Stern?
RT Hydrae	,,	255	2 ,2	γ_2?	RV Tauri-Stern?
R Pictoris	,,	333	2 ,8	γ_2	RV Tauri-Stern?
Z Ursae maj.	,,	198	2 ,4	γ_2	RV Tauri-Stern.
R Normae	Mb	487	6 ,2	γ_2	
RU Cygni	Mc	462	2 ,3	γ_2 pec	
R Ursae min.	Mc	332	1 ,7	$\beta_4-\gamma_2$	
V Hydrae	N	530	4 ,0	γ_2?	
RZ Cygni	pec	556	4 ,6	γ_2	RV Tauri-Stern?

Die ausgesprochensten und zugleich in ihrem Lichtwechsel regelmäßigsten Vertreter dieser Gruppe von Mira-Sternen sind R Centauri und R Normae, und man kann diese Unterklasse der Mira-Sterne daher als R Centauri-Sterne bezeichnen. Mehrere von ihnen sind der Zugehörigkeit zur RV Tauri-Klasse verdächtig, bei welcher sich die Haupt- und die Nebenminima gelegentlich vertauschen, Z Ursae majoris gehört sogar sicher zu dieser Klasse. Diese Sterne

[1]) J B A A 26, S. 162 (1916). [2]) M N 67, S. 489 (1907).

sind aber hier in die Tabellen der Mira-Sterne mit aufgenommen worden, da sich eine reinliche Scheidung zwischen Mira- und RV Tauri-Sternen nicht in allen Fällen vornehmen läßt wegen des geringen Umfanges der für manche dieser Sterne zur Verfügung stehenden Beobachtungsreihen. Die RV Tauri-Sterne werden später eingehend behandelt werden.

Im Abstiege der Lichtkurve treten Buckel oder Stufen höchstens ganz gelegentlich einmal auf, bei keinem Mira-Stern ist eine solche Erscheinung als die Regel zu bezeichnen. Die betreffenden Fälle sind in den Tabellen gekennzeichnet (z. B. R und V Bootis, S Coronae borealis, RY Ophiuchi).

Bei den glatten Lichtkurven ist der Abstieg durchschnittlich fast ausnahmslos langsamer als der Anstieg oder ebenso steil wie dieser, allenfalls auch kaum merklich steiler. Bei S Cygni (Spektr. Me) scheint der Abstieg in der Regel bedeutend steiler als der Anstieg zu sein, in einigen andern, in den Tabellen gekennzeichneten Fällen treten steile Abstiege häufig oder wenigstens manchmal auf. Bei den nicht glatten Lichtkurven (γ) sind dagegen Abstiege, die merklich steiler als die Anstiege sind, nicht selten.

Unter den Me-Sternen kommt, wie bereits erwähnt, die Kurvenform β_4 (lang ausgedehnte konstante Phase im Maximum) nicht vor, und von den Mira-Sternen der übrigen Spektralklassen sind nur die Lichtkurven von η Geminorum (Ma) und S Aurigae (N) mit β_4 bezeichnet, während mehrere andere Lichtkurven sich dieser Form nähern (β_3—β_4, β_4—γ_2). Man hat vielfach eine „η Geminorum-Klasse" von Veränderlichen unterscheiden zu müssen geglaubt. Dies ist weder notwendig noch zulässig, denn offenbar ist die Lichtkurve von η Geminorum nur ein extremer Fall einer β-Kurve, und im übrigen scheint sich η Geminorum nicht grundsätzlich von manchen anderen Angehörigen der Mira-Klasse zu unterscheiden. Ob die Veränderlichkeit der Radialgeschwindigkeit von η Geminorum (um etwa 10 km) in irgendeinem Zusammenhang mit den Helligkeitsänderungen steht, wissen wir noch nicht.

27. Die Spektra der Mira-Sterne der Spektralklasse Me. α) Vorbemerkungen. Unsere Tabellen der Mira-Sterne mit bekanntem Spektrum enthalten:

321	Sterne der Spektralklasse	Me
18	,, ,, ,,	Se
39	,, ,, ,,	K, Ma, Mb, Mc
24	,, ,, ,,	N, R
3	Sterne mit Spektrum pec,	

wobei nur die Sterne mit bekannten Perioden berücksichtigt sind. Unter den Mira-Sternen überwiegen also diejenigen mit Me-Spektrum die übrigen weit. Allerdings ist zu bemerken, daß die Mira-Sterne mit Spektrum K, Ma, Mb, Mc, N, R durchschnittlich kleinere Amplituden haben als die mit Spektrum Me, so daß ihre Perioden schwerer festzustellen sind als die der letzteren. In Wirklichkeit mögen sich also die Zahlenverhältnisse etwas verschieben, aber an der obigen Tatsache wird sich dadurch nichts ändern. Wir wollen uns also zunächst mit den Me-Spektren näher beschäftigen.

Das Me-Spektrum ist dadurch charakterisiert, daß sich über ein Absorptionsspektrum der Klasse M Emissionslinien lagern. Der neue Draper-Katalog unterscheidet bei der Klasse M drei Unterabteilungen Ma, Mb, Mc, die er wie folgt beschreibt:

"Class Ma. The spectrum is banded. The bands extending from 4762 to 4954 and from 5168 to 5445 are well marked. The change in light at $H\beta$ is much less conspicuous than at 4762. Several bright spaces are seen, such as from 4556 to 4586, and from 4657 to 4668. The lines of the G band are well separated, and line 4315,2 is very faint. Line 4226,9 is the most conspicuous absorption line. The spectrum is faint towards the end of shorter wave length, so that bands H and K are generally barely seen.

Class Mb. The edges of the absorption bands, at wave lengths 4762, 4954, 5168 and 5445 are strong and appear somewhat like bright bands. These bands fade gradually towards the edge of shorter wave length. Line 4226,9 is very wide and sometimes appears to be as intense as $H\delta$ in the spectrum of α Canis majoris. Conspicuous bright bands of equal intensity are seen from 4556 to 4586 and from 4614 to 4626. Lines 4299,4, 4300,7, and the compound line 4305,6, 4308,0 and 4309,5 are the only well marked lines remaining of the band G. On isochromatic plates, absorption bands are also seen having edges at the wave lengths 5763, 5816 and 5857, approximately.

Class Mc. The continuous spectrum is fainter, and the bright edged bands are stronger, than in Classes Ma and Mb, so that the spectrum appears to be of a fluted character, and on plates of small dispersion many of the dark lines seem to have disappeared."

Neuerdings ist man dazu übergegangen, die Unterabteilungen Ma, Mb, Mc aufzugeben und, wie bei den anderen Spektralklassen, dezimale Unterabteilungen der Klasse M einzuführen; man ist aber damit noch nicht weit gediehen. Der Draper-Katalog gibt für die Me-Sterne an, welcher der drei Klassen Ma, Mb, Mc ihr kontinuierliches Spektrum angehört.

Die Spektralklasse Me wird im Draper-Katalog, so wie es früher allgemein üblich war, mit Md bezeichnet. Diese Klasse beschreibt der Draper-Katalog mit folgenden Worten:

"Class Md. This designation is used for spectra of any division of Class M, in which at least one hydrogen line is bright. The spectra differ widely. Either $H\beta$, $H\gamma$, or $H\delta$ may be the strongest bright line, while the underlying spectrum may belong to Class Ma, Mb, or Mc. The subject is further complicated by changes in the relative intensity of the hydrogen lines and probably in the class of spectrum, connected with the variation in the light of the star . . . It is evident that no accurate subdivision of these spectra can be made until observations have been obtained at different points on the light curve."

Auf die gewöhnlichen Spektra der Klasse M näher einzugehen, ist hier nicht der Ort, da dies in einem anderen Abschnitte des ,,Handbuches der Astrophysik" geschieht. [Eine übersichtliche Zusammenstellung über das Verhalten der Linien verschiedener Elemente in den Spektren der Klassen M, N und S hat kürzlich P. W. MERRILL gegeben[1].] Wir haben uns hier zunächst nur mit den Spektren der Klasse Me zu beschäftigen.

β) Das Spektrum von o Ceti. Am eingehendsten ist von allen Me-Spektren dasjenige von Mira Ceti untersucht worden. Wir übergehen hier die ältesten Beobachtungen. Die erste genauere Untersuchung rührt von H. C. VOGEL her[2]; er stellt fest, daß, wie damals übrigens schon bekannt war, das kontinuierliche Spektrum der VOGELschen Klasse IIIa angehört, und daß die Wasserstofflinien $H\gamma$ bis $H\iota$ mit Ausnahme von $H\varepsilon$ als helle Linien erscheinen ($H\alpha$ und $H\beta$ lagen außerhalb des Bereiches der von VOGEL benutzten Aufnahmen). Er gibt ferner eine Tabelle der Wellenlängen der Absorptionslinien nebst einem Vergleich mit dem Sonnenspektrum, der natürlich im weniger brechbaren Teil des Spektrums versagt. Eine Beschreibung des Spektrums von o Ceti findet sich auch in Harv Ann 28, 1897[3]. W. W. CAMPBELL[4] hat dann das Spektrum mit einem Spektrographen von starker Dispersion aufgenommen. Die Ab-

[1] Ap J 63, S. 13 (1926) = Mt Wilson Contr 306.
[2] Sitzungsber. d. Kgl. Preuß. Akad. d. Wiss. 1896, S. 395.
[3] S. 45, 98, 108. [4] Ap J 9, S. 31 (1899).

sorptionslinien ergaben eine konstante Radialgeschwindigkeit von +62 km, die Emissionslinie $H\gamma$ ergab auf verschiedenen Platten Werte zwischen +42 km und +59 km, deren Differenzen CAMPBELL hauptsächlich wirklichen Änderungen in der Lage und dem Charakter der Linie zuschreibt. Die Linie ist nämlich keineswegs scharf und symmetrisch, sondern ziemlich breit und nach Rot hin verwaschen, nach Violett hin scharf abgegrenzt. Auf einer Aufnahme erschien die Linie dreifach. Auf allen Aufnahmen war die $H\gamma$-Emissionslinie gegen die entsprechende Absorptionslinie nach Violett verschoben. Andere Aufnahmen ergaben für die Emissionslinie $H\delta$ ganz ähnliche Resultate. Bei Abnahme der Helligkeit des kontinuierlichen Spektrums infolge der Helligkeitsabnahme des Sternes tauchten noch mehrere helle Linien auf, besonders die Fe-Linien λ 4376 und 4308. Gleichzeitig beschäftigte sich auch W. SIDGREAVES mit dem Spektrum von Mira Ceti; seine Resultate hat er in zwei Arbeiten veröffentlicht[1].

Einen großen Fortschritt in unserer Kenntnis des Spektrums von Mira Ceti brachte dann eine eingehende Untersuchung von J. STEBBINS[2]. Die von ihm benutzten Spektralaufnahmen sind mit einem am großen Lick-Refraktor angebrachten Einprismenspektrographen gewonnen worden, und zwar von Juni 1902 bis Januar 1903, während der Veränderliche vom Maximum (3^m,0) bis zum Minimum (9^m,0) abnahm. STEBBINS gibt ein ausführliches Verzeichnis der Linien des Absorptionsspektrums von λ 3936 bis λ 5568; dabei befinden sich die Köpfe von 22 Banden. Die Identifizierung der Linien mit solchen im Spektrum der Sonne erwies sich als sehr schwierig und war nur in wenigen Fällen zweifelsfrei. Die Radialgeschwindigkeit ergab sich, in guter Übereinstimmung mit CAMPBELL, zu +66 km.

Um die Absorptionslinien besser mit Linien im Sonnenspektrum identifizieren zu können, hat STEBBINS eine ältere Aufnahme des Spektrums von Mira vermessen, die mit starker Dispersion erhalten worden war. Als sicher nachgewiesen können auf Grund dieser Messungen gelten Linien des Ca, Fe, Cr, V.

Änderungen der Intensität und des Charakters konnte STEBBINS nur bei g (λ 4227) sicher nachweisen, doch waren solche Änderungen bei vielen anderen Linien wahrscheinlich.

Bei den Banden war STEBBINS zweifelhaft, ob sie als helle Banden mit scharfer Begrenzung nach Rot oder als dunkle mit scharfem Ende nach Violett aufzufassen seien (wir wissen heute, daß letztere Auffassung die richtige ist). STEBBINS maß die Wellenlängen der scharfen Grenzen, wobei er nicht zwischen beiden Auffassungen zu entscheiden brauchte. Er gibt die Wellenlängen von 21 Bandenköpfen an (der mit der kleinsten Wellenlänge liegt bei λ 4313) und vergleicht die Banden mit denen im Spektrum von fünf gewöhnlichen M-Sternen, wobei er gute Übereinstimmung der Wellenlängen findet. Identifizieren konnte er die Banden nicht, ihr Ursprung blieb vorläufig noch in Dunkel gehüllt.

Von den Emissionslinien der Balmer-Serie des H konnten auch, im Gegensatz zu früheren Beobachtungen, $H\beta$ und $H\varepsilon$ wenigstens zeitweise, wenn auch schwach, festgestellt werden. Ein Versuch, $H\alpha$ durch visuelle Beobachtungen nachzuweisen, mißlang dagegen. Außer den hellen Linien $H\beta$ bis $H\varkappa$ wurde noch eine Anzahl anderer heller Linien wahrgenommen, unter denen sich solche des Fe, Si (λ 3906), Mg (λ 4571) ziemlich sicher identifizieren ließen. Die Erscheinung, daß die Emissionslinien relativ zu den Absorptionslinien nach Violett verschoben sind, wurde bestätigt, und zwar nicht nur für die des H, sondern auch für die übrigen.

[1] M N 58, S. 344 (1898) u. 59, S. 509 (1899).

[2] Lick Bull 2, S. 78 (1903); Ap J 18, S. 341 (1903).

Die Intensitäten der hellen Linien änderten sich mit dem Lichtwechsel des Sternes; wir werden später an Hand der Beobachtungen von A. H. JOY diese Änderungen näher schildern. Besonders weist STEBBINS auf die ungewöhnlichen Intensitätsverhältnisse der hellen Linien des H hin. Namentlich zur Zeit des Maximums waren $H\delta$ und $H\gamma$ ungemein hell ($H\delta$ noch heller als $H\gamma$), während relativ zu ihnen $H\beta$ und $H\varepsilon$ sehr schwach waren.

Durch STEBBINS' Arbeit waren höchst merkwürdige Tatsachen betreffs des Spektrums von Mira Ceti festgestellt. Eine weitere Untersuchung erfolgte durch J. S. PLASKETT[1] während des sehr hellen Maximums der Mira um die Jahreswende 1906/07, bei dem der Veränderliche die zweite Größe erreichte. $H\beta$ war diesmal bedeutend heller als auf STEBBINS' Aufnahmen, während von $H\varepsilon$ keine Spur wahrzunehmen war. Aber wiederum war $H\delta$ die hellste, $H\gamma$ die zweithellste Emissionslinie. Außer den Linien des H konnten noch mehrere andere helle Linien festgestellt werden, die aber nicht sicher zu identifizieren waren.

Gleichzeitig (Dezember 1906) hat V. M. SLIPHER[2] das Spektrum von Mira Ceti auf farbenempfindlichen Platten aufgenommen; auf diesen war auch $H\alpha$ als Emissionslinie sichtbar, wenn auch schwächer als $H\beta$ bis $H\delta$. W. SIDGREAVES[3] machte Aufnahmen mit Objektivprisma. Das Linienabsorptionsspektrum erschien ihm ebenso wie 1897/98, die Banden aber waren viel schwächer als damals, was wohl der schon erwähnten großen Helligkeit des Maximums im Dezember 1906 zuzuschreiben ist. Die Emissionslinie $H\gamma$ war relativ zu $H\delta$ entschieden stärker als 1897/98, und die Serie der Linien des H war bis $H\sigma$ erkennbar ($H\varepsilon$ fehlte). A. L. CORTIE[4] hat das Spektrum während des Maximums von 1906 Dezember visuell beobachtet und hebt namentlich die große Helligkeit im roten Teil hervor.

Über das Verhalten der hellen Linien des H während der Maxima im Februar 1905 und Dezember 1906 liegen Beobachtungen des Harvard-Observatoriums vor[5].

Während so die Erforschung des Spektrums der Mira immer weitere Fortschritte machte, war es auch gelungen, den Ursprung der Banden im Spektrum dieses Sternes und der M-Sterne überhaupt festzustellen, und zwar auf Grund der Untersuchungen A. FOWLERS über das Bandenspektrum des Titanoxyds[6]. Diese Untersuchungen entschieden zunächst die noch von STEBBINS offengelassene Frage, ob es sich im Spektrum von o Ceti um helle oder um dunkle Banden handle, im Sinne der letzteren Auffassung. Es zeigte sich, daß die Banden des Titanoxyds ihrer Lage und ihrem Aussehen nach genau mit den Absorptionsbanden der Sterne der Spektralklasse M übereinstimmten. Eine spezielle Vergleichung der Banden des Titanoxyds mit denen im Spektrum von Mira Ceti hat FOWLER etwas später gegeben[7]. A. S. KING hat nachgewiesen, daß die Titanoxydbanden im elektrischen Ofen bei einer Temperatur von 1900° auftauchen[8].

Bei dem Maximum von Mira Ceti im August 1909 machte W. H. WRIGHT auf dem Lick-Observatorium den Versuch, an der hellen $H\gamma$-Linie Polarisation festzustellen[9], wie man sie erwarten konnte, falls die von CAMPBELL beobachtete Verdreifachung dieser Linie einem Zeeman-Effekt zuzuschreiben war. WRIGHT kam aber nicht zu bestimmten Resultaten.

[1] J Can R A S 1, S. 45 (1907). [2] Ap J 25, S. 66 u. 235 (1907).
[3] M N 67, S. 534 (1907). [4] M N 67, S. 537 (1907). [5] Harv Ann 56, S. 104 (1912).
[6] London R S Proc 73, S. 219 (1904); M N 64, App. S. 16 (1904); London R S Proc 79 (Series A), S. 509 (1907).
[7] M N 69, S. 508 (1909). [8] Publ A S P 34, S. 348 (1922).
[9] Lick Bull 6, S. 60 (1910).

W. S. ADAMS und A. H. JOY[1] fanden, daß bei Abnahme der Helligkeit des Veränderlichen unter den Absorptionslinien die Funkenlinien an Intensität abnahmen. Auch gaben sie[2] ein Verzeichnis der von ihnen auf einer Aufnahme vom 2. März 1918 (als der Stern schwach war) gemessenen hellen Linien.

Während des Maximums, das zu Anfang August 1919 stattfand, gelangte C. D. SHANE auf dem Lick-Observatorium zu wichtigen Ergebnissen[3]. Über seine Untersuchung der Absorptionsbanden berichtet er, wie folgt:

"The most conspicuous features in the visual portion of the spectrum are the many and prominent absorption bands, over fifty having been measured between wavelengths 5550 *A* and 6700 *A*. All of these are sharp toward the violet and fade away gradually toward the red. Most of the bands in *o* Ceti of shorter wavelength than 5840 *A*, i. e., in the yellow and blue, have been identified by FOWLER with bands of titanium oxide. In the visual region, between 5500 *A* and 6700 *A*, as observed in the present investigation, there are no fewer than thirty coincidences with the band spectrum of titanium oxide. There remain twenty bands in *o* Ceti which apparently do not belong to this substance. Six of these have been found to coincide with bands which ordinarily occur in the arc spectrum of yttrium. Of these six, three coincide respectively with the first band in each of three known groups in the yttrium band spectrum. While the origin of these bands cannot be definitely attributed to yttrium, it is quite probable that bands due to yttrium in some form are represented in *o* Ceti. The remaining fourteen bands have not been identified."

Mit der Helligkeitsabnahme wurden die Absorptionsbanden stärker, das kontinuierliche Spektrum in seinem roten Teil relativ zum brechbareren Teil intensiver. Von den hellen Wasserstofflinien (es konnten alle von $H\alpha$ bis $H\varrho$ festgestellt werden) erwiesen sich gerade die als besonders schwach ($H\varepsilon$, $H\varkappa$, $H\lambda$, $H\mu$, $H\xi$), die sehr nahe mit starken Absorptionslinien des Bogenspektrums anderer Elemente (Ca, Fe, V) zusammenfallen. SHANE erachtet es hiernach für wahrscheinlich, daß die abnorme Schwäche jener Linien des H durch die Absorption übergelagerter metallischer Dämpfe erzeugt wird, wie man es für $H\varepsilon$ schon früher vermutet hatte. Die helle $H\alpha$-Linie war gegen die übrigen hellen Linien des H um 1,2 AE nach Violett verschoben, während $H\varkappa$ im Juli und August 1919 um mehr als 1 AE gegen die übrigen Linien des H nach Rot verschoben war, sich dagegen im September und Oktober in normaler Lage befand.

W. S. ADAMS und A. H. JOY[4] beschrieben auf Grund von Aufnahmen auf dem Mt. Wilson-Observatorium, die mit den besprochenen von SHANE gleichzeitig waren, die Änderungen im Spektrum von *o* Ceti während der Helligkeitsabnahme.

F. E. BAXANDALL[5] hat die Frage aufgeworfen, ob die Identifizierung der hellen Linien, die beim Schwächerwerden des Veränderlichen auftreten, mit den Linien niedriger Temperatur des Fe, Mg und Si stichhaltig sei, und er macht Bedenken dagegen geltend. W. S. ADAMS und A. H. JOY[6] gaben darauf eine genaue Bestimmung der Wellenlängen dieser Linien und zeigten, daß diese Wellenlängen tatsächlich mit denen der erwähnten Linien des Fe, Mg und Si sehr genau übereinstimmten.

---

[1] Publ A S P 29, S. 112 (1917). [2] Publ A S P 30, S. 193 (1918).
[3] Publ A S P 32, S. 234 (1920), ausführlicher Lick Bull 10, S. 131 (1921).
[4] Publ A S P 32, S. 163 (1920). [5] Obs 46, S. 82, 226 (1923).
[6] Publ A S P 35, S. 168 (1923). Vgl. auch MERRILL, Ap J 58, S. 195 (1923) = Mt WILSON Contr 265.

Auch E. B. FROST und Miss FR. LOWATER[1] haben auf dem Yerkes-Observatorium eine Untersuchung über das Spektrum von o Ceti angestellt.

Während des Maximums zu Beginn des Jahres 1924 wurde das Spektrum von o Ceti von W. J. S. LOCKYER[2] mit Objektivprismen aufgenommen. Er fand, daß die Reihenfolge der photographischen Helligkeiten der Emissionslinien die folgende war, wenn man mit den größten Helligkeiten beginnt ($H\varepsilon$ war nicht sichtbar):

$$H\delta,\ H\gamma,\ H\zeta,\ H\beta,\ H\eta,\ H\iota,\ H\vartheta.$$

Die Diskussion anderweitiger Beobachtungen während früherer Maxima führte LOCKYER zu der Überzeugung, daß diese Reihenfolge während aller Maxima innegehalten wird, welche Helligkeit der Veränderliche auch im Maximum erreicht.

Alle von FOWLER im Laboratorium gefundenen Titanoxydbanden, die in den Bereich der LOCKYERschen Spektralaufnahmen fallen, konnten im Spektrum von Mira Ceti festgestellt werden.

D. H. MENZEL hat auf Grund der vorliegenden Beobachtungen darauf aufmerksam gemacht[3], daß bei der Helligkeitsabnahme von o Ceti die Emissionslinien in der Reihenfolge wachsenden Ionisationspotentials erscheinen. Kein Element, welches Emissionslinien im Spektrum von o Ceti gibt, hat ein weniger als 7,4 Volt betragendes Ionisationspotential.

Zusammenfassend hat A. H. JOY[4] im Jahre 1926 über seine eingehenden Untersuchungen über das Spektrum von Mira Ceti berichtet. Seine Beobachtungen sind besonders dadurch wertvoll, daß sie sich über alle Phasen der Lichtkurve erstrecken und so ein vollständiges Bild von den Änderungen des Spektrums geben. Wir wollen die Hauptergebnisse JOYS hier kurz anführen.

Das kontinuierliche Spektrum von Mira Ceti ist im Helligkeitsmaximum durchschnittlich M5, im Minimum M9, und zwar hängt es durchschnittlich in folgender Weise von der Helligkeit des Sternes ab:

Mittlere Größe	Mittlerer Spektraltypus
$3^m,1$	M 5,2
3 ,7	M 6,3
4 ,6	M 6,9
5 ,5	M 7,6
6 ,4	M 8,0
7 ,7	M 8,4
8 ,9	M 9,0

Mehrere neue Absorptionsbanden wurden von JOY aufgefunden, die sich, mit Ausnahme von zweien, mit solchen des Titanoxyds identifizieren ließen. Mit abnehmender Helligkeit des Sternes nehmen die Absorptionsbanden an Intensität zu. Es konnten in den Banden einzelne Linien gemessen werden, die dieselbe Radialgeschwindigkeit ergaben wie die übrigen Absorptionslinien.

Die Absorptionslinien, soweit sie bei der angewandten ziemlich geringen Dispersion gut meßbar waren, sind mit wenigen Ausnahmen Linien niedriger Temperatur des Fe, V, Cr, Mn, Ca, Mg. Die Linien des V sind bemerkenswert durch ihre Zahl und ihre Beständigkeit; sie sind die einzigen Absorptionslinien mäßiger Intensität, welche nahe dem Helligkeitsminimum gemessen werden können. Die Linien des Ti sind auffallend schwach. JOY gibt ein Verzeichnis von 44 identifizierten Absorptionslinien, die den obengenannten Elementen sowie dem Ti, Sr und Ba angehören; doch ließen sich auch noch Linien des Co und Na nachweisen. Wenn der Stern an Helligkeit abnimmt, verschwinden die schwächeren Absorptionslinien, und einige verwandeln sich in Emissionslinien. Die Ca-Linie λ 4226 wird im Helligkeitsminimum 30 AE breit, und auch die Cr-Linien λ 4254 und λ 4274 verbreitern sich außerordentlich.

---

[1] Ap J 58, S. 265 (1923). [2] M N 84, S. 558 (1924). [3] Harv Circ 258 (1924).
[4] Ap J 63, S. 281 (1926) = Mt Wilson Contr 311.

Die Absorptionslinien ergeben eine veränderliche Radialgeschwindigkeit. Die Geschwindigkeitskurve hat dieselbe Form wie die Lichtkurve, das positive Maximum der Geschwindigkeitskurve fällt mit dem Helligkeitsmaximum, das negative Maximum mit dem Helligkeitsminimum zusammen, so daß die Beziehungen zwischen Geschwindigkeits- und Lichtkurve gerade umgekehrt wie bei den δ Cephei-Sternen sind. Die Geschwindigkeitskurve läßt sich durch folgende spektroskopische Bahnelemente darstellen:

$$P = 330^{\mathrm{d}} \qquad e = 0{,}20$$
$$K = 5{,}9\ \mathrm{km} \qquad a \sin i = 26200000\ \mathrm{km}$$
$$\gamma = +58{,}2\ \mathrm{km} \qquad \frac{m_2^3 \sin^3 i}{(m_1 + m_2)^2} = 0{,}007\odot.$$
$$\omega = 265^\circ{,}2$$

Daß den früheren Beobachtern diese immerhin nicht sehr große Veränderlichkeit der Radialgeschwindigkeit entgangen ist, erklärt sich dadurch, daß ihre Messungen ganz vorwiegend nur in der Nähe des Helligkeitsmaximums angestellt sind.

Von den Wasserstofflinien wurden $H\alpha$ bis $H\iota$ als Emissionslinien beobachtet. Im Helligkeitsminimum ist keine der Wasserstofflinien hell. Wenn der Stern im Helligkeitsaufstieg ungefähr die siebente Größe erreicht, tritt plötzlich $H\delta$ und kurz darauf $H\gamma$ in Erscheinung; ihre größte Helligkeit erreichen diese Linien etwa einen Monat nach dem Helligkeitsmaximum, und zwar ist $H\delta$ während dieser ganzen Zeit heller als $H\gamma$. Dann nehmen sie an Intensität ab, wobei $H\gamma$ etwas heller als $H\delta$ wird. Wenn der Stern in der Abnahme die achte Größe erreicht, verschwinden die Emissionslinien $H\gamma$ und $H\delta$, und nahe dem Minimum sind $H\gamma$ und $H\delta$ als Absorptionslinien sichtbar.

$H\beta$ und $H\zeta$ erscheinen um die Zeit des Maximums und verschwinden nach 3 oder 4 Monaten, wenn der Veränderliche die Größe 6,5 hat. $H\varepsilon$ und $H\eta$ erscheinen etwa einen Monat nach dem Maximum und verschwinden gleichzeitig mit $H\beta$ und $H\zeta$. Für weitere Einzelheiten müssen wir auf die Abhandlung von JOY verweisen.

Auf Platten mit starker Dispersion weisen $H\gamma$ und $H\delta$ eine komplizierte Struktur auf, indem sie in 3 Komponenten zerfallen (vgl. W. W. CAMPBELLS früher besprochene Beobachtungen).

Außer den hellen Linien des H treten noch viele andere Emissionslinien auf; im ganzen konnten, einschließlich der des H, 49 Emissionslinien zweifelsfrei festgestellt werden, die, soweit sie sich sicher identifizieren ließen, dem H, Fe, Si, Mn, Mg, In (?) angehören. (Dazu kommt noch eine Anzahl von anderen Beobachtern gesehener Linien.) Man kann unter ihnen zwei Gruppen unterscheiden. Die erste Gruppe umfaßt die Linien des H und die Funkenlinien des Fe; diese sind am stärksten bald nach dem Helligkeitsmaximum. Die zweite Gruppe umfaßt Linien niedriger Temperatur des Fe, Mg, Mn, In; sie haben ihre größte Intensität 4 bis 6 Monate nach dem Helligkeitsmaximum. Die Linien des Si zeigen kein ausgesprochenes Maximum der Helligkeit.

Im Minimum oder gleich danach verschwinden alle Emissionslinien, so daß während eines Zeitraumes von 20^{d} bis 40^{d} bald nach dem Minimum keine Emissionslinien vorhanden sind (abgesehen von denen, die dem Spektrum des Begleiters angehören; s. unten).

Aus den Verschiebungen der Emissionslinien des H und einiger Emissionslinien niedriger Temperatur des Fe und Mg hat JOY die entsprechenden Radialgeschwindigkeiten bestimmt. Auch diese ergeben sich als veränderlich. Die Emissionslinien sind gegen die Absorptionslinien nach Violett verschoben, nur zur Zeit des Helligkeitsminimums ergeben beide Arten von Linien dieselbe Radialgeschwindigkeit. Die Differenzen Δ der Radialgeschwindigkeiten (im Sinne Absorptions- minus Emissionslinien) für die verschiedenen Phasen (Helligkeitsmaximum bei Phase 0^d, Minimum bei Phase 215^d) sind in folgender Tabelle gegeben:

Phase	Δ	Phase	Δ
16^d	$+16,7$ km	148^d	$+6,1$ km
49	19,1	163	0,9
78	17,9	194	0,3
96	17,4	297	14,0
116	12,2	327	16,1
134	7,7		

Die Tabelle hat eine Lücke zwischen den Phasen 194^d und 297^d, da zu dieser Zeit die Emissionslinien ganz fehlen oder schwach sind.

Die Kurve, die den Verlauf der den Emissionslinien entsprechenden Radialgeschwindigkeiten darstellt, hat einen wesentlich anderen Verlauf als die Geschwindigkeitskurve, die den Absorptionslinien entspricht. Ihr positives Maximum hat sie ungefähr zur Zeit des Helligkeitsminimums, ihr negatives Maximum ungefähr bei der Phase 90^d. Die Geschwindigkeitskurven (sowohl die den Absorptions-, wie die den Emissionslinien entsprechenden) scheinen in verschiedenen Perioden unverändert zu sein.

Auf weitere von JOY behandelte Einzelheiten über das Spektrum von Mira Ceti können wir hier nicht eingehen. Wir erwähnen nur noch, daß die Intensitäten der Emissionslinien zur Zeit der Helligkeitsmaxima des Veränderlichen von der Helligkeit der einzelnen Maxima abhängen, und ferner, daß in dem ungewöhnlich schwachen Maximum von 1924 Februar sonst nicht beobachtete helle Linien oder Banden erschienen und ebenso auch unbekannte Absorptionsbanden, die vielleicht dem Magnesiumhydrid angehörten.

Eine genauere Untersuchung führt JOY zu der Überzeugung, daß die großen visuellen Helligkeitsschwankungen von Mira Ceti zu erklären sind durch die geringen, etwa eine Größenklasse (vgl. Ziff. 31) betragenden radiometrischen Änderungen in Verbindung mit den Änderungen der Intensität der Absorptionsbanden des Titanoxyds.

Wir haben, wie bereits erwähnt, durch die soeben skizzierten Untersuchungen von JOY ein recht vollständiges Bild der Änderungen des Spektrums von Mira Ceti während des Lichtwechsels gewonnen. Die wichtigste Folgerung, die wir aus diesen und den früheren Untersuchungen des Spektrums dieses Sternes ziehen können, ist die, daß die Helligkeitsabnahme des Veränderlichen mit einer Abnahme der Temperatur verbunden ist.

Im Minimum 1920 Januar von Mira Ceti beobachteten ADAMS und JOY[1] das Auftreten breiter, heller Linien des H und He, und dieselben Erscheinungen nahmen sie auch 1921[2] und später wahr. A. H. JOY[3] kam zu der Vermutung, daß Mira einen Begleiter haben müsse, dessen Spektrum in der Regel von dem des Veränderlichen überstrahlt wird und nur dann hervortritt, wenn der Veränderliche sich im Minimum seiner Helligkeit befindet. E. E. BARNARD konnte 1921 mit dem großen Refraktor des Yerkes-Observatoriums diesen Begleiter nicht auffinden. Dies gelang erst R. G. AITKEN[4] mit dem großen Lick-Refraktor im Oktober 1923. Der Begleiter war damals eine halbe bis dreiviertel Größenklasse schwächer als der Veränderliche, der sich im Minimum ($9^m,2$) befand. AITKENS Messungen ergaben für Positionswinkel und Distanz des Begleiters:

$$\begin{array}{lll} 1923,84 & \vartheta = 130^\circ,3 & \varrho = 0'',91, \\ 1924,69 & 131,6 & 0,85. \end{array}$$

[1] Publ A S P 32, S. 163 (1920). [2] Publ A S P 33, S. 107 (1921).
[3] Pop Astr 31, S. 237 (1923).
[4] Publ A S P 35, S. 323 (1923); Harv Bull 792 (1923); Publ A S P 36, S. 296 (1924).

1925 war der Begleiter merklich schwächer und 1926 konnte AITKEN[1] ihn nicht mehr mit Sicherheit feststellen. JOY glaubt als sicher annehmen zu dürfen, daß der Begleiter gleichfalls veränderlich ist. Vielleicht besitzt er außerdem eine rasche Bewegung relativ zum Hauptstern.

Das Spektrum des Begleiters hat JOY 1924[2] und in seiner oben besprochenen großen Abhandlung über Mira Ceti beschrieben. Die Wasserstofflinien ($H\alpha$ bis $H\varepsilon$ wurden beobachtet) bestehen aus je zwei hellen Komponenten mit einer Absorptionslinie dazwischen. Die brechbareren Komponenten haben veränderliche Helligkeit. Außerdem sind breite helle Banden des He und des Ca (H und K) vorhanden, sowie scharfe, schwache Emissionslinien des ionisierten Fe. Absorptionslinien sind nicht sichtbar außer den erwähnten des H. Das starke kontinuierliche Spektrum gleicht in seiner Intensitätsverteilung dem eines Sternes der Spektralklasse B8.

Aus den Fe-Linien ergibt sich eine Radialgeschwindigkeit von $+51$ km, die der Radialgeschwindigkeit von Mira Ceti im Helligkeitsminimum sehr nahe gleich ist.

Wenn man die absolute Größe von Mira Ceti im normalen Maximum zu $-0^{m},3$ annimmt, so folgt für die absolute Größe des Begleiters $+6^{m}$; diese Helligkeit ist für einen B-Stern außerordentlich gering.

γ) Weitere Spezialuntersuchungen über die Spektra einzelner Me-Sterne. Eine Untersuchung über das Spektrum von χ Cygni auf Grund von in Potsdam gewonnenen Aufnahmen wurde von G. EBERHARD bereits 1902 veröffentlicht[3]. Die Aufnahmen erstrecken sich von Anfang August (etwas vor dem Maximum) bis Ende November 1901, als der Stern etwa die Größe 9,0 erreicht hatte. Die hellen Linien $H\gamma$ und $H\delta$ waren, wie bei o Ceti, nach Violett scharf begrenzt, nach Rot verwaschen. Aus den Messungen von $H\gamma$ ergab sich während des ganzen Zeitintervalls eine konstante Radialgeschwindigkeit von etwa -20 km, in Übereinstimmung mit der hellen Fe-Linie λ 4308, die anfangs unsichtbar war und erst allmählich auftauchte. Die Emissionslinien zeigten, wie bei o Ceti, eine Verschiebung nach Violett gegen die entsprechenden Absorptionslinien. Aus letzteren, die übrigens nur auf zwei Platten gemessen werden konnten, folgte eine Radialgeschwindigkeit von $+2,5$ km. Es ließen sich Absorptionslinien des Fe, Ca, Ti, Cr, V, Si und H identifizieren. Auf früheren Aufnahmen mit schwacher Dispersion waren die hellen Wasserstofflinien von $H\gamma$ bis $H\iota$ (außer $H\varepsilon$) sichtbar, ferner noch die Si-Linie λ 3906.

In einer zweiten Abhandlung[4] hat EBERHARD seine Untersuchungen weitergeführt, indem er noch Aufnahmen aus der Zeit von September bis Dezember 1902 verwertet. Die helle Fe-Linie λ 4308 erschien diesmal erst gegen Schluß der Beobachtungsreihe. $H\gamma$ ergab dieselbe Radialgeschwindigkeit wie 1901. Die Absorptionslinien ergaben auf zwei Platten die Radialgeschwindigkeit $-2,3$ km.

Von diesen beiden Arbeiten EBERHARDS über χ Cygni liegt die erste vor der von J. STEBBINS über Mira Ceti, während die zweite mit letzterer gleichzeitig ist. Die Übereinstimmung der Resultate für χ Cygni und o Ceti ließ schon damals schließen, daß die beobachteten Erscheinungen für die Veränderlichen der Klasse Me typisch seien.

Nach P. W. MERRILL[5] weist das Spektrum von χ Cygni gewisse Einzelheiten auf, die an die Spektren der Klasse Se erinnern.

---

[1] Publ A S P 38, S. 334 (1926). [2] Publ A S P 36, S. 290 (1924).
[3] A N 157, S. 341 (1902). [4] Ap J 18, S. 198 (1903).
[5] Publ A S P 38, S. 329 (1926).

R. H. CURTISS[1] beobachtete 1903 bei W Cygni das Schwächerwerden der hellen Linien des H während der Helligkeitsabnahme des Sternes; gleichzeitig nahm die Absorptionslinie λ 4226 des Ca an Breite zu. W Cygni zeigte also dasselbe Verhalten wie *o* Ceti. Dies ist um so interessanter, als dieser Veränderliche einen für einen Me-Stern ziemlich abnormen Lichtwechsel hat und im Verdacht der Zugehörigkeit zur RV Tauri-Klasse steht. Übrigens scheinen bei diesem Stern die Emissionslinien nicht in jedem Maximum aufzutreten[2].

δ) Die Nebellinien im Spektrum von R Aquarii. Im Oktober 1919 fand P. W. MERRILL[3] auf dem Mt. Wilson-Observatorium gelegentlich seiner systematischen Aufnahmen der Spektren der Klasse Me, daß im Spektrum von R Aquarii, der sich damals im schnellen Helligkeitsaufstieg befand, nicht nur die Linien des Wasserstoffs, sondern auch die charakteristischen Linien des Spektrums der Gasnebel hell waren. Diese Entdeckung wurde an anderen Orten bestätigt[4]. Später hat MERRILL über seine nähere Untersuchung des Spektrums ausführliche Mitteilungen gemacht[5].

Das Spektrum von R Aquarii stellt sich dar als ein gewöhnliches Me-Spektrum, über welches das charakteristische Spektrum der Gasnebel gelagert ist. Um das Helligkeitsmaximum waren, wie es bei den Mira-Sternen die Regel ist, die Wasserstofflinien $H\beta$, $H\gamma$ und $H\delta$ hell; während der Helligkeitsabnahme des Sternes wurden sie, wie es gleichfalls die Regel ist, immer schwächer. Außer den hellen Linien $H\beta$, $H\gamma$, $H\delta$ des Me-Spektrums sind auch dieselben hellen H-Linien des Nebelspektrums vorhanden, und erstere sind, wie gegen das Absorptionsspektrum, so auch gegen letztere etwas nach Violett verschoben. Im Maximum der Helligkeit sind die H-Linien des Nebelspektrums viel schwächer als die des Me-Spektrums und werden von diesen verdeckt; sie nehmen aber nicht mit dem Stern zugleich an Helligkeit ab, so daß sie nahe dem Helligkeitsminimum des Sterns schließlich viel heller als die des Me-Spektrums sind. Überhaupt scheint die Helligkeit des Nebelspektrums nicht von der des Sternes abhängig zu sein. Außer $H\beta$, $H\gamma$ und $H\delta$ wurden noch folgende Linien in dem Nebelspektrum beobachtet: λ 4363,2 (Nebellinie), 4471,5 (He), 4658,2 (Nebellinie), 4958,9 (Nebellinie N_2), 5006,8 (Nebellinie N_1). Auf Aufnahmen des Lick-Observatoriums war auch noch die Nebellinie λ 3869 sichtbar.

Für die Radialgeschwindigkeit ergaben sich folgende Werte:

Aus den Emissions-Linien des Me-Spektrums	−33 km
„ „ Absorptions- „ „ „ „	−19 „
„ der Nebellinie λ 4363	−25 „
„ den übrigen Nebellinien	−10 „

Die hellen Linien des Me-Spektrums sind also gegen die Absorptionslinien nach Violett verschoben, wie es die Regel ist. Die Nebellinien, außer λ 4363, sind dagegen etwas nach Rot verschoben.

Später[6] hat MERRILL auch noch das Vorhandensein der Nebellinien λ 3967 und λ 4068 im Spektrum von R Aquarii festgestellt; beide waren aber sehr schwach. Ende 1926 hat der Veränderliche dann überraschenderweise ein Spektrum wie das von P Cygni (vgl. Ziff. 10) entwickelt[7].

Es lag nahe zu untersuchen, ob sich nicht auf photographischen Aufnahmen eine Nebelhülle von R Aquarii nachweisen ließe. Es gelang in der Tat LAMPLAND[8]

[1] Lick Bull 3, S. 41 (1904).
[2] Publ. of the Astron. Obs. of the Univ. of Michigan (Detroit Obs.) 2, S. 60 (1916).
[3] Harv Bull 697 (1919); Publ A S P 31, S. 305 (1919); 32, S. 247 (1920).
[4] Publ A S P 31, S. 309 (1919); A N 210, S. 61 (1919).
[5] Ap J 53, S. 375 (1921) = Mt Wilson Contr 206.
[6] Publ A S P 34, S. 134 (1922).
[7] Harv Bull 842 (1927); Publ A S P 39, S. 48 (1927).
[8] Pop Astr 30, S. 162, 618 (1922).

auf dem Lowell-Observatorium, eine solche festzustellen, in der der Veränderliche zentrisch liegt. Irgendwelche Änderungen in dieser Hülle mit dem Lichtwechsel des Sternes konnte LAMPLAND[1] nicht wahrnehmen.

R Aquarii gehört zu den Mira-Sternen mit Lichtkurven der Klasse α_1, d. h. mit langandauernder konstanter Phase im Minimum ($P = 387^d$); das Minimum im Jahre 1924 hatte allerdings keine konstante Phase, war aber immerhin auch sehr flach.

ε) Allgemeine Untersuchungen über die Me-Spektren. Die bis vor kurzem übliche Bezeichnung Md für die Spektralklasse Me wurde von Miss A. J. CANNON eingeführt[2]. In den beiden vom Harvard-Observatorium herausgegebenen Katalogen veränderlicher Sterne[3] ist aus der Kolumne für Angabe des Spektrums ersichtlich, welche Veränderliche nach dem damaligen Stande der Kenntnis der Klasse Md angehören. Ein Verzeichnis der Md-Sterne findet sich auch in der Abhandlung „Stars with Peculiar Spectra" von WILLIAMINA P. FLEMING[4]. Hier ist die Klasse Md noch in Unterklassen Md1 bis Md10 eingeteilt, welche nicht ganz zureichend wie folgt definiert werden (S. 196): "A classification was made from an examination of the continuous spectrum, the comparative brightness of the hydrogen lines being also carefully estimated, always assuming the brightness of $H\gamma$ as 10. The first class, of which R Lyncis is the typical star, shows a spectrum resembling α Tauri, and having also $H\beta$ and $H\gamma$ strongly bright and nearly equal, while $H\delta$ is barely visible. The last group, of which R Leonis is the typical star, shows a continuous spectrum, similar to $-2°3653$, or a little beyond that star in the classification. Of the bright hydrogen lines in R Leonis $H\beta$ is not seen, $H\gamma$ is barely visible, and $H\delta$ is strongly marked. The other classes form a nearly continuous sequence between extremes." S. 225 der zitierten Abhandlung werden noch einige weitere knappe Erläuterungen gegeben. Es ist daraus ersichtlich, daß bei

Md1 das kontinuierliche Spektrum Ma ist, $H\beta$ und $H\gamma$ hell, und zwar $H\beta$ heller als $H\gamma$,

Md4 $H\beta$ unsichtbar, $H\delta$ sichtbar, aber nicht so hell wie $H\gamma$,

Md6 das kontinuierliche Spektrum Mb, $H\delta$ und $H\gamma$ hell, und zwar $H\delta = H\gamma$, $H\zeta$ auch hell,

Md9 $H\delta$ heller als $H\gamma$; $H\zeta$, $H\eta$, $H\vartheta$, $H\iota$ hell,

Md10 das kontinuierliche Spektrum Mc, $H\delta$ viel heller als $H\gamma$.

(R Lyncis, dessen Spektrum hier als Md1 bezeichnet wird, wird jetzt zur Spektralklasse Se gerechnet.)

Die hier skizzierte Einteilung der Klasse Md ist nicht einwandfrei, da, wie wir gesehen haben, sich bei demselben Stern die relativen Helligkeiten der Linien des H ändern. Immerhin ist aber auch dieser Einteilung eine gewisse Bedeutung nicht abzusprechen. Die Angaben über die Spektra der Me-Sterne im Katalog der G. u. L. beruhen hauptsächlich auf der Tabelle von WILLIAMINA P. FLEMING.

Im neuen Draper-Katalog ist von den Unterklassen Md1 bis Md10 nicht mehr Gebrauch gemacht. Es ist vielmehr für jeden Md-Stern, soweit es möglich war, angegeben, welcher der drei Klassen Ma, Mb, Mc sein kontinuierliches Spektrum angehört, und außerdem sind Angaben über die Intensitätsverhältnisse der Linien des H auf den für den neuen Draper-Katalog verwandten Aufnahmen gemacht. Hauptsächlich auf den Draper-Katalog stützen sich die Angaben über die Spektra der Me-Sterne in unserer Tabelle I.

[1] Publ A S P 34, S. 218 (1922). [2] Harv Ann 28, Part II (1901).
[3] Harv Ann 48, No. III (1903); 55, Part I (1907).
[4] Harv Ann 56, No. VI (1912).

Unter Benutzung der FLEMINGschen Unterabteilungen Md1 bis Md10 hat W. GYLLENBERG[1] eine Abhängigkeit des Spektraltypus von der Periodenlänge nachgewiesen. Er findet für verschiedene Periodenintervalle die folgenden mittleren Spektraltypen (n=Zahl der Sterne): (s. nebenstehende Tabelle).

P	Mittl. Spektrum	n
$<200^d$	Md 4,5	22
$200^d - 249^d$	Md 5,1	39
250 −299	Md 5,7	53
300 −349	Md 6,6	53
$>350^d$	Md 7,1	62

Für eine ähnliche Statistik hat LUDENDORFF[2] die Angaben des neuen Draper-Katalogs über das Helligkeitsverhältnis h von $H\delta$ zu $H\gamma$ benutzt. Es ergibt sich folgende Tabelle:

h	Mae P_m	Mae n	Mbe P_m	Mbe n	Mce P_m	Mce n
<1	235^d	22	203^d	9	272^d (für <1 und =1)	5
=1	229	27	275	37		
1,1−2,9	344	10	278	23	386	6
3,0−5,0	329	9	319	31	336	16
>5,0	341	3	357	9	396	14

Man sieht, daß für jede der drei Spektralklassen die Periodenlängen mit wachsendem h zunehmen; der Gang der Zahlen ist allerdings nicht überall glatt, aber es ist auch zu bedenken, daß die h für die einzelnen Sterne veränderlich und überhaupt recht unsicher sind, und daß die einzelnen Mittelwerte P_m zum Teil nur auf wenigen Einzelwerten beruhen.

Daß im Mittel die Mae-Sterne die kürzesten, die Mce-Sterne die längsten Perioden haben, wissen wir bereits aus früheren Betrachtungen (Ziff. 23).

Neuerdings ist die Einteilung der Spektralklasse M in die drei Unterklassen Ma, Mb, Mc aufgegeben worden und man hat dafür eine dezimale Unterteilung M1 bis M10 eingeführt. Nach diesem System hat P. W. MERRILL[3] die von ihm auf ihre Radialgeschwindigkeit hin beobachteten Me-Sterne klassifiziert. Es zeigt sich natürlich auch bei dieser Art der Klassifikation eine starke Korrelation zwischen Periode und Spektralklasse.

Die Arbeiten MERRILLS über die Radialgeschwindigkeiten der Me-Sterne werden später eingehend besprochen werden. Sie enthalten auch wertvolle Angaben über die Spektren der einzelnen Veränderlichen.

Miss CANNON[4] hat die Änderungen im Spektrum langperiodischer Veränderlicher mit Hilfe der Objektivprismenaufnahmen des Harvard-Observatoriums studiert. Sie kommt hinsichtlich der Me-Sterne zu folgenden Resultaten:

1. Es ist eher die Ausnahme als die Regel, daß die $H\beta$-Linie hell ist, und wenn sie hell ist, so ist sie nie die hellste der Linien des H. (Die nähere Untersuchung hat nämlich gezeigt, daß, wenn $H\beta$ die hellste Emissionslinie ist, das kontinuierliche Spektrum nicht der Klasse M angehört, sondern der jetzt mit S bezeichneten Spektralklasse.)

2. Im Spektrum gehen parallel dem Lichtwechsel Veränderungen vor sich. In einigen Fällen erscheint $H\delta$ zuerst und wird 10- bis 20mal so hell als $H\gamma$. Im Maximum erreicht $H\gamma$ oft die Helligkeit von $H\delta$. Die relativen Intensitäten dieser beiden Linien variieren in verschiedenen Maxima desselben Sternes.

3. Die Verteilung des Lichtes im kontinuierlichen Spektrum ändert sich mit der Helligkeit, ebenso die Intensität gewisser Absorptionslinien. Z. B. war das kontinuierliche Spektrum von S Carinae im Maximum K5, im Minimum Mc.

Über die Intensitätsänderungen der Emissionslinien im Spektrum der Me-Sterne hat MERRILL eine genauere Untersuchung ausgeführt[5]. Von den Sternen R Leonis, R Hydrae, R Serpentis, X Ophiuchi, χ Cygni und T Cephei

[1] Lund Medd Série I, Nr. 90 (1918). [2] A N 228, S. 369 (1926).
[3] Ap J 58, S. 215 (1923) = Mt Wilson Contr 264. [4] Pop Astr 27, S. 527 (1919).
[5] Ap J 53, S. 185 (1921) = Mt Wilson Contr 200.

wurden Spektrogramme hergestellt, die der Zeit nach über einen beträchtlichen Teil der Periode ihres Lichtwechsels verstreut liegen, und die Intensitäten der Linien *Hδ*, *Hγ*, λ 4202 (Fe), 4308 (Fe), 4571 (Mg) wurden näher untersucht. Das Ergebnis stellt MERRILL in Abb. 16 bildlich dar. Die starke Kurve ist die mittlere Lichtkurve für *o* Ceti, R Leonis, R Hydrae, χ Cygni und T Cephei; die übrigen Kurven stellen den zeitlichen Verlauf der Intensitäten der genannten fünf Emissionslinien in einer willkürlichen Skala dar. Die Abszissen sind die

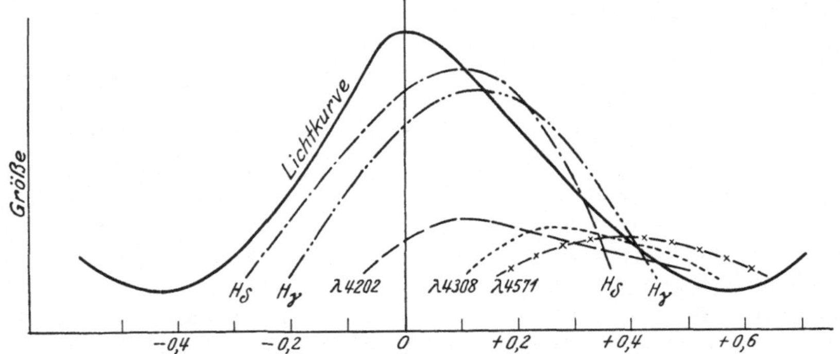

Abb. 16. Intensitätsänderungen der hellen Linien bei den Me-Sternen nach MERRILL.

vom Helligkeitsmaximum an gezählten Phasen des Lichtwechsels, ausgedrückt in Bruchteilen der Perioden. Einer weiteren Erläuterung bedarf die Abbildung nicht. Man ersieht aus ihr, wie die Linien ihre absoluten und ihre relativen Intensitäten ändern.

MERRILL macht in dieser Abhandlung noch einige interessante Nebenbemerkungen. Er hebt hervor, daß um das Helligkeitsmaximum herum die Si-Linie λ 3905,5 nächst den Linien des H die hellste Emissionslinie ist. Die Mg-Linie λ 4571 ist besonders für niedrige Temperaturen charakteristisch. Linien niedriger Temperatur sind auch die Fe-Linien λ 4202 und 4308; es ist aber eigentümlich, daß die übrigen der gleichen Temperaturklasse angehörenden Linien des Fe nicht auftreten. Auf diese merkwürdige Tatsache hat für *o* Ceti später auch BAXANDALL in einer schon zitierten Arbeit[1] hingewiesen und auf Grund derselben die Richtigkeit der Identifizierung dieser Linien bezweifelt (vgl. diese Ziffer unter β).

Ferner ist zu erwähnen, daß, im Gegensatz zu ihrem Verhalten bei *o* Ceti, die Absorptionsbanden bei X Ophiuchi bei der Helligkeitsabnahme des Sternes schwächer werden.

Später[2] hat MERRILL ein Verzeichnis von allen hellen Linien gegeben, die er auf fünf oder mehr Spektrogrammen von Me-Sternen gemessen hat. Sie gehören außer dem Wasserstoff dem Fe (λ 4202, 4308, 4376), Si (3905, 4103), Mg (4571), Mn (4031) an. Eine Anzahl weiterer heller Linien läßt sich nicht mit Sicherheit identifizieren.

Ferner hat MERRILL auch für eine größere Zahl (fast 100) Absorptionslinien in den Me-Spektren die Wellenlängen abgeleitet und die Linien nach Möglichkeit identifiziert. Es sind darunter vor allem die Linien niedriger Temperatur des Fe, V, Mn, Cr, Ca, Sr vertreten. Wie ein Vergleich mit A. S. KINGS Studien über das Verhalten dieser Linien im elektrischen Ofen lehrt, ist anzunehmen, daß die Temperatur in der umkehrenden Schicht der Me-Sterne zur Zeit des

[1] Obs 46, S. 82 (1923). [2] Ap J 58, S. 195 (1923) = Mt Wilson Contr 265.

Helligkeitsmaximums etwa 2200° ist. Wir haben früher erwähnt, daß nach KING die Titanoxydbanden bei 1900° auftreten.

Wir haben schon mehrfach hervorgehoben, daß im Spektrum der Me-Sterne die hellen Linien gegen die Absorptionslinien nach Violett verschoben sind. Näher werden wir darauf bei der Besprechung der Untersuchungen über die Radialgeschwindigkeiten der Me-Sterne eingehen.

Es geht aus den obigen Ausführungen hervor, daß das früher eingehend besprochene Spektrum von *o* Ceti bzw. sein Verhalten zu den Helligkeitsänderungen des Sternes im ganzen durchaus typisch für die Me-Sterne ist.

Nur sehr wenige M-Sterne mit hellen Wasserstofflinien sind bekannt, die nicht zur Klasse der Mira-Sterne (bzw. in einigen Fällen zur Klasse der RV Tauri-Sterne) zu rechnen sind. Eine Zusammenstellung derselben (mit Literaturangaben) liefert A. H. JOY[1]. Es handelt sich dabei um 6 Zwerg- und 6 Riesensterne. Die 6 Zwergsterne scheinen keine Veränderungen der Helligkeit aufzuweisen, von den 6 Riesensternen sind einige, darunter W Cephei, gering und unregelmäßig veränderlich, also wahrscheinlich zur Klasse der μ Cephei-Sterne zu rechnen. In den Spektren fast aller dieser Sterne zeigen sich aber doch wesentliche Verschiedenheiten gegenüber den Me-Spektren der Mira-Sterne. Eine Ausnahme bildet nur C. D. M. $-33°16843$, der ein Riesenstern konstanter Helligkeit mit typischem Me-Spektrum ist. Auf den Veränderlichen S Persei, der ein typisches Me-Spektrum besitzt und unregelmäßig veränderlich ist, haben wir schon hingewiesen.

28. Die Spektra der Mira-Sterne der Klasse Se. Schon seit einiger Zeit ist man darauf aufmerksam geworden, daß die Spektren einiger Veränderlicher, die man bis dahin zur Klasse Me rechnete, besondere Eigentümlichkeiten zeigen, z. B. das Spektrum von R Cygni[2]. Man erkannte schließlich, daß das kontinuierliche Spektrum dieser Veränderlichen dem einer kleinen Zahl von nichtveränderlichen Sternen gleicht, für welche π^1 Gruis das typische Beispiel ist. Man hat für diese Spektralklasse die Bezeichnung S eingeführt.

Die Spektren der Klasse S, und speziell die der Klasse Se, sind von P. W. MERRILL[3] eingehend untersucht worden. Die von ihm gegebene Tabelle der S-Sterne umfaßt 16 Veränderliche (alle Se) und 6 nichtveränderliche Sterne (alle S). Die Veränderlichen sind auch in unserer Tabelle II zusammengestellt; es sind dort, gegenüber MERRILLS Tabelle, noch drei Objekte dieser Art hinzugekommen.

Nach MERRILL ist die Spektralklasse S der Klasse Ma näher verwandt als irgendeiner anderen Spektralklasse. Besonders kompliziert ist das Spektrum zwischen λ 4630 und λ 4660, wo augenscheinlich Absorptionslinien und -banden sowie Emissionslinien auftreten. Der Abfall der Intensität des kontinuierlichen Spektrums ist von λ 4500 nach kleineren Wellenlängen hin stärker als bei dem Ma-Spektrum. Unter den Absorptionslinien fallen besonders λ 4554 (Ba) und λ 4607 (Sr) durch ihre Intensität auf.

Die Emissionslinien sind, wie bei den Me-Sternen, gegen die Absorptionslinien nach Violett verschoben. $H\alpha$ ist hell. $H\beta$ ist heller als $H\gamma$ (was bei den Me-Sternen nie vorzukommen scheint) und $H\gamma$ heller als $H\delta$ oder in einigen Fällen gleich $H\delta$. $H\varepsilon$ ist stets unsichtbar, $H\zeta$ war bei einem oder zwei Sternen nachzuweisen. Merkwürdig ist das Vorkommen der Funkenlinien des Fe λ 4584, 4924, 5018, die als breite Emissionen auftreten. Vielleicht ist aber diese Identifi-

[1] Ap J 63, S. 301 (1926) = Mt Wilson Contr 311, S. 21.

[2] ESPIN, M N 72, S. 546 (1912); W. H. WRIGHT, M N 72, S. 548 (1912).

[3] Ap J 56, S. 457 (1922) = Mt Wilson Contr 252. Vgl. auch Ap J 63, S. 13 (1926) = Mt Wilson Contr 306.

zierung nicht zutreffend. Auch die sonst noch vorkommenden Emissionslinien sind breit, außer den Linien des H und den beiden auch bei den Me-Sternen vorkommenden Linien λ 4511,4 und 4521,4. Diese breiten Emissionslinien lassen sich nicht identifizieren, ebensowenig wie die beiden letztgenannten Linien.

Die im weniger brechbaren Teile des Spektrums auftretenden Absorptionsbanden lassen sich mit den im Bogenspektrum des Zirkons auftretenden, wahrscheinlich dem Zirkonoxyd zuzuschreibenden Banden identifizieren, wie MERRILL nachweisen konnte[1], und zwar mit großer Gewißheit. Auch im brechbareren Teil des Spektrums stimmen zahlreiche Linien mit Bogenlinien des Zirkons überein. Die Titanoxydbanden der M-Spektren sind nicht vorhanden.

A. S. KING[2] gibt als untere Grenze der Temperatur für das Auftreten der Zirkonoxydbanden im elektrischen Ofen die Zahl 2400° bis 2500° an, während die Banden des Titanoxyds schon bei 1900° erscheinen. D. H. MENZEL[3] ist der Ansicht, daß die Temperatur der S-Sterne besonders niedrig sein müsse.

Nach MERRILLS Meinung spalten sich die S-Spektren von der gewöhnlichen Spektralfolge B A F G K M zwischen K und Ma ab.

29. Die Spektra der Mira-Sterne der Spektralklassen R und N. Für die allgemeine Beschreibung der Spektralklassen R und N muß auf den Abschnitt über die Sternspektra im „Handbuch der Astrophysik“ sowie auf den neuen Draper-Katalog verwiesen werden, für die R-Spektra ferner auf die Untersuchungen von W. C. RUFUS[4] und von R. F. SANFORD[5], für die N-Spektra auf die von C. D. SHANE[6], von J. H. MOORE[7], von R. F. SANFORD[8] und von P. W. MERRILL[9]. Es sei hier nur daran erinnert, daß die Spektralklasse R einen Übergang von G zu N bildet, und daß für N die Absorptionsbanden des Kohlenstoffes (Swan-Banden und Zyan-Banden) charakteristisch sind.

Speziell über die Spektra der wenigen R-Sterne (vgl. Tabelle IV), von denen wir wissen, daß sie zur Mira-Klasse gehören, ist nur wenig bekannt. Im Spektrum von RU Virginis sind nach dem neuen Draper-Katalog $H\beta$ und $H\gamma$ hell, und zwar an Intensität ungefähr gleich. Bei VX Geminorum (Spektrum Rp, $P = 377^d$; der Stern ist in unserer Tabelle IV nicht enthalten, da er im Hauptkatalog der G. u. L. noch nicht vorkommt) sind nach SANFORD $H\beta$, $H\gamma$ und $H\delta$ hell[10], und zwar ist $H\beta$ heller als $H\gamma$, $H\gamma$ heller als $H\delta$[11]; diese Linien sind gegen die Absorptionslinien, wie bei den Me-Sternen, nach Violett verschoben, und zwar um einen Betrag, der einer Differenz der Radialgeschwindigkeiten von 13 km entspricht. Das Spektrum von VX Geminorum hat Ähnlichkeit mit einem solchen der Klasse Se.

Das Spektrum von U Cygni (R8) ist von SHANE näher untersucht worden zusammen mit den Spektren von N-Sternen, zu denen er diesen Stern rechnet. Es handelt sich dabei um R Leporis sowie ferner um mehrere andere N-Sterne, deren Lichtwechsel gegenwärtig als unregelmäßig angesehen wird; es ist aber keineswegs ausgeschlossen, daß der Lichtwechsel dieser Sterne doch bestimmte Perioden innehält, die nur wegen der Kleinheit der Lichtschwankungen noch nicht erkannt worden sind. SHANE kommt zu folgenden Schlüssen:

---

[1] Publ A S P 35, S. 217 (1923). [2] Publ A S P 36, S. 140 (1924).
[3] Harv Circ 258 (1924).
[4] Publ. of the Astr. Obs. of the Univ. of Michigan (Detroit Obs.) 2, S. 103 (1916).
[5] Ap J 59, S. 339 (1924) = Mt Wilson Contr 276. [6] Lick Bull 10, S. 79 (1920).
[7] Lick Bull 10, S. 160 (1922). [8] Publ A S P 38, S. 177 (1926).
[9] Ap J 63, S. 13 (1926) = Mt Wilson Contr 306.
[10] Vgl. auch Harv Circ Nr. 184 (1914).
[11] Publ A S P 36, S. 351 (1924).

1. Die hellen Linien des H, die sich übrigens keineswegs bei allen beobachteten Sternen feststellen ließen, fehlen oder sind sehr schwach zur Zeit des Helligkeitsminimums des Veränderlichen. Bei der Helligkeitszunahme tauchen sie auf, zuerst $H\alpha$ und $H\beta$, dann $H\gamma$ und $H\delta$. Kurz vor dem Maximum (bei den Me-Sternen nach dem Maximum) erreichen sie ihre größte Intensität und verschwinden dann, wahrscheinlich in umgekehrter Reihenfolge, wenn der Stern wieder um eine bis zwei Größenklassen abgenommen hat.

2. Die Kohlenstoffabsorptionsbanden scheinen um das Maximum an Intensität abzunehmen. Das rote Ende des Spektrums wird bei der Helligkeitsabnahme stärker relativ zum brechbareren Ende.

3. Gewisse helle Linien unbekannten Ursprungs, namentlich eine solche bei λ 5845,3, scheinen mit dem Lichtwechsel an Intensität veränderlich zu sein.

4. Die Absorptionslinien D des Na sind um das Helligkeitsminimum sehr breit, zur Zeit des Maximums schmaler.

Die unter 1. bis 3. angeführten Tatsachen zeigen, daß die Änderungen im Spektrum der N-Sterne ganz ähnlich verlaufen wie bei den Me-Sternen. Zudem sind bei U Cygni die hellen Wasserstofflinien gegen die Absorptionslinien nach Violett um einen Betrag verschoben, der einer Differenz der Radialgeschwindigkeiten von 14 km entspricht. ($H\beta$ war heller als $H\gamma$, $H\gamma$ heller als $H\delta$; auf den Aufnahmen von SHANE war manchmal $H\gamma$ heller als $H\beta$.) Bei den anderen von MOORE untersuchten Veränderlichen der Klasse N — ganz vorwiegend solche, die man gegenwärtig als unregelmäßig ansieht, oder deren Periode man wenigstens nicht kennt — ließen sich helle Wasserstofflinien nicht wahrnehmen. Bei UV Aurigae hat P. W. MERRILL[1] sie neuerdings festgestellt.

Es kann nach den hier gemachten Ausführungen kaum zweifelhaft sein, daß die Ursache des Lichtwechsels bei den Mira-Sternen der Spektralklassen R und N dieselbe ist wie bei denen der Spektralklasse Me. Dasselbe trifft natürlich auch für die Mira-Sterne der Spektralklassen Se, sowie K und M (ohne helle Linien) zu.

30. Die Farben der Mira-Sterne. Da die Mira-Sterne Spektraltypen angehören, die tiefen Temperaturen entsprechen, so muß ihre Farbe rötlich oder rot sein. Daß dies zutrifft, ist eine längst bekannte Tatsache. Näheres über die Entwicklung der Erkenntnis, daß die Mira-Sterne rot seien, mag man in Bd. I von HAGENS Werk „Die veränderlichen Sterne", S. 727ff., nachlesen. Ebenso ist es auch schon längere Zeit bekannt, daß bei den Mira-Sternen längere Perioden mit stärkerer Rotfärbung der Sterne verbunden sind (vgl. Ziff. 7).

Im Generalkatalog der G. u. L. sind die Farben der Mira-Sterne in der OSTHOFFschen Skala gegeben (vgl. dazu Vorwort zu Bd. I der G. u. L., S. XIV). Dieses Material hat H. THOMAS in seiner schon mehrfach erwähnten Dissertation einer Bearbeitung unterzogen. Er weist zunächst nach, daß die scheinbaren Maximalhelligkeiten der Mira-Sterne einen deutlichen Zusammenhang mit der Farbe zeigen, und zwar gelangt er zu nebenstehender Tabelle:

Maximalhelligkeit	Mittl. Farbe	Anzahl
$\geqq 5^m,9$	$8^c,0$	10
$6^m,0-6^m,9$	7,8	24
7,0–7,9	7,0	74
8,0–8,9	6,4	147
9,0–9,9	5,3	83
$\leqq 10^m,0$	5,0	29

Diese Korrelation schreibt THOMAS wohl mit Recht einem physiologischen Effekt zu. Indem er nun alle Farbenangaben der G. u. L. auf die achte Größenklasse reduziert (mit Hilfe obiger Tabelle), findet er schließlich folgenden Zusammenhang zwischen Periode und Farbe:

P	Mittl. Farbe	Anzahl
$\leqq 250^d$	$5^c,8$	130
250^d-300^d	6,1	61
300–350	6,6	78
350–400	7,1	56
$>400^d$	7,8	60

[1] Publ A S P 38, S. 176 (1926).

Auf den Spektraltypus ist dabei keine Rücksicht genommen. Das Gesetz, das sich hier widerspiegelt, war bei unseren früheren Betrachtungen schon dadurch zum Ausdruck gekommen, daß die Mae-Sterne die kürzesten, die Mce-Sterne die längsten Perioden haben, und daß auch die sehr roten N-Sterne (Farbe meist 9^c oder 10^c) sehr lange Perioden besitzen.

Der Farbenindex der Spektralklasse M ist nach E. S. KING $+1^m,35$. In guter Übereinstimmung damit ergibt sich als Farbenindex von 10 Me-Sternen im Maximum ihrer Helligkeit im Mittel $+1^m,2$ (nach Table VI in Harv Ann 84, No. 4 unter Hinzuziehung von T Andromedae nach Harv Ann 80, No. 8), während man als Farbenindex für 4 Se-Sterne im Mittel $+1^m,7$ findet (P. W. MERRILL gibt als Farbenindex der S-Sterne $+1^m,8$ an). Diese Mittelwerte sind aber keineswegs sehr sicher. Für die Sterne der Spektralklasse R gibt W. C. RUFUS in seiner früher zitierten Arbeit $+1^m,7$ an, für den Farbenindex der N-Sterne fand J. A. PARKHURST $+2^m,5$. Für die extrem roten N-Sterne (Unterklasse Nc) ist der Farbenindex augenscheinlich noch viel größer; Miss H. S. LEAVITT[1] fand für S Cephei im Maximum seiner Helligkeit einen solchen von rund $+5^m$.

Gleichzeitig mit dem Lichtwechsel der Mira-Sterne gehen, wie wir gesehen haben, im Spektrum dieser Objekte Veränderungen vor sich, auf Grund derer man ein Anwachsen des Farbenindex mit Abnehmen der Helligkeit erwarten könnte. In der Tat konnte Miss LEAVITT nachweisen, daß der Farbenindex des eben erwähnten N-Sternes S Cephei mit Abnahme der Helligkeit des Veränderlichen auf mehr als $+6^m,5$ zunahm. Auch bei Mira-Sternen der Klasse Me wollen einige Beobachter ein Röterwerden mit der Helligkeitsabnahme bemerkt haben[2]. Es ist aber über die Änderung der Farbe bzw. des Farbenindex bei den Me-Sternen erst sehr wenig bekannt. Bei T Andromedae beträgt nach H. C. WILSON[3] der Farbenindex im Helligkeitsmaximum $+1^m,35$, im Helligkeitsminimum $+0^m,70$, der Stern wäre hiernach also im Maximum röter als im Minimum. Bei den in Harv Ann 84, No. 4, bearbeiteten 13 zirkumpolaren Mira-Sternen (Spektra Me und Se) sind die in Table V angegebenen Farbenindices im Helligkeitsminimum teils größer, teils kleiner als die in Table VI gegebenen Farbenindices im Maximum, und meist beträgt der Unterschied nur wenige Zehntel der Größenklasse. (Am größten, $0^m,70$, ist der Unterschied bei T Cassiopeiae, und zwar liegt er hier im selben Sinne wie bei T Andromedae.) Im Durchschnitt scheint sich also der Farbenindex bei den Mira-Sternen der Spektralklassen Me und Se nicht sehr zu ändern. Da dieses Resultat ziemlich unerwartet ist, so wären weitere und genaue Untersuchungen über diese Frage sehr erwünscht.

Da sich der Farbenindex nicht sehr zu ändern scheint, so müssen die photographischen und die visuellen Lichtkurven der Me- und Se-Sterne sehr nahe parallel verlaufen; daß dies der Fall ist, lehrt die graphische Darstellung solcher Kurven in Harv Ann 84, No. 1. (Vgl. ferner z. B. auch die photographische und die visuelle Lichtkurve von T Columbae in Harv Bull 837, von S Sculptoris in Harv Bull 838 und von R Chamaeleontis in Harv Bull 842.)

31. Die Temperaturen der Mira-Sterne. Aus den Eigenschaften der Spektra der Mira-Sterne muß der Schluß gezogen werden, daß die Temperaturen dieser Objekte ziemlich tief sind. In der Tat nimmt man ja für gewöhnliche M-Sterne eine effektive Temperatur von etwa $3000°$ an.

---

[1] Harv Circ 188 (1915).

[2] Näheres darüber bei HAGEN, Die veränderlichen Sterne, I, S. 742.

[3] Harv Ann 80, No. 8 (1917).

Zu sehr interessanten Resultaten über die Temperaturen der Me-Sterne sind S. B. NICHOLSON und E. PETTIT auf dem Mt. Wilson-Observatorium auf Grund von Messungen mit der Vakuumthermosäule gelangt[1]. Ihre wichtigsten Ergebnisse sind in folgender Tabelle zusammengestellt:

Sämtliche Zahlen in dieser Tabelle außer den Temperaturen sind in Größenklassen ausgedrückt. Der Wärmeindex ist die Differenz zwischen visueller und radiometrischer Größe, wobei letztere definiert ist als Größe eines A0-Sternes, der dieselbe Gesamtstrahlung ergibt wie der in Rede stehende Stern. In der Kolumne „Visuelle Größe" ist diese für die Zeit der Beobachtung gegeben; die nächste Kolumne enthält die genäherten Helligkeitsgrenzen für die betreffenden Veränderlichen. Die Temperaturen sind aus den Wärmeindices unter gewissen Annahmen berechnet, auf die hier nicht näher eingegangen werden kann, und zu denen auch die recht zweifelhafte gehört, daß diese Sterne wie schwarze Körper strahlen. Es zeigt sich nun auf Grund der Tabelle, daß nahe dem Minimum R Leonis, R Hydrae und *o* Ceti effektive Temperaturen von 1700° haben, während bei den nahe dem Maximum beobachteten Sternen R Cancri, X Ophiuchi und *o* Ceti die Temperaturen etwa 1900° bis 2000°, bei RT Cygni sogar nahezu 4000° betragen. (Das Studium der Spektra hatte [vgl. Ziff. 27] zur Annahme einer Temperatur von 2200 °im Helligkeitsmaximum geführt.) Bei R Aquilae entspricht einer Helligkeitsänderung von $3^m,3$ eine Temperaturänderung von 450°, während die entsprechenden Zahlen bei *o* Ceti $2^m,4$ und 130° sind. Der aus den Änderungen in den Spektren gezogene Schluß, daß bei den Me-Sternen die Temperatur im Helligkeitsminimum niedriger ist als im Helligkeitsmaximum, wird hier bestätigt. Im allgemeinen scheinen die Temperaturen auch im Helligkeitsmaximum niedriger zu sein als bei den gewöhnlichen M-Sternen; allerdings darf man die Unsicherheiten, denen solche Temperaturberechnungen unterliegen, nicht vergessen.

	Wärmeindex	Visuelle Größe	Visuelle Amplitude	Temperatur
RT Cygni	1,5	6,8	7—12	3960°
χ ,,	8,0	9,1	4—13	1690
R Cancri	6,2	8,3	7—12	1910
X Ophiuchi	5,6	7,3	7— 9	2040
,,	6,1	8,0	—	1940
R Leonis	8,3	9,4	6—10	1650
R Hydrae	8,0	9,0	5—10	1710
R Aquilae	6,8	9,8	6—12	1810
,,	4,4	6,5	—	2260
o Ceti	7,6	9,0	2— 9	1720
,,	6,6	6,6	—	1850

Man sieht ferner, daß sich die radiometrische Größe, also die Gesamtstrahlung, weit weniger ändert als die visuelle Helligkeit. Für die zweimal beobachteten Sterne ergibt sich:

	Radiometr. Größe	Visuelle Größe
X Ophiuchi	1,7	7,3
,,	1,9	8,0
R Aquilae	3,0	9,8
,,	2,1	6,5
o Ceti	1,4	9,0
,,	0,0	6,6

Spätere Messungen von E. PETTIT und S. B. NICHOLSON[2] an χ Cygni ergaben, daß mit Anwachsen der visuellen Größe des Sternes von 12^m auf $4^m,3$, also auf ungefähr das 1300fache, die Gesamtstrahlung nur auf das 1,7fache anwuchs. Für *o* Ceti fanden sie[3] als durchschnittliche Temperaturen im Maximum und Minimum 2300° und 1800° und eine Änderung der Gesamtstrahlung um $1^m,1$.

Daß die roten Sterne radiometrisch überraschend hell sind, ist eine bekannte Tatsache. Die Me-Sterne scheinen besonders große Wärmeindices zu haben,

[1] Publ A S P 34, S. 132, 181, 290 (1920); Pop Astr 31, S. 18 (1923).
[2] Annual Report of the Director of the Mt Wilson Obs., 1923, S. 205.
[3] Ap J 63, S. 330 (1926) = Mt Wilson Contr 311, S. 50.

denn für gewöhnliche M-Sterne liegen letztere nach NICHOLSON und PETTIT zwischen $2^m,4$ und $4^m,2$. Die Gesamtstrahlung von χ Cygni ist, wenn dieser Stern die visuelle Größe 12^m hat, ebenso stark wie die von Regulus (B8, Größe $1^m,3$).

Für zwei N-Sterne (X Cancri und U Hydrae) haben NICHOLSON und PETTIT Wärmeindices gefunden ($3^m,2$ bzw. $2^m,9$), die auffallend klein sind.

Kolorimetrisch sind χ Cygni und o Ceti von J. HOPMANN in Bonn beobachtet worden[1]. Bei beiden Sternen kommt er zu dem Schluß, daß die großen Helligkeitsschwankungen durch Änderungen der effektiven Oberflächentemperatur hervorgerufen werden, während die Gesamtstrahlung nur geringen Schwankungen unterliegt, die einen anderen zeitlichen Verlauf haben, etwa so, wie sich die visuellen Lichtkurven von Periode zu Periode ändern. Als Maximaltemperatur bzw. Minimaltemperatur ergaben sich für o Ceti die Werte etwa 3800° bzw. $<2000°$, für χ Cygni 2600° bzw. $<1800°$. HOPMANNS Ergebnisse stimmen also mit denen von PETTIT und NICHOLSON leidlich überein, wenn auch die Maximaltemperaturen höher sind als die von letzteren gefundenen. HOPMANN[2] hat auch Veränderliche der Spektralklasse N (Mira-Sterne und unregelmäßige) kolorimetrisch beobachtet; die von ihm aus diesen Messungen berechneten Temperaturen liegen zwischen 2000° und 3000°, für S Cephei (Nc) findet er 1400°.

32. Die Radialgeschwindigkeiten der Mira-Sterne. Die Radialgeschwindigkeiten der den Spektralklassen K, M (ohne helle Linien), R und N angehörenden Mira-Sterne sind nur in ganz vereinzelten Fällen bekannt, und es ist ganz unmöglich, aus diesen vereinzelten Werten irgendwelche Schlüsse zu ziehen. Erwähnt soll nur werden, daß die Radialgeschwindigkeit von η Geminorum (Spektrum Ma) veränderlich ist[3], und zwar, soweit wir bisher wissen, zwischen den Grenzen +14 km und +25 km; es entzieht sich aber noch unserer Kenntnis, ob die Änderungen der Radialgeschwindigkeit in irgendeiner Beziehung zu dem Lichtwechsel dieses Sternes stehen.

Im folgenden haben wir uns also ausschließlich mit den Radialgeschwindigkeiten der Mira-Sterne der Spektralklassen Me und Se zu beschäftigen. Schon gelegentlich der Besprechung der Untersuchungen über die Spektra der Me-Sterne haben wir einige Untersuchungen über die Radialgeschwindigkeiten solcher Sterne, namentlich von o Ceti, erwähnt, und wir haben vor allem gesehen, daß die Emissionslinien relativ zu den Absorptionslinien nach Violett verschoben sind. Des weiteren brauchen wir hier solche vereinzelten Bestimmungen von Radialgeschwindigkeiten nicht anzuführen, da sie von P. W. MERRILL bei seiner Gesamtdiskussion über die Radialbewegung der Mira-Sterne zusammengestellt und benutzt worden sind. MERRILL hat jahrelang, zuerst auf dem Detroit Observatory, dann auf dem Mt. Wilson Observatory, die Radialgeschwindigkeiten dieser Sterne beobachtet, und unsere heutigen, schon ziemlich eingehenden Kenntnisse darüber sind in allererster Linie ihm zu verdanken.

MERRILL hat die Resultate seiner Forschungen in drei Arbeiten veröffentlicht[4]. In der ersten sind Radialgeschwindigkeiten von 24 Mira-Sternen gegeben, die aber meist nur auf Messungen der Emissionslinien des H beruhen. Er weist darauf hin, daß, wie es für o Ceti und χ Cygni schon aus den Untersuchungen von W. W. CAMPBELL und von G. EBERHARD bekannt war, auch bei den anderen Sternen, bei denen Absorptionslinien gemessen werden konnten, nämlich bei R Leonis, R Serpentis und R Cassiopeiae, die hellen Linien gegen die dunklen

---

[1] A N 222, S. 237 (1924); 226, S. 1 (1925); 227, S. 257 (1926).
[2] A N 226, S. 225 (1925). [3] Ap J 16, S. 114 (1902).
[4] Ap J 41, S. 247 (1915); Publ. of the Astr. Obs. of the Univ. of Michigan (Detroit Obs.) 2, S. 45 (1916); Ap J 58, S. 215 (1923) = Mt Wilson Contr 264.

nach Violett verschoben sind. In der zweiten zitierten Arbeit bringt MERRILL bereits ein viel größeres Material bei, nämlich Messungen von Emissionslinien an 43 Mira-Sternen und von Absorptionslinien an 13 von diesen. Die Aufnahmen sind fast ausnahmslos in der Nähe der Helligkeitsmaxima gewonnen. Änderungen der Radialgeschwindigkeit ließen sich in keinem Falle nachweisen. Die Emissionslinien sind in allen beobachteten Fällen gegen die Absorptionslinien nach Violett verschoben. Die von der Sonnenbewegung befreiten Radialbewegungen sind sehr groß (Näheres darüber später). Die weiteren Schlüsse, die MERRILL damals zog, können hier übergangen werden, da seine dritte Arbeit ein festeres Fundament für solche Folgerungen liefert und diese daher besser bei Besprechung dieser dritten Arbeit erörtert werden.

Im Anschluß an die zweite MERRILLsche Abhandlung machte H. LUDENDORFF[1] darauf aufmerksam, daß die von MERRILL gemessenen Verschiebungen der Emissionslinien gegen die Absorptionslinien eine starke Korrelation mit der Periodenlänge zeigen; je größer diese ist, umso größer ist auch die Verschiebung. Aus den (im ganzen 14) damals bekannten Werten dieser Verschiebung ließ sich eine provisorische Formel ableiten, welche aus der Periode die Verschiebung zu berechnen gestattet. Bringt man nun bei den Mira-Sternen, bei welchen nur die hellen Linien gemessen sind, an die aus diesen folgenden Radialgeschwindigkeiten diese errechneten Verschiebungen an, so erhält man, wenigstens genähert, die den Absorptionslinien entsprechenden Radialgeschwindigkeiten. Nach W. HEISKANEN und H. LUDENDORFF[2] zeigen auch diese, nachdem man sie von der Sonnenbewegung befreit hat, eine Korrelation mit der Periode, und zwar haben die Me-Sterne kurzer Periode durchschnittlich größere Radialgeschwindigkeiten als die längerer Periode. (Zum mindesten ist zu konstatieren, daß mit wachsender Periode die Streuung der Werte der Radialgeschwindigkeit abnimmt.)

Eine große Vertiefung unserer Kenntnis von den Radialgeschwindigkeiten der Mira-Sterne brachte dann die dritte der oben zitierten Arbeiten von MERRILL. Sie gibt uns die Radialgeschwindigkeiten von im ganzen 133 Veränderlichen der Spektralklassen Me und Se. Bei 47 von ihnen sind sowohl die Absorptions- wie die Emissionslinien gemessen, bei den übrigen nur letztere. Von den Emissionslinien kamen hauptsächlich $H\beta$, $H\gamma$ und $H\delta$ in Betracht, doch wurden in manchen Fällen auch noch weitere Linien des H, sowie Emissionslinien anderer Elemente gemessen. Was zunächst die aus den Verschiebungen der Emissionslinien folgenden Radialgeschwindigkeiten V_e angeht, so scheinen diese bei demselben Stern in verschiedenen Maxima konstant zu sein. Dagegen zeigen sich bei einigen Sternen Änderungen von V_e, die von der Phase des Lichtwechsels abhängig sind. In Abb. 17 sind die auffälligsten Beispiele dieser Art graphisch dargestellt. Die Ordinaten sind die V_e, die Abszissen die Abstände vom Helligkeitsmaximum in Tagen; Kreise stellen Beobachtungen von halbem Gewicht dar. Bei R Hydrae scheint keine Abhängigkeit vorhanden zu sein, bei R Leonis, X Ophiuchi, χ Cygni und T Cephei dagegen haben die V_e bald nach dem Helligkeitsmaximum des Sternes ein negatives Maximum. Ebenso wie bei diesen Sternen scheint die Sachlage bei R Virginis und R Canis minoris zu sein, und auch bei anderen Sternen sind Andeutungen für dieselbe Erscheinung vorhanden.

Die kontinuierlichen Spektra der Sterne wurden meist nur zu Zeiten, die den Helligkeitsmaxima nahe lagen, auf den Platten erhalten. Änderungen in den aus den Messungen der Absorptionslinien folgenden Radialgeschwindigkeiten V_a ließen sich nicht feststellen.

---

[1] A N 212, S. 483 (1921). [2] A N 213, S. 297 (1921).

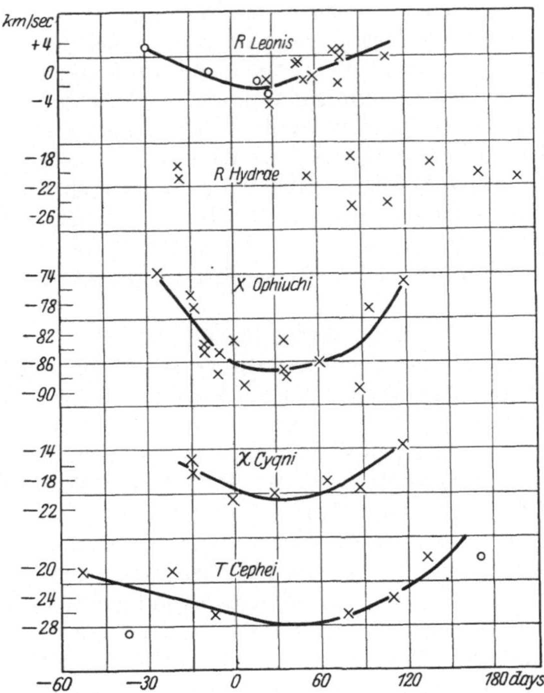

Abb. 17. Abhängigkeit der Radialgeschwindigkeit V_e von der Phase des Lichtwechsels nach MERRILL.

Die Verschiebungen der Emissionslinien gegen die Absorptionslinien, also die Differenzen $V_a - V_e$, hat MERRILL, wie erwähnt, für 47 Sterne (darunter drei der Spektralklasse Se, nämlich R Andromedae, U Cassiopeiae und R Geminorum) messen können. Wir geben in der folgenden Tabelle diese Werte an, auf ganze Kilometer abgerundet. Bei der Berechnung derselben wurden von MERRILL auch die V_e als konstant betrachtet, d. h. die Mittel aus den beobachteten V_e genommen. In der Tabelle geben wir aber nicht nur die von MERRILL ermittelten Werte von $V_a - V_e$, sondern auch noch die für 16 Me-Sterne des südlichen Himmels, welche später von LEAH B. ALLEN[1] veröffentlicht worden sind, und die MERRILL noch nicht zur Verfügung standen. Die betreffenden Veränderlichen sind in der Tabelle durch einen beigesetzten Stern gekennzeichnet.

Stern	$V_a - V_e$	Stern	$V_a - V_e$	Stern	$V_a - V_e$
R Andromedae	+28 km	R Ceti	+11 km	T Normae*	+10 km
T	+ 5	U	+12	R Octantis*	+15
W	+16	T Columbae*	+ 8	X Ophiuchi	+13
R Aquarii	+14	S Coronae bor.	+21	U Orionis	+20
T	+17	χ Cygni	+19	S Pavonis*	+ 4
R Aquilae	+ 9	W	− 1	T *	+ 5
R Arietis	+12	X Delphini	+ 7	R Persei	+11
R Bootis	+11	R Draconis	+ 6	U	+ 5
V	+11	R Geminorum	+21	R Phoenicis*	+28
R Cancri	+14	V	+11	S *	+ 5
R Canum ven.	+15	S Gruis*	+ 2	R Pictoris*	+ 8
R Carinae*	+11	RY Herculis	+11	L_2 Puppis	+ 2
S *	+15	R Horologii*	+11	RR Scorpii*	+ 7
R Cassiopeiae	+22	R Hydrae	+18	RS *	+ 4
T	+14	W	+16	S Sculptoris*	+21
U	+12	RT	+ 5	R Serpentis	+20
V	+17	S Lacertae	+ 6	R Trianguli	+ 7
R Centauri*	+ 8	R Leonis	+17	R Ursae maj.	+11
T	+ 3	R Leonis min.	+13	Z	+ 6
T Cephei	+15	W Lyrae	+ 9	R Virginis	− 1
o Ceti	+16	X Monocerotis	+10	U	+16

Es ist zu bemerken, daß S Pavonis bisher zur Spektralklasse Mc gerechnet wurde und demnach auch in unsere Tabelle III (Ziff. 22) aufgenommen ist. Vielleicht handelt es sich bei ihm um einen Stern, bei dem, wie bei W Cygni, die Emissionslinien nicht in jedem Maximum auftauchen.

[1] Lick Bull 12, S. 71 (1925).

Aus der Tabelle geht nun folgendes hervor: Nur bei zwei Sternen, W Cygni und R Virginis, ist $V_a - V_e$ negativ, nämlich $= -1$ km. W Cygni ist ein Stern mit abnormer Lichtkurve (γ_2) und gehört wohl sicher zur RV Tauri-Klasse, die hellen Linien sind bei diesem Stern nicht in jedem Maximum vorhanden; R Virginis hat eine sehr kurze Periode. Bei allen anderen Sternen der Tabelle ist $V_a - V_e$ positiv, d. h. die Emissionslinien sind gegen die Absorptionslinien nach Violett verschoben. Es zeigt sich, in Übereinstimmung mit der erwähnten Bemerkung von LUDENDORFF, eine starke Korrelation zwischen den $V_a - V_e$ und der Periodenlänge, und zwar findet MERRILL, wenn er die von ihm beobachteten Sterne nach der Größe von $V_a - V_e$ ordnet und in drei Gruppen teilt, folgende Mittelwerte für $V_a - V_e$ und die Perioden:

$V_a - V_e =$	$P =$
$+ 5{,}7$ km	219^d
12,7	314
19,0	356

Die Beziehung zwischen $V_a - V_e$ und der Periode stellt MERRILL des näheren durch die nebenstehende Abb. 18 dar, in der die $V_a - V_e$ die Ordinaten, die P die Abszissen sind. (Die kleinen Kreise bedeuten Bestimmungen von halbem

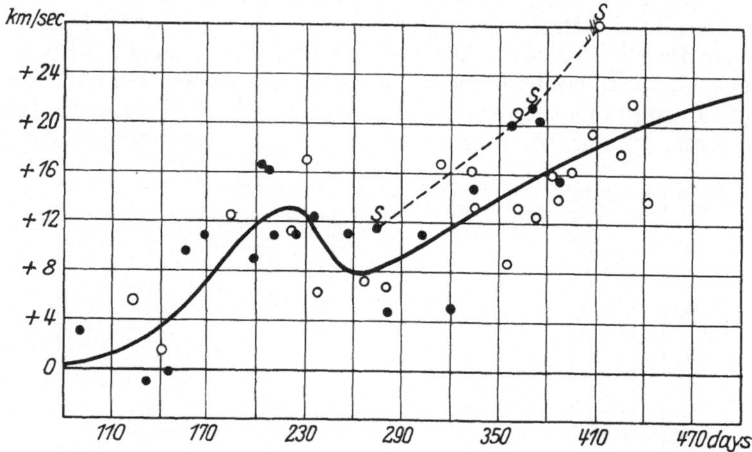

Abb. 18. Beziehung zwischen der Verschiebung der Emissionslinien und der Periode nach MERRILL.

Gewicht.) Die Kurve gibt den mittleren Verlauf der $V_a - V_e$ an. Die drei vorkommenden Sterne der Klasse Se sind durch ein beigeschriebenes S gekennzeichnet; es scheint, als ob bei ihnen $V_a - V_e$ größer ist als bei den übrigen Sternen. Eine gerade Linie scheint nicht zur Darstellung der $V_a - V_e$ als Funktion der Periode zu genügen.

Auch zu der Amplitude der Helligkeitsschwankungen und zu der spektralen Einteilung in die Unterklassen M1e, ..., M10e stehen die $V_a - V_e$ in Korrelation; dies ist selbstverständlich, da ja, wie wir wissen, diese Größen ihrerseits mit der Periode in starker Korrelation stehen. Am deutlichsten ist jedenfalls die Beziehung der $V_a - V_e$ zur Periode, und die von LEAH B. ALLEN später bestimmten $V_a - V_e$ für südliche Mira-Sterne bestätigen diese Korrelation durchaus, wenn auch einige größere Abweichungen vorkommen.

Eine unpublizierte Untersuchung von LUDENDORFF über etwaige Beziehungen der $V_a - V_e$ zur Gestalt der Lichtkurve führte nicht zu klaren Ergebnissen. Es zeigte sich indessen, daß bei Sternen mit γ-Kurven $V_a - V_e$ oft auffällig klein ist (eine Ausnahme bildet S Carinae); in besonders starkem Grade ist dies bei

R Centauri ($V_a - V_e = +8$ km bei $P = 561^d$) der Fall. Ferner treten bei Sternen mit sehr großen Helligkeitsamplituden meist größere Werte von $V_a - V_e$ auf, als man nach der MERRILLschen Kurve erwarten sollte.

Mit Hilfe der in Abb. 18 wiedergegebenen Kurve ermittelt nun MERRILL für diejenigen Sterne, bei denen nur die V_e gemessen sind, die zugehörigen V_a, so daß diese ihm schließlich für 133 Mira-Sterne zur Verfügung standen. Die V_a betrachtet er als die wirklichen Radialgeschwindigkeiten und berechnet aus ihnen die Koordinaten des Apex, die Sonnengeschwindigkeit V_0 und die Größe des K-Effektes; er findet:

$$A_0 = 287°{,}1 \quad D_0 = +41°{,}0 \quad V_0 = 55{,}0 \text{ km} \quad K = +3{,}9 \text{ km}.$$

Wenn man den sehr rasch bewegten Stern S Librae ausschließt, so wird $K = -0{,}2$ km, und auch die anderen Zahlen ändern sich. Auffällig ist der sehr große Betrag von V_0. Er läßt darauf schließen, daß die Me- und Se-Sterne eine Gruppenbewegung haben. Wie MERRILLS nähere Untersuchung zeigt, ist diese Gruppenbewegung für die stärker bewegten Sterne noch größer als für die schwächer bewegten.

Bringt man nun die nach den oben angegebenen Elementen berechnete Sonnenbewegung an die V_a an, so erhält man die „residual velocities" V'_a. Das Mittel der absoluten Beträge der V'_a ist 33,4 km, ein ungewöhnlich großer Wert, der mit dem für die planetarischen Nebel gefundenen ungefähr übereinstimmt. Die V'_a zeigen, in Übereinstimmung mit dem erwähnten Befunde von HEISKANEN und LUDENDORFF, eine starke Korrelation mit der Periode in dem Sinne, daß sehr große V'_a nur bei Sternen mit relativ kurzen Perioden vorkommen.

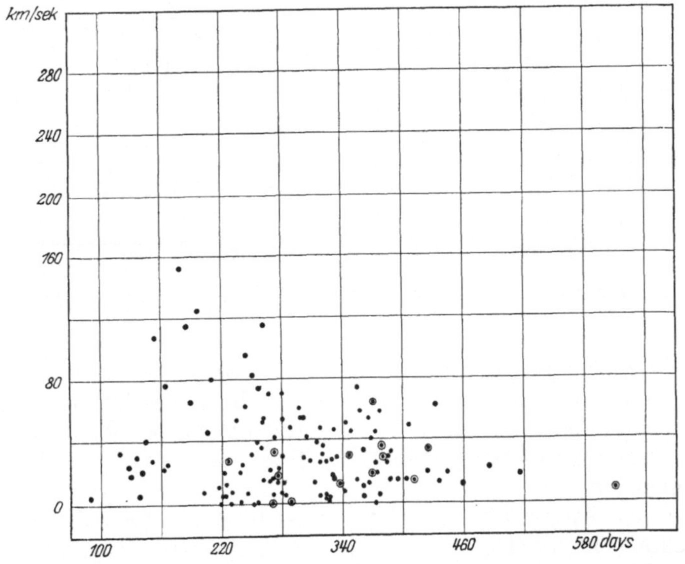

Abb. 19. Beziehung zwischen den V'_a und den Perioden nach MERRILL.

In der in Abb. 19 wiedergegebenen graphischen Darstellung MERRILLS sind die V'_a als Ordinaten eingezeichnet, während die Abszissen die Perioden sind (die kleinen Kreise beziehen sich auf die Se-Sterne). Die Abbildung läßt die erwähnte Regel deutlichst erkennen, und die späteren ergänzenden Beobachtungen von LEAH B. ALLEN stehen mit diesem Ergebnis vollkommen in Übereinstimmung.

Ferner zeigt sich nach MERRILL auch eine deutliche Beziehung zwischen den Werten V'_a und den spektralen Unterabteilungen M1e bis M10e: Je niedriger die

Ziffer dieser Unterabteilungen, desto größer ist der zugehörige Mittelwert der V'_a. Diese Korrelation ist auf Grund der zwischen Periode und Spektralklasse einerseits und Periode und V'_a andererseits bestehenden Beziehungen zu erwarten.

Wir geben hier noch ein Verzeichnis derjenigen Mira-Sterne, deren (auf die Sonne bezogene) Radialgeschwindigkeiten $V_a > 100$ km sind:

Stern	P	V_a	Stern	P	V_a
RR Aquarii	181^d	-182 km	T Herculis	165^d	-125 km
R Arietis	186	$+114$	S Librae	192	$+295$
S Carinae	149	$+289$	W Lyrae	197	-174
Z Cygni	263	-165	X Monocerotis	155	$+157$
RT	192	-115	R Pictoris	333	$+208$
R Draconis	245	-138			

Mit Ausnahme von R Pictoris sind, entsprechend der Korrelation zwischen Periode und Radialgeschwindigkeit, die Perioden aller dieser Sterne kurz. R Pictoris ist höchst wahrscheinlich ein RV Tauri-Stern[1], und es ist möglich, daß man bei diesen Objekten nicht, wie es hier geschehen ist, den Abstand von Hauptminimum zu Hauptminimum als Periode anzusehen hat, sondern die Hälfte dieses Betrages (Abstand von Haupt- zu Nebenminimum), für R Pictoris also 166^d.

Wir haben bisher die Frage unerörtert gelassen, ob denn nun in der Tat die V_a und nicht vielmehr die V_e den wirklichen Radialgeschwindigkeiten der Me- und Se-Sterne entsprechen. Man wird von vornherein ersteres für wahrscheinlicher halten. Für die erstere Anschauung spricht auch, daß, wenn man die V_e als Grundlage für eine Berechnung der Sonnenbewegung benutzt, sich ein sehr großer negativer K-Effekt ergibt ($K =$ etwa -12 km), und ferner eine nähere Untersuchung von Merrill über den Me-Stern X Ophiuchi, die in einer besonderen Abhandlung enthalten ist[2].

X Ophiuchi ist ein Doppelstern, dessen Komponenten $0'',22$ voneinander entfernt sind. Die eine Komponente hat ein Me-Spektrum und ist in einer Periode von 339^d veränderlich. Die zweite Komponente hat ein K0-Spektrum, das photographiert werden kann, wenn die erste Komponente im Minimum ihrer Helligkeit ist. Es ergeben sich nun folgende Radialgeschwindigkeiten:

aus den Absorptionslinien der ersten Komponente $V_a = -70,6$ km,
aus den Emissionslinien der ersten Komponente $V_e = -83,4$ km,
aus den (Absorptions-)Linien der zweiten Komponente $V = -70,8$ km.

Die sehr langsame Bahnbewegung kann auf die Radialgeschwindigkeiten keinen erheblichen Einfluß haben, letztere müssen also für beide Komponenten nahezu gleich sein. Diese Bedingung ist erfüllt, wenn wir als wirkliche Radialgeschwindigkeit der Me-Komponente die aus der Messung der Absorptionslinien folgende ansehen, dagegen nicht, wenn wir V_e als Radialgeschwindigkeit betrachten (V_e ist übrigens, wie schon erwähnt, etwas veränderlich, der oben angesetzte Wert ist ein Mittelwert). Der Befund bei X Ophiuchi spricht also entschieden dafür, daß die V_a als die wahren Radialgeschwindigkeiten der Me-Sterne anzusehen sind.

Nebenbei sei hier erwähnt, daß Adams auf spektroskopischem Wege die absolute Größe der K0-Komponente zu $+2^m,4$ bestimmt hat; daraus ergibt sich als absolute Größe der Me-Komponente im Maximum der Helligkeit $+0^m,3$.

Nach den Arbeiten Merrills über die Radialgeschwindigkeiten der Mira-Sterne sind noch einige andere über dieses Gebiet erschienen, nämlich eine solche

[1] A N 225, S. 249 (1925). [2] Ap J 57, S. 251 (1923) = Mt Wilson Contr 261.

von E. B. FROST und Miss FR. LOWATER[1] über die Radialgeschwindigkeiten von *o* Ceti, R Leonis, T Cephei und R Serpentis, deren Resultate aber MERRILL schon zur Verfügung gestanden haben und von ihm benutzt worden sind, und die schon zitierte von LEAH B. ALLEN, die sich auf 20 südliche Veränderliche bezieht und deren Erörterung wir schon in die der MERRILLschen Arbeiten einbezogen haben. Miss ALLEN hat, wie hier noch erwähnt werden möge, die Elemente der Sonnenbewegung aus 146 Werten von V_a neu berechnet und findet

$$A_0 = 278° \pm 7° \quad D_0 = 35° \pm 6° \quad V_0 = 57\,\text{km} \pm 6\,\text{km} \quad K = +2\,\text{km},$$

die mit den MERRILLschen Werten in befriedigender Übereinstimmung stehen.

Schließlich sind noch zwei Spezialuntersuchungen zu erwähnen. Die erste rührt von A. H. JOY[2] her und bezieht sich auf *o* Ceti; die Resultate dieser Arbeit sind bereits in Ziff. 27β) ausführlich besprochen worden. Danach sind bei *o* Ceti nicht nur die V_e, sondern auch die V_a veränderlich. Man hatte bis dahin Änderungen in den aus den Messungen der Absorptionslinien folgenden Radialgeschwindigkeiten V_a bei keinem Mira-Stern nachweisen können. Zum Teil mag dies daran gelegen haben, daß Messungen der Absorptionslinien in der Regel nur auf Spektrogrammen möglich waren, die nahe den Helligkeitsmaxima der Sterne aufgenommen worden sind, da in größerer Entfernung vom Maximum das Absorptionsspektrum sich nicht mehr mit genügender Schwärzung auf den Platten abbildet. Durch die Veränderlichkeit der V_a bei Mira Ceti wird die Frage, welchen Wert wir als wirkliche Radialgeschwindigkeit zu betrachten haben, kompliziert. Einstweilen wird man vielleicht am besten tun, den sich aus der spektroskopischen Bahn ergebenden Wert γ der Schwerpunktsgeschwindigkeit als eigentliche Radialgeschwindigkeit des Sternes anzusehen.

Die zweite der erwähnten Spezialuntersuchungen verdanken wir LEAH B. ALLEN[3], die den Stern T Centauri zum Gegenstand des Studiums gemacht hat. Dieser Stern ist dadurch besonders interessant, daß er unter allen bekannten Me-Veränderlichen die kürzeste Periode (91^d) hat. Auch in seinem spektroskopischen Verhalten ist er abnorm. Die Emissionslinien des H erreichen ihre größte Helligkeit kurz vor dem Maximum und nehmen dann sehr rasch an Helligkeit ab, so daß sie im Helligkeitsmaximum des Sternes schon schwach sind und wenige Tage später verschwinden, in einer Phase, wo sie in der Regel bei den Me-Sternen am stärksten sind. Das kontinuierliche Spektrum ist im Maximum K9 oder M0, im Minimum M4.

Die Differenz $V_a - V_e$ ist, entsprechend der kurzen Periode des Sternes, sehr klein, im Mittel +3 km. Sowohl V_a wie V_e scheinen veränderlich zu sein. Es ist nämlich

nahe dem Helligkeitsmaximum	$V_a = +26$ km	$V_e = +25$ km
zu anderen Zeiten	+34	+32

Es zeigt sich also hier ein anderes Verhalten wie bei den V_a von Mira Ceti. Es ist nämlich bei T Centauri im Maximum der Helligkeit die vom Beobachter fort gerichtete Geschwindigkeit am kleinsten, wie bei den δ Cephei-Sternen. Diese Änderungen scheinen sich aber nicht in jeder Periode des Lichtwechsels zu wiederholen.

33. Die lateralen Eigenbewegungen und die absoluten Helligkeiten der Mira-Sterne. Obwohl Auseinandersetzungen über die lateralen Eigenbewegungen der Mira-Sterne nicht eigentlich in den Rahmen dieses Werkes gehören, muß doch kurz darauf eingegangen werden, da sich aus diesen Bewegungen Schlüsse

[1] Ap J 58, S. 265 (1923).
[2] Pop Astr 31, S. 645 (1923). Ausführlich in Ap J 63, S. 281 (1926) = Mt Wilson Contr 311.
[3] Lick Bull 12, S. 73 (1925).

auf die absoluten Helligkeiten dieser Sterne ziehen lassen. Zuerst hat sich W. GYLLENBERG[1] speziell mit den Eigenbewegungen der Mira-Sterne beschäftigt und auf Grund des geringen Materials, das ihm zur Verfügung stand, die Bewegung der Sonne aus diesen Eigenbewegungen sowie die mittlere absolute Helligkeit der Mira-Sterne bestimmt. 1923 hat dann R. E. WILSON vollständige Verzeichnisse der bekannten Eigenbewegungen der Mira-Sterne veröffentlicht, und zwar in zwei Abhandlungen[2]. Das erste Verzeichnis enthält die Eigenbewegungen von 315 roten Sternen der Spektralklassen Mc, Me, N, R und Pec; weitaus die meisten dieser Sterne sind Veränderliche, allerdings nicht nur Mira-Veränderliche, sondern auch unregelmäßige. Das zweite Verzeichnis gibt die Eigenbewegungen von 86 Veränderlichen, meist der Spektralklassen Ma und Mb, aber auch früherer (A bis K). Es handelt sich teils um Mira-Sterne, zum größeren Teil aber um unregelmäßige Veränderliche, darunter auch Sterne wie U Geminorum, R Coronae u. a. m. Die Eigenbewegungen in den beiden WILSONschen Verzeichnissen sind alle klein, nur wenige übersteigen 0″,1 im Jahr. Unter den Mira-Sternen hat o Ceti die größte Eigenbewegung (0″,23). Als Mittelwerte der Eigenbewegungen μ gibt WILSON in der ersten Arbeit

für die Me-Sterne	$\mu =$ 0″,0455	121	Sterne
R ,,	0 ,0299	34	
N ,,	0 ,0252	92	

(Bei den R- und N-Sternen sind die Mira-Sterne zusammen mit den übrigen Sternen dieser Klassen behandelt.) Aus den Eigenbewegungen von 119 Me-Sternen ergab sich der folgende Zielpunkt der Sonnenbewegung und folgende parallaktische Bewegung M:

$$A_0 = 276^\circ{,}0 \quad D_0 = +34^\circ{,}0 \quad M = 0'',0204.$$

A_0 und D_0 stimmen sehr nahe mit den von Miss ALLEN (cf. vorige Ziffer) aus Radialgeschwindigkeiten bestimmten Werten (278°, 35°) überein. Nach weiteren Untersuchungen WILSONS zeigen diese Sterne auch die Strombewegung nach dem KAPTEYNschen Vertex in befriedigender Übereinstimmung mit der Gesamtheit der Sterne des Boss-Katalogs. Unter Heranziehung der älteren Radialgeschwindigkeitsbestimmungen MERRILLS bzw. der daraus folgenden Geschwindigkeit der Sonne findet WILSON schließlich als mittlere absolute Helligkeit der Me-Sterne in ihrem Helligkeitsmaximum den Wert $0^m{,}0$.

Auf WILSONS Rechnungen über die Sterne der Spektraltypen Ma, Mb, Mc, N und R gehen wir hier nicht ein, da er die Mira-Sterne, um die es sich für uns hier allein handelt, nicht von den anderen Sternen dieser Spektralklassen (unregelmäßige Veränderliche und nichtveränderliche Sterne) sondert. Es sei nur erwähnt, daß sich auch alle diese Sterne, wie die Me-Sterne, als Riesensterne ergeben.

P. W. MERRILL und G. STRÖMBERG haben der Bestimmung der absoluten Größe der Me- und Se-Sterne eine besondere Abhandlung gewidmet[3]. Sie berechnen die absolute Größe sowohl mit Hilfe der wenigen trigonometrisch gemessenen Parallaxen solcher Sterne als auch durch Vergleichung der Eigenbewegungen mit den Resultaten der Messungen der Radialgeschwindigkeiten, indem sie aus diesen letzteren die Geschwindigkeit der Sonne in bezug auf die Mira-Sterne ermitteln. Das Endresultat ist die absolute Größe $+0^m{,}1$ als Mittelwert für die Me-Sterne im Maximum ihrer Helligkeit, ein Wert, der mit dem von WILSON nahezu übereinstimmt. Für die Sterne der verschiedenen Unter-

---

[1] Lund Medd Série I, Nr. 90 (1918).
[2] A J 34, S. 183 (1923); 35, S. 125 (1923).
[3] Ap J 59, S. 97 (1924) = Mt Wilson Contr 267.

abteilungen der Klasse Me und für die Se-Sterne finden sich folgende absolute Größen:

M2e, M3e	$-0^m,6$
M4e	$-0\ ,5$
M5e	$0\ ,0$
M6e	$+0\ ,5$
M7e, M8e	$+0\ ,6$
Se	$+0\ ,4$

Da die durchschnittliche Periode mit dem Index der Spektralklasse wächst, so folgt aus dieser Tabelle, daß die Me-Sterne mit kürzeren Perioden absolut heller sind als die mit längeren, im Gegensatz zu dem Verhalten der δ Cephei-Sterne.

Darauf, daß die Me-Sterne im Maximum ihrer Helligkeit Riesensterne sind, deuten, wie MERRILL und STRÖMBERG bemerken, auch Beobachtungen von ADAMS und JOY hin. Diese haben in derselben Weise wie für gewöhnliche M-Sterne die absoluten Helligkeiten für 17 Me-Sterne bestimmt und als Mittelwert dafür $-0^m,5$ gefunden.

P. DOIG[1] hat die Frage erörtert, ob nicht, entsprechend der Korrelation zwischen Radialgeschwindigkeit und Periode, auch eine solche zwischen lateraler Eigenbewegung und Periode besteht. Er gelangt zu nebenstehender Tabelle, in der die lateralen jährlichen Eigenbewegungen μ auf die scheinbare Größe 8,0 reduziert sind.

P	Sterne	Mittlere Radialgeschwindigkeit	μ
$<260^d$	27	67,5 km	$0'',032$
260^d-360^d	32	34,7	$0\ ,025$
$>360^d$	22	15,0	$0\ ,029$

In den μ zeigt sich also keine Abhängigkeit von der Periode. Es folgt, daß die Sterne mit kurzen Perioden, da sie die größeren Radialgeschwindigkeiten besitzen, weiter entfernt und absolut heller sein müssen als die mit langen. Dies deckt sich mit dem Befunde von MERRILL und STRÖMBERG.

Endlich soll hier nur noch ganz kurz auf eine Untersuchung von G. STRÖMBERG und P. W. MERRILL[2] über die Raumgeschwindigkeiten der Me- und Se-Sterne hingewiesen werden. Sie haben unter bestimmten Annahmen über die absoluten Helligkeiten der einzelnen Veränderlichen für 80 Me- und Se-Sterne die Parallaxen und alsdann die drei rechtwinkligen Geschwindigkeitskomponenten (in bezug auf die Sonne) berechnet. Dann wurden für zwei Gruppen von Sternen, nämlich erstens für die mit M1e- bis M6e-Spektren und zweitens für die mit M7e-, M8e- und Se-Spektren die Geschwindigkeitsellipsoide berechnet. Die Richtung der großen Achse ist bei beiden Ellipsoiden ungefähr dieselbe wie für andere Klassen von Sternen, aber die Streuung der Geschwindigkeiten ist viel größer. Die Asymmetrie in der Verteilung der Geschwindigkeiten ist sehr groß, namentlich für die erste Gruppe, welche die am schnellsten bewegten Sterne enthält.

Nach allem scheinen die Me- und Se-Sterne besondere Bewegungsverhältnisse zu besitzen. Für die Einordnung dieser Sterne in das allgemeine Entwicklungsschema erwachsen aus diesem Umstande große, einstweilen noch nicht zu überwindende Schwierigkeiten.

Neuerdings hat W. J. LUYTEN[3] eine Liste der Eigenbewegungen von 25 Mira-Sternen veröffentlicht.

34. Parallaxen der Mira-Sterne. Bei den in der vorigen Ziffer skizzierten Untersuchungen über die lateralen Eigenbewegungen und die absoluten Helligkeiten der Mira-Sterne hat es sich ergeben, daß ihre mittleren Parallaxen sehr

[1] J B A A 33, S. 326 (1923); 34, S. 106 (1924).
[2] Ap J 59, S. 148 (1924) = Mt Wilson Contr 268.
[3] Harv Circ 293 (1926).

klein sind, ungefähr von der Größenordnung eines Hundertstels der Bogensekunde. Diese Kleinheit der Parallaxen wird bestätigt durch direkte trigonometrische Messungen der Parallaxen von einigen Me-Sternen durch A. VAN MAANEN auf dem Mt. Wilson-Observatorium, die wir hier unter Hinzufügung einiger später bestimmten Werte nach der Zusammenstellung von MERRILL und STRÖMBERG[1] wiedergeben:

Stern	Absolute Parallaxe	Wahrscheinlicher Fehler
R Cancri	−0″,002	±0″,005
R Canum ven.	−0 ,011	0 ,003
T Cassiopeiae	+0 ,026	0 ,003
T Cephei	+0 ,001	0 ,003
U Herculis	+0 ,006	0 ,006
R Leonis	+0 ,003	0 ,009
X Ophiuchi	−0 ,001	0 ,005
R Trianguli	+0 ,004	0 ,006
R Virginis	+0 ,009	0 ,004

Die Bestimmung der Parallaxen dieser Veränderlichen wird durch die Änderungen ihrer Helligkeit und die gleichzeitigen Änderungen im Spektrum erschwert und unsicher gemacht, worauf hier nicht näher eingegangen werden kann. Speziell die Parallaxe von *o* Ceti, für die mehrere Bestimmungen vorliegen, wird noch verfälscht durch den erst neuerdings bekanntgewordenen Umstand, daß dieser Veränderliche ein enger Doppelstern ist; eine unter Beobachtung besonderer Vorsichtsmaßregeln ausgeführte Messungsreihe von S. A. MITCHELL[2] ergibt für die relative Parallaxe desselben den Wert +0″,009 ± 0″,010, woraus eine absolute Parallaxe von +0″,015 folgt.

35. Durchmesser der Mira-Sterne. Wenn wir nach MERRILL und STRÖMBERG die absolute Größe eines typischen Me-Sternes im Maximum seiner Helligkeit zu $+0^m,1$ annehmen, so ist es möglich, den Durchmesser angenähert zu berechnen. Die genannten beiden Autoren stellen folgende Überlegung an[3]: Der Größe $+0^m,1$ entspricht eine Helligkeit, die rund 100mal so groß ist wie die der Sonne. Die Flächenhelligkeit andererseits wird nach dem Spektrum auf etwa 0,01 von der der Sonne zu schätzen sein, die Oberfläche muß also 10000mal, der Durchmesser demnach 100mal so groß sein wie bei der Sonne. Auf Grund der radiometrischen Messungen kommt man zu einem Durchmesser der Mira-Sterne, der gleich 160 Sonnendurchmessern ist. Man wird hiernach also sagen können, daß normale Me-Sterne im Helligkeitsmaximum Durchmesser haben werden, die gleich 100 bis 200 Sonnendurchmessern sind. Auch H. THOMAS stellt in seiner Dissertation Rechnungen über die Durchmesser der Me-Sterne an und gelangt gleichfalls zu hohen Werten (etwa 80 Sonnendurchmesser). Wenn wir nicht für die Mira-Sterne ganz außerordentlich große Massen annehmen wollen, so müssen ihre Dichten enorm gering sein.

Nach dem obigen Befunde über die Durchmesser der Mira-Sterne mußte eine interferometrische Messung derselben aussichtsreich erscheinen, und F. G. PEASE[4] hat auf dem Mt. Wilson-Observatorium eine solche an *o* Ceti im Helligkeitsmaximum von Anfang 1925 versucht. Er fand einen Durchmesser von 0″,056, etwa 1,3mal so groß wie bei α Orionis. Nehmen wir für *o* Ceti eine Parallaxe von rund 0″,02 an, so ergibt sich damit ein Durchmesser gleich 300 Sonnendurchmessern. Jedenfalls kann man sagen, daß der Durchmesser von *o* Ceti von derselben Größenordnung wie der von α Orionis ist.

36. Galaktische Verteilung der Mira-Sterne. Die Verteilung der Me-Sterne in galaktischer Breite und Länge ist von H. SHAPLEY und Miss A. J. CANNON näher untersucht worden[5]. Wir geben die von ihnen gefundenen Zahlen in

[1] Ap J 59, S. 98 (1924) = Mt Wilson Contr 267.
[2] Pop Astr 35, S. 140 (1927).
[3] Ap J 59, S. 106 (1924) = Mt Wilson Contr 267.
[4] Publ A S P 37, S. 89 (1925).
[5] Harv Circ Nr. 245 (1923).

der folgenden Tabelle wieder. n bedeutet die Zahl der in den betreffenden Zonen gelegenen Me-Sterne, (n) die auf gleiche Flächengröße der Zonen reduzierte Zahl.

Galaktische Breite	n	(n)	Galaktische Länge	n
+90° bis +70°	3	17	0° bis 30°	38
+70 „ +50	20	40	30 „ 60	36
+50 „ +30	34	44	60 „ 90	26
+30 „ +10	49	52	90 „ 120	29
+10 „ −10	66	66	120 „ 150	11
−10 „ −30	81	86	150 „ 180	20
−30 „ −50	56	73	180 „ 210	17
−50 „ −70	25	50	210 „ 240	19
−70 „ −90	7	40	240 „ 270	29
	341		270 „ 300	31
			300 „ 330	54
			330 „ 360	31
				341

Sehr stark ist die Ungleichmäßigkeit der Verteilung in galaktischer Länge. In den dem Sternbilde Sagittarius entsprechenden galaktischen Längen liegen doppelt soviel Me-Sterne als in den dem Taurus entsprechenden. In Breite zeigt sich eine Konzentration nach −20° hin. Für die Gesamtheit der langperiodischen Veränderlichen (ohne Rücksicht auf das Spektrum und einschließlich derer mit unbekanntem Spektrum) ist nach H. THOMAS diese Konzentration um −20° Breite nicht vorhanden, sondern die Verteilung ist so wie für die Gesamtheit der M-Sterne (Konzentration um den galaktischen Äquator). Die N-Sterne sind, wie bekannt, sehr stark nach dem galaktischen Äquator hin angehäuft. Wie hier gleich bemerkt werden möge, verhalten sich nach THOMAS die unregelmäßig veränderlichen M-Sterne in bezug auf ihre Verteilung in galaktischer Breite ebenso wie die Gesamtheit der M-Sterne, in bezug auf ihre Verteilung in Länge ungefähr so wie die Me-Sterne; auch die schwächeren, nichtveränderlichen M-Sterne zeigen in galaktischer Länge dieselbe merkwürdige Verteilung.

Schon H. H. TURNER hat darauf hingewiesen[1], daß die Mira-Sterne mit sehr langen Perioden die Milchstraße bevorzugen. Dies wird durch eine eingehendere Statistik von H. LUDENDORFF[2] für die Me-Sterne durchaus bestätigt. Es ergeben sich nämlich für verschiedene Periodenintervalle folgende mittlere galaktische Breiten g_m:

P	g_m	n
≦150^d	25°	14
151^d −180^d	19	12
181 −210	33	16
211 −240	29	34
241 −270	36	42
271 −300	32	36
301 −330	28	45
331 −360	29	37
361 −390	33	32
391 −420	23	20
421 −450	15	12
451 −500	22	8
>500^d	12	6

Auch die Me-Sterne mit $P < 180^d$ scheinen hiernach durchschnittlich etwas kleinere galaktische Breiten zu haben als die mit Perioden mittlerer Länge. Ferner scheinen nach LUDENDORFF diejenigen Me-Sterne, die galaktische Breiten unter 40° haben, durchschnittlich um etwa $0^m,7$ kleinere Amplituden zu besitzen als die mit galaktischen Breiten >40°.

Über die räumliche Verteilung der langperiodischen Veränderlichen hat W. GYLLENBERG[3] Untersuchungen angestellt, bei denen er gleiche absolute Helligkeit aller dieser Veränderlichen voraussetzt. Nach den früher besprochenen Untersuchungen von MERRILL und STRÖMBERG über die absoluten Helligkeiten der Me-Sterne scheint es aber nicht so, als ob diese Voraussetzung erfüllt ist, es scheint vielmehr eine Korrelation zwischen absoluter Helligkeit und der Spektralklasse zu bestehen.

37. Die Zahl der Mira-Sterne. In unseren früher gegebenen Tabellen I bis V (Ziff. 22) sind 399 Mira-Sterne mit bekanntem Spektrum und bekannter Periode

[1] M N 79, S. 373 (1919). [2] A N 220, S. 148 (1924).
[3] Lund Medd Série I, Nr. 90 (1918).

enthalten, wenn wir nur diejenigen Sterne zur Miraklasse rechnen, deren Perioden $> 90^d$ sind. (Unter diesen 399 Sternen befinden sich ohne Frage einige RV Tauri-Sterne; es ist aber schwer, die Grenze zwischen beiden Klassen von Veränderlichen zu ziehen.) Die erwähnten Tabellen enthalten nur Sterne, die im Hauptkatalog der G. u. L. vorkommen; dieser zählt ferner noch 183 Mira-Sterne mit bekannter Periode, aber unbekanntem Spektrum auf. Unter den 1687 Sternen des Hauptkatalogs befinden sich also 582 = 34,5 % Mira-Sterne. Nun sind aber in dem Katalog noch viele wahrscheinlich zur Miraklasse gehörige Sterne enthalten, deren Perioden noch nicht gewährleistet werden können, und auch unter den Veränderlichen des Katalogs, über deren Lichtwechsel wir noch so gut wie nichts wissen, werden sich noch Mira-Sterne befinden. Man wird hiernach also sagen können, daß gegen 40% der Sterne des Hauptkatalogs der G. u. L. Mira-Sterne sind.

Einen Versuch, die Zahl der Mira-Sterne, die im Maximum die 9. Größe überschreiten, abzuschätzen, hat H. THOMAS in seiner Dissertation gemacht. Wir wollen hier seine Überlegungen nicht näher wiedergeben, sondern nur deren Ergebnis mitteilen. Er meint, daß es im ganzen etwa 1500 solcher Mira-Sterne geben wird, von denen etwa 1200 der Spektralklasse Me angehören werden. Etwa 14% aller Sterne der Spektralklasse M bis zur 9. Größe dürften Mira-Sterne sein.

38. Entwicklungsgang der Mira-Sterne. Sofern man überhaupt Hypothesen über die zeitliche Entwicklung der Mira-Sterne schon zur Diskussion zulassen will, ergeben sich dafür aus den vorangehenden Ausführungen gewisse Anhaltspunkte, auf die H. LUDENDORFF[1] hingewiesen hat. Es ist im folgenden stets von Mira-Sternen der Spektralklasse Me die Rede.

Als absolute Größe eines normalen Me-Sternes im Maximum seiner Helligkeit hatte sich der abgerundete Betrag $0^m,0$ ergeben. Diese Sterne sind also im Helligkeitsmaximum ohne Zweifel Riesen. Nehmen wir als normale Amplitude den Betrag $5^m,5$ an, so haben also die Me-Sterne im Minimum der Helligkeit absolute Helligkeiten von $+5^m,5$, während die Zwerge der Spektralklasse M absolute Größen von $+8^m,5$ und schwächer haben. Im Minimum stehen die Me-Sterne also zwischen Riesen und Zwergen. Es ist aber zu bemerken, daß die Gesamtstrahlung der Me-Sterne im Minimum nur verhältnismäßig wenig kleiner ist als im Maximum, im Gegensatz zu dem Verhalten der visuellen Strahlung. Die Me-Sterne werden also auch im Helligkeitsminimum als Riesen zu betrachten sein. Stellen wir uns nun auf den Boden der RITTER-RUSSELLschen Entwicklungstheorie, so müssen wir also annehmen, daß es sich um junge Sterne handelt, und daß die zeitliche spektrale Entwicklung in der Richtung Mce, Mbe, Mae bzw. M10e, M9e, M8e ... vor sich geht.

Nun sahen wir, daß die Mce-Sterne, also die jüngsten, durchschnittlich die längsten Perioden haben, und daß andererseits bei den Me-Sternen mit den längsten Perioden besonders häufig Lichtkurven der Form α_1 (sehr breites Minimum mit langandauernder konstanter Phase, steiler Helligkeitsanstieg) auftreten. Wir können also vielleicht annehmen, daß wir in den Me-Sternen mit Lichtkurven der Form α_1 die Urform der Mira-Sterne vor uns haben; das Minimum wäre der Normalzustand des Sternes, und dieser Normalzustand wird durch plötzliche Aufhellungen unterbrochen, wie bei den neuen Sternen und den U Geminorum-Sternen. Für diese an sich schon ganz plausible Vorstellung spricht noch der Umstand, daß die Minimalhelligkeiten bei diesen Sternen nur wenig, die Maximalhelligkeiten aber stark veränderlich sind. Im Laufe der Zeit ver-

[1] A N 220, S. 145 (1924).

wandelt sich das Spektrum Mce in ein solches der Klasse Mbe, dann Mae, die Lichtkurven gehen in die Formen α_2, α_3, α_4, β über, und gleichzeitig nehmen die Periode und die Amplitude der Lichtschwankung ab (vgl. die Korrelationen zwischen Spektrum und Periodenlänge, zwischen Periodenlänge und Form der Lichtkurve und zwischen Periodenlänge und Amplitude). Während der Entwicklung wird die absolute Helligkeit zunächst noch etwas größer, denn wir sahen, daß die Sterne der Spektralklasse Mne bei kleinem n größere absolute Helligkeiten haben als bei größerem n. Auch diese Zunahme ist ganz plausibel, denn in der Anfangsentwicklung eines Riesensternes muß es ein Stadium geben, wo seine Helligkeit noch wächst.

Die hellen Linien im Spektrum werden allmählich verlöschen, der Me-Stern wird in einen M-Stern übergehen. Dies braucht übrigens keineswegs erst zu geschehen, nachdem das Stadium Mae erreicht ist, denn es gibt ja Mira-Sterne mit Mc- und Mb-Spektren ohne helle Linien.

Was schließlich aus einem Mira-Stern wird, ist schwer zu sagen. Einige wenige scheinen in RV Tauri-Sterne überzugehen, die Mehrzahl könnte sich vielleicht in rote, unregelmäßige Veränderliche mit kleinen Amplituden und schließlich in gewöhnliche M-Sterne konstanter Helligkeit verwandeln.

Natürlich können alle diese Ansichten bestenfalls als Arbeitshypothese gelten. Man kann ihnen vor allem das entgegenhalten, daß die Me-Sterne besondere Bewegungsverhältnisse aufweisen, die sie von anderen M-Sternen unterscheiden, — auf ähnliche Schwierigkeiten stößt man ja auch sonst bei Betrachtungen über die Entwicklung der Sterne. Auch im übrigen läßt jene Arbeitshypothese noch mancherlei zu wünschen übrig, z. B. bleibt die Rolle der Mira-Sterne mit Lichtkurven der Form γ ungeklärt, die ja allerdings eine Ausnahme bilden.

Betrachtungen ähnlicher Art über die Mira-Sterne der Spektralklassen S, N und R lassen sich nicht anstellen, da das Material an derartigen Sternen zu klein ist.

39. Hypothesen zur Erklärung des Lichtwechsels der Mira-Sterne. Eine auch nur einigermaßen befriedigende Erklärung des Lichtwechsels der Mira-Sterne gibt es bisher nicht; wir können uns daher bei der Besprechung der betreffenden Hypothesen sehr kurz fassen, um so mehr, als sie J. STEIN im zweiten Bande (1924) des von HAGEN und ihm herausgegebenen Werkes „Die veränderlichen Sterne“ mit der größten Ausführlichkeit behandelt hat. Das folgende ist im wesentlichen nur ein Auszug aus STEINS Darstellung.

Die frühesten Versuche, periodische Helligkeitsschwankungen der Sterne (nicht nur der Mira-Sterne, sondern auch der übrigen periodischen Veränderlichen) zu erklären, gingen von der Annahme aus, daß ein Stern, dessen Oberflächenteile verschiedene Helligkeiten besitzen, sich um seine Achse drehe. ZÖLLNER hat zuerst in seinen „Photometrischen Untersuchungen“[1] dieser Hypothese eingehende Betrachtungen gewidmet. Er nimmt zur Erklärung des Lichtwechsels der Mira-Sterne auf der Oberfläche derselben Schlackenbildung an (die Sonnenflecke betrachtet er als Anfang einer solchen), und unter Einführung weiterer Hilfshypothesen erklärt er auch die Tatsache, daß bei den meisten Mira-Sternen die Helligkeitszunahme rascher vor sich geht als die Abnahme. Wir wissen heute, daß die Mira-Sterne Riesensterne sind, und daß infolgedessen von Schlackenbildung auf ihnen keine Rede sein kann.

H. BRUNS[2] hat das Problem mathematisch behandelt. Er zeigt, daß jede beliebige stetige und periodische Lichtkurve durch ungleichförmige Helligkeits-

[1] Leipzig 1865.
[2] Monatsberichte d. Kgl. Preuß. Akad. d. Wiss. 1881, S. 48.

verteilung auf der Oberfläche einer um eine feste Achse rotierenden Kugel dargestellt werden kann, und zwar auf unendlich viele verschiedene Arten. Er hat dabei aber die wichtige physikalische Bedingung übersehen, daß die aus der Lichtkurve abgeleitete Helligkeitsfunktion auf der Oberfläche des Sternes nirgends negativ werden darf, und es ist somit nicht erwiesen, daß jede periodische Lichtkurve durch die Fleckentheorie erklärt werden kann. H. N. RUSSELL[1] ist zu ähnlichen Resultaten wie BRUNS gelangt, doch hebt er die eben erwähnte Schwierigkeit hervor.

ZÖLLNER hatte die oft beobachteten Veränderungen der Periodenlänge bei den Mira-Sternen durch Verschiebungen der Schlackenfelder erklärt. H. GYLDÉN nimmt in seiner Abhandlung „Versuch einer mathematischen Theorie zur Erklärung des Lichtwechsels der veränderlichen Sterne"[2] an, daß diese Änderungen wiederum periodisch seien und erklärt sie durch die Hypothese, daß die Umdrehungsachse nicht mit einer Hauptträgheitsachse zusammenfalle. Er betrachtet also bei der von ihm entwickelten Theorie die Sterne als starre Körper, was natürlich im vorliegenden Falle ganz unzulässig ist.

P. GUTHNICK[3] hat das Problem, ob jede periodische Lichtkurve durch geeignete Fleckenverteilung auf der Oberfläche einer rotierenden Kugel und durch Randverdunklung erklärt werden könne, experimentell behandelt durch photometrische Messungen an einer dunklen Kugel, auf der Kreideflecken angebracht waren. Es gelang ihm in der Tat, alle vorkommenden Typen von Lichtkurven zu erzeugen. K. F. BOTTLINGER[4] hat dann aber gezeigt, daß man zur Erklärung der Lichtkurven mit steilem Helligkeitsanstieg ganz unwahrscheinlich hohe Beträge für die Randverdunklung annehmen muß.

Eine zweite Gruppe von Theorien über die periodischen Helligkeitsschwankungen der Mira-Sterne sieht diese als durch periodische Fleckenbildungen (entsprechend der elfjährigen Periode der Sonnenflecke) hervorgerufen an. Die Rotation spielt hier also keine oder höchstens eine sekundäre Rolle. R. WOLF hat schon 1852 auf die Ähnlichkeit der Fleckenhäufigkeitskurve der Sonne mit der typischen Lichtkurve der Mira-Sterne hingewiesen, und er wie auch E. SCHÖNFELD glaubten daher, daß vielleicht der Lichtwechsel der letzteren durch periodische Fleckenbildungen zu erklären sei. Schwierig war dabei die Entscheidung der Frage, ob das Sonnenfleckenmaximum dem Maximum oder dem Minimum der Helligkeit des Veränderlichen entspräche.

Handelte es sich bei WOLF und bei SCHÖNFELD mehr um gelegentliche Bemerkungen als um wirkliche Untersuchungen über den Gegenstand, so hat hingegen H. H. TURNER sich eingehend mit der Vergleichung der Sonnenfleckenkurve mit den Lichtkurven beschäftigt. In der ersten von zwei schon in Ziff. 21 teilweise besprochenen Abhandlungen[5] stellt er sowohl die Lichtkurven wie die Sonnenfleckenkurve durch FOURIERsche Reihen dar und findet zwischen diesen FOURIER-Entwicklungen Ähnlichkeiten, wenn er annimmt, daß das Fleckenmaximum dem Helligkeitsmaximum entspricht. In der zweiten Abhandlung verfolgt er den Gegenstand weiter und erklärt die Verschiedenheiten der Lichtkurven dadurch, daß er für die Veränderlichen verschiedene Neigungen der Umdrehungsachse gegen die Gesichtslinie annimmt. So geistvoll diese Betrachtungen auch sind, muß ihnen doch entgegengehalten werden, daß die Mira-Sterne nach der heutigen Anschauung Riesen und demnach von ganz anderer Beschaffenheit als die Sonne sind.

Eine weitere Theorie zur Erklärung der periodischen Veränderlichen ist die KLINKERFUESsche Fluthypothese. Da sie hauptsächlich auf die δ Cephei-

[1] Ap J 24, S. 1 (1906). [2] Helsingfors 1879. [3] A N 209, S. 1 (1919).
[4] A N 210, S. 33 (1919). [5] M N 64, S. 543 (1904) u. 67, S. 332 (1907).

Sterne angewandt wird, soll sie erst später näher besprochen werden. Sie erfordert, daß die Veränderlichen enge Doppelsternsysteme sind. Nun sind ja neuerdings von A. H. JOY periodische Verschiebungen der Absorptionslinien im Spektrum von Mira Ceti festgestellt worden, und man ist daher vielleicht berechtigt, diesen Veränderlichen als spektroskopischen Doppelstern mit einer Periode, die gleich der des Lichtwechsels ist, aufzufassen. Aber A. H. JOY hat darauf aufmerksam gemacht, daß selbst unter den günstigsten Annahmen über Neigung und Massen die Bahn des Begleiters von 330^d Umlaufszeit innerhalb des Hauptsternes verlaufen müßte, da dieser, wie wir gesehen haben, enorm groß ist. Ob alle Mira-Veränderlichen ähnliche Veränderungen der Radialgeschwindigkeit zeigen, wie *o* Ceti, wissen wir noch nicht.

Alle bisher angeführten Theorien über die Mira-Veränderlichen lassen die Eigentümlichkeiten der Spektren dieser Sterne fast ganz außer Betracht. Sie wurden zuerst von A. BRESTER in seinem „Essai d'une explication du mécanisme de la périodicité dans le soleil et les étoiles rouges variables“[1] etwas mehr berücksichtigt. Nach seiner komplizierten Theorie wird der Lichtwechsel hervorgerufen durch das periodische Entstehen und Verschwinden von Öffnungen in einer den Stern umgebenden kühleren Hülle. Nach dem heutigen Stande der Wissenschaft haben solche Hypothese auf Hypothese häufenden Theorien nur geringen Wert, und wir verweisen für eine ausführliche Darstellung der Ansichten BRESTERS auf STEINS zitiertes Werk. Auch eine von W. W. CAMPBELL in seinem Buche „Stellar Motions“ (New Haven 1913) entwickelte Hypothese genügt dem heutigen Stande der Forschung nicht.

In vieler Hinsicht ähnlich der BRESTERschen Theorie ist MERRILLS[2] Schleiertheorie (veil theory); sie geht indessen nicht so ins einzelne und will wohl mehr als allgemeine Anregung gelten. Er meint, daß sich die Mira-Sterne wohl in Wirklichkeit nicht so sehr an Helligkeit ändern, sondern daß zur Zeit des Minimums ein Schirm kondensierter Gase (vielleicht Ca) in der oberen Atmosphäre des Sterns sich zwischen diesem und dem Beobachter befindet. „The cloud formed by condensation would conserve the heat radiated from the photosphere to space so that the temperature of the materials immediately above the photosphere would increase until the overlying veil is vaporized and the star shines out brightly. It is easy to conceive that these phenomena would be periodic and would cause variations in the spectrum, particularly in chromospheric emission. Possibly electric effects having their origin in the evaporation of the cloud are effective in stimulating the hydrogen and other gaseous emission. If so the source would be at a high level, which seems to accord with observation especially if we assume that the relative displacement of bright and dark lines is due to pressure.“ Diese letztere Annahme ist jedenfalls ganz unzulässig und überhaupt wird man wohl die ganze Hypothese kaum als den Tatsachen entsprechend ansehen dürfen. J. HOPMANN[3] hat MERRILLS Schleiertheorie modifiziert und einige Mängel derselben beseitigt. Eine thermodynamische Begründung für seine Anschauungen vermag er aber, wie er selbst hervorhebt, nicht zu geben.

A. S. EDDINGTON[4] ist der Ansicht, daß seine Pulsationstheorie, auf die wir bei der Besprechung der Theorien der δ Cephei-Sterne zurückkommen werden, auch auf die Mira-Sterne anwendbar sei, und er findet, daß die Dimensionen von *o* Ceti mit einer Pulsationsperiode von rund 300^d gut vereinbar sind.

---

[1] Verhandl. Kon. Akad. van Wetensch. te Amsterdam, 1. Sectie 9, Nr. 6 (1908).

[2] Publ. of the Astr. Obs. of the Univ. of Michigan (Detroit Obs.) 2, S. 70 (1916).

[3] A N 228, S. 105 (1926).

[4] The Internal Constitution of the Stars. Cambridge 1926, S. 206.

Eine bemerkenswerte Theorie hat J. H. JEANS[1] in seiner Arbeit „On Cepheid and Long-period Variation and the Formation of Binary Stars" aufgestellt. Er macht zunächst Bedenken gegen EDDINGTONS Pulsationstheorie der δ Cephei-Sterne geltend und zeigt dann, daß große Schwierigkeiten dieser Theorie wegfallen unter der Annahme, daß der Stern nicht, wie EDDINGTON voraussetzt, kugelförmig ist und nicht rotiert, sondern daß er eine Rotation besitzt und eine birnenförmige Gestalt hat, also sich im Vorstadium der Trennung in ein Doppelsternsystem befindet. Vorher hat der Stern die Gestalt eines dreiachsigen Ellipsoids besessen, und der Übergang in die birnenförmige Figur ist nach JEANS wahrscheinlich oszillatorischen Charakters mit einer bestimmten Periode, so daß auch eine Oszillation der ganzen Gasmasse angenommen werden kann. Der Lichtwechsel kommt zustande durch diese Oszillation und dadurch, daß der Stern uns infolge seiner Rotation verschieden große Teile seiner Oberfläche zuwendet. Zuerst brauchen die Perioden der Oszillation und der Rotation nicht gleich und auch nicht kommensurabel zu sein: wir haben einen unregelmäßigen Veränderlichen vor uns. Allmählich, wenn die eigentliche Trennung des Sternes in zwei Komponenten näher rückt, gleichen sich die Perioden aneinander an, der Stern geht in einen langperiodischen bzw. δ Cephei-Stern über. Schließlich tritt die Trennung ein, es entsteht ein spektroskopischer Doppelstern. Dabei ändert sich das Spektrum nach JEANS' Theorie grundlegend; z. B. wird aus einem in zwei gleiche Teile zerfallenden M-Stern ein spektroskopischer Doppelstern der Spektralklasse B. Es gelingt JEANS, auf Grund dieser Theorie die typischen Lichtkurven der δ Cephei- und der Mira-Sterne darzustellen.

Nach JEANS' Theorie müßten ganz allmähliche Übergänge zwischen den Mira-Sternen und den δ Cephei-Sternen, sowie zwischen diesen und den spektroskopischen Doppelsternen vorhanden sein. Daß ersteres der Fall ist, kann man wohl kaum behaupten, wenigstens sind die Übergangsformen sehr selten. JEANS leitet aus seiner Theorie das Vorhandensein der „Period-Luminosity Curve" der δ Cephei-Sterne ab. Eine entsprechende Kurve müßte auch für die Mira-Sterne bestehen. In Wirklichkeit scheinen aber, wie wir früher gesehen haben, bei diesen gerade die mit kürzeren Perioden größere absolute Helligkeit zu besitzen als die mit längeren. Auch erörtert JEANS nicht, inwieweit seine Theorie den Eigentümlichkeiten und den Änderungen des Spektrums der Mira-Sterne gerecht wird. Immerhin erscheinen JEANS' Ausführungen sehr beachtenswert. Eine restlose Erklärung des Lichtwechsels der Mira-Sterne aber ist bei dem gegenwärtigen Stande der Wissenschaft kaum innerhalb der Grenzen des Möglichen.

Die Verschiebungen der Emissionslinien gegen die Absorptionslinien in den Spektren der Mira-Sterne werden von CH. E. ST. JOHN und W. S. ADAMS[2] durch die Annahme aufsteigender Ströme von Gasen, die die hellen Linien hervorbringen, erklärt; die Absorptionslinien werden nach dieser Vorstellung durch die langsam absteigenden Ströme von Gasen erzeugt. Diese Ansicht ist in Übereinstimmung mit dem Befund von ST. JOHN, daß die hellen Linien H und K im Sonnenspektrum eine aufwärts gerichtete Bewegung relativ zu den Absorptionslinien anzeigen. Bei den Me-Sternen müßten die in Betracht kommenden Geschwindigkeiten größer sein als auf der Sonne.

Das Vorhandensein heller Linien und speziell der hellen Wasserstofflinien in M-Spektren ist ziemlich schwer zu erklären. Über diesen Punkt hat SV. ROSSELAND[3] Betrachtungen angestellt, auf die wir hier verweisen müssen.

40. Die Veränderlichen mit Perioden von 45^d bis 90^d. Besondere Schwierigkeiten in bezug auf ihre Einordnung in die Klassen der Veränderlichen bereiten

[1] M N 85, S. 797 (1925). [2] Ap J 60, S. 43 (1924) = Mt Wilson Contr 279.
[3] Ap J 63, S. 218 (1926) = Mt Wilson Contr 309.

die Veränderlichen, deren Perioden 45^d bis 90^d betragen; sie sind von H. LUDENDORFF[1] eingehend behandelt worden, und wir geben hier seine Ausführungen mit einigen nachträglichen Ergänzungen wieder. Es sei zunächst daran erinnert, daß T Centauri, ein unzweifelhafter Mira-Stern, eine Periode von 91^d hat; man wird daher periodische Veränderliche mit mehr als 90^d Periode (abgesehen von den RV Tauri-Sternen und natürlich den Bedeckungsveränderlichen) zur Miraklasse rechnen. Schwieriger ist es, die obere Grenze der Periode für die δ Cephei-Sterne festzusetzen. Hier ist l Carinae ($P = 35^d,5$) der die längste Periode besitzende Stern, der unzweifelhaft zur δ Cephei-Klasse gehört, wie Lichtkurve und spektrale Eigenschaften lehren. Es folgen dann aber der Periodenlänge nach noch vier Sterne, die man mit großer Wahrscheinlichkeit in diese Klasse einordnen darf, nämlich (g = Abstand vom galaktischen Äquator, $\varepsilon = M-m/P$ = Abstand des Minimums vom folgenden Maximum, ausgedrückt in Einheiten der Periode):

Nr.	Sterne	Spektrum	g	P	ε	A
634	U Carinae	K0	0°	$38^d,7$	0,14	$1^m,2$
481	RS Puppis	K0 bis K5 oder Ma	0	41 ,3	0,29	1 ,5
—	SV Vulpeculae	K0 bis K5 oder Ma	1	44 ,5	0,36	1 ,1
367	SS Geminorum	G5	3	44 ,9	0,37	0 ,8

Der Spektralcharakter (bei RS Puppis und SV Vulpeculae sind die Änderungen des Spektrums vom Maximum bis zum Minimum angegeben), die geringe galaktische Breite, die Werte von ε und A, alles spricht dafür, daß wir hier δ Cephei-Sterne vor uns haben. Bei SV Vulpeculae scheinen zwar kleine Änderungen der Periode oder Epochensprünge vorzukommen[2], und bei SS Geminorum scheinen P und A sowie die Form der Lichtkurve ebenfalls Veränderungen zu erleiden[3], aber auch bei l Carinae, bei dem die Änderungen der Radialgeschwindigkeit durchaus die für δ Cephei-Sterne charakteristischen sind, und der daher unzweifelhaft zu dieser Klasse gehört, will man Ähnliches beobachtet haben. Es ist sehr wohl denkbar, daß bei δ Cephei-Sternen mit ungewöhnlich langen Perioden der Lichtwechsel stärkere Änderungen zeigt, als bei denen mit normalen Perioden.

Wir wollen daher $P = 45^d,0$ als obere Grenze für die δ Cephei-Sterne annehmen. Die folgende Tabelle enthält nun eine Zusammenstellung der Sterne, die Perioden von 45^d bis 90^d haben. Es sind hier nicht nur die Sterne des Hauptkatalogs der G. u. L. berücksichtigt, sondern auch die später entdeckten. Ausgeschlossen sind (außer Bedeckungsveränderlichen) die Sterne, die zur RV Tauri-Klasse gehören oder der Zugehörigkeit zu dieser verdächtig sind, ferner solche, deren Perioden noch stark angezweifelt werden können, wie SZ Cassiopeiae, TT Persei, TW Pegasi, RU Aquarii u. a. m. Soweit möglich, ist die Form der Lichtkurve in der früheren Bezeichnungsweise angegeben.

Die meisten von diesen Sternen sind nur unzureichend beobachtet worden, und namentlich die ε sind wohl in vielen Fällen noch unsicher. Außer den in der letzten Kolumne gemachten Bemerkungen ist noch folgendes über die einzelnen Sterne zu erwähnen:

Z Leonis. Die Maxima und Minima werden durch eine konstante Periode gut dargestellt, $M-m$ scheint konstant zu sein. Die Helligkeiten im Maximum und Minimum sind wohl etwas veränderlich. Die hohe galaktische Breite bei

[1] A N 214, S. 77 (1921).
[2] E. LEINER, A N 221, S. 137 (1924); Lichtkurve von G. ZACHAROV, A N 222, S. 293 (1924).
[3] P. AHNERT, A N 226, S. 87 (1925).

Nr.	Stern	Spektrum	g	P	ε	A	Lk	Bemerkungen
979	UU Herculis	—	41°	$45^d,4$	0,43	$0^m,5$	α_4	Max. breiter als Min. (Mem S A It 2, S. 184).
556	Z Leonis	Mb	51	56,4	0,42	1,7	α?	
—	CG Sagittarii	—	18	64,1	—	1,5ph	?	Harv Bull 803. Nahe bei Kugelsternhaufen NGC 6723.
1350	S Vulpeculae	—	0	67,5	0,37	1,4	α_3	A N 214, S. 61; 216, S. 383.
489	Z Cancri	Mc	27	70	0,51	0,8	?	Wenig bekannt, Periode veränderl.
241	UX Aurigae	Mb	7	72	0,31	0,8	α_3?	Lk. und P sehr veränderlich (A N 223, S. 191).
773	V Ursae min.	Ma	43	73	0,45	1,2	pec	Lk. sehr veränderlich.
824	UV Draconis	Mb	53	77	0,48	1,0	?	
389	V Lyncis	—	21	82	0,35	0,8	?	
—	TZ Cephei	—	11	84	—	1,8	?	
1326	AF Cygni	Mb	12	88	0,46	1,6	α	Lk. und P sehr veränderlich.
—	CE Sagittarii	—	18	90	—	>2,0ph	?	Harv Bull 803. Nahe bei Kugelsternhaufen NGC 6723.

diesem Sterne wie bei UU Herculis spricht aber gegen den gewöhnlichen δ Cephei-Charakter.

S Vulpeculae. Von W. HEISKANEN in den oben zitierten Arbeiten eingehend untersucht. Die Periode ist stark veränderlich, die Form der mittleren Lichtkurve ist die für δ Cephei-Sterne typische. Auch die Lichtkurve ist veränderlich.

Z Cancri. Wenig bekannt, P veränderlich.

UV Draconis. Wenig bekannt. Die Maxima und Minima weichen von den mit konstanter Periode gerechneten Zeiten bis zu 15^d ab, doch scheinen diese Abweichungen ganz regellos zu erfolgen. Die Maximal- und die Minimalhelligkeiten sind um einige Zehntel einer Größenklasse veränderlich.

V Lyncis. Wenig bekannt. Die Maxima und Minima weichen bis zu 24^d von der mit konstanter Periode gerechneten Ephemeride ab. Bisweilen bleiben Maxima oder Minima aus. M. L. LUIZET hält es für wahrscheinlich, daß die wahre Periode doppelt so groß ist wie angegeben. RV Tauri-Typus liegt augenscheinlich nicht vor.

AF Cygni. Angaben über neuere Literatur siehe Tabelle III in Ziff. 22. Über die starken, anscheinend periodischen Änderungen der Periode cf. Ziff. 24. Lichtkurve sehr veränderlich.

Was die Spektra der hier betrachteten Sterne angeht, so gehören sie, soweit sie bekannt sind, der Klasse M an und schließen sich damit gut an die δ Cephei-Sterne einerseits und die Mira-Sterne andererseits an. Die galaktische Breite ist bei einigen groß, so daß diese Sterne, als Einheit aufgefaßt, nicht mehr als galaktische Objekte, wie es die δ Cephei-Sterne sind, betrachtet werden können. Die mittlere galaktische Breite der 12 Sterne ist 25°, genau übereinstimmend mit der galaktischen Breite derjenigen Me-Sterne, deren $P < 150^d$ sind (vgl. Ziff. 36) und kleiner als die mittlere galaktische Breite der Me-Sterne, deren P zwischen 180^d und 390^d liegen.

Bei den Werten von P ist auffallend, daß zwischen $P = 46^d,0$ und $P = 64^d,0$ nur ein Stern vorkommt, während sonst die Verteilung der P recht gleichmäßig ist. Falls also eine scharfe Grenze zwischen den δ Cephei- und den Mira-Sternen zu ziehen ist, so ist es wahrscheinlich, daß sie bei Periodenlängen von 50^d bis 60^d liegt. Was die Amplituden angeht, so ähneln die Sterne mehr den δ Cephei- als den Mira-Sternen; die mittlere Amplitude der Me-Sterne mit Perioden zwischen 90^d und 150^d ist $3^m,6$, aber es ist zu bedenken, daß für Me-Sterne mit kurzen Perioden die Amplitude rasch mit der Periode abnimmt, und ferner, daß überhaupt die Mira-Sterne mit M-Spektrum ohne helle

Linien durchschnittlich kleinere Amplituden haben als Me-Sterne gleicher Periode. Man kann also nicht behaupten, daß die kleinen Amplituden stark gegen die Zugehörigkeit der Sterne zur Mira-Klasse sprechen.

Die Lichtkurven und Perioden der hier betrachteten Sterne zeichnen sich zum Teil durch starke Veränderlichkeit aus, so als ob die Sterne sich in einem sehr instabilen Zustande befänden. Im ganzen sind die Lichtkurven wohl denen der δ Cephei-Sterne ähnlicher als denen der Me-Sterne kurzer Periode, welche fast ausschließlich nahezu symmetrische Lichtkurven besitzen.

Zusammenfassend können wir sagen, daß die angedeutete Periodenlücke bei 50^d bis 60^d und das gleichzeitige Auftreten höherer galaktischer Breiten auf eine nahe Verwandtschaft dieser Sterne mit den Mira-Sternen hindeutet, während andererseits die Form der Lichtkurven wieder auf eine solche mit den δ Cephei-Sternen hinweist. Es ist natürlich auch möglich, daß wir hier wirkliche Übergangstypen vor uns haben, oder daß einige der Sterne zur Mira-Klasse, andere zur δ Cephei-Klasse gehören. In der Abgrenzung beider Klassen gegeneinander wird uns erst die Untersuchung der Radialgeschwindigkeiten dieser Sterne weiterbringen. Im Jahresbericht für 1924 des Mt. Wilson-Observatoriums findet sich die Bemerkung, daß A. H. JOY die Radialgeschwindigkeiten mehrerer der Veränderlichen gemessen hat, deren Perioden zwischen denen der δ Cephei-Sterne und der langperiodischen Veränderlichen liegen, und daß sich sehr große Werte dafür ergeben haben, entsprechend wie bei den Me-Sternen. Dies scheint für die Zugehörigkeit der betreffenden Sterne zur Mira-Klasse entscheidend zu sein, und man darf auf die Veröffentlichung der Ergebnisse gespannt sein.

f) Die veränderlichen Sterne der μ Cephei-Klasse.

41. Definition und Vorbemerkungen. Als μ Cephei-Sterne bezeichnen wir rötlich oder rot gefärbte Veränderliche (Spektrum N, R, M und auch K, sowie in einigen Fällen besonderes Spektrum), die unregelmäßige Lichtschwankungen aufweisen. Die Gesamtamplitude der Schwankungen ist in den meisten Fällen gering, der Betrag von zwei Größenklassen wird nur in wenigen Ausnahmefällen überschritten.

Infolge der roten Färbung dieser Sterne und der Kleinheit der Helligkeitsschwankungen ist die nähere Untersuchung des Lichtwechsels derselben schwierig. Subjektive Beobachtungsfehler verdecken häufig die wirklichen Lichtschwankungen nahezu oder völlig. Diese Schwierigkeiten und die Regellosigkeit der Änderungen haben bewirkt, daß die Beobachter diese Sterne sehr vernachlässigt haben. In der Tat hat auch die Verfolgung vieler dieser Objekte mit Hilfe der gewöhnlichen Schätzungsmethoden kaum Zweck; die Eigentümlichkeiten des Lichtwechsels lassen sich nur durch sorgfältige photometrische Messungen (möglichst unter Benutzung roter Vergleichsterne) oder besser noch mit Hilfe objektiver photometrischer Meßmethoden ergründen.

Unter diesen Umständen sind unsere Kenntnisse über die μ Cephei-Sterne zur Zeit noch sehr gering, und die Resultate, die man für einige von ihnen abgeleitet hat, verdienen in der Regel wenig Vertrauen. Die Ansichten darüber, inwieweit die beobachteten Helligkeitsschwankungen als reell zu betrachten sind, gehen sehr auseinander. T. W. BACKHOUSE[1] glaubt auf Grund eigener Beobachtungen behaupten zu dürfen, daß die ausgesprochen rot gefärbten Sterne, insbesondere die der Spektralklasse N, fast ausnahmslos veränderlich seien. H. OSTHOFF[2] ist dagegen der Ansicht, daß die bei den roten Sternen

[1] J B A A 27, S. 382 (1913). [2] A N 212, S. 97 (1920).

beobachteten Helligkeitsschwankungen in weitestgehendem Maße nur scheinbar, durch Änderungen in der Auffassung des Beobachters bedingt seien. Er begründet diese Ansicht sehr ausführlich, u. a. durch Vergleichung gleichzeitiger Beobachtungen eines Beobachters an verschiedenen Sternen. So zeigen z. B. H. E. LAUS gleichzeitige Schätzungen der Helligkeiten von R Lyrae (Mb) und g Herculis (Mb) eine ausgesprochene Parallelität, und die aus den Schätzungen eines so erfahrenen Beobachters wie J. PLASSMANN abgeleiteten Lichtkurven

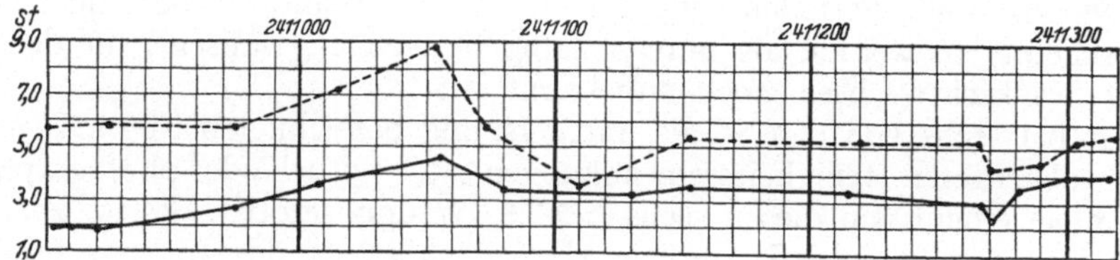

Abb. 20. Lichtkurven von α Cassiopeiae (oben) und μ Cephei (unten) nach PLASSMANN.

von α Cassiopeiae (K0) und μ Cephei (Ma) stimmen für dasselbe Zeitintervall ganz erstaunlich überein (Abb. 20). Es gibt nach OSTHOFF sowohl Auffassungsänderungen von sehr langsamem, jahrelangem Verlauf, als auch solche kurzer Periode.

Man wird auf Grund der Ausführungen von OSTHOFF und auf Grund der vielen Widersprüche, auf die man bei den Beobachtungen desselben roten Sternes durch verschiedene Beobachter ständig stößt, jedenfalls die kleinen beobachteten Lichtschwankungen sehr kritisch beurteilen müssen. Indessen kann kein Zweifel bestehen, daß viele dieser Sterne wirklich veränderlich sind. Das geht schon daraus hervor, daß man für manche rote Sterne mit kleinen Lichtschwankungen, die zuerst als unregelmäßig galten, schließlich einwandfreie Perioden festgestellt hat, wodurch diese Sterne dann freilich in die Klasse der Mira-Sterne übergeführt wurden. Auch photographische Beobachtungen oder z. B. bei α Orionis solche mit der Selenzelle haben die Veränderlichkeit roter Sterne bewiesen.

Eine scharfe Grenze läßt sich zwischen den μ Cephei-Sternen und den Mira-Sternen nicht ziehen. Namentlich unter den in unserer früher gegebenen Tab. III aufgeführten, den Spektralklassen K, Ma, Mb, Mc angehörenden Mira-Sternen sind einige enthalten, deren Lichtwechsel so große Unregelmäßigkeiten aufweist, daß man sie schon fast als μ Cephei-Sterne betrachten könnte. Z. B. hält RU Cephei (K 8) nach den Untersuchungen von W. HASSENSTEIN[1] zwar die Periode von 110^d ganz gut inne, aber der Verlauf der Lichtkurve von Minimum zu Minimum ist in verschiedenen Perioden ganz verschieden. Bei W Persei[2] liegen die Abstände der Minima zwischen 400^d und 600^d, und auch bei anderen Sternen sind starke Unregelmäßigkeiten vorhanden.

Im folgenden betrachten wir die μ Cephei-Sterne gesondert nach den Spektralklassen, zu denen sie gehören; wir beginnen mit den am stärksten rot gefärbten, d. h. denen der Klasse N, welchen wir die der verwandten Spektralklasse R gleich angliedern.

42. Die μ Cephei-Sterne der Spektralklassen N und R. In „Katalog und Ephemeriden veränderlicher Sterne für 1926" sind 137 Sterne der Spektralklassen N und R enthalten, wenn man vier Sterne, deren Spektren als „Pec" bezeichnet sind, mit einbegreift, nämlich VX Andromedae und die Sterne mit ähnlichen, heute als Nc bezeichneten Spektren (AF Aurigae, R Leporis, SS Virginis). Von den 137 Sternen sind nach dem jetzigen Stande der Kenntnis 24 als

[1] Potsd Publ Nr. 83 (1926). [2] Harv Bull Nr. 784 (1923).

Mira-Sterne anzusehen und schon in unserer Tabelle IV angeführt worden, einer ist ein R Coronae-Stern (S Apodis), einer eine ehemalige Nova (Z Centauri), und vier werden als δ Cephei-Sterne klassifiziert (RU Camelopardalis, VX und VZ Cygni mit R-Spektren, S Scuti mit Nb-Spektrum; hierzu ist zu bemerken, daß S Scuti wahrscheinlich unregelmäßig veränderlich ist, VZ Cygni in Wirklichkeit ein cF8-Spektrum hat, und daß das Spektrum von VX Cygni bei M. GÜSSOW und in der G. u. L. als K angegeben ist). Für neun von den übrigbleibenden 107 Sternen sind Perioden angegeben, deren Realität aber noch nicht sicher scheint, 36 werden als unregelmäßig bezeichnet, während für den Rest die Art des Lichtwechsels noch unbekannt ist.

Nur bei sechs von den 36 als unregelmäßig bezeichneten Sternen sind die bisher beobachteten Helligkeitsschwankungen $>2^m$ (visuell oder photographisch), und nur bei einem von diesen sechs $>2^m,5$ (W Monocerotis). Es ist aber wohl noch bei keinem der 36 Sterne die Unregelmäßigkeit des Lichtwechsels mit Sicherheit zu garantieren. Gerade bei den sehr roten N-Sternen ist die Unsicherheit der Helligkeitsschätzungen sehr groß im Verhältnis zu dem kleinen Umfang der Helligkeitsschwankungen, und man darf wohl sagen, daß wir über den Lichtwechsel dieser „unregelmäßigen" Veränderlichen noch so gut wie nichts Sicheres wissen.

Über die Spektra der veränderlichen N- und R-Sterne haben wir bereits in Ziff. 29 einige Bemerkungen gemacht und die nötigen Literaturnachweise gegeben. Die Radialgeschwindigkeiten von 25 N-Sternen (größtenteils Veränderliche mit unregelmäßigem Lichtwechsel oder solchem unbekannter Art) sind von J. H. MOORE[1] bestimmt worden. Als mittlere, von der Sonnenbewegung befreite Radialgeschwindigkeit ergab sich der Wert 18 km, der bedeutend kleiner ist als bei den Me-Sternen. Während also die Radialgeschwindigkeiten der von MOORE beobachteten N-Sterne durchaus mäßig sind, fand später R. F. SANFORD[2] für V Arietis die sehr große Radialgeschwindigkeit von 175 km. Man glaubte früher, daß V Arietis eine eintägige Periode habe, doch trifft dies nicht zu[3].

Erwähnt sei hier noch, daß das Spektrum von TU Tauri, bei dem die Art des Lichtwechsels noch unbekannt ist, zusammengesetzt ist, und zwar Nb + A0[4]. Es liegt hier offenbar, wie im Falle von Mira Ceti, ein enges Doppelsternsystem vor, dessen visuelle Trennung aber noch nicht möglich gewesen ist.

Endlich sei noch daran erinnert, daß die N- und die R-Sterne Riesensterne sind.

43. Die μ Cephei-Sterne der Spektralklasse M. Wir haben bereits früher erwähnt, daß von den Me-Sternen nur ein einziger mit Sicherheit zur Klasse der unregelmäßigen Veränderlichen gerechnet werden kann, nämlich S Persei. H. LUDENDORFF[5] hat eine Zusammenstellung der von 1881 bis 1918 beobachteten Maxima und Minima dieses Sternes gegeben. Der mittlere Abstand zweier aufeinanderfolgenden Minima ist 811^d, und im allgemeinen wird dieser Abstand auch einigermaßen innegehalten, es kommen aber auch sehr große Abweichungen vor. Der erwähnte Abstand ist übrigens auffallend groß, denn die längsten bisher bei Me-Sternen festgestellten Perioden liegen unterhalb von 600^d, und nur der S-Stern S Cassiopeiae hat eine Periode von 613^d.

Die Form der Lichtkurve und die Helligkeitsgrenzen wechseln bei S Persei außerordentlich. Besonders klein waren die Helligkeitsänderungen 1894 bis 1896. Das schwächste Minimum war $12^m,2$, das hellste Maximum $7^m,2$, so daß $A = 5^m,0$ wird, während die G. u. L. nur $A = 3^m,5$ angibt. Auch der zeitliche Abstand des Minimums von dem darauf folgenden Maximum ist sehr variabel.

[1] Lick Bull 10, S. 160 (1922). [2] Publ A S P 36, S. 351 (1924).
[3] A N 226, S. 313 (1926); 227, S. 285 (1926). [4] Publ A S P 37, S. 35 (1925).
[5] A N 220, S. 241 (1924).

Die Helligkeitsamplitude von S Persei ist für einen μ Cephei-Stern sehr groß, und es ist daher fraglich, ob man diesen Stern noch zu dieser Klasse von veränderlichen Sternen rechnen darf oder ihn nicht vielmehr besser als einen entarteten Mira-Stern betrachten sollte. H. H. TURNER[1] hat die Lichtkurve von S Persei harmonisch analysiert und Analogien zur Kurve der Sonnenfleckenhäufigkeit zu finden geglaubt.

In „Katalog und Ephemeriden veränderlicher Sterne für 1926" sind 194 Sterne der Spektralklassen Ma, Mb, Mc enthalten, von denen 44 als unregelmäßig bezeichnet sind. Bei der Mehrzahl der übrigen ist infolge der Kleinheit der Schwankungen die Art des Lichtwechsels noch unbekannt oder wenigstens die Realität der angegebenen Perioden noch nicht gewährleistet; die 35 mit einigermaßen sichergestellten Perioden sind in der früher gegebenen Tabelle III zusammengestellt. Bei einigen wenigen wird eine kurze Periode, also Zugehörigkeit zur δ Cephei-Klasse, vermutet.

Unter den als unregelmäßig bezeichneten Veränderlichen der Klassen Ma, Mb, Mc haben nur zwei eine Helligkeitsschwankung von 2^m oder mehr, nämlich S Draconis ($A = 2^m,5$) und TZ Cygni ($A = 2^m,0$). Bei S Draconis scheinen die Lichtschwankungen manchmal lange Zeit ganz auszusetzen und dann wieder rasch zu verlaufen; auch TZ Cygni scheint manchmal lange Zeit konstant zu sein. Es ist auch bei diesen Sternen schwer, die reellen Lichtschwankungen von den durch Beobachtungsfehler vorgetäuschten zu unterscheiden; jedenfalls wäre eine sorgfältige Diskussion der Beobachtungen dazu nötig.

Die am öftesten genannten unter den im allgemeinen als unregelmäßig angesehenen Veränderlichen der Klasse M sind μ Cephei, α Orionis, α und g Herculis, ϱ Persei und R Lyrae. Die Helligkeitsschwankungen aller dieser Sterne halten sich innerhalb der Größenklasse.

μ Cephei (Ma) ist von J. PLASSMANN zum Gegenstand einer eingehenden Arbeit gemacht worden (Untersuchungen über den Lichtwechsel des Granatsterns μ Cephei, Münster i. W., 1904). Danach erlitt dieser Stern zur Zeit ARGELANDERS Lichtschwankungen in einer Periode von 400^d bis 460^d, die aber bis zu Ende des Jahrhunderts auf etwa 1000^d anwuchs. Die Amplitude dieses langsamen Lichtwechsels ist etwa $0^m,5$. Außerdem glaubt PLASSMANN noch eine kurze Periode feststellen zu können, deren Länge um 90^d beträgt, während die zugehörige Helligkeitsschwankung von der Größenordnung $0^m,1$ ist. Es ist aber für die richtige Beurteilung der Ergebnisse auf die oben erwähnten Bemerkungen von OSTHOFF zu verweisen. Die Radialgeschwindigkeit von μ Cephei ist nach W. W. CAMPBELL[2] um etwa 13 km veränderlich; ob irgendein Zusammenhang zwischen den Änderungen der Radialgeschwindigkeit und denen der Helligkeit besteht, wissen wir nicht.

α Orionis (Ma) ist sehr viel beobachtet worden, aber die Resultate der verschiedenen Beobachtungsreihen sind voll von Widersprüchen. Die Helligkeitsschätzungen sind bei diesem Stern wegen seiner großen Helligkeit besonders schwierig. Nach Messungen von J. STEBBINS[3] mit dem Selenphotometer sind langsame Helligkeitsänderungen vorhanden, über die sich rascher verlaufende (Periode etwa 250^d, Umfang $0^m,2$) lagern. Auch lichtelektrische Messungen von P. GUTHNICK und R. PRAGER[4] deuten auf langsame und gleichzeitige raschere Schwankungen hin. Schließlich glaubte H. OSTHOFF[5] aus seinen eigenen, aus den Jahren 1908 bis 1922 stammenden Beobachtungen eine Periode von sechs Jahren nachweisen zu können. Diese Periode stimmt überein mit

[1] M N 73, S. 116 (1912). [2] Lick Bull 7, S. 102 (1912). [3] Pop Astr 21, S. 5 (1913).
[4] Veröffentl. d. Stw. Babelsberg 2, Heft 3, S. 110 (1918).
[5] A N 216, S. 187 (1922).

derjenigen, die K. F. BOTTLINGER[1] vorher in den Radialgeschwindigkeiten von α Orionis gefunden hatte, die aber nicht ganz sicher schien, da die Änderungen der Radialgeschwindigkeit nur etwa 5 km betragen. J. LUNT[2] hat in der Tat später gezeigt, daß BOTTLINGERS Bahnelemente von α Orionis jedenfalls noch erheblicher Verbesserungen bedürfen.

Der Durchmesser von α Orionis ist auf dem Mt. Wilson-Observatorium von A. A. MICHELSON und F. G. PEASE[3] interferometrisch gemessen worden; er ergab sich zu $0'',047$, und wenn man die Parallaxe zu $0'',018$ annimmt, so folgt daraus ein linearer Durchmesser, der nur wenig kleiner ist als der der Marsbahn. Es scheint nach neueren Beobachtungen von PEASE[4] fast so, als ob der Durchmesser veränderlich ist. Ob Beziehungen zwischen den Änderungen der Helligkeit, der Radialgeschwindigkeit und des Durchmessers bestehen, ist eine noch ungeklärte Frage.

α Herculis (Mb), über dessen Helligkeitsschwankungen ebenfalls äußerst widerspruchsvolle Beobachtungen vorliegen, ist neuerdings von J. STEBBINS[5] photoelektrisch gemessen worden; es hat sich eine Periode von 120^d und eine Gesamtschwankung von $0^m,3$ ergeben, doch hat der Lichtwechsel Unregelmäßigkeiten aufzuweisen. (Näheres hat STEBBINS noch nicht veröffentlicht.) α Herculis ist ein Doppelstern; der Begleiter steht etwa $5''$ von dem Veränderlichen entfernt und hat bisher noch keine Bahnbewegung um diesen erkennen lassen. Der Begleiter (Spektrum F9) ist ein spektroskopischer Doppelstern, dessen Bahnelemente von R. F. SANFORD[6] bestimmt worden sind. Die Periode beträgt $51^d,6$, die Radialgeschwindigkeit des Schwerpunktes $-37,2$ km, während die Radialgeschwindigkeit des Veränderlichen $-32,2$ km, also um 5 km kleiner ist. Obwohl die beiden Sterne, nach ihrer spektroskopischen Parallaxe zu schließen, mindestens 250 astronomische Einheiten voneinander entfernt sind, ist diese Differenz doch immerhin auffallend; als Einsteineffekt kann sie wohl nicht gedeutet werden.

Der Lichtwechsel von R Lyrae (Mb) ist von B. OKOUNEFF[7] eingehend untersucht worden. Er ist sehr verwickelt, und wir müssen für Einzelheiten auf die zitierte Abhandlung verweisen. Die Grenzen der Lichtschwankung sind $4^m,0$ und $4^m,5$. Die Radialgeschwindigkeit ist um etwa 15 km veränderlich[8].

Über den Lichtwechsel der übrigen zwei oben besonders genannten unregelmäßigen Veränderlichen der Spektralklasse M, nämlich g Herculis (Mb) und ϱ Persei (Mb) wissen wir noch nichts Positives.

Als Beispiel für die Lichtkurve eines μ Cephei-Sternes geben wir die von X Herculis nach den Beobachtungen von P. M. RYVES[9] hier teilweise wieder.

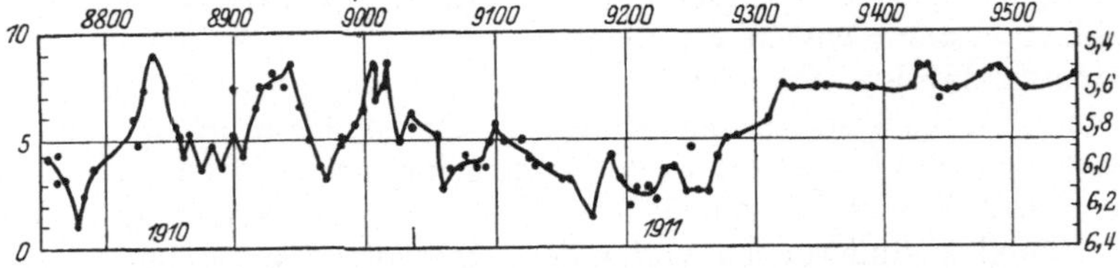

Abb. 21. Lichtkurve von X Herculis.

Inwieweit die Kurve durch persönliche Auffassungsfehler verfälscht ist, läßt sich nicht sagen.

[1] A N 187, S. 33 (1911). [2] Ap J 44, S. 250 (1916).
[3] Ap J 53, S. 249 (1921) = Mt Wilson Contr 203. [4] Publ A S P 34, S. 346 (1922)
[5] Pop Astr 27, S. 677 (1919). [6] Ap J 53, S. 212 (1921).
[7] A N 228, S. 153 (1926). [8] Publ A S P 36, S. 137 (1924).
[9] M N 84, S. 171 (1924).

Die μ Cephei-Sterne der Spektralklasse M sind ohne Zweifel im allgemeinen Riesen, wie die der Spektralklassen N und R. Wir kennen aber auch einen Zwergstern unter ihnen, nämlich SZ Ursae majoris[1]. Die absolute Größe dieses eine jährliche Eigenbewegung von 3″ besitzenden Sternes ist $10^m,9$, er ist also ein ausgesprochener Zwergstern; das Spektrum ist Ma. Miss E. F. LELAND hat mit Hilfe von photographischen Aufnahmen gefunden, daß er manchmal $0^m,5$ schwächer ist, als seiner normalen Helligkeit entspricht. Es ist wohl anzunehmen, daß bei solchen Zwergsternen die Ursache des Lichtwechsels eine ganz andere ist als bei den Riesen. Vielleicht besteht aber noch die Möglichkeit, daß es sich bei SZ Ursae majoris um einen Bedeckungsveränderlichen handelt.

44. μ Cephei-Sterne der Spektralklasse K. In „Katalog und Ephemeriden veränderlicher Sterne für 1926“ sind nur sechs Sterne der Spektralklasse K als unregelmäßig bezeichnet. Die größten Helligkeitsamplituden unter diesen haben: TW Aquilae ($A = 2^m,1$ phot.), bei dem die Unregelmäßigkeit aber noch zweifelhaft ist; V Pyxidis ($A = 2^m,4$ phot.), bei dem aber nach neueren Feststellungen[2] in Wirklichkeit die Lichtschwankungen bedeutend kleiner zu sein scheinen; SS Scorpii ($A = 2^m,0$), auf den die gleiche Bemerkung zutrifft. Größer als in den „Ephemeriden“ angegeben, sind nach Harv Bull 831 die Lichtschwankungen bei BM Scorpii, nämlich gleich $2^m,2$ phot.; dieser Stern scheint aber ein RV Tauri-Stern zu sein[3]. In Harv Bull 831 wird noch für mehrere weitere K-Sterne die Unregelmäßigkeit des Lichtwechsels festgestellt; die Helligkeitsschwankungen sind aber in allen diesen Fällen gering.

Das bekannteste Beispiel für die unregelmäßig veränderlichen K-Sterne ist α Cassiopeiae. (In den „Ephemeriden“ ist als Spektrum G8 angegeben, im neuen Draper-Katalog aber K0). H. PLATE[4] fand aus Beobachtungen von J. PLASSMANN für den Lichtwechsel dieses Sternes Perioden von $22^d,6$ bis $30^d,6$ (vgl. aber die oben erwähnten Ausführungen von OSTHOFF); H. GROUILLER und M. BLOCH[5] fanden aus Beobachtungen von M. L. LUIZET keine kurzen Perioden, wohl aber eine lange, deren Dauer im Mittel gerade ein Jahr beträgt. Die Lichtschwankung ist überhaupt nur klein ($0^m,5$). Es fragt sich aber, ob diesen Ergebnissen irgendeine reelle Bedeutung zuzuschreiben ist. OSTHOFF vermutet, daß der Stern unveränderlich sei[6].

Sehr merkwürdig ist nach K. BOHLIN[7] der Lichtwechsel des Sternes UZ Tauri, der nach BOHLIN ein K-Spektrum zu besitzen scheint. Dieser Veränderliche hatte im Oktober 1921 die Größe 9,2; dann nahm er bis Januar 1923 unter unregelmäßigen Schwankungen von 10^d bis 20^d Dauer etwa bis zur Größe $13^m,5$ ab. Es scheinen hier ähnliche Verhältnisse vorzuliegen wie bei α Orionis, nur daß die Amplitude des Lichtwechsels größer ist. Immerhin ist es wohl noch fraglich, ob der Stern zu den μ Cephei-Sternen gerechnet werden darf; die von BOHLIN gegebene Lichtkurve erinnert in mancher Beziehung an die einer Nova.

45. μ Cephei-Sterne mit besonderem Spektrum. Unter den Sternen, deren Spektrum in den „Ephemeriden für 1926“ als „Pec“ angegeben ist, können folgende wahrscheinlich oder vielleicht zur μ Cephei-Klasse gerechnet werden (diejenigen unter ihnen, die ein Spektrum ähnlich wie VX Andromedae besitzen, haben wir schon unter den N-Sternen mit betrachtet):

SU Monocerotis. Spektrum ähnlich denen der Klasse Mc (Draper-Katalog). $A = 0^m,7$. Über den Lichtwechsel ist noch nichts Näheres bekannt.

[1] Harv Bull 772 (1922). [2] Harv Bull 831 (1926). [3] Harv Circ 221 (1920).
[4] Mitt V A P 21, S. 132 (1911). [5] Lyon Bull 5, S. 74 (1923).
[6] A N 219, S. 135 (1923). [7] A N 218, S. 203 (1923).

W Cephei. Das Spektrum ist denen der Klasse Me verwandt[1]. Die weniger brechbaren Linien des H sind hell, und $H\alpha$ ist die hellste von ihnen. Außerdem sind noch andere helle Linien vorhanden. Das kontinuierliche Spektrum scheint M zu sein. Der Stern ist unregelmäßig veränderlich zwischen den photographischen Größen $8^m,6$ und $9^m,3$[2]. Ein ähnliches Spektrum wie W Cephei scheinen die geringe Lichtschwankungen aufweisenden Veränderlichen VV Cephei[3] und WY Geminorum[4] zu besitzen.

DI Carinae[5]. Hellster Teil des Spektrums zwischen $H\beta$ und $H\gamma$; zwei starke dunkle Banden sind sichtbar. Größe $11^m,8$ bis $13^m,2$ phot. Wohl unregelmäßig veränderlich.

RU Lupi. $H\beta$, $H\gamma$, $H\delta$, H und K hell, kontinuierliches Spektrum unbekannt. (Neuer Draper-Katalog.) Veränderlich zwischen $9^m,0$ und $11^m,0$ phot.

RY Telescopii. $H\beta$, $H\gamma$, $H\delta$ hell, kontinuierliches Spektrum unbekannt (Draper-Katalog). Veränderlich zwischen 11^m und $<15^m$ phot. Dürfte wohl nach der Größe des Lichtwechsels eher ein Mira-Stern als unregelmäßig veränderlich sein.

46. Abnorme unregelmäßige Veränderliche. In ,,Katalog und Ephemeriden veränderlicher Sterne für 1926" ist noch eine Reihe von unregelmäßigen Veränderlichen enthalten, über deren Spektrum und Farbe wir nichts wissen. Wenn die Amplituden klein sind, werden wir diese Sterne einstweilen ruhig zu den μ Cephei-Sternen zählen können. Es kommen nun aber in dem genannten Verzeichnis auch unregelmäßige Veränderliche vor, deren Spektrum oder deren Farbe gegen die Zugehörigkeit zur μ Cephei-Klasse sprechen, oder bei denen die Größe der Amplitude die Zugehörigkeit zu dieser Klasse zweifelhaft erscheinen läßt. Wir sehen hier natürlich ab von denjenigen Objekten, die man mit Sicherheit oder Wahrscheinlichkeit zur Klasse der Neuen Sterne oder der Novaähnlichen Veränderlichen, der R Coronae-, der U Geminorum- oder der RV Tauri-Sterne zählen kann. Sondert man alle diese aus, so bleiben schließlich 15 unregelmäßige Veränderliche übrig, deren Einordnung in unsere Klassen der veränderlichen Sterne Schwierigkeiten bereitet. Nun befinden sich unter diesen Objekten einige, die zwar als unregelmäßig bezeichnet sind, deren Lichtwechsel aber noch so unbekannt ist, daß diese Klassifikation immerhin als zweifelhaft bezeichnet werden muß; es sind dies: X Canum ven. (Farbe 3 in OSTHOFFS Skala, $A = 0^m,8$), RU Centauri (G0, $A = 0^m,7$), RV Librae (G5, $A = 0^m,7$), d Serpentis (A0p, $A = 0^m,7$), Z Serpentis (Farbe 3, $A = 0^m,8$), X Tauri (F5, $A = 1^m,5$), RT Vulpeculae (A0, $A = 1^m,5$). Wenn man diese Sterne beiseite läßt, so bleiben die folgenden übrig, die man also zunächst als abnorm bezeichnen muß und in keine der hier aufgestellten Klassen von Veränderlichen einordnen kann (die Sterne sind nach der Rektaszension geordnet und ihre Nummer in der G. u. L. ist angegeben):

Nr. 28. T Piscium (Farbe 0, $A = 1^m,5$). Lichtwechsel wenig bekannt. Es liegt nur eine einzige größere Beobachtungsreihe von E. SCHÖNFELD vor, deren Reduktion von Interesse wäre.

Nr. 52. RX Andromedae (Spektrum und Farbe unbekannt, $A = 3^m,0$). Gehört vielleicht zu den U Geminorum-Sternen mit raschem Lichtwechsel, wie X Leonis.

Nr. 234. RW Aurigae (Farbe 1, $A = 2^m,9$). Ein sehr merkwürdiger Veränderlicher, der hauptsächlich von S. ENEBO beobachtet ist. Die Lichtschwankungen verlaufen meist rasch und völlig regellos.

[1] Publ A S P 34, S. 58, 175 (1922); Pop Astr 30, S. 103 (1922).
[2] Harv Bull 764 (1922). [3] Publ A S P 33, S. 263 (1921).
[4] Publ A S P 34, S. 133 (1922); Harv Bull 767 (1922). [5] Harv Bull 780 (1922).

Nr. 1192. SS Scuti (Spektrum F8 bis K0 nach dem Draper-Katalog, $A = 0^m{,}8$). Nach E. ZINNER[1] verlaufen die Lichtschwankungen rasch und unregelmäßig. Die Zu- und die Abnahme erfolgen in weniger als 1^d.

Nr. 1283. X Lyrae (Farbe 4, $A = 1^m{,}3$). Nach den Beobachtungen der American Association of Variable Star Observers ist der Stern von 1913 bis Ende 1924 jedenfalls sehr nahe konstant gewesen. Auch W. DOBERCK[2] betrachtet seine Helligkeit als unveränderlich.

Nr. 1378. S Telescopii (Farbe und Spektrum unbekannt, $A = 3^m{,}5$). Könnte eine Nova gewesen sein, die im September 1887 hell (9^m) war. Der Stern war in den letzten Jahren jedenfalls nur sehr wenig veränderlich (nach Beobachtungen der American Association of Variable Star Observers). Gegen die Annahme, daß es sich um eine Nova handelt, spricht allerdings die erhebliche galaktische Breite.

Nr. 1607. RZ Lacertae (Spektrum A0, $A = 0^m{,}8$). Nach E. ZINNER[3] verlaufen die Helligkeitsänderungen ohne Regel innerhalb eines Tages.

Nr. 1670. ϱ Cassiopeiae (Spektrum cF8, $A = 0^m{,}7$). Auffällig ist die rötlichgelbe, zu dem Spektralcharakter des Sternes in Widerspruch stehende Farbe[4]. Die Helligkeitsbeobachtungen dieses Sternes geben zum Teil widerspruchsvolle Resultate. B. OKOUNEFF[5] glaubt aus einem größeren Beobachtungsmaterial, das von verschiedenen Beobachtern herrührt, eine Periode von 1100^d feststellen zu können. Es scheinen aber starke Unregelmäßigkeiten vorzukommen. W. S. ADAMS und A. H. JOY[6] bezeichnen das Spektrum als G5 und weisen auf die große Ähnlichkeit desselben mit dem der δ Cephei-Sterne hin.

Am merkwürdigsten unter diesen Sternen sind RW Aurigae und ϱ Cassiopeiae. Bei den übrigen besteht wohl noch die Möglichkeit, daß sie auf Grund eines sichereren Beobachtungsmaterials in eine der bekannten Klassen von Veränderlichen werden eingeordnet werden können.

g) Die Veränderlichen der RV Tauri-Klasse.

47. Definition und Vorbemerkungen. Ebensowenig wie für die meisten anderen Klassen von Variabeln läßt sich auch für die RV Tauri-Klasse bei dem heutigen Stande der Kenntnis eine scharfe Definition geben. Das am meisten charakteristische Kennzeichen dieser Veränderlichen ist, daß zwischen zwei Hauptminima in der Regel ein sekundäres Minimum eintritt, ähnlich wie bei dem Bedeckungsveränderlichen β Lyrae oder bei dem Mira-Veränderlichen R Centauri. Die sekundären Minima bleiben aber zuweilen aus, sie vertauschen sich auch zuweilen mit den Hauptminima, und die Lichtkurven sowie zum Teil auch die Abstände der Hauptminima sind stark veränderlich; diese Sterne können daher eine wirkliche Verwandtschaft mit den β Lyrae-Sternen nicht haben, wenn auch zeitweise auf Grund kürzerer Beobachtungsreihen Zweifel bestehen können, ob ein bestimmter Stern ein RV Tauri- oder ein β Lyrae-Stern ist.

Daß die RV Tauri-Sterne eine Klasse für sich bilden, haben schon S. ENEBO[7], der wohl auch zuerst den Namen eingeführt hat, und J. VAN DER BILT[8] ausdrücklich hervorgehoben. Eine systematische Aussonderung der RV Tauri-

[1] Ergänzungshefte zu den A N Bd. 4, Nr. 3 (1922). [2] A J 32, S. 32 (1919).
[3] Ergänzungshefte zu den AN Bd. 4, Nr. 3 (1922).
[4] P. GUTHNICK in A N 199, S. 177 (1914).
[5] A N 221, S. 225 (1924). [6] Publ A S P 31, S. 184 (1919).
[7] Beobachtungen veränderlicher Sterne VI u. VIII. Kristiania 1912 u. 1914.
[8] Recherches astron. de l'Obs. d'Utrecht VI (1916).

Sterne von den übrigen Veränderlichen hat zuerst H. LUDENDORFF[1] vorgenommen. In der Seeliger-Festschrift[2] ist er dann nochmals näher auf sie eingegangen.

Bei der Angabe der Perioden der RV Tauri-Sterne hat bisher oft eine gewisse Inkonsequenz obgewaltet. Man hat nämlich in manchen Fällen die Zeit vom Hauptminimum (Hm) bis zum folgenden Nebenminimum (Nm), meist aber die Zeit von einem Hm bis zum folgenden Hm als Periode bezeichnet. Stützt man sich allein auf die Lichtkurven der Sterne, so ist es keine Frage, daß die letztere Art der Periodenrechnung vorzuziehen ist. Die Nm bleiben nämlich manchmal aus, z. B. bei R Sagittae und R Scuti, oder sie sind nur schwach angedeutet. Niemand würde in solchen Fällen auf den Gedanken kommen, die Periode anders zu rechnen, also von Hm zu Hm. Sehr merkwürdig ist es nun, daß bei R Scuti die Periode der Änderung der Radialgeschwindigkeit gleich der Hälfte der von Hm zu Hm gerechneten Periode ist. Einstweilen wollen wir aber bei der Periodenzählung von Hm zu Hm bleiben, die wir daher in der folgenden Tabelle der RV Tauri-Sterne durchgängig zur Anwendung gebracht haben.

48. Die einzelnen RV Tauri-Sterne. Die folgende Tabelle enthält alle Sterne, die nach dem augenblicklichen Stande der Kenntnis zur RV Tauri-Klasse gehören (diese sind durch einen Stern bezeichnet), oder die in starkem Verdachte der Zugehörigkeit zu dieser Klasse stehen. Die erste Kolumne gibt den Namen des Sternes, die zweite die Periode P (von Hm zu Hm gerechnet), die dritte die Gesamtamplitude A der Lichtschwankung, die vierte den Abstand g vom galaktischen Äquator, die letzte endlich, soweit möglich, die Spektralklasse. Auf die Tabelle folgen Bemerkungen über die einzelnen Sterne. Hervorgehoben sei hier noch, daß auch BM Scorpii[3] und nach K. GRAFF RR Tauri[4] in dem Verdachte stehen, RV Tauri-Sterne zu sein, und daß ferner E. ZINNER[5] einige weitere Sterne namhaft gemacht hat, die er für RV Tauri-Sterne hält. Nach W. ZESSEWITSCH[6] scheint auch CX Cygni ein RV Tauri-Stern zu sein. Bei all diesen Objekten kann aber eine endgültige Entscheidung noch nicht getroffen werden.

Stern	P	A	g	Spektrum
TT Ophiuchi	61^d	$1^m,5$	27°	cF5e
*R Sagittae	71	1 ,8	11	cG1
*V Vulpeculae	75	1 ,0	9	cG7p
AC Herculis	76	1 ,0	13	F8
*RV Tauri	78	2 ,5	12	
UZ Ophiuchi	88	3 ,2	22	
*U Monocerotis	92	1 ,5	5	cG0p
TX Persei	95	1 ,4	21	
*TV Andromedae	127	2 ,2	16	
*R Scuti	140	4 ,5	3	G5e bis K2ep
S Aquilae	146	2 ,3	10	
RV Andromedae	172	2 ,7	11	Me?
U Bootis	190	3 ,9	59	
*RS Camelopardalis	(190)	1 ,0ph	34	Mb
*Z Ursae maj.	198	2 ,4	58	Mce
RT Hydrae	255	2 ,2	19	Me
*W Cygni	259	1 ,1	6	Me
*R Pictoris	333	2 ,8	39	Mae
RZ Cygni	556	4 ,6	1	Pec

[1] A N 214, S. 217 (1921). [2] S. 83.
[3] Harv Circ 221 (1920); Harv Bull 831 (1926).
[4] A N 213, S. 165 (1921); B Z 4, S. 27 (1922).
[5] A N 224, S. 269 (1925). [6] A N 227, S. 55 (1926).

Anmerkungen. TT Ophiuchi. Nach E. LEINER[1] ist die Lichtkurve β Lyrae-artig, doch sind die Hm und Nm in der Helligkeit im allgemeinen nicht allzu verschieden. Sie scheinen sich manchmal zu vertauschen. Die Lichtkurve ist etwas veränderlich, mehr, als es für einen β Lyrae-Stern statthaft wäre; namentlich ist der Abstand der beiden Maxima variabel. Die Zeiten der Minima werden sehr genau innegehalten, entsprechend der Periode von $61^d,12$. Für einen RV Tauri-Stern ist der Lichtwechsel auffallend regelmäßig, und der letzte Zweifel, ob hier nicht vielleicht doch ein Bedeckungsveränderlicher vorliegt, ist noch nicht behoben. Nach W. S. ADAMS und A. H. JOY[2] ist das Spektrum cF5e; $H\beta$ und $H\gamma$ sind hell, ihre Intensität ändert sich mit der Phase des Lichtwechsels, und zwar sind sie im Maximum heller als im Minimum[3]. Die Radialgeschwindigkeit ist etwas veränderlich.

R Sagittae. Der Lichtwechsel ist von P. GUTHNICK in der G. u. L. sehr genau beschrieben worden. Die Periode von ungefähr 71^d bleibt erhalten, Haupt- und Nebenminima vertauschen sich nach längeren Intervallen miteinander. Die Lichtkurve ist außerordentlich veränderlich, sie ähnelte z. B. 1861 der von β Lyrae, 1880 glich sie mehr der eines Algol-Sternes (zeitweise ohne sekundäres Minimum, zeitweise mit einem solchen), Ende 1862 der eines ζ Geminorum-Sternes (d. h. eines δ Cephei-Sternes mit angenähert symmetrischer Lichtkurve) von $^1/_2\, P = 35^d$ Periode.

Das Spektrum ist nach W. S. ADAMS, A. H. JOY und R. F. SANFORD[4] cG1. Für die Radialgeschwindigkeit ergaben sich aus acht Platten Werte von -28 km bis $+32$ km; sie ist also stark veränderlich.

V Vulpeculae hatte 1903 bis 1904 eine β Lyrae-artige Lichtkurve, 1905 bis 1906 dagegen eine δ Cephei-artige von der halben Periode. Die Beobachtungen der American Association of Variable Star Observers 1915 bis 1924 sind zu wenig zahlreich, um viel daraus folgern zu können, doch scheint die Lichtkurve während dieses Zeitintervalles teils δ Cephei-artig, teils β Lyrae-artig gewesen zu sein. Zeitweise sind die Änderungen überhaupt nur sehr klein. Die Periode von etwa 75^d scheint sich zu erhalten. Auch Beobachtungen von E. LEINER[5] aus den Jahren 1919 bis 1926 bestätigen die Zugehörigkeit zur RV Tauri-Klasse.

Das Spektrum ist nach W. S. ADAMS, A. H. JOY und R. F. SANFORD[4] cG7. Die Radialgeschwindigkeit ist veränderlich, da sich aus 20 Platten dafür Werte zwischen -29 km und $+4$ km ergeben. Die Wasserstoffabsorptionslinien sind auffallend schwach[3], „a condition almost certainly due to partial balancing of absorption and emission in these lines".

AC Herculis. Nach G. ZACHAROV[6] und E. LEINER[7] ähnelt die Lichtkurve der von β Lyrae, doch ist das Hauptminimum stark unsymmetrisch (Helligkeitsabnahme vor dem Hm weit langsamer als die Zunahme nach demselben). LEINER betrachtet die Lichtkurve als veränderlich und nennt AC Herculis einen typischen R Sagittae-Stern, ZACHAROV sieht ihn dagegen als β Lyrae-Stern an. Die Hm lassen sich mit einer konstanten Periode bisher sehr gut darstellen. Die Zugehörigkeit zur RV Tauri-Klasse ist noch zweifelhaft. Das Spektrum ist F8[8].

RV Tauri. Der Lichtwechsel dieses Sternes ist von J. VAN DER BILT[9] ausführlich untersucht worden; dortselbst ist auch eine Lichtkurve wiedergegeben.

[1] A N 210, S. 275 (1920); 213, S. 213 (1921); 218, S. 153 (1923).
[2] Pop Astr 28, S. 513 (1920.)
[3] Report of the Director of the Mount Wilson Obs. 1922, S. 234.
[4] Publ A S P 36, S. 139 (1924). [5] A N 229, S. 27 (1926).
[6] A N 221, S. 136 (1924); 222, S. 295 (1924). [7] A N 221, S. 247 (1924).
[8] Harv Circ 225 (1921). [9] Recherches astr. de l'Obs. d'Utrecht VI (1916).

Sie ist sehr veränderlich, bald β Lyrae-, bald δ Cephei-artig, aber niemals Algol-artig. Die Nebenminima sind zuweilen ebenso tief oder tiefer als die Hauptminima. Die Maxima und Minima lassen sich mit einer Periode von 39^d leidlich darstellen, wenn man aber die β Lyrae-artigen Lichtkurven als charakteristisch ansieht, so ist natürlich $P = 78^d$ zu setzen. Wegen der Einzelheiten ist auf die genannte Abhandlung zu verweisen.

Ein Stück der Lichtkurve nach J. VAN DER BILT geben wir in Abb. 22 wieder. (Die Ordinaten in dieser Abbildung sind Intensitäten.) Die Farb-

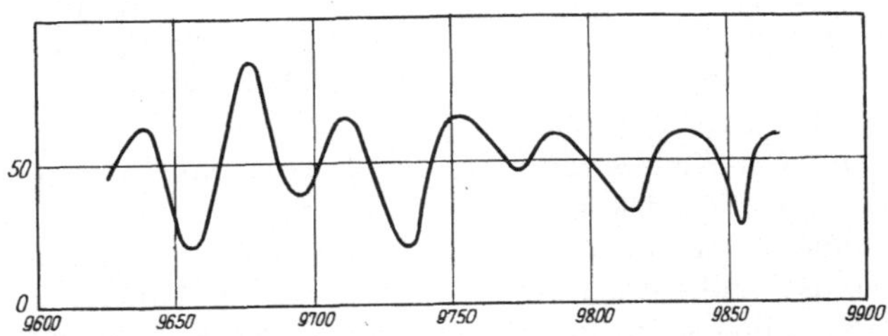

Abb. 22. Lichtkurve von RV Tauri nach VAN DER BILT.

klasse des Sternes ist *f*, das direkt nicht beobachtete Spektrum dürfte also ungefähr F sein.

UZ Ophiuchi. Nach E. LEINER[1] ist die β Lyrae-artige Lichtkurve von Periode zu Periode merklichen Änderungen unterworfen. Die Hm wie die Nm besitzen wechselnde Tiefen, die Helligkeit im Hm variiert um $^1/_2{}^m$. Die Hm treten sehr regelmäßig, der Länge der Periode entsprechend, ein. Die Zugehörigkeit zur RV Tauri-Klasse ist wohl noch sehr zweifelhaft. Die Farbe ist gelb.

U Monocerotis. Für diesen Stern sind nur wenige Beobachtungen veröffentlicht[2]. E. F. SAWYER, der ihn lange beobachtet hat, hat nur die Zeiten und Helligkeiten der Maxima und Minima angegeben. Danach treten häufig abwechselnd flache und tiefe Minima auf, so daß die Lichtkurve β Lyrae-artig wird ($P = 92^d$), während zu andern Zeiten der Lichtwechsel δ Cephei-artig ist ($P = 46^d$). Die Haupt- und Nebenminima vertauschen sich bisweilen, die Periode scheint sich zu erhalten. Es ist kein Zweifel, daß der Stern zum RV Tauri-Typus gehört.

Das Spektrum[3] ist cG0p; die Absorptionslinien des Wasserstoffs zeigen dieselben Eigentümlichkeiten wie bei V Vulpeculae, aber in noch stärkerem Grade. Die Radialgeschwindigkeit ist veränderlich.

TX Persei. Nach den Beobachtungen von S. ENEBO[4] und M. L. LUIZET[5] dürfte dieser Stern wohl zur RV Tauri-Klasse gehören, wenn auch noch weitere Beobachtungen zur endgültigen Entscheidung dieser Frage nötig sind.

TV Andromedae. S. ENEBO hat in Nr. VI seiner „Beobachtungen veränderlicher Sterne" eine graphische Darstellung des Lichtwechsels für die Zeit von 1908 bis 1912 gegeben. Es herrschten damals β Lyrae-ähnliche Lichtkurven vor; im übrigen war der Lichtwechsel sehr unregelmäßig. Nach späteren, in Nr. IX veröffentlichten Beobachtungen ENEBOS aus der Zeit von 1912 bis 1916 sind dann die Nebenminima augenscheinlich ganz weggefallen, die Form der Lichtkurve ist auch in dieser Zeit sehr wechselnd. Auch die Periode scheint erheblichen Änderungen unterworfen. Es ist kein Zweifel, daß hier ein RV Tauri-Stern vorliegt. Die Farbe ist ungefähr 6 in der OSTHOFFschen Skala.

[1] A N 216, S. 295 (1922). [2] G. u. L 1, S. 227.
[3] Report of the Director of the Mt. Wilson Obs. 1922, S. 234.
[4] Beobachtungen veränderlicher Sterne VIII (1914); A N 217, S. 442 (1923).
[5] Lyon Bull. 6, S. 178 (1924).

R Scuti. Für diesen Stern hat sich ein sehr großes Beobachtungsmaterial angesammelt, dessen einheitliche Bearbeitung erwünscht wäre; ohne Zweifel gehört er zu der hier besprochenen Klasse von Veränderlichen. Die Lichtkurve ist außerordentlich veränderlich, immer wieder aber treten β Lyrae-artige Kurven mit ungefähr 140^d Periode hervor. Zeitweise (z. B. 1876 bis 1877) war der Lichtwechsel überhaupt nur gering, manchmal erreicht er in kurzer Zeit 3 bis 4 Größenklassen. Ausgeprägt Algol-artige Kurven scheinen nur selten vorzukommen. Wenn der Lichtwechsel beträchtlich ist, überwiegen stets die β Lyrae-artigen Kurven; um diese Zeiten scheinen sich die Haupt- und Nebenminima auch nur ziemlich selten zu vertauschen.

Das Spektrum von R Scuti ist zuerst von R. H. CURTISS näher untersucht worden[1]. Nach A. H. JOY[2] variiert es zwischen G5e im Maximum der Helligkeit und K2ep (K2 mit hellen Linien des Wasserstoffs und Absorptionsbanden des Titanoxyds, also eng mit Me verwandt). Die Radialgeschwindigkeit ist nach R. H. CURTISS[3] um etwa 10 km variabel und besitzt eine veränderliche Periode von 65^d bis 75^d (also offenbar gleich der Hälfte der mittleren Periode des Lichtwechsels, die von Hauptminimum zu Hauptminimum gerechnet ist). Die Radialgeschwindigkeit des Schwerpunktes, befreit von der Sonnenbewegung, ist etwa +58 km, also, wie bei den Me-Sternen, beträchtlich. Die hellen Linien des H sind gegen die Absorptionslinien nach Violett verschoben, und zwar im Helligkeitsmaximum mehr als bald nach dem Minimum. Auch hier zeigt R Scuti also ein ähnliches Verhalten wie die Me-Sterne. Die Absorptionslinien des H werden schwächer, wenn die Helligkeit zunimmt.

Kolorimetrische Messungen von R Scuti hat H. VOGT[4] in Heidelberg angestellt. Nach denselben hat der Stern im Helligkeitsmaximum eine effektive Temperatur von etwa 4000°, im Minimum eine solche von 2500°. Der effektive Durchmesser ergibt sich am größten im Helligkeitsminimum, am kleinsten im Maximum. Die bolometrische Helligkeit ändert sich, wie bei den Mira-Sternen, viel weniger als die visuelle.

S Aquilae. Der Lichtwechsel dieses Sternes ist von R. MÜLLER[5] diskutiert worden. Die Form der Lichtkurve ist sehr veränderlich. Häufig treten Nebenminima auf, die sich allerdings nicht mit den Hauptminima vertauschen. Wahrscheinlich ist der Stern trotzdem zur RV Tauri-Klasse zu rechnen.

RV Andromedae ist von 1904 bis 1912 ziemlich viel beobachtet worden; es zeigen sich zuweilen ausgeprägte Nebenminima ungefähr in der Mitte zwischen den Hauptminima, die Lichtkurve scheint stark veränderlich. Es ist aber doch wohl noch fraglich, ob der Stern zur RV Tauri-Klasse zu rechnen ist.

U Bootis. Die bisherigen Beobachtungen dieses Sternes sind ebenfalls von R. MÜLLER[6] diskutiert worden. Die Periode und die Form der Lichtkurve sind sehr veränderlich, manchmal sind Nebenminima angedeutet, so daß der Stern vielleicht als eine Übergangsform zwischen der Mira- und der RV Tauri-Klasse angesehen werden darf. Um einen typischen RV Tauri-Stern handelt es sich jedenfalls nicht. Die Farbe ist 5 in der OSTHOFFschen Skala, das Spektrum ist unbekannt.

RS Camelopardalis. Nach der von C. MARTIN und H. C. PLUMMER[7] gegebenen und in Abb. 23 teilweise reproduzierten photographischen Lichtkurve für 1909 bis 1914 ist sicher anzunehmen, daß der Stern zum RV Tauri-Typus gehört. Die Periode scheint sehr veränderlich zu sein. Die genannten Autoren

[1] Ap J 20, S. 232 (1904).
[2] Publ A S P 34, S. 349 (1922).
[3] Pop Astr 32, S. 220 (1924).
[4] A N 228, S. 89 (1926).
[5] A N 229, S. 179 (1927).
[6] A N 223, S. 284 (1925).
[7] M N 76, S. 612 (1916); s. auch M N 77, S. 118 (1916).

haben die Lichtkurve genau analysiert, jedoch kann hier auf ihre Ergebnisse nicht näher eingegangen werden.

Z Ursae majoris. Nach der eingehenden Bearbeitung der bisherigen Beobachtungen dieses Sternes durch R. MÜLLER[1] liegt wohl sicher Zugehörig-

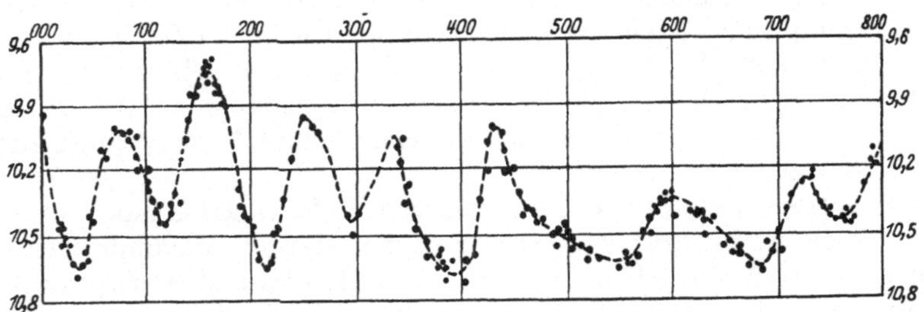

Abb. 23. Lichtkurve von RS Camelopardalis nach MARTIN und PLUMMER.

keit zur RV Tauri-Klasse vor. Die Lichtkurve ist oft β Lyrae-artig, die Hm und Nm vertauschen sich manchmal, es kommt auch vor, daß das Nm ausbleibt. Zeitweise ist der Lichtwechsel sehr unregelmäßig.

RT Hydrae. Die Lichtkurve scheint sehr veränderlich zu sein; das vorliegende Beobachtungsmaterial ist aber wenig befriedigend, und die Zugehörigkeit des Sternes zur RV Tauri-Klasse ist noch sehr zweifelhaft. H. LUDENDORFF hat den Lichtwechsel näher diskutiert[2].

W Cygni. Diesem Stern haben H. H. TURNER und MARY A. BLAGG[3] ausführliche Untersuchungen gewidmet, und sie haben versucht, die komplizierte Lichtkurve durch eine verwickelte Formel darzustellen. Die Lichtkurve setzt sich danach im wesentlichen zusammen aus zwei übereinander lagernden Wellen von $129^d,6$ (Amplitude $0^m,72$) und 243^d (Amplitude $0^m,52$) Periode. Nach $15 \cdot 129^d,6 = 8 \cdot 243^d = 1944^d$ wiederholt sich der Lichtwechsel. Die Lichtkurve, die für dieses Intervall von 1944^d gegeben wird, ähnelt durchaus der eines RV Tauri-Sternes (sekundäres Minimum zwischen zwei Hauptminima, allmählich vertauschen sich Haupt- und Nebenminimum). Betrachtet man W Cygni als RV Tauri-Stern, so ist $P = 2 \cdot 129^d,6 = 259^d$ zu setzen. Wegen der auftretenden Unregelmäßigkeiten muß auf die zitierten Arbeiten verwiesen werden. Es bedürfte einer näheren Untersuchung, ob der weitere Verlauf des Lichtwechsels die abgeleiteten Gesetze, die übrigens schon in der zweiten zitierten Abhandlung gewisse Einschränkungen erfahren, bestätigt hat. 1922 bis 1924 ist der Lichtwechsel jedenfalls kleiner gewesen, als es die Lichtkurve von TURNER und Miss BLAGG erfordert. Es ist wohl sicher, daß der Stern zur RV Tauri-Klasse gehört. Über das Spektrum ist schon in Ziff. 27γ berichtet worden.

R Pictoris. Die bisherigen Beobachtungen dieses Veränderlichen sind von R. MÜLLER[4] zusammenfassend bearbeitet worden. In der Mitte zwischen zwei Hm liegt ein Nm, und die Hm und Nm haben sich mehrmals miteinander vertauscht. Die Periode ist Änderungen unterworfen. Es ist wohl kein Zweifel, daß der Stern zur RV Tauri-Klasse gehört. Die Radialgeschwindigkeit ist nach LEAH B. ALLEN[5] sehr groß (+208 km); dieser Umstand wäre weniger auffällig, wenn man die Periode vom Hm bis zum nächsten Nm rechnet, also $= 166^d$ ansetzt, da nur Me-Sterne mit kurzen Perioden so große Radialgeschwindigkeiten haben.

---

[1] A N 223, S. 281 (1925). [2] A N 219, S. 15 (1923).
[3] M N 80, S. 41 (1919); 81, S. 144 (1920). [4] A N 225, S. 249 (1925).
[5] Lick Bull 12, S. 71 (1926).

RZ Cygni. Auch diesen Veränderlichen hat R. MÜLLER[1] bearbeitet. Die Periode unterliegt langsamen Änderungen. Die Lichtkurve ist sehr veränderlich, häufig ist sie β Lyrae-artig. Manchmal ist die Helligkeit des Sternes im Hm längere Zeit (etwa 100^d) konstant. Der Stern dürfte ebenfalls zur RV Tauri-Klasse zu rechnen sein, wenn auch die Länge der Periode (556^d) sehr auffällig ist, so daß noch einige Zweifel bestehen bleiben. (Betreffs des Spektrums vgl. Ziff. 22, Tab. V.)

Die Radialgeschwindigkeiten der hier vorkommenden Me-Sterne sind schon in Ziff. 32 behandelt worden.

49. Allgemeines über die RV Tauri-Sterne. Die RV Tauri-Sterne mit kurzen Perioden haben manches mit den δ Cephei-Sternen gemeinsam; z. B. gleicht die Lichtkurve von R Sagittae zeitweise der eines δ Cephei-Sternes mit symmetrischer Lichtkurve und mit der Periode von 35^d, die von V Vulpeculae zeitweise der Lichtkurve eines δ Cephei-Sternes mit einer Periode von 37^d. Die Spektren haben, wie die der δ Cephei-Sterne, c-Charakter, und die Radialgeschwindigkeiten sind variabel wie die dieser Sterne. (Allerdings ist es noch nicht bekannt, ob der Zusammenhang zwischen den Änderungen der Radialgeschwindigkeit und denen der Helligkeit ähnlich ist wie bei den δ Cephei-Sternen.) Die Amplituden des Lichtwechsels und die galaktischen Breiten sind im allgemeinen klein, wie bei den δ Cephei-Sternen. Ein grundlegender Unterschied aber besteht darin, daß die Lichtkurven bei den typischen RV Tauri-Sternen sehr veränderlich, bei den δ Cephei-Sternen dagegen nahezu unveränderlich sind.

Noch enger sind die Beziehungen zwischen den RV Tauri-Sternen mit langen Perioden und den Mira-Sternen, die Lichtkurven der Form γ_2 (mit sekundärem Minimum) besitzen. So hat z. B. Z Ursae majoris zeitweise einen sehr ähnlichen Lichtwechsel wie R Ursae minoris. Wie die Mira-Sterne, sind die langperiodischen RV Tauri-Sterne nicht auf die Nähe der Milchstraße beschränkt, die Perioden haben bei beiden Klassen von Sternen dieselbe Länge. Auch in den Spektren sind Unterschiede nicht vorhanden. Allerdings ist bei den der Spektralklasse Me angehörenden RV Tauri-Sternen die Violettverschiebung der Emissionslinien geringer, als man es nach der MERRILLschen Kurve (vgl. Ziff. 32) für die betreffenden Periodenwerte erwarten sollte. Diese Verschiebung $V_a - V_e$ ist für vier Sterne unserer Tabelle der RV Tauri-Sterne gemessen, und zwar für:

	P	$V_a - V_e$	MERRILLS Kurve
Z Ursae majoris	198^d	+6 km	+12 km
RT Hydrae	255	+5	+ 8
W Cygni	259	−1	+ 8
R Pictoris	333	+8	+13

Eine bessere Übereinstimmung zwischen den gemessenen Werten $V_a - V_e$ und den aus MERRILLS Kurve entnommenen würde man erhalten, wenn man P vom Hm zum Nm rechnete, also für P die Hälfte der oben angebenen Zahlen einsetzte. Aber auch bei R Centauri, der zweifellos ein Mira-Stern mit Lichtkurve γ_2 ist, ist $V_a - V_e$ klein, nämlich $= +8$ km, während aus der (für die Periode $P = 561^d$ von R Centauri zu extrapolierenden) Kurve von MERRILL ein Wert von etwa $+27$ km folgen würde. Ein Unterschied zwischen den langperiodischen RV Tauri-Sternen und den Mira-Sternen besteht insofern, als die Helligkeitsamplituden bei den meisten der ersteren relativ klein und die Lichtkurven veränderlicher sind als bei den Mira-Sternen. Immerhin ist es fraglich, ob man berechtigt ist, die langperiodischen RV Tauri-Sterne als besondere Klasse von den Mira-Sternen abzutrennen, oder ob man sie nur als Mira-Sterne mit besonders stark veränderlichen Lichtkurven zu betrachten hat. Es ist das schließlich eine Sache der Definition.

[1] A N 225, S. 249 (1925).

Es wäre nach diesen Ausführungen denkbar, daß die in der Tabelle aufgezählten RV Tauri-Sterne in zwei wesensverschiedene Gruppen zerfallen, nämlich in kurzperiodische, die mit den δ Cephei-Sternen verwandt sind, und langperiodische, die in sehr engen Beziehungen zu den Mira-Sternen stehen. Aber der Lichtwechsel dieser beiden Gruppen hat doch so viel Gemeinsames, daß man sich zunächst noch gegen eine solche Zweiteilung aussprechen möchte. Auf die Einheit der RV Tauri-Sterne deutet das gleichmäßige Fortschreiten der Periodenlängen in der Tabelle hin (es ist höchstens etwa bei $P = 100^d$ bis 120^d eine Lücke angedeutet), das gleichzeitige Anwachsen der durchschnittlichen galaktischen Breite und das gleichzeitige Fortschreiten des Spektraltypus von F bis M. Hinsichtlich der eben erwähnten Andeutung einer Periodenlücke ist allerdings bemerkenswert, daß die Lücke, wenn man die halben Periodenwerte als maßgebend ansieht, zusammenfällt mit der angedeuteten Lücke in der Periodenverteilung der Veränderlichen mit Perioden von 45^d bis 90^d (vgl. Ziff. 40).

Auch mit den μ Cephei-Sternen haben die langperiodischen RV Tauri-Sterne manche Ähnlichkeit, wenn auch unter den ersteren solche mit Me-Spektren (mit Ausnahme des für sich allein dastehenden Sternes S Persei) nicht vorkommen. In der Tat sind die Grenzen zwischen den beiden eben genannten Klassen und den Mira-Sternen sehr schwer zu ziehen. Z. B. kann der überaus interessante, von W. HASSENSTEIN[1] sorgfältig untersuchte Stern RU Cephei (Spektrum K8) als Mira-Stern angesehen werden, weil er seine Periode von 110^d leidlich gut innehält (Epochensprünge wie bei RU Cephei kommen auch bei anderen Mira-Sternen vor), als μ Cephei-Stern, weil sich die Gestalt seiner Lichtkurve von Periode zu Periode oft gänzlich ändert, und schließlich als Übergangsform zwischen einer dieser beiden Klassen und den RV Tauri-Sternen, weil manchmal sekundäre Minima angedeutet oder auch deutlich ausgeprägt sind.

Aus dem Spektralcharakter der RV Tauri-Sterne müssen wir schließen, daß sie Riesensterne sind. Befriedigende Theorien zur Erklärung ihres Lichtwechsels fehlen natürlich noch gänzlich.

h) Die Veränderlichen der δ Cephei-Klasse.

50. Definition, Unterabteilungen und Bezeichnungen. Unter δ Cephei-Sternen, die nach ihrem typischen Vertreter, δ Cephei, so benannt werden, versteht man diejenigen veränderlichen Sterne, deren Helligkeit sich in Perioden bis zu 45^d ändert. Die Lichtschwankungen gehen sehr regelmäßig vor sich, d. h. die Lichtkurve bleibt von Periode zu Periode unverändert oder ändert sich doch nur außerordentlich wenig. Auch die Änderungen der Periode sind, sofern überhaupt solche festgestellt werden können, sehr klein und verlaufen sehr langsam. Die visuellen Helligkeitsschwankungen sind nur von geringem Umfang; in der Mehrzahl der Fälle übersteigen sie nicht $1^m,0$, nur in wenigen Fällen erreichen oder übersteigen sie $1^m,5$.

Die δ Cephei-Sterne zerfallen deutlich in zwei Gruppen, nämlich in solche mit Perioden bis zu 1^d (kurzperiodische δ Cephei-Sterne) und solche mit Perioden von mehr als 1^d (langperiodische δ Cephei-Sterne). Diese beiden Gruppen zeigen viele Verschiedenheiten untereinander, es kann aber kein Zweifel daran bestehen, daß sie, was die Ursache des Lichtwechsels angeht, sehr eng miteinander verwandt sind. Eine dritte Gruppe endlich, mit Perioden von Bruchteilen des Tages und äußerst kleinen, nur durch genaueste Messungen feststellbaren Lichtschwankungen, bilden die heute noch wenig zahlreichen Veränderlichen, deren Hauptvertreter β Cephei ist, und die wir daher kurz

[1] Potsd Publ Nr. 83 (1926).

als β Cephei-Sterne bezeichnen wollen. Ob sie wirklich zu den δ Cephei-Sternen zu rechnen sind, steht noch dahin; ihre Lichtschwankungen sind im allgemeinen viel unregelmäßiger als die der eigentlichen δ Cephei-Sterne. Sie sollen mit in diesem Kapitel behandelt werden, zunächst sind sie aber nur dort in die Betrachtung eingeschlossen, wo es besonders erwähnt wird.

In den kugelförmigen Sternhaufen, den MAGELLANschen Wolken und den Spiralnebeln kommen viele δ Cephei-Sterne vor. Diese „Sternhaufen-Veränderlichen“ werden, obwohl es sich um δ Cephei-Sterne handelt, erst im nächsten Kapitel näher besprochen. Soweit es nicht besonders hervorgehoben wird, haben wir es also zunächst nur mit den isolierten, zum galaktischen System gehörigen δ Cephei-Sternen zu tun.

Die scheinbaren Helligkeiten der δ Cephei-Sterne sind meist ziemlich gering. Immerhin erreicht von den langperiodischen α Ursae minoris im Maximum die Größe 2,3, l Carinae und δ Cephei die Größe 3,6, η Aquilae und ζ Geminorum die Größe 3,7. Die kurzperiodischen sind durchweg sehr schwach; der hellste von ihnen, RR Lyrae, hat die Maximalgröße 7,1.

Über die namentliche Bezeichnung der δ Cephei-Sterne besteht große Uneinigkeit. Häufig nennt man sie statt „δ Cephei-Sterne“ einfach „Cepheiden“. Diese Bezeichnung ist trotz ihrer größeren Einfachheit nicht empfehlenswert, denn nach sonstigem astronomischen Sprachgebrauch hat man unter „Cepheiden“ Sternschnuppen zu verstehen, deren Radiationspunkt im Sternbilde Cepheus liegt. E. HARTWIG hat für diejenigen δ Cephei-Sterne, deren Helligkeitszunahme rasch erfolgt, den Namen „Blinksterne“ vorgeschlagen, der aber keinen Anklang gefunden hat. δ Cephei-Sterne mit angenähert symmetrischen Lichtkurven nennt man häufig nach ihrem bekanntesten Vertreter „ζ Geminorum-Sterne“. Diejenigen kurzperiodischen δ Cephei-Sterne, die eine konstante Phase im Minimum und einen steilen Helligkeitsanstieg besitzen, nennt man nach einem Vorschlage von HARTWIG oft „Antalgol-Sterne“, weil ihre Lichtkurve sozusagen entgegengesetzt verläuft wie die des Bedeckungsveränderlichen Algol. Angelsächsische Autoren bezeichnen die Antalgol-Sterne häufig als „Cluster Type“, weil sie hauptsächlich in Sternhaufen vorkommen; in weiterem Sinne rechnet man zum „Cluster Type“ dann auch die übrigen kurzperiodischen δ Cephei-Sterne, deren Lichtkurven nicht Antalgol-Charakter tragen.

51. Vorbemerkungen über die δ Cephei-Sterne. Wie bei den Mira-Sternen wollen wir an den Eingang unserer Ausführungen über die δ Cephei-Sterne eine kurze Schilderung der hauptsächlichen Eigenschaften dieser Veränderlichen stellen, da hierdurch das Verständnis des Folgenden erleichtert wird. Wir stützen uns dabei auf eine Darstellung von MARGARETE GÜSSOW[1].

Die kurzperiodischen δ Cephei-Sterne ($P < 1^d$) haben ein Maximum der Häufigkeit bei Perioden von $0^d,5$, die langperiodischen bei solchen von 5^d. Ein deutliches Minimum in der Häufigkeitsfunktion der Perioden findet sich bei $P = 1^d$, denn nur zwei δ Cephei-Sterne haben Perioden zwischen $0^d,75$ und $1^d,5$, nämlich UZ Cassiopeiae ($P = 0^d,81$) und TX Scorpii ($0^d,94$). Durch dieses Minimum in der Häufigkeitsverteilung der Perioden wird zunächst die Zweiteilung in kurz- und in langperiodische δ Cephei-Sterne bedingt. Die kurzperiodischen δ Cephei-Sterne haben scheinbar (und auch absolut) bedeutend geringere Helligkeiten als die langperiodischen. Für letztere besteht eine enge Korrelation zwischen absoluter Helligkeit und Periode, in dem Sinne, daß längeren Perioden eine größere absolute Helligkeit entspricht (SHAPLEYS „Period-Luminosity Curve“); die kurzperiodischen haben alle dieselbe absolute Helligkeit ($-0^m,3$). Alle δ Cephei-Sterne sind Riesensterne, ihre Parallaxen sind durchweg sehr klein. Größere

[1] V J S 59, S. 171 (1924).

Eigenbewegungen kommen nur bei den kurzperiodischen vor; diese haben auch durchschnittlich große Radialgeschwindigkeiten, während die der langperiodischen klein sind. Die kurzperiodischen δ Cephei-Sterne verteilen sich auf alle galaktischen Breiten, die langperiodischen zeigen eine sehr starke Konzentration nach der Milchstraße.

Die Spektra der δ Cephei-Sterne gehören fast ausschließlich den Klassen A bis K an; die kurzperiodischen haben durchschnittlich „früheren" Spektraltypus als die langperiodischen. Die Spektra tragen c-Charakter. Mit dem Lichtwechsel ändert sich auch das Spektrum der δ Cephei-Sterne; der Spektraltypus im Helligkeitsmaximum entspricht einer höheren Temperatur als der im Helligkeitsminimum.

Die visuelle Helligkeitsamplitude übersteigt nicht wesentlich $1^m,5$. Die photographische Helligkeitsamplitude ist größer als die visuelle.

Die Lichtkurven sind zumeist stark unsymmetrisch, d. h. der Helligkeitsanstieg ist steiler als die Helligkeitsabnahme. Sekundäre Wellen kommen in der Lichtkurve hauptsächlich auf dem absteigenden Aste vor. Im allgemeinen sind die Änderungen der Lichtkurve nur gering. Veränderliche Perioden scheinen hauptsächlich bei den kurzperiodischen δ Cephei-Sternen vorzukommen.

Die Radialgeschwindigkeiten der δ Cephei-Sterne sind veränderlich, und zwar synchron mit den Änderungen der Helligkeit. Im großen ganzen sind Licht- und Geschwindigkeitskurve einander spiegelbildlich ähnlich. Durchschnittlich entsprechen größeren Helligkeitsamplituden auch größere Amplituden der Radialgeschwindigkeit.

Die Literatur über die δ Cephei-Sterne ist ganz außerordentlich umfangreich. Zur vorläufigen Orientierung darüber sollen hier die Titel einiger Abhandlungen allgemeineren Inhalts über diese Veränderlichen angeführt werden; die Beobachtungssammlungen (s. Ziff. 3), die Erklärungsversuche und die Arbeiten über die Veränderlichen in Sternhaufen sind an dieser Stelle nicht berücksichtigt. Zahlreiche weitere Arbeiten werden bei Gelegenheit der folgenden Ausführungen zitiert werden.

CHANG, Y., Monographie préliminaire des Céphéides. Lyon Bull 8, S. 135 (1926).

GÜSSOW, MARGARETE, Kritische Zusammenstellung sämtlicher Beobachtungsergebnisse der Veränderlichen vom δ Cephei-Typus und Kritik der EDDINGTONschen Pulsationstheorie. Inaug.-Diss. Berlin 1924.

(Diese leider nur in wenigen Exemplaren vervielfältigte Abhandlung war für die Darlegungen dieses Kapitels von großem Nutzen und wird im folgenden kurz als GÜSSOW, Dissertation, zitiert werden; sie berücksichtigt die Literatur bis Ende 1922.)

HELLERICH, J., Neue Bearbeitung der photometrischen und spektroskopischen Beobachtungen der Veränderlichen vom δ Cephei-Typus. Inaug.-Diss. Berlin 1913; Ergänzungen dazu A N 215, S. 291 (1922).

HENROTEAU, F., The Cepheid Problem. Publ. of the Dominion Observatory at Ottawa 9, Nr. 1 (1925).

HOFFMEISTER, C., Über die Lichtkurven der δ Cephei-Sterne und ihre statistische Verwertung. A N 225, S. 201 (1925).

LUDENDORFF, H., Untersuchungen über die δ Cephei- und ζ Geminorum-Sterne. A N 203, S. 361 (1916).

LUIZET, M., Les Céphéides considérées comme étoiles doubles avec une monographie de l'étoile variable δ Céphée. Annales de l'Université de Lyon I, Fasc. 33 (1912).

SHAPLEY, H., On the Nature and Cause of Cepheid Variation. Ap J 40, S. 448 (1914) = Mt Wilson Contr 92. — On the Determination of the Distances of the Globular Clusters. Ap J 48, S. 89 (1918) = Mt Wilson Contr 151. — The Luminosities and Distances of 139 Cepheid Variables. Ap J 48, S. 279 (1918) = Mt Wilson Contr 153. — Three Notes on Cepheid Variation. Ap J 49, S. 24 (1919) = Mt Wilson Contr 154.

Von größeren Monographien über einzelne δ Cephei-Sterne seien außer der schon oben zitierten von LUIZET über δ Cephei noch folgende angeführt:

BECKER, FR., Der veränderliche Stern ζ Geminorum. Berlin 1924.

GROUILLER, H., Monographie de l'étoile variable η Aigle. B A Deuxième Sér. Première Partie, 1, S. 331 (1920).

KRON, E., Über den Lichtwechsel von XX Cygni. Potsd Publ Nr. 65 (1912).

LOCKYER, W. J. S., Resultate aus den Beobachtungen des veränderlichen Sternes η Aquilae. Inaug.-Diss. Göttingen 1897.

MEYERMANN, B., Resultate aus den Beobachtungen des veränderlichen Sternes δ Cephei. Inaug.-Diss. Göttingen 1902.

SHAPLEY, H., On the Changes in the Spectrum, Period, and Light-Curve of the Cepheid Variable RR Lyrae. Ap J 43, S. 217 (1916) = Mt Wilson Contr 112.

SHAPLEY, H., and SHAPLEY, MARTHA BETZ, A Study of the Light-Curve of XX Cygni. Ap J 42, S. 148 (1915) = Mt Wilson Contr 104.

52. Die rechnerische Darstellung der Lichtkurven der δ Cephei-Sterne. Bei der rechnerischen Darstellung der Lichtkurven der δ Cephei-Sterne fällt eine große Schwierigkeit fort, die bei derjenigen der Lichtkurven der Mira-Sterne auftritt, nämlich die Veränderlichkeit der Kurve von Periode zu Periode. In der Tat sind die Lichtkurven der δ Cephei-Sterne im allgemeinen nur sehr wenig veränderlich (wir werden darauf später noch näher eingehen), so daß die für eine rechnerische Darstellung der Lichtkurve eines Mira-Sternes erforderliche umständliche Ableitung einer mittleren Lichtkurve nicht nötig ist. Es kommt noch hinzu, daß, außer etwa bei kurzperiodischen δ Cephei-Sternen, die Lichtkurven in der Regel schon auf Beobachtungen aus verschiedenen Perioden beruhen, die mit Hilfe des bekannten Periodenwertes auf eine Periode reduziert worden sind. Die erhaltenen Lichtkurven sind also eigentlich schon mittlere.

Wie bei den Mira-Sternen hat man es auch bei den δ Cephei-Sternen unternommen, die Lichtkurven durch Fourier-Entwicklungen darzustellen, aber, wie bei den ersteren, so sind auch bei den δ Cephei-Sternen die greifbaren Resultate, die man auf diese Weise erhalten hat, recht gering. E. C. PICKERING hat solche Rechnungen schon 1880 bis 1881 angestellt[1] und ist später nochmals auf seine Ergebnisse zurückgekommen[2]. Er setzt die Helligkeit L des Veränderlichen im Maximum = 100 und findet für die bekannten δ Cephei-Sterne ζ Geminorum, η Aquilae und δ Cephei folgende genäherte Formeln:

ζ Geminorum: $L = 89{,}6 + 10{,}2 \sin(v - 11°{,}3)$,

η Aquilae: $L = 74{,}6 + 20{,}0 \sin(v - 60°) + 6{,}0 \sin(2v - 120°)$,

δ Cephei: $L = 72{,}1 + 20{,}0 \sin(v - 45°) + 7{,}0 \sin(2v - 120°)$.

v wird der Zeit proportional vom Minimum an gerechnet und durchläuft während der Periode die Werte von 0° bis 360°. Zu einer genauen Darstellung der Lichtkurven mußten noch Sinusglieder mit $3v$ im Argument mitgenommen werden, doch betrachtete PICKERING deren Realität als fraglich.

Es kann hier nicht auf alle die für einzelne δ Cephei-Sterne unternommenen Darstellungen durch Fourier-Reihen eingegangen werden, vielmehr soll nur auf einige größere Untersuchungen dieser Art hingewiesen werden.

[1] Proceed. of the American Acad. of Arts and Sciences 16, S. 257 (1881).

[2] Harv Circ 190 (1916).

H. C. PLUMMER[1] gab die Analyse der Lichtkurven einiger kurzperiodischer δ Cephei-Sterne. In Gemeinschaft mit C. MARTIN hat er dann die Lichtkurven zahlreicher δ Cephei-Sterne auf photographischem Wege bestimmt und in den Monthly Notices Bd. 73 (1913) bis Bd. 81 (1921) veröffentlicht. Auf dem absteigenden Aste der Lichtkurven ergaben sich stets mehr oder weniger zahlreiche sekundäre Wellen, die von MARTIN und PLUMMER als reell betrachtet werden. Bei der für alle diese Lichtkurven durchgeführten rechnerischen Darstellung durch Fourier-Entwicklungen war daher die Mitnahme einer großen Zahl von Gliedern (bis zu elf) notwendig. F. C. JORDAN[2] hat aber mit Hilfe eigener Beobachtungen von einigen der Sterne gezeigt, daß jene Wellen nicht reell, sondern Ungenauigkeiten der Beobachtungen zuzuschreiben sind, und auch bei weiteren der von MARTIN und PLUMMER untersuchten Sterne haben anderweitige Beobachtungen die sekundären Wellen nicht bestätigt.

Schon MARTIN und PLUMMER[3] wiesen darauf hin, daß die ersten Glieder der Fourier-Entwicklungen für die Lichtkurven der δ Cephei-Sterne Analogien mit den entsprechenden Entwicklungen von T. E. R. PHILLIPS für die Lichtkurven der Mira-Sterne (vgl. Ziff. 21) zeigen. PHILLIPS selbst hat dann[4] die Lichtkurven von 18 Veränderlichen mit Perioden von $0^d,3$ bis 27^d analysiert und in derselben Form dargestellt, die er für die zahlenmäßige Wiedergabe der Lichtkurven der Mira-Sterne gewählt hatte, nämlich in der Form

$$M + k_1 \cos(\vartheta - 180°) + k_2 \cos(2\vartheta - \varphi_2) + k_3 \cos(3\vartheta - \varphi_3).$$

Er findet, daß sich, wie bei den Mira-Sternen, die Lichtkurven nach den Werten von φ_2 und φ_3 in zwei Gruppen einordnen lassen. (Hierbei ist zu bemerken, daß von den fünf Sternen, die PHILLIPS zu Gruppe I rechnet, drei Bedeckungsveränderliche sind, nämlich S Antliae, RS Sagittarii und β Lyrae.) Das Resultat bietet nichts Überraschendes, denn schon der Augenschein lehrt, daß die Formen der Lichtkurven bei den δ Cephei- und den Mira-Sternen im großen ganzen übereinstimmen. MARTIN und PLUMMER sind später[5] nochmals auf die Bedeutung von PHILLIPS' beiden Gruppen für die δ Cephei-Sterne zurückgekommen und haben dabei auch die kurzperiodischen δ Cephei-Sterne in den Sternhaufen in den Kreis der Betrachtung gezogen, deren Lichtkurven zuvor von PLUMMER[6] analysiert worden waren.

Auf wesentlich anderem Wege gelangt M. L. LUIZET in seiner schon zitierten Abhandlung „Les Céphéides" zu einer numerischen Darstellung des Verlaufes der Lichtkurven. Allerdings beschränkt er sich bei seinen Untersuchungen auf glatte, also keine sekundären Wellen aufweisende Lichtkurven. Dabei geht er nicht von den in Größenklassen ausgedrückten Helligkeiten der Veränderlichen aus, sondern rechnet die Größenklassen in Helligkeiten um, legt also Intensitätskurven zugrunde. Er nimmt an, daß die δ Cephei-Sterne Doppelsterne seien, deren eine Komponente dunkel ist, ferner daß, wie es wenigstens angenähert auch der Fall ist, das Helligkeitsmaximum zusammenfällt mit dem Minimum (negativen Maximum) der Radialgeschwindigkeit, das Helligkeitsminimum dagegen mit dem positiven Maximum der Radialgeschwindigkeit; weiter macht er die Annahme, daß die Halbkugel, die uns der (sphärische) Stern im Helligkeitsmaximum zukehrt, heller ist als die von uns abgekehrte. Die Trennungsebene beider Halbkugeln geht stets durch den Schwerpunkt des Systems, und die Helligkeitsverteilung ist symmetrisch zu den Polen dieser Halbkugeln. Es sei $2S$ die Helligkeit der helleren, $2S'$ die der dunkleren Hemisphäre, v_1 die wahre Anomalie des Sternes zur Zeit des Helligkeitsmaximums, v diejenige zur

[1] M N. 73, S. 652 (1913). [2] M N 82, S. 38 (1921). [3] M N 78, S. 156 (1917). [4] M N 78, S. 185 (1918). [5] M N 80, S. 33 (1919). [6] M N 79, S. 639 (1919).

Zeit t. Dann wird, wie einfache Betrachtungen zeigen, die Helligkeit E zur Zeit t sein:

$$E = S + S' + (S - S') \cos(v - v_1).$$

Man kann $2S$ als Einheit der Helligkeit wählen; dann kann man den Wert von $2S'$ sofort aus der Intensitätskurve des Sternes ablesen. LUIZET zeigt nun, wie man aus dieser Kurve die Exzentrizität e der Bahnellipse, ferner den Winkel $\lambda = 360° - v_1$, d. h. die Länge des Periastrons vom Knoten, und schließlich die Zeit T des Periastrons (gezählt von der Zeit des Helligkeitsmaximums an) bestimmen kann. Berechnet man rückwärts aus S, S', e, λ, T und der bekannten Periode P die Intensitätskurve, so zeigt sich bei den von LUIZET untersuchten 12 Sternen (unter denen sich übrigens die beiden jetzt als Bedeckungsveränderliche erkannten Veränderlichen S Antliae und W Ursae majoris befinden) eine recht gute Übereinstimmung zwischen Beobachtung und Rechnung. Eine Vereinfachung der von LUIZET befolgten Methode der „Bahnbestimmung" haben später S. SCHARBE[1] und F. HENROTEAU[2] gegeben. Es sei auch auf die bezüglichen Bemerkungen von J. HELLERICH[3] hingewiesen.

Es ist sicher anzunehmen, daß die oben skizzierten primitiven Grundanschauungen, von denen LUIZET ausgeht, nicht richtig sind, und daß die zwischen Beobachtung und Rechnung erzielte Übereinstimmung rein formaler Natur ist. (Nebenbei sei bemerkt, daß die Rechnungen, die LUIZET über die Massen der δ Cephei-Sterne anstellt, auf schweren Mißverständnissen beruhen und gänzlich hinfällig sind.) Daß sich zwischen den von LUIZET berechneten photometrischen Bahnelementen und den aus den Änderungen der Radialgeschwindigkeiten ermittelten spektroskopischen Bahnelementen eine gewisse Übereinstimmung herausstellt, erklärt sich dadurch, daß die Lichtkurven und Geschwindigkeitskurven angenähert Spiegelbilder voneinander sind.

Eine weitere Art der Darstellung der Lichtkurven von δ Cephei-Sternen hat A. A. NIJLAND[4] angegeben, und zwar durch die Formel

$$y = \frac{\sin\vartheta}{1 - k\cos(\vartheta - \psi)}.$$

Hierin sind k und ψ passend gewählte Konstanten, und ϑ durchläuft während einer Periode der Zeit proportional die Werte von 0° bis 360°. Nähere Resultate über die Anwendung dieser Formel hat NIJLAND noch nicht veröffentlicht.

Für die statistische Diskussion der Lichtkurven der δ Cephei-Sterne ist die numerische Charakterisierung ihrer Form äußerst wichtig. Man hat sich bis vor kurzem damit begnügt, die Größe $M-m/P$, d. h. die in Einheiten der Periodenlänge ausgedrückte Zeit zwischen Minimum und folgendem Maximum, in den Kreis der Betrachtungen zu ziehen. Diese Größe charakterisiert die Form der Lichtkurve natürlich keineswegs vollständig, und C. HOFFMEISTER[5] ist daher dazu übergegangen, eine andere, diesen Zweck besser erfüllende Größe als Charakteristik der Lichtkurve einzuführen, nachdem er anfangs mit Fourier-Entwicklungen operiert hatte. Er sagt darüber folgendes: „Es sei G_a das Absolutglied der Fourierentwicklung, d. i. der Mittelwert der Ordinaten aller der Entwicklung zugrunde gelegten Kurvenpunkte. Sodann werde die Größe

$$C = 4 \cdot G_a \cdot \frac{M - m}{P}$$

als die Charakteristik der Lichtkurve bezeichnet. Mit dem Faktor 4 soll dabei erreicht werden, daß C für ideale ζ Geminorum-Kurven, d. h. für Sinus-

[1] A N 198, S. 225 (1914). [2] A J 32, S. 57 (1919). [3] A N 219, S. 115 (1923).
[4] A N 220, S. 96 (1923). [5] V J S 59, S. 213 (1924); A N 225, S. 201 (1925).

kurven, den Wert 10 annimmt, wenn man dem Maximum die Ordinate 10 gibt. Mittels dieses Verfahrens wird sowohl der Steilheit des Aufstiegs als auch der allgemeinen Höhenlage der Kurve und damit besonders der Gestalt des Maximums Rechnung getragen. Das Verfahren genügt den hier in Frage stehenden Zwecken vollkommen und kennzeichnet die Gestalt der Lichtkurven eindeutig, da die an sich möglichen Fälle, die eine Zweideutigkeit ergeben würden, in der Natur nicht auftreten. Die Charakteristiken der Kurven liegen in einem ziemlich weiten Bereich, für Antalgolsterne vorwiegend zwischen 1 und 2, für ζ Geminorum-Sterne zwischen 8 und 10. Die Fälle, in denen $C > 10$ ist, sind selten. Es kann dies eintreten, wenn entweder $M - m/P > 0{,}5$ ist oder der Stern ein sehr breites Maximum hat."

Zur Ermittlung von C zeichnete HOFFMEISTER zunächst alle Kurven auf einheitlichen Maßstab um, indem er die Ordinaten für jedes Zwölftel der Periode bestimmte, der Ordinate des Maximums den Wert 10, der des Minimums den Wert 0 beilegte und die anderen Ordinaten entsprechend umrechnete. „Zeichnet man die Kurven dann so, daß die Länge der Periode bei allen Sternen durch gleiche Strecken dargestellt wird, so erhält man die reine Kurvengestalt in einer Form, die die Kurven der einzelnen Sterne unmittelbar miteinander vergleichen läßt." C ließ sich alsdann ohne weiteres berechnen. Für einige Sterne wurde C auf Grund von Lichtkurven, die von verschiedenen Beobachtern herrührten, getrennt berechnet; es ergab sich dann eine gute Übereinstimmung, z. B. für δ Cephei, aus einer Lichtkurve von

GUTHNICK	$C = 4{,}0$
HORNIG	4,9
SCHMIDT	4,2
VOGELENZANG	4,4

Es ist hier zu erwähnen, daß die von HOFFMEISTER eingeführte Größe G_a in enger Beziehung zu der von H. THOMAS (vgl. Ziff. 21) für die Mira-Sterne eingeführten Größe S_x steht.

53. Die Lichtkurven der δ Cephei-Sterne. Nachdem wir uns in der vorigen Ziffer mit der Darstellung der Lichtkurven der δ Cephei-Sterne durch Formeln bzw. durch numerische Angaben beschäftigt haben, gehen wir nunmehr zur näheren Betrachtung der Lichtkurven selbst über. Wir geben zunächst in einer Reihe von Abbildungen (Nr. 24 bis 32) einige besonders gut festgelegte Lichtkurven wieder. Die angegebenen Werte von C sind der Tabelle von C. HOFFMEISTER entnommen und brauchen den wiedergegebenen Lichtkurven nicht immer ganz genau zu entsprechen.

Eine große Zahl von Lichtkurven, nämlich 115, von δ Cephei-Sternen hat M. GÜSSOW in ihrer Dissertation auf Grund der besten vorliegenden Beobachtungsreihen sowohl durch numerische Angabe der Helligkeiten für jedes Vierundzwanzigstel der Periode als auch durch graphische Wiedergabe zur Darstellung gebracht. Betrachtet man die von ihr reproduzierten Lichtkurven, so erkennt man, daß sich, wenn man von den sekundären Wellen absieht, die meisten von ihnen in das für die Mira-Sterne angewandte Klassifikationsschema (vgl. Ziff. 21) einordnen lassen würden. Es treten aber doch auch Unterschiede hervor. Lichtkurven der Form γ_1 (Buckel im Aufstieg) kommen nur sehr selten vor, und der Buckel ist in diesen wenigen Fällen in der Regel sehr klein; er kann unter den von M. GÜSSOW wiedergegebenen Lichtkurven wohl nur bei Z Lacertae (Abb. 26) als sicher reell verbürgt werden. Ferner fehlen Kurven der Form γ_2 (Doppelmaximum); nur die Lichtkurve von S Sagittae (Abb. 29) könnte man allenfalls als γ_2 bezeichnen, wenn man das zweite Maximum auch wohl besser als Buckel im Abstieg der Kurve aufzufassen hat. Wellen im Abstieg kommen — wenn auch in der Regel nur schwach ausgeprägt — bei vielen der von M. GÜSSOW zusammengestellten Lichtkurven vor; mit der Frage ihrer Realität werden wir uns noch weiter zu beschäftigen haben. Daß sie in manchen Fällen, z. B. bei η Aquilae (Abb. 28) und W Virginis (Abb. 30) reell sind, steht außer Frage. Im Gegensatz dazu hatten wir früher gesehen, daß bei den Mira-Sternen Wellen im Abstiege der Lichtkurve so gut wie niemals auftreten.

Sehen wir einstweilen von diesen Wellen ab, so können wir sagen, daß bei den δ Cephei-Sternen wie bei den Mira-Sternen der Spektralklasse Me Lichtkurven der Form α (Aufstieg merklich steiler als Abstieg) die der Form β (im wesentlichen symmetrische Kurven) an Zahl übertreffen, und zwar in noch stärkerem Grade als bei den Me-Mira-Sternen. Das Maximum ist bei den α-Kurven in der Regel spitzer als das Minimum, was auch bei den Me-Mira-Sternen mit Lichtkurven der Form α fast ausnahmslos zutrifft. Lichtkurven der Form α_1 (Antalgolkurven) kommen im wesentlichen nur bei den kurzperiodischen δ Cephei-Sternen vor, während bei den Me-Mira-Sternen gerade die mit sehr langen Perioden diese Form bevorzugen.

Wir wollen indessen hier die Lichtkurven der δ Cephei-Sterne nicht in das für die Mira-Sterne angewandte Schema einordnen, sondern diese Kurven durch die Größe $M-m/P$ und die HOFFMEISTERsche Zahl C (s. vorige Ziffer) charakterisieren. HOFFMEISTER gibt in seiner schon zitierten Abhandlung[1] diese Werte für 118 δ Cephei-Sterne an. [Ein weiterer Stern der Liste von HOFFMEISTER, S Antliae, der durch den extremen Wert $C = 11{,}9$ auffällt, ist kein δ Cephei-Stern, sondern hat sich nach den Beobachtungen seiner Radialgeschwindigkeit von A. H. JOY[2] als Bedeckungsveränderlicher erwiesen; dasselbe gilt von SS Hydrae[3], SZ Tauri[4] und RV Canum ven[5]. In der Liste ist ferner X Vulpeculae für V Vulpeculae zu lesen.] Unter diesen Sternen sind die enthalten, für welche M. GÜSSOW Lichtkurven gegeben hat, und ferner noch einige weitere. Alle diese Sterne sind in der folgenden Tabelle alphabetisch geordnet zusammengestellt. Die Kolumnen geben der Reihe nach die Nummer des Sternes in der G. u. L. (Z bedeutet Zusatzkatalog), den Namen des Sternes, die Periode P, den Wert $M-m/P$, die HOFFMEISTERsche Charakteristik C, die Amplitude A [nicht eingeklammerte Werte von A sind auf die photometrische Harvard-Skala oder, soweit photographische Werte (ph) angeführt sind, auf die internationale Skala reduziert und der Dissertation von M. GÜSSOW entnommen, eingeklammerte Werte beziehen sich nicht auf diese Skalen und sind im allgemeinen derselben Quelle entlehnt] und das Spektrum [meist der Dissertation von M. GÜSSOW und einer Zusammenstellung von W. S. ADAMS, A. H. JOY und R. F. SANFORD[6] entnommen. Die Spektren der δ Cephei-Sterne sind synchron mit dem Lichtwechsel veränderlich; soweit möglich, sind die Grenzklassen angegeben].

Die Tabelle umfaßt keineswegs alle δ Cephei-Sterne, sondern nur die, für welche HOFFMEISTER einigermaßen brauchbare Lichtkurven zur Verfügung standen. Die Dissertation von M. GÜSSOW enthält einen Katalog von 174 Sternen, die als sicher zur δ Cephei-Klasse gehörig angesehen werden, und im „Katalog und Ephemeriden veränderlicher Sterne für 1926“ sind 239 Veränderliche mit Perioden von weniger als 50^d aufgezählt, die sicher oder wahrscheinlich zur δ Cephei-Klasse gehören (s. Abteilung II des eben zitierten Katalogs; die RV Tauri-Sterne sind vor der Zählung ausgesondert). Dazu kommen dann noch die

[1] A N 225, S. 201 (1925).
[2] Pop Astr 29, S. 629 (1921). Ausführlich in Ap J 64, S. 287 (1926) = Mt Wilson Contr 322.
[3] Publ A S P 36, S. 139 (1924). [4] B A N 3, S. 11 (1925).
[5] Ap J 65, S. 124 (1927) = Mt Wilson Contr 330.
[6] Publ A S P 36, S. 139 (1924).

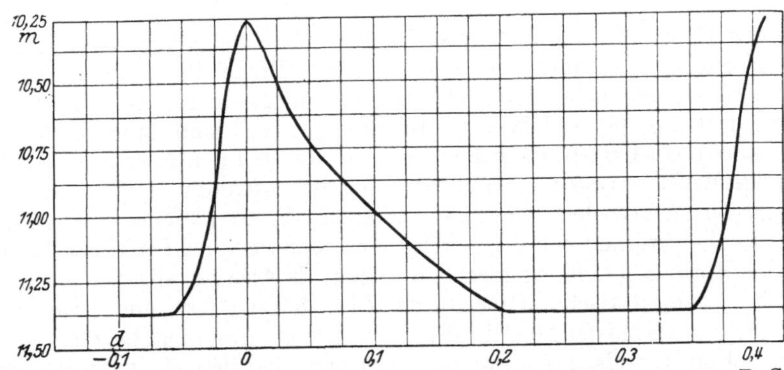

Abb. 24. Lichtkurve von ST Virginis nach photometrischen Messungen von P. GUTHNICK (A N 179, S. 181). $P = 0^d{,}411$, $C = 1{,}5$.

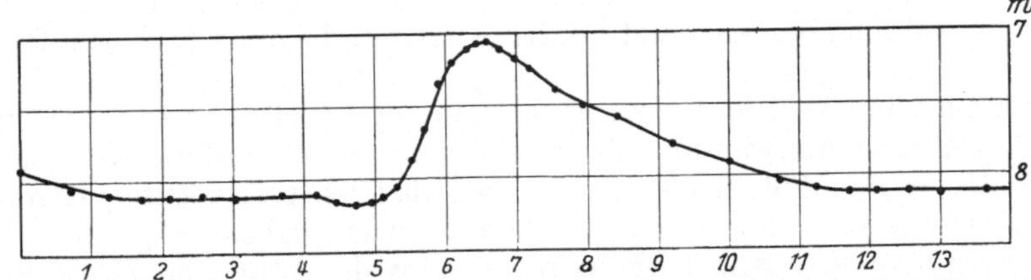

Abb. 25. Lichtkurve von RR Lyrae nach photographisch-photometrischen Messungen von E. HERTZSPRUNG (B A N 1, S. 143). $P = 0^d{,}567$, $C = 2{,}1$.

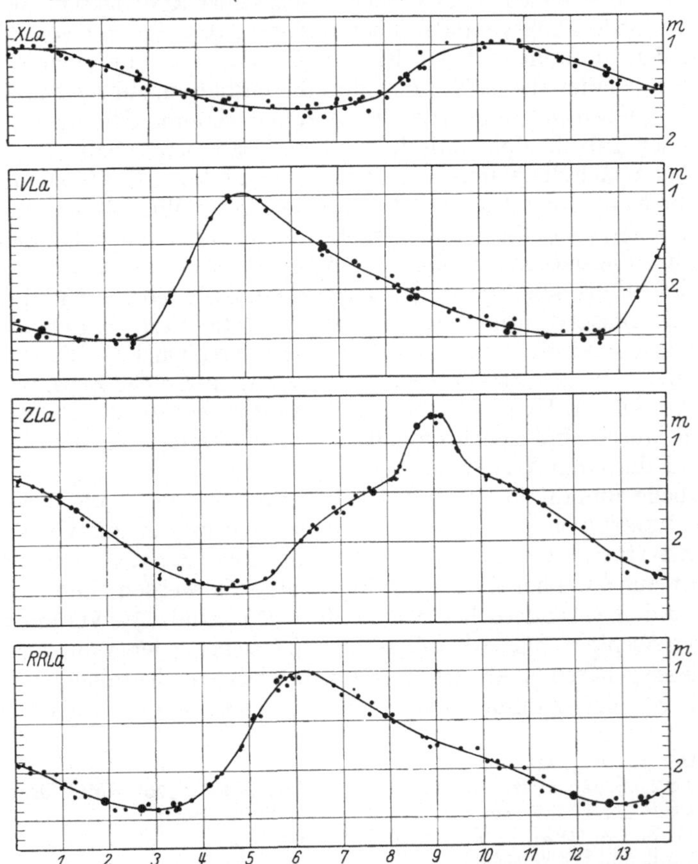

Abb. 26. Lichtkurven von X, V, Z, RR Lacertae nach photographisch-photometrischen Beobachtungen von E. HERTZSPRUNG (B A N 1, S. 67). Perioden: $X = 5^d{,}444$, $V = 4^d{,}983$, $Z = 10^d{,}886$, $RR = 6^d{,}415$. $C = 6{,}9$, $4{,}5$, $6{,}7$, $4{,}2$.

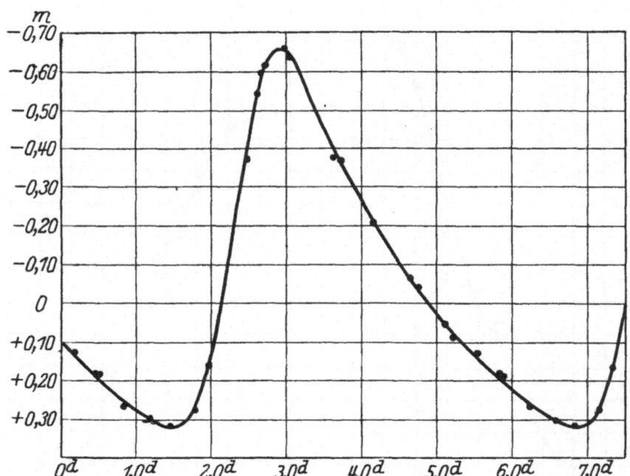

Abb. 27. Lichtkurve von δ Cephei nach lichtelektrischen Messungen von P. GUTHNICK (A N 208, S. 171). $P = 5^d,366$, $C = 4,4$.

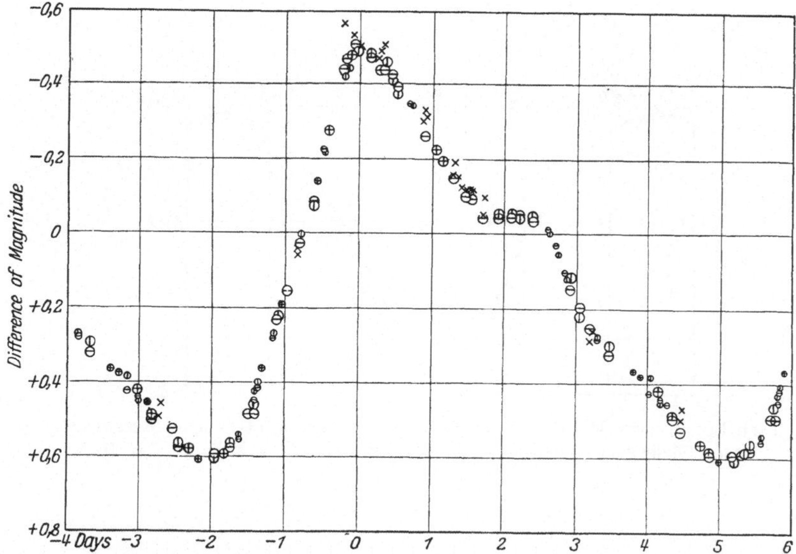

Abb. 28. Lichtkurve von η Aquilae nach lichtelektrischen Beobachtungen von CH. C. WYLIE (Ap J 56, S. 225). $P = 7^d,176$, $C = 5,8$.

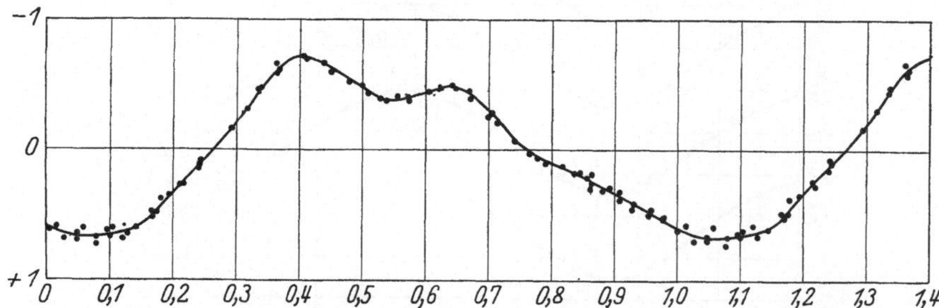

Abb. 29. Lichtkurve von S Sagittae nach photographisch-photometrischen Messungen von E. HERTZSPRUNG (A N 205, S. 287). $P = 8^d,382$, $C = 6,1$.

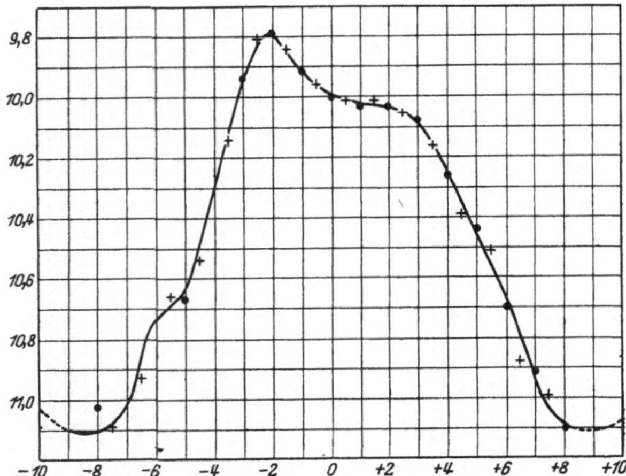

Abb. 30. Photographische Lichtkurve von W Virginis nach C. A. CHANT (Harv Ann 80. No. 12). $P = 17^d{,}271$, $C = 7{,}7$.

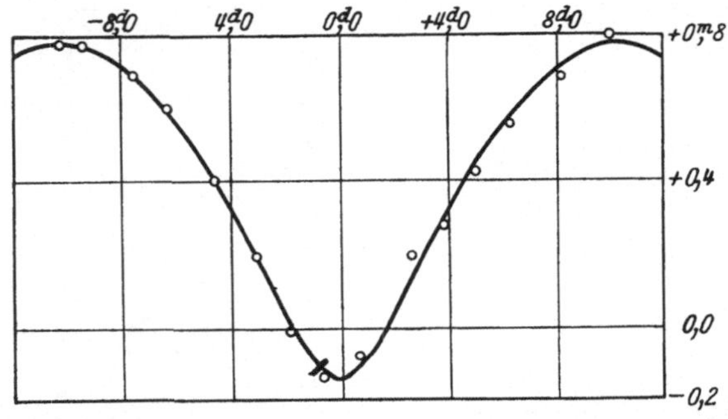

Abb. 31. Lichtkurve von RU Camelopardalis nach photometrischen Messungen auf dem Laws Observatory (Laws Obs Bull Nr. 21). $P = 22^d{,}17$, $C = 10{,}5$.

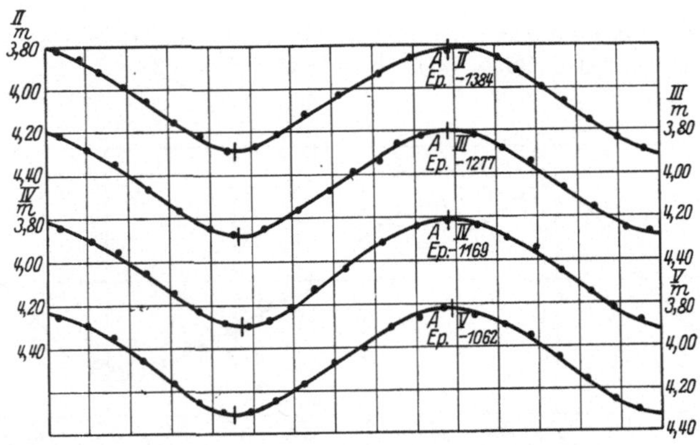

Abb. 32. Lichtkurven von ζ Geminorum nach Beobachtungen von FR. W. A. ARGELANDER (nach FR. BECKER, ζ Geminorum). $P = 10^d{,}155$, $C = 11{,}0$.

Verzeichnis von δ Cephei-Sternen.

Nr.	Stern	P	$\frac{M-m}{P}$	C	A	Spektrum
21	SW Andromedae	0^d,442	0,23	3,3	0^m,94	F1
1360	η Aquilae	7 ,176	,29	5,8	0 ,51	A8—G5
1320	U	7 ,024	,31	5,8	0 ,7	F6—G2
1259	SZ	17 ,135	.38	6,7	1 ,4	G1
1270	TT	13 ,753	,34	5,9	0 ,83	G2
1095	S Arae	0 ,452	,16	1,7	(1 ,28)	A
140	X Arietis	0 ,651	,11	1,7	(1 ,2)	A0
253	Y Aurigae	3 ,859	,28	3,9	0 ,54	
391	RT	3 ,728	,28	4,4	0 ,80	A7—G1
226	RX	11 ,626	,33	4,2	0 ,6	F9
220	SU	0 ,470	,48	9,2	0 ,64	F5
239	SY	10 ,138	,36	6,9	0 ,46	G5
433	TZ	0 ,392	,15	1,5	(1 ,19)	
Z 31	YZ	18 ,356	,38	6,0	0 ,75	
—	γ Bootis	0 ,290	,41	7,3	0 ,05±	F0
817	RS	0 ,377	,19	2,2	0 ,85	B8—F0
828	RY	9 ,007	,17	2,9	(0 ,30)	F5
Z 189	SV	0 ,365	,16	3,4	(0 ,5ph)	
Z 186	SW	0 ,513	,10	1,2	(1 ,5)	
438	RU Camelopardalis	22 ,17	,39	10,5	0 ,96	Rp
163	RW	16 ,402	,41	7,5	0 ,70	K5
173	RX	7 ,908	,26	4,6	(0 ,7)	K2
522	RW Cancri	0 ,548	,18	2,1	(1 ,3)	
Z 75	SS	0 ,367	,16	2,3	1 ,10	
787	RU Canum ven.	0 ,364	,17	2,6	(1 ,0)	
1658	RS Cassiopeiae	6 ,295	,28	6,1	0 ,77	G5
75	RW	14 ,80	,38	7,9	1 ,01	F
1668	RY	12 ,133	,45	7,0	1 ,28	G5
125	SU	1 ,950	,25	4,2	0 ,33	A9.6—F4.5
1639	SW	5 ,440	,28	5,0	0 ,97	G2
10	SY	4 ,071	,31	4,7	0 ,9	G5
24	TU	2 ,139	,25	2,9	1 ,26	F1—F8
Z 7	UZ	0 ,810	,32	5,2	1 ,3ph	
Z 11	VV	6 ,208	,19	3,0	1 ,5ph	
73	RR Ceti	0 ,553	,16	2,2	0 ,70	F0
—	β Cephei	0 ,191	,47	7,0	0 ,05±	B1
1603	δ	5 ,366	,27	4,4	0 ,61	F0—G2
1611	RZ	0 ,309	,32	6,1	0 ,60	A
698	S Comae	0 ,587	,10	1,4	(0 ,56)	
Z 165	V	0 ,469	,13	1,8	(0 ,8)	
1451	X Cygni	16 ,390	,32	5,7	0 ,69	F5p
1344	SU	3 ,846	,28	4,3	0 ,74	A6—F7
1437	SZ	15 ,113	,36	6,8	0 ,80	K
1485	TX	14 ,709	,36	5,4	1 ,17	
1478	UY	0 ,561	,14	2,0	0 ,85	F0
1481	VX	20 ,131	,33	4,1	1 ,26	K
1492	VY	7 ,859	,33	7,7	0 ,54	K
1554	VZ	4 ,864	,23	4,1	0 ,41	F8
1386	XX	0 ,135	,30	4,4	0 ,71	A5
1331	XZ	0 ,467	,17	2,3	1 ,04	A0—A6
—	CD	17 ,023	,36	7,5	(1 ,1)	K5?
982	RW Draconis	0 ,443	,19	2,8	0 ,70	A5
651	SU	0 ,660	,18	2,8	0 ,75	A5
683	SW	0 ,570	,32	4,5:	1 ,0	F4
423	ζ Geminorum	10 ,155	,52	11,0	0 ,42	G0p
399	W	7 ,916	,32	6,3	0 ,76	F2—G0
442	RR	0 ,397	,13	1,6	1 ,17	A8
363	RZ	5 ,529	,27	4,7	0 ,75	F2?
367	SS	44 ,87	,42	9,2	(0 ,51)	G5
1029	SW Herculis	0 ,493	,34	6,4:	(1 ,94ph)	

Verzeichnis von δ Cephei-Sternen.

Nr.	Stern	P	$\frac{M-m}{P}$	C	A	Spektrum
1092	TW Herculis	$0^d,400$	0,10	1,3	(1^m,0)	
Z 204	VX	0 ,455	,11	1,2	(2 ,0ph)	A3
Z 213	VZ	0 ,440	,10	1,3	(1 ,3)	
—	12 Lacertae	0 ,193	,47	9,3	0 ,125±	B2
1622	V	4 ,983	,30	4,5	0 ,96	G2
1623	X	5 ,444	,39	6,9	0 ,35	G2
1579	Y	4 ,324	,28	5,3	0 ,50	F
1612	Z	10 ,886	,42	6,7	(0 ,8)	G0
1614	RR	6 ,415	,24	4,2	(0 ,7)	F
569	RR Leonis	0 ,452	,19	2,1	1 ,00	A9
Z 99	RV	0 ,515	,14	1,7	(0 ,9)	
Z 151	RX	0 ,653	,24	3,8	(0 ,5)	
587	V Leonis min.	0 ,544	,12	1,3	(1 ,2)	
223	U Leporis	0 ,581	,20	2,5	(0 ,90)	A4
1183	Y Lyrae	0 ,503	,14	1,6	(1 ,03)	
1317	RR	0 ,567	,18	2,1	0 ,85	B9—F2
1197	RZ	0 ,511	,10	0,8	(1 ,2)	A2
388	T Monocerotis	27 ,003	,29	4,9	1 ,09	F5—G2
Z 46	SV	15 ,228	,33	5,9	(0 ,6)	G5
870	U Normae	12 ,641	,45	7,9	(0 ,64)	G5
1084	Y Ophiuchi	17 ,121	,37	6,4	0 ,61	F5—G3
1062	ST	0 ,450	,10	1,2	1 ,55	
385	RS Orionis	7 ,567	,33	7,3	0 ,80	F8
210	SV Persei	11 ,128	,41	6,7	1 ,00	F5
184	SX	4 ,290	,26	4,5	0 ,58	
86	VX	10 ,89	,50	10,1	(0 ,4)	
Z 15	VY	5 ,531	,25	3,9	1 ,4ph	G5
454	X Puppis	25 ,953	,19	4,5	(0 ,9)	G5
481	RS	41 ,313	,34	7,0	(1 ,1)	K0—K5 oder Ma
1366	S Sagittae	8 ,382	,30	6,1	0 ,50	F4—G3
1172	U Sagittarii	6 ,745	,43	9,6	0 ,79	G0
1112	W	7 ,595	,29	5,0	0 ,85	A8—G2
1075	X	7 ,012	,42	8,7	0 ,67	F1—G5
1146	Y	5 ,773	,35	5,6	0 ,74	F4—G4
1210	YZ	9 ,553	,51	9,5	0 ,55	G1
1132	AP	5 ,057	,32	6,8	(0 ,6)	F5
1113	AV	15 ,394	,27	5,0	(0 ,8)	
1150	AY	6 ,744	,33	6,4	(0 ,8)	
1214	BB	6 ,754	,45	8,9	(0 ,3)	G0
1027	RV Scorpii	6 ,062	,30	5,4	0 ,63	F5—G5
1171	X Scuti	4 ,200	,39	6,7	(1 ,0)	K?
1179	Y	10 ,347	,41	7,4	(0 ,5)	G5
1191	Z	12 ,900	,37	6,6	(0 ,9)	G0
1189	RU	19 ,700	,25	3,7	1 ,47	G5
—	TY	11 ,056	,46	8,3	0 ,86	
154	SS Tauri	0 ,370	,20	3,0	(0 ,9)	
337	ST	4 ,035	,28	4,8	(0 ,5)	G0
190	SW	1 ,584	,29	5,5	(0 ,8)	F
200	SZ	3 ,149	,37	5,1	0 ,58ph	A9—G
764	RV Ursae maj.	0 ,468	,28	3,5	0 ,9	F0
755	SX	0 ,307	,47	9,4	(0 ,5)	
68	α Ursae min.	3 ,968	,46	9,1	0 ,08	F8
754	W Virginis	17 ,271	,35	7,7	1 ,06	Pec
809	ST	0 ,411	,15	1,5	1 ,11	
1464	T Vulpeculae	4 ,436	,28	4,7	0 ,71	A9—G1
1333	U	7 ,990	,40	7,7	0 ,79	F7—G5
1372	X	6 ,319	,32	5,0	0 ,85	K?
—	SV	44 ,730	,35	7,0	1 ,15	K0—K5 oder Ma

zahlreichen δ Cephei-Sterne in Sternhaufen, den MAGELLANschen Wolken, Spiralnebeln und ähnlichen Gebilden.

Es ist ferner noch zu bemerken, daß die auf Grund der HOFFMEISTERschen Liste angefertigte Tabelle auch drei Sterne, nämlich γ Bootis, β Cephei und 12 Lacertae, enthält, die sowohl von M. GÜSSOW wie von C. HOFFMEISTER ohne weiteres zu den kurzperiodischen δ Cephei-Sternen gerechnet werden, obwohl sich dagegen gewichtige Einwände erheben lassen. Wir rechnen diese Sterne zu der mit den δ Cephei-Sternen verwandten β Cephei-Klasse (vgl. Ziff. 50).

54. Statistik der Perioden und Lichtkurven der δ Cephei-Sterne. Ausführliche statistische Untersuchungen über die Perioden und Lichtkurven der δ Cephei-Sterne haben zuletzt M. GÜSSOW und C. HOFFMEISTER in ihren schon zitierten Abhandlungen angestellt, wodurch die älteren Arbeiten auf diesem Gebiet, auf die wir im folgenden gelegentlich hinweisen werden, überholt worden sind.

α) Statistik der Perioden. M. GÜSSOW legt der Statistik der Perioden die 174 Sterne ihres Verzeichnisses von δ Cephei-Veränderlichen zugrunde. Schließen wir von diesen die drei schon erwähnten β Cephei-Sterne aus, so ergibt sich folgende Übersicht über die Verteilung der Periodenlängen (n = Zahl der Sterne in jedem Periodenintervall):

P	n	P	n	P	n	P	n	P	n
$0^d,0-0^d,1$	0	$1^d,0-1^d,5$	0	$6^d,0-6^d,5$	7	$12^d,0-13^d,0$	4	$30^d,0-35^d,0$	0
0 ,1—0 ,2	3	1 ,5—2 ,0	3	6 ,5— 7 ,0	4	13 ,0—14 ,0	1	35 ,0—40 ,0	2
0 ,2—0 ,3	0	2 ,0—2 ,5	1	7 ,0— 7 ,5	3	14 ,0—15 ,0	4	40 ,0—45 ,0	3
0 ,3—0 ,4	9	2 ,5—3 ,0	1	7 ,5— 8 ,0	5	15 ,0—16 ,0	5	45 ,0—50 ,0	0
0 ,4—0 ,5	20	3 ,0—3 ,5	3	8 ,0— 8 ,5	2	16 ,0—17 ,0	4		
0 ,5—0 ,6	17	3 ,5—4 ,0	5	8 ,5— 9 ,0	0	17 ,0—18 ,0	4		
0 ,6—0 ,7	8	4 ,0—4 ,5	8	9 ,0— 9 ,5	2	18 ,0—20 ,0	4		
0 ,7—0 ,8	0	4 ,5—5 ,0	6	9 ,5—10 ,0	3	20 ,0—22 ,0	3		
0 ,8—0 ,9	2	5 ,0—5 ,5	8	10 ,0—11 ,0	5	22 ,0—25 ,0	2		
0 ,9—1 ,0	0	5 ,5—6 ,0	5	11 ,0—12 ,0	3	25 ,0—30 ,0	2		

Zusammenfassung: $P < 1^d,0$ $n = 59$
$= 1^d,0-30^d,0$ 107
$> 30^d,0$ 5

Auffällig ist die Lücke in der Verteilung der Perioden zwischen $P = 0^d,7$ und $P=1^d,5$. Zwei deutliche Maxima in der Verteilung der Periodenwerte treten hervor bei $P = 0^d,5$ und bei $P = 5^d$. Diese Eigentümlichkeiten der Häufigkeitsverteilung der Periodenwerte lassen die Einteilung der δ Cephei-Sterne in kurz- und langperiodische berechtigt erscheinen. Die kürzeste bisher bekannte Periode hat XX Cygni ($P = 0^d,134865$).

M. GÜSSOW macht auch auf die Lücke in der Periodenverteilung bei $P = 30^d$ bis 35^d aufmerksam und unterscheidet eine dritte Klasse von δ Cephei-Sternen, die sie „außenstehende" nennt. Die geringe Zahl von δ Cephei-Sternen mit Perioden von mehr als 20^d läßt es aber zweifelhaft erscheinen, ob diese Lücke irgendeine wirkliche Bedeutung hat. Wir wollen daher die δ Cephei-Sterne mit $P > 35^d$ mit zu den langperiodischen rechnen.

β) Statistik der Helligkeitsamplituden. Für die Verteilung der Werte der visuellen Amplituden gibt M. GÜSSOW umstehende Übersicht (die drei β Cephei-Sterne sind hier wiederum ausgeschlossen; zwischen den auf die Harvardskala reduzierten A und den auf andere Skalen bezogenen ist hier nicht unterschieden). Es ist bei der Beurteilung der Zahlen natürlich nicht zu vergessen, daß Veränderliche mit kleineren Amplituden schwerer aufgefunden werden als solche mit größeren. Bemerkenswert ist, daß visuelle Amplituden von mehr als $1^m,6$ bei den hier betrachteten δ Cephei-Sternen nicht vorkommen. Eine deutliche Beziehung zwischen Periode und Amplitude, wie sie sich bei den Mira-

Sternen der Spektralklasse Me ergeben hatte, ist weder bei den kurz- noch bei den langperiodischen δ Cephei-Sternen vorhanden, wenn auch bei letzteren vielleicht ein kleines Anwachsen der Amplitude mit der Periode angedeutet ist. Die kurzperiodischen δ Cephei-Sterne haben durchschnittlich ein wenig größere Amplituden als die langperiodischen.

A	$P<1^d$	$P>1^d$
$<0^m,3$	$n=0$	$n=1$
0 ,3	0	4
0 ,4	1	3
0 ,5	2	9
0 ,6	4	8
0 ,7	4	13
0 ,8	3	16
0 ,9	7	6
1 ,0	9	9
1 ,1	5	5
1 ,2	5	4
1 ,3	1	5
1 ,4	0	2
1 ,5	5	4
1 ,6	1	0
>1 ,6	0	0

Veränderliche Amplituden sind bei einer größeren Anzahl von δ Cephei-Sternen, und zwar hauptsächlich bei kurzperiodischen, vermutet worden. Da es sich aber dabei nur um kleine Änderungen von wenigen Zehnteln der Größenklasse handelt, so ist es in der Regel sehr schwer zu entscheiden, ob sie reell sind.

Die photographischen Amplituden sind größer als die visuellen, und zwar ist nach M. GÜSSOW der Durchschnittswert des Verhältnisses photographische Amplitude : visuelle Amplitude für die kurzperiodischen δ Cephei-Sterne 1,33, für die langperiodischen 1,73. Wir werden hierauf später noch zurückkommen.

γ) Statistik der Form der Lichtkurven. Es ist zuerst von H. LUDENDORFF[1] darauf hingewiesen worden, daß bei den langperiodischen δ Cephei-Sternen die Häufigkeit der einzelnen Werte von $\varepsilon = M-m/P$ in sehr eigentümlicher Weise von der Periodenlänge abhängt. Während nämlich für Sterne mit $P<6^d$ und für solche mit $P>13^d$ die Werte von ε stark gestreut (etwa zwischen 0,14 und 0,50) sind, fehlen bei Sternen mit $P=6^d$ bis $P=13^d$ die Werte von $\varepsilon<0,33$. LUDENDORFF benutzte bei seiner Untersuchung die P und ε von 91 langperiodischen δ Cephei-Sternen. Das von M. GÜSSOW benutzte, etwas größere Material ergibt im wesentlichen dieselbe Erscheinung, und auch E. HERTZSPRUNG[2] fand sie später an den zahlreichen von ihm am südlichen Himmel entdeckten δ Cephei-Sternen bestätigt, wenn auch der Beginn jenes Intervalles etwas nach der Seite der größeren Perioden hin verschoben erscheint. Er macht auch darauf aufmerksam, daß sekundäre Wellen gerade in den Lichtkurven solcher δ Cephei-Sterne auftreten, deren Perioden ungefähr in jenes Intervall fallen, und zwar bei den Sternen mit kürzeren Perioden auf dem absteigenden, bei denen mit längeren Perioden auf dem aufsteigenden Aste der Lichtkurve.

Schon aus dem Gesagten geht hervor, daß der Zusammenhang zwischen Form der Lichtkurve und Periode bei den δ Cephei-Sternen nicht so einfach ist, wie bei den Mira-Sternen der Spektralklasse Me. Dies wird auch bestätigt durch die Untersuchung, die C. HOFFMEISTER in seiner schon zitierten Abhandlung[3] über die Beziehungen zwischen P und der Charakteristik C angestellt hat. Er gibt eine graphische Darstellung, in der die log P als Abszissen, die C als Ordinaten eingetragen sind. Für die kurzperiodischen δ Cephei-Sterne ergibt sich eine starke Häufung bei log $P = -0,3$ und $C = 2$, d. h. diese Sterne haben ganz vorwiegend Lichtkurven mit steilem Aufstieg, spitzem Maximum und breitem Minimum; nur selten kommen bei den kurzperiodischen δ Cephei-Sternen große Werte von C (symmetrische Lichtkurven) vor, und diesen entsprechen meist außerordentlich kurze Perioden. Bei den langperiodischen δ Cephei-Sternen kommen dagegen kleine Werte von C (etwa unter 2,5 bis 3,0) nicht vor; die Werte von C zeigen bei diesen Sternen keine Abhängigkeit von der Periode, doch ist die von LUDENDORFF und HERTZSPRUNG in der Verteilung

[1] A N 209, S. 217 (1919). [2] B A N 3, S. 115 (1926). [3] A N 225, S. 201 (1925).

der ε auf die verschiedenen Perioden gefundene Regel auch entsprechend in der Verteilung der Werte von C vorhanden.

Auch eine Untersuchung HOFFMEISTERS, ob eine Abhängigkeit zwischen C und der visuellen Helligkeitsamplitude besteht, hat keine besonders bemerkenswerten Ergebnisse gezeitigt. Nur bei den kurzperiodischen δ Cephei-Sternen nimmt die Amplitude mit wachsendem C ab.

Schließlich lassen wir noch eine Statistik der in unserer Tabelle der Lichtkurven enthaltenen Werte von C folgen:

C	$P<1^d$	$P>1^d$
< 1,0	1	0
1,0–1,9	15	0
2,0–2,9	12	2
3,0–3,9	5	4
4,0–4,9	2	17
5,0–5,9	1	15
6,0–6,9	2	15
7,0–7,9	2	12
8,0–8,9	0	3
9,0–9,9	3	4
≧ 10,0	0	3
	43	75

Die Verschiedenheit der Lichtkurven bei den kurz- und den langperiodischen δ Cephei-Sternen geht aus dieser Tabelle klar hervor.

55. Die sekundären Wellen in den Lichtkurven der δ Cephei-Sterne. Wir haben schon mehrfach erwähnt, daß die Beobachtungen der δ Cephei-Sterne häufig sekundäre Wellen, namentlich in dem absteigenden Aste der Lichtkurven, ergeben haben. Da die δ Cephei-Sterne nur geringe Helligkeitsamplituden besitzen, so stellen diese sekundären Wellen in der Regel nur sehr kleine Helligkeitsschwankungen dar, und es liegt daher von vornherein in vielen Fällen der Verdacht nahe, daß sie nur durch Beobachtungsfehler vorgetäuscht werden, zumal viele Beobachter geneigt sind, ihren Beobachtungen eine größere Genauigkeit zuzutrauen, als sie wirklich besitzen. So sahen wir schon (Ziff. 52), daß die von C. MARTIN und H. C. PLUMMER bei vielen δ Cephei-Sternen gefundenen sekundären Wellen sich nicht als reell erwiesen haben.

K. GRAFF[1] hat mit einem Keilphotometer die drei Sterne RU Scuti, RR Geminorum und RV Canum venaticorum, welch' letzterer sich inzwischen als Bedeckungsveränderlicher erwiesen hat, in der Absicht beobachtet, die Frage der Existenz sekundärer Wellen zu prüfen; er findet in den Lichtkurven dieser drei Sterne kleine, unregelmäßige Wellen, an deren Realität seiner Ansicht nach „ein Zweifel nicht mehr möglich ist". Diese Wellen wiederholen sich aber nicht periodisch, sondern ändern sich von Periode zu Periode.

M. GÜSSOW äußert sich in ihrer Dissertation sehr skeptisch über die sekundären Wellen. Sie macht darauf aufmerksam, daß solche Wellen vorgetäuscht werden können, wenn die Lichtkurve von Periode zu Periode veränderlich ist und bei der Ableitung der Lichtkurve Beobachtungen aus verschiedenen Perioden vereinigt werden, was ja in der Regel geschieht. Als sicher festgestellt betrachtet sie die sekundären Wellen nur bei folgenden sechs Sternen:

RR Ceti	$P = 0^d,55$	SV Vulpeculae	$P = 44^d,7$
ζ Geminorum	10 ,2	S Sagittae	8 ,4
SZ Aquilae	17 ,1	RZ Cephei	0 ,31

Selbst bei diesen Sternen dürften die Wellen wohl nicht in allen Fällen sicher reell sein; bei S Sagittae ist die Welle verbürgt. Auch gibt es noch einige andere Fälle, bei denen dies zutrifft, und auf die wir weiter unten zurückkommen.

Auf einem ganz anderen Standpunkt gegenüber der Frage der sekundären Wellen steht C. HOFFMEISTER in seiner schon zitierten Abhandlung[2]. Er neigt offenbar dazu, die beobachteten Wellen als vorwiegend reell anzusehen, und spricht, wenn auch mit einiger Reserve, die Ansicht aus, daß bei 69% aller δ Cephei-Sterne solche Wellen vorhanden seien. Er ordnet die Wellen in

[1] A N 217, S. 305 (1922). [2] A N 225, S. 201 (1925).

drei Klassen ein. Insbesondere weist er (wie auch M. GÜSSOW) auf das häufige Vorkommen einer kleinen sekundären Welle kurz vor Beginn der Helligkeitszunahme hin, die in der Tat bei manchen Sternen durch genaue Beobachtungen verbürgt zu sein scheint (z. B. bei RR Lyrae, vgl. Abb. 25). Bei der Unsicherheit der ganzen Frage können wir hier auf HOFFMEISTERS Ausführungen nicht näher eingehen.

Auf sehr sinnreiche Weise hat A. A. NIJLAND die Frage der Realität der sekundären Wellen untersucht[1]. Er hat 17 δ Cephei-Sterne und 36 Bedeckungsveränderliche nach der Schätzungsmethode eingehend beobachtet. Dieses Beobachtungsmaterial diskutiert er nun und leitet sowohl für die δ Cephei-Sterne wie für die Bedeckungsveränderlichen durch Vergleichung der Beobachtungen mit den daraus abgeleiteten Lichtkurven den mittleren Fehler seiner Schätzungen ab. Dabei macht er die Voraussetzung, daß die Lichtkurven der δ Cephei-Sterne (außer denen von S Sagittae und η Aquilae) glatt, ohne sekundäre Wellen, verlaufen, wie es für die Bedeckungsveränderlichen als selbstverständlich angenommen werden darf. Es ergibt sich nun, daß unter dieser Voraussetzung die Beobachtungen der δ Cephei-Sterne ebenso genau, ja sogar noch etwas genauer sind als die der Bedeckungsveränderlichen. Die bei Annahme glatter Kurven übrigbleibenden Fehler verhalten sich ganz wie zufällige. Es liegt also kein Grund vor, bei den δ Cephei-Sternen (von den beiden genannten Fällen abgesehen) sekundäre Wellen anzunehmen, um die Übereinstimmung zwischen den Beobachtungen und der Lichtkurve zu steigern. Nur zwei unter den 17 δ Cephei-Sternen besitzen also sekundäre Wellen; dieser Befund steht in starkem Gegensatz zu HOFFMEISTERS Annahme, daß 69% der δ Cephei-Sterne sekundäre Wellen in ihren Lichtkurven aufweisen.

Wir wollen nun noch einige Fälle aufzählen, in denen das Vorhandensein einer sekundären Welle keinem Zweifel unterliegen kann:

Die Lichtkurve von η Aquilae (Abb. 28) hat einen Buckel im Helligkeitsabstieg, der in den verschiedenen Beobachtungsreihen immer wieder hervortritt (vgl. W. J. S. LOCKYER, Resultate aus den Beobachtungen des veränderlichen Sternes η Aquilae, Diss., Göttingen 1897).

Die Lichtkurve von S Sagittae (Abb. 29) hat einen starken Buckel im Helligkeitsabstieg; man kann fast sagen, daß sie ein sekundäres Maximum besitzt.

Die Lichtkurve von W Virginis (Abb. 30) hat ebenfalls einen starken Buckel im Abstieg. Der in der Abbildung angedeutete Buckel im Aufstieg ist wohl noch nicht verbürgt.

Z Lacertae hat in der Lichtkurve (Abb. 26) vor und nach dem auffallend spitzen Maximum je einen flachen Buckel.

VY Cygni hat einen starken Buckel im Abstieg[2].

Sekundäre Wellen sind wohl auch bei einer Anzahl von den durch E. HERTZSPRUNG am südlichen Himmel entdeckten δ Cephei-Sternen verbürgt[3]. Auf die Regel, die HERTZSPRUNG über den Zusammenhang zwischen dem Auftreten sekundärer Wellen und der Periodenlänge gefunden hat, haben wir schon in der vorigen Ziffer hingewiesen.

Wie man auch in den einzelnen Fällen über die Realität der Wellen in den Lichtkurven der δ Cephei-Sterne denken mag, es steht jedenfalls fest, daß einigermaßen deutliche Wellen im aufsteigenden Aste der Kurven weit seltener auftreten als im absteigenden. Bei den Mira-Sternen ist, wie wir gesehen haben, das Gegenteil der Fall.

[1] Amsterdam Proc 28, S. 142 (1925). [2] Publ. Dominion Obs. Ottawa 9, S. 79 (1925).
[3] B A N 3, S. 115 u. 204 (1926).

56. Die Veränderungen der Form der Lichtkurven bei den δ Cephei-Sternen. Die Frage, ob Änderungen der Form der Lichtkurve vorkommen, ist für die langperiodischen δ Cephei-Sterne noch ziemlich ungeklärt. M. GÜSSOW betrachtet die Lichtkurven der folgenden Sterne dieser Art als veränderlich:

l Carinae ($P = 35^d,5$). Es sollen sich Unregelmäßigkeiten im Helligkeitsanstieg zeigen. Auch Änderungen der Helligkeitsamplitude will man wahrgenommen haben.

TY Scuti ($P = 11^d,1$) Auch bei diesem Stern soll der Helligkeitsanstieg Veränderungen aufweisen.

W Virginis ($P = 17^d,3$). Die Lichtkurve soll langsamen Änderungen in der Weise unterliegen, daß sich Form und Lage des Buckels der Kurve auf dem Helligkeitsabstieg ändern.

η Aquilae ($P = 7^d,2$). Ebenso (vgl. die schon zitierte Dissertation von W. J. S. LOCKYER und FR. BECKERS neuere Untersuchungen[1]).

Ein fünfter Stern, dem M. GÜSSOW eine veränderliche Lichtkurve zuschreibt, ist SX Aurigae; es handelt sich aber bei diesem um einen Bedeckungsveränderlichen.

Die Zahl der Beispiele von langperiodischen δ Cephei-Sternen, bei denen man solche Änderungen der Lichtkurve vermutet hat, ließe sich leicht vermehren. Im allgemeinen wird man aber solchen Vermutungen ziemlich skeptisch gegenüberstehen; nur Messungen nach den genauesten Methoden können die Änderungen sicher feststellen, und es ist auch zu beachten, daß die Beobachtungen, aus denen die Lichtkurven langperiodischer δ Cephei-Sterne abgeleitet werden, naturgemäß in der Regel verschiedenen Perioden angehören, so daß die resultierenden Kurven in gewissem Sinne schon mittlere Kurven sind.

Etwas günstiger ist die Sachlage bei den kurzperiodischen δ Cephei-Sternen, bei denen sich die Beobachtungen aus einer Nacht über die ganze oder wenigstens einen sehr großen Teil der Periode erstrecken können. H. SHAPLEY hat im Jahre 1914[2] das damals zur Entscheidung der vorliegenden Frage vorhandene Material erörtert, und besonders eingehend hat er sich mit den Veränderlichen XX Cygni[3] und RR Lyrae[4] beschäftigt. Bei diesen beiden Sternen findet er, daß die Helligkeitsmaxima in verschiedenen Perioden verschieden verlaufen. Für RR Lyrae haben E. HERTZSPRUNG[5] und andere diese Erscheinung bestätigt.

M. GÜSSOW hält die Lichtkurven der folgenden kurzperiodischen δ Cephei-Sterne für veränderlich: SU und SW Draconis ($P = 0^d,66$ bzw. $0^d,57$), S Comae ($0^d,59$), XZ Cygni ($0^d,47$), SS Cancri ($0^d,37$), XX Cygni ($0^d,13$), RR Lyrae ($0^d,57$) und RS Bootis ($0^d,38$). Bei den ersten vier von diesen Sternen soll der Helligkeitsanstieg veränderlich sein, bei den folgenden drei sich die Lichtkurve von Periode zu Periode ändern und bei dem letzten langsam veränderlich sein. Ob alle diese Fälle einer schärferen Kritik standhalten, muß dahingestellt bleiben. Am wahrscheinlichsten ist die Realität der Änderungen außer bei den schon oben erwähnten Sternen XX Cygni und RR Lyrae wohl bei XZ Cygni nach S. BLAŽKO[6] und RS Bootis nach FR. H. SEARES und H. SHAPLEY[7].

Den von M. GÜSSOW aufgezählten kurzperiodischen δ Cephei-Sternen mit veränderlichen Lichtkurven kann man mit demselben Recht wohl auch noch

---

[1] A N 225, S. 1 (1925). [2] Ap J 40, S. 452 (1914) = Mt Wilson Contr 92.
[3] Ap J 42, S. 148 (1915) = Mt Wilson Contr 104.
[4] Ap J 43, S. 217 (1916) = Mt Wilson Contr 112. [5] B A N 1, S. 139 (1922).
[6] Annales de l'Obs. Astr. de Moscou, Sér. 2, Vol. VIII, Livr. 2, No. 2 (1926).
[7] Ap J 48, S. 214 (1918) = Mt Wilson Contr 159.

beizählen SW Andromedae ($P = 0^d,44$) nach H. SHAPLEY[1], RW Draconis ($P = 0^d,44$) nach S. BLAŽKO[2] und AA Aquilae ($P = 0^d,36$) nach N. IVANOV[3]. Weitere fragliche Fälle wollen wir hier nicht anführen.

57. Änderungen der Perioden bei den δ Cephei-Sternen. Wir betrachten zunächst die Periodenänderungen bei den langperiodischen δ Cephei-Sternen. Eine säkulare Änderung ist in der Periode von δ Cephei festgestellt worden. E. HERTZSPRUNG[4] hat auf Grund einer Bearbeitung des gesamten Beobachtungsmaterials von 1785 an für sie folgende Formel aufgestellt:

$$P = 5^d,3663770 - 0^d,000000916\,(t - 1883),$$

wo t die Jahreszahl ist. Die Periode verkürzt sich also jährlich um $0^s,079$. Eine nachträgliche Bearbeitung der 1823 bis 1826 von FR. M. SCHWERD angestellten Beobachtungen durch H. LUDENDORFF[5] hat die obige Formel bestätigt.

Bei η Aquilae scheint sich um das Jahr 1890 die Periode sprungweise verlängert zu haben; sie war nach J. HELLERICH[6] vorher $7^d,176401$, nachher $7^d,176678$, also um 24^s länger. Die Konstanz der Periode in den letzten Jahrzehnten wird durch eine Untersuchung von FR. BECKER[7] bestätigt.

Bei T Monocerotis zeigt sich nach A. BEMPORAD[8] eine säkulare Zunahme der Periode. Die Periodenlänge ergibt sich zu

$$P = 27^d,00313 + 0^d,00004168\,E,$$

wo E die Epoche ist, gezählt vom julianischen Tage 2410011 an; das zur Ableitung dieser Formel benutzte Beobachtungsmaterial erstreckt sich über die Jahre 1881 bis 1920.

Bei einigen anderen langperiodischen δ Cephei-Sternen stellt man die Periodenänderungen durch ein periodisches Glied in der Formel dar, welche zur Berechnung der Maxima der Helligkeit dient. Dieses periodische Glied lautet bei

ζ Geminorum	($P = 10^d,15$):	$1^d,05 \sin(0°,070\,E + 112°)$[9],
ϰ Pavonis	($P = 9,09$):	$0,43 \sin(1,13\,E + 36)$[10],
AY Sagittarii	($P = 6,74$):	$1,1 \sin(18,71\,E + 321)$[11].

(Auf die Angabe der Nullepochen kann hier verzichtet werden.)

Diese Formeln können höchst wahrscheinlich nur interpolatorischen Charakter beanspruchen; die für AY Sagittarii und ϰ Pavonis bedürfen wohl auch noch der Bestätigung.

Bei BI Orionis hat man ebenfalls ein solches periodisches Glied eingeführt. Es scheint sich aber bei diesem Stern um eine Übergangsform zwischen δ Cephei- und U Geminorum-Sternen zu handeln; wir haben ihn bereits in Ziff. 17 näher besprochen. Auch noch bei einigen anderen langperiodischen δ Cephei-Sternen hat man Änderungen der Periode vermutet, so z. B. bei SZ Tauri und RU Camelopardalis.

Bei den kurzperiodischen δ Cephei-Sternen scheinen Änderungen der Periode relativ häufiger zu sein als bei den langperiodischen. Man hat die beobachteten Änderungen teils durch säkulare, teils durch periodische Glieder, teils durch

---

[1] M N 81, S. 208 (1921).
[2] Russisches Astronomisches Journal 1, Heft 2, S. 27 (1924).
[3] A N 228, S. 143 (1926). [4] A N 210, S. 17 (1919). [5] A N 212, S. 185 (1920).
[6] A N 222, S. 25 (1924). [7] A N 225, S. 1 (1925).
[8] Mem. S. A. It. 1, S. 229 (1921).
[9] FR. BECKER, Der veränderliche Stern ζ Geminorum. Berlin 1924.
[10] A. W. ROBERTS, A J 24, S. 91 (1904).
[11] C. HOFFMEISTER, A N 218, S. 325 (1923).

beide zugleich dargestellt. So stellt z. B. S. BLAŽKO[1] die Maxima von XZ Cygni durch folgende komplizierte Formel dar:

$$\text{Max} = 2417201^d{,}235 \text{ t. m. Gr.} + 0^d{,}4665892\,E - 0^d{,}0002\left(\frac{E}{1000}\right)^2 + 0^d{,}0079 \sin 2°{,}9268\,(E + 41{,}5) + 0^d{,}0024 \sin 5°{,}8536\,(E + 1).$$

Es ist wohl ohne weiteres klar, daß diese Formel nur interpolatorische Bedeutung besitzt und wahrscheinlich späteren Beobachtungen nicht genügen wird. Es hat keinen Zweck, die entsprechenden, allerdings einfacheren Formeln für die anderen hier in Betracht kommenden Objekte an dieser Stelle anzuführen, zumal sie häufig geändert werden. Näheres ist aus den Elementenverzeichnissen und Ephemeriden, die alljährlich veröffentlicht werden, zu entnehmen. Hervorgehoben sei, daß bei SW Andromedae ($P = 0^d{,}44$) der Koeffizient des periodischen Gliedes in der Formel für die Maxima $0^d{,}10$, also fast $^1/_4$ der Periode beträgt[2] und eine Periode von ungefähr 1900^d besitzt.

Etwas näher wollen wir uns noch mit den Periodenänderungen von RR Lyrae befassen, die mehrfach diskutiert worden sind, zuletzt von J. HELLERICH[3] und von R. PRAGER[4]. Letzterer stellt die Zeiten T, zu denen der Veränderliche im Helligkeitsaufstieg seine mittlere Größe erreicht, durch folgende Formel dar:

$$T = 2414856^d{,}4083 + 0^d{,}56683735\,E - 0^d{,}0693 \sin 0°{,}0155(E - 1200) + 0^d{,}0086 \sin 0°{,}0554(E - 325).$$

(Außerdem erleiden diese Zeiten noch eine Schwankung kurzer Periode, die in ungefähr 72 Lichtwechselperioden verläuft, aber augenscheinlich Veränderungen unterworfen ist.) J. HELLERICH glaubt die langsamen Änderungen der Periode durch die Annahme einer einmaligen sprunghaften Änderung der Periode ersetzen zu können. Daß die Form der Lichtkurve von RR Lyrae veränderlich ist, haben wir schon in Ziff. 56 gesehen.

Über die interessanten Periodenänderungen von RZ Cephei hat Miss H. S. LEAVITT[5] eingehende Untersuchungen angestellt.

An das Vorhandensein periodischer Änderungen der Periode bei gewissen δ Cephei-Sternen hat B. GERASIMOVIČ[6] statistische und theoretische Folgerungen geknüpft. Das zur Verfügung stehende Material ist aber noch so unsicher und spärlich, daß wir uns hier mit dem Hinweis auf diese Arbeit begnügen können.

58. Die Spektren der δ Cephei-Sterne. Mit den Spektren von δ Cephei-Sternen hat sich zuerst A. BELOPOLSKI eingehend beschäftigt, und zwar zunächst mit demjenigen von δ Cephei selbst und dann auch mit dem von η Aquilae. Als Spektralklasse von δ Cephei gibt BELOPOLSKI[7] die Klasse IIa VOGELS an, doch fiel es ihm auf, daß gewisse Linien, die im Spektrum der Sonne schmal und schwach sind, in dem des Sternes breit und stark erscheinen. Wesentliche Änderungen im Spektrum, die im Zusammenhang mit dem Lichtwechsel stehen, konnte er nicht wahrnehmen. Die Radialgeschwindigkeit erwies sich als veränderlich, und BELOPOLSKI berechnete eine spektroskopische Bahn für den Stern. (Mit den Radialgeschwindigkeiten der δ Cephei-Sterne werden wir uns indessen erst in Ziffer 60 ausführlicher beschäftigen.) Zu ganz ähnlichen Ergebnissen kam BELOPOLSKI kurze Zeit darauf für η Aquilae[8].

---

[1] Ann. de l'Obs. Astr. de Moscou, Sér. 2, Vol. VIII, Livr. 2, No. 2 (1926).
[2] A. H. JOY, Publ A S P 36, S. 82 (1924). [3] A N 227, S. 133 (1926).
[4] Veröffentl. d. Univ.-Sternw. Berlin-Babelsberg 5, Heft 4 (1926).
[5] Harv Circ 261 (1924). [6] A N 221, S. 167 (1924).
[7] Bull. de l'Acad. Imp. des Sc. de St. Pétersbourg, V. Serie, Vol. 1, S. 267 (1894). A N 136, S. 281 (1894); 140, S. 17 (1896); Ap J 1, S. 160 (1895)
[8] Bull. de l'Acad. Imp. des Sc. de St. Pétersbourg, V. Sér., Vol. 7, S. 367 (1897); Ap J 6, S. 393 (1897).

Miss A. C. MAURY[1] rechnete die δ Cephei-Sterne η Aquilae, δ Cephei, T Monocerotis und ζ Geminorum zu ihrer „Division ac" und hatte damit eine wichtige Eigenschaft der Spektren dieser Sterne erkannt. Gegenwärtig hat man bekanntlich die „Division ac" fallen lassen und rechnet die ihr angehörigen Sterne mit zu den c-Sternen.

Eine weitere höchst wichtige Eigenschaft der δ Cephei-Sterne entdeckte sodann S. ALBRECHT[2] an mehreren dieser Objekte, nämlich die, daß das Intensitätsmaximum des Spektrums sich mit wachsender Helligkeit des Sterns nach Violett und mit abnehmender wieder nach Rot verschiebt. Diese Eigenschaft kehrt bei allen daraufhin untersuchten δ Cephei-Sternen wieder.

Es ergaben sich aber auch sonst noch merkwürdige Änderungen in den Spektren dieser Veränderlichen. INA LEHMANN[3] fand, daß sich im Spektrum von δ Cephei die Intensitäten einiger Linien stark ändern, und zwar in dem Sinne, daß sie zur Zeit des Helligkeitsmaximums am schwächsten, zur Zeit des Minimums am stärksten sind. Bei ζ Geminorum sind nach derselben Autorin[4] die auftretenden Änderungen in ihrem Verlauf komplizierter. Auch andere Beobachter haben, zum Teil schon früher, kleine Änderungen in den Spektren dieser Klasse von Veränderlichen wahrgenommen. Schließlich aber hat es sich erwiesen, daß diese Änderungen weit größer sind, als man anfangs glaubte. F. G. PEASE[5] wies 1914 darauf hin, daß der kurzperiodische δ Cephei-Stern RS Bootis im Helligkeitsmaximum ein B8-, im Minimum ein F0-Spektrum besitzt. H. SHAPLEY veröffentlichte 1916 verschiedene Arbeiten über die Veränderlichkeit der Spektren der δ Cephei-Sterne und faßte seine Resultate schließlich in einer größeren Abhandlung[6] zusammen. Alle daraufhin untersuchten δ Cephei-Sterne zeigten eine mit der Änderung der Helligkeit synchrone Veränderlichkeit des Spektrums in dem Sinne, daß das Spektrum im Helligkeitsmaximum einer höheren Temperatur entspricht als das im Helligkeitsminimum. Es wurden bei diesen Untersuchungen hauptsächlich die Wasserstofflinien beachtet.

Die von SHAPLEY zusammengestellten Fälle sind, nach der Periode geordnet, die folgenden:

Stern	P	Spektrum	Stern	P	Spektrum
RS Bootis	$0^d,38$	B8 bis F0	Y Sagittarii	$5^d,77$	F4 bis G4
XZ Cygni	0,47	A0 „ A6	X	7,01	F1 „ G5
RR Lyrae	0,57	B9 „ F2	U Aquilae	7,02	F6 „ G2
SU Cassiopeiae	1,95	A8 „ F5	η	7,18	A8 „ G5
TU	2,14	F1 „ F8	W Sagittarii	7,59	A8 „ G2
SZ Tauri	3,15	A9 „ F7	W Geminorum	7,92	F2 „ G1
RT Aurigae	3,73	A7 „ G1	U Vulpeculae	7,99	F7 „ G5
SU Cygni	3,85	A6 „ F7	S Sagittae	8,38	F4 „ G3
T Vulpeculae	4,44	A9 „ G1	Y Ophiuchi	17,11	F5 „ G3
δ Cephei	5,37	F0 „ G2	T Monocerotis	27,01	F5 „ G2

Man erkennt schon an dieser geringen Zahl von δ Cephei-Sternen, daß eine Beziehung zwischen Periode und Spektralklasse besteht, worauf wir noch des Näheren zurückkommen.

In Abb. 33 geben wir als Beispiel eine Darstellung der Änderungen des Spektrums von SU Cygni. Die Kurve ist die Lichtkurve, und die in verschiedenen Phasen beobachteten Spektraltypen sind beigeschrieben.

---

[1] Harv Ann 28, Part I (1897). [2] Lick Bull 4, S. 131 (1907).
[3] Mitteil. d. Russ. Hauptsternwarte Pulkowo 5, S. 176 (1911).
[4] Bull. de l'Acad. Imp. des Sc. de St. Pétersbourg, VI. Sér., Vol 8, S. 423 (1914).
[5] Publ A S P 26, S. 256 (1914). [6] Ap J 44, S. 273 (1916) = Mt Wilson Contr 124.

Auch für einige weitere δ Cephei-Sterne hat man die Veränderlichkeit des Spektrums konstatiert. Die obigen Beispiele mögen indessen hier genügen; in der Tabelle in Ziff. 53 findet man noch einige mehr. Es besteht wohl kein Zweifel, daß die Spektra sämtlicher δ Cephei-Sterne veränderlich sind.

W. S. ADAMS und A. H. JOY haben die Änderungen im Spektraltypus einiger δ Cephei-Sterne noch näher untersucht[1]. Sie kommen zu dem Schluß, daß diese Änderungen hauptsächlich in Intensitätsänderungen der Wasserstofflinien bestehen; diese Linien sind abnorm stark, und zwar im Maximum stärker als im Minimum. Das übrige Spektrum zeigt nach den genannten Autoren nur geringe Änderungen mit dem Lichtwechsel, namentlich ist die Intensität einiger Funkenlinien im gleichen Sinne wie die der Wasserstofflinien veränderlich. ADAMS und JOY machen ferner darauf aufmerksam, daß, während im Sonnenspektrum $H\beta$ erheblich stärker als $H\gamma$ ist, bei den δ Cephei-Sternen das Umgekehrte der Fall ist, und zwar im Maximum in noch etwas stärkerem Grade als im Minimum. ADAMS und SHAPLEY fanden ferner bei δ Cephei eine geringe allgemeine Verbreiterung der Spektrallinien im Minimum[2].

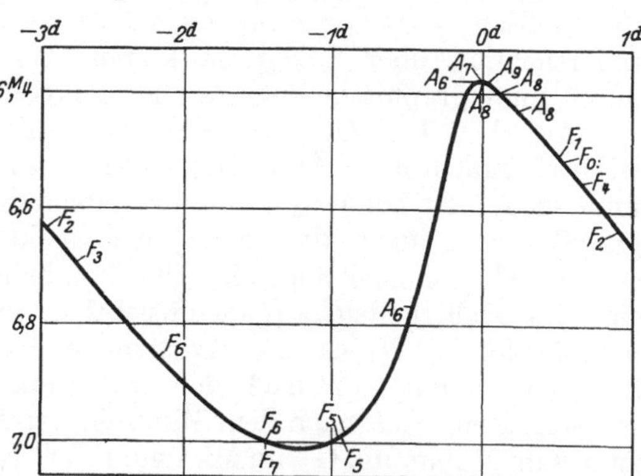

Abb. 33. Änderungen des Spektrums von SU Cygni nach SHAPLEY.

F. HENROTEAU[3] fand, daß bei δ Cephei und η Aquilae das Maximum der Ionisation, also das Maximum der Intensität der Funkenlinien, ungefähr mit dem Maximum der Helligkeit, das Minimum der Ionisation ungefähr mit dem Minimum der Helligkeit zusammenfällt. Bei ζ Geminorum und α Ursae minoris tritt dagegen das Maximum der Ionisation ungefähr um $^1/_4$ der Periode vor, das Minimum ungefähr $^1/_4$ der Periode nach dem Helligkeitsmaximum auf.

Im Widerspruch zu der Ansicht von ADAMS und JOY, daß die spektralen Änderungen der δ Cephei-Sterne in erster Linie nur in Änderungen der Intensitäten der Wasserstofflinien bestehen, befinden sich gewisse Ergebnisse von S. ALBRECHT. Dieser hatte schon 1906 gefunden[4], daß die Wellenlängen gewisser Linien in den Spektren nicht veränderlicher Sterne sich mit fortschreitendem Spektraltypus um geringe Beträge ändern, und er schlug vor, diese Verschiedenheiten der Wellenlängen zur Klassifizierung der Sternspektren zu benutzen. Er fand nun weiter[5] starke Anzeichen dafür, daß jene Linien im Spektrum von η Aquilae mit dem Lichtwechsel ganz entsprechende Änderungen der Wellenlänge erleiden, was auf eine Änderung des Spektraltypus hindeutet. Näher hat ALBRECHT dann diese Erscheinung an l Carinae untersucht[6], und er fand sie voll bestätigt. Diejenigen Linien nämlich, welche bei den Sternen konstanter Helligkeit ihre Wellenlängen mit fortschreitendem Spektraltypus ändern, zeigen bei l Carinae ganz entsprechende Änderungen mit dem Lichtwechsel. Das könnte nicht der Fall sein, wenn, wie ADAMS und JOY annehmen, sich bei den δ Cephei-Sternen im wesentlichen nur die Intensitäten der Wasserstofflinien ändern, nicht aber der sonstige Spektralcharakter. Auch A. PANNEKOEK und

[1] Wash Nat Ac Proc 4, S. 129 (1918). [2] Wash Nat Ac Proc 2, S. 136 (1916).
[3] J Can R A S 19, S. 81 (1925); Publ. Dom. Obs. Ottawa 9, S. 52 u. 115 (1925).
[4] Ap J 24, S. 333 (1906). [5] Lick Bull 4, S. 131 (1907).
[6] Pop Astr 25, S. 519 (1917); 27, S. 520 (1919); ausführlich Ap J 54, S. 161 (1921).

J. J. M. REESINCK[1] kommen, indem sie die Änderungen der Linienintensitäten im Spektrum der δ Cephei-Sterne untersuchen, zu dem Schluß, daß die Ansicht von ADAMS und JOY nicht haltbar sei. Die schon erwähnten Verschiebungen des Intensitätsmaximums im Spektrum der δ Cephei-Sterne scheinen ebenfalls für die Richtigkeit von ALBRECHTS Anschauungen zu sprechen. In einer Spezialuntersuchung über S Sagittae konnte W. GYLLENBERG[2] diese Verschiebung durch photographische spektralphotometrische Messungen nachweisen.

Auf dem Lick-Observatorium hat CH'ING-SUNG YÜ[3] die mit dem Lichtwechsel synchronen Änderungen des Spektrums von ζ Geminorum genauer untersucht. Er benutzte dabei mit einem spaltlosen Quarzspektrographen hergestellte Aufnahmen, die einen großen Spektralbereich umfassen. Seine wichtigsten Ergebnisse sind folgende: Die Zyanbanden (Köpfe bei λ 3590, 3883, 4217) sind im Helligkeitsmaximum unsichtbar, im Minimum am stärksten. Ebenso verhält sich *G*. $H\gamma$ ist am stärksten um $^1/_4\, P = 2^d,5$ nach dem Maximum, am schwächsten um $^1/_4\, P$ nach dem Minimum. *H* und *K* scheinen im Maximum am stärksten, $^1/_4\, P$ nach dem Minimum am schwächsten zu sein. Das Spektrum ist F8 im Maximum, G5 im Minimum. Die Intensitätskurve des kontinuierlichen Spektrums ergibt für das Helligkeitsmaximum eine effektive Temperatur von 7200°; im Minimum weicht sie erheblich von der Intensitätskurve des schwarzen Körpers ab, und die dem Minimum entsprechende Temperatur kann nur unsicher auf etwa 5000° geschätzt werden.

Von wenigen Ausnahmen abgesehen, gehören die δ Cephei-Sterne den Spektraltypen A bis K an (vgl. die Tabelle in Ziff. 53). M. GÜSSOW konnte in ihrer Dissertation (mit Fortlassung von γ Bootis, β Cephei und 12 Lacertae, die keine eigentlichen δ Cephei-Sterne sind) die Spektren von 103 δ Cephei-Sternen angeben. Von diesen besitzen 14 ein A-, 33 ein F-, 42 ein G-, 11 ein K-, 1 ein R- (M. GÜSSOW rechnet diesen Stern, RU Camelopardalis, zur Spektralklasse K) und 2 (W Serpentis und W Virginis) ein besonderes Spektrum. (Augenscheinlich hat M. GÜSSOW bei der Zählung für diejenigen δ Cephei-Sterne, bei denen die Änderungen des Spektrums schon näher bekannt sind, die mittlere Spektralklasse angesetzt.) Auf W Serpentis, W Virginis und RU Camelopardalis gehen wir später noch näher ein. Soweit bekannt, scheinen die Spektren aller δ Cephei-Sterne ac- bzw. c-Charakter zu besitzen; das deutet auf sehr große absolute Helligkeit dieser Sterne hin.

Ein Zusammenhang zwischen Periode und Spektraltypus ist von W. W. CAMPBELL[4], H. SHAPLEY[5] und S. ALBRECHT[6] festgestellt und zuletzt von M. GÜSSOW und von Y. CHANG[7] untersucht worden. Nach M. GÜSSOW sind die mittleren Spektralklassen für die einzelnen Periodenintervalle folgende:

P	Spektrum
$<1^d$	A2
$1^d - 3^d$	F1
3 — 10	G0
10 — 30	G4
30 — 45	G8

Hierbei sind in der ersten Zeile β Cephei (B1) und 12 Lacertae (B2) mitgenommen worden, dagegen sind ausgeschlossen W Serpentis ($P = 14^d$, pec.) und W Virginis ($P = 17^d$, pec).

Im einzelnen ist noch folgendes zu bemerken: Die kurzperiodischen δ Cephei-Sterne haben mit Ausnahme von β Cephei und 12 Lacertae (die keine eigentlichen δ Cephei-Sterne sind) A- bis F-Spektren. Die Sterne mit $P > 5^d$ haben keine früheren Spektren als F5, abgesehen von RW Aquilae (F0). Im Minimum ihrer Helligkeit haben RS Puppis ($P = 41^d,3$) und SV Vulpeculae

[1] B A N 3, S. 47 (1926). [2] Lund Medd Sér. II, Nr. 24 (1920).
[3] Publ A S P 38, S. 357 (1926). [4] Lick Bull 6, S. 51 (1910).
[5] Ap J 40, S. 451 (1914) = Mt Wilson Contr 92; Ap J 48, S. 106 (1918) = Mt Wilson Contr 151.
[6] Ap J 54, S. 188 (1921). [7] Lyon Bull. 8, S. 161 (1926).

($P = 44^d,7$) Spektra der Klasse K5 oder Ma. Es zeigt sich hier also schon ein Übergang zu den Veränderlichen der Mira-Klasse.

Es verdient erwähnt zu werden, daß es eine Anzahl von Sternen gibt, deren Spektrum die größte Ähnlichkeit mit dem der δ Cephei-Sterne besitzt, die aber keine Änderungen der Helligkeit erkennen lassen[1]. Als Beispiele nennen wir α Persei und γ Cygni.

59. δ Cephei-Sterne mit besonderem Spektrum. Der erste Fall, in welchem im Spektrum eines δ Cephei-Sternes helle Linien gefunden wurden, ist der von W Serpentis ($P = 14^d$). Das Spektrum dieses Sternes ist F9, hat aber nach W. S. Adams und A. H. Joy[2] doppelt umgekehrte helle Wasserstofflinien Die zentrale dunkle Komponente derselben liegt nahezu symmetrisch, $H\beta$ ist besonders hell.

W Virginis ($P = 17^d$) hat nach A. H. Joy[3] ein cG0-Spektrum, doch sind die Linien des Wasserstoffs hell, wenn die Helligkeit des Sternes zum Maximum ansteigt. Die Radialgeschwindigkeit ist, wie bei allen δ Cephei-Sternen, veränderlich (um 74 km), die Schwerpunktsgeschwindigkeit ungewöhnlich groß (−62 km). Die galaktische Breite ist ebenfalls für einen langperiodischen δ Cephei-Stern abnorm groß (57°). Nach weiteren Beobachtungen von Joy[4] nehmen die hellen Linien nicht an den periodischen Verschiebungen der Absorptionslinien teil. Das Auftreten heller Linien bei diesem Stern deutet auf eine spektrale Verwandtschaft mit den Mira-Veränderlichen hin.

Besonders interessant ist RU Camelopardalis ($P = 22^d$). Wiederum nach Joy[5] ist das Spektrum dieses Sternes R; im Helligkeitsminimum sind $H\beta$, $H\gamma$ und $H\delta$ hell, und zwar ist $H\beta$ die hellste, $H\delta$ die schwächste unter diesen Linien; im Maximum sind die Wasserstofflinien nur als Absorptionslinien sichtbar. Die hellen Linien sind, wie bei den Me-Sternen, gegen die dunkeln nach Violett verschoben. Die Radialgeschwindigkeit ist um 45 km veränderlich, und, wie bei allen δ Cephei-Sternen, tritt die größte Annäherungsgeschwindigkeit ungefähr zur Zeit des Helligkeitsmaximums auf. Die Absorptionslinien ähneln in ihrem Charakter denen der normalen δ Cephei-Sterne, die Funkenlinien sind wie bei diesen sehr stark.

60. Die periodischen Linienverschiebungen im Spektrum der δ Cephei-Sterne. Wie wir zu Anfang von Ziff. 58 gesehen haben, fand Belopolski, daß δ Cephei und η Aquilae veränderliche Radialgeschwindigkeit besitzen. Die Periode derselben ergab sich in beiden Fällen gleich der des Lichtwechsels, und Belopolski konnte für beide Sterne ihre Bahnen als spektroskopische Doppelsterne berechnen. Dabei zeigte es sich, daß die Zeit der Minimalhelligkeit auch nicht angenähert mit der Zeit zusammenfällt, zu welcher die Radialgeschwindigkeit gleich der des Schwerpunkts ist; hieraus mußte man schließen, daß der Lichtwechsel nicht durch eine Bedeckung zu erklären ist.

Im Laufe der letzten Jahrzehnte sind nun die Radialgeschwindigkeiten einer größeren Zahl von δ Cephei-Sternen untersucht worden. In allen Fällen fand man, daß, wie bei δ Cephei und η Aquilae, die Radialgeschwindigkeit synchron mit dem Lichtwechsel veränderlich ist. Es ergab sich ferner, daß das positive Maximum der Radialgeschwindigkeit sehr nahe mit dem Minimum der Helligkeit, das Minimum (negative Maximum) der Radialgeschwindigkeit sehr nahe mit dem Maximum der Helligkeit zusammenfällt, d. h. im Helligkeitsminimum hat der Stern die größte Geschwindigkeit von uns fort, im Helligkeits-

---

[1] Publ A S P 31, S. 184 (1919); 32, S. 165 (1920).
[2] Publ A S P 30, S. 306 (1918). [3] Publ A S P 37, S. 156 (1925).
[4] Annual Report of the Director of the Mount Wilson Obs. 1925, S. 113.
[5] Publ A S P 31, S. 180 (1919).

maximum dagegen besitzt er die größte Annäherungsgeschwindigkeit. Zeichnet man die Lichtkurve und die Kurve der Radialgeschwindigkeiten in passendem Maßstabe, so sind beide nahezu Spiegelbilder voneinander. Es ergibt sich also ganz allgemein derselbe Schluß wie bei δ Cephei und η Aquilae, daß es nämlich unmöglich ist, den Lichtwechsel der δ Cephei-Sterne durch Bedeckungserscheinungen zu erklären.

Infolge der Veränderlichkeit der Radialgeschwindigkeit der δ Cephei-Sterne sah man diese zunächst als spektroskopische Doppelsterne an und berechnete deren Bahnen nach den üblichen Methoden. Auch nachdem es höchst zweifelhaft geworden ist, ob man jene periodischen Linienverschiebungen wirklich als durch Bewegung des Sternes um den Schwerpunkt eines Doppelsternsystems hervorgerufen ansehen darf, ist es üblich geblieben, sie in Form von Bahnelementen eines solchen Systems darzustellen. Wir wollen uns hier zunächst auf den Standpunkt stellen, daß es sich bei den δ Cephei-Sternen tatsächlich um spektroskopische Doppelsterne handelt, und zusehen, was das Studium ihrer Bahnelemente ergibt.

In der folgenden Tabelle sind die bisher bekannten spektroskopischen Bahnelemente von δ Cephei-Sternen zusammengestellt. Es bedeutet:

P die Periode in Tagen,
ω den Abstand des Periastrons vom Knoten,
e die Exzentrizität,
K die halbe Gesamtamplitude der Radialgeschwindigkeiten (in km),
i die Neigung,
a die große Halbachse der Bahn (in km),
γ die Radialgeschwindigkeit des Schwerpunktes des Systems (in km),
m_1 die Masse der hellen (sichtbaren) Komponente, m_2 die der dunkeln (unsichtbaren) Komponente (in Einheiten der Sonnenmasse).

Die Zeit T des Periastrons lassen wir, als hier ohne wesentliches Interesse, fort; die Bahnen sind, wie schon erwähnt, so orientiert, daß um die Zeit des Helligkeitsmaximums die Annäherungsgeschwindigkeit am größten ist. Auch sind in der Tabelle von jedem Stern nur die zuverlässigsten Bahnbestimmungen angeführt, und SU Cygni, ein Veränderlicher, dessen Bahn erst unsicher bestimmt ist, ist fortgelassen. Bei RT Aurigae, Y Sagittarii und α Ursae minoris scheinen die Bahnelemente stark veränderlich zu sein, und es sind daher zwei Bahnbestimmungen, die verschiedenen Zeiten entsprechen, angeführt. Bei S Sagittae ist die Radialgeschwindigkeit γ des Schwerpunktes veränderlich, und die Elemente seiner Bahn stehen in der zweiten auf diesen Stern bezüglichen Zeile. Unter den in der Tabelle enthaltenen Sternen befindet sich nur ein kurzperiodischer δ Cephei-Stern, RR Lyrae, der sich, was seine Bahnelemente angeht, in nichts von den langperiodischen unterscheidet, außer dadurch, daß γ einen sehr großen Wert besitzt.

Zu den einzelnen Sternen sind noch folgende Bemerkungen zu machen:

η Aquilae. Neuere Beobachtungen der Radialgeschwindigkeit 1923 bis 1925 durch TH. S. JACOBSEN[1] werden durch die in der Tabelle gegebene Bahn gut dargestellt, wenn man annimmt, daß sich die Schwerpunktsgeschwindigkeit γ seit WRIGHTS Beobachtungen um etwa 1 km geändert hat.

RT Aurigae. Die Bahnen nach DUNCAN und KIESS sind sehr verschieden; die Elemente scheinen stark veränderlich zu sein.

δ Cephei. Die von BELOPOLSKI angenommene starke Veränderlichkeit von γ scheint nicht reell zu sein. TH. S. JACOBSEN[1] hat die Radialgeschwindigkeit

[1] Lick Bull 12, S. 138 (1926).

Tabelle der spektroskopischen Bahnelemente der δ Cephei-Sterne.

Stern	P	ω	e	K	$a \sin i$	$\frac{m_2^3 \sin^3 i}{(m_1+m_2)^2}$	γ	Quelle
η Aquilae	7^d,176	68°,9	0,49	20,6	1773000	0,0043	−14,2	W. H. WRIGHT, Ap J 9, S. 59 (1899).
RT Aurigae	3 ,7282	95 ,0	0,37	18,0	856500	0,0018	+21,4	J. C. DUNCAN, Lick Bull 5, S. 120 (1909).
	3 ,72806	115 ,3	0,43	12,0	554700	0,0005	+20,0	C. C. KIESS, Publ Detroit Obs 3, S. 131 (1917).
l Carinae	35 ,523	99 ,5	0,36	18,9	8593000	0,0210	+ 4,1	R. E. WILSON u. C. M. HUFFER, Pop Astr 29, S. 85 (1921).
SU Cassiopeiae	1 ,9495	—	0,0	11,0	295000	0,0003	− 7,0	Genäherte Bahn. W. S. ADAMS u. H. SHAPLEY, Ap J 47, S. 46 (1918) = Mt Wilson Contr 145.
δ Cephei	5 ,36640	85 ,4	0,48	19,7	1270600	0,0028	−16,8	J. H. MOORE, Lick Bull 7, S. 153 (1913).
X Cygni	16 ,38543	101 ,1	0,25	28,0	6121000	0,034	+ 9,3	J. C. DUNCAN, Ap J 53, S. 95 (1921) = Mt Wilson Contr 196.
ζ Geminorum	10 ,154	333	0,22	13,2	1797800	0,0023	+ 6,8	W. W. CAMPBELL, Ap J 13, S. 90 (1901).
RR Lyrae	0 ,566848	110 ,8	0,255	22,1	166600	0,00057	−68,3	R. PRAGER, Sitzungsber. d. Preuß. Akad. d. Wiss. 1916, S. 216.
Y Ophiuchi	17 ,1207	201 ,7	0,16	7,70	1790000	0,0011	− 5,1	Miss ST. UDICK, Publ Allegheny Obs 2, S. 153 (1912).
S Sagittae	8 ,381589	52 ,8	0,34	15,1	1635000	0,0025	var.	J. A. ALDRICH, Pop Astr 32, S. 218 (1924).
	682	195	0,16	15	136000000		− 3	[Elemente der Bewegung des Schwerpunkts.]
W Sagittarii	7 ,59460	70 ,0	0,32	19,5	1930000	0,0050	−28,6	R. H. CURTISS, Ap J 20, S. 172 (1904).
X	7 ,01185	93 ,6	0,40	15,2	1334000	0,0016	−13,5	J. H. MOORE, Lick Bull 5, S. 111 (1909).
Y	5 ,773268	43 ,0	0,21	19,3	1500000	0,004	+ 3,6	Elemente für 1908 } J. C. DUNCAN, Ap J 56, S. 340 (1922)
		74 ,5	0,42	20,6	1354000	0,003	− 5,9	„ „ 1921 } = Mt Wilson Contr 248.
SZ Tauri	3 ,1484	76 ,7	0,24	10,9	460000	0,0004	−3,15	E. S. HAYNES, Lick Bull 8, S. 85 (1913).
R Trianguli austr.	3 ,3891	58	0,17	16,0	735000	0,0014	−18.9	Genäherte Bahn. J. HELLERICH, A N 219, S. 169 (1923).
α Ursae min.	3 ,96809	69 ,8	0,09	3,03	164700	0,000011	−11,5	Elemente für 1899 } J. H. MOORE u. E. A. KHOLODOVSKY,
		47 ,3	0,19	2,74	146900	0,000008	−17,5	„ „ 1923 } Lick Bull 11, S. 166 (1924).
T Vulpeculae	4 ,43578	104 ,0	0,44	17,6	966000	0,0018	− 1,4	A. F. BEAL, Publ Allegheny Obs 3, S. 198 (1915).

1923 bis 1924 von neuem gemessen; seine Ergebnisse werden durch die Bahnelemente von MOORE gut dargestellt, wenn man ω um 3°,8 vergrößert und γ um +1,8 km ändert. Auch Beobachtungen von F. HENROTEAU[1] scheinen auf kleine Änderungen der spektroskopischen Bahnelemente hinzudeuten.

ζ Geminorum. Nach CAMPBELL zeigen die Radialgeschwindigkeiten kleine systematische Abweichungen von der einfachen elliptischen Bewegung. Diese Abweichungen lassen sich durch eine sekundäre Welle der Geschwindigkeitskurve mit einer Gesamtamplitude von 4,5 km und einer Periode, die = $^1/_3$ der Periode des Lichtwechsels ist, darstellen. JACOBSENS[2] Beobachtungen aus dem Jahre 1923 bestätigen CAMPBELLS Resultate; auch Beobachtungen von F. HENROTEAU[3] scheinen die sekundäre Welle zu bestätigen. JACOBSEN fand ferner, daß die Spektrallinien zur Zeit des Helligkeitsminimums etwas breiter sind als zur Zeit des Maximums; dasselbe haben, wie wir sahen, W. S. ADAMS und H. SHAPLEY für δ Cephei

[1] Publ. Dominion Obs. Ottawa 9, S. 59 (1925).

[2] Lick Bull 12. S. 138 (1926).

[3] Publ. Dominion Obs. Ottawa 9, S. 107 (1925).

festgestellt. Es sei hier schließlich noch auf die Untersuchungen von H. C. PLUMMER über die Geschwindigkeitskurve dieses Sternes hingewiesen[1].

Y Ophiuchi. Nach H. LUDENDORFF[2] ist es wahrscheinlich, daß die Geschwindigkeitskurve raschen, wenn auch nicht sehr großen Änderungen unterliegt.

S Sagittae. Nach ALDRICH beschreibt der Schwerpunkt des Systems eine Bahn, deren Elemente in der Tabelle angegeben sind. Die Geschwindigkeitskurve, die der Periode des Lichtwechsels ($8^d,38$) entspricht, hat außerdem eine sekundäre Welle mit einer Gesamtamplitude von 7 km und einer Periode von $4^d,19$. Die Verhältnisse sind also bei diesem Stern sehr kompliziert.

W Sagittarii. Nach CURTISS hat die Geschwindigkeitskurve eine sekundäre Welle mit einer Gesamtamplitude von 9,7 km und einer Periode von $3^d,80$ ($= 1/_2\, P$). Nach JACOBSENS[3] Beobachtungen aus den Jahren 1923 bis 1925 hat sich die Geschwindigkeitskurve gegen früher erheblich geändert.

Y Sagittarii. Die von J. C. DUNCAN für 1908 und für 1921 bestimmten Elemente sind so verschieden, daß man Änderungen der Bahn annehmen muß. Namentlich ist auch γ veränderlich. Schon LUDENDORFF[4] hatte eine Veränderlichkeit der Geschwindigkeitskurve vermutet. H. SHAPLEY und A. D. WALKER[5] haben, angeregt durch DUNCANS Befund, mit Hilfe des ihnen zur Verfügung stehenden Plattenmaterials untersucht, ob die Lichtkurve Änderungen erfahren hat, die denen der Geschwindigkeitskurve entsprechen. Sie konnten aber solche Änderungen nicht nachweisen.

α Ursae minoris. Die Verschiedenheit der für 1899 und der für 1923 abgeleiteten Bahnelemente deutet auf Änderungen derselben hin. Ganz sicher ist γ veränderlich, und die Periode dieser Veränderlichkeit beträgt mehr als 30 Jahre. Von verschiedenen Seiten ist versucht worden, diese langperiodische Bewegung des Sternes auch in seiner lateralen Eigenbewegung zu erkennen. B. GERASIMOVIČ[6] hat indessen nachgewiesen, daß dies nicht möglich ist. — Daß α Ursae minoris veränderliche Helligkeit besitzt, wurde erst von E. HERTZSPRUNG endgültig festgestellt, nachdem die viertägige spektroskopische Periode bekannt geworden war. HERTZSPRUNG wurde nämlich durch den Charakter der Bahnelemente und durch den ac-Charakter des Spektrums auf die Vermutung geführt, daß α Ursae minoris ein δ Cephei-Stern sei, und seine genauen photographisch-photometrischen Messungen bestätigten dies[7]. Die photographische Helligkeitsamplitude fand HERTZSPRUNG zu nur $0^m,17$, die Lichtkurve ist sehr nahe eine Sinuskurve. Die Veränderlichkeit war auch schon von anderen vermutet worden.

Betrachten wir nun das in der Tabelle vorliegende Material genauer, so erkennen wir folgende Eigentümlichkeiten der spektroskopischen Bahnen der δ Cephei-Sterne:

1. Die ω sind nicht über alle Quadranten gleichmäßig verteilt, sondern beschränken sich auf das Intervall von 333° bis 202°, und, wenn wir von ζ Geminorum und Y Ophiuchi (sowie natürlich auch von der Bahn des Schwerpunktes von S Sagittae) absehen, sogar auf das Intervall von 43° bis 115°. Eine solche ungleichmäßige Verteilung läßt sich schon von vornherein auf Grund der Tatsache erwarten, daß die Geschwindigkeitskurven angenähert Spiegelbilder der Lichtkurven sind und bei letzteren in der Regel der Anstieg merklich steiler als der Abstieg ist.

---

[1] M N 73, S. 661 (1913). [2] A N 203, S. 367 (1916). [3] Lick Bull 12, S. 138 (1926).
[4] A N 203, S. 370 (1916). [5] Harv Circ 236 (1922).
[6] A J 35, S. 181 (1924). [7] A N 189, S. 90 (1911).

2. Während sonst die Exzentrizitäten von spektroskopischen Doppelsternen kurzer Periode in der Regel klein sind, sind die der δ Cephei-Sterne durchschnittlich recht groß.

3. Die $a \sin i$ und entsprechend die Massenfunktionen $f = \frac{m_2^3 \sin^3 i}{(m_1 + m_2)^2}$ sind sehr klein. (Diese beiden Größen sind durch die Relation $a^3 \sin^3 i = \text{Const} \cdot f P^2$ verbunden.) Hieraus folgt, daß, sofern man nicht wider alle Wahrscheinlichkeit sehr kleine Massen für die δ Cephei-Sterne annehmen will, die Masse des Begleiters im Verhältnis zu der des Hauptsterns sehr klein sein muß.

4. Die Schwerpunktsgeschwindigkeit γ ist in auffallend vielen Fällen veränderlich, nämlich bei S Sagittae, Y Sagittarii, α Ursae minoris und auch bei SU Cygni, dessen Bahn erst ganz genähert bekannt ist. In einigen weiteren Fällen besteht der Verdacht einer geringeren Veränderlichkeit von γ.

5. Die Kurven der Radialgeschwindigkeit und damit die Bahnelemente sind bei mehreren von den Sternen veränderlich. Diese Frage ist zuerst von H. LUDENDORFF[1] untersucht worden, der hauptsächlich bei Y Ophiuchi und Y Sagittarii solche Änderungen wahrscheinlich fand. Für letzteren Stern sind sie durch die Resultate von DUNCAN bestätigt worden, und auch bei RT Aurigae, W Sagittarii und α Ursae minoris hat man derartige Änderungen festgestellt.

6. Bei mehreren Sternen verrät die Kurve der Radialgeschwindigkeiten, auch abgesehen von der Veränderlichkeit von γ, systematische Abweichungen von der elliptischen Bewegung (ζ Geminorum, S Sagittae, W Sagittarii).

Alle diese Erscheinungen, durch die sich die δ Cephei-Sterne zum Teil sehr wesentlich von den gewöhnlichen spektroskopischen Doppelsternen unterscheiden, lassen den Verdacht rege werden, daß die mit den Helligkeitsänderungen synchronen Linienverschiebungen vielleicht doch nicht durch die Bewegung der Lichtquelle um den Schwerpunkt eines Doppelsternsystems zu erklären seien, oder daß dies wenigstens nicht in vollem Umfange der Fall sei.

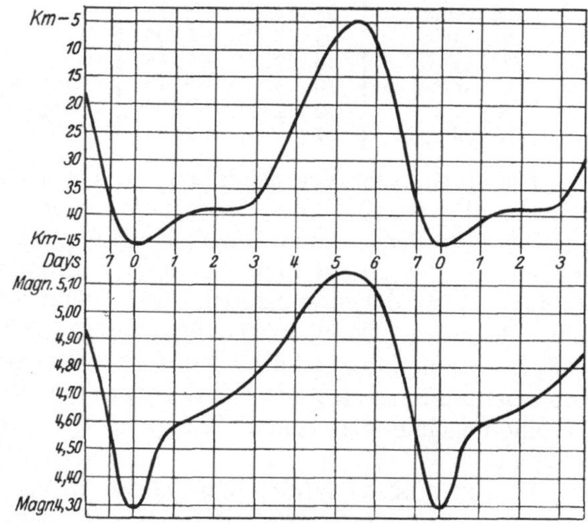

Abb. 34. Die Geschwindigkeits- und die Lichtkurve von W Sagittarii nach R. H. CURTISS, Lick Bull 3, S. 167.

Wir haben bereits erwähnt, daß die Lichtkurven und die Geschwindigkeitskurven der δ Cephei-Sterne, in passendem Maßstabe gezeichnet, mit einer gewissen Annäherung Spiegelbilder voneinander sind. Wie auffallend diese Erscheinung in einigen Fällen ist, lehrt das Beispiel von W Sagittarii. In Abb. 34 sehen wir oben die Geschwindigkeitskurve, unten die Lichtkurve von W Sagittarii, und zwar ist letztere umgekehrt wie sonst üblich gezeichnet (kleinere Helligkeiten oben). Der Buckel im Abstieg der Lichtkurve kehrt an der entsprechenden Stelle der Geschwindigkeitskurve wieder.

[1] A N 203, S. 367 (1916).

In anderen Fällen ist die Übereinstimmung zwischen beiden Kurven nicht so gut, z. B. spiegeln sich, wie wir sahen, die Änderungen der Geschwindigkeitskurve von Y Sagittarii nicht in der Lichtkurve wieder.

Auch zwischen den Amplituden der Geschwindigkeits- und der Lichtkurven scheint ein gewisser Zusammenhang zu bestehen. H. LUDENDORFF[1] machte darauf aufmerksam, daß für 12 δ Cephei-Sterne, deren Helligkeitsamplituden photometrisch gemessen waren, genähert die Relation

$$2K = 47{,}3\,A$$

besteht, und später[2] fand er, daß auch drei weitere Objekte, darunter auch der kurzperiodische δ Cephei-Stern RR Lyrae, sich zwanglos dieser Beziehung anpaßten. Wenn man nicht photometrisch gemessene, sondern auf Schätzungen beruhende Werte von A verwendet, so ist die Übereinstimmung mit der Formel weit schlechter.

Wie schon erwähnt, fällt bei den δ Cephei-Sternen das Helligkeitsmaximum zeitlich genähert mit der größten Annäherungsgeschwindigkeit zusammen, das Helligkeitsminimum dagegen mit dem positiven Maximum der Radialgeschwindigkeit. Diese Frage ist von J. HELLERICH zuerst in seiner Dissertation[3] und später in einer zweiten Arbeit[4] näher untersucht worden. Da die Maxima der Lichtkurven wegen ihrer geringeren Breite genauer festzulegen sind, als die Minima, so beschränken wir uns hier darauf, die für sie geltenden Ergebnisse HELLERICHS wiederzugeben. Es sei Δ = Zeit der größten Annäherungsgeschwindigkeit minus Zeit des Helligkeitsmaximums, so findet HELLERICH für die einzelnen Sterne folgende Werte dieser Differenz:

Stern	P	Δ	Stern	P	Δ
η Aquilae	$7^d{,}18$	$+0^d{,}03$	S Sagittae	$8^d{,}38$	$-0^d{,}28$
RT Aurigae	3 ,73	+0 ,15	W Sagittarii	7 ,59	+0 ,51
l Carinae	35 ,52	+3 ,23	X	7 ,01	−0 ,05
δ Cephei	5 ,37	+0 ,04	Y	5 ,77	+0 ,77
X Cygni	16 ,39	+1 ,45	SZ Tauri	3 ,15	+0 ,32
ζ Geminorum	10 ,15	+0 ,78	α Ursae min.	3 ,97	+0 ,18
RR Lyrae	0 ,57	+0 ,06	T Vulpeculae	4 ,44	+0 ,06
Y Ophiuchi	17 ,12	+2 ,64			

Die Δ sind im Verhältnis zu den Perioden alle klein und mit zwei Ausnahmen alle positiv. Im allgemeinen treten also die Helligkeitsmaxima kurz vor der Zeit der größten Annäherungsgeschwindigkeit ein, und entsprechend ergibt sich, daß die Helligkeitsminima kurz vor dem positiven Maximum der Radialgeschwindigkeit eintreten. Die Δ sind natürlich mit gewissen Unsicherheiten behaftet. Durch weitere Rechnungen auf Grund der Annahme, daß die δ Cephei-Sterne Doppelsterne sind, führt HELLERICH den Nachweis, daß in allen untersuchten Systemen zur Zeit der Extremwerte der Helligkeit die Stellung der Verbindungslinie der beiden Komponenten zur Gesichtslinie annähernd dieselbe ist.

H. LUDENDORFF fand[5], daß sich für 10 Veränderliche von ausgesprochenem δ Cephei-Charakter die Amplitude K der spektroskopischen Bahn mit großer Genauigkeit durch eine empirische Formel als Funktion von e, ω und P darstellen läßt. Für diejenigen Sterne, welche mehr ζ Geminorum-artige Lichtkurven haben, stimmte die Formel nicht mehr, und auch später hinzugekommene

[1] A N 193, S. 301 (1913). [2] A N 203, S. 373 (1916).

[3] Neue Bearbeitung der photometrischen und spektroskopischen Beobachtungen der Veränderlichen vom δ Cephei-Typus. Berlin 1913.

[4] A N 215, S. 291 (1922). [5] A N 203, S. 361 (1916).

Bahnbestimmungen von Sternen mit ausgesprochen δ Cephei-artigen Lichtkurven haben sie nicht bestätigt.

Die bei einigen der δ Cephei-Sterne, z. B. bei ζ Geminorum und W Sagittarii, beobachteten sekundären Wellen in der Geschwindigkeitskurve will A. W. ROBERTS[1] wenigstens teilweise durch Annahme einer ellipsoidischen Gestalt der Sterne erklären. Über den Einfluß der Rotation auf die Geschwindigkeitskurven hat R. H. CURTISS Betrachtungen angestellt[2].

Wie wir schon erwähnt haben, entstehen beim Studium der periodischen Linienverschiebungen in den Spektren der δ Cephei-Sterne Bedenken, ob diese Verschiebungen in vollem Umfange durch die Bewegung des betreffenden Sternes um den Schwerpunkt eines Doppelsternsystems zu erklären sind, oder ob dies vielleicht überhaupt ganz unzulässig ist. Diese Bedenken werden noch wesentlich verstärkt durch einige Spezialuntersuchungen, auf die wir jetzt eingehen wollen.

W. C. RUFUS[3] hat im Spektrum von η Aquilae Gruppen von Linien voneinander gesondert, die auf der Sonne den höchsten, den mittleren und den tiefen Schichten der Atmosphäre angehören. Die aus diesen einzelnen Gruppen von Linien sich ergebenden Radialgeschwindigkeiten zeigen nun ganz erhebliche Verschiedenheiten. Bildet man z. B. die Differenz zwischen den Radialgeschwindigkeiten, die aus den Wasserstofflinien (höchste Schicht), und denen, die aus den Linien der mittleren Schicht folgen, so erhält man für diese Differenz eine mit dem Lichtwechsel synchrone Kurve mit einer Amplitude von mehr als 20 km, während die mittlere Geschwindigkeitskurve eine Amplitude von 40 km hat. Die Geschwindigkeitskurve für die Wasserstofflinien bleibt in der Phase um 1^d hinter der mittleren Geschwindigkeitskurve zurück. Auch D .W. LEE hat in der Atmosphäre von η Aquilae komplizierte Verhältnisse feststellen können, aber es ist erst ein kurzes Referat über seine Arbeit veröffentlicht worden[4]. J. A. ALDRICHS[5] ebenfalls noch nicht ausführlich veröffentlichte Untersuchungen über S Sagittae ergeben ein fortschreitendes Zurückbleiben der Phase, wenn man von den niederen zu den höheren Schichten fortschreitet. Nach R. H. CURTISS[6] herrschen auch bei W Sagittarii ähnlich verwickelte Verhältnisse; die Amplitude der Radialgeschwindigkeiten nimmt mit der Höhe in der Atmosphäre ab und auch die Form der Geschwindigkeitskurve ist für die einzelnen Schichten verschieden. Bei ζ Geminorum konnte F. HENROTEAU[7] keine Verschiedenheiten der aus verschiedenen Gruppen von Linien abgeleiteten Radialgeschwindigkeiten wahrnehmen, während nach W. C. RUFUS[8] auch bei diesem Stern derartige Erscheinungen auftreten.

CH. E. ST. JOHN und W. S. ADAMS[9] fanden, daß im Spektrum von δ Cephei die Funkenlinien etwas andere Radialgeschwindigkeiten als die Bogenlinien ergeben, und zwar ist die Differenz (im Sinne Funkenlinien minus Bogenlinien)

im Helligkeitsmaximum $+2{,}4$ km,
im Helligkeitsminimum $+0{,}9$ km.

Die beiden genannten Autoren erklären diese Differenz durch Konvektionsströme, die im Maximum stärker sind als im Minimum.

Wenn man sich auf den Boden der Pulsationstheorie der δ Cephei-Sterne stellt, so kann man sich nach Ansicht von R. H. CURTISS auf Grund der soeben

[1] M N 66, S. 329 (1906). [2] Publ Detroit Obs 1, S. 104 (1915).
[3] Wash Nat Ac Proc 10, S. 264 (1924). [4] Pop Astr 34, S. 622 (1926).
[5] Pop Astr 32, S. 218, 471 (1924).
[6] Pop Astr 32, S. 547 (1924); Publ A S P 38, S. 148 (1926).
[7] Publ. Dominion Obs. Ottawa 9, S. 111 (1925). [8] Pop Astr 34, S. 242 (1926).
[9] Ap J 60, S. 43 (1924) = Mt Wilson Contr 279.

geschilderten Tatsachen nicht dem Schlusse entziehen, daß die Pulsationen unterhalb der von uns beobachteten strahlenden Schichten entstehen und sich letzteren von unten nach oben mitteilen. Die Relativbewegungen benachbarter Schichten werden zu Kompressionen und Expansionen in der Atmosphäre führen, denen Änderungen der Temperatur und der Strahlung entsprechen werden. Für die Erklärung des Lichtwechsels dürften diese Erscheinungen von größter Bedeutung sein.

Jedenfalls ist es nach dem Gesagten klar, daß die Linienverschiebungen im Spektrum eines δ Cephei-Sternes nicht als allein durch Radialbewegungen des ganzen Sternes verursacht angesehen werden dürfen. In der Tat würden sich ja die spektroskopischen Bahnelemente sehr verschieden ergeben je nach der Auswahl der Linien, die man zur Messung benutzt. Die in der Tabelle angegebenen Bahnelemente sind jedenfalls mittlere, wie man sie erhält, wenn man Linien aus verschiedenen Schichten der Sternatmosphären zur Bestimmung der Radialgeschwindigkeiten heranzieht.

Die aus den spektroskopischen Bahnen folgende Radialgeschwindigkeit γ des Schwerpunktes wird, wie die Dinge auch liegen mögen, höchst wahrscheinlich der wirklichen, von den mit dem Lichtwechsel synchronen Bewegungen befreiten Radialgeschwindigkeit des betreffenden δ Cephei-Sternes nahezu entsprechen, wenngleich auch hier die so auffällig oft beobachtete Veränderlichkeit von γ ein Gefühl der Unsicherheit hervorruft. Jedenfalls bleibt zunächst nichts anderes übrig, als die Größe γ für die Radialgeschwindigkeit dieser Sterne anzusehen. Mit diesen Radialgeschwindigkeiten werden wir uns später zu beschäftigen haben.

61. Spektralphotometrische Eigenschaften der δ Cephei-Sterne. Bereits im Jahre 1899 bewies K. SCHWARZSCHILD[1] durch photographisch-photometrische Messungen, daß die photographische Helligkeitsamplitude von η Aquilae doppelt so groß ist wie die visuelle, während im übrigen der photographische und der visuelle Lichtwechsel dieses Sternes vollkommen parallel verlaufen. Ähnliche Untersuchungen, wie sie SCHWARZSCHILD über η Aquilae angestellt hat, sind später von anderen Beobachtern, z. B. von C. W. WIRTZ[2] und A. WILKENS[3], für eine Reihe weiterer δ Cephei-Sterne ausgeführt worden. Es hat sich dabei die allgemeine Regel ergeben, daß die photographische Amplitude der δ Cephei-Sterne stets größer ist als die visuelle. M. GÜSSOW gibt in ihrer Dissertation das Verhältnis $p =$ photographische Amplitude : visuelle Amplitude für 32 δ Cephei-Sterne an, und zwar für 11 kurzperiodische und 21 langperiodische. Für die kurzperiodischen liegt p zwischen 1,11 (XX Cygni) und 1,78 (RS Bootis), für die langperiodischen zwischen 1,08 (UX Persei) und 2,65 (Y Aurigae). Im Mittel ist für erstere $p = 1{,}33$, für letztere $p = 1{,}73$; es ist also p für die kurzperiodischen δ Cephei-Sterne durchschnittlich kleiner als für die langperiodischen. Eine Beziehung zwischen p und der Periode scheint sonst nicht zu bestehen. Die Werte von p für die einzelnen Sterne dürften zum Teil noch etwas unsicher sein.

In engstem Zusammenhang mit der soeben besprochenen Erscheinung stehen die Änderungen der Farbe der δ Cephei-Sterne während ihres Lichtwechsels. Wir haben bereits erwähnt, daß sich das Intensitätsmaximum in den Spektren der δ Cephei-Sterne während des Lichtwechsels verschiebt, und zwar liegt es im Helligkeitsmaximum am weitesten nach Violett. Mit diesen Verschiebungen ist natürlich eine Veränderung der Farbe bzw. des Farbenindex

---

[1] Publ. der v. Kuffnerschen Sternwarte, 5 C, S. 100 (1900).
[2] A N 154, S. 317 (1901). [3] A N 172, S. 305 (1906).

verbunden. Sehr eingehend haben F. H. SEARES und H. SHAPLEY[1] die Änderungen des Farbenindex bei dem kurzperiodischen δ Cephei-Stern RS Bootis ($P = 0^d,377$) untersucht. Ihre Resultate werden durch Abb. 35 veranschaulicht; es zeigt sich, daß die Änderungen der Helligkeit und die des Farbenindex genau parallel verlaufen, in dem Sinne jedoch, daß im Helligkeitsmaximum der Farbenindex am kleinsten ($-0^m,15$), im Minimum am größten ($+0^m,52$) ist. Die Änderungen in der Farbe entsprechen durchaus denen, die man nach den bei RS Bootis konstatierten Änderungen des Spektrums (vgl. Ziff. 58) erwarten darf. Ähnliche Untersuchungen, wenn auch meist nicht so eingehend, sind auch für einige andere Sterne dieser Art angestellt worden. Wir erwähnen von diesen die über XX Cygni (bei diesem Stern ist der Farbenindex auffällig wenig veränderlich) von H. SHAPLEY und MARTHA BETZ SHAPLEY[2], über XZ Cygni von MARTHA BETZ SHAPLEY[3] und die über X Cygni, S Sagittae, TT Aquilae und δ Cephei von F. C. JORDAN[4]). Über S Sagittae hat W. GYLLENBERG[5] eine eingehende spektralphotometrische Studie ausgeführt und Lichtkurven für verschiedene Wellenlängen abgeleitet. Lichtelektrische Messungen des Farbenindex von δ Cephei und ζ Geminorum rühren von P. GUTHNICK[6] her. Visuelle Messungen in verschiedenen Teilen des sichtbaren Spektrums mit Hilfe von Farbfiltern haben CH. NORDMANN[7] für δ Cephei und CH. GALISSOT[8] für T Monocerotis und ζ Geminorum angestellt. Letzterer findet, im Gegensatz zu GUTHNICK, keine Änderungen des Farbenindex für diesen letzteren Stern.

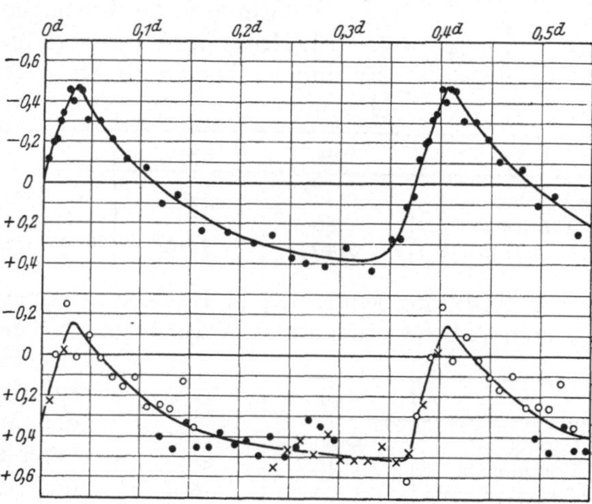

Abb. 35. Visuelle Lichtkurve (oben) und Kurve des Farbenindex (unten) für RS Bootis nach SEARES und SHAPLEY.

M. GÜSSOW gibt nebenstehende Farbenindizes von δ Cephei-Sternen als gut bestimmt an:

	Farbenindex	
	im Maximum	im Minimum
TT Aquilae	$+0^m,30$	$+0^m,84$
RS Bootis	− ,15	,52
SS Cancri	+ ,28	,82
SU Cassiopeiae	,59	,73
X Cygni	,85	1 ,32
XX	,24	0 ,32
XZ	,00	,58
δ Cephei	,45	,92
S Sagittae	,56	,92
Y Sagittarii	,46	,85

Die Änderung des Farbenindex ist, wie leicht ersichtlich, gleich der Differenz zwischen der photographischen und der visuellen Helligkeitsamplitude des betreffenden Sternes.

[1] Ap J 48, S. 214 (1918) = Mt Wilson Contr 159.
[2] Ap J 42, S. 148 (1915) = Mt Wilson Contr 104.
[3] Ap J 45, S. 182 (1917) = Mt Wilson Contr 128.
[4] Ap J 50, S. 174 (1919).
[5] Lund Medd, Sér. II, Nr. 24 (1920).
[6] A N Jubiläumsnummer S. 10 (1921).
[7] C R 146, S. 518 (1819).
[8] B A, Deuxième Sér., Première Partie, 3, S. 207 (1923).

P	Farbenindex
$0^d,55$	$+0^m,15$
6	+0 ,6
11	+1 ,0
18	+1 ,4
33	+1 ,6

Auf Grund der Beziehungen zwischen Periode und Spektrum nimmt H. SHAPLEY[1] nebenstehende Beziehungen zwischen Periode und mittlerem Farbenindex an.

Kolorimetrische Messungen und Temperaturbestimmungen hat J. HOPMANN[2] für verschiedene δ Cephei-Sterne geliefert. Er findet folgende effektive Temperaturen im Maximum und Minimum:

	T_{max}	T_{min}		T_{max}	T_{min}
η Aquilae	5240°	3960°	δ Cephei	6700°	4780°
R T Aurigae	5880	4160	ζ Geminorum	4570	3580
SU Cassiopeiae	6440	5110	T Vulpeculae	4670	3290

A. PANNEKOEK und J. J. M. REESINCK haben dagegen in einer schon früher zitierten Arbeit[3] auf andere Weise folgende Temperaturen abgeleitet:

	T_{max}	T_{min}
δ Cephei	6400°	5400°
ζ Geminorum	5310	5030
α Ursae minoris	5770	5660

Man ersieht hieraus, daß die Resultate HOPMANNS einerseits und die der niederländischen Forscher andererseits recht wenig übereinstimmen, und CH'ING-SUNG YÜ hat in seiner bereits in Ziff. 58 besprochenen spektralphotometrischen Untersuchung über ζ Geminorum noch weit mehr abweichende Zahlen gefunden. Doch ist kein Zweifel, daß die Temperaturen im Maximum merklich höher sind als im Minimum, wie es ja auch nach dem früher Gesagten gar nicht anders möglich ist. Auch R. A. SAMPSONS[4] spektralphotometrische Messungen von δ Cephei bestätigen dies.

Nach HOPMANN laufen die Temperatur- und die Lichtkurven im ganzen parallel, wenn auch gewisse Abweichungen vorkommen. Die Temperaturen scheinen niedriger zu sein, als sie sonst den betreffenden Spektralklassen entsprechen; es scheint dies überhaupt eine Eigenschaft der „Übergiganten" zu sein, zu denen die δ Cephei-Sterne gehören. Die bolometrischen Gesamtamplituden der δ Cephei-Sterne sind nach HOPMANN nur gering; wir sahen früher, daß auch bei den Mira-Sternen die Gesamtstrahlung weit weniger veränderlich ist als die visuelle Helligkeit.

62. Die galaktische Verteilung der δ Cephei-Sterne. Es ist eine schon lange bekannte Tatsache, daß die langperiodischen δ Cephei-Sterne sehr stark nach der galaktischen Ebene konzentriert sind, während sich im Gegensatz dazu die kurzperiodischen gleichmäßig über alle galaktischen Breiten verteilen. Wie gut sich die langperiodischen δ Cephei-Sterne der galaktischen Ebene anschmiegen, geht daraus hervor, daß, wenn man aus ihrer Verteilung an der Sphäre die bevorzugte Ebene berechnet, diese nach E. HERTZSPRUNG[5] sehr genau mit der galaktischen Ebene zusammenfällt.

Die galaktische Verteilung der δ Cephei-Sterne ist von M. GÜSSOW in ihrer Dissertation ausführlich behandelt worden. Sie gelangt zu folgenden Ergebnissen:

In galaktischer Länge zeigt die Häufigkeit der langperiodischen δ Cephei-Sterne ein Minimum zwischen 170° und 240°, dem sich unmittelbar ein Maximum zwischen 250° und 260° anschließt. Ein zweites Minimum zeigt sich zwischen 300° und 330°, ein zweites Maximum zwischen 330° und 360°. Die kurzperiodischen δ Cephei-Sterne sind ziemlich gleichmäßig auf die Längen von

[1] Ap J 48, S. 106 (1918) = Mt Wilson Contr 151.
[2] A N 221, S. 337 (1924); 222, S. 1, 233 (1924); 226, S. 1 (1925); 227, S. 257 (1926).
[3] B A N 3, S. 47 (1925). [4] M N 85, S. 240 (1925).
[5] A N 192, S. 261 (1912).

0° bis 200° verteilt, nur fünf liegen außerhalb dieses Intervalls. Für die Verteilung in galaktischer Breite ergeben sich folgende Zahlen:

Die vier langperiodischen δ Cephei-Sterne mit galaktischen Breiten g von mehr als 30° sind nach M. GÜSSOW SS Hydrae ($g = 38°$, $P = 8^d{,}25$), RS Ceti ($g = 52°$, $P = 17^d{,}4$), W Virginis ($g = 58°$, $P = 17^d{,}3$) und RY Bootis ($g = 62°$, $P = 9^d{,}01$). Von diesen Sternen hat sich aber SS Hydrae durch neuere Beobachtungen der Radialgeschwindigkeit als Bedeckungsveränderlicher erwiesen[1]. RS Ceti und RY Bootis sind in ihrem Lichtwechsel wenig bekannt. W Virginis hat ein ungewöhnliches Spektrum (Ziff. 59).

Breite	Zahl der Sterne	
	langperiodische	kurzperiodische
0°–10°	95	6
10 –20	9	10
20 –30	4	7
30 –40	1	14
40 –50	0	6
50 –60	2	7
60 –70	1	6
70 –80	0	3
80 –90	0	3
	112	62

H. SHAPLEY[2] und H. LUDENDORFF[3] fanden für die kurzperiodischen δ Cephei-Sterne eine Abhängigkeit zwischen scheinbarer Helligkeit und galaktischer Breite in dem Sinne, daß die scheinbar helleren unter diesen Sternen durchschnittlich geringere galaktische Breiten haben als die scheinbar schwächeren. Als ganz sichergestellt darf man diese Beziehung indessen vielleicht noch nicht betrachten.

Über die Beziehungen zwischen der Häufigkeitsfunktion der Perioden der langperiodischen δ Cephei-Sterne zu der galaktischen Länge hat J. SCHILT[4] Untersuchungen angestellt.

63. Bewegungen der δ Cephei-Sterne. Mit den Bewegungen der δ Cephei-Sterne haben sich früher namentlich H. SHAPLEY[5] sowie J. C. KAPTEYN und P. J. VAN RHIJN[6]) und W. J. LUYTEN[7] beschäftigt. Von den Ergebnissen dieser Arbeiten wollen wir hier nur diejenigen SHAPLEYS anführen, daß, während die langperiodischen δ Cephei-Sterne nur parallaktische und keine Strombewegung zeigen, die Bewegungen der kurzperiodischen einen Apex ergeben, der mit dem von STRÖMBERG für die schnellbewegten Sterne gefundenen nahezu übereinstimmt. Wir werden in der nächsten Ziffer noch auf diese Arbeiten zurückzukommen Gelegenheit haben. Hier wollen wir uns etwas eingehender mit der Abhandlung von R. E. WILSON „The Proper-Motions and Mean Parallax of the Cepheid Variables“[8] beschäftigen.

WILSON gibt zunächst für 84 Veränderliche eine Zusammenstellung der Eigenbewegungen, die er auf Grund alles erreichbaren Beobachtungsmaterials abgeleitet hat. Es befinden sich unter diesen 84 Sternen auch einige wenige, die wir nicht zu den δ Cephei-Sternen rechnen, z. B. die RV Tauri-Sterne R Sagittae und V Vulpeculae, doch wird die Mitnahme dieser Sterne bei der Diskussion die Ergebnisse nicht merklich beeinträchtigt haben.

Unter den 84 Sternen kommen nur vier vor, die eine Eigenbewegung μ von mehr als 10″ in 100 Jahren haben. Es sind dies RR Lyrae ($\mu = 22''$), RZ Cephei (20″), SU Draconis (13″) und XZ Cygni (11″). Alle diese vier Sterne sind kurzperiodische δ Cephei-Sterne. Überhaupt ergibt sich, was auch schon auf Grund der früheren Arbeiten auf diesem Gebiete bekannt war, daß die kurzperiodischen δ Cephei-Sterne viel größere Eigenbewegungen besitzen als die

[1] Publ A S P 36, S. 139 (1924).
[2] Ap J 48, S. 291 (1918) = Mt Wilson Contr 153.
[3] A N 209, S. 217 (1919).
[4] Ap J 64, S. 149 (1926) = Mt Wilson Contr 315.
[5] Ap J 48, S. 93 (1918) = Mt Wilson Contr 151; Harv Circ 237 (1922).
[6] B A N 1, S. 37 (1922).
[7] Publ A S P 34, S. 166 (1922).
[8] A J 35, S. 35 (1923).

langperiodischen, unter denen der sich auch durch große galaktische Breite und abnormes Spektrum auszeichnende Stern W Virginis die größte Eigenbewegung (8″) hat. Ist M_0 das Mittel der Größen (wobei als Helligkeit jedes einzelnen Sternes das Mittel aus Maximal- und Minimalhelligkeit angesetzt ist), μ_0 das Mittel der μ, τ_0 das der τ (der Komponenten der Eigenbewegungen senkrecht zur Richtung zum Apex), q_0 das der parallaktischen Bewegungen, so finden sich nach WILSON folgende Werte:

	Zahl der Sterne	M_0	μ_0	τ_0	q_0
Gruppe I (kurzper. Sterne)	19	$9^m,77$	5″,65	+0″,79	+1″,69
„ II (langper. „)	51	6 ,77	2 ,18	−0 ,09	+1 ,32

wo die μ_0, τ_0, q_0 für 100 Jahre gelten. Sterne mit Perioden von mehr als 40^d und fragliche Objekte sind ausgeschlossen. Trotzdem nun M_0 für Gruppe I um 3^m schwächer ist als für Gruppe II, ist doch μ_0 für Gruppe I $2^1/_2$mal so groß als für Gruppe II. Der große Wert τ_0 für Gruppe I deutet auf große Pekuliarbewegungen, der im Verhältnis zu q_0 sehr kleine Wert τ_0 für Gruppe II auf kleine Pekuliarbewegungen hin.

Auch die Radialgeschwindigkeiten der δ Cephei-Sterne sind von WILSON diskutiert worden. Wie wir schon früher sahen, erleiden die Linien im Spektrum dieser Sterne mit dem Lichtwechsel synchrone periodische Verschiebungen, und man betrachtet als Radialgeschwindigkeit des betreffenden Sternes den Wert γ, den man erhält, wenn man jene Linienverschiebungen als durch Bewegungen in einem Doppelsternsystem hervorgerufen ansieht und die Radialgeschwindigkeit des Schwerpunktes dieses Systems bestimmt.

Betrachten wir nun diese Werte γ in der Tabelle der spektroskopischen Bahnelemente der δ Cephei-Sterne in Ziff. 60, so fällt sofort auf, daß die γ für alle langperiodischen δ Cephei-Sterne klein sind, daß dagegen der einzige in der Tabelle vorkommende kurzperiodische Stern RR Lyrae einen großen Wert von γ (68 km) aufweist. Man hat nun noch für eine Reihe weiterer δ Cephei-Sterne, für die spektroskopische Bahnbestimmungen noch nicht vorliegen, die Werte von γ auf Grund mehr oder weniger zahlreicher Beobachtungen der Radialgeschwindigkeit abgeschätzt und hat obige Regel bestätigt gefunden: die langperiodischen δ Cephei-Sterne haben kleine, die kurzperiodischen im allgemeinen große Radialgeschwindigkeiten. Es zeigt sich also bei den Radialgeschwindigkeiten das entsprechende Verhalten wie bei den lateralen Eigenbewegungen.

Mit Hilfe der lateralen Eigenbewegungen und der Radialgeschwindigkeiten der langperiodischen δ Cephei-Sterne berechnet nun WILSON in seiner Abhandlung den Apex der Sonnenbewegung und die Geschwindigkeit V_0 der Sonne. Er findet

$$A = 275^\circ,1 \quad D = +34^\circ,2 \quad V_0 = 21,4\,\text{km}.$$

Die Koordinaten des Apex stimmen sehr gut mit den gewöhnlich angenommenen Werten (rund 270°, +30°) überein, und V_0 ist fast genau gleich dem Wert, den G. STRÖMBERG aus den Radialgeschwindigkeiten von 1400 F- bis M-Sternen gefunden hat (21,5 km). Für die kurzperiodischen δ Cephei-Sterne führt WILSON die entsprechenden Rechnungen wegen der zu geringen Zahl der Objekte, für die die nötigen Daten vorliegen, nicht aus.

Mit dem STRÖMBERGschen Wert der Sonnengeschwindigkeit befreit nun WILSON die beobachteten Radialgeschwindigkeiten γ der δ Cephei-Sterne von dem Einfluß der Sonnenbewegung und findet so die „absoluten“ Radialgeschwindigkeiten V, die wir nebst den von WILSON angenommenen Werten γ

in der folgenden Tabelle wiedergeben. Die hier verwandten γ weichen zum Teil etwas von den in der Tabelle in Ziff. 60 angebenen ab, doch ist dies ohne Belang. Wir kürzen alle Zahlen auf volle Kilometer ab. Die Sterne sind nach ihrer Rektaszension geordnet.

Kurzperiodische δ Cephei-Sterne.

	γ	V		γ	V
SU Draconis	−202 km	−193 km	RR Lyrae	− 69 km	− 50 km
SW	− 83	− 74	XZ Cygni	−215	−196
RS Bootis	− 66	− 51			

Langperiodische δ Cephei-Sterne.

	γ	V		γ	V
TU Cassiopeiae	−23 km	−19 km	RV Scorpii	−28 km	−19 km
α Ursae minoris	−17	− 6	X Sagittarii	−14	− 2
SU Cassiopeiae	− 7	− 2	Y Ophiuchi	− 5	+12
SZ Tauri	− 3	−16	W Sagittarii	−29	−18
T Monocerotis	+12	− 5	Y	+ 4	+18
RT Aurigae	+21	+11	$\varkappa$ Pavonis	+36	+33
ζ Geminorum	+ 7	− 6	SU Cygni	−36	−16
l Carinae	+ 4	−10	η Aquilae	−14	+ 3
S Muscae	+ 3	− 7	S Sagittae	− 9	+10
R Trianguli austr.	−20	−24	X Cygni	+ 9	+27
S	+ 2	− 1	T Vulpeculae	− 1	+16
S Normae	− 8	− 8	δ Cephei	−17	− 4

Die Richtigkeit der früher erwähnten Regel, daß die kurzperiodischen δ Cephei-Sterne große, die langperiodischen kleine Radialgeschwindigkeiten haben, geht aus dieser Tabelle klar hervor. Daß erst so wenige Radialgeschwindigkeiten kurzperiodischer δ Cephei-Sterne bekannt sind, hat seinen Grund in der Lichtschwäche dieser Objekte.

Obwohl wir uns mit den Parallaxen der δ Cephei-Sterne erst später zu beschäftigen haben, sei hier doch schon erwähnt, daß WILSON aus den Eigenbewegungen und Radialgeschwindigkeiten der δ Cephei-Sterne auch deren mittlere Parallaxen π_m bestimmt. Er findet für die kurzperiodischen δ Cephei-Sterne $\pi_m = 0'',0013$ mit einem mittleren Fehler von genau demselben Betrage, und für die langperiodischen $\pi_m = 0'',0033$ mit einem mittleren Fehler von $0'',0004$.

Es sei hier noch bemerkt, daß der kurzperiodische δ Cephei-Stern RZ Cephei zwar eine große Eigenbewegung ($19'',7$ in 100 Jahren), aber nach Untersuchungen von W. J. LUYTEN[1] eine kleine Radialgeschwindigkeit (veränderlich zwischen den Grenzen −42 km und +14 km, soweit die Beobachtungen reichen) besitzt. Die Parallaxe dieses Sternes läßt sich mit Hilfe von SHAPLEYS „Period-Luminosity Curve" (vgl. nächste Ziffer) ermitteln, und es ergibt sich eine Entfernung von 3800 Lichtjahren[2]. In diesem Abstand entspricht die hundertjährige Eigenbewegung von $19'',7$ einer Transversalgeschwindigkeit von etwa 1100 km pro Sekunde. RZ Cephei ist damit der am raschesten bewegte einzelne Stern, den wir kennen.

Einzelne Beobachtungen der Radialgeschwindigkeit einer größeren Zahl von δ Cephei-Sternen haben W. S. ADAMS, A. H. JOY und R. F. SANFORD[3] angestellt. Danach lassen sich noch einige weitere kurzperiodische δ Cephei-Sterne

[1] Publ A S P 35, S. 69 (1923). [2] Harv Bull 773 (1922).
[3] Publ A S P 36, S. 139 (1924).

mit großen Radialgeschwindigkeiten anführen; z. B. liegen die beobachteten Werte der Radialgeschwindigkeit bei

RR Ceti	zwischen	−144 km	und	− 59 km
U Leporis		+ 90		+139
RV Ursae maj.		−209		−148
VX Herculis		−405		−374
RW Draconis		−124		− 94
RZ Lyrae		−281		−161
RV Capricorni		−121		− 51

während die von den genannten Autoren beobachteten langperiodischen δ Cephei-Sterne alle ziemlich geringe Radialgeschwindigkeiten zu besitzen scheinen außer W Virginis (beobachtete Werte zwischen −95 km und −26 km). Dieser Stern ist uns schon häufiger wegen seiner Ausnahmestellung aufgefallen. Vielleicht besitzt auch WZ Sagittarii eine für einen langperiodischen δ Cephei-Stern ungewöhnlich große Radialgeschwindigkeit.

G. STRÖMBERG[1] hat neuerdings mit Hilfe eines größeren, zum Teil noch nicht veröffentlichten Materials an Radialgeschwindigkeiten von δ Cephei-Sternen die Elemente der Sonnenbewegung abgeleitet. Aus den Radialgeschwindigkeiten von 26 δ Cephei-Sternen mit Perioden von weniger als $0^d,7$ findet er:

$$A = 306° \quad D = +47° \quad V_0 = 109 \text{ km}.$$

A und D weichen also von den gewöhnlich angenommenen Werten nicht allzu weit ab (die Bestimmung ist wegen der geringen Zahl von Sternen natürlich unsicher), dagegen ist V_0 sehr groß. Aus den Radialgeschwindigkeiten von 37 langperiodischen δ Cephei-Sternen mit $P > 2^d$ findet STRÖMBERG:

$$A = 283° \quad D = +24° \quad V_0 = 11,5 \text{ km}.$$

Nimmt man $A = 270°$, $D = +30°$, so ergibt sich nach STRÖMBERG $V_0 = 12,3$ km. V_0 ist hiernach also viel kleiner, als man nach WILSONS Rechnungen angenommen hatte.

Einige weitere Betrachtungen über die Eigenbewegungen der langperiodischen δ Cephei-Sterne rühren von J. SCHILT[2] her.

64. Shapleys „Period-Luminosity Curve". Im Jahre 1912 veröffentlichte E. C. PICKERING[3] eine von Miss H. S. LEAVITT gemachte Entdeckung, die sich in der Folgezeit als außerordentlich wichtig erweisen sollte. Für 25 von den zahlreichen Veränderlichen, die in der Kleinen MAGELLANschen Wolke aufgefunden worden waren, und auf die wir später noch zurückkommen, hatte Miss LEAVITT die Perioden bestimmt; der Lichtwechsel dieser Veränderlichen zeigte denselben Charakter wie der der δ Cephei-Sterne. Miss LEAVITT fand nun, daß, je heller diese Sterne im Maximum bzw. im Minimum sind, desto länger ihre Periode ist. Nebenstehend geben wir die Tabelle von Miss LEAVITT in abgekürzter Form wieder. Die 25 Sterne sind nach der Periodenlänge geordnet, unter Max steht die photographische Maximal-, unter Min die photographische Minimalhelligkeit.

Max	Min	P	Max	Min	P
$14^m,8$	$16^m,1$	$1^d,25$	$14^m,1$	$14^m,8$	$6^d,65$
14 ,8	16 ,4	1 ,66	14 ,0	14 ,8	7 ,48
14 ,8	16 ,4	1 ,76	13 ,9	15 ,2	8 ,40
15 ,1	16 ,3	1 ,88	13 ,6	14 ,7	10 ,34
14 ,7	15 ,6	2 ,17	13 ,4	14 ,6	11 ,64
14 ,4	15 ,7	2 ,91	13 ,8	14 ,8	12 ,42
14 ,7	15 ,9	3 ,50	13 ,4	14 ,4	13 ,08
14 ,6	16 ,1	4 ,29	13 ,4	14 ,3	13 ,47
14 ,3	15 ,3	4 ,55	13 ,0	14 ,6	16 ,75
14 ,3	15 ,5	4 ,99	12 ,2	14 ,1	31 ,94
14 ,4	15 ,4	5 ,31	11 ,4	12 ,8	65 ,8
14 ,3	15 ,2	5 ,32	11 ,2	12 ,1	127 ,0
13 ,8	14 ,8	6 ,29			

[1] Ap J 61, S. 363 (1925) = Mt Wilson Contr 293.
[2] Ap J 64, S. 149 (1926) = Mt Wilson Contr 315.
[3] Harv Circ 173 (1912).

Die oben erwähnte Regel geht aus diesen Zahlen klar hervor. Nimmt man statt der Perioden selbst deren Logarithmen, so ergibt sich, daß die Helligkeit im Maximum bzw. im Minimum eine lineare Funktion von $\log P$ ist. Die einzelnen Sterne weichen nur sehr wenig von dieser Regel ab. Da wir annehmen können, daß die Sterne in der sehr weit entfernten Kleinen MAGELLANschen Wolke alle ungefähr dieselbe Entfernung von uns haben, so folgt, daß nicht nur die scheinbaren, sondern auch die absoluten Helligkeiten jener Veränderlichen eine Funktion der Periodenlänge sind.

Es lag nun der Gedanke sehr nahe, eine solche Abhängigkeit zwischen absoluter Helligkeit und Periode auch für die δ Cephei-Sterne unseres engeren, galaktischen Sternsystems anzunehmen. Ob dieser Gedanke richtig ist, läßt sich wegen der ungenauen Kenntnis der Entfernungen dieser Sterne nicht unmittelbar feststellen. Ist er aber richtig, so ist uns ein Mittel gegeben, aus den scheinbaren Helligkeiten und den Perioden die absoluten Helligkeiten und die Entfernungen der δ Cephei-Sterne zu bestimmen, sobald nur für einige von ihnen die Parallaxen genau bekannt sind.

Zuerst ist E. HERTZSPRUNG[1] diesem Gedankengange nähergetreten. Er berechnete für 13 δ Cephei-Sterne, für die in Boss' Preliminary General Catalogue die Eigenbewegungen gegeben sind, aus letzteren die mittlere Parallaxe und aus dieser die mittlere absolute Helligkeit. Indem er dann annimmt, daß das LEAVITTsche Gesetz auch für die δ Cephei-Sterne unseres Systems gilt, kann er Untersuchungen über die räumliche Verteilung dieser Sterne anstellen. Außerdem leitet er auch die Entfernung der Kleinen MAGELLANschen Wolke ab. Wir brauchen uns hier indessen mit HERTZSPRUNGS Resultaten nicht näher zu beschäftigen, da sie durch die Untersuchungen von H. SHAPLEY[2] überholt worden sind, die wir nun ausführlich darlegen wollen.

SHAPLEY geht bei seinen Rechnungen ebenfalls von den im Boss-Kataloge für 13 langperiodische δ Cephei-Sterne gegebenen Eigenbewegungen aus, schließt aber von diesen 13 Sternen zwei, nämlich l Carinae und $\varkappa$ Pavonis, aus gewissen Gründen aus. Für die übrigbleibenden 11 Sterne zieht er auch die Radialgeschwindigkeiten heran. Mit Hilfe dieses recht knappen Materiales berechnet er nun unter der wohl (vgl. die in der vorigen Ziffer besprochenen späteren Untersuchungen WILSONS) nahe zutreffenden Voraussetzung, daß die langperiodischen δ Cephei-Sterne keine gemeinsame Strombewegung haben, die mittlere Parallaxe p_0 und die mittlere absolute Helligkeit M_0 der 11 Sterne. Es ergibt sich:

$$p_0 = 0'',0034,$$
$$M_0 = -2^m,35.$$

Die gemessenen trigonometrischen Parallaxen einiger der Sterne haben im Verhältnis zu ihren kleinen Werten zu große wahrscheinliche Fehler, als daß sie für diese Untersuchung von Nutzen sein könnten. Unter der Voraussetzung, daß die nach dem Antiapex der Sonnenbewegung gerichtete Komponente der Eigenbewegung eines jeden der Sterne rein parallaktischer Natur ist, wird alsdann für jeden Stern die individuelle Parallaxe und die absolute Helligkeit M berechnet. Die folgende Tabelle gibt letztere nebst der Periode des Lichtwechsels:

Stern	M	P	Stern	M	P
α Ursae minoris	$-2^m,7$	$3^d,97$	Y Sagittarii	$-1^m,6$	$5^d,77$
SU Cassiopeiae	$-1\ ,6$	$1\ ,95$	η Aquilae	$-2\ ,7$	$7\ ,18$
RT Aurigae	$-0\ ,7$	$3\ ,73$	S Sagittae	$-2\ ,9$	$8\ ,38$
ζ Geminorum	$-4\ ,1$	$10\ ,15$	T Vulpeculae	$-1\ ,4$	$4\ ,44$
X Sagittarii	$-1\ ,5$	$7\ ,01$	δ Cephei	$-3\ ,3$	$5\ ,37$
W	$-2\ ,7$	$7\ ,59$			

[1] A N 196, S. 201 (1913). [2] Ap J 48, S. 89 (1918) = Mt Wilson Contr 151.

M ist diejenige absolute Größe, die dem Mittel aus Maximal- und Minimalgröße entspricht. In den Zahlen läßt sich eine Korrelation zwischen M und P erkennen, die noch deutlicher hervortritt, wenn wir die Sterne nach der Länge der Periode gruppenweise zu Mitteln zusammenfassen:

Zahl der Sterne	M	P
1	$-4^m,1$	$10^d,15$
3	$-2\ ,8$	$7\ ,72$
3	$-2\ ,1$	$6\ ,05$
3	$-1\ ,6$	$4\ ,05$
1	$-1\ ,6$	$1\ ,95$

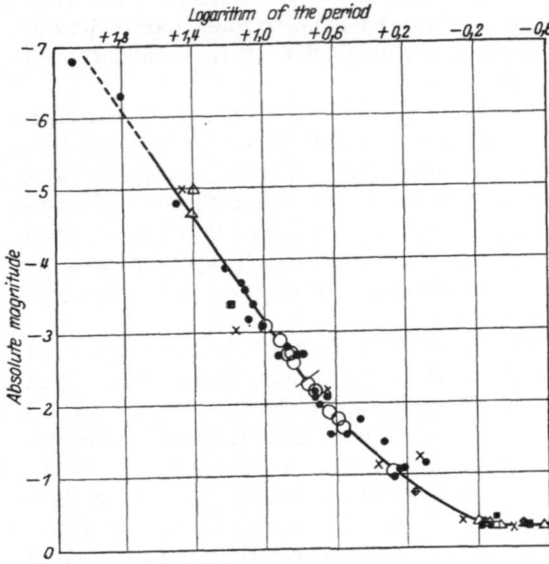

Abb. 36. SHAPLEYS Period-Luminosity Curve.

Es ist also hier ein ähnlicher Zusammenhang vorhanden wie bei den Veränderlichen in der Kleinen MAGELLANschen Wolke; dabei ist noch zu beachten, daß die Herleitung der M infolge der dabei gemachten, sicher nicht streng zutreffenden Voraussetzungen recht unsicher ist.

Auf Grund der Zahlen in den beiden letzten Tabellen zeichnet nun SHAPLEY eine ausgleichende Kurve, indem er die log P als Abszissen, die M als Ordinaten verwendet. Aus dieser Kurve liest er alsdann die zu jedem Werte von P bzw. log P gehörigen Werte M ab; diese ausgeglichenen Werte von M sind in Abb. 36 als Kreise eingezeichnet.

Um nun den Verlauf der Kurve sicherer festzulegen und namentlich um sie auf ein größeres Periodenintervall auszudehnen, benutzt SHAPLEY zunächst die von Miss LEAVITT beobachteten Veränderlichen in der Kleinen MAGELLANschen Wolke. Hierbei ist vor allem die Schwierigkeit zu überwinden, daß Miss LEAVITT photographische Größen angibt, während die Kurve sich auf visuelle Größen bezieht. SHAPLEY bestimmt daher den Zusammenhang zwischen Farbenindex und Periodenlänge (vgl. Ziff. 61) und kann nun die photographischen Größen von Miss LEAVITT in visuelle umwandeln. Er vergleicht alsdann diese scheinbaren Größen mit den für die betreffenden Periodenwerte sich aus Abb. 36 ergebenden Werten der absoluten Größen und findet, daß an die scheinbaren Größen eine Korrektion von $-16^m,4$ anzubringen ist, um sie auf die letzteren zu reduzieren. Mit anderen Worten: die Kurve, die den Zusammenhang zwischen scheinbaren Größen und log P für die Veränderlichen der Kleinen MAGELLANschen Wolke darstellt, ist im Sinne der Ordinaten um $-16^m,4$ zu verschieben, um mit der Kurve, die die Beziehung zwischen absoluter Größe und log P für die galaktischen δ Cephei-Sterne wiedergibt, möglichst nahe zur Deckung gebracht zu werden. Da nun die LEAVITTschen Veränderlichen ein größeres Periodenintervall ($1^d,25$ bis 127^d) umfassen als die von SHAPLEY benutzten galaktischen δ Cephei-Sterne ($1^d,95$ bis $10^d,15$), so wird durch die Heranziehung der ersteren nicht nur der Verlauf der Kurve in dem letzteren Intervall sicherer festgelegt, sondern zugleich auch das Periodenintervall, das die Kurve umfaßt, bedeutend erweitert. Vorausgesetzt ist natürlich bei der ganzen Operation, daß in der Kleinen MAGELLANschen Wolke genau derselbe Zusammenhang zwischen absoluter Größe und log P besteht, wie in unserem engeren System. Die den Veränderlichen in der Kleinen MAGELLANschen Wolke entsprechenden Punkte sind in die Abbildung als kleine schwarze Kreise eingezeichnet.

Um die Kurve auch nach der Seite der kleineren Perioden hin weiter auszudehnen, benutzt SHAPLEY in ganz ähnlicher Weise die Veränderlichen in dem kugelförmigen Sternhaufen ω Centauri. Es standen ihm hier auf Grund der Arbeiten von S. I. BAILEY die Helligkeiten und Perioden von 5 langperiodischen und 90 kurzperiodischen δ Cephei-Sternen zur Verfügung, welch' letztere er nach ihrer Periodenlänge zu drei Mittelwerten zusammenfaßt. Die den Veränderlichen in ω Centauri entsprechenden Punkte sind in der Abbildung durch Kreuze dargestellt.

Ferner benutzt SHAPLEY auch noch die Veränderlichen in Messier 5 (Dreiecke), Messier 3 (schwarze Quadrate), Messier 13 (offene Quadrate) und Messier 15 (ganz kleine Kreise). Auf Grund des Gesamtmaterials wird dann die in der Abbildung wiedergegebene „Period-Luminosity Curve“ gezogen, der folgende Zahlen entsprechen:

log P	Absolute visuelle Größe	log P	Absolute visuelle Größe	log P	Absolute visuelle Größe	log P	Absolute visuelle Größe
−0,6	$-0^{m},34$	+0,1	$-0^{m},81$	+0,8	$-2^{m},43$	+1,5	$-4^{m},95$
−0,5	−0 ,33	+0,2	−0 ,99	+0,9	−2 ,79	+1,6	−5 ,31
−0,4	−0 ,33	+0,3	−1 ,17	+1,0	−3 ,15	+1,7	−5 ,67
−0,3	−0 ,34	+0,4	−1 ,37	+1,1	−3 ,51	+1,8	−6 ,02
−0,2	−0 ,38	+0,5	−1 ,58	+1,2	−3 ,87	+1,9	−6 ,38
−0,1	−0 ,50	+0,6	−1 ,81	+1,3	−4 ,23	+2,0	−6 ,74
0,0	−0 ,64	+0,7	−2 ,10	+1,4	−4 ,59	+2,1	−7 ,1

Der Verlauf der Kurve ist recht sicher festgelegt, unsicher ist nur ihre Höhenlage, die ganz auf der Bestimmung der absoluten Helligkeiten der 11 von SHAPLEY benutzten galaktischen δ Cephei-Sterne beruht. R. H. WILSON hält es auf Grund seiner in der vorigen Ziffer behandelten Untersuchungen für möglich, daß die aus SHAPLEYS Kurve folgenden Parallaxen höchstens um etwa 40% vergrößert werden müssen, während eine Verkleinerung unwahrscheinlich sei. Wenn man die von G. STRÖMBERG ermittelte, auf die langperiodischen δ Cephei-Sterne bezogene Sonnengeschwindigkeit $V_0 = 12{,}3$ km (vgl. vorige Ziffer) den Rechnungen zugrunde legt, so ergibt sich in der Tat eine Verringerung der Entfernungen um 38% und eine Korrektion der absoluten Größen um $+1^{m},06$. Auch Untersuchungen von J. SCHILT[1] deuten auf eine Korrektion im gleichen Sinne hin, und ebenso solche von K. G. MALMQUIST[2], der eine Korrektion der absoluten Größen SHAPLEYS von $+0^{m},8$ findet; dieser Korrektion entsprechend wären die aus SHAPLEYS Kurve folgenden Entfernungen mit 0,7 zu multiplizieren. Es ist also wahrscheinlich, daß die absoluten Größen in obiger Tabelle noch eine erhebliche konstante Korrektion erfordern. Es sei nochmals erwähnt, daß diese Größen sich auf das Mittel aus Maximal- und Minimalhelligkeit der Veränderlichen beziehen („median magnitudes“).

Aus der Kurve und aus der Tabelle geht hervor, daß die kurzperiodischen δ Cephei-Sterne kaum noch eine Abhängigkeit zwischen absoluter Helligkeit und Periodenlänge aufweisen; sie scheinen vielmehr alle nahezu dieselbe absolute Helligkeit zu haben. Dafür spricht auch die Tatsache, daß innerhalb eines jeden Sternhaufens die vorkommenden kurzperiodischen Veränderlichen sehr nahe dieselbe „median magnitude“ besitzen, obwohl die Maximalgrößen, die

[1] Ap J 64, S. 149 (1926) = Mt Wilson Contr 315.
[2] Lund Medd Sér I, Nr. 113 (1926).

Amplituden, die Perioden und die Lichtkurven mitunter merkliche Verschiedenheiten aufweisen.

Auf Grund von SHAPLEYS Period-Luminosity Curve kann man nun, sobald man die Periode eines δ Cephei-Sternes kennt, unmittelbar seine absolute mittlere Größe angeben, und wenn auch seine scheinbare mittlere Größe bekannt ist, kann man seine Parallaxe berechnen. Es lassen sich daher auch die Parallaxen derjenigen Sternhaufen, Spiralnebel und ähnlichen Gebilde, die δ Cephei-Sterne enthalten, bestimmen. Über die Resultate derartiger Untersuchungen wird in den betreffenden Kapiteln dieses Handbuches berichtet. Mit den Entfernungen und der räumlichen Verteilung der δ Cephei-Sterne unseres galaktischen Systems werden wir uns dagegen in der nächsten Ziffer beschäftigen.

Aus der Period-Luminosity Curve folgt, daß die kurzperiodischen δ Cephei-Sterne die absolute Größe $-0^m,3$ haben, also, ebenso wie die langperiodischen, Riesensterne sind. Hiergegen erhoben J. C. KAPTEYN und P. J. VAN RHIJN[1] Bedenken. Sie leiteten die Eigenbewegungen einer Anzahl von kurzperiodischen δ Cephei-Sternen ab und fanden dafür verhältnismäßig große Werte; daraus schlossen sie auf verhältnismäßig große Parallaxen und geringe absolute Helligkeiten. H. SHAPLEY[2] machte aber darauf aufmerksam, daß sich für diese Sterne sehr beträchtliche Radialgeschwindigkeiten ergeben haben. Die Größe der scheinbaren Eigenbewegungen ist demnach nicht geringer Entfernung, sondern großer Geschwindigkeit zuzuschreiben. Zudem haben sich in der Kleinen MAGELLANschen Wolke (wie auch in einigen Sternhaufen) kurzperiodische Veränderliche gefunden[3], welche gegen die dort vorhandenen langperiodischen gerade die Helligkeitsdifferenz zeigen, die man auf Grund von SHAPLEYS Kurve erwarten muß. Auch B. LINDBLAD[4] ist durch eine Diskussion der Eigenbewegungen der kurzperiodischen δ Cephei-Sterne zu dem Schlusse gekommen, daß SHAPLEYS Ansichten über die Entfernung derselben im wesentlichen richtig sind.

Die oben angegebene Period-Luminosity Curve bezieht sich auf visuelle Größen. Die entsprechende Kurve, bezogen auf photographische Größen, hat SHAPLEY[5] später allein aus den Veränderlichen der Kleinen MAGELLANschen Wolke abgeleitet, nachdem für 107 von diesen Objekten die Perioden und photographischen Helligkeiten bestimmt worden waren. Da die Messungen der Helligkeiten der schwächsten, d. h. der kurzperiodischen Veränderlichen noch nicht genau sind, so beschränkt sich SHAPLEY auf die langperiodischen. Wir geben hier diese photographische Period-Luminosity Curve in Zahlen wieder:

log P	Absolute Größe	log P	Absolute Größe
0,0	$-0^m,51$	1,2	$-2^m,48$
0,2	$-0\ ,82$	1,4	$-2\ ,89$
0,4	$-1\ ,14$	1,6	$-3\ ,37$
0,6	$-1\ ,45$	1,8	$-3\ ,94$
0,8	$-1\ ,78$	2,0	$-4\ ,81$
1,0	$-2\ ,12$		

Angesichts der großen Bedeutung, die SHAPLEYS Kurve für die ganze Stellarastronomie besitzt, muß man nun die Frage erheben, ob irgendwelche Anzeichen dafür vorhanden sind, daß vielleicht nicht alle δ Cephei-Sterne sich dem durch diese Kurve ausgedrückten Gesetze fügen. SHAPLEY[6] selbst hat darauf aufmerksam gemacht, daß im dichtesten Teil der Kleinen MAGELLANschen Wolke einige Veränderliche um $0^m,5$ schwächer sind, als man auf Grund ihrer Perioden erwarten sollte. Man kann dies aber vielleicht durch Absorption des Lichtes in den umgebenden Nebelmassen oder auch durch einen photographischen Effekt

[1] B A N 1, S. 37 (1922). [2] Harv Circ 237 (1922). [3] Harv Bull 765 (1922). [4] Ap J 59, S. 37 (1924). [5] Harv Circ 280 (1925). [6] Harv Circ 280 (1925).

erklären. Ferner hat E. HERTZSPRUNG[1] bei den δ Cephei-Sternen in der Umgebung von η Carinae keinen Zusammenhang zwischen scheinbarer Helligkeit und Periode gefunden. Man muß also annehmen, daß diese Veränderlichen sehr verschiedene Entfernungen von uns haben, oder daß SHAPLEYS Gesetz dort nicht gilt. Ähnlich scheinen nach F. HENROTEAU[2] die Dinge bei den δ Cephei-Sternen in den Sternwolken im Sagittarius zu liegen. Weitere Untersuchungen über diese Frage wären dringend erwünscht.

65. Die Entfernungen und die räumliche Verteilung der δ Cephei-Sterne. Im folgenden nehmen wir an, daß SHAPLEYS Period-Luminosity Curve so, wie er sie abgeleitet hat, richtig ist. Es können danach also für alle δ Cephei-Sterne aus der Periode die mittlere absolute Helligkeit, und, wenn auch die mittlere scheinbare Helligkeit bekannt ist, die Parallaxe oder Entfernung berechnet werden. Unter Benutzung der sphärischen Koordinaten der Sterne kann man alsdann auch ihre Lage im Raume angeben. Eine gewisse Ungenauigkeit in den Rechnungen entsteht dadurch, daß die scheinbaren Helligkeiten der Veränderlichen meist nicht auf ein einheitliches und einwandfreies photometrisches System bezogen sind.

SHAPLEY[3] hat derartige Rechnungen für 139 δ Cephei-Sterne ausgeführt. Er schließt dabei alle Sterne mit Perioden von mehr als 40^d aus, da er ihre Zugehörigkeit zur δ Cephei-Klasse als zweifelhaft betrachtet. (Wir haben diese Grenze früher bei $P = 45^d$ gezogen.) Die sich ergebenden Parallaxen sind außerordentlich klein und erreichen nur in verhältnismäßig wenigen Fällen die Größe von einigen tausendsteln Bogensekunden. Weniger als ein Drittel der Sterne hat Parallaxen von mehr als 0″,001. Die größte Parallaxe hat α Ursae minoris mit 0″,016, dann folgt δ Cephei mit 0″,0054 und η Aquilae mit 0″,0046. (Für β Cephei folgt $\pi = 0'',018$, aber dieser Stern ist kein eigentlicher δ Cephei-Stern, und es ist durchaus fraglich, ob er in SHAPLEYS Kurve hineinpaßt.) Der mittlere Abstand von der galaktischen Ebene ergibt sich für die 45 kurzperiodischen δ Cephei-Sterne zu 960 Parsecs, für die 94 langperiodischen zu nur 150 Parsecs; in dieser letzteren Zahl spricht sich die starke galaktische Konzentration der langperiodischen δ Cephei-Sterne aus. Für die kurzperiodischen gibt SHAPLEY folgende Übersicht über die in Parsecs ausgedrückten Abstände $R \sin\beta$ von der galaktischen Ebene:

Sterne nördlich von der galaktischen Ebene		Sterne südlich von der galaktischen Ebene	
$R \sin\beta$	Zahl	$R \sin\beta$	Zahl
$> +1500$	7	> -1000	3
$+1500$ bis $+1000$	7	-1000 bis -500	5
$+1000$ „ $+500$	9	-500 „ 0	8
$+500$ „ 0	6		

Für die räumliche Verteilung der langperiodischen δ Cephei-Sterne dagegen ergibt sich folgende Übersicht ($R \cos\beta$ ist der auf die galaktische Ebene projizierte Abstand des Sternes von der Sonne):

		Grenzen von $R \cos\beta$			
		0—1000	1000—2000	2000—4000	>4000
Nördlich der Milchstraße	Zahl der Sterne	14	6	11	2
	Mittl. $R \sin\beta$	+ 220	+220	+160	+100
	Größtes $R \sin\beta$	+1740	+810	+640	+210
Südlich der Milchstraße	Zahl der Sterne	17	17	23	4
	Mittl. $R \sin\beta$	− 80	−100	−140	−260
	Größtes $R \sin\beta$	− 160	−240	−400	−510

[1] B A N 2, S. 113 (1924). [2] J Can R A S 18, S. 343 (1924).
[3] Ap J 48, S. 279 (1918) = Mt Wilson Contr 153.

Später hat SHAPLEY[1] darauf aufmerksam gemacht, daß der langperiodische δ Cephei-Stern W Virginis ($P = 17^d,1$), der uns schon durch seine hohe galaktische Breite, sein abnormes Spektrum und seine große Eigenbewegung aufgefallen ist, und der in SHAPLEYS Liste von 139 δ Cephei-Sternen nicht enthalten war, unter allen δ Cephei-Sternen die größte Entfernung hat, nämlich $R = 7100$ Parsecs. Für diesen Stern ist $R \cos \beta = 3900$, $R \sin \beta = +6000$ Parsecs; er hat damit auch den größten Abstand von der galaktischen Ebene. — Einen Nachtrag zu seiner Untersuchung gab SHAPLEY[2] im Jahre 1920; es ist in dieser Liste W Virginis statt W Serpentis zu lesen.

Man hat die Parallaxen einer Anzahl von δ Cephei-Sternen auch trigonometrisch gemessen[3]; bei der außerordentlich großen Entfernung dieser Objekte sind die wahrscheinlichen Fehler dieser Parallaxen, trotz der großen Genauigkeit der Messungen, durchaus von derselben Größenordnung wie die Parallaxen selbst. SHAPLEY[4] hat nun solche trigonometrischen Parallaxen mit den auf Grund seiner Kurve abgeleiteten „photometrischen" Parallaxen verglichen und gezeigt, daß sich Widersprüche nicht ergeben. Zu demselben Schlusse ist neuerdings A. VAN MAANEN[3] in bezug auf die 11 von ihm auf dem Mt. Wilson-Observatorium gemessenen Parallaxen von δ Cephei-Sternen gelangt. Weiter hat SHAPLEY[4] gefunden, daß die photometrischen Parallaxen recht gut mit den auf dem Mt. Wilson Observatory[5] bestimmten spektroskopischen übereinstimmen; daß sie im Mittel übereinstimmen, ist von vornherein zu erwarten, da die Nullpunkte der beiden Methoden der Parallaxenbestimmung auf den Eigenbewegungen von Sternen beruhen, welche dieselben spektroskopischen Charakteristiken haben.

66. Die veränderlichen Sterne der β Cephei-Klasse und andere Veränderliche mit äußerst kleinen Helligkeitsschwankungen. Wir haben bereits früher (Ziff. 50) erwähnt, daß es einige Veränderliche mit sehr kleinen Helligkeitsschwankungen gibt, die zwar mit den δ Cephei-Sternen verwandt sind, aber sich doch in vieler Beziehung von ihnen unterscheiden. Der am längsten bekannte Vertreter dieser Art von Sternen ist β Cephei, und wir benennen daher diese Klasse von Veränderlichen nach ihm; ähnlich wie β Cephei verhalten sich vor allem 12 Lacertae und γ Bootis.

β Cephei hat ein Spektrum der Klasse B1. E. B. FROST stellte fest, daß die Radialgeschwindigkeit dieses Sternes veränderlich sei und fand[6] für die Periode der Änderungen den sehr kleinen Wert von $0^d,1904$. Später ergaben lichtelektrische Messungen von P. GUTHNICK[7], daß die Helligkeit von β Cephei synchron mit den Änderungen der Radialgeschwindigkeit um $0^m,07$ veränderlich sei. Die Beobachtungen von P. GUTHNICK und R. PRAGER sind in den „Veröffentlichungen der Sternwarte Berlin-Babelsberg"[8] ausführlich mitgeteilt. Die Form und die Amplitude der Lichtkurve ergeben sich als veränderlich; im allgemeinen ist die Lichtkurve ziemlich symmetrisch. Dieses Ergebnis wurde durch lichtelektrische Messungen von Miss E. E. CUMMINGS[9] auf dem Lick-Observatorium bestätigt. Miss CUMMINGS fand ein sekundäres Maximum im Aufstieg der Lichtkurve, das auch GUTHNICK und PRAGER zeitweise festgestellt hatten.

---

[1] Publ A S P 32, S. 156 (1920). [2] Publ A S P 32, S. 162 (1920).
[3] Listen z. B. bei F. HENROTEAU, Publ. Dominion Obs. Ottawa 9, S. 14 (1925); A. v. MAANEN, Publ A S P 38, S. 327 (1926).
[4] Harv Circ 237 (1922).
[5] Ap J 53, S. 13 (1921) = Mt Wilson Contr 199.
[6] Ap J 24, S. 259 (1906). [7] A N 196, S. 357 (1913).
[8] Bd. 1, Heft 1 (1914); Bd. 2, Heft 3 (1918).
[9] Lick Bull 11, S. 115 (1923).

Die Radialgeschwindigkeit von β Cephei ist oft untersucht worden. CL. C. CRUMP[1] fand, daß die Amplitude der Geschwindigkeitskurve und ebenso, daß die Radialgeschwindigkeit des Schwerpunktes veränderlich ist; im allgemeinen ist K = etwa 16 km. Es ergab sich ferner $e = 0{,}04$; die Bahn ist also nahezu kreisförmig. Der der Zeit des Helligkeitsmaximums entsprechende Punkt der Geschwindigkeitskurve liegt um ungefähr $^1/_7$ der Periode vor der Zeit der größten Annäherungsgeschwindigkeit (nach der sorgfältigen Diskussion von Miss CUMMINGS um 0,17 P). Die Periode ist $P = 0^d{,}1904795$. — Auch F. HENROTEAU[2] hat Beobachtungen der Radialgeschwindigkeit von β Cephei angestellt und die Veränderlichkeit von K und γ bestätigt gefunden; die Periode erwies sich dagegen als konstant. Nach neueren lichtelektrischen Messungen von Frl. M. GÜSSOW[3] scheint sie sich aber etwas verlängert zu haben.

Ganz ähnlich ist die Sachlage bei 12 Lacertae (Spektrum B2). Daß dieser Stern veränderliche Radialgeschwindigkeit besitzt, war bereits seit längerer Zeit bekannt; R. K. YOUNG[4] hatte die spektroskopischen Bahnelemente bestimmt und $P = 0^d{,}193089$, $K = 17$ km gefunden, wobei er bereits eine veränderliche Amplitude vermutete. J. KUNZ und J. STEBBINS[5] stellten darauf durch lichtelektrische Messungen eine geringe Veränderlichkeit der Helligkeit fest, und diese ist dann von GUTHNICK[6] näher untersucht worden. Die Schwankungen der Helligkeit besitzen dieselbe Periode wie die der Radialgeschwindigkeit; die Lichtkurve ist stark veränderlich, meist ist die Amplitude etwa $0^m{,}12$, geht aber mitunter auf $0^m{,}04$ herunter. Die von GUTHNICK in der Jubiläumsnummer der AN wiedergegebenen Lichtkurven sind ziemlich symmetrisch.

Weitere Untersuchungen der Radialgeschwindigkeit von 12 Lacertae sind von R. K. YOUNG[7] und W. H. CHRISTIE[8] veröffentlicht worden. Die wichtigsten Schlüsse, zu denen YOUNG kam, sind:

1. Die Amplitude der Geschwindigkeitskurve ist rasch veränderlich; ihre Werte liegen zwischen 15 bis 20 km und 70 km.
2. Auch die Form der Kurve ist veränderlich, die Periode dagegen konstant.
3. Die Breiten der Spektrallinien sind veränderlich, synchron mit dem Lichtwechsel; im Periastron sind die Linien breiter als im Apastron.
4. Die Schwerpunktsgeschwindigkeit ist langsam veränderlich.
5. Die periodischen Verschiebungen der Ca-Linien H und K sind viel kleiner als die der anderen Linien.

Nach W. H. CHRISTIE scheint die Schwerpunktsgeschwindigkeit, die sich aus den Linien H und K ergibt, nahezu konstant zu sein.

Im ganzen liegen die Verhältnisse bei β Cephei und 12 Lacertae also sehr ähnlich. Auch γ Bootis (Spektrum F), dessen Lichtkurve ($P = 0^d{,}290$) GUTHNICK und PRAGER[9] untersucht haben, und dessen Radialgeschwindigkeit von A. A. BELOPOLSKI[10] beobachtet worden ist, gehört wohl zu dieser Klasse von Sternen (wenn auch nach M. GÜSSOW[11] in den Jahren 1925 bis 1926 ein Lichtwechsel mit der genannten Periode nicht mehr nachweisbar war), und ebenso ν Eridani

[1] Publ. Detroit Obs. 2, S. 144 (1916).
[2] Publ. Dominion Obs. Ottawa 5, S. 77 (1921); 9, S. 36 (1924).
[3] A N 229, S. 197 (1927).
[4] Publ. Dominion Obs. Ottawa 3, S. 65 (1915).
[5] Pop Astr 25, S. 657 (1917).
[6] A N 208, S. 219 (1919) u. Jubiläums-Nr. der A N (1921).
[7] Publ. Dominion Astrophys. Obs. Victoria 1, S. 105 (1919).
[8] Ebenda 3, S. 210 (1926).
[9] Veröffentl. d. Sternw. Berlin-Babelsberg 1, Heft 1 (1914); 2, Heft 3 (1918).
[10] Bull. de l'Acad. Imp. des Sc. Petrograd 1917, S. 27.
[11] A N 229, S. 197 (1927).

(Lichtkurve von R. H. BAKER[1], $P = 0^d,15$, Spektrum B2), wenngleich bei letzterem Stern die Periode der Änderungen der Radialgeschwindigkeit nach Angabe in J. H. MOORES Third Catalogue of Spectroscopic Binary Stars[2] $0^d,19$ beträgt. Wahrscheinlich gehören noch mehr von den spektroskopischen Doppelsternen, die man als zum β Canis majoris-Typus gehörig bezeichnet, und die in dem Abschnitt über spektroskopische Doppelsterne näher behandelt werden, zu der hier in Rede stehenden Gruppe von Veränderlichen, so z. B. β Canis majoris selbst, β Ursae majoris, γ Ursae minoris, σ Scorpii u. a. m. Der jedenfalls äußerst kleine Lichtwechsel dieser Objekte ist aber noch nicht hinreichend erforscht worden, um ein Urteil in dieser Frage fällen zu können.

Daß Sterne wie β Cephei und 12 Lacertae mit den δ Cephei-Sternen, und zwar mit den kurzperiodischen, ziemlich eng verwandt sind, geht aus der synchronen Veränderlichkeit der Helligkeit und der Radialgeschwindigkeit hervor; Lichtkurve und Geschwindigkeitskurve scheinen in ungefähr denselben Beziehungen zueinander zu stehen, wie bei den δ Cephei-Sternen. Aber es sind doch auch Unterschiede vorhanden, die es verbieten, jene beiden Sterne kurzerhand zu den δ Cephei-Sternen zu rechnen. Lichtkurve und Geschwindigkeitskurve sind bei ihnen viel stärker veränderlich als bei letzteren, der Spektraltypus (B) ist ein früherer, die scheinbare Helligkeit ($4^m,9$ bzw. $5^m,2$) weit größer und die Amplitude des Lichtwechsels weit kleiner als bei den kurzperiodischen δ Cephei-Sternen. Ebensowenig wie man bei den δ Cephei-Sternen alle Eigentümlichkeiten der periodischen Linienverschiebungen durch wirkliche Änderungen der Radialgeschwindigkeit erklären kann, wird dies übrigens bei β Cephei und 12 Lacertae der Fall sein, ja es liegen hier die Dinge offenbar noch viel schwieriger als bei den δ Cephei-Sternen.

Außer bei den schon erwähnten Sternen sind nun noch bei einer großen Anzahl anderer Sterne durch lichtelektrische Messungen sehr kleine Helligkeitsschwankungen, meist nur von einigen Hundertsteln der Größenklasse, konstatiert worden. Im wesentlichen handelt es sich dabei um Beobachtungen von P. GUTHNICK und R. PRAGER[3]. Als Beispiele seien genannt α Aurigae, α Canum venaticorum, α Cygni, α Geminorum, α und γ Lyrae, α, o und φ Persei, ε und ζ Ursae majoris. Meist ist der Lichtwechsel sehr kompliziert und der Zusammenhang mit den bei fast all diesen Sternen vorhandenen Änderungen der Radialgeschwindigkeit noch nicht genügend klargestellt. Wir können hier auf die einzelnen Sterne nicht näher eingehen, sondern müssen auf die zitierten Abhandlungen und auf die von GUTHNICK gegebene zusammenfassende Übersicht[4], sowie auf eine Arbeit von R. H. BAKER[5] verweisen. Ob und inwieweit in diesen Fällen Beziehungen zu den δ Cephei-Sternen vorliegen, ist eine wohl noch offene Frage.

67. Theorien zur Erklärung des Lichtwechsels der δ Cephei-Sterne. Namentlich in den letzten zwei Dezennien ist eine große Anzahl von Arbeiten erschienen, in denen eine Erklärung des Lichtwechsels der δ Cephei-Sterne versucht wird. Der Erfolg dieser Bemühungen ist ziemlich bescheiden geblieben. Keine der vielen Theorien ist allgemein angenommen worden, und wir müssen gestehen, daß wir heute von einer wirklichen Erklärung der beobachteten Erscheinungen noch recht weit entfernt sind. Eine sehr ausführliche Besprechung aller bis etwa 1923 erschienenen einschlägigen Abhandlungen ist im zweiten Bande des schon öfters zitierten Werkes „Die veränderlichen Sterne. Heraus-

[1] Publ A S P 38, S. 86 (1926). [2] Lick Bull 11, S. 141 (1924).
[3] Veröffentl. d. Sternw. Berlin-Babelsberg 1, Heft 1 (1914); 2, Heft 3 (1918).
[4] Seeliger-Festschrift, S. 391 (1924).
[5] Publ A S P 38, S. 86 (1926).

gegeben von J. G. HAGEN und J. STEIN" zu finden. Wir wollen uns daher hier möglichst kurz fassen, indem wir den spezieller interessierten Leser auf dieses Werk verweisen.

Bevor die δ Cephei-Sterne spektroskopisch näher untersucht waren, und bevor man namentlich die periodischen, mit dem Lichtwechsel synchronen Linienverschiebungen in ihren Spektren erkannt hatte, erklärte man ihren Lichtwechsel in der Regel durch Verfinsterungserscheinungen oder vor allem durch die Annahme eines mit ungleichmäßig verteilten Flecken (Schlacken) bedeckten, rotierenden Sternes. Die erste Erklärung scheidet, wie wir schon früher (Ziff. 60) gesehen haben, auf Grund des spektroskopischen Befundes aus. Auch die zweite Erklärung ist nicht haltbar; wir haben über diesen Punkt schon bei der Besprechung der Theorien zur Erklärung des Lichtwechsels der Mira-Sterne (Ziff. 39) die nötigen Ausführungen gemacht. Bei den δ Cephei-Sternen, die wie die Mira-Sterne Riesensterne sind, liegen die Dinge genau ebenso.

Als man die periodischen Linienverschiebungen in den Spektren der δ Cephei-Sterne aufgefunden hatte, glaubte man zunächst als sichere Tatsache annehmen zu müssen, daß diese Veränderlichen Doppelsterne seien. Der Lichtwechsel war dadurch natürlich noch in keiner Weise erklärt, aber man glaubte doch, bei jedem Erklärungsversuch von dieser Tatsache ausgehen zu müssen. Dieser Anschauung sind manche Astrophysiker treu geblieben, und man kann die auf ihr fußenden Theorien kurz als „Doppelsterntheorien der δ Cephei-Sterne" bezeichnen. Andere Astrophysiker haben sich indessen von dieser Anschauung frei gemacht; sie erklären den Lichtwechsel der δ Cephei-Sterne durch Pulsationen der Gasmassen, als welche diese Himmelskörper anzusehen sind („Pulsationstheorie"). Diese beiden Theorien stehen sich gegenwärtig noch gegenüber (wenn auch neuerdings eine Vereinigung beider versucht wird), ohne daß eine wirklich endgültige Entscheidung getroffen werden kann. Man darf aber wohl sagen, daß, sofern überhaupt eine der beiden Theorien in ihren Grundlagen richtig ist, sich die Wagschale zur Zeit mehr zugunsten der Pulsationstheorie neigt. Freilich wird auch sie in ihrer gegenwärtigen Form noch nicht allen Anforderungen gerecht.

Wir wenden uns zunächst der Besprechung der Doppelsterntheorien zu. Zu diesen ist schon die sogenannte Fluthypothese zu rechnen, die W. KLINKERFUES aufstellte, um damit den Lichtwechsel sämtlicher Veränderlicher zu erklären. Er sieht die Veränderlichen als Doppelsterne mit sehr exzentrischen Bahnen an; der Lichtwechsel kommt dadurch zustande, daß durch die Anziehung des Begleiters in den Atmosphären der Sterne gewaltige Flutwellen entstehen, und daß die scheinbare Helligkeit durch die Absorption des Lichtes in der Flutwelle geändert wird. In dieser primitiven Form ist die Theorie nicht haltbar. Dasselbe gilt auch von den Theorien, die A. W. ROBERTS[1], L. A. EDDIE[2] und F. H. LOUD[3] aufgestellt haben. Einer gewissen Beliebtheit erfreute sich eine Zeitlang die Hypothese von J. C. DUNCAN[4], und wir wollen ihr daher einige Worte widmen. (Übersichten über diese älteren Theorien haben P. R. SUTTON[5] und D. BRUNT[6] gegeben.)

Da es bisher nicht gelungen ist, in den Spektren der δ Cephei-Sterne Spuren der zweiten Komponente nachzuweisen, so nimmt DUNCAN an, daß diese dunkel sei. Die helle Komponente ist von einer dichten, stark absorbierenden Atmosphäre umgeben, die dunkle dagegen von einer sehr dünnen, die soweit reicht, daß die Bahn des hellen Sternes noch ganz von ihr umschlossen wird. Durch

[1] Ap J 2, S. 283 (1895). [2] Ap J 3, S. 227 (1896). [3] Ap J 26, S. 369 (1907).
[4] Lick Bull 5, S. 82 (1909); 6, S. 154 (1911).
[5] Pop Astr 19, S. 408 (1911). [6] Obs 36, S. 59 (1913).

den Widerstand, den die Atmosphäre des hellen Sternes in diesem Mittel erleidet, wird sie bei der Vorwärtsbewegung des Sternes zurückgedrängt, so daß sie auf der bei der Bewegung vorangehenden Seite des Sternes weniger tief ist als auf der nachfolgenden und daher mehr Licht durchläßt. Unter diesen Verhältnissen wird uns in der Tat der Stern am hellsten erscheinen, wenn er sich gerade auf uns zu bewegt, und am schwächsten im entgegengesetzten Teile der Bahn, wie es auch nach den Beobachtungen der Radialgeschwindigkeiten der Fall ist. H. LUDENDORFF[1] hat indessen darauf hingewiesen, daß, da die Größe $\frac{m_2^3 \sin^3 i}{(m_1 + m_2)^2}$ bei den δ Cephei-Sternen sehr klein, die absolute Helligkeit dieser Sterne aber sehr groß ist, m_2 im Verhältnis zu m_1 ziemlich klein sein muß, da man sonst eine sehr kleine Masse für diese Sterne annehmen müßte. Die von DUNCAN angenommene Konstitution des Systems, bei der m_2 von einer ausgedehnten Atmosphäre umgeben ist, die diejenige von m_1 zurückzudrängen vermag, ist daher in hohem Maße unwahrscheinlich.

J. HELLERICH erklärt in seiner schon zitierten Dissertation (1913) und auch in einer späteren Abhandlung[2] den Lichtwechsel durch ungleichmäßige Verteilung der Helligkeit auf der Oberfläche der hellen Komponente, ohne indessen zu erörtern, wie diese merkwürdige Verteilung zustande kommt. Über den Einfluß der Rotation auf die beobachteten Geschwindigkeitskurven bei ungleicher Oberflächenhelligkeit hat R. H. CURTISS[3] Betrachtungen angestellt.

Auf den Boden der Doppelsterntheorie der δ Cephei-Sterne stellt sich auch P. GUTHNICK[4]. Er hat bei einigen Sternen mit veränderlicher Radialgeschwindigkeit, die zweifellos spektroskopische Doppelsterne sind, sehr kleine Lichtschwankungen gefunden, die mit denen der δ Cephei-Sterne eine gewisse Ähnlichkeit besitzen. Dies scheint ihm dafür zu sprechen, daß auch die δ Cephei-Sterne Doppelsterne sind. Wie er nun den Lichtwechsel erklärt, geben wir am besten mit seinen eigenen Worten wieder: „Unter dem gegenseitigen Einfluß der Komponenten des Systems bilden sich auf deren Oberflächen mehr oder weniger beständige Gebiete verschiedener Leuchtkraft, die in Verbindung mit der im Mittel in der Umlaufszeit sich vollziehenden Rotation den Lichtwechsel verursachen. Die ungleichförmige Helligkeitsverteilung ist wahrscheinlich auf meteorologische Vorgänge zurückzuführen: über den helleren Gebieten ist die Atmosphäre des Sternes durchlässiger als über den dunkleren Gebieten. Infolge ungleicher Rotationsgeschwindigkeit der verschiedenen Schichten und Zonen der Atmosphäre haben die hellen Gebiete, die als die eigentlichen Störungsgebiete zu betrachten sind, die Neigung, nach der der Rotationsrichtung entgegengesetzten Seite abzufließen, ähnlich wie es an den größeren Flecken des Jupiter direkt beobachtet wird. Es entsteht dadurch eine Helligkeitsverteilung mit einem in der Richtung der Rotation sehr starken, in der entgegengesetzten Richtung dagegen kleineren Gradienten. Dies verursacht die gewöhnlich vorhandene Asymmetrie der Lichtkurven. Die beobachteten periodischen Veränderungen des Spektrums der δ Cephei-Sterne stehen mit der Vorstellung im Einklang. Da das Licht der helleren Gebiete aus tieferen und heißeren Schichten der Atmosphäre stammt als das Licht der dunkleren Gebiete, so muß, wie es tatsächlich der Fall ist, der Spektraltypus des Sterns im Helligkeitsmaximum ein früherer sein als im Helligkeitsminimum. Ob die Schwankungen der Spektrallinien nur von der Bahnbewegung herrühren, ist unter diesen Um-

[1] A N 184, S. 373 (1910). [2] A N 215, S. 291 (1922).

[3] Publ. Detroit Obs. 1, S. 104 (1915).

[4] Veröffentl. d. Sternw. Berlin-Babelsberg 2, Heft 3 (1918); Die Naturwissenschaften 6, S. 716 (1918); A N 208, S. 171 (1919).

ständen höchst zweifelhaft, und es liegen auch manche Anzeichen des Gegenteils vor. Es ist daher vorderhand nicht statthaft, aus ihnen Schlüsse mechanischer Natur zu ziehen, wenn das Vorhandensein eines starken Lichtwechsels die Gefahr erheblicher Beeinflussung der Linienverschiebungen durch andere Effekte als den von der Bewegung in der Bahn herrührenden Dopplereffekt naherückt. Es scheint z. B. nicht ausgeschlossen, daß die starken Exzentrizitäten der δ Cephei-Bahnen wenigstens zum Teil von solchen Effekten herrühren und in Wirklichkeit viel kleiner sind als die beobachteten Linienverschiebungen sie ergeben. Der Umstand, daß mit sehr kleinem Wert der Massenfunktion in der Regel — jedoch nicht immer — Helligkeitsschwankungen nach δ Cephei-Art von beträchtlichem Umfang verbunden sind, dagegen mit normalem oder großem Wert der Funktion nur geringe Helligkeitsschwankungen dieser Art, scheint anzudeuten, daß die große Masse des Begleiters dem Zustandekommen eines δ Cephei-Lichtwechsels nicht günstig ist. Von einem meteorologischen Standpunkt aus kann dies nicht als unplausibel angesehen werden."

Wie wir schon früher erwähnt haben (Ziff. 39), hat GUTHNICK die Möglichkeit des Zustandekommens der charakteristischen Lichtkurven der δ Cephei-Sterne durch geeignete Verteilung von Flecken auf rotierenden Kugeln bei gleichzeitiger Randverdunkelung experimentell geprüft. Auf die Bedenken BOTTLINGERS haben wir gleichfalls hingewiesen.

Es würde zu weit führen, wenn wir alle die Abhandlungen, die sich in günstigem Sinne mit der Doppelsterntheorie der δ Cephei-Sterne beschäftigen, und die teilweise neue Modifikationen derselben bringen, hier einzeln besprechen wollten. Wir beschränken uns darauf, Literaturnachweise für diese Abhandlungen zu geben:

B. FESSENKOFF, A N 199, S. 387 (1914); Publ. de l'Obs. de Kharkow Nr. 8 (1914).

C. D. PERRINE, Ap J 41, S. 307 (1915); Ap J 50, S. 81, 148 (1919); M N 81, S. 442 (1921).

J. G. HAGEN, A N 209, S. 33 (1919); Ap J 51, S. 62 (1920); AN 211, S. 247, 413 (1920); Mem. S A It 1, S. 173 (1920); M N 81, S. 226 (1921); A N 225, S. 175 (1925). [Diese Abhandlungen HAGENS, außer der letzten, sind abgedruckt in Specola Astr. Vaticana, Miscellanea Astronomica, Parte II (1922).]

H. C. PLUMMER, M N 80, S. 496 (1920).

A. A. NIJLAND, Jubil.-Nr. der A N, S. 28 (1921).

H. VOGT, A N 212, S. 473 (1921).

T. J. J. SEE, A N 216, S. 193 (1922).

FR. NÖLKE, A N 217, S. 65 (1922).

J. HELLERICH, A N 218, S. 33 (1923).

Aus dem Inhalt dieser Abhandlungen sei nur erwähnt, daß nach HAGENS Ansicht die Annäherung des Begleiters an den Hauptstern auf letzterem Lichteruptionen auslöst. Diese Hypothese nähert sich also bereits der Pulsationstheorie, erklärt aber die Pulsation eben durch den Einfluß des Begleiters. HAGEN gelingt es auch, eine Erklärung für die eigentümliche Verteilung der Werte von ω bei den spektroskopischen Bahnen der δ Cephei-Sterne aufzustellen. — SEE wendet in seiner zitierten Abhandlung seine „Resisted Tidal Wave-theory" auf die δ Cephei-Sterne an; er ist der Meinung, daß sich diese Theorie auch für die Erklärung des Lichtwechsels der Mira-Sterne verwenden lasse.

Gegen die Doppelsterntheorie überhaupt und speziell gegen NIJLANDS oben zitierte Arbeit macht A. PANNEKOEK[1] Bedenken geltend, indem er durch theore-

[1] A N 215, S. 227 (1922).

tische Betrachtungen nachweist, daß jede Theorie, die die zusätzliche, für die Maximalhelligkeit des Veränderlichen notwendige Wärme in jeder Periode aus mechanischer Quelle neu erzeugen und dann wieder verschwinden läßt, unhaltbar ist. J. STEIN[1] pflichtet PANNEKOEK im Prinzip bei, macht aber auf einige Mängel von PANNEKOEKS Beweisführung aufmerksam.

Sehr eigenartig ist die Theorie von M. LA ROSA[2], mit der er die Lichtschwankungen aller Veränderlichen, also auch die der δ Cephei-Sterne, erklären will. Er nimmt an, daß die Veränderlichen Doppelsterne sind, und daß die Geschwindigkeit des Lichtes abhängt von der Geschwindigkeit der Lichtquelle (RITZsche Theorie). Die Lichtschwankungen sind nur scheinbar und kommen dadurch zustande, daß infolge der Bahnbewegung des Sternes dessen Geschwindigkeit relativ zum Beobachter sich ändert, so daß die Lichtstrahlen, die den Beobachter in einem bestimmten Augenblicke erreichen, zu verschiedenen Zeiten von dem Stern ausgegangen sind und sich in wechselnder Stärke superponieren. Diese Theorie hat eine lebhafte Diskussion hervorgerufen, und W. DE SITTER[3], H. J. GRAMATZKI[4], P. SALET[5] sowie W. E. BERNHEIMER[6] haben sie widerlegt. In mancher Hinsicht mit LA ROSAs Theorie verwandt ist die von J. PEREZ DEL PULGAR[7]. Dieser folgert ohne Zuhilfenahme neuer Hypothesen aus den MAXWELLschen Gleichungen, daß sich für einen Doppelstern Lichtschwankungen ergeben, wenn der Beobachter sehr weit entfernt ist, und diese Lichtschwankungen sollen denen der δ Cephei-Sterne entsprechen. Aber auch diese Theorie scheint nicht einwandfrei zu sein.

Wenn die Doppelsterntheorie der δ Cephei-Sterne richtig ist, so ist es zu erwarten, daß bei diesen Sternen Bedeckungserscheinungen vorkommen, daß sich also über die eigentliche Lichtkurve, die dem δ Cephei-artigen Lichtwechsel entspricht, eine Bedeckungslichtkurve lagert. Schreibt man den spektroskopischen Bahnelementen dieser Sterne reelle Bedeutung zu, so muß die Bedeckung eintreten, wenn der Veränderliche sich in der Helligkeitszunahme befindet. Diese Bedeckungserscheinungen werden aber wenig auffällig sein, da, wie wir gesehen haben, der Begleiter im Verhältnis zum Hauptstern ziemlich klein ist.

Anzeichen von Bedeckungserscheinungen hat P. GUTHNICK[8] bei ST Ophiuchi gefunden, doch würde hier die Bedeckung ziemlich genau mit dem Maximum der Helligkeit des δ Cephei-artigen Lichtwechsels zusammenfallen. A. W. ROBERTS[9] glaubt, daß bei dem Lichtwechsel von S Arae Verfinsterungen eine Rolle spielen, doch nimmt er hier eine kleine helle und eine große dunkle Komponente an, so daß das System eine ganz andere Konstitution, als oben angedeutet, besitzen würde. Bei $\varkappa$ Pavonis findet A. W. ROBERTS[10] eine kleine Einsenkung im aufsteigenden Aste der Lichtkurve, und bei RR Lyrae findet R. PRAGER[11] Anzeichen von Bedeckungserscheinungen. C. HOFFMEISTER[12] glaubt in den sekundären Wellen der Lichtkurven einer größeren Anzahl von δ Cephei-Sternen Bedeckungserscheinungen angedeutet zu sehen. Wie wir indessen früher dargelegt haben (Ziff. 55), ist die Frage, ob solche Wellen in den Lichtkurven reell sind, recht schwer zu beantworten.

---

[1] A N 217, S. 59 (1922).

[2] Z f Phys 21, S. 333 (1924); A N 222, S. 249 (1924); A N 223, S. 293, 359 (1925); Z f Phys 34, S. 698 (1925); Lyon Bull 7, S. 121 (1925).

[3] B A N 2, S. 121, 163 (1924).

[4] A N 223, S. 135 (1924); 224, S. 247 (1925).

[5] Lyon Bull 7, S. 42, 124, 147 (1925).

[6] Z f Phys 36, S. 302 (1926). [7] Z f Phys 32, S. 730 (1925).

[8] A N 179, S. 181 (1908). [9] Ap J 33, S. 197 (1911). [10] Ap J 34, S. 164 (1911

[11] Veröffentl. d. Sternw. Berlin-Babelsberg 5, Heft 4 (1926).

[12] A N 225, S. 201 (1925).

Auch J. HELLERICH[1] hat sich mit dem Problem der Verfinsterungen bei δ Cephei-Sternen beschäftigt. Er meint, daß, wenn man auch dem Begleiter eine relativ zum Hauptstern geringe Masse zuschreiben müsse, deswegen doch keineswegs seine Dimensionen relativ zu denen des Hauptsternes klein zu sein brauchen. Es werden unter dieser Annahme dann nicht nur kleine Einsenkungen in der Lichtkurve infolge der Bedeckung eintreten, sondern die Form der Lichtkurve wird durch die Bedeckungen wesentlich umgestaltet werden. Es gelingt ihm, unter einfachen Annahmen (Kreisbahn, kugelförmige Gestalt der Komponenten, Berührung derselben usw.), die Lichtkurve von δ Cephei ziemlich genau darzustellen, indem er der hellen Komponente einen „physischen" Lichtwechsel von symmetrischem Verlauf beilegt. Das Verhältnis der Radien von dunkler und heller Komponente ergibt sich zu 0,575, die Helligkeit der hellen Komponente ist $L_1=0{,}990$, die der dunklen $L_2=0{,}010$ (es wird $L_1+L_2=1$ angenommen). Es muß natürlich vorausgesetzt werden, daß die beobachteten Radialgeschwindigkeiten durch Rotationseffekte und dergleichen verfälscht sind, und es bedarf auch noch weiterer Hypothesen, um diese Theorie mit dem spektroskopischen Befunde vereinbar zu machen. Auch ist die angenommene Konstitution des Systems nicht gerade wahrscheinlich. Daß bei Annahme eines engen Doppelsternsystems, dessen Komponenten sich berühren, die für die δ Cephei-Sterne gefundene Beziehung zwischen absoluter Helligkeit und Periode nicht unplausibel ist, zeigt HELLERICH am Schlusse seiner Abhandlung.

Für die Doppelsterntheorie der δ Cephei-Sterne spricht der Umstand, daß die Häufigkeitskurven der Perioden bei ihnen und bei den spektroskopischen Doppelsternen und Bedeckungsveränderlichen einen sehr ähnlichen Verlauf besitzen. Darauf haben J. HELLERICH[2] und O. STRUVE[3] aufmerksam gemacht. Andererseits weist O. STRUVE darauf hin, daß bei den δ Cephei-Sternen die Amplitude der Radialgeschwindigkeiten nicht, wie bei den eigentlichen spektroskopischen Doppelsternen, von der Periode abhängt, sondern für alle Perioden im Mittel konstant und kleiner als bei spektroskopischen Doppelsternen gleicher Periode ist.

A. S. EDDINGTON[4] führt als Hauptbedenken gegen die Doppelsterntheorie der δ Cephei-Sterne folgendes an: Aus der von ihm gefundenen Beziehung zwischen absoluter Helligkeit und Masse der Sterne kann er, da die absoluten Helligkeiten der einzelnen δ Cephei-Sterne aus der Period-Luminosity Curve bekannt sind, die Masse m_1 berechnen und mit Hilfe der wenigstens angenähert bekannten effektiven Temperaturen auch den Radius R. Er findet so für die R sehr große Werte, z. B. (in Einheiten von 10^6 km) für l Carinae 145, ζ Geminorum 43, δ Cephei 20, RR Lyrae 4,3. Die Masse m_2 des Begleiters muß, wie wir gesehen haben, relativ zu m_1 klein sein; EDDINGTON nimmt als oberen Grenzwert von $m_2:m_1$ etwa $^1/_{12}$ an. Praktisch kann man also m_1 als die Masse des ganzen Systems betrachten. Dann ist die große Halbachse der relativen Bahn gleich dem Radius a' der Bahn, die ein hypothetischer Satellit in der Periode P um die Masse m_1 beschreibt. Es zeigt sich nun, daß bei den meisten δ Cephei-Sternen $a' < R$ ist, so daß der Begleiter sich innerhalb des Hauptsternes bewegen würde. Das ist unmöglich, und deshalb muß man nach EDDINGTONS Ansicht die Doppelsterntheorie der δ Cephei-Sterne fallen lassen.

Nun liegen EDDINGTONS Betrachtungen manche noch unsichere Elemente zugrunde, und als einen zwingenden Beweis gegen die Doppelsterntheorie wird man sie daher nicht betrachten können, so bemerkenswert sie auch sind.

---

[1] A N 224, S. 277 (1925). [2] A N 218, S. 33 (1923); 224, S. 277 (1925).
[3] M N 86, S. 63 (1925).
[4] The Internal Constitution of the Stars. Cambridge 1926, S. 184.

Erhebliche Bedenken gegen die Doppelsterntheorie erwachsen aus den Untersuchungen über die Vorgänge in den Atmosphären der δ Cephei-Sterne, die W. C. RUFUS, J. A. ALDRICH, R. H. CURTISS und D. W. LEE angestellt und die wir am Schlusse von Ziff. 60 besprochen haben. Bei der großen Dehnbarkeit, die die Doppelsterntheorie der δ Cephei-Sterne besitzt, wird es vielleicht unter Zuhilfenahme neuer Hypothesen gelingen, auch den soeben erwähnten Erscheinungen gerecht zu werden. Im ganzen aber muß man diese Dehnbarkeit eher für einen Mangel als für einen Vorzug der Doppelsterntheorie halten.

Mathematisch weit besser faßbar als die Doppelsterntheorie ist die Pulsationstheorie der δ Cephei-Sterne, deren Ausbau wir hauptsächlich A. S. EDDINGTON verdanken. Zuerst ist wohl A. RITTER[1] auf den Gedanken gekommen, daß Pulsationen für die Erklärung des Lichtwechsels der δ Cephei-Sterne eine Rolle spielen können. Später hat F. R. MOULTON diesen Gegenstand berührt[2]. Die Untersuchungen der beiden Genannten sind aber nach dem heutigen Stande der Kenntnis nicht auf die δ Cephei-Sterne anwendbar. H. C. PLUMMER[3] hat dann darauf hingewiesen, daß die periodischen Linienverschiebungen im Spektrum von ζ Geminorum vielleicht ganz oder teilweise durch radiale Pulsationen der Atmosphäre zu erklären seien und ist bald darauf unter spezieller Anwendung auf RR Lyrae noch etwas näher auf dieses Thema eingegangen[4].

Vor PLUMMERS zuletzt zitierter Arbeit war aber bereits SHAPLEYS bekannte Abhandlung „On the Nature and Cause of Cepheid Variation“[5] erschienen, in der er sich entschieden gegen die Doppelsterntheorie äußert und die Vorstellung verficht, daß der Lichtwechsel der δ Cephei-Sterne durch periodische Pulsationen in den Massen isolierter Sterne zu erklären sei. A. S. EDDINGTON[6] hat dann diese Theorie ausführlich mathematisch behandelt und neuerdings mit einigen Verbesserungen nochmals in seinem Buche „The Internal Constitution of the Stars“ (Cambridge, 1926) auseinandergesetzt.

In dem „Handbuch der Astrophysik“ ist eine Darlegung der Pulsationstheorie in dem Abschnitt über die Thermodynamik der Gestirne gegeben, so daß wir hier nicht darauf zurückzukommen brauchen. Eine eingehende Darstellung der Pulsationstheorie gibt auch M. GÜSSOW in ihrer früher zitierten Dissertation.

Von Arbeiten, die sich mit der Pulsationstheorie befassen, führen wir hier noch folgende an:

H. SHAPLEY [Ap J 49, S. 25 (1919) = Mt Wilson Contr 154; Ap J 52, S. 79 (1920) = Mt Wilson Contr 190]. Es wird gezeigt, daß, wenn die Pulsationstheorie zutrifft, die Abhängigkeit der Periode von der absoluten Helligkeit durchaus plausibel ist.

H. SHAPLEY und S. B. NICHOLSON [Mt Wilson Comm Nr. 63 = Wash Nat Acad Proc 5, S. 417 (1919)]. Wenn die Pulsationstheorie richtig ist, so müssen infolge der Verteilung der Werte der Radialgeschwindigkeit über die Sternscheibe die Spektrallinien eine periodische Verbreiterung und eine kleine Asymmetrie aufweisen. Es wird dargelegt, daß es kaum möglich sein wird, diese Erscheinungen mit den jetzigen Beobachtungsmethoden nachzuweisen.

G. LEMAÎTRE, Harv Circ 282 (1925).

E. PERSICO, M N 86, S. 98 (1926).

---

[1] Ann d Phys u Chem, Neue Folge, 8, S. 179 (1879); 13, S. 366 (1881).
[2] Ap J 29, S. 257 (1909).
[3] MN 73, S. 665 (1913). [4] MN 75, S. 566 (1915).
[5] Ap J 40, S. 448 (1914) = Mt Wilson Contr 92.
[6] M N 79, S. 2 (1918), S. 177 (1919).

W. BAADE, A N 228, S. 359 (1926). Es wird eine Methode zur Prüfung der Richtigkeit der Pulsationstheorie angegeben.

J. WOLTJER, B A N 3, S. 235 (1927).

Die Pulsationstheorie vermag in recht befriedigender Weise zu erklären, daß bei Sternen von der Beschaffenheit der δ Cephei-Sterne periodische Änderungen der Helligkeit, der Radialgeschwindigkeit und des Spektrums vorkommen, wie wir sie an diesen Veränderlichen wahrnehmen. Auch die Periodenlängen sind mit der Pulsationstheorie vereinbar. Für eine Pulsation in den Atmosphären der δ Cephei-Sterne sprechen auch die in Ziff. 60 besprochenen Beobachtungen von W. C. RUFUS, J. A. ALDRICH, R. H. CURTISS und D. W. LEE. Aber die Pulsationstheorie weist doch einige Mängel auf, die J. H. JEANS in folgenden Worten[1] darlegt (er bezeichnet dabei mit r den Radius):

"I. The luminosity of the star would be in phase with r, instead with $\dot{r}\left[=\frac{dr}{dt}\right]$ as observed.

II. The light-curve would show fore-and-aft symmetry, contrary to observation, and it is difficult to suppose that the inclusion of terms of second and higher orders in the displacement could modify this.

III. The pulsation would disappear altogether in a period of about 100000 years, and this is too small to reconcile with the observed frequency of Cepheids in the sky.

IV. We should expect to find oscillations of all possible amplitudes down to zero, whereas no Cepheids are known whose amplitude of oscillation is less than a certain limiting value.

V. We should expect to find oscillations of different and incommensurable frequencies in progress simultaneously in the same star, whereas the observed light-curves of Cepheids show a single definite period."

JEANS hatte schon früher[2] die Ansicht geäußert, daß die Lichtaufhellungen bei den δ Cephei-Sternen wohl eher auf erzwungene als auf freie Schwingungen zurückzuführen seien, und schließlich[3] hat er dann eine neue Theorie der δ Cephei-Sterne veröffentlicht, die in gewissem Sinne die Doppelsterntheorie mit der Pulsationstheorie vereinigt. Wir haben diese Theorie bereits in Ziff. 39 bei der Besprechung der Theorien zur Erklärung der Mira-Veränderlichen kurz skizziert und verweisen auf die dortigen Ausführungen. Die Period-Luminosity Curve bleibt auch bei der Annahme von JEANS' Hypothese verständlich.

In einer späteren Abhandlung[4] hat JEANS untersucht, inwieweit seine Theorie den oben unter I bis V aufgezählten Mängeln der reinen Pulsationstheorie nicht unterliegt. Er kommt zu folgenden Resultaten:

I. Der richtige Zusammenhang zwischen der Phase des Lichtwechsels und der von $\frac{dr}{dt}$ läßt sich unter gewissen Annahmen herstellen. (Hiergegen sind indessen Bedenken von J. J. M. REESINCK[5] erhoben worden.)

II. Diese Schwierigkeit bleibt in gewissem Umfange bestehen.

III. Die Zeit, während der die Pulsationen andauern, wird einfach bestimmt durch die Länge der Zeit, die für die Vollendung des Trennungsprozesses der beiden Komponenten notwendig ist. Diese Zeit dürfte lang genug sein, um die beobachtete Häufigkeit der δ Cephei-Sterne zu erklären.

IV. Nach JEANS' Theorie ist es nicht zu erwarten, daß die Pulsationen abklingen. Sie bleiben von endlicher Amplitude, bis sie in Bahnbewegungen von Doppelsternsystemen umgewandelt werden. So besteht die Schwierigkeit IV nicht mehr.

---

[1] M N 86, S. 90 (1926).
[2] Obs 42, S. 88 (1919).
[3] M N 85, S. 797 (1925).
[4] M N 86, S. 86 (1926).
[5] M N 86, S. 379 (1926).

V. Nach JEANS' Theorie sind die Oszillationen anfänglich alle von derselben Periode. Zuerst müssen sich diese Periode und die der Rotation des Sternes in der Lichtkurve superponieren, aber später werden diese beiden Perioden gleich, und der Lichtwechsel zeigt eine bestimmte Periode.

JEANS selbst gibt zu, daß seine Beweisführung nicht streng ist, daß es sich vielmehr zunächst nur um „a rough preliminary survey" handelt. Am bedenklichsten scheint ihm die unter II genannte Schwierigkeit. Auch scheint es nicht so, als ob der Lichtwechsel der roten unregelmäßigen Veränderlichen den Anforderungen der Theorie entspricht. Ferner dürfte zwischen den δ Cephei-Sternen und den Mira-Sternen nicht eine so relativ deutliche Grenze bestehen, wie es tatsächlich der Fall ist.

P. TEN BRUGGENCATE[1] hat versucht, die JEANSsche Theorie der δ Cephei-Sterne in einigen Punkten zu modifizieren und gewisse Schwierigkeiten derselben zu beheben. Seine Schlüsse sind aber, worauf H. VOGT[2] aufmerksam gemacht hat, zum Teil nicht einwandfrei.

A. S. EDDINGTON[3] erhebt gegen die Theorie von JEANS einen Einwand, der zunächst sehr schwerwiegend erscheint. Während EDDINGTON selbst nämlich in seiner Pulsationstheorie annimmt, daß die δ Cephei-Sterne nicht rotierende Gasmassen seien, ist nach JEANS die Periode des Lichtwechsels gleich der Periode der Rotation des Sternes. Bei der ungeheuren Größe der δ Cephei-Sterne — EDDINGTON berechnet für δ Cephei einen Radius von 20000000 km — müssen daher verschiedene Teile der Oberfläche enorme Unterschiede in der Radialgeschwindigkeit aufweisen. Diese müßten sich durch starke Verbreiterungen der Linien in den Spektren bemerkbar machen; solche sind aber keineswegs vorhanden. JEANS[4] behebt diese Schwierigkeit durch den Hinweis, daß Gasmassen, wie die δ Cephei-Sterne, nicht wie starre Körper rotieren. Die Periode des Lichtwechsels entspricht der Rotation eines inneren Kernes, während die äußeren Schichten langsamer rotieren.

Schließlich wollen wir noch einige Theorien der δ Cephei-Sterne erwähnen, die sich nicht in die bisher besprochenen einordnen lassen. S. BLAŽKO[5]) erklärt den Lichtwechsel mit Hilfe von dunkeln Ringen um den Stern. SHINZO SHINJO[6] hält die δ Cephei-Sterne für ausgedehnte Kugeln meteorischer Beschaffenheit mit einem exzentrischen Kern. Beide Theorien dürften nicht haltbar sein. K. BOTTLINGER[7] erwägt die Möglichkeit, daß die δ Cephei-Sterne JACOBIsche Ellipsoide seien, die von den Enden ihrer großen Achsen Materie abschleudern; er sieht aber selbst diese Theorie nicht für einwandfrei an.

Früher hat man mehrfach versucht, bei solchen δ Cephei-Sternen, deren Lichtkurven nicht zu unsymmetrisch sind, den Lichtwechsel durch die Annahme zu erklären, daß die Sterne rotierende Ellipsoide seien. H. SHAPLEY hat unter Zugrundelegung dieser Hypothese die Lichtkurven von RU Camelopardalis[8] und SZ Tauri[9] befriedigend darstellen können. Die Messungen der Radialgeschwindigkeiten dieser Sterne zeigen aber, daß sie sich in bezug auf die periodischen Linienverschiebungen ebenso verhalten wie die übrigen δ Cephei-Sterne, und daß daher die genannte Erklärung ihres Lichtwechsels nicht zutreffend sein kann.

---

[1] M N 86, S. 335 (1926); A N 228, S. 217 (1926); Die Naturwissenschaften 14, S. 905 (1926).
[2] A N 229, S. 125 (1927).
[3] Obs 49, S. 88 (1926).
[4] Obs 49, S. 125 (1926).
[5] A N 218, S. 69 (1923).
[6] Jap. Journ. of Astr. and Geoph. 1, S. 7 (1922).
[7] A N 210, S. 33 (1919).
[8] Laws Obs Bull 2, S. 71 (1913).
[9] A N 194, S. 353 (1913).

i) Veränderliche in Sternhaufen, Nebelflecken und ähnlichen Gebilden.

68. Veränderliche in kugelförmigen Sternhaufen. In offenen Sternhaufen hat man nur in ganz vereinzelten Fällen Veränderliche gefunden, und diese wenigen Objekte bieten kein besonderes Interesse. In kugelförmigen Sternhaufen, Nebeln und ähnlichen Gebilden kommen dagegen zahlreiche Veränderliche vor, und zwar ganz vorwiegend δ Cephei-Sterne. Besonders häufig sind solche Veränderliche in den kugelförmigen Sternhaufen; die Eigenschaften und die kosmische Stellung dieser Sternhaufen werden in einem anderen Teile dieses Werkes behandelt, hier interessieren uns lediglich die veränderlichen Sterne in ihnen. Eingehende Untersuchungen über dieselben liegen vor für die kugelförmigen Sternhaufen NGC 5139 = ω Centauri, NGC 5272 = M3 Canum venaticorum, NGC 5904 = M5 Serpentis und NGC 7078 = M15 Pegasi, und wir wollen uns zunächst mit diesen beschäftigen. Für die Entdeckungsgeschichte verweisen wir auf die Angaben in der „Geschichte und Literatur der veränderlichen Sterne", 2. Band.

NGC 5139 = ω Centauri. Die veränderlichen Sterne in diesem kugelförmigen Sternhaufen, der ungemein reich an Sternen ist, sind von S. I. Bailey[1] untersucht worden. Er konnte bei 128 Sternen des Haufens Veränderlichkeit der Helligkeit nachweisen, und für 95 von diesen konnte er die Lichtkurven und die Elemente des Lichtwechsels ableiten. 90 von diesen 95 Veränderlichen sind kurzperiodische δ Cephei-Sterne. Nach der Form der Lichtkurven teilt Bailey sie in drei Unterklassen a, b, c ein. Diese definiert er in folgender Weise:

a) Aufstieg sehr rasch, Abstieg ebenfalls, aber doch viel langsamer als der Aufstieg. Im Minimum sind die Änderungen der Helligkeit während längerer Zeit (etwa der Hälfte der Periode) sehr klein (sogenannte „Antalgol"-Lichtkurven).

b) Aufstieg nicht so rasch wie bei a), Abstieg langsam und bis zum Beginn des neuen Aufstieges andauernd (Lichtkurven wie bei δ Cephei).

c) Aufstieg nicht wesentlich rascher als Abstieg, in einigen Fällen sogar etwas langsamer, Lichtkurve also im wesentlichen symmetrisch, etwa wie bei ζ Geminorum.

Die drei Unterklassen sind keineswegs scharf voneinander getrennt, sondern gehen allmählich ineinander über. Die wichtigsten Eigenschaften der Sterne der einzelnen Unterklassen ersehen wir aus der folgenden Tabelle, in der n die Zahl der betreffenden Sterne bezeichnet:

Unterklasse	a	b	c
n	37	19	34
Mittel der Größen im Maximum	$12^m,99$	$13^m,10$	$13^m,33$
„ „ „ „ Minimum	14 ,11	13 ,97	13 ,89
„ „ Helligkeitsamplituden	1 ,12	0 ,87	0 ,56
„ „ mittleren Größen	13 ,55	13 ,54	13 ,61
Grenzwerte der Periode	$0^d,50-0^d,66$	$0^d,66-0^d,90$	$0^d,30-0^d,62$
Mittelwert der Perioden	$0^d,586$	$0^d,752$	$0^d,395$

Unter „mittlerer Größe" wird der Mittelwert aus den Größen im Maximum und im Minimum der Helligkeit verstanden. Das Mittel dieser mittleren Größen ist für die drei Unterklassen praktisch dasselbe, während sich in den Amplituden und Perioden deutliche Unterschiede zeigen.

[1] Harv Ann 38 (1902). Ausführliche Referate in VJS 38, S. 28 (1903) und G. u. L. 2, S. 455 (1920).

Nur bei 5 von den 95 Veränderlichen, deren Elemente bekannt sind, ist $P > 1^d$. Einer von diesen 5 Sternen hat eine Periode von $1^d,35$ und eine Lichtkurve der Klasse a, ein zweiter hat $P = 14^d,75$ und eine Lichtkurve der Klasse c, ein dritter, der hellste Veränderliche des Haufens ($9^m,80$ bis $11^m,10$), hat $P = 29^d,34$ und eine Lichtkurve der Klasse b, von der indessen die beobachteten Helligkeiten mitunter stark abweichen. Diese drei Sterne sind also langperiodische δ Cephei-Sterne. Die beiden übrigen Veränderlichen scheinen Mira-Sterne mit Perioden von 148^d und 242^d zu sein; beide scheinen abwechselnd hellere und schwächere Maxima zu haben.

Wir haben früher gesehen (Ziff. 64), welch' wichtige Rolle die Veränderlichen in ω Centauri und anderen Sternhaufen für die Festlegung von SHAPLEYS Period-Luminosity Curve spielen, und wie sie die Möglichkeit liefern, die Parallaxen der Sternhaufen zu berechnen. Die Period-Luminosity Curve für ω Centauri hat SHAPLEY besonders abgeleitet[1], nachdem er auf Grund von BAILEYS Angaben noch die Perioden von zwei weiteren langperiodischen δ Cephei-Sternen ($2^d,3$ und $4^d,6$) bestimmt hatte[2].

Auf die statistischen Rechnungen von K. PEARSON und JULIA BELL[3] über die veränderlichen Sterne in ω Centauri kann hier nur kurz hingewiesen werden und ebenso auf die Betrachtungen, die H. C. PLUMMER[4] sowie C. MARTIN und H. C. PLUMMER[5] auf Grund von Darstellungen der Lichtkurven von Veränderlichen in ω Centauri und anderen Sternhaufen durch Fourier-Entwicklungen angestellt haben.

NGC 5272 = M3 Canum venaticorum. Auch die Veränderlichen in diesem Sternhaufen hat S. I. BAILEY[6] näher untersucht. Er konnte 137 Veränderliche nachweisen; bei einigen von ihnen ist allerdings die Veränderlichkeit zweifelhaft.

Für 110 Veränderliche hat BAILEY die Elemente des Lichtwechsels feststellen können; sie alle sind kurzperiodische δ Cephei-Sterne und haben Lichtkurven, die denen der Unterklasse a in ω Centauri gleichen („Antalgol"-Kurven). Die Perioden liegen zwischen $0^d,4117$ und $0^d,7077$, und zwar zeigt die Häufigkeit der Perioden ein Maximum bei $0^d,50$ bis $0^d,55$. Im Maximum haben die Sterne die Größen $14^m,20$ bis $15^m,60$, im Minimum $16^m,02$ bis $16^m,93$, die Amplituden liegen zwischen $0^m,95$ und $2^m,55$. Die Größen im Maximum sind indessen erheblich verfälscht durch den Umstand, daß die Platten weit länger exponiert sind, als die meist sehr spitzen Maxima dauern. [Die mittleren Größen (Mittel aus Größe im Maximum und Minimum) sind nach SHAPLEY[7] in M3 wie in ω Centauri und in anderen kugelförmigen Sternhaufen im wesentlichen konstant.] Eine Abhängigkeit der Maximalhelligkeit von der Länge der Periode P zeigt sich in dem Sinne, daß mit wachsendem P die Maximalhelligkeit durchschnittlich etwas abnimmt; dasselbe ist mit der Helligkeitsamplitude der Fall, sowie mit der Geschwindigkeit der Helligkeitszunahme. Die Perioden sind bei einigen von den Sternen nicht völlig konstant.

Einer von den Veränderlichen, für die sich noch keine Periode feststellen ließ, zeigt ein besonderes Verhalten. Er ist in der Regel im Maximum seiner Helligkeit und alsdann einer der hellsten Sterne des Haufens; mitunter aber ist er bedeutend schwächer. Nach H. SHAPLEY[8] ist dieser Stern (Nr. 95 in BAILEYS Katalog) sehr rot, während SHAPLEY als mittleren Farbenindex für 15 andere der BAILEYschen Veränderlichen den Wert $+0^m,05$, entsprechend

[1] Harv Circ 237 (1922). [2] Ap J 48, S. 109 (1918) = Mt Wilson Contr 151.
[3] M N 69, S. 128 (1908). [4] M N 79, S. 639 (1919). [5] M N 80, S. 33 (1919).
[6] Harv Ann 78, Part 1 (1913). [7] Ap J 48, S. 117 (1918) = Mt Wilson Contr 151.
[8] Publ A S P 28, S. 81 (1916).

der Spektralklasse A1, findet. Nach R. F. SANFORD[1] zeigt das Spektrum des erwähnten Sternes auf einer Platte die Linien $H\gamma$ und $H\delta$ hell auf kontinuierlichem Untergrunde, der augenscheinlich einer späten Spektralklasse entspricht. Auf einer zweiten Platte, die zur Zeit geringerer Helligkeit des Sternes aufgenommen ist, zeigten sich keine Emissionslinien.

Durch Vergleichung der photographischen und der photovisuellen Größen der Veränderlichen in M3 hat SHAPLEY[2] die Abhängigkeit des Farbenindex von dem Lichtwechsel festgestellt. Im Maximum der Helligkeit ist der Farbenindex im Mittel $= -0^m,07$, im Minimum $+0^m,43$. Die photovisuelle und die photographische Lichtkurve verlaufen ganz so wie bei den galaktischen δ Cephei-Sternen.

E. E. BARNARD hatte schon 1899 einen Veränderlichen in M3 entdeckt, der nicht in BAILEYS Verzeichnis enthalten ist, und der nach Beobachtungen seines Entdeckers[3] ein langperiodischer δ Cephei-Stern ist ($P = 15^d,78$). Entsprechend seiner langen Periode ist er der hellste Veränderliche des Sternhaufens. Weitere 40 neue Veränderliche nebst einer Anzahl der Veränderlichkeit verdächtiger Sterne fand SHAPLEY[4], ohne indessen die Elemente des Lichtwechsels abzuleiten.

Außer der Abhandlung von BAILEY liegt noch eine zweite umfangreiche Untersuchung über die Veränderlichen in M3 vor, nämlich von J. LARINK[5]. Er beschäftigt sich mit dem Lichtwechsel von 133 Sternen, von denen 129 mit BAILEYschen, einer mit einem SHAPLEYschen Veränderlichen identisch sind. Es gelingt ihm, für eine Anzahl von Sternen, für die BAILEY noch keine Perioden gegeben hat, solche abzuleiten. Bei 18 Sternen werden Änderungen der Periode festgestellt, und zwar hat diese in 11 Fällen ab-, in 7 Fällen zugenommen. Die Änderungen der Periode scheinen meist sprungweise zu erfolgen. Es scheint ferner, als ob der Lichtwechsel der Sterne in verschiedenen Perioden nicht immer völlig gleichartig verläuft. Die Zusammenhänge, die sich zwischen Periodenlänge, Maximalhelligkeit und Amplitude ergeben, sind nicht so einfach, wie BAILEY sie gefunden hatte.

Einer der Sterne, deren Veränderlichkeit neu von LARINK entdeckt worden ist, und zwar Nr. 141 seines Verzeichnisses liegt 25′ von der Mitte des Sternhaufens entfernt und hat den Namen RV Canum venaticorum erhalten. Man hielt ihn zuerst für einen δ Cephei-Stern von außerordentlich kurzer Periode ($0^d,135$); neuerdings hat aber J. SCHILT[6] nachgewiesen, daß dieser Stern ein Bedeckungsveränderlicher vom W Ursae majoris-Typus ist und eine Periode von $0^d,270$ besitzt. — P. GUTHNICK und R. PRAGER[7] haben noch 19 weitere Veränderliche in M3 entdeckt.

NGC 5904 = M5 Serpentis. Für diesen Sternhaufen hat BAILEY[8] ebenfalls eine eingehende Untersuchung der Veränderlichen angestellt. Sein Katalog von Veränderlichen in M5 umfaßt 92 Sterne; bei 8 von ihnen ist aber die Veränderlichkeit noch zweifelhaft. Die Elemente des Lichtwechsels konnten für 72 Veränderliche bestimmt werden. 8 Sterne haben auffallend kurze Perioden von $0^d,23$ bis $0^d,35$, vier besitzen auffallend lange Perioden, nämlich von $0^d,85$,

[1] Pop Astr 27, S. 99 (1917).
[2] Ap J 49, S. 30 (1919) = Mt Wilson Contr 154.
[3] A N 172, S. 345 (1906).
[4] Ap J 40, S. 443 (1914) = Mt Wilson Contr 91 und Ap J 51, S. 140 (1920) = Mt Wilson Contr 176.
[5] Astr. Abh. d. Hamburger Sternwarte in Bergedorf Bd. 2, Nr. 6 (1922).
[6] Ap J 65, S. 124 (1927) = Mt Wilson Contr 330.
[7] Sitzungsber. d. Preuß. Akad. d. Wiss. 1925, S. 508.
[8] Harv Ann 78, Part 2 (1917).

$25^d,7$, $26^d,5$, $106^d,0$ (dieser letztere Stern paßt nicht in SHAPLEYs Period-Luminosity Curve); die Perioden der übrigen 60 liegen zwischen $0^d,45$ und $0^d,74$ mit Anhäufung um etwa $0^d,50$. Für die 8 Sterne mit besonders kurzen Perioden setzt BAILEY, um bessere Übereinstimmung mit der Periodenlänge der übrigen Sterne zu erzielen, den doppelten Periodenwert an und bezeichnet diese Veränderlichen als solche mit „doppeltem Maximum“. Abgesehen von der Kürze der Periode scheint aber kein zwingender Grund für eine solche Betrachtungsweise vorzuliegen, da es noch fraglich ist, ob die beiden Maxima sich irgendwie unterscheiden. Die Lichtkurven dieser 8 Veränderlichen haben Ähnlichkeit mit der von ζ Geminorum (Unterklasse c in ω Centauri). Auch ein Veränderlicher in M3 zeigt, wie BAILEY hier nachträglich bemerkt, ein ähnliches Verhalten.

Da die Lichtkurven der einzelnen Veränderlichen zum Teil unsicher sind, so wählt BAILEY 30 von den Sternen zur Ableitung mittlerer Lichtkurven aus. Diese 30 Objekte ordnet er nach der Periodenlänge in drei Gruppen von je 10 und bildet für jede Gruppe die mittlere Lichtkurve. Er erhält so folgende Resultate:

Gruppe	Mittlere Periode	Mittleres Maximum	Mittleres Minimum	Amplitude
1	$0^d,484$	$14^m,32$	$15^m,63$	$1^m,31$
2	0 ,543	14 ,40	15 ,57	1 ,17
3	0 ,643	14 ,45	15 ,48	1 ,03

In den Zahlen lassen sich gewisse Gesetzmäßigkeiten erkennen, die mit den bei M3 gefundenen übereinstimmen. Die Form der mittleren Lichtkurve ist bei Gruppe 1 und 2 „Antalgol“-artig, bei Gruppe 3 dauert die Helligkeitsabnahme bis zum neuen Helligkeitsanstieg.

E. E. BARNARD hat verschiedene der BAILEYschen Veränderlichen in M5 beobachtet und hat sich besonders mit dem Veränderlichen Nr. 33 des Kataloges von BAILEY beschäftigt[1], bei dem er geringe Änderungen der Periode feststellen konnte. H. H. TURNER[2] hat auf Grund des Beobachtungsmaterials von BARNARD Untersuchungen über diesen Stern angestellt.

NGC 7078 = M15 Pegasi. Auch für die Veränderlichen in diesem kugelförmigen Sternhaufen besitzen wir eine eingehende Untersuchung von BAILEY[3]. Danach sind 66 Veränderliche darin bekannt, und für 61 von diesen konnten die Elemente des Lichtwechsels bestimmt werden. Bis auf einen, der eine Periode von $1^d,44$ besitzt, sind diese Sterne sämtlich kurzperiodische δ Cephei-Sterne. Von der einen Ausnahme abgesehen, zerfallen sie deutlich in zwei Gruppen:

1. Gruppe: 29 Sterne, $P = 0^d,30 - 0^d,44$. Lichtkurven wie in Unterklasse c in ω Centauri, wenn auch Anstieg steiler als Abstieg.

2. Gruppe: 31 Sterne, $P = 0^d,57 - 0^d,76$. Lichtkurven wie in Unterklassen a und b.

Der Mittelwert der Perioden ist für die 1. Gruppe $0^d,38$, für die 2. $0^d,64$; die mittleren Amplituden für die beiden Gruppen sind $0^m,67$ bzw. $0^m,84$. Die mittlere Größe der Variablen beider Gruppen ist identisch, etwa $15^m,7$. Von den statistischen Ergebnissen BAILEYs ist das bemerkenswerteste, daß die mittlere Helligkeit der Veränderlichen von der Mitte nach dem Rande des Haufens zu um $0^m,2$ bis $0^m,3$ wächst, und zwar zuerst rasch und dann sehr langsam. SHAPLEY[4] hat nachgewiesen, daß die beiden Gruppen von Veränderlichen keinen Unterschied in der Farbe, die ihrer mittleren Helligkeit entspricht, besitzen. Der

[1] Pop Astr 27, S. 522 (1919) u. 30, S. 548 (1922).
[2] M N 80, S. 640 u. 81, S. 74 (1920). [3] Harv Ann 78, Part 3 (1919).
[4] Ap J 49, S. 37 (1919) = Mt Wilson Contr 154.

Farbenindex hängt in derselben Weise von der Phase des Lichtwechsels ab, wie bei den Veränderlichen in M3. Für sehr kleine Wellenlängen des Lichtes scheint nach SHAPLEY die Amplitude des Lichtwechsels bei diesen Veränderlichen kleiner zu sein als für die photographisch hauptsächlich wirksamen Wellenlängen. — P. GUTHNICK und R. PRAGER[1] haben noch 8 weitere Veränderliche in M15 entdeckt.

Für die übrigen kugelförmigen Sternhaufen, die Veränderliche enthalten, liegen so ausführliche Untersuchungen wie für die vier bisher behandelten nicht vor. In folgenden Fällen ist noch Näheres über die Veränderlichen bekannt.

NGC 104 = 47 Tucanae (in der Kleinen MAGELLANschen Wolke). Vier von den 7 Veränderlichen in diesem Haufen sind in ihrem Maximum die hellsten unter den 20000 Sternen des Haufens. Die Perioden und die photographischen Grenzgrößen für die drei hellsten sind bestimmt[2] worden:

1.	$P = 211^d$	Max = $11^m,0$	Min = $14^m,4$
2.	203	11 ,0	14 ,2
3.	192	11 ,0	14 ,3

Es handelt sich hier also auffallenderweise offenbar um typische Mira-Sterne. Es ergibt sich für diese Objekte eine große absolute Helligkeit, wie sie auch die galaktischen Mira-Sterne besitzen.

NGC 6656 = M22. Von den 16 bekannten Veränderlichen in diesem Haufen haben nach BAILEY[3] 14 Perioden von etwa ${}^2/{}_3{}^d$, einer eine Periode von ${}^1/{}_3{}^d$ und einer eine solche von 200^d. Dieser letztere ist schwächer als die übrigen Veränderlichen, paßt also sicher nicht in die Period-Luminosity Curve.

ST. CHEVALIER[4] hat 17 Veränderliche in diesem Sternhaufen entdeckt; die Frage der Identität dieser Objekte mit den von BAILEY behandelten ist noch zu prüfen.

NGC 6723. BAILEY[5] hat die Elemente des Lichtwechsels der 16 in diesem Sternhaufen aufgefundenen Veränderlichen bestimmt. Die Perioden liegen zwischen $0^d,338$ und $0^d,619$, die Amplituden zwischen $0^m,50$ und $1^m,60$, die mittlere Helligkeit ist $15^m,3$. Die Form der Lichtkurven ist die für Sternhaufenveränderliche typische (Unterklasse a oder b). Innerhalb eines Kreises von $1^1/_2{}^\circ$ Radius um die Mitte des Haufens liegen 9 Veränderliche mit folgenden Perioden und photographischen Grenzgrößen nach BAILEY[6]:

1.	$P = 0^d,49$	Max = $14^m,5$	Min = $16^m,5$	6.	$P = 64^d,1$	Max = $12^m,0$	Min = $13^m,5$
2.	8 ,03	14 ,5	15 ,5	7.	28 ,0	11 ,0	17 ,0
3.	90 ,4	14 ,0	<16 ,0	8.	7 ,13	14 ,5	16 ,0
4.	347 ,5	10 ,0	<17 ,0	9.	0 ,52	13 ,5	15 ,5
5.	0 ,57	14 ,0	15 ,5				

Besonders auffällig ist angesichts der Periode von 28^d die große Amplitude von 7, falls es sich hier nicht um einen Druckfehler handelt.

NGC 6981 = M72. H. SHAPLEY und MARY RITCHIE[7] haben die Lichtkurven von 26 Veränderlichen in diesem Haufen bestimmt, während dies für 3 weitere noch nicht gelungen ist. Von den Lichtkurven gehören 3 zu BAILEYS Unterklasse b und 4 zu c, die übrigen zu a. Die Perioden liegen zwischen $0^d,33$ und $0^d,66$, die Amplituden zwischen $0^m,35$ und $1^m,43$. Der Stern mit der längsten Periode ($0^d,66$) liegt am weitesten ($4',6$) von der Mitte des Haufens entfernt und hat die größte Amplitude ($1^m,43$) und verhältnismäßig große mittlere Helligkeit ($16^m,43$); die mittlere Helligkeit der anderen Veränderlichen ist $16^m,8$ im Mittel.

[1] Sitzungsber. d. Preuß. Akad. d. Wiss. 1925, S. 508. [2] Harv Bull 783 (1923).
[3] Pop Astr 28, S. 518 (1920). [4] Ann de l'Obs de Zô-Sè 10 (1918).
[5] Harv Circ 266 (1924). [6] Harv Bull 803 (1924).
[7] Ap J 52, S. 232 (1920) = Mt Wilson Contr 195.

In einer größeren Anzahl anderer kugelförmiger Sternhaufen sind ebenfalls Veränderliche entdeckt worden, über deren Lichtwechsel wir aber noch nichts oder höchstens sehr wenig wissen. Wir können hier diese Fälle nicht alle einzeln anführen, sondern verweisen auf die Angaben im 2. Bande der „Geschichte und Literatur der veränderlichen Sterne". Folgende Hinweise auf neuere Literatur auf diesem Gebiete, die sich an der eben genannten Stelle noch nicht verzeichnet findet, werden indessen von Nutzen sein [es sind dabei die Listen von HELEN DAVIS in Publ A S P 29, S. 260 (1917) und H. SHAPLEY in Ap J 52, S. 76 (1920) = Mt Wilson Contr 190, einfach mit „DAVIS" bzw. „SHAPLEY" bezeichnet]:

NGC 1851. S. I. BAILEY, Harv Bull 802 (1924).
3201. IDA E. WOODS, Harv Circ 216 (1919); S. I. BAILEY, Harv Circ 234 (1922).
4147. DAVIS.
4590 = M 68. H. SHAPLEY, Publ A S P 31, S. 226 (1919); Ap J 51, S. 49 (1920) = Mt Wilson Contr 175.
5024 = M 53. SHAPLEY; W. BAADE, Mitt. d. Hamburger Stw. in Bergedorf 6, S. 67 (1926).
5466. W. BAADE, ebenda, S. 61.
6205 = M 13. H. SHAPLEY, Ap J 48, S 112 (1918) = Mt Wilson Contr 151; P. GUTHNICK und R. PRAGER, Sitzungsber. d. Preuß. Akad. d. Wiss. 1925, S. 508.
6229. DAVIS.
6293. SHAPLEY.
6341 = M 92. P. GUTHNICK und R. PRAGER, Sitzungsber. d. Preuß. Akad. d. Wiss. 1925, S. 508.
6362. IDA E. WOODS, Harv Circ 217 (1919).
6541. IDA E. WOODS, Harv Bull 764 (1922).
6553. SHAPLEY.
6712. DAVIS.
6779 = M 56. H. DAVIS, Publ A S P 29, S. 210 (1917); SHAPLEY.
6864 = M 75. SHAPLEY.
7006. SHAPLEY; H. SHAPLEY a. BEATRICE W. MAYBERRY, Wash Nat Ac Proc 7, S. 152 (1921).
7492. SHAPLEY.

Es geht aus den vorangehenden Ausführungen hervor, daß in den kugelförmigen Sternhaufen von Veränderlichen ganz vorwiegend kurzperiodische δ Cephei-Sterne vorkommen, die sich in allem genau so verhalten wie die galaktischen Objekte dieser Art. Die mittlere Helligkeit dieser Veränderlichen ist in jedem Sternhaufen eine Konstante, wie SHAPLEY[1] nachgewiesen hat, oder es kommen jedenfalls nur geringe Abweichungen von diesem Gesetz vor.

Außer den zahlreichen kurzperiodischen δ Cephei-Sternen hat man aber, wie wir gesehen haben, in kugelförmigen Sternhaufen auch einige langperiodische δ Cephei-Sterne und sogar einige Mira-Sterne aufgefunden.

69. Veränderliche in der nächsten Umgebung von kugelförmigen Sternhaufen. Wir haben bereits oben gesehen, daß sich in der Umgebung von NGC 6723 9 Veränderliche mit Perioden von $^1/_2{}^d$ bis 347^d vorfinden. Auch in der Umgebung anderer kugelförmiger Sternhaufen hat man Veränderliche entdeckt. Am bemerkenswertesten ist die Sachlage bei NGC 5024 = M 53. In diesem Haufen sind nach H. SHAPLEY[2] und W. BAADE[3] 40 Veränderliche bekannt.

[1] Ap J 48, S. 115 (1918) = Mt Wilson Contr 151.
[2] Ap J 52, S. 76 (1920) = Mt Wilson Contr 190.
[3] Mitt. d. Hamburger Stw. in Bergedorf 6, S. 67 (1926).

In der Umgebung des Haufens, auf einer Fläche von 8 Quadratgrad, fand nun BAADE[1] 7 Veränderliche, von denen 5 kurzperiodische δ Cephei-Sterne sind und einer eine Periode von mehr als 20^d hat, während sich über die Periode des letzten noch nichts aussagen läßt. Bis auf einen passen diese Veränderlichen ihrer Helligkeit nach nicht zu den Veränderlichen des Haufens, wenn man die Gültigkeit von SHAPLEYS Period-Luminosity Curve voraussetzt; sie sind vielmehr viel heller als die Haufenveränderlichen. Man muß also zunächst annehmen, daß jene Veränderlichen nichts mit dem Haufen zu tun haben und nur scheinbar in seiner Nähe liegen. Nun hat NGC 5024 die hohe galaktische Breite von 79°, und es war zu folgern, daß kurzperiodische δ Cephei-Sterne in unerwartet großer Zahl in hohen galaktischen Breiten vorkommen. (Wie es scheint, ist ja sogar auch ein langperiodischer δ Cephei-Stern unter jenen Veränderlichen, und Sterne dieser Art sind in so hohen galaktischen Breiten sonst überhaupt unbekannt.) Zur Klärung dieser Frage hat BAADE[2] darauf zwei Gegenden in der Nähe des galaktischen Poles auf Veränderliche durchsucht, aber keinerlei solche Objekte gefunden. Wir stehen hier also vor einem noch ungelösten Rätsel. Man wird die Annahme nicht von der Hand weisen können, daß jene Veränderlichen doch zu dem Haufen in Beziehung stehen, und daß die SHAPLEYsche Kurve in diesem Falle nicht gilt. Noch auffälliger wird die Sachlage dadurch, daß BAADE auch nahe bei dem gleichfalls hohe galaktische Breite besitzenden Kugelhaufen NGC 5466 fünf kurzperiodische δ Cephei-Sterne gefunden hat, die gleichfalls heller sind als die Veränderlichen des Haufens.

Angesichts des Interesses, das hiernach das Vorkommen von veränderlichen Sternen in unmittelbarer Nähe von kugelförmigen Sternhaufen gewinnt, seien hier die weiteren Fälle zusammengestellt, in denen Veränderliche in solcher Lage gefunden worden sind:

NGC 4833. BAILEY [Harv Bull 792 (1923)] fand 11 Veränderliche innerhalb 90′ Entfernung von diesem Haufen. 8 davon scheinen Mira-Sterne zu sein, deren Perioden mit 129^d, 151^d, 170^d, 200^d, 235^d, 255^d, 260^d, 360^d angegeben sind, zum Teil aber noch falsch sein können. Für 2 ist nur angegeben, daß die Periode kurz ist, für den letzten ist die Periode unbekannt.

NGC 6093. Nach BAILEY [Harv Bull 798, (1924)] liegen nahe bei diesem Haufen 5 Veränderliche: TW Scorpii $P = 199^d$, UY Scorpii $P = 151^d$ (?), S Scorpii $P = 177^d$, VW Scorpii $P = 258^d$, AR Scorpii $P < 1^d$.

NGC 6397. Nach BAILEY [Harv Bull 796 (1923)] liegen nahe dem Haufen 4 Veränderliche mit Perioden von 170^d, 220^d, 255^d(?) 320^d.

NGC 6541. Nach BAILEY [Harv Bull 799 (1924)] liegen nahe dem Haufen 3 kurzperiodische δ Cephei-Sterne und 9 Veränderliche mit Perioden von 186^d, 207^d, 212^d, 216^d, 230^d, 244^d, 252^d, 268^d und 353^d(?).

NGC 6584. Nach BAILEY [Harv Bull 801 (1924)] liegen innerhalb $1^1/_2°$ Entfernung von der Mitte des Haufens 9 Veränderliche. Bei 5 von diesen ist über die Periode nur gesagt, daß sie kurz (einige Tage oder weniger) sei, die 4 andern haben Perioden von 112^d, 238^d, 247^d, 361^d.

NGC 6809. Nach BAILEY [Harv Bull 813 (1925)] liegen innerhalb einer Entfernung von $1^1/_2°$ von dem Haufen 8 veränderliche Sterne. Zwei davon haben wahrscheinlich Perioden von 217^d und 286^d; die Perioden der andern sind unbekannt. In derselben Nummer des Harv Bull sind noch 5 weitere, von J. S. PARASKÉVOPOULOS entdeckte Veränderliche unbekannter Periode angezeigt, die in der Nähe des Sternhaufens liegen.

In fast allen den zuletzt aufgezählten Fällen überwiegen unter den nahe bei den Kugelhaufen gefundenen Veränderlichen die Mira-Sterne. Merkwürdiger-

[1] Mitt. d. Hamburger Sternw. in Bergedorf 5, S. 35 (1922). [2] Ebenda 6, S. 66 (1926).

weise besitzen diese Mira-Sterne ganz vorwiegend Perioden von weniger als 300^d, während bei den Mira-Sternen im allgemeinen das Maximum der Periodenhäufigkeit bei etwa 300^d liegt. S. BELIAWSKY[1] hat aber darauf aufmerksam gemacht, daß überhaupt bei den Mira-Sternen von geringer scheinbarer Maximalhelligkeit (um solche handelt es sich im vorliegenden Falle) relativ kurze Perioden vorzuwiegen scheinen, während H. LUDENDORFF[2] nur eine ganz geringe Abhängigkeit zwischen Periodenlänge und Maximalhelligkeit gefunden hatte. Letzteres erklärt sich aber vielleicht dadurch, daß das von LUDENDORFF benutzte Material — die Mira-Sterne des Hauptkataloges der G. u. L. mit bekannten Spektren — nur wenige besonders schwache Objekte umfaßt. Auch die schwachen Mira-Veränderlichen in den Milchstraßenwolken im Scutum und Sagittarius scheinen vorwiegend kleine Perioden zu besitzen[3]. Es ist einstweilen durchaus möglich, anzunehmen, daß die in der Nähe der Kugelhaufen gefundenen Mira-Sterne nichts mit ihnen zu tun haben, sondern nur scheinbar in ihrer Nähe liegen.

70. Die Veränderlichen in den MAGELLANschen Wolken und in NGC 6822. Auf photographischen Aufnahmen der beiden MAGELLANschen Wolken wurden von HENRIETTA S. LEAVITT zahlreiche Veränderliche entdeckt. Sie veröffentlichte im Jahre 1908 einen Katalog[4] von 969 Veränderlichen in der Kleinen und 808 in der Großen MAGELLANschen Wolke. Für 16 von den ersteren leitete sie die Lichtkurven ab. Die Perioden dieser 16 Sterne liegen zwischen $1^d,25$ und 127^d; die Lichtkurven der meisten haben die charakteristische Form, die bei Sternhaufen-Variablen überwiegt, nämlich steilen Aufstieg, spitzes Maximum, langsameren Abstieg und breites Minimum.

Wir haben bereits in Ziff. 64 gesehen, daß Miss LEAVITT[5], nachdem sie die Lichtkurven von noch 9 weiteren Veränderlichen bestimmt hatte, die Entdeckung machte, daß Periode und Helligkeit der Veränderlichen in einfacher Beziehung zueinander stehen. Die Elemente der 25 Veränderlichen haben wir an der soeben zitierten Stelle wiedergegeben. Die Entdeckung von Miss LEAVITT gab dann Anlaß zu SHAPLEYS Untersuchungen über die Period-Luminosity Curve, die wir eingehend besprochen haben.

Von den erwähnten 25 Veränderlichen haben 23 Perioden von $1^d,25$ bis 32^d und zeigen in ihren Lichtkurven die Eigentümlichkeiten der δ Cephei-Sterne. Es besteht also keinerlei Bedenken, sie als solche zu betrachten. Aber auch die beiden übrigen mit Perioden von 66^d und 127^d sind ohne Zweifel langperiodische δ Cephei-Sterne. Sie fügen sich zwanglos in die Period-Luminosity Curve ein, ihre Amplituden sind gering ($1^m,4$ bzw. $0^m,9$ photographisch), und die Form ihrer Lichtkurven spricht auch nicht gegen ihre Zugehörigkeit zu dieser Klasse. Die Lichtkurve des Veränderlichen mit 127^d Periode wird von SHAPLEY in Harv Circ 237 (1922) wiedergegeben; dieser macht darauf aufmerksam, daß die Lichtkurve dieses Sternes genau dieselbe Form hat, wie die des Veränderlichen SX Herculis, den wir zu den Mira-Sternen rechnen.

Die galaktischen δ Cephei-Sterne haben, soweit bekannt, nicht längere Perioden als etwa 45^d, und es ist merkwürdig, daß in der Kleinen MAGELLANschen Wolke nach dem Gesagten solche Veränderliche mit längeren Perioden vorkommen. Vielleicht kann man sie als Übergänge zur Mira-Klasse auffassen.

Die hellsten Veränderlichen in der Kleinen Wolke haben Spektra[6] etwa der Klassen K 5 und M; eine genauere Klassifizierung war nicht möglich.

H. SHAPLEY, I. YAMAMOTO und HARVIA H. WILSON[7] haben später die Perioden, Amplituden und mittleren photographischen Größen von 107 Ver-

---

[1] A N 227, S. 277 (1926). [2] A N 220, S. 145 (1924).
[3] Harv Circ 265 (1924). [4] Harv Ann 60, No. 4 (1908).
[5] Harv Circ 173 (1912). [6] Harv Circ 255 (1924). [7] Harv Circ 280 (1925).

änderlichen der Kleinen Wolke veröffentlicht. Perioden von mehr als 45^d kommen, von den beiden genannten Fällen und von einer Periode von 620^d abgesehen, nicht vor; der Stern mit dieser auch für galaktische Veränderliche ganz ungewöhnlichen Periode von 620^d steht wahrscheinlich nur scheinbar in der Wolke. Eine Periode von weniger als 1^d hat nur einer von den 107 Veränderlichen. Die aus diesem Material abgeleitete photographische Period-Luminosity Curve für $P > 1^d$ haben wir in Ziff. 64 zahlenmäßig wiedergegeben.

Daß in der Kleinen MAGELLANschen Wolke auch kurzperiodische δ Cephei-Sterne in größerer Zahl vorkommen, hat SHAPLEY[1] nachgewiesen. Für 13 von diesen Objekten gibt er die Perioden an, die zwischen $0^d,40$ und $0^d,81$ liegen. Die mittlere Helligkeit ist für diese Sterne $16^m,1$.

Über die Veränderlichen in der Großen MAGELLANschen Wolke ist nur wenig bekannt. Einige knappe Angaben darüber hat SHAPLEY[2] gemacht. Für 8 besonders helle unter ihnen hat Miss A. J. CANNON[3] die Spektra klassifiziert; sie gehören den Klassen K 5 und M an. Auf dem Union Observatory in Johannesburg ist eine Anzahl Veränderlicher in der Großen Wolke entdeckt worden[4], die in der Liste von Miss LEAVITT nicht enthalten sind; ferner hat man daselbst auch in der Nähe der Großen Wolke eine Anzahl von Veränderlichen festgestellt[5] und für 6 von ihnen die Lichtkurven abgeleitet; darunter ist ein kurzperiodischer δ Cephei-Stern, die andern sind langperiodische Sterne dieser Klasse.

Die Häufigkeitsfunktion der Perioden scheint bei den langperiodischen δ Cephei-Sternen in den MAGELLANschen Wolken anders zu verlaufen als bei den galaktischen δ Cephei-Sternen. Darüber hat J. HELLERICH[6] Untersuchungen angestellt, indem er nach den veröffentlichten Harvard-Beobachtungen die scheinbaren mittleren Helligkeiten der Veränderlichen in den Wolken und aus diesen mittleren Helligkeiten mit Hilfe der Period-Luminosity Curve die Perioden bestimmte.

Ein Objekt, daß mit den MAGELLANschen Wolken Ähnlichkeit hat, obwohl es eine viel geringere Ausdehnung besitzt, ist NGC 6822. In dieser Wolke hat E. HUBBLE[7] 15 Veränderliche aufgefunden, von denen 11 sicher δ Cephei-Sterne sind. Die Perioden dieser 11 Veränderlichen liegen zwischen 11^d und 64^d, ihre Helligkeiten zeigen die zu erwartende Abhängigkeit von der Periodenlänge. Drei von den 4 übrigen Veränderlichen zeigen kleine, vielleicht unregelmäßige Schwankungen; es ist aber möglich, daß auch sie δ Cephei-Sterne sind. Der letzte Veränderliche muß, wenn er nicht unregelmäßig ist, eine Periode von mehr als 150^d besitzen.

Aus den vorangehenden Ausführungen geht hervor, daß, während in kugelförmigen Sternhaufen fast ausschließlich kurzperiodische δ Cephei-Sterne vorkommen und langperiodische sehr selten sind, letzteres für die MAGELLANschen Wolken und NGC 6822 keineswegs zutrifft. Eher scheinen in diesen letztgenannten Gebilden die langperiodischen δ Cephei-Sterne die kurzperiodischen an Zahl zu überwiegen. Doch ist eine gewisse Vorsicht bei der Beurteilung der Sachlage geboten, da die kurzperiodischen δ Cephei-Sterne schwächer sind als die langperiodischen.

71. Veränderliche Sterne in Spiralnebeln. Auf das Vorkommen neuer Sterne in Spiralnebeln war man schon vor längerer Zeit aufmerksam geworden, während

[1] Wash Nat Ac Proc 8, S. 69 (1922); Harv Bull 765 (1922).
[2] Harv Circ 268 (1924) u. 271 (1925).
[3] Harv Bull 754 (1921).
[4] Union Circ 43 (1918). [5] Ebenda 61 (1924).
[6] A N 218, S. 33 (1923).
[7] Ap J 62, S. 409 (1925) = Mt Wilson Contr 304.

man veränderliche Sterne nicht in ihnen gefunden hatte. Dies gelang indessen J. C. DUNCAN[1], der in dem bekannten Spiralnebel NGC 598 = M33 Trianguli 3 Veränderliche nachwies, ohne indessen den Charakter ihrer Lichtschwankungen näher festzulegen. M. WOLF[2] stellte unabhängig die Veränderlichkeit eines dieser Sterne fest. Später hat dann E. HUBBLE[3] diesen Nebel auf Veränderliche näher erforscht und noch 42 weitere aufgefunden. 35 von diesen 45 Veränderlichen sind δ Cephei-Sterne, 4 sind unregelmäßig und einer wahrscheinlich ein Bedeckungs-Veränderlicher. Der Lichtwechsel der übrigen 5 ist seiner Natur nach noch unbekannt, doch dürften die meisten von ihnen δ Cephei-Sterne sein. Die 35 als δ Cephei-Sterne erkannten Veränderlichen haben Perioden zwischen 13^d und 70^d (nur 3 von den Perioden sind $>45^d$, nämlich 46^d, 55^d und 70^d), die Amplituden übersteigen nicht $1^m,2$, die Helligkeiten im Maximum liegen zwischen $18^m,0$ und $19^m,1$. Die Lichtkurven haben die typische Form der Lichtkurven von Sternhaufen-Veränderlichen, und die Helligkeit wächst mit der Periodenlänge. Mit Hilfe der Period-Luminosity Curve konnte HUBBLE die Entfernung von M33 zu 263000 Parsecs = 850000 Lichtjahren bestimmen.

Die 4 unregelmäßigen Veränderlichen in M33 scheinen nur sehr langsame Veränderungen zu erleiden; sie scheinen wenigstens teilweise zur Klasse der Nova-ähnlichen Veränderlichen zu gehören. Das Spektrum des hellsten von ihnen war im Helligkeitsmaximum kontinuierlich mit darüber gelagerten Emissionslinien des Wasserstoffes und vielleicht noch einigen anderen hellen Linien. Auch zwei Novae konnte HUBBLE in M 33 nachweisen.

Im großen Andromedae-Nebel M31 hat HUBBLE[4] 36 Veränderliche gefunden, während die Gesamtzahl der in diesem Spiralnebel gefundenen neuen Sterne sich schon auf 66 beläuft. Für 12 von den 36 Veränderlichen hat HUBBLE die Lichtkurven bestimmt. Es handelt sich um langperiodische δ Cephei-Sterne mit Perioden von 18^d bis 50^d. Die Entfernung des Andromeda-Nebels ergibt sich mit Hilfe der Helligkeiten dieser Sterne zu 285000 Parsecs.

Auch in den Spiralnebeln NGC 3031 = M81, NGC 5457 = M101 und NGC 2403 hat HUBBLE veränderliche Sterne entdeckt, doch liegen nähere Veröffentlichungen darüber noch nicht vor.

72. Veränderliche Sterne in Nebelflecken. Wir haben bereits in früheren Abschnitten auf den Zusammenhang zwischen veränderlichen Sternen und Nebelflecken hingewiesen, namentlich bei der Besprechung der Nova-ähnlichen Veränderlichen und der veränderlichen Sterne der R Coronae-Klasse. Von ersteren liegt η Carinae in einem hellen, ausgedehnten Nebel, und in der Umgebung dieses interessanten Sternes sind auf dem Harvard-Observatorium (vgl. Geschichte und Literatur der veränderlichen Sterne) und auf dem Union Observatory in Johannesburg[5], dort namentlich von E. HERTZSPRUNG[6], zahlreiche Veränderliche entdeckt worden, von denen viele δ Cephei-Sterne und Bedeckungs-Veränderliche sind. Ob sie aber mit dem Nebel in Verbindung stehen, ist wohl noch zweifelhaft, da es sich hier um eine Gegend der Milchstraße handelt und in dieser solche Veränderliche häufig sind.

Von den Sternen der R Coronae-Klasse liegen T Orionis, T Tauri, R Monocerotis, R Coronae austr. in Nebeln, und zwar die drei letztgenannten in solchen, welche selbst auch veränderlich sind. In diesen Fällen darf ein Zusammenhang zwischen Stern und Nebel als sicher angenommen werden, wenn

[1] Publ A S P 34, S. 290 (1922).
[2] A N 217, S. 476 (1923).
[3] Ap J 63, S. 236 (1926) = Mt Wilson Contr 310.
[4] Pop Astr 33, S. 252 (1925); Obs 48, S. 139 (1925).
[5] Union Circ 18 (1914); 26 (1915); 46 (1919).
[6] An verschiedenen Stellen in B A N 2 u. 3.

auch eine Abhängigkeit zwischen den Lichtschwankungen von Stern und Nebel in keinem der drei Fälle festgestellt ist. Wir verweisen wegen dieser Objekte auf unsere Ausführungen in Ziff. 13 und 14.

Der oben erwähnte R Coronae-Stern T Orionis liegt im Orionnebel, der auch sonst noch viele Veränderliche enthält, von denen bereits in Ziff. 15 die Rede gewesen ist.

Der einzige Mira-Stern, der eine Nebelhülle hat, ist, soweit man bis jetzt weiß, R Aquarii (vgl. Ziff. 27).

Andere bisher noch nicht erwähnte Fälle, in denen Veränderliche in Nebeln liegen, sind folgende:

NGC 2264. Höhlennebel bei S Monocerotis. M. Wolf [A N 221, S. 379 (1924)] fand in diesem 20 Veränderliche, von denen der hellste im Maximum die 14. Größe erreicht. Näheres über die Art des Lichtwechsels ist noch nicht bekannt.

NGC 6514. Trifidnebel. C. O. Lampland fand in diesem Nebel 3 Veränderliche [Pop Astr 27, S. 32 (1919); Publ A S P 33, S. 207 (1921)].

NGC 6523. „Lagoon-Nebula" im Sagittarius. C. O. Lampland hat in diesem Nebel 18 Veränderliche entdeckt [Publ A S P 28, S. 192 (1916); Pop Astr 27, S. 32 (1919)].

NGC 6727. Dieser Nebel umhüllt den Stern CPD $-37°8450$. Nach Innes [Union Circ 33 (1916)] sind sowohl der Stern wie der Nebel veränderlich. Nach Harv Circ 780 (1922) ändert sich die Helligkeit des Sternes zwischen den photographischen Größen 8,6 und 10,5, und die Änderungen scheinen einen ungewöhnlichen Charakter zu besitzen.

NGC 7023. Nach C. D. Perrine [Lick Bull 1, S. 187 (1902)] enthält dieser Nebel zwei veränderliche Sterne.

IC 405. Dieser Nebel enthält nach Harv Bull 786 (1923) einen veränderlichen Stern, AE Aurigae, der ein Spektrum der Klasse B0p besitzt.

In den bisher aufgezählten Fällen handelt es sich bei den Nebeln, in denen die Veränderlichen liegen, um solche, die zur Klasse der diffusen Nebel gehören. Um einen planetarischen Nebel, der einen veränderlichen Zentralstern besitzt, handelt es sich dagegen bei NGC 7662. Dieses Objekt ist von E. E. Barnard[1] viel beobachtet worden. H. H. Turner[2] vermutet eine Periode von $27^1/_3{}^d$ in den Lichtschwankungen des Sternes.

k) Über die Beziehungen zwischen den verschiedenen Klassen von veränderlichen Sternen.

73. Vorbemerkungen. Wir haben in den vorangehenden Ausführungen die veränderlichen Sterne in verschiedene Klassen eingeteilt, und es ergibt sich nun die Frage, inwieweit zwischen diesen verschiedenen Klassen verwandtschaftliche Beziehungen bestehen. Diese Frage ist deshalb so ungemein schwierig zu beantworten, weil wir über die Ursachen des Lichtwechsels der Veränderlichen nichts Sicheres wissen. Man muß sich daher bei der Untersuchung der Beziehungen zwischen den einzelnen Klassen auf äußerliche Anzeichen stützen, und das macht die ganzen Überlegungen unsicher und wenig befriedigend.

Betrachtungen über Ähnlichkeiten und Verschiedenheiten einzelner Klassen von Veränderlichen und daran sich anschließende Vermutungen über eine physische Verwandtschaft oder Nichtverwandtschaft finden sich in der Literatur ziemlich häufig. Systematisch — unter Ausdehnung auf alle Klassen der Veränderlichen — ist die Frage zuerst von H. Ludendorff[3] behandelt worden

[1] M N 68, S. 465 (1908). [2] Ebenda S. 481 (1908). [3] Seeliger-Festschrift S. 80 (1924).

Eine etwaige Verwandtschaft zwischen Mira- und δ Cephei-Sternen hat dann besonders eingehend A. V. NIELSEN[1] erörtert. Wir folgen hier im ganzen den Darlegungen des ersteren, indem wir dieselben in einigen Punkten ergänzen und verbessern.

Im voraus sei noch bemerkt, daß alle Veränderlichen (wir sehen, wie immer, von den Bedeckungs-Veränderlichen ab) Riesensterne sind, vielleicht mit Ausnahme einiger μ Cephei-Sterne (bisher ist nur ein solcher Fall bekannt, vgl. Ziff. 43) und der Veränderlichen im Orionnebel (vgl. Ziff. 15).

74. Die Mira-Sterne. Der Lichtwechsel der Mira-Sterne weist, wie wir gesehen haben, je nach der Spektralklasse, der sie angehören, gewisse Unterschiede auf. Nun sind die Helligkeitsschwankungen der Mira-Sterne der Spektralklassen K, M (ohne helle Linien) und Se denen der Mira-Sterne der Klasse Me im ganzen so ähnlich, daß wohl niemand hier verschiedene Ursachen annehmen wird. Größer sind schon die Unterschiede zwischen den Mira-Sternen der Klasse Me einerseits und denen der Klassen N und R andererseits. Aber auch hier überwiegen doch die Ähnlichkeiten, namentlich auch in der Abhängigkeit der Änderungen des Spektrums von der Phase des Lichtwechsels. Man wird daher ziemlich sicher sein können, daß bei den Mira-Sternen der Spektralklassen N und R die Ursache des Lichtwechsels dieselbe ist wie bei den übrigen Mira-Sternen, wenn auch der Verlauf der Lichtkurve durch die abweichende physikalische Beschaffenheit etwas modifiziert wird. Wir sind daher nach dem heutigen Stande der Kenntnis berechtigt, die Mira-Sterne als eine einheitliche Klasse von Veränderlichen zu betrachten.

75. Beziehungen zwischen den Mira-Sternen und den μ Cephei-Sternen. Daß die Mira-Sterne mit den μ Cephei-Sternen eng verwandt sind, wird wohl nicht bezweifelt werden. Unter den Me-Sternen gibt es zwar, abgesehen von einigen RV Tauri-Sternen, nur einen, nämlich S Persei, von dem mit Sicherheit behauptet werden kann, daß er unregelmäßig veränderlich ist. Aber die Veränderlichen der Spektralklasse M (ohne helle Linien) gehören zum Teil zu den Mira-, zum Teil zu den μ Cephei-Sternen, und der Unterschied scheint nur graduell zu sein. Z. B. bildet W Persei (Mc) einen Übergang. Die Periode ist bei diesem Stern im Mittel 496^d, die Einzelwerte schwanken aber zwischen 400^d und 600^d, die Form der Lichtkurve ist stark veränderlich. Im allgemeinen scheint die Sache so zu liegen, daß M-Sterne mit größeren Amplituden eine regelmäßige Periode haben, solche mit kleinen Amplituden dagegen meist oder mindestens sehr häufig unregelmäßig sind; ganz ähnlich ist es mit den veränderlichen N-Sternen. Leider sind die roten unregelmäßigen Veränderlichen, wie wir schon erwähnt haben, meist nur wenig beobachtet worden, und die Beobachtungen sind im Verhältnis zu der kleinen Helligkeitsamplitude zu ungenau, so daß wir erst sehr wenig über den Lichtwechsel dieser Sterne wissen. Es ist keineswegs ausgeschlossen, daß wir bei ihnen noch Unterklassen unterscheiden müssen.

76. Beziehungen der RV Tauri-Sterne zu den Mira-Sternen und den langperiodischen δ Cephei-Sternen. Diese Beziehungen sind schon in Ziff. 49 erörtert worden; es kann danach kaum ein Zweifel bestehen, daß die RV Tauri-Sterne mit kurzen Perioden mit den δ Cephei-Sternen, die mit langen Perioden dagegen mit den Mira-Sternen wesensverwandt sind. Auch zu den μ Cephei-Sternen liegen vielleicht, wie wir an der zitierten Stelle gesehen haben, Beziehungen vor.

77. Beziehungen der U Geminorum-Sterne zu den Mira-Sternen und den langperiodischen δ Cephei-Sternen. Die U Geminorum-Sterne haben wir im

[1] A N 227, S. 177 (1926).

Abschnitt d) eingehend besprochen. Es geht aus dem dort Gesagten hervor, daß diejenigen U Geminorum-Sterne, bei denen längere Pausen zwischen den Maxima vorhanden sind, Lichtkurven besitzen, die den Lichtkurven der Form α_1 der Mira-Sterne (lange Konstanz im Minimum, sehr rascher Aufstieg) ähnlich sind, nur treten bei letzteren die Aufhellungen regelmäßig ein. Es ist aber zu erwähnen, daß auch bei den Mira-Sternen mit α_1-Kurven gelegentlich starke Unregelmäßigkeiten auftreten. So scheinen bei Z Tauri die Maxima manchmal auszubleiben oder nur schwach angedeutet zu sein. Bei RW Lyrae erheben sich die Maxima manchmal bis fast zur 9. Größe, manchmal aber auch nur bis zur 13., d. h. nur etwa 1 bis $1^1/_2$ Größenklassen über die Minimalhelligkeit. Besonders die Ähnlichkeit der Lichtkurven einiger dieser Me-Sterne mit der von SS Cygni ist verblüffend, nur der zeitliche Maßstab der Kurven ist bei SS Cygni ein anderer. In der Regel ist allerdings die Dauer des konstanten Minimums relativ zur Dauer der Erhebung über dieses bei SS Cygni größer als bei den Mira-Sternen; immerhin dauert bei V Camelopardalis ($P = 515^d$) der konstante Teil des Minimums mitunter 300^d und mehr, bei Z Tauri ($P = 500^d$) etwa 320^d, und es kommen andererseits bei SS Cygni Fälle vor, wo die Dauer der Minima ziemlich kurz ist.

Der Me-Stern kürzester Periode, der eine Lichtkurve der Form α_1 hat, ist wohl Z Cygni mit $P = 263^d$. Der Periodenlänge nach schließen sich also die Me-Sterne mit α_1-Kurven an die U Geminorum-Sterne leidlich an, denn bei UV Persei beträgt das kürzeste beobachtete Intervall zwischen zwei Aufhellungen zwar nur 142^d, aber im allgemeinen scheint dieses Intervall größer zu sein. (Die Maxima bei UV Persei sind allerdings viel spitzer als bei den Me-Sternen mit α_1-Lichtkurven.)

Nach allem kann man sich des Eindrucks nicht erwehren, daß eine Verwandtschaft zwischen den Me-Sternen mit Lichtkurven der Form α_1 und den U Geminorum-Sternen vorliegt. Dagegen spricht freilich der Umstand, daß SS Cygni und U Geminorum Spektra haben, die von denen der Mira-Sterne grundverschieden sind.

Eine Verwandtschaft derjenigen U Geminorum-Sterne, bei denen die Aufhellungen ziemlich rasch aufeinanderfolgen, mit den langperiodischen δ Cephei-Sternen ist gleichfalls nicht unwahrscheinlich. Wir wissen, daß TZ Persei und Z Camelopardalis keine konstante Phase im Minimum haben, ebenso wie es eine solche bei den langperiodischen δ Cephei-Sternen höchstens ganz ausnahmsweise gibt. Die durchschnittlichen Intervalle zwischen den Maxima sind bei TZ Persei, Z Camelopardalis, SU Ursae majoris und X Leonis nicht größer als die Perioden mancher langperiodischen δ Cephei-Sterne. Allerdings besitzen die drei letztgenannten Sterne erhebliche galaktische Breiten, im Gegensatz zu den langperiodischen δ Cephei-Sternen.

78. Beziehungen zwischen den Mira-Sternen und den langperiodischen δ Cephei-Sternen. Gegen die Verwandtschaft dieser beiden Klassen von Veränderlichen sprechen zunächst folgende Umstände:

1. Die verschiedene galaktische Verteilung (die Mira-Sterne sind, abgesehen von den der Spektralklasse N angehörigen, über den ganzen Himmel verteilt, die langperiodischen δ Cephei-Sterne zeigen eine sehr starke galaktische Konzentration).

2. Die von der Sonnenbewegung befreiten Radialgeschwindigkeiten der Mira-Sterne sind durchschnittlich ziemlich groß, die der langperiodischen δ Cephei-Sterne dagegen nur klein.

3. Die Mira-Sterne gehören vorwiegend den Spektralklassen M und N, die langperiodischen δ Cephei-Sterne den Klassen F, G und K an.

4. Die Radialgeschwindigkeiten verhalten sich bei beiden Klassen von Sternen zum Lichtwechsel ganz verschieden. Die Radialgeschwindigkeiten der δ Cephei-Sterne sind stark veränderlich, das Maximum bzw. Minimum der Helligkeit fällt ungefähr mit dem negativen bzw. positiven Maximum der Radialgeschwindigkeit zusammen. Bei den Mira-Sternen ist die aus den Verschiebungen der Absorptionslinien folgende Radialgeschwindigkeit nicht oder, wie es neuerdings scheint, gering variabel, und bei Mira Ceti selbst ist das Verhalten dieser Radialgeschwindigkeiten zur Lichtkurve gerade umgekehrt wie bei den δ Cephei-Sternen. Bei T Centauri scheint dagegen in den Radialgeschwindigkeiten ein ähnliches Verhalten wie bei den δ Cephei-Sternen angedeutet zu sein (vgl. Ziff. 32). Die Emissionslinien der Mira-Sterne scheinen bei einigen Sternen, ganz wie die Absorptionslinien der δ Cephei-Sterne, sehr bald nach dem Maximum der Helligkeit ein Maximum der Annäherungsgeschwindigkeit zu ergeben. Bei Mira Ceti selbst ist die Sachlage komplizierter, doch ergeben bei diesem Stern die hellen Linien im Minimum ein positives Maximum der Radialgeschwindigkeit.

5. Bei den δ Cephei-Sternen wächst die absolute Helligkeit mit der Periode; bei den Me-Mira-Sternen scheint eher das Gegenteil der Fall zu sein. Auch sind die absoluten visuellen Helligkeiten der Mira-Sterne augenscheinlich viel geringer als die der δ Cephei-Sterne mit sehr langen Perioden. Dafür sind aber erstere radiometrisch sehr hell.

Von diesen Gründen, die eine Verwandtschaft zwischen den Mira- und δ Cephei-Sternen zunächst unwahrscheinlich machen, sind aber 1. bis 3. wenig stichhaltig, da ganz ähnliche Unterschiede auch zwischen den lang- und den kurzperiodischen δ Cephei-Sternen bestehen, an deren Verwandtschaft wohl niemand Zweifel hegt.

Die Verschiedenheit der Spektra könnte sogar vielleicht als Grund für die Verwandtschaft angeführt werden. Ordnet man nämlich die kurzperiodischen und die langperiodischen δ Cephei- sowie die Mira-Sterne nach ihrer Periodenlänge in eine fortlaufende Reihe und bildet (unter Auslassung der N-, R- und S-Sterne) die mittleren Spektren für gewisse Periodenintervalle, so erhält man für die Spektra eine stetig fortlaufende Reihe von etwa A bis M, und auch innerhalb der Klasse der Mira-Sterne werden die Sterne noch immer röter, je länger die Periode ist.

Stärker als 1. bis 3. spricht 4. gegen die Verwandtschaft; aber auch dieser Gegengrund wird wenigstens etwas abgeschwächt durch das oben geschilderte Verhalten der Emissionslinien bei einigen der Mira-Sterne. 5. Dagegen scheint einen wesentlichen Unterschied darzustellen.

Für die Verwandtschaft der δ Cephei- und der Mira-Sterne sprechen folgende Umstände:

1. Mira-Sterne und δ Cephei-Sterne sind Riesensterne; im Minimum der Helligkeit ist ihre Temperatur niedriger als im Maximum, wie aus den Änderungen im Spektrum hervorgeht.

2. Die Lichtkurven beider Klassen haben ihrer Gestalt nach in vielen Fällen die größte Ähnlichkeit. (Allerdings sind auch manche Unterschiede vorhanden: So kommen ausgesprochene sekundäre Wellen bei den Mira-Sternen hauptsächlich im aufsteigenden, bei den δ Cephei-Sternen hauptsächlich im absteigenden Aste der Lichtkurve vor.)

3. Bei den Mira-Sternen mit verhältnismäßig kurzen Perioden ist die Helligkeitsamplitude durchschnittlich kleiner als bei denen mit längeren Perioden; die ersteren nähern sich dadurch den δ Cephei-Sternen, die kleine Helligkeitsamplituden besitzen.

4. Wenn auch Sterne mit Perioden zwischen 45^d (der ungefähren oberen Periodengrenze für die unzweifelhaft der δ Cephei-Klasse angehörigen Sterne) und 90^d (der unteren Periodengrenze der unzweifelhaften Mira-Sterne) selten sind, so gibt es doch einige Sterne in dieser Lücke, die dem Spektraltypus M angehören (soweit die Spektra bekannt sind), und die einen Übergang zwischen den beiden Klassen zu bilden scheinen. Ganz ähnlich liegen die Dinge bei dem Übergang von den kurz- zu den langperiodischen δ Cephei-Sternen.

5. In der Kleinen MAGELLANschen Wolke kommt neben δ Cephei-Sternen mit normalen Perioden auch ein Veränderlicher mit einer Periode von 127^d vor. Dieser paßt seiner Helligkeit nach zwanglos in die Kurve, welche die Beziehung zwischen Helligkeit und Periodenlänge für die übrigen δ Cephei-Sterne der Kleinen MAGELLANschen Wolke darstellt. Man hat also hier einen δ Cephei-Stern von 127^d Periode vor sich, den man sonst auf Grund der langen Periode als Mira-Stern betrachten würde. Die Lichtkurve ist die typische für Mira-Sterne; die photographische Amplitude ist aber nur 0,9 Größenklassen. Ein anderer δ Cephei-Stern in der Kleinen MAGELLANschen Wolke hat die lange Periode von 66^d. Auch sonst kommen in Sternhaufen und Nebeln einige wenige δ Cephei-Sterne vor, die die obere Periodengrenze der galaktischen δ Cephei-Sterne (45^d) überschreiten.

6. Die Mira-Sterne besitzen eine deutliche Verwandtschaft mit den RV Tauri-Sternen. Andererseits läßt sich, wie wir gesehen haben, auch eine solche zwischen den RV Tauri-Sternen und den langperiodischen δ Cephei-Sternen feststellen.

7. Der δ Cephei-Stern RU Camelopardalis ($P = 22^d$) hat ein Spektrum der Klasse R, im Minimum treten merkwürdigerweise Emissionslinien des Wasserstoffs auf, die, wie bei den Me-Mira-Sternen, gegen die Absorptionslinien nach Violett verschoben sind.

Auch der δ Cephei-Stern W Virginis ($P = 17^d$) hat in gewissen Phasen des Lichtwechsels helle Wasserstofflinien, die aber an den periodischen Verschiebungen der Absorptionslinien nicht teilnehmen. Endlich hat W Serpentis ($P = 14^d$) doppelt umgekehrte helle Wasserstofflinien. Wenn sich nun auch bei diesen Objekten die Emissionslinien des H in mancher Hinsicht anders verhalten als bei den Mira-Sternen, so deutet ihr Vorhandensein doch immerhin auf eine Verwandtschaft dieser langperiodischen δ Cephei-Sterne mit den Mira-Sternen hin.

Zusammenfassend wird man sagen dürfen, daß die Frage einer etwaigen Verwandtschaft zwischen den langperiodischen δ Cephei-Sternen und den Mira-Sternen noch offenbleiben muß. Es scheint, daß mehr für als gegen eine solche Verwandtschaft spricht, doch ist das eine subjektive Ansicht. Auch NIELSEN (in seiner zitierten Abhandlung) und EDDINGTON (in seinem Buch „The Internal Constitution of the Stars“) sprechen sich für das Vorhandensein der Verwandtschaft beider Gruppen von Veränderlichen aus. Letzterer erklärt den Lichtwechsel der Mira-Ceti-Sterne, wie den der δ Cephei-Sterne, durch Pulsationen.

79. Beziehungen zwischen den übrigen Klassen von veränderlichen Sternen. Daß die langperiodischen δ Cephei-Sterne mit den kurzperiodischen nahe verwandt sind, wenngleich auch manche Verschiedenheiten zwischen ihnen bestehen, geht aus unsern früheren Ausführungen über die δ Cephei-Sterne klar hervor und braucht hier nicht nochmals dargelegt zu werden. Auch zwischen den kurzperiodischen δ Cephei-Sternen und den β Cephei-Sternen bestehen enge Beziehungen; werden doch letztere von vielen Autoren einfach zu den ersteren gerechnet.

Ob etwa auch eine Verwandtschaft zwischen den δ Cephei- und den unregelmäßigen μ Cephei-Sternen besteht, ist eine Frage, die wir noch ganz unentschieden lassen müssen; wir kennen den Lichtwechsel der letzteren noch zu wenig.

Wenden wir uns nun den beiden hier noch nicht behandelten Klassen von Veränderlichen zu, den Nova-ähnlichen und den R Coronae-Sternen. Es scheint, als ob zwischen den Nova-ähnlichen Veränderlichen und den U Geminorum-Sternen eine gewisse Ähnlichkeit vorhanden ist. Der U Geminorum-Stern SS Cygni hat, wie wir sahen, ein Spektrum, das an das der neuen Sterne in gewissen Phasen ihrer Erscheinung erinnert. Das Aufleuchten geschieht bei manchen U Geminorum-Sternen, z. B. bei UV Persei, außerordentlich plötzlich; das Maximum dauert bei diesem Sterne nur wenige Tage. Für die Nova-ähnlichen Veränderlichen ist T Pyxidis ein typischer Fall. Dieser Stern erhob sich aus einem nahezu konstanten Minimum plötzlich in den Jahren 1890, 1902 und 1920 zu größerer Helligkeit und zeigte dabei 1920 ein typisches Nova-Spektrum. Man könnte ihn einen U Geminorum-Stern nennen, der nur in sehr langen Intervallen aufleuchtet, ebensogut aber auch eine Nova, die mehrmals erschienen ist. Daß auch sonst die Nova-ähnlichen Veränderlichen mit den neuen Sternen selbst in engster Verwandtschaft stehen, haben wir früher gesehen.

Die R Coronae-Sterne endlich liegen mit sehr wenigen Ausnahmen in geringen galaktischen Breiten, wie die Nova-ähnlichen Veränderlichen und die neuen Sterne selbst; sie besitzen sehr verschiedene Spektren, aber auffallend zahlreiche unter ihnen haben zeitweise oder dauernd helle Linien, und einige haben im Spektrum Eigentümlichkeiten, die an die neuen Sterne erinnern. Es ist die Hypothese aufgestellt worden, daß der Lichtwechsel durch Staub- und Nebelmassen zustande kommt, die sich an dem Stern vorbeibewegen oder durch die der Stern sich hindurchbewegt. Um einfache Verfinsterungsvorgänge kann es sich aber wegen der Eigentümlichkeiten der Spektren vieler dieser Sterne sicherlich nicht in allen Fällen handeln. Diese hier angedeutete Hypothese, die an SEELIGERS Theorie der neuen Sterne anklingt, gewinnt noch dadurch an Wahrscheinlichkeit, daß manche der R Coronae-Sterne in Nebeln liegen. So scheint alles in allem eine Verwandtschaft mit den Nova-ähnlichen Veränderlichen und den neuen Sternen selbst vorhanden zu sein.

80. Übersicht über die Beziehungen zwischen den verschiedenen Klassen der veränderlichen Sterne. Eine solche Übersicht geben wir in folgendem Schema:

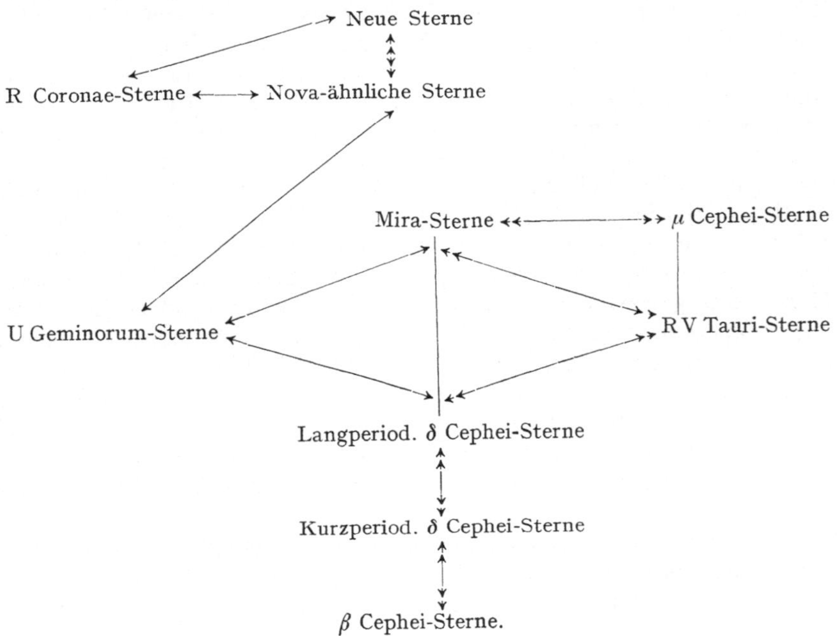

Die Stärke der Verwandtschaft zwischen den einzelnen Klassen wird durch die Art der Verbindungslinien zwischen den Bezeichnungen der Klassen gekennzeichnet. Linien mit zwei Spitzen an jedem Ende bedeuten eine nach dem heutigen Stande der Kenntnis unzweifelhafte Verwandtschaft, solche mit einer Spitze eine stark zu vermutende, und solche ohne Spitze eine zweifelhafte Verwandtschaft. Wo keine Verwandtschaft erkennbar ist, sind keine Verbindungslinien gezogen.

Nicht in das Schema aufgenommen sind die am Ende von Ziff. 66 erwähnten Veränderlichen mit äußerst kleinen Lichtschwankungen, über die wir erst sehr wenig wissen, und die möglicherweise mit den δ Cephei-Sternen verwandt sind.

Wir haben schon früher erwähnt, daß es nur sehr wenige Veränderliche gibt, die sich nicht in die oben aufgeführten Klassen einordnen lassen, und haben diese Objekte bei der Besprechung derjenigen Klassen angeführt, mit denen sie einige Ähnlichkeit besitzen.

Es scheint nach dem Schema, als ob die veränderlichen Sterne (von den Verfinsterungs-Veränderlichen abgesehen) in gewissem Sinne eine Gesamtheit bilden, deren Glieder durch allmähliche Übergänge verbunden sind, so daß die Sachlage ähnlich, wenn auch weniger einfach, wie bei den Spektren der Sterne ist. (Vor allem sind bei den veränderlichen Sternen die Übergangsformen seltener, als dies bei den Spektren im allgemeinen der Fall ist.)

Freilich dürfen wir nicht vergessen, daß wir uns bei der Erörterung der Verwandtschaftsverhältnisse zwischen den verschiedenen Klassen von Veränderlichen auf sehr unsicherem Boden bewegt haben, so daß die Betrachtungen über diesen Gegenstand nur als vorläufiger Versuch zu werten sind.

Nachtrag.

Während des Druckes des vorliegenden Kapitels ist eine Anzahl von Arbeiten über veränderliche Sterne erschienen; auf die wichtigsten von ihnen wird in den folgenden Nachträgen kurz hingewiesen:

[Zu Ziff. 4, S. 57.] R. PRAGER hat „Tabellen zur Nomenklatur der veränderlichen Sterne“ veröffentlicht, die sehr nützlich sind [Kleinere Veröffentl. d. Stw. Babelsberg, Nr. 2].

[Zu Ziff. 10, S. 69.] Über das Spektrum von P Cygni ist eine Arbeit von CH'ING-SUNG YÜ erschienen [Publ A S P 39, S. 112].

[Zu Ziff. 27, S. 143.] Über einen absolut äußerst schwachen, nicht veränderlichen M-Stern mit heller $H\beta$- und $H\gamma$-Linie vgl. Publ A S P 39, S. 173.

[Zu Ziff. 28, S. 143.] Über die Spektra der Klasse S hat P. W. MERRILL eine zweite ausführliche Abhandlung veröffentlicht [Ap J 65, S. 23 = Mt Wilson Contr 325]. Die neue von MERRILL aufgestellte Liste von S-Sternen enthält an Veränderlichen außer den in Tabelle II, Ziff. 22, aufgeführten Objekten noch AA Cygni, S Cygni (der bisher zur Spektralklasse Me gerechnet wurde und sich durch steilen Helligkeitsabstieg auszeichnet, vgl. Tabelle I, Ziff. 22) und Z Delphini (bisher Spektrum pec, vgl. Tabelle V, Ziff. 22).

[Zu Ziff. 40, S. 164.] RS Puppis ist [Harv Bull 848] seinem Lichtwechsel nach ein typischer δ Cephei-Stern, ebenso SV Vulpeculae [Harv Bull 846]; SS Geminorum hat sich dagegen als RV Tauri-Stern erwiesen [Harv Bull 846]. Der in der Tabelle auf S. 165 angeführte Stern CG Sagittarii ist seinem Lichtwechsel nach ein typischer δ Cephei-Stern mit einer Periode von 64^d [Harv Bull 846]. Ein neu entdeckter Veränderlicher im Aquarius hat eine Periode von $72^d,7$ [Harv Bull 848].

[Zu Ziff. 43, S. 168.] Einen sehr eigentümlichen Lichtwechsel hat RU Aquarii [Harv Bull 848]; das Spektrum ist Mc. Es ist schwierig, diesen Stern in eine bestimmte Klasse von Veränderlichen einzuordnen.

[Zu Ziff. 48, S. 174.] Eine Untersuchung über TT Ophiuchi findet sich in Harv Bull 847, eine solche über AC Herculis in Harv Bull 845; danach sind beide Sterne sicher zur RV Tauri-Klasse zu rechnen. Die Liste der RV Tauri-Sterne ist ferner nach Harv Bull 846 noch um SS Geminorum ($P = 89^d,3$) und nach Harv Bull 847 um TX Ophiuchi ($P = 136^d$) zu bereichern. Ein weiterer, neu entdeckter RV Tauri-Stern wird in Harv Bull 847 angezeigt ($P = 49^d$). Der Lichtwechsel der genannten Sterne hat meist Ähnlichkeit mit dem von β Lyrae, nur ist die Lichtkurve stark veränderlich; die Harvard-Astronomen bezeichnen diesen Typus der Veränderlichkeit daher als „quasi β Lyrae type".

Das Spektrum von SS Geminorum ändert sich zwischen Maximum und Hauptminimum von G0 zu K0, das von AC Herculis von F8 zu K5. In dieser Hinsicht verhalten sich also die genannten Veränderlichen ganz so wie die δ Cephei-Sterne.

[Zu Ziff. 56, S. 197.] Bei den kurzperiodischen δ Cephei-Sternen V Leonis minoris und TW Herculis hat C. HOFFMEISTER Änderungen der Lichtkurven in ihrem Abstieg festgestellt [A N 230, S. 113].

[Zu Ziff. 58, S. 202.] In Harv Circ 313 teilen H. SHAPLEY und MARGARET L. WALTON die Spektra von 70 δ Cephei-Sternen nebst den beobachteten Änderungen während des Lichtwechsels mit und untersuchen aufs neue die Abhängigkeit des mittleren Spektrums von der Periode. Für die kurzperiodischen δ Cephei-Sterne, von denen nur sechs in der Liste vorkommen, ergibt sich keine solche Abhängigkeit, ihr mittleres Spektrum ist durchschnittlich A6. Bei den langperiodischen δ Cephei-Sternen ist dagegen die Abhängigkeit sehr ausgesprochen, und SHAPLEY stellt sie in Harv Circ 314 durch folgende Tabelle dar:

Mittleres Spektrum	log P	Mittleres Spektrum	log P
F4	(0,16)	G2	0,79
F6	0,30	G4	1,04
F8	0,43	G6	1,38
G0	0,59	G8	(1,70)

Die einzelnen Sterne weichen im allgemeinen nur wenig von der durch diese Zahlen definierten Kurve ab. SHAPLEY zeigt, wie aus dieser Beziehung zwischen Spektrum und Periode seine Period-Luminosity Curve (vgl. Ziff. 64) hergeleitet werden kann.

[Zu Ziff. 68, S. 237.] Über die Veränderlichen in NGC 6656 = M 22 enthält Harv Bull 848 einige Mitteilungen.

[Zu Ziff. 70, S. 241.] Einer der Veränderlichen in der Großen MAGELLANschen Wolke hat sich als R Coronae-Stern erwiesen [Harv Bull 846].

Chapter 3.

Novae

By

F. J. M. STRATTON-Cambridge.

With 15 illustrations.

a) Early History.

1. Ancient Records. The earliest record of the phenomenon of a bright star appearing in the sky is to be found in the Chinese cyclopaedia T'u shu chi ch'-êng for the year B. C. 2679. The phenomenon described may not, however, refer to a nova as in many of the earlier records the Chinese do not distinguish between comets and stars newly visible. The appearance of a bright star in Scorpio in B. C. 2255, followed by auspicious events, of an inauspicious yellow star in B. C. 2238, followed by the death of the Emperor, of a new star in Aquarius in B. C. 532 complete the list of observations of likely novae before the days when European astronomers became active observers of the Heavens. The first new star recorded in China and in Europe also was one that appeared in Scorpio in the year 134 B. C.[1]. The elder PLINY suggested[2] that the appearance of this star was the factor that led HIPPARCHUS to catalogue the stars; it was probably too late in date for this though it may well have influenced the direction of his later work. If we may take low galactic latitude and the absence of independent evidence as to the presence of comets in the heavens in the same year as factors in deciding which of the vague references to bright stars without tails are to be ascribed to novae then we may consider the following list of possible novae[3]: stars observed in China in September 107 A. D. in Gemini, in December 123 A. D. in Ophiuchus, in June 304 in the Hyades, in March 369 in Cassiopeia, from March to October 393, in Scorpio, in February 1011 in Sagittarius, in July 1054 in Taurus, and in December 1230 in Ophiuchus.

Western observations during this period are not well evidenced. Leaving aside HADRIAN's nova of 130 A. D., as uncertain, and one or two early records where there is independent evidence of a comet visible in the same year we come to a statement of HEPIDANNUS[4] that a bright new star appeared in the South in 1012 and was visible for three months. The point of interest here is that the star oscillated in brightness; the record "aliquando contractior aliquando diffusior" may refer to a comet with varying tail, but if not it may be noted as the first reference to oscillations in the magnitude curve of a nova.

2. Novae from 1572 to 1866. Passing from the vague and uncertain to the well attested we may consider the first nova which definitely contributed to

[1] See E. BIOT, Connaissance des Temps, p. 61 (1846). This star was probably a comet.

[2] Historia Naturalis, Bk. II, Ch. 26.

[3] For an account of these early doubtful Novae see three articles by ZINNER, in Sirius (1919). See also MÜLLER u. HARTWIG, Geschichte und Literatur des Lichtwechsels, II, S. 449 (1920) (ZINNER).

[4] Annales Breves Rerum in Alamannia Gestarum, HEPIDANNUS, c. 1050.

astronomical history, the famous nova of 1572 which restored TYCHO BRAHE to astronomy[1]. First seen as a conspicuous object on November 11, 1572, this star, N Cassiopeiae 1572, was the first of which it was definitely ascertained from measures for parallax that it belonged to the stellar universe and not to the solar system. At discovery it was much brighter than any other star or even than Jupiter or Venus at their brightest; it was easily visible at midday. It took three months to drop to the first magnitude and then faded steadily away remaining visible to the naked eye until March 1574. During its period of visibility TYCHO noted among other things that it scintillated like the stars and unlike the planets; also he records its change of colour from white to red and back to a whitish colour again, the colour of Saturn.

A possible nova in Scorpio observed in China in 1584 was followed by the discovery of Mira Ceti in 1596 by DAVID FABRICIUS[2] and of a new star of the third magnitude in Cygnus observed by WILLIAM JANZOON BLAEU in 1600. This latter star remained variable for some years with a second bright maximum in 1657/9 and it is a naked eye object to the present day as P Cygni. Though denied the title of nova by some writers its spectrum at the present day bears striking resemblance to a fairly early stage in the evolution of the typical nova and justifies its being classed with the novae as N Cygni, 1600.

Shortly afterwards followed another famous nova, N Ophiuchi, 1604, first observed on October 10, 1604 by BRUNOWSKI, as a first magnitude star. This star remained visible for 12 months; white in colour throughout it was remarkable for the strength of its scintillations. It too is linked with a name famous in astronomy, KEPLER[3] having discussed its visible history and its possible origin. He pointed out that like TYCHO BRAHE's nova, the star appeared in the Milky Way and confirmed TYCHO's suggestion, "stellas hujus modi procreari ex via lactea".

Naturally it not infrequently happened that, as in the case of Mira Ceti, variables observed at maximum were hailed as novae. One of these was a star first observed in Vulpecula near β Cygni on 1669 Dec. 20 by the CARTHUSIAN MONK ANTHELME[4] at Dijon. This star when first seen was of the third magnitude; it disappeared for a time but was again observed by ANTHELME on 1671 March 17 as of the fourth magnitude. Careful observation showed that the star was fluctuating considerably in brightness; it again disappeared in August reappearing on 1672 March 29, this time only of the sixth magnitude[5]. It has not since been definitely observed. If a nova it was of an unusual type with its three well-defined bright maxima. Another possible nova first observed by HEVELIUS in 1667 between χ^1 and χ^3 Orionis was re-observed seventy years later by BEVIS[6]. It is mentioned here as a possible type to be considered in formulating a theory of novae.

Save for a star of the sixth magnitude observed three nights running in Sagitta by D'AGELET in 1783[7]—a case of an early catalogued star now missing

[1] For a short account and reference to the literature on the magnitude changes of this and 31 subsequent novae see an article by ZINNER in MÜLLER u. HARTWIG, Geschichte und Literatur des Lichtwechsels, II, S. 415; III, S. 98 (1920).

[2] This star was re-discovered as a Nova in 1637 by PHOCYLIDES HOLWARDA.

[3] De Stella Nova in Pede Serpentarii (1606).

[4] Phil Trans No. 165, p. 2087 (1670).

[5] M N 21, p. 231 (1861). [6] J B A A 4, p. 96 (1893).

[7] Wash Nat Ac Mem. 1, p. 237 (1866) — star entered as Anonyma. Another star of this nature is a star in Aries observed by KRUEGER at Bonn twice in 1854 and entered in the Bonn Durchmusterung as of magnitude 9,5, catalogued somewhere between 1854—62 in the Atlas Ecliptique of M. CHACORNAC as of the 11th or 12th magnitude and subsequently lost.

that may represent the observation of a faint nova near its maximum—the eighteenth century passed without an observation of a new star and it was not until 1848 April 28 that another discovery of a nova was announced by HIND in Ophiuchus[1]. This star when first seen was between the fourth and fifth magnitude and was described as having "a very intense reddish-yellow light". It soon faded away. Next in date comes the discovery by AUWERS on 1860 May 21 of a seventh magnitude star in the globular cluster N G C 6093 in Scorpio[2]. This is the first definite case of a nova discovered in a cluster or nebula, a phenomenon since not infrequently observed.

TEBBUTT's discovery of N Arae 1862 not recognised as a variable until 1878[3] was followed by BIRMINGHAM's discovery of a second magnitude star in Corona Borealis on 1866 July 7[4]. This star now known as T Coronae Borealis, was the first nova to be examined through the spectroscope. HUGGINS[5] found the spectrum to consist of bright hydrogen lines superposed on a continuous spectrum which was crossed by many absorption lines. At this point, where the study of the novae enters upon a new and very important stage and where observations of novae become more frequent, we can conveniently close the introductory historical section and develop the observational results with proper reference to the stars concerned in each section. We should however refer the student of early novae to an article by LUNDMARK [Publ A S P 33, p. 225 (1921)] which gives reference to many of the statements made by mediaeval and ancient writers.

One word should be added here about the definition of novae. Generally speaking the term is limited to a star which has had one observed outburst of radiation giving it for a short period a magnitude far brighter than it has been known to exhibit at any other time. In a few cases, where spectroscopic evidence supports the claim for the star to be classed as a nova, rather than as an irregular variable of more normal type, a star has been included which has shewn several fairly equal maxima at wide intervals of time. Such a star is η Carinae, a star with several recorded bright maxima; its claim to appear in the list of novae has recently received strong support from the spectral development of N Pictoris 1925, whose spectrum some months after maximum developed into the unusual type of η Carinae. In border line cases it must naturally depend upon individual choice whether a star is included in the list of novae or not[6].

b) Distribution of Novae.

3. Galactic Novae. The following tables give the right ascensions and declinations for the epoch 1925,0 of novae observed in the last four hundred years, the names of the respective discoverers and the dates of the principal maxima. Table I contains those novae which do not lie in nebulae or clusters and for them the galactic longitude, G, and the galactic latitude, g, have been added. The galactic co-ordinates have been derived from the table in Harv Ann

[1] M N 8, p. 146 (1848). Meanwhile the interesting star, η Carinae, frequently classed as a nova, had been observed at several bright maxima. Observed by HALLEY as of magnitude 4 in 1677, it had been seen as of magnitude 2 in 1685, 1751, 1826, of magnitude 1 in 1827, 2 from 1828—34, slowly rising with fluctuations to -1^{m} in 1843. It dropped to $0^{m},2$ by 1856 and is now about $8^{m},0$. Its remarkable spectrum would indicate a close relationship with novae.

[2] A N 53, p. 293 (1848). [3] M N 38, p. 330 (1878). [4] M N 26, p. 310 (1866).

[5] London R S Proc 15, p. 146. (1866)

[6] HERTZSPRUNG has described [Harv Bull 845 (1927)] a peculiar object of short duration apparently a star of about 6^{m} visible on two photographic plates of 13^{m} and 29^{m} 20^{s} exposure on Dec. 15, 1900 near RX Camelopardalis. There is no evidence of any comet or other member of the solar system being responsible for the images.

Table I. Galactic Novae.

	Star and date of maximum	Discoverer	α (1925,0)	δ (1925,0)	G	g
1	N Cassiopeiae 1572 (B)	SCHULER	0^h 20^m 37^s	+63° 43′,9[1]	88°	+ 2°
2	N Piscium 1907	ERNST	0 30 54	+ 9 53,2	86	−52
3	N Persei 1887 (V)	FLEMING	1 56 44	+56 22,2[1]	101	− 4
4	N Persei 1912 (UW)	D'ESTERRE	2 7 9	+56 45,0	102	− 3
5	N Arietis 1854	KRUEGER	2 44 19	+17 3,3	128	−36
6	N Arietis 1855 (W)	SCHÖNFELD	3 16 7	+28 43,9	127	−22
7	N Arietis 1905	WOLF	3 20 45	+19 35,3	134	−28
8	N Persei 1901	ANDERSON	3 26 6	+43 38,9	119	− 9
9	N Orionis 1916	THIELE	5 17 43	+ 1 5,7	169	−17
10	N Aurigae 1891	ANDERSON	5 27 10	+30 23,3	145	+ 0
11	N Pictoris 1925 (RR)	WATSON	6 34 57	−62 34,6	239	−25
12	N Geminorum 1903	TURNER	6 39 25	+30 1,3	153	+13
13	N Geminorum 1912	ENEBO	6 50 3	+32 14,2	151	+16
14	N Geminorum 1892	BARNARD	6 54 24	+17 8,5	144	+11
15	N Monocerotis 1918	WOLF	7 23 7	− 6 31,4	191	+ 6
16	N Puppis 1902	WOODS	8 10 39	−26 20,2	214	+ 5
17	N Pyxidis 1890 (T)	LEAVITT	9 1 34	−32 4,6	225	+10
18	N Leonis 1855 (U)	SCHÖNFELD	10 20 3	+14 23,0	196	+54
19	N Velorum 1905	LEAVITT	10 59 24	−53 59,0	255	+ 5
20	N Carinae 1895 (RS)	FLEMING	11 4 59	−61 31,8	259	− 1
21	N Leonis 1918 (RZ)	BELJAWSKY	11 33 32	+ 2 13,9	236	+60
22	N Virginis 1871 (X)	PETERS	11 58 1	+ 9 29,4	240	+69
23	N Comae Berenices 1877	SCHWAB	14 3 33	+21 6,1	345	+70
24	N Bootis 1860 (T)	BAXENDELL	14 10 35	+19 24,8	342	+67
25	N Circini 1906	LEAVITT	14 42 24	−59 41,8	285	− 1
26	N Normae 1893 (R)	FLEMING	15 24 2	−50 19,3	295	+ 4
27	N Normae 1920	WOODS	15 33 52	−52 4,9	295	+ 2
28	N Coronae Borealis 1866 (T)	BIRMINGHAM	15 56 22	+26 7,9	10	+46
29	N Serpentis 1903 (X)	LEAVITT	16 15 23	− 2 19,1	338	+30
30	N Scorpii 1863 (U)	POGSON	16 18 11	−17 42,2	326	+20
31	N Arae 1910	FLEMING	16 34 57	−52 16,7	301	− 5
32	N Scorpii 1917 } N Ophiuchi 1917 }	WOODS	16 49 59	−29 30,5	321	+ 7
33	N Ophiuchi 1848	HIND	16 55 18	−12 46,7	336	+15
34	N Ophiuchi 1897	WOODS	17 19 54	−24 26,0	328	+ 4
35	N Ophiuchi 1604	BRUNOWSKI	17 26 8	−21 25,1[1]	333	− 4
36	N Arae 1862	TEBBUTT	17 33 41	−45 26,4	313	− 9
37	N Scorpii 1922	CANNON	17 43 22	−36 36,6	321	− 6
38	N Ophiuchi 1901 (RS)	FLEMING	17 46 11	− 6 41,2	348	+ 8
39	N Scorpii 1906	CANNON	17 49 7	−34 20,2	324	− 6
40	N Sagittarii 1910	FLEMING	17 55 23	−27 33,1	331	− 4
41	N Sagittarii 1914,6	WOODS	18 1 22	−31 45,0	328	− 7
42	N Sagittarii 1901	CANNON	18 2 1	−27 26,6	331	− 5
43	N Sagittarii 1905	WOODS	18 4 9	−32 29,0	327	− 8
44	N Ophiuchi 1919	MACKIE	18 10 37	+11 35,2	7	+12
45	N Sagittarii 1926	CANNON	18 12 43	−23 39,9	333	− 5
46	N Sagittarii 1899	CANNON	18 15 20	−25 13,1	335	− 7
47	N Sagittarii 1917 (BS)	INNES	18 22 10	−27 9,3	334	− 9
48	N Sagittarii 1919	WOODS	18 27 17	−29 27,1	233	−11
49	N Sagittarii 1914	INNES	18 29 43	−26 58,0	335	−10
50	N Sagittarii 1912	INNES	18 29 59	−29 37,9	332	−11
51	N Sagittarii 1914,6	INNES	18 36 21	−28 13,5	331	−12
52	N Aquilae 1918	BOWER	18 45 5	+ 0 30,0	1	− 1
53	N Lyrae 1919	MACKIE	18 50 29	+29 7,8	28	+11
54	N Lyrae 1905 (SU)	WOLF	18 50 59	+36 24,8	34	+15
55	N Sagittarii 1898	FLEMING	18 57 37	−13 16,2	350	−10
56	N Aquilae 1905	FLEMING	18 58 8	− 4 33,2	359	− 6
57	N Aquilae 1919	WOLF	19 14 33	+ 1 38,7	5	− 6

[1] Uncertain identifications.

Continuation of Table I.

	Star and date of maximum	Discovered	α (1925,0)	δ (1925,0)	G	g
58	N Aquilae 1899 (DO)	FLEMING	$19^h\ 16^m\ 33^s$	− 0° 16′,3	4°	− 8°
59	N Aquilae 1925	WOLF	19 27 24	− 6 35,2	0	−13
60	N Sagittae 1783	D'AGELET	19 29 23	+17 35,1	21	− 2
61	N Vulpeculae 1670	ANTHELME	19 44 29	+27 7,7	31	− 0
62	N Cygni 1909	HINKS	19 50 50	+36 50,8	40	+ 4
63	N Cygni 1920	DENNING	19 56 32	+53 24,8	55	+12
64	N Vulpeculae 1923 (SW)[1]	WOLF	19 56 51	+22 43,7	29	− 5
65	N Sagittae 1913	MACKIE	20 4 12	+ 7 28,4	26	− 9
66	N Cygni 1600 (P)	BLAEU	20 15 1	+37 47,9	44	+ 0
67	N Aquarii 1907	ROSS	21 8 9	− 9 8,2	10	−37
68	N Cygni 1876 (Q)	SCHMIDT	21 38 46	+42 29,9	58	− 8
69	N Lacertae 1910	ESPIN	22 32 45	+52 19,6	71	− 4
70	N Andromedae 1901 (Z)	FLEMING	23 30 3	+48 24,2	78	−12

56, p. 5, the origin of the coordinates being taken at $\alpha = 18^h 40^m$, $\delta = 0°$ and the northern pole of the galaxy at $\alpha = 12^h 40^m$, $\delta = +28°$ (1900,0).

Table II contains the novae observed in the nebulae or clusters, the number of the nebula in the N G C and its position for the epoch 1925,0, the date of the maximum of the nova, the name of the discoverer and the distance of the nova from the nucleus of the nebula.

4. Earlier Nomenclature. The following list gives the numerical descriptions of the novae commonly used in the earlier literature of the subject. As will be seen

3	N Persei	1887 (V)	Nova Persei No. 1.
5	N Arietis	1854	Nova Arietis No. 1.
7	N Arietis	1905	Nova Arietis No. 2.
8	N Persei	1901	Nova Persei No. 2.
12	N Geminorum	1903	Nova Geminorum No. 1.
13	N Geminorum	1912	Nova Geminorum No. 2.
26	N Normae	1893 (R)	Nova Normae No. 1.
27	N Normae	1920	Nova Normae No. 2.
84[2]	N Scorpii	1860 (T)	Nova Scorpii No. 1.
32 {	N Scorpii	1917	Nova Scorpii No. 3, or 2.
	N Ophiuchi	1917	Nova Ophiuchi No. 6, or 5.
33	N Ophiuchi	1848	Nova Ophiuchi No. 2.
34	N Ophiuchi	1897	Nova Ophiuchi No. 3.
35	N Ophiuchi	1604	Nova Ophiuchi No. 1.
37	N Scorpii	1922	Nova Scorpii No. 2.
38	N Ophiuchi	1901 (RS)	Nova Ophiuchi No. 4, or 3.
39	N Scorpii	1906	Nova Scorpii No. 2, or 1.
40	N Sagittarii	1910	Nova Sagittarii No. 2, or 4.
41	N Sagittarii	1914,6 (WOODS)	Nova Sagittarii No. 7.
42	N Sagittarii	1901	Nova Sagittarii No. 4, or 2.
43	N Sagittarii	1905	Nova Sagittarii No. 6.
44	N Ophiuchi	1919	Nova Ophiuchi No. 5, or 4.
46	N Sagittarii	1899	Nova Sagittarii No. 3.
47	N Sagittarii	1917 (BS)	Nova Sagittarii No. 8.
48	N Sagittarii	1919	Nova Sagittarii No. 5.
52	N Aquilae	1918	Nova Aquilae No. 3.
55	N Sagittarii	1898	Nova Sagittarii No. 1.
56	N Aquilae	1905	Nova Aquilae No. 2.
57	N Aquilae	1919	Nova Aquilae No. 4.
58	N Aquilae	1899 (DO)	Nova Aquilae No. 1.
60	N Sagittae	1783	Nova Sagittae No. 1.
63	N Cygni	1920	Nova Cygni No. 3.
65	N Sagittae	1913	Nova Sagittae No. 2.
66	N Cygni	1600 (P)	Nova Cygni No. 1.
68	N Cygni	1876 (Q)	Nova Cygni No. 2.

[1] Near Dumb-bell Nebula. [2] Table II.

different writers have not all used the same numbers, owing to discoveries of novae some years after their outburst; the notation is now being abandoned.

5. Novae in Nebulae.

Table II. Novae observed in Nebulae or in Clusters.

	NGC	α (1925,0)	δ (1925,0)	Star and date of maximum	Discoverer	Position relative to centre of nebula Δα	Δδ
1	224	$0^h\ 38^m\ 39^s$	+40° 51′,6	N Andromedae 1885 (S)	GULLY	− 16″	− 4″
2				,, 1909	RITCHEY	−191	−160
3				,, 1909	RITCHEY	−194	− 42
4				,, 1917	SHAPLEY	+360	+480
5				,, 1917	RITCHEY	− 26	−225
6				,, 1917	RITCHEY	+165	+275
7				,, 1918	RITCHEY	−143	+ 11
8				,, 1918	RITCHEY	−115	− 46
9				,, 1918	DUNCAN	+440	+330
10				,, 1918	SANFORD	+120	−450
11				,, 1918	SANFORD	− 15	−380
12				,, 1919	SANFORD	− 85	+235
13				,, 1919	SANFORD	−220	−275
14				,, 1919	RITCHIE	+290	−180
15				,, 1919	DUNCAN	−160	+170
16				,, 1919	SHAPLEY	−200	−190
17				,, 1919	HUMASON	+ 15	+150
18				,, 1920	DUNCAN	− 50	+ 50
19				,, 1920	DUNCAN	− 10	−180
20				,, 1920	DUNCAN	+140	+100
21				,, 1922	HUMASON	+250	+168
22				,, 1923	HUMASON	+ 67	+ 71
—	—	— — —	—	—	—	—	—
—	—	— — —	—	—	—	—	—
67	598	1 29 36	+30 16,4	N Piscium 1919	HUBBLE	+110	+ 12
68				N Piscium 1925	HUBBLE	+360	+540
69	2403	7 29 35	+65 45,8	N Camelopardalis 1910	RITCHEY	—	—
70	2355	8 6 43	+25 25,8	N Cancri 1901	REINMUTH	+ 21	+ 7
71	2608	8 30 42	+28 43,3	N Cancri 1920	WOLF	{ − 20 − 23	+ 2[1] + 8
72	2841	9 16 51	+51 17,7	N Ursae Majoris 1912	PEASE	− 50	+ 20
73	3147	10 10 30	+73 46,5	N Draconis 1904	MRS. ISAAC ROBERTS	−1140	+118
74	3372	10 42 8	−59 17,1	N Carinae 1843 (η)	HALLEY etc.	+ 19	+ 18
75	4303	12 18 5	+ 4 53,3	N Virginis 1926	WOLF and REINMUTH	−450	+ 6
76	4321	12 19 8	+16 14,3	N Leonis 1901	CURTIS	−110	+ 4
77				N Leonis 1914	CURTIS	+ 24	−111
78	4424	12 23 21	+ 9 50,0	N Virginis 1895 (VW)	WOLF	+105	0
79	4486	12 27 1	+12 48,3	N Virginis 1919	BALANOWSKY	− 15	+100
80				N Virginis 1922	WOLF	0	+120
81	4527	12 30 19	+ 3 4,0	N Virginis 1915	CURTIS	+ 44	+ 8
82	5236	13 32 47	−29 29,1	N Centauri 1923	LAMPLAND	+109	+ 58
83	5253	13 35 40	−31 15,6	N Centauri 1895 (Z)	FLEMING	+ 19	+ 23
84	6093	16 12 34	−22 47,6	N Scorpii 1860 (T)	AUWERS	+ 9	+ 6
85	6946	20 33 24	+59 53,1	N Cephei 1917	RITCHEY	− 37	−105

Date	Novae
1923−4	11
1924−5	13
1925−6	15
1926−7	5

N Andromedae 23 to N Andromedae 66 have been found by HUBBLE at MT. WILSON in the observing seasons of the nebula here given:

Of the whole 66 novae observed in the great Andromeda nebula 34 have appeared in the central region of

[1] Two nuclei in this nebula.

unresolved nebulosity and 32 in the extreme outer arms[1]. HUBBLE states that the high concentration in the central region is real (Fig. 1).

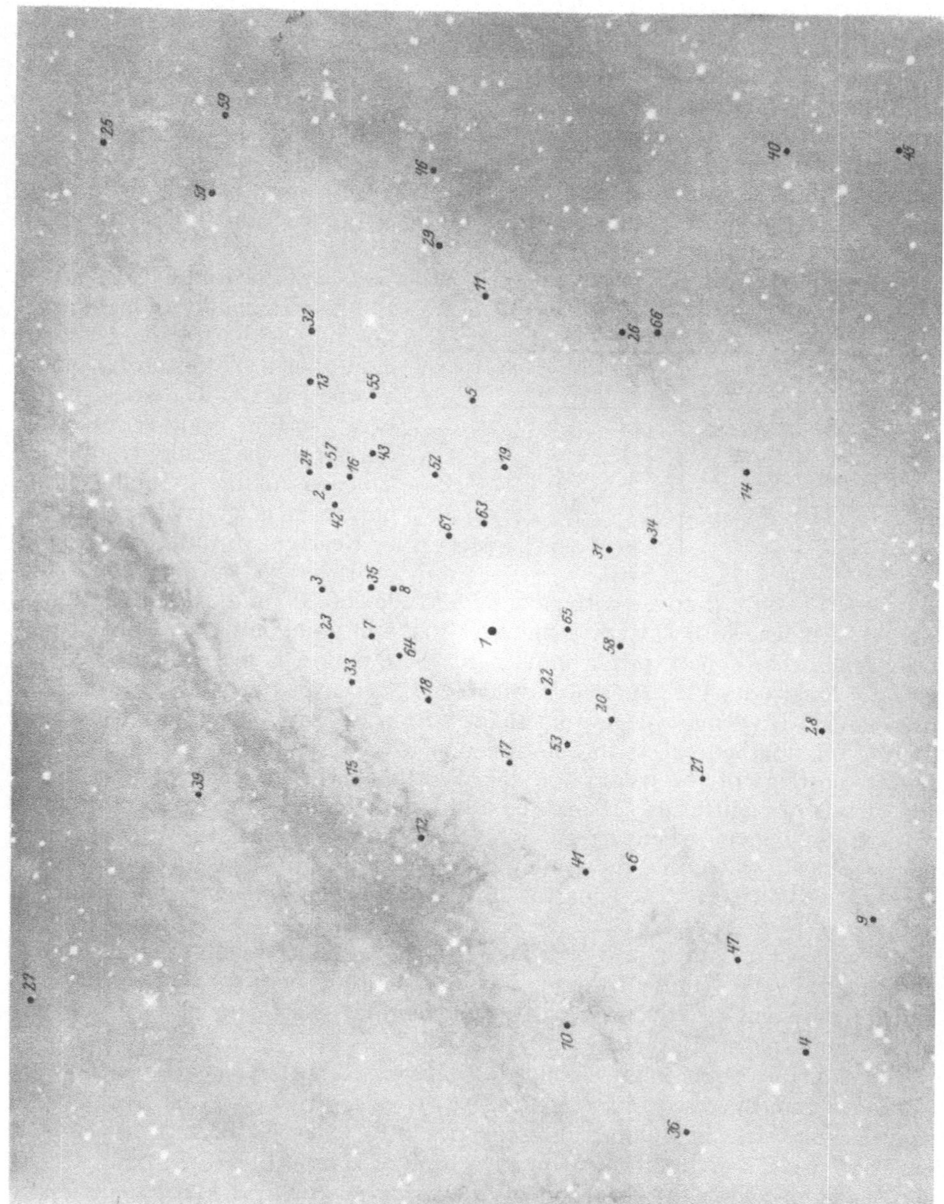

Fig. 1. Novae in central portion of the Andromeda nebula (HUBBLE, Mt. Wilson).

6. Galactic Concentration. The most striking point in the distribution of the novae, which are not intimately connected with nebulae, is the strong concentration in the plane of the galaxy. The 70 novae in Table I have the following distribution in galactic latitude:

[1] Ap J 63, p. 258 (1926) and from information privately supplied by HUBBLE.

Galactic latitude	Number of novae	Galactic latitude	Number of novae	Galactic latitude	Number of novae
0°—10°	39	30°—40°	3	60°—70°	3
10 —20	17	40 —50	1	70 —80	1
20 —30	4	50 —60	2	80 —90	0

Of the fourteen novae with galactic latitude over 20° information as to the spectrum is only available in three cases—N Piscium 1907, N Pictoris 1925 (RR) and N Coronae 1866 (T). Eight of them date from over 50 years ago when opportunity of following their development was much less than at present; material is in consequence lacking for drawing any conclusions as to differences between these novae and the more common galactic type. All that can be said is that the two of which most is known, T Coronae and RR Pictoris, are both of rather unusual types.

A preponderance of novae in low southern galactic latitudes (26 between 0° and −10° as against 13 between 0° and +10°) is presumably to be ascribed to the position of the solar system, north of the median plane of the galaxy. The effect of the solar system being eccentrically placed in the galaxy is also conspicuous in the grouping of the novae in galactic longitude.

Galactic longitude	Number of novae	Galactic longitude	Number of novae
0°— 30°	11	180°—210°	2
30 — 60	6	210 —240	4
60 — 90	4	240 —270	3
90 —120	3	270 —300	3
120 —150	4	300 —330	9
150 —180	4	330 —360	17

Nearly two-thirds of the novae lie in galactic longitudes 300°—360°—60°; if novae with galactic latitudes of 20° and over are excluded, we get 37 novae with galactic longitudes 300°—60° and only 19 with longitudes 60°—300°. The concentration implied by these figures is most intense between galactic longitudes 320° and 340°, where one quarter of the galactic novae have been found. If we may judge from analogy from the Andromeda nebula we may ascribe this marked excess in one direction to a concentration of novae in the central portions of the galaxy—in the direction from the solar system to the star-clouds of Sagittarius. It may be added that there is a tendency for novae to occur in regions where irregular nebulae, bright or dark, are present, and not necessarily where the star density is at maximum. In the Andromeda nebula there is a similar tendency for the novae to appear in the neighbourhood of dark lanes (Fig. 2).

The general data of the distribution of novae indicate that spiral nebulae and the Milky Way present the most favourable conditions for their occurrence, and are consistent with the view that the nebular and galactic novae are essentially of the same nature.

One further point ought perhaps to be discussed in this connection. How are we to compare the discovery in recent years of over 60 novae in the great Andromeda nebula with the number of discoveries in the galaxy? BAILEY[1], as a result of a systematic search for novae on 1049 pairs of Harvard plates, estimated that in a series of bi-weekly photographs of the type secured at Harvard as many as 9 novae would be found in each year and that the number of apparent magnitude 9 or brighter could be estimated at 25 a year and the number of magnitude 6 or brighter at one or two a year. HUBBLE estimates the number of novae in the Andromeda nebula at 30 a year and he adds that the 65 novae actually discovered in the Andromeda nebula in the past 9 years have a maximum frequency about apparent photographic magnitude $17^{m},2$. Assuming that half

[1] Publ A S P 24, p. 554 (1921).

these stars are brighter than magnitude 17,2 and that the Andromeda nebula is 950000 light years away we find that this frequency is equivalent to one new star a year of apparent magnitude brighter than 6 at a distance of 10000 light years. To allow for the fact that all the novae in the nebula are unlikely to

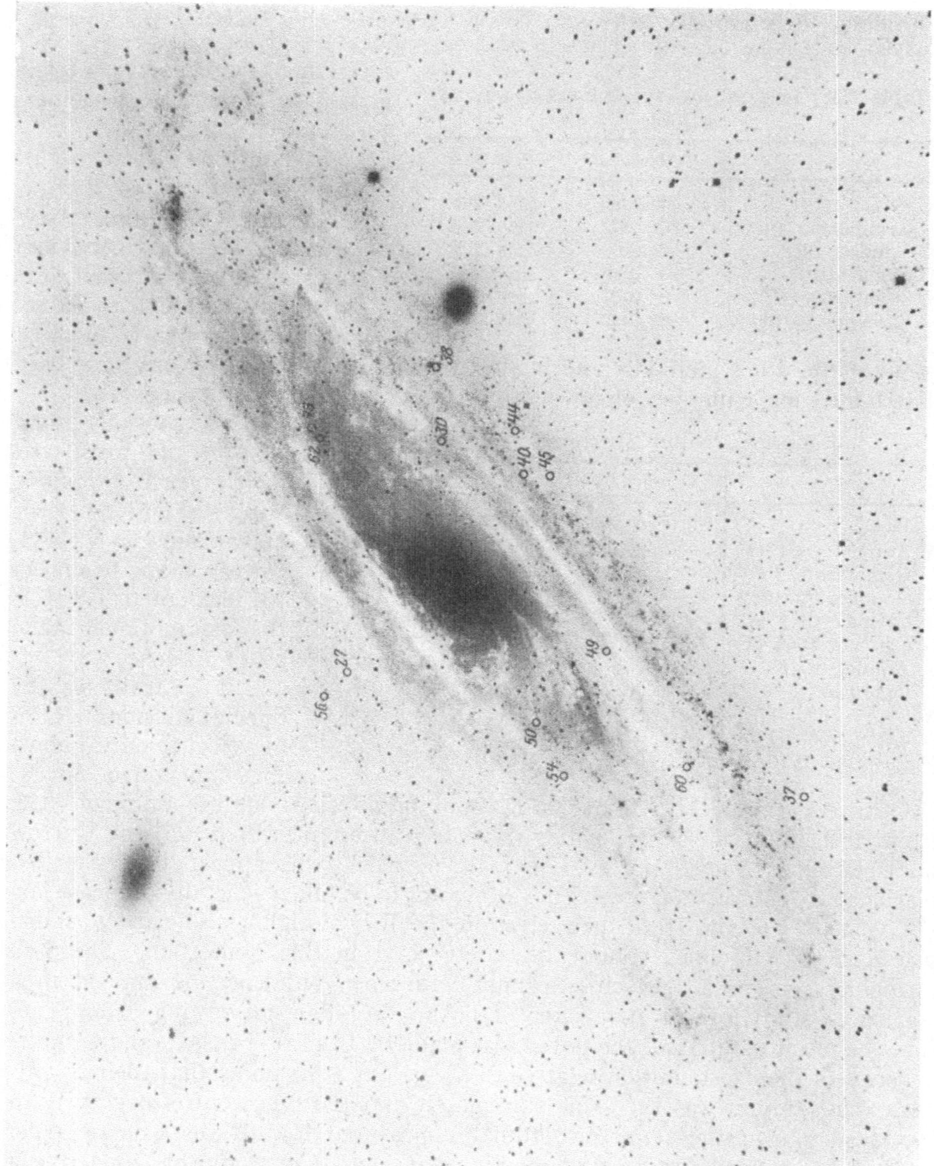

Fig. 2. Novae in the outer portion of the Andromeda nebula (Hubble, Mt. Wilson)

have been caught and in many of them the maximum may have been missed in the photographs examined, this number must be multiplied by some factor which at present can only be guessed. But Shapley's estimate of 50000 light years for the distance of the centre of the galaxy, where the great concentration of the galactic novae is to be found indicates that the frequency of the phenomenon in the Andromeda nebula and in the galaxy is of the same order. More than that can hardly be stated.

c) Parallaxes and Proper Motions of Novae.

7. Trigonometrical Parallaxes. Trigonometrical parallaxes have been determined for a number of novae. Six of these figure in SCHLESINGER's General Catalogue of Parallaxes and his values are given below:

Table III. Trigonometrical Parallaxes of Novae.

	$\pi_{abs.}$	Prob. error
N Persei 1901	+0″,011	±0″,003
N Geminorum 1912	+0 ,002	±0 ,008
N Aquilae 1918	−0 ,003	±0 ,004
N Cygni 1920	+0 ,026	±0 ,008
N Cygni 1600 (P)	−0 ,021	±0 ,009
N Lacertae 1910	+0 ,012	±0 ,014

For one other nova a trigonometrical parallax has been measured by VAN MAANEN:

	$\pi_{rel.}$	
N Ophiuchi 1919	−0″,005	±0″,009

8. Proper Motions. Assuming that in general behaviour novae are not very different from other stars it is possible to deduce parallaxes of groups of them by some of the other methods used in stellar statistics. Proper motions have been determined for a number of novae; LUNDMARK[1] gives the following table:

Table IV. Proper Motions of Novae.

	μ_α	μ_δ	μ
N Persei 1901	−0″,009	−0″,015	0″,016
N Aurigae 1891 (T)	−0 ,002	−0 ,003	0 ,004
N Geminorum 1903	−0 ,007	−0 ,003	0 ,007
N Geminorum 1912	−0 ,014	+0 ,017	0 ,021
N Carinae 1843 (η)	+0 ,007	+0 ,002	0 ,004
N Coronae 1866 (T)	+0 ,008	+0 ,011	0 ,013
N Aquilae 1918	−0 ,005	−0 ,018	0 ,019
N Ophiuchi 1848	+0 ,008	−0 ,016	0 ,018
N Cygni 1876 (Q)	+0 ,004	+0 ,003	0 ,004
N Cygni 1600 (P)	−0 ,015	−0 ,011	0 ,016
N Lacertae 1910	0 ,000	+0 ,001	0 ,001

LUNDMARK's solution for the parallactic drift given by applying AIRY's equations to the above stars is 0″,0051, corresponding to a mean parallax of 0″,0012, if the solar motion is taken as 17,7 km/sec.

9. Parallaxes by Special Methods. Radial velocities are known for only three novae: N Carinae 1843 and N Aurigae 1897 from bright lines, and N Coronae 1866, from absorption lines. Velocities have also been measured for the narrow undisplaced *H* and *K* lines. For reasons which will appear later the displacements of absorption lines in novae are not to be taken generally as measuring the velocities of the stars themselves in the line of sight; also narrow undisplaced *H* and *K* lines cannot be safely used in this connection. The data available from radial velocities cannot with any confidence be applied in a statistical study of the parallaxes of novae. Another interesting attempt to find the mean parallax of the novae was made by LUPLAU-JANSSEN and HAARH[2] in terms of their distribution relative to the galaxy. Assuming that their brightness at maximum was the same throughout and that the centre of gravity of the system had the same Z-coordinate (measured toward the galactic pole) as that determined for the B-stars by CHARLIER, they were able to derive a scale for the distribution of the nova system. A second closely allied method was to consider the dispersion of the novae in the Z-coordinate relative to the galaxy and to compare it again with CHARLIER's figures for the B-stars. The mean of the two solutions gave for the parallax of N Persei = 0″,048, a value considerably larger than that given by the trigonometrical method; probably

[1] Publ A S P 35, p. 102 (1923). [2] A N 211, p. 89 (1920).

the proper deduction from this is that the distribution of novae and B-stars, though both are galactic objects, is different.

Other solutions due to LUNDMARK are based upon the well known concentration in the Sagittarius region and the agreement of the small negative value for the mean galactic latitude of the novae with that of the Milky Way star clouds. Either of these implies that novae follow the Milky Way structure and leads to views as to the mean parallax of novae and their absolute brightness which must vary with changing views as to the dimensions of the galaxy and the position of the solar system in the galaxy. LUNDMARK's solution gives from the galactic latitudes a mean distance of 11000 light years corresponding to a parallax of 0″,0003 and a mean absolute magnitude of −6,5. For the stars in the Sagittarius region he accepts SHAPLEY's distance of 68000 light years and deduces a mean absolute magnitude of −9,5.

We may now consider a few cases of the determination of parallax of individual stars by special methods. For N Coronae 1866, now apparently a giant star of type Mb and of magnitude 9,6, we get, from the mean absolute magnitude of M giant stars, a parallax of 0″,0014. From the proper motion given in Table IV above, 0″,013, and the magnitude of the star, VAN RHIJN's relationship between parallax, spectrum and proper motion (Ap J 43, p. 36) would suggest a parallax of approximately double the above value, or 0″,003. LUNDMARK taking into account the giant nature of the star has derived a value 0″,0016. For N Persei 1901, nebulosity was photographed surrounding the star eight months after the star reached maximum brightness. The nebulosity was found to be moving outwards at an incredibly high rate for moving matter and a study of the spectrum of its light confirmed the supposition that what was being observed was a spreading out of radiation from the central disturbance, which was being reflected from surrounding matter. W. E. WILSON shewed[1] that this theory was feasible in terms of the amount of light required and available to illuminate the nebulosity, and several writers (KAPTEYN, TURNER, etc.) determined the parallax of the nova from the rate of spread of the illumination. There are necessarily some uncertain factors, such as foreshortening, but the mean result arrived at, 0″,010, is in good agreement with the trigonometrical parallax, 0″,011.

In the case of N Aquilae 1918 a hypothetical parallax has been deduced from the spread of the measurable disc into which the nova has developed and the velocity with which matter is flowing out from the centre of disturbance in the line of sight, as evidenced by displaced elements of the bright spectral bands due to hydrogen and nebulium. There are difficulties of interpretation due to the fact that while hydrogen and nebulium show the same radial velocities the discs given in a slitless spectrogram for $H\beta$ and the principal nebular line are growing at different rates. Again as the spectral bands are complex there is an element of uncertainty as to which of the alternative DOPPLER velocities that are available should be selected to correlate with the growth of the disc. The two published estimates 0″,003, 0″,004 are both possible and show that the suggestion that the two phenomena are connected is quite reasonable. For N Aquilae 1919, spectrum R0, LUNDMARK has given a spectroscopic parallax of 0″,0003; for N Scorpii 1860, which appeared in the globular cluster NGC 6093, we have a parallax by SHAPLEY of 0″,00005[2] and for N Carinae 1843, assumed involved in the surrounding nebulosity, we have a hypothetical parallax by LUNDMARK of 0″,0015[3]. Lastly for the novae in the Andromeda nebula we have a parallax given by HUBBLE of 0″,000003. This contrasts markedly with the

[1] Nature 65, p. 298 (1902). [2] Ap J 48, p. 161 (1918).
[3] M N 85, p. 880 (1925).

value in SCHLESINGER's table, also derived from Mt. Wilson, of 0″,005 ± 0″,008 but it helps to emphasize the view that the parallaxes given above must not be regarded with too great confidence, nor must too much be built upon them.

DAVIDOVICH calculated the absolute magnitude of N Pictoris 1925 by the spectroscopic method, comparing the star before maximum with ι_1 Scorpii. He obtained a value for M of -5^m at maximum corresponding to a parallax of 0″,006. The validity of the method in the case of a nova is open to question but the result is quite reasonable.

d) Light Curves and Absolute Magnitudes.

10. Absolute Magnitudes. Taking the parallaxes given in the above section and the apparent magnitudes of the stars at maximum we get the following table for the corresponding absolute magnitudes.

Table V. Magnitudes at Maximum.

		π	m_{max}	M_{max}
	N Andromedae 1885	0″,000003	7,2	−15,4
	N Persei 1901	0 ,011	0,0	− 4,8
	N Pictoris 1925	0 ,006	0,15	− 5
	N Geminorum 1912	0 ,002	3,7	− 4,8
	N Carinae 1843	0 ,0015	− 1,0	−10,1
	N Scorpii 1863	0 ,00005	7,0	− 7,5
	N Serpentis 1918	0 ,003	no recorded maximum	
	N Aquilae 1918	0 ,003	− 1,0	− 8,6
	N Aquilae 1919	0 ,0003	10,4	− 2,2
	N Lacertae 1910	0 ,012	4,5	− 0,1
	N Cygni 1920	0 ,026	1,6	− 1,2
Mean parallax	Other Novae in NGC 224	0 ,000003	17,0	− 5,2
Mean parallax	Novae in Sagittarius region	0 ,00005	8,2	− 8,3

The range of M_{max} (15 magnitudes for 10 objects) does not support the view of some writers that we can assume a constant magnitude at maximum for all novae. Different spectral development and light curves of varied shapes indicating a wide range in the forces at work in different cases join in the warning against a too ready unifying of diverse data.

One comment should, however, be made. The novae in the Andromeda nebula show a strong concentration about an absolute magnitude at maximum of $-5^m,2$. (There is an exceptional case in N Andromedae 1885, which is clearly separate from the main body of these novae.) The mean value of the absolute magnitude for the other novae of reasonably well known parallax is $-4^m,9$, indicating that the phenomena in the nebulae and in the galaxy are comparable. LUNDMARK from additional statistical considerations has obtained as a mean value for the absolute magnitude of the novae at maximum $-7^m,1$[1].

11. Light Curves around Maximum. We now come to the study of the light curves themselves (Fig. 3) and we commence with some details of the rise in brightness to the principal maximum. Save for a certain number of novae which were known as faint stars before their outburst we are mainly dependent upon sporadic photographic information for the early portions of the light curves. Naturally we have the fullest information in the case of the stars whose rise in brightness was more gradual. The first star of whose history previous to its great brightening we have some knowledge is T Coronae.

[1] Publ A S P 35, p. 106 (1923).

N Coronae 1866 (T) was observed as a star of magnitude 9,5 in the Bonner Durchmusterung and seems to have escaped notice as a variable star[1] until on the evening of May 12, 1866, it was observed at Tuam by BIRMINGHAM as

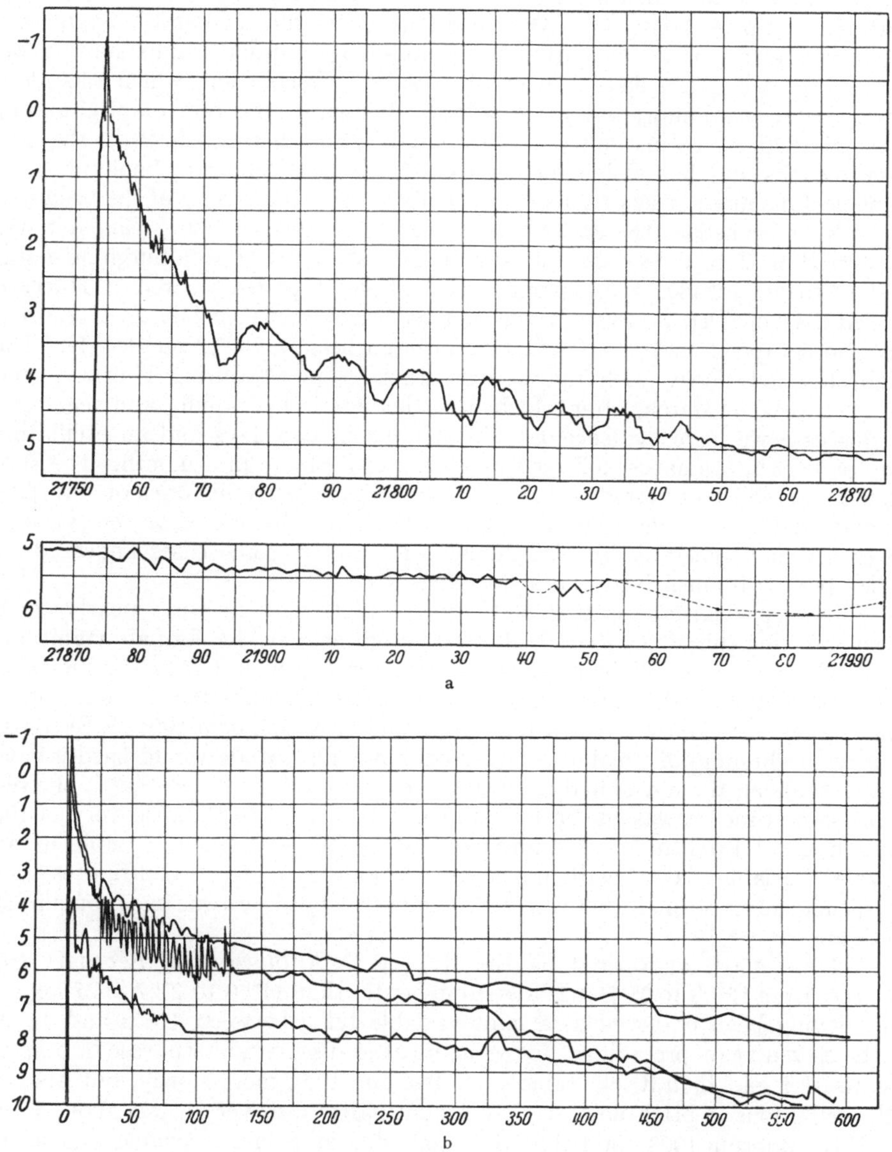

Fig. 3. a) Light curve of N Aquilae 1918. b) Comparison of light curves of N Aquilae 1918, N Persei 1901 and N Geminorum 1912. (L. CAMPBELL, Harvard Annals, 81 [1923].)

being "very brilliant, of about the second magnitude". All we know of its rise is that SCHMIDT observing at Athens on the same evening did not notice the star and was convinced that it could not have been as bright as $4^m,0$ two hours before BIRMINGHAM discovered it. Several astronomers working on May 11

[1] An observation by HERSCHEL on June 9, 1842, of a star of the 6th magnitude in the same neighbourhood cannot with any certainty be attributed to this star.

and earlier days expressed the belief that it could not have been conspicuous to the naked eye on those dates. Two astronomers discovered it independently on May 13, one on May 14 and one more on May 15. It seems likely that the rise was very rapid, much more so than the fall which was for the first ten days at the rate of $0^m,5$ a day. Of N Ophiuchi 1848, N Scorpii 1860 and N Cygni 1876 we have evidence that the outburst was rapid, no unusual bright star having been recognized a few days previous to the first observation, which was at or shortly after maximum; but no details are known of the rate of brightening. N Cygni 1876, 3^m at maximum in November 1876, fell through 4^m in the first three months and had fallen further to $10^m,5$ by the following September. A number of disputed observations throw doubt on the final stages of the brightening of N Andromedae 1885 and we come next to N Aurigae 1891. This star was discovered on Feb. 1, 1892 by the Rev. T. D. ANDERSON at Edinburgh as a star of the 5th magnitude. Subsequent examination of plates secured at Harvard shewed that on Dec. 8, 1891, it was not as bright as $13^m,2$ but on Dec. 10 it was photographed as $5^m,4$. Its maximum brightness was $4^m,2$ on Dec. 17. The brightness fell slowly with fluctuations from 5^m to 6^m during February and the first week in March, after which date the star faded rapidly and regularly. In this case we have evidence of a rise of $7^m,8$ in two days and an oscillating drop of $1^m,5$ in 3 months, followed by a drop of 8^m more in a month. The subsequent rise will be referred to below. Details of the early histories of stars like N Persei 1887, N Normae 1893, N Ophiuchi 1897 etc., discovered on the Harvard plates, are not generally known, though the stars in some cases appeared on earlier plates, frequently as faint variables.

An exceptional case is that of N Pyxidis 1890, with its three nearly equal maxima during thirty years, each about 6^m brighter than the normal magnitude, about which it varies slightly. Spectroscopic reasons have led to the inclusion of this star among the novae rather than to its classification as some other type of irregular variable. The same can be said of N Ophiuchi 1901 (R S) whose solitary brightening of 2^m above its normal mean magnitude would hardly have placed it among the novae had it not been for the nova spectrum that it shewed at the same time; it was the spectrum in fact which led to the discovery of the star's unusual brightness. N Andromedae 1901 (Z) is another example of an irregular variable classed with novae on account of its spectral affinities. Its magnitude ranges normally between $10^m,0$ and $11^m,5$ but there was one pronounced maximum at $9^m,2$.

N Persei 1901, discovered by the Rev. T. D. ANDERSON on February 21, 1901 rose from $12^m,8$ to $2^m,7$ in 2 days (from fainter than $11^m,0$ to $2^m,7$ in 27 hours) after having been observed as a faint variable lying between $11^m,0$ and $14^m,0$ for the eleven years preceding. For 38 hours after discovery it still rose in brightness to $0^m,1$ and then it fell, rapidly at first and then more slowly, but always with oscillations in brightness, so that a month after discovery it was about $4^m,0$.

N Geminorum 1903, probably the same star as a faint variable known as ranging between $12^m,0$ and $14^m,4$ over a period of 15 years is shown by the Harvard photographers to have risen from $12^m,0$ to $4^m,97$ in 5 days, and then it dropped 3^m in the following three weeks, at the end of which time it was discovered by TURNER on an Oxford plate. In the case of N Aquilae 1905 the rise from the earlier value about $13^m,0$ to the maximum $8^m,0$ took 26 days; after half that interval of time again the nova settled down to a new value about which it oscillated slightly for some time, namely $10^m,5$. N Circini 1906 and N Scorpii 1906 had both a somewhat different history to any of the foregoing: the former took 6 months, the later 10 months to rise through 4^m to its solitary pronounced maximum and then fell back to its initial mean brightness at a slightly slower rate: in each case before and after the chief outburst the star was a faint irregular variable with no great range of magnitude. N Sagittarii 1910 in a similar way took 8 months to rise 5^m and 17 months to fall, with many marked oscillations through the same range of magnitude. N Lacertae 1910 after brightening slowly through 2^m suddenly brightened up through 6^m in 5 days and fell back nearly to its original brightness in about a year. Like N Geminorum 1903 this star was not caught until past maximum, in this case more than a month past, and the early history is again based on the Harvard plates.

In the case of N Geminorum 1912 we return to the usual behaviour of the brighter novae, a rise from 12^m to $4^m,0$ in a day, a slight brightening to a maximum of $3^m,7$ during the following three days and then a fall with oscillations in brightness at an average rate of 1^m in 10 days. In this case Enebo discovered it 2 days before maximum but after its chief rapid rise had been completed. N Sagittae 1913 after rising through $2^m,5$ in 21 days rose further through $2^m,8$ in 1 day, remained at maximum for a couple of days, fell through $0^m,9$ in 3 days, 2^m further in 10 days and 4^m further in 6 months. Of N Sagittarii 1914,6 (Woods) we only know that it rose more than 3^m in 4 days and fell 4^m again in 2 months, while N Sagittarii 1926 rose 5^m inside a fortnight to maximum $8^m,6$, and fell 5^m again in 2 months.

N Aquilae 1918, known as a variable star lying between 10^m and 11^m for some thirty years, was photographed at $10^m,5$ on June 5, 1918. A photograph taken on June 7 at Harvard showed its brightness to be $6^m,6$; the following day—in fact 9^h10^m later—it was observed in India by Bower to be a first magnitude star, and was during the following hours independently discovered by hundreds of observers. It rose on June 9th—after a slight halt—to a magnitude of $-1^m,08$, rising through 1^m in about $0^d,5$: it then fell through 1^m in about 8 hours and then dropped with fluctuations through 2^m in 6 days, 2^m more in about 10 days after which it reached a periodically oscillating stage to be discussed below. The chief point of interest in the light curve of N Ophiuchi 1919 is its flat maximum which with fluctuations covers a period of two months: it might be best perhaps to describe it as having two maxima, at $7^m,5$ on September 13 and at $7^m,2$ on November 5, with a 2^m drop in between.

Of N Normae 1920 we only know that it rose through more than 3^m to its maximum in 3 days, but more is known of N Cygni 1920, which is one of the few novae discovered before maximum, by Denning. Rising at the rate of 1^m a day for the last five days before maximum ($2^m,2$) it fell at first at the rate of 2^m in 6 days, followed by a slower but steady drop with only minor fluctuations at the rate of 1^m in 9 days. More interesting still is N Pictoris 1925, a nova which rose only gradually to maximum and was discovered on May 25, 1925, 15 days before reaching its final maximum. A faint uncatalogued star of magnitude 12,7 its rise in brightness was missed, save for one photograph 42 days before discovery, when it had a magnitude 3,0; at discovery by Watson it was a naked eye object of magnitude $2^m,4$ and it reached a maximum of $1^m,7$ on the following day. Two days later it had dropped to $2^m,37$ rising to its final maximum, $1^m,15$, 15 days after discovery. Then with many oscillations in brightness it dropped 2^m in 3 months and another $1^m,5$ in the following three months.

We can summarize the preceding accounts of the behaviour of novae round about their maxima as follows. The final rise to maximum is more rapid than the first drop after maximum and the difference is more marked the more rapid the rise. In fact the ratio of the rate of change of magnitude before to that after maximum may change from 50 or more, for such stars as N Persei 1901, N Geminorum 1912, down to 10 to 15 for such stars as N Aurigae 1891, N Sagittae 1913, N Sagittarii 1914, N Aquilae 1918 with smaller values for N Geminorum 1903, N Cygni 1920 and values approaching unity for such novae as N Aquilae 1905, N Sagittarii 1910, N Pictoris 1925. The mean value for the number of magnitudes that a star rises during a day just before maximum for the four groups above is 7,6, 3,0, 1,2 and 0,16 respectively. It seems then fair to say that the more violent the outburst the greater the contrast just before and just after maximum. It is probably not without significance that the absolute magnitude at maximum of the stars in the above groups which figure in Table V are $-4^m,8$ for N Persei 1901 and for N Geminorum 1912 in group 1 and $-1^m,2$ for N Cygni 1920 in group 3. N Aquilae 1918, however, whose parallax is much less reliable, probably had a higher absolute magnitude (−8,6) at maximum than either of the stars in group 1 so that not too much stress must be laid on the relationship between absolute magnitude at maximum and the relative rates of rise and fall.

12. Later Stages of Light Curves. As we get further away from the maximum the contrast between the two sides of the magnitude curve gets less striking; also the rate of change is less. There is not much evidence available for the rising part of the curve except just before maximum but what evidence there is points to the rise being always faster than the corresponding drop in brightness on the other side of the maximum. The disturbance develops more rapidly than it settles down.

A marked feature of the settling down process is frequently a distinct set of oscillations in brightness almost periodic in nature. This is very different from the slight drop noticed occasionally on the rising branch of the curve. It is accompanied by spectroscopic changes which will be discussed later. The periodic changes have been studied in detail for three stars. In the case of N Persei 1901 these oscillations set in about a month after maximum, had a period of about 4 days (ranging from 2 to 5 days) and lasted for about 3 months. The range of magnitude was about 2^m. In N Geminorum 1912 the period was about 3,5 days, appearing very shortly after maximum and being followed for about a month; at the end of that time the period shewed signs of a slight increase in length. The range in magnitude dropped from about $1^m,5$ at the start to about $0^m,5$ at a later stage. The average interval of time from minimum to maximum was 1,8 days, the average from maximum to minimum was 2,3 days—the difference may have been partly due to the effect of the falling brightness on the periodic oscillation. In the case of N Aquilae 1918 there were two periods. For the first ten days after maximum periodic fluctuations with a period of about one day were superposed on the steeply dropping light curve. After that a marked period of 11,2 days set in, clearly marked for about three months with the range of magnitude falling from about $0^m,8$ to $0^m,2$. The fluctuations then became irregular and the apparent period rather shorter becoming about 7 days in 1919. During the oscillating stage the interval from minimum to maximum was on the average 2 days shorter than the interval from maximum to minimum. In the case of N Pictoris 1925 well marked oscillations in brightness occurred and the light curve suggests a combination of several terms of different periods. The most marked period after discovery was one of about 8 days and a range of a little more than 1^m. After 2 months the period lengthened to about 13 days with evidence of a shorter period of about 5 days superposed. BARABASCHEFF has pointed[1] out that this increase of period with time is a

[1] A N 222, p. 389 (1924).

common feature with novae and he has coupled it with a progressive change in the colour index of a nova and a pulsation accompanying the growth of its diameter. The evidence as to periods, however, is very far from being clear.

Before passing on to the discussion of the later stages of the light curves of novae reference must be made to the special case of N Aurigae 1891. As already pointed out this star remained for the unusually long time of three months at about 5^m (near its maximum brightness $4^m,2$) and then fell through 8^m in a month. Five months later when the nova came into view again after conjunction with the sun it had risen through 2^m to $11^m,5$, at which brightness it remained very fairly constant for three years. Its spectrum had also changed in the interval to that of a planetary nebula. After three years it commenced to fade slowly to its present magnitude between 14^m and 15^m.

13. Persistent Variability. Observations of novae years after their discovery have been made by Barnard and Steavenson[1]. They show that with few exceptions—N Lacertae 1910 is almost the only exception—the brighter novae remain variable stars, returning steadily, if somewhat haltingly to magnitudes but little different to those that they had before the outburst. The more recent novae are in most cases still fading, Barnard estimating the average time taken by a nova to return to its normal condition (prior to the outburst) at about 12 years. Right throughout the period of decline and even after reaching a fairly settled condition the fluctuations which occur are irregular. As an illustration of the difficulty of fixing periods from the observations of magnitude we may note that periods of 10, 15, 30, 55, 363 and 396 days have been suggested for N Persei 1901 and periods of 27, 40 and 50 days for N Ophiuchi 1848. The last named star now varies between $12^m,1$ and $12^m,7$. N Coronae 1866, before discovery $9^m,5$, was varying between $9^m,8$ and $10^m,2$ in the last recorded observation. N Cygni 1876 has settled at about $14^m,9$ having fallen to 14^m in the first 6 years, N Aurigae 1891 lies between 14^m and 15^m. N Persei 1901 has had in some ways the most interesting late career of all the novae. In addition to the nebulosity discovered a few months after the great outburst to be surrounding it a much smaller ring of nebulosity spreading out from the stellar centre was discovered by Barnard in 1916. The star's magnitude now varying between 13^m and 14^m is the same as before the outburst. N Geminorum 1903, now fainter than 16^m is one of the few stars which has ended up definitely fainter than it was for many years before its outburst. N Lacertae 1910 now steady at $14^m,1$ is notable for being no longer recognisably variable and for the speed at which it fell back to its original brightness, which it reached finally about three years after maximum. N Geminorum 1912 has fallen now to about 14^m and N Aquilae 1918 by the year 1925 reached its earlier magnitude $10^m,4$—$10^m,6$. N Cygni 1920 is still falling in brightness at the rate of about 1^m a year, and in 1925 it reached $13^m,28$. Before 1920 it was fainter than 15^m.

Table VI. Range of Variation.

	M_{max}	Range
N Persei 1901	$-4^m,8$	14^m
N Geminorum 1912	$-$ 4,8	10
N Cygni 1920	$-$ 1,2	13
N Lacertae 1910	$-$ 0,1	9,5
N Carinae 1843	$-$10,1	9

Lundmark has made an interesting suggestion that the range of magnitude between maximum and the undisturbed (or final) magnitude increases with the maximum apparent magnitude. Applying this to the stars with known absolute magnitudes at maximum (see Table V) we get the figures given in Table VI.

[1] M N 80 and later volumes.

No real physical connection is indicated by the above figures and it would look as though the connection traced by LUNDMARK is accidental.

14. Light Curves of Novae in Spirals. Figure 3 gives the light curves of N Persei 1901, N Geminorum 1912 and N Aquilae 1918 and illustrates the general resemblance that the light curves of novae bear to each other. It may be added that the light curves of the novae in spirals though naturally not followed so closely as those of the brighter novae share the general features of the very rapid rise and the somewhat less rapid fall, with oscillations or checks. Figure 4 gives the light curve of N Andromedae 1924,7 (No. 34), as determined by HUBBLE—the best determined light curve as yet of a nova in a spiral nebula. In general outline it resembles the typical light curves of the galactic novae.

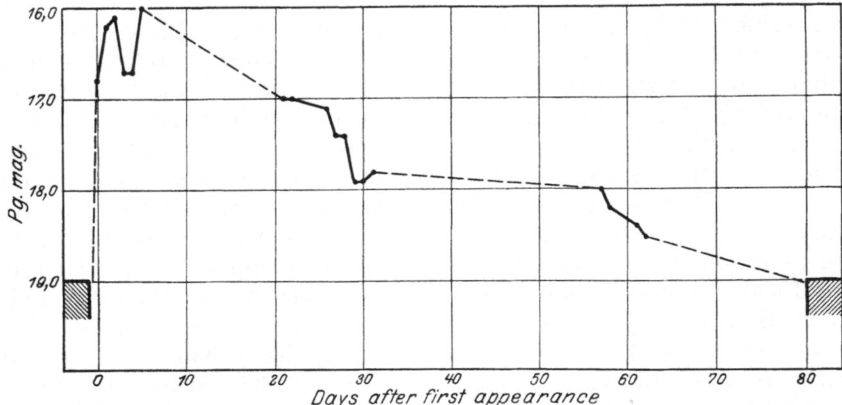

Fig. 4. Light Curve of Nova No. 34 in Andromeda nebula (HUBBLE, Mt. Wilson).

15. Abnormal Light Curves. A few of the abnormal light curves must now be discussed. Like N Aurigae 1891 we find that N Coronae 1866 after its first rapid drop from 2^m to $9^m,5$ rose about 4 months after maximum and remained steadily brighter than 8^m for over a month before once again fading. This is a fluctuation of a type common at later stages of some novae but so pronounced a fluctuation at a fairly early stage only occurs in these two stars. In N Pyxidis 1890, with its three sharp maxima, $7^m,5$, at irregular intervals of 12 to 18 years and with fairly constant brightness, $13^m,5$, in between, in N Ophiuchi 1901 with its one sharp maximum at 9^m, and its irregular variations about $11^m,5$, we have two stars which are only classed with the novae because of the presence of a typical nova spectrum at maximum. N Vulpeculae 1670 shares with N Pyxidis 1890 the unusual feature of the treble maximum though with much shorter intervals between the maxima. N Carinae 1843 has, with much longer intervals still, had several marked maxima and its spectrum affiliates it to so definite a nova as N Pictoris 1925.

The other unusual type of light curve is best exemplified by some of the novae in the Andromeda nebula. N Andromedae 1921 No. 26 retained a constant brightness for four years after its first appearance before it began to fade away. While novae Nos. 36 and 39 which appeared between October 1915 and September 1917 and in February 1924 respectively have both maintained a constant brightness ever since discovery. In this they resemble RT Serpentis 1918 which appeared in 1909 first at $13^m,9$ and had risen to $9^m,2$ in 1918 now remaining a variable about 10^m. Its spectrum A8eβ does not place it in the ordinary group of novae and its absolute magnitude, $+2^m,9$, also suggests that here we are dealing with a different class of phenomenon. (In the chapter on variable stars, it is considered as belonging to the R Coronae-class of variables. Unless we regard such stars as having emerged from behind a dark cloud, it would seem that they are more entitled to the title "new star" than the main class of objects to which the title is generally applied.)

e) Spectrophotometry and Temperature.

16. Spectrophotometry of the Continuous Spectrum. The spectrophotometry of novae is complicated by the fact that in the course of development of a nova there is in general a change from a strong continuous spectrum crossed by narrow absorption lines to a weak continuous spectrum crossed by broad bright bands, varying in relative strength. No method that does not take account of this change can give results of value as to the temperature and physical changes which accompany the development of the star. The most complete investigation of the subject made so far has been by BRILL with the star N Geminorum 1912[1]. The chief results of interest which he obtains for the six weeks after the star's maximum are the following:

(1) The maximum intensity of the continuous spectrum shifted towards the violet as the star faded—this involves a much more rapid fall in intensity in the red than in the violet. Compared in the distribution of its continuous radiation with the A-type star, B D $+32°1433$, the nova showed strong maxima in the red-yellow and the ultra-violet with a very pronounced minimum in the blue. These irregularities give cause for hesitation in taking the above mentioned shift to be due to a mere temperature change, to be interpreted in terms of the formula for black body radiation.

(2) The intensity of the continuous spectrum varied with the magnitude of the star and its changes are the chief factor in determining the integrated magnitude. The fluctuations were more pronounced in the photographic than in the visual region of the spectrum. But the relative energy curve changed with oscillations in the star's brightness, different stretches of the continuous spectrum behaving differently—thus for the region $\lambda\lambda$ 4700—5500 a maximum in the star's light curve accompanied a shift in the maximum radiation towards the violet, indicating a higher temperature; on the other hand for the region $\lambda\lambda$ 3400—4700 the shift was in the other direction and a fall in temperature was indicated. The explanation may be found in the selective absorption of a surrounding atmosphere of varying thickness or in the behaviour of bright bands, not sufficiently pronounced to stand out from the continuous spectrum (while strong enough to cause apparent irregularities in the behaviour of the continuous spectrum), or in the presence of such stretches of radiation as the ultra-violet continuous radiation due to hydrogen which begins at the head of the BALMER series and extends into the extreme ultra-violet.

(3) The bright emission bands after their first rise to maximum joined in the general changes in the star's magnitude though they did not reach their maximum strength until the star was well past maximum. When they began to fade, they faded faster than the continuous but in their places fresh bright bands were rising to maximum.

BRILL also examined some spectrograms of the same star secured nine to thirteen months after maximum.

The light curve retained still the marked features of two maxima in the yellow-red and the ultra-violet compared with the A-type comparison star.

[1] Publ. d. Astrophys. Obs. zu Potsdam 23, No. 70 (1914); A N 211, p. 1 (1920); A N 212, p. 457 (1921).

The latter might well be due to hydrogen ultra-violet emission replacing a corresponding absorption in the A-type star. In the changes which accompanied the oscillations of the light curve greater fluctuations were shown, as previously, in the ultra-violet continuous. The photographic continuous spectrum at this later stage showed a slight shift in its maximum intensity back towards the red, but at the same time the red continuous had dropped so much that the colour index of the star changed from $+0^m,22$ to $0^m,0$: the effective wave-length changed from 3820 to 3780. These contradictory results illustrate the difficulty of drawing conclusions about changes in the effective temperature of the novae at different stages. In the interval of seven months since the earlier series of observations the hydrogen emission bands had fallen in brightness at about the same rate as the continuous spectrum, but those bands which involved nebular lines had fallen more slowly and as before new bands had in the meantime come into prominence. Further discussion of these changes belongs to a later section of this article. They emphasize, however, the need to consider with caution conclusions derived from the more crude methods of examining spectral changes in terms of colour index or effective wave-length.

17. Colour Index. In view of the complex changes in the spectra of the novae at different stages it is only to be expected that the observed colour should vary irregularly. In the first three months after maximum several observers found colour oscillations in N Aquilae 1918. It had a negative colour index at maximum on June 9 which gradually increased negatively for two months, then oscillated, having maximum values (close to that which it had at maximum brightness) on August 9, 15, 30 and September 7, which showed no close relation with colour estimates or with magnitudes. Visual colour estimates made over the same interval of time by other observers shewed a change of colour with fluctuations from white at maximum to red; this is in agreement with similar observations in the case of the early stages of N Persei 1901 and of N Cygni 1920 but in disagreement with the decrease of the colour index and the shift of the maximum intensity of the continuous spectrum in the direction of the shorter wave-lengths.

It seems pretty clear that the varying strength of certain bright bands, such as $H\alpha$, must play a large part in visual colour estimates and in photo-visual magnitudes while other bright bands affect the photographic or photo-electric magnitudes.

18. Effective Wave-lengths. Naturally the fluctuations in colour estimates and in colour index are reflected in the effective wave-lengths obtained by the use of a coarse wire grating placed over the object glass of the telescope. A comparison of the effective wave-length obtained for N Aquilae 1918 with the corresponding colour index shows the same fluctuating changes, but the general agreement is crossed by sufficient cases of marked disagreement to shew that some factor, presumably variation in the bright bands, is affecting the two criteria differently. The following table gives the effective wave-length obtained by EDDINGTON and GREEN at Cambridge[1] and the colour index obtained by GUTHNICK and HÜGELER[2] at Berlin-Babelsberg on the days when the star was observed at both stations.

Date	Colour index	Effective w.l.
1918, June 30	$-0^m,17$	4395
July 11	$-$ 0,19	4346
July 18	$-$ 0,22	4426
July 25	$-$ 0,28	4315
Aug. 9	$-$ 0,11	4328
Aug. 14	$-$ 0,32	4293

For N Cygni 1920 the effective wave-lengths increased irregularly

[1] M N 79, p. 361 (1919). [2] A N 210, p. 345 (1920).

from 4230 (type F5) to 4330 (type K) in the first month after discovery, fell again to a value smaller than at discovery and subsequently increased again. The variations in spectral type usually associated with the different effective wave-lengths were not such as were observed in the star during the same period; it must be recognized that variations in the bright bands also affected these results and robbed this method of examining the changes in the distribution of the continuous spectrum of the star of its full value. The method may, however, be fairly applied to a nova before the bright bands make their appearance. HERTZSPRUNG did this for N Aquilae 1918 on June 8[1]. He found from the effective wave-length a temperature of 7300°, and from theoretical considerations as to the radiation from the star an angular radius of 0″,0020. If the nova had been of the same size as the Sun this would have given for its parallax, 0″,44. This is much too large to be possible, hence the surface of the nova must have been much larger than that of the Sun even at this early date. The value of the parallax assumed for this star in an earlier section would indicate on June 8 a diameter about 150 times the length of the Sun's diameter.

Expansion with an accompanying drop in the pressure of the radiating gases is probably much more important than high temperature in producing a change of spectrum toward an earlier spectral class, even while the star is dropping in magnitude. If, however, the distribution of the continuous spectrum is to be trusted as an indication of temperature, even after the bright bands have emerged, we can say that N Aquilae 1918 rose to the temperature of a B-type star while it was dropping through three magnitudes.

One point should be mentioned in connection with the changes in the colour of a nova. BARNARD and other observers have frequently stated that the focus required for a nova is generally longer than for ordinary stars and is the same as that used for a nebula. This is a natural accompaniment of the development of the nova into a nebula but in some instances the difficulty in focussing a nova in the early stages has been due to the presence of out-of-focus images corresponding to such bright bands as the red band at $H\alpha$. BARNARD's colour estimate of N Persei 1901 and of N Aquilae 1918, after they had developed the nebular spectrum, was palish-white; the latter star he described in 1919 as shining through a blue nebular envelope and as no longer possessing its original red and out-of-focus image. The development of the bright nebular spectrum in partial replacement of the bright hydrogen spectrum had changed the visual estimates of colour, which had altered initially from white to red, back towards white.

f) Spectroscopic History.

19. Spectra before Maximum Brightness. It is not to be expected that much should be known of its spectrum before a nova's principal outburst. There are a few cases recorded only. N Aquilae 1918, while a faint variable before its great brightening, showed on Harvard plates a spectrum of no very definite type, but one near to A-type in the distribution of the continuous radiation. On the night of maximum light June 9 it had a spectrum of type cA2eα, closely resembling that of α Cygni or of the chromosphere. The narrow absorption lines were displaced to the violet by an amount $40 \cdot 10^{-4}\lambda$; bright companions were not certainly present. N Pictoris 1925, as photographed on the Harvard plates during its protracted rise to maximum, was described as having a spectrum of type cF5. The absorption lines before maximum were displaced slightly to the violet (about 1 Å) and emission lines accompanied many

[1] A N 207, p. 75 (1918).

of the absorption lines on their red edge, their centres being about 1 Å to the red of the corresponding undisplaced line. DAVIDOVICH states that the change in the fortnight before maximum was that from a dwarf to a giant of class F 5. N Ophiuchi 1919 although observed a week before its brightest maximum, $7^m,2$ on November 5, showed a typical post-maximum spectrum of bright bands. As however a previous maximum of nearly equal intensity, $7^m,5$, had occured in the September preceding, this observation is probably not to be classed with that of novae examined before their first principal maximum. N Cygni 1920 was observed for two days before its maximum. Again there was a displaced α Cygni absorption spectrum, but this time the evidence for bright lines before maximum was more definite: bright lines were observed at Sidmouth, two nights before maximum, flanking on the red side the displaced absorption lines $H\beta$, $H\gamma$, $H\delta$, Fe^+4924, Fe^+5018. Nearly every nova, when observed at or shortly after maximum, has shewn an A-type spectrum, closely resembling that of α Cygni. The one striking exception is N Persei 1901. The day before its maximum, when observed at Harvard, its displaced absorption spectrum with its hydrogen, oxygen, carbon and helium lines[1], was definitely of B-type rather than of A-type. The hydrogen lines were accompanied by weak emission lines on their red edge. In this star too the A-type spectrum very rapidly developed shortly after maximum. DAVIDOVICH's statement that the pre-maximum absorption spectra of novae are generally characteristic of super-giant stars may be accepted.

20. Spectra at and immediately after Maximum Brightness. The spectrum at maximum is very nearly always an absorption spectrum, of type closely resembling that of the chromosphere or better that of α Cygni but with all the lines broadened and displaced to the violet by amounts proportional to their wave-lengths; nearly always a few of the lines, at least the first few hydrogen lines of the BALMER Series, have weak bright companions not much displaced from the normal wave-length of the corresponding line: $H\alpha$ is frequently abnormally strong in the earlier stages. The H and K lines of ionized calcium are always strongly present as absorption lines, both as broad lines displaced to the violet with the same displacement factor as the other lines and also as fine absorption lines displaced from their normal position by a small amount, generally very closely that due to the solar motion. As a typical absorption spectrum of the maximum stage we can take a portion of the measured spectrum of N Geminorum 1912 on March 15 the day after maximum (Figure 5); we add the wave-lengths given by displacing the measured lines by the factor $19 \cdot 10^{-4}\lambda$ and a comparison with the stronger lines of α Cygni[2]. It will be noticed how completely the lines are due to ionized atoms,—the enhanced spark lines of LOCKYER, FOWLER and NEWALL. LOCKYER and FOWLER first pointed this out in a study of N Aurigae 1891 compared with α Cygni. It should be added that CORNU noticed the close resemblance of the early nova spectrum to that of the chromosphere as long ago as 1876, in his discussion of the bright line spectrum of N Cygni 1876[3].

The agreement of the reduced nova lines with the stellar lines is not perfect but it must be remembered that the spectrum is coarse and complex; part of the failure to secure complete agreement lies in the slight systematic differences in the displacements of different elements. Thus in this case a factor $20 \cdot 10^{-4}\lambda$ would fit the displacements of the Ti^+ lines better than the factor $19 \cdot 10^{-4}\lambda$

[1] Harv Circ 56 (1901).
[2] Annals of the Solar Physics Observatory Cambridge 4, Part I, p. 36 (1920).
[3] M N 37, p. 201 (1877).

given by the hydrogen lines. The displacement for the Fe^{+} lines is intermediate between these two. But, though there are minor discrepancies of this nature the method of increasing the wave-lengths of the absorption lines by a quantity

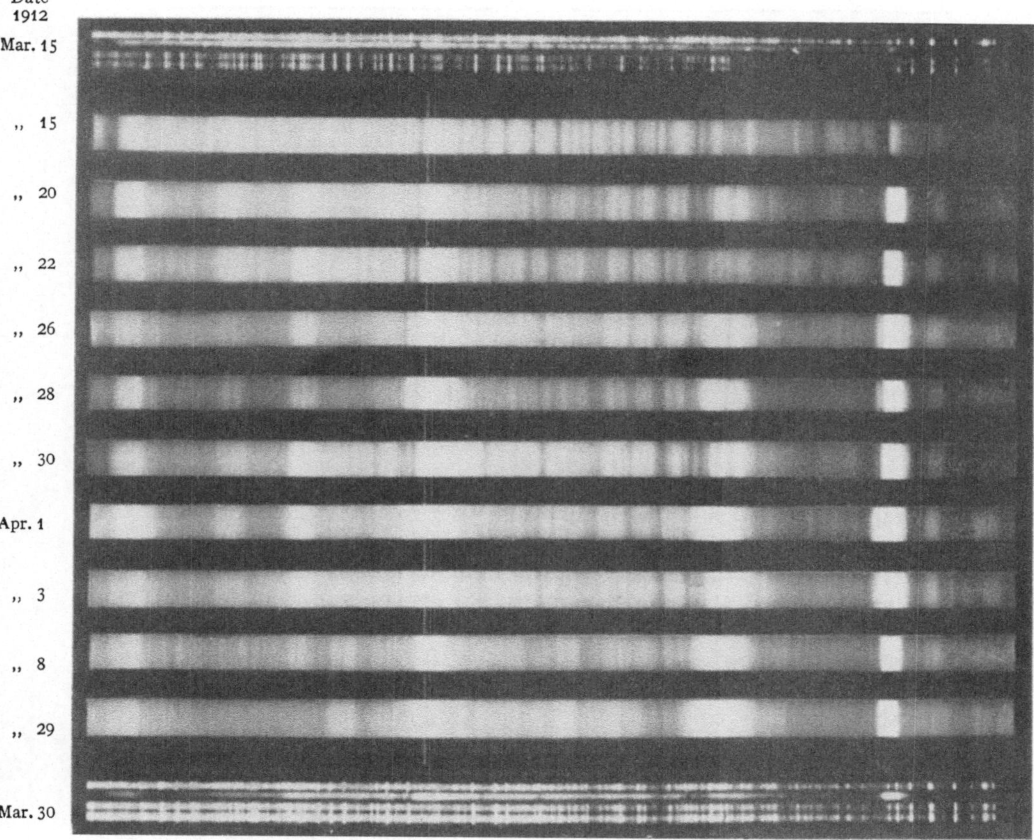

Fig. 5. Early spectrum of Nova Geminorum 1912 (STRATTON, Cambridge).

varying directly as λ does lead to a simple identification of the lines in a Nova spectrum near the period of maximum brightness. The factor varies from day to day and at this stage of the star's development it generally increases with the time. However, in the case of the first set of absorption lines in N Geminorum 1912—the lines given in Table VII—the displacement steadily decreased for several days until the absorption lines were lost in the broad bright bands that emerged on their red side[1].

The next step in the complication of the nova spectrum after the emergence of the bright companions on the red side of the displaced absorption spectrum is the widening of the bright bands and the development generally of structural details in these bands; several maxima appear which may vary in position from day to day. Crossed as the spectrum is by a large number of broad displaced absorption lines it is a difficult matter to disentangle the details of the complex

[1] Annals of the Solar Physics Observatory Cambridge 4, p. 45 (1920).

Table VII. Comparison of N Geminorum 1912 on March 15, near Maximum, with α Cygni.

Measured λ	Reduced λ	α Cygni λ	Element	Measured λ	Reduced λ	α Cygni λ	Element
4332,5	40,7	40,7	$H\gamma$	4499,7	08,2	08,5	Fe^+
50,6	58,9	57,8	—	4506,5	15,1	15,5	Fe^+
59,7	68,0	69,9	Fe^+	14,1	22,7	22,7	Fe^+
67,6	75,9	74,9	Ti^+	25,7	34,3	34,1	Ti^+
77,6	85,9	85,5	Fe^+	34,3	42,9	41,4	Fe^+
86,8	95,1	{95,0	Mg^+	41,2	49,8	{49,6	Fe^+
		{95,2	Ti^+			{49,8	Ti^+
4391,9	00,2	99,9	Ti^+	50,2	58,8	58,8	Cr^+
4408,8	17,2	{17,0	Fe^+	56,5	65,2	64,0	Ti^+
		{17,9	Ti^+	64,1	72,8	72,2	Ti^+
35,7	44,1	44,0	Ti^+	68,8	77,5	76,5	Fe^+
43,1	51,5	50,6	Ti^+	75,5	84,2	84,0	Fe^+
48,1	56,6	55,3	Fe^+	80,9	89,6	88,4	Cr^+
61,1	69,6	68,7	Ti^+	4583,5	92,2	92,5	Cr^+
72,2	80,7	81,3	Mg^+	4607,1	15,9	16,8	Cr^+
79,8	88,3	88,5	Ti^+	11,0	19,8	19,0	Cr^+
82,2	90,7	91,6	Fe^+	4620,8	29,6	29,5	Fe^+
93,7	02,2	01,4	Ti^+				

bright bands. WRIGHT has shewn[1] that the widths of the bright bands immediately after maximum in the case of N Geminorum 1912 varied as the square of the wave-lengths of the corresponding lines and not directly as the wave-lengths (Fig. 6). Within a few days their widths became proportional to the wave-lengths. The centres of these bright bands are generally but little displaced either way from the normal wave-lengths of the presumed origins and the complex maxima may be displaced either to the red or to the violet. These maxima vary in relative strength for short intervals. The continuous spectrum due to hydrogen which extends to the violet of the head of the BALMER series has also been seen bright in novae at this stage[2].

Accompanying the increased complexity in the bright bands or rapidly following it comes a doubling of many lines in the absorption spectrum. This doubling nearly always occurs for the hydrogen lines of the BALMER series and sometimes for some of the ionized iron lines as well as for such lines as H and K of ionized calcium. We can indicate the displacement to the violet in ÅNGSTRÖM units by the quantity $\varkappa \cdot 10^{-4} \lambda$ ($\varkappa$ is called the displacement factor; the corresponding velocity of approach, if the displacement is regarded as a DOPPLER effect is $30\,\varkappa$ km/sec). Such pairs of values for $\varkappa$ have been found as 15 and 46 for N Geminorum 1912 three days after maximum and 55, 82 for N Aquilae 1918 four days after maximum. On the same day the two extreme maxima of the bright bands of N Aquilae 1918 had displacement factors $+44$ and -40. ADAMS has suggested that not only the different $\varkappa$'s in the same star but those in different stars are related to one another by very simple numerical ratios. Two general statements can be made about the displacement factors. On the whole they are larger for the stars of brighter absolute magnitude—high velocities, if the displacements may be so interpreted, accompanying greater luminosities and more violent outbursts. The other point of general application is that the groups of absorption lines which develop later have larger displacement factors than those which develop earlier. The bright bands spread out beyond the absorption lines, frequently covering the first and least displaced absorption lines; the later

[1] Lick Obs Publ 14, p. 36.

[2] See for example M N 81, p. 141 (1920).

and more displaced absorptions, though weakening at times and varying in position, remain visible until the continuous spectrum fades and they lose their background. In some cases the hydrogen absorption lines show a still more

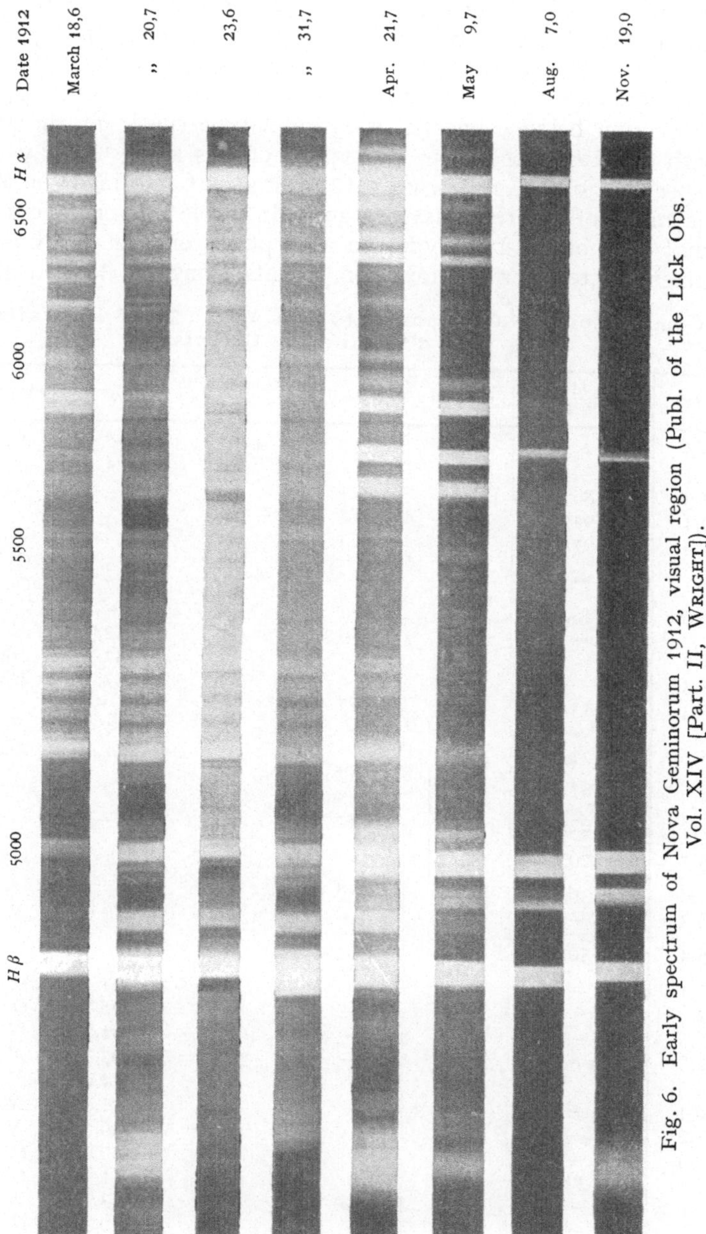

Fig. 6. Early spectrum of Nova Geminorum 1912, visual region (Publ. of the Lick Obs. Vol. XIV [Part. II, WRIGHT]).

complex structure, thus in N Cygni 1920 we find six absorption lines for the BALMER lines, the corresponding values of $\varkappa$ being 20,5, 30,0, 35,9, 44, 51,8, 59,7. Some of the Fe^{+} lines also shew multiple absorptions, but generally only a portion of the hydrogen absorptions are repeated for the iron lines.

21. Complex Spectra involving more than one Type. A partial explanation of the failure of the metallic lines to shew the same number of displacement as the hydrogen lines is that in some cases a new set of hydrogen absorptions more displaced than the earlier sets may appear accompanied by a set of absorption lines with the same displacement factor and due not to iron, calcium and titanium but to helium, oxygen, nitrogen and carbon, the typical lines of B-type stars. N Geminorum 1912, N Aquilae 1918, N Lyrae 1919 may be mentioned as all shewing this feature of double hydrogen absorptions, the less displaced absorption being connected with an A-type spectrum and the more displaced with a B-type spectrum. As typical of this stage we may take the absorption spectrum of N Geminorum 1912 secured at Cambridge [1] on March 30 for the same range of spectrum as that given in Table VII on March 15. We may note that each of the two hydrogen absorptions on this day was double, the displacement factors for the four components being 28, 35 and 48, 60.

Table VIII. Comparison of N Geminorum 1912 on March 30, 16 Days after Maximum, with α Cygni and γ Orionis.

Measured	Reduced with $\varkappa = 32$	α Cygni	Element	Reduced with $\varkappa = 54$	γ Orionis	Element
4355,6	—	—	—	4379,1	79,7	*N*
65,1	—	—	—	88,6	88,1	*He*
72,3	86,3	85,5	Fe^+	—	—	—
80,4	94,4	95,2	Ti^+	—	—	—
83,5	97,5	99,9	Ti^+	—	—	—
4391,1	—	—	—	4414,8	15,1 17,1	*O* *O*
4401,6	15,7	17,0	Fe^+	—	—	—
07,7	—	—	—	31,5	32,9	*N*
11,8	—	—	—	35,6	37,7	*He*
21,8	—	—	—	45,7	47,2	*He*
29,3	43,5	44,0	Ti^+	—	—	—
46,6	—	—	—	70,6	71,6	*He*
56,9	71,2	71,6	*He*	81,0	81,4	Mg^+
73,6	87,9	89,3 89,5	Fe^+ Ti^+	—	—	—
4487,7	4502,1	01,4	Ti^+	—	—	—
4505,7	20,1	20,4 22,7	Fe^+	—	—	—
21,6	36,1	34,1	Ti^+	—	—	—
28,3	—	—	—	4552,8	52,6	*N*
36,4	50,9	49,6 49,8	Fe^+ Ti^+	—	—	—
66,1	—	—	—	90,8	91,1	*O*
72,0	—	—	—	96,7	96,3	*O*
76,5	—	—	—	4601,2	01,7	*N*
82,8	—	—	—	07,5	07,3	*N*
88,5	—	—	—	13,3	14,0	*N*
4596,4	—	—	—	21,2	21,6	*N*
4605,2	—	—	—	30,1	30,7	*N*
13,3	—	—	—	38,2	39,0	*O*
15,6	30,4	29,5	Fe^+	40,5	41,9	*O*
17,6	—	—	—	42,5	43,3	*N*
24,7	—	—	—	49,7	49,3	*O*
37,1	—	—	—	62,1	61,8	*O*
51,4	—	—	—	76,5	76,3	*O*

Although the agreement of the coarse and complex spectrum with the composite spectrum built up of an A-type spectrum and a B-type spectrum

[1] Annals of the Solar Physics Observatory Cambridge 4, p. 41 (1920).

with different displacement factors is sufficiently close to make it pretty clear that the sources of the lines have been rightly identified, closer inspection reveals certain discrepancies. Thus as pointed out above for different elements in the same group the displacement factors may vary slightly among themselves. There seems to be a slight tendency for the hydrogen lines to be more displaced than the lines of the heavier elements, though the reverse is sometimes the case. The matter is complicated somewhat by the tendency of the hydrogen absorptions to become double, with components of varying strength. Very puzzling and so far unexplained is the fact pointed out by WRIGHT[1] that the displacements of the absorption lines do not vary directly as λ but can be better represented by the formula $\delta\lambda = A\lambda + B\lambda^2$ (Fig. 7). So long as the widths of the emission bands are conditioned by the absorption companions at their violet ends (and also very occasionally by absorptions at the red ends) a somewhat similar formula best fits their widths but with the fading of the continuous spectrum and the dark companions the bright bands settle down and remain fairly constant at widths varying with the wave-length of their line of origin for each star. The actual widths of the bands vary greatly from star to star from a few ÅNGSTRÖM units up to such values as 50 Å, at $H\gamma$.

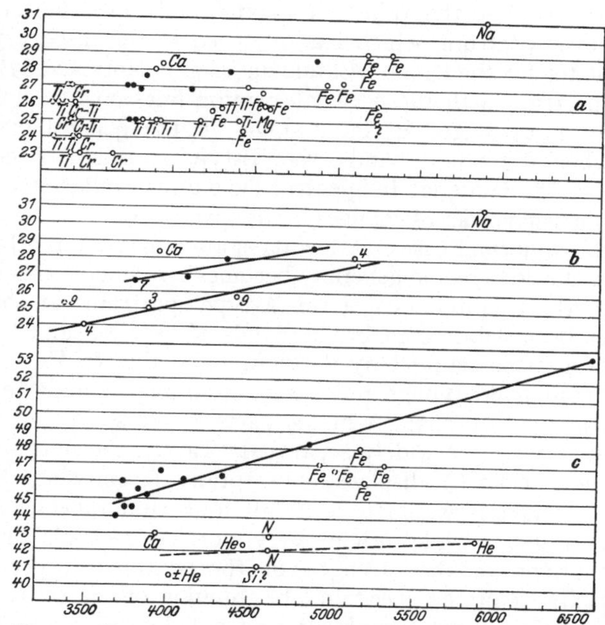

Fig. 7. Change of displacement factor with wave-length for Nova Geminorum 1912 (WRIGHT, Lick).
(b is mean of a. Black dots represent H lines.)

A striking feature in the broad bands right from the earliest stages consists in the multiple maxima in the bands. The displacements of these maxima from the normal position of the presumed line of origin varies generally with the wave-length. The different maxima vary considerably in relative brightness though after some initial fluctuations they generally settle down to fairly constant displacements. Even after the change from an A- or B-type spectrum to a nebular spectrum is complete, the displacement of the maxima in the nebular bands is frequently the same multiple of the wave-length as for the hydrogen and other bands in the early stages.

Several observers, notably BELOPOLSKY, have reported the presence in the spectra of fine undisplaced metallic lines. These have not been generally recorded, possibly on account of the dispersion generally used for nova spectrograms, save in the case of H and K, where narrow undisplaced absorption lines are found on the bright bands for practically every nova. The velocity given by these lines is in nearly every case the solar velocity reversed e. g. for N Persei 1901, N Geminorum 1912, N Aquilae 1918. The one notable exception is N Ophiuchi

[1] Lick Obs Publ 14, p. 51, and M N 81, p. 191, 501 (1921).

1919 for which there was a difference of 24 km/sec between the velocity (+ 6 km/sec) given by the fine H and K absorption lines and the reversed velocity of the Sun. For this star these narrow undisplaced lines were rather stronger than in most novae. This star was exceptional in having two maxima of about the same magnitude at an interval of about 2 months. The spectra found six or seven weeks after the first maximum were of the ordinary A-type with bright bands and absorption companions on the violet side, but the strength of the continuous was varying considerably from day to day.

The emergence of the B-type absorption spectrum, along with the A-type absorption spectrum, may come a few days after the principal maximum in stars where the disturbance is of exceptional strength but as a rule it is several weeks after the principal maximum before this stage develops. The typical nova spectrum, which has often led to the discovery of novae on the Harvard plates for the DRAPER Catalogue, e. g. N Normae 1893, is the A-type bright band spectrum with the usual absorption lines on the violet side: some of the novae were discovered at later stages in the general evolutionary scheme, namely when the bright bands were B-type in origin or nebular.

22. Sequence in Spectral Type and Oscillations. We may follow still further the usual order of changes in the spectra of novae, reserving for separate discussion the abnormal cases. A few days after maximum a broad composite bright band in the 4640 region generally becomes prominent; this is only in part to be ascribed to the bright bands of the A-type spectrum; at some later stage it is mainly to be attributed to bright bands of O^+, N^+ and C^+ and at a later stage still it may be flanked by a bright band due to He^+, 4686. Fluctuations in this broad band though not easily disentangled may be in part explained on comparison with changes in the parts of the spectrum where strong A- and B-type lines are more widely separated. Taking the four spectra in the order in which they occur (A displaced absorption, A emission, B displaced absorption and B emission) we may have all four simultaneously present or some of them prominent while others may not be recognisably present. Naturally fluctuations in the strength of the continuous spectrum affect the general appearance of the spectrum: brightening of the continuous makes the absorption spectrum more prominent and a fading of the continuous throws the bright bands into strong relief. On the whole there is in general a progressive tendency, with fluctuations, in which the B emission gradually replaces the A absorption, only to give place as we shall see later to a nebular and WOLF-RAYET stage.

In some stars, e. g. N Geminorum 1912, the nitrogen lines in the B-type spectrum are mainly the ordinary spark lines such as 3995, 4447, 4602, 4607, 4614, 4621, 4631, whereas in other novae such as N Aquilae 1918 the lines of N^{++} become prominent e. g. 4097, 4104, 4634, 4641. In this star the absorption lines of N^{++} 4097, 4104 had 24 days after maximum a larger displacement factor (114) than that given 17 days earlier (92) by the same lines when present along with the lines of N^+, O^+, C^+ and He familiar in the spectrum of γ Orionis[1]. The velocity 3420 km/sec required by the explanation in terms of a DOPPLER effect is the largest recorded in a nova spectrum. N Cygni 1920 showed similarly a larger displacement factor (80) for the two lines 4097, 4104 than for any other set of lines[2].

Accompanying these prominent absorption lines in N Aquilae 1918 during July and part of August 1918 there were two very prominent broad absorptions

[1] BAXANDALL, M N 81, p. 70 (1920).
[2] WRIGHT, Publ A S P 32, p. 340 (1920).

some 16 Å apart which oscillated in position between 4545, 4561 and 4557, 4573. These have not yet been identified. WRIGHT from a pair of absorptions at 4572, 4588 in N Geminorum 1912 on May 2, 1912 suggests that, if the absorptions are displaced with the nitrogen spectrum, origins should be looked for at 4605, 4621[1]; PADDOCK comparing the varying displacement of this pair with that of 4097, 4103 suggests origins at 4601, 4616[2].

The oscillations in position of these two pairs of lines accompanied periodic changes in the light curve of the Nova. At minimum the lines moved to the red or at a later stage vanished, at maximum they moved out to the violet and were much strengthened. In nearly all novae which have passed through the B-type, this stage is marked by periodicity (with a period as a rule of several days) in the light curve. The general rule appears to be that when the star brightens its spectrum returns to that of an earlier and brighter stage, the spectrum being in fact a simple function of the magnitude. Thus in the earlier stages where fluctuation in brightness appears to be mainly a question of a strong or faint continuous spectrum, a brightening means a return to an A- or B-absorption spectrum or to a blend of both; a fading in brightness generally means an A or B bright band spectrum in which as the star fades A gives place to B, while B in due course gives place to P or O. As a rule the WOLF-RAYET bands due to He^{+}, 4686, or to N^{++}, 4634, 4641, become prominent before the emergence of the bright nebular bands; N Pictoris 1925 is an exception in this as in many other details of its spectral development. The nebular line 4363 was seen on Jan. 27, 1926 but 4686 He^{+} did not appear till March 14 or 15 when it suddenly came up very strongly as the brightest band in the spectrum.

Although many of the bright bands in this stage of the star's development appear to be featureless, a like structure seems common to the hydrogen, helium and some of the metallic bands. This structure may vary in the early stages particularly, but after a time the bright maxima in the bands settle down to a stable position—displaced by constant displacement factors. As fresh bands due to lighter elements emerge, they generally repeat the structure found in the earlier bands, though frequently with more detailed structure. Thus in N Persei 1901, the early Cambridge plates and the late Cambridge plates gave respectively, for the bright bands the following table for the ratios (observed frequency): (undisplaced frequency):

	Early Plates	Late Plates
	H	4363 (Neb.)
Absorption	1,0051	—
Max. bright	22	1,0021
2nd bright max.	07	05
3rd bright max.	—	0,9990
4th bright max.	—	78
5th bright max.	—	69

ADAMS and JOY record the same continuity of structure of the bright bands for N Aquilae 1918[3].

It may be added that the more violent the outburst, apparently, the greater the width of the emission bands and the displacements of the absorption lines, and the more quickly does the star run through the sequence of spectral changes.

23. The Nebular Spectrum. Bright bands of unknown origin but referred by their structure to lines found in the spectra of planetary nebulae emerge in gradual or oscillatory replacement of the A- and B-type bright bands. They very rarely appear along with the A-type bands but are frequently present with the bright B-type bands of O^{+}, N^{+}, C^{+} and He. These B-type bands

[1] Lick Obs Publ 14, p. 70. [2] Publ A S P 30, p. 244 (1918).
[3] Publ A S P 31, p. 182 (1919).

give place as a rule not only to nebular bands but also to O-type bands corresponding to more highly ionized atoms of the same elements. The nebular lines may begin to appear, as in the case of N Aquilae 1918, only 11 days after maximum[1], but as a rule it is several months, often as many as six, after the principal maximum before these lines are seen. Save for one or two exceptional cases, where no bright bands have been recorded at all, the nebular stage always occurs after the initial A- or B-type bright band stage, continues for a considerably longer period and gradually gives place—again with occasional fluctuations in the spectrum—to a WOLF-RAYET or O-type spectrum. As in many planetary nebulae the nebular lines proper and the ionized nitrogen and other lines more closely connected with WOLF-RAYET stars are present simultaneously as bright bands. Although there are occasional differences in structure between the two groups of bright bands there is no such clean-cut division as between the spectrum of a nucleus and of an outer envelope common among the planetary nebulae.

The first nova in which the nebular spectrum was detected was N Cygni 1876, which in September 1877, 10 months after its maximum, was observed by COPELAND at Dun Echt to show a spectrum closely resembling that of an ordinary planetary nebula. Lord LINDSAY added in his account of the star that anyone observing it for the first time through the spectroscope would have no doubt about its being a planetary nebula. In the case of N Aurigae 1891 the appearance of the nebular spectrum in the autumn of 1892 on the emergence of the star from close proximity to the Sun was accompanied by an unusual feature: the star had brightened through 2^m since it had last been seen and the spectrum had simultaneously reached a later stage in the normal sequence. The complex structure of the broad bright bands corresponding to the nebular lines was also noticed. This was but a continuance of complexity in the structure of the bright bands shewn by the star in its earlier stage as an A-type star.

The nebular lines vary in brightness not only relatively to the hydrogen and nitrogen lines but also relatively to each other. The range of variation in one star is much wider than among the planetary nebulae. For instance 4363 is frequently at an early stage much brighter than 4959 or 5007. Only one of the nebulae (I C 4997) whose spectra are tabulated in Lick Publ 13, p. 246, shows 4363 as strong as this pair. Again N Persei 1901 shows on plates taken during the autumn of 1901 a bright complex band, which from its structure should be referred to a source at 4724,2 (Fig. 8). This line has only been found so far in one nebula (N G C 7027) as a weak line at wave-length 4725,5. Other nebular lines which appear in Novae are 5755, 6302, 6364, 6730 in the visual region, and 3342, 3426, 3445 and 3869 in the ultra-violet. Of these 3445 and 3869 were both very strong in N Persei 1901, a further point of resemblance to N G C 7027. As in nebulae generally, 3869 differs considerably in strength from star to star. The line 4363 which strengthens early also weakens early the nova generally tending towards the more normal nebular type. In the case of N Geminorum 1912 the 4640 stage was several times recovered on a sudden brightening of the star (e. g. May and September 1912 and February 1913), after the nebular stage had been reached: on these occasions bright bands due to N^{++} came up strongly at 3412, 3484, 4104, 4640 while bands at 3445, 4686 (He^+) faded away as also the abnormal N^{++} line at 4379. The bands of He at 4472 and N^+ at 3995 remained unaffected[2].

[1] M N 79, p. 279 (1919).
[2] WRIGHT, M N 81, p. 181 (1921).

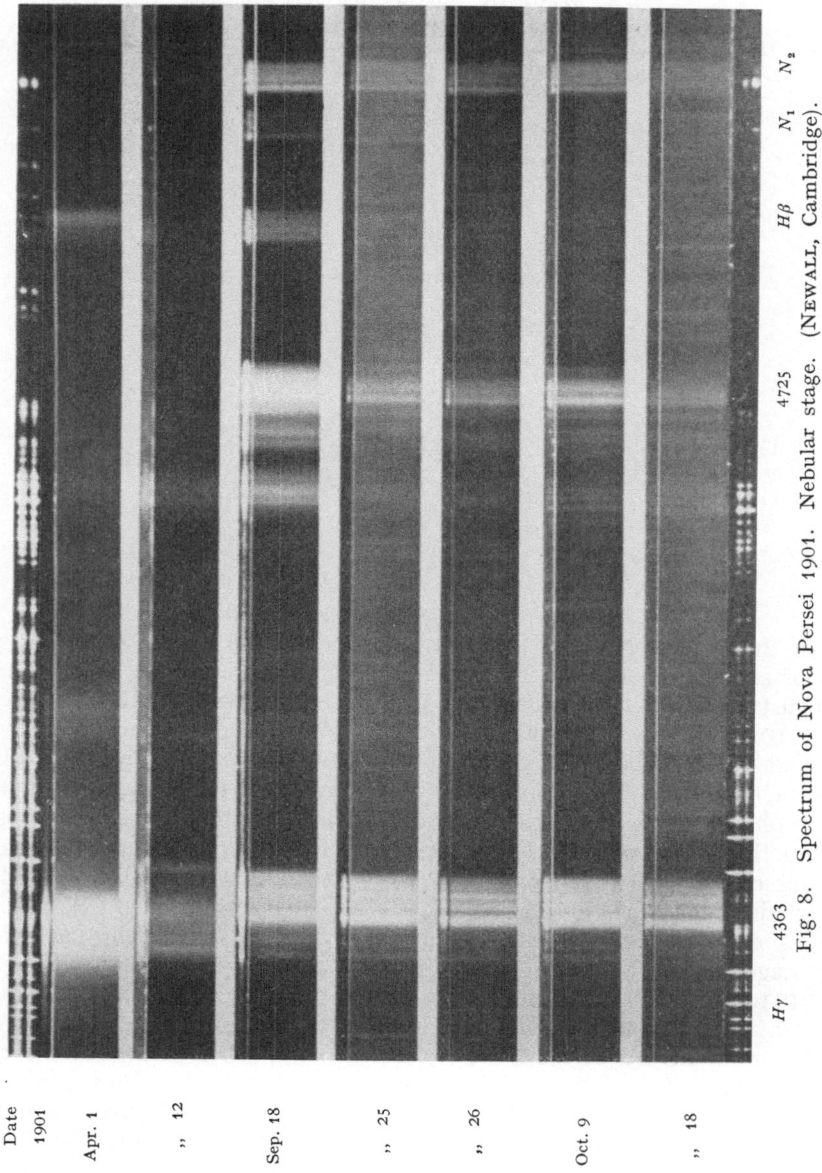

Fig. 8. Spectrum of Nova Persei 1901. Nebular stage. (NEWALL, Cambridge).

The following table due to WRIGHT gives the emission spectrum of N Geminorum 1912, eight months after maximum.

Table IX. Emission Spectrum of N Geminorum, 1912[1].

(a) 1912 Sep. 23.

3342	Neb.	4146	*He*
3414	N^{++}	4187	O^+
3444	Neb.	4201	N^+, He^+
3482	N^{++}	4238	N^+
3696—3737		4267,5	C^+

[1] Lick Obs Publ 14, p. 78.

Table IX. (Continuation.)

(a) 1912 Sep. 23.

3747–3773	O^{++} and H	4315	O^{+}
3797	$H\vartheta$	4338	$H\gamma$
3835	$H\eta$	4364	Neb.
3869	Neb.	4378	N^{++}
3890	$H\zeta$	4415	O^{+}
3926	He^{+}	4444	N^{+}
3969	$H\varepsilon$	4471	He
3996	N^{+}	4511	N^{++}
4027	He, He^{+}	4543	He^{+}
4041	N^{+}	4591–4732	N^{++}, He^{+}, C^{+}, O^{+}, Neb.
4071	O^{+}		
4089	Si^{+++}	4777	N^{+}
4097	N^{++}	4801	N^{+}
4102	$H\delta$	4862	$H\beta$
4103	N^{++}	4960	Neb.
4119	Si^{+++}, He	5007	Neb.

(b) 1912 Oct. 6.

5045	N^{+}	5876	He
5180	—	5944	N^{+}
5411	He^{+}	6300	Neb.
5667	N^{+}	6369	Neb.
5681	N^{+}	6479	N^{+}
5755	Neb.	6562	$H\alpha$

In the case of N Persei 1901 the presence of fine undisplaced absorption lines of sodium, D_1 and D_2, has been noticed on the bright D_3 band of helium in August 1901, well after the nebular stage had developed. Similarly in N Geminorum 1912 the fine undisplaced H line is still to be measured on the bright $H\varepsilon$ band in December 1912, nine months after maximum. The bright bands in the nebular stage still vary in structure from time to time though the maxima remain on the whole more fixed in position relative to the undisplaced position than in the earlier more unsettled stages. The displacements of like maxima vary directly as the wave-length and the same displacement factor holds for lines of different elements e. g. for H, He and the nebular lines. Some of the features remain unchanged in the bright hydrogen bands right back to an early stage e.g. in N Aquilae 1918 LUNT has traced back details from September 1919 to August 1918 almost without change, similarly in N Persei 1901 maxima have remained unchanged from March 1901 to November 1902.

Attempts have been made to explain the complex structures of the bright bands in terms of the ZEEMAN effect but without success. The most illuminating observations came from the Lick Observatory when WRIGHT first pointed out that in N Aquilae 1918, which was growing into a visible disc, the different maxima of the bright bands came from different portions of the disc and further different elements were strong in different portions of the disc. Thus two maxima in He^{+} 4686 came from outer portions of the expanding nebula while outbursts of N^{+} radiation came from the central portions. These N^{+} outbursts were accompanied by a temporary fading of the 4686 radiation in the centre of the disc (compare the account of N Geminorum 1912 above) while the maxima in the outer portions were unaffected. The hydrogen and nebular bands showed maxima agreeing in position with both sets of bands. When N Aquilae 1918 was photographed during 1919 and later with the slit across the disc in position angle 202° and the nova kept stationary on the slit Messrs. MOORE and SHANE found a number of very interesting phenomena[1]. In the bright nebular and

[1] Lick Bull No. 322 (1919).

hydrogen bands maxima occured in pairs displaced equally to the red and violet (the latter being the stronger) and each corresponding pair of maxima came from opposite sides of the centre of the disc (Fig. 9). Rotation at once suggests itself as the explanation but while the least displaced maxima suggest rotation in one sense and come mainly from the top and bottom of the disc the most displaced maxima which come from inner parts of the disc suggest rotation in the opposite sense. No simple scheme of expansions and rotations has been devised to fit all the facts. The magnitude of the growing disc could be measured from slitless spectrograms and for the nebular bands the rate of growth of the diameter, 2″,0 a year, gave a reasonable parallax, 0″,0025, if the velocity of

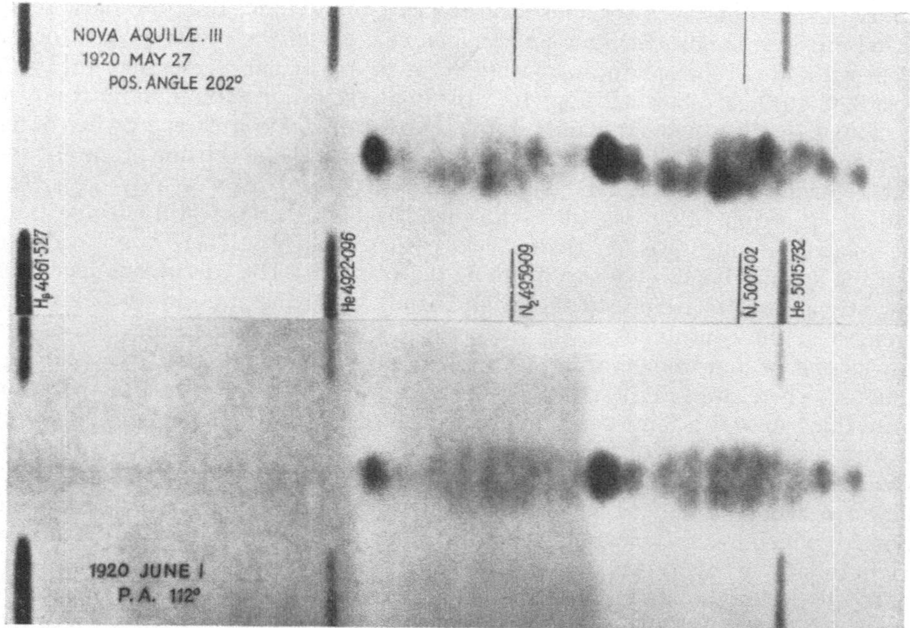

Fig. 9. Complex bands in Nova Aquilae 1918 (Moore and Shane, Lick).

growth was supposed to be the same as that indicated by the maxima in the bands most displaced from the normal wave-length. But difficulties in the way of this simple application arose from the fact that the $H\beta$ bands had maxima with the same displacements as those of the nebular bands while the corresponding disc only grew outwards at less than half the rate of the disc given by the nebular lines[1].

Other phenomena noted by Messrs. Moore and Shane in late 1920 and during 1921 included the growth of a dark streak along the spectrum dividing the upper and the lower maxima suggesting a dark space separating the nucleus from the growing outer envelope. At the same time the less displaced maxima, coming from the outer layers of the disc, weakened perceptibly, compared with the more displaced maxima coming from the centre of the disc, and the displacements of those latter maxima slightly decreased. Other bright bands such as He^+ 4686 continued as a bright band in the continuous spectrum of the central star. That the nebulous envelope grew outwards and dissipated itself

[1] Publ A S P 32, p. 322 (1920); 33, p. 218 (1921).

leaving a WOLF-RAYET nuclear star seems clearly indicated; for an examination of the star and its spectrum in 1926 by AITKEN and MENZEL[1] showed no trace of the visible outer envelope and only very slight traces of emission from the principal nebular lines and 4686. Photographs taken at Mt. Wilson in 1926 by HUBBLE and DUNCAN showed the disc to be still expanding, the diameter still growing about 2″,0 a year. The brightest disc was at He^+ 4686; the hydrogen lines also showed as discs but there were only faint traces of images at the nebular lines. The slower growth of the hydrogen discs was not confirmed.

24. WOLF-RAYET Stage and Late Observations of Spectra. At about a year after maximum most novae are still in the nebular stage, very little continuous spectrum seen but broad bright nebular—and frequently also WOLF-RAYET—bands. Oscillations in brightness still occur, a brightening bringing back some of the nitrogen bands. Striking and sudden changes in the spectrum may occur even without a change in magnitude (see in particular N Geminorum 1912, Dec. 8, 9, 1912, as shewn in Fig. 10). By two years after maximum most novae seem to lose their nebular bands—probably by dissipation of a growing outer envelope—and a nucleus is left which has generally the spectrum of an O-type star[2]. Bright bands of H, He^+, and N^{++} (4610, 4641) show weakly on a continuous spectrum. In cases where the nebular bands proper still survive they are relatively much weaker than a year earlier (Fig. 11). There is no evidence that the nebular bands ever survive more than 9 years. The continuous spectrum now strengthens relatively to the bright bands while the star slowly fades: the strength of the continuous in the ultraviolet supports the identification of the star as one of high temperature even where no bands stand out from the continuous. There are only a very few cases known where the ultimate spectrum is anything except a faint continuous spectrum, either without detailed features or with weak bright WOLF-RAYET bands. Two of these, N Cygni 1600 (P) and N Carinae 1843 (η), have halted at a typical stage in the evolution of a nova: their case will be discussed below. The only other outstanding case is N Coronae 1866 (T) which has a typical Ma-spectrum along with certain broad and complex bright bands of hydrogen and helium and a suggestion of nebular bands at 4658, 4363. On the strength of this star's spectrum LUNDMARK has suggested that a faint star of $13^m,7$ and of Mb-spectrum close to TYCHO BRAHE's position may be the famous N Cassiopeiae 1572.

25. Spectra of Novae in Nebulae. The evidence as to the spectra of the novae that have appeared in the nebulae is somewhat meagre and conflicting. Thus most observers described the spectrum of N Andromedae 1885 as continuous, but HUGGINS, VOGEL and VON KONKOLY all suspected bright bands and noted variations in intensity along the spectrum. SHANE described the spectrum of N Virginis 1926 (in N G C 4303) as continuous with no bright bands in it, possibly with absorption bands breaking the continuous spectrum, while HUBBLE described the same star as having very wide diffuse emission bands suggestive of N Aquilae 1918 a month after maximum. Again ADAMS found N Cephei 1917 (in NGC 6946) to have a continuous spectrum with bright lines, while N Centauri 1895 (Z) (in NGC 5253) described on discovery at Harvard in July 1895 as having dark lines and resembling class R was examined visually in December 1895 and found to be monochromatic and nebular; two years later when examined by CAMPBELL it was found to have a continuous spectrum with no bright lines but with a distribution of intensity pointing to an unusually high temperature. Though the evidence is not enough to be decisive

[1] Publ A S P 38, p. 391 (1926).
[2] See for example ADAMS and PEASE, Ap J 40, p. 294 (1914).

Date	Magn.
Titanium	
1912 May 9,6	7,3
December 8,8	,, 7,8
,, 9,8	,, 7,8
,, 13,8	,, 8,0
,, 22,8	,, 8,2
1913 January 8,7	,, 8,4

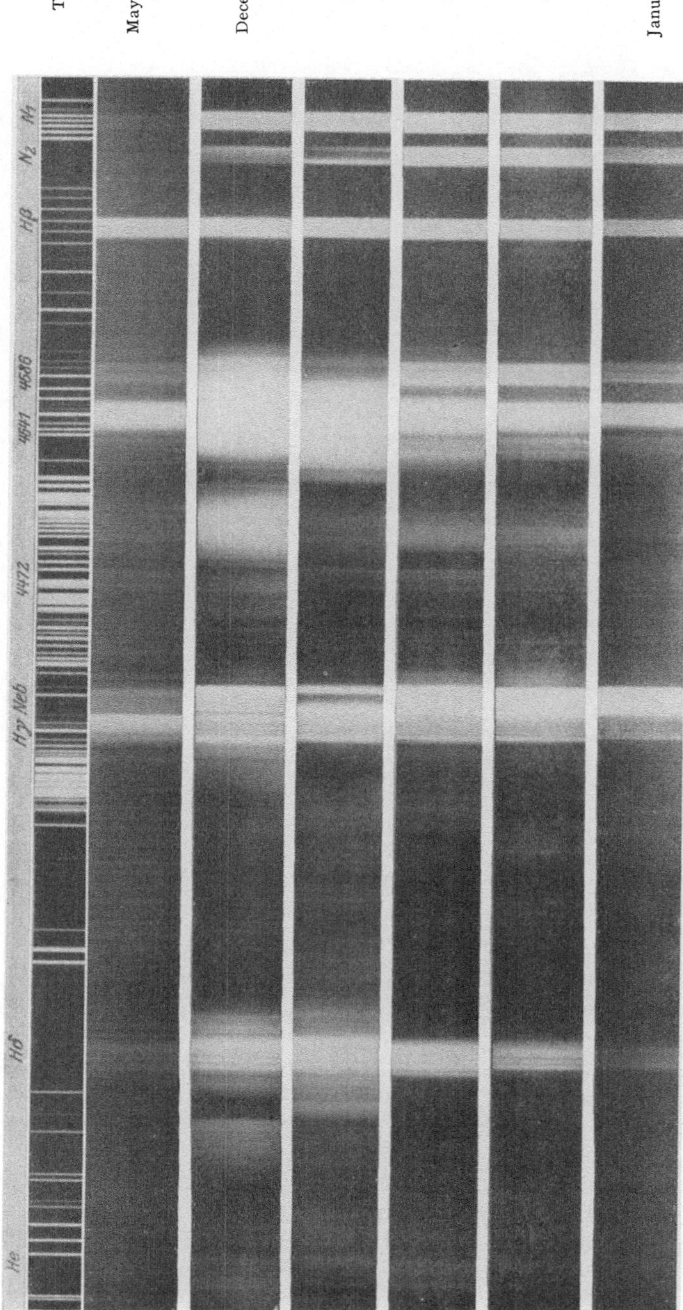

Fig. 10. Spectrum of Nova Geminorum 1912. Wolf-Rayet stage (Curtiss, Michigan).

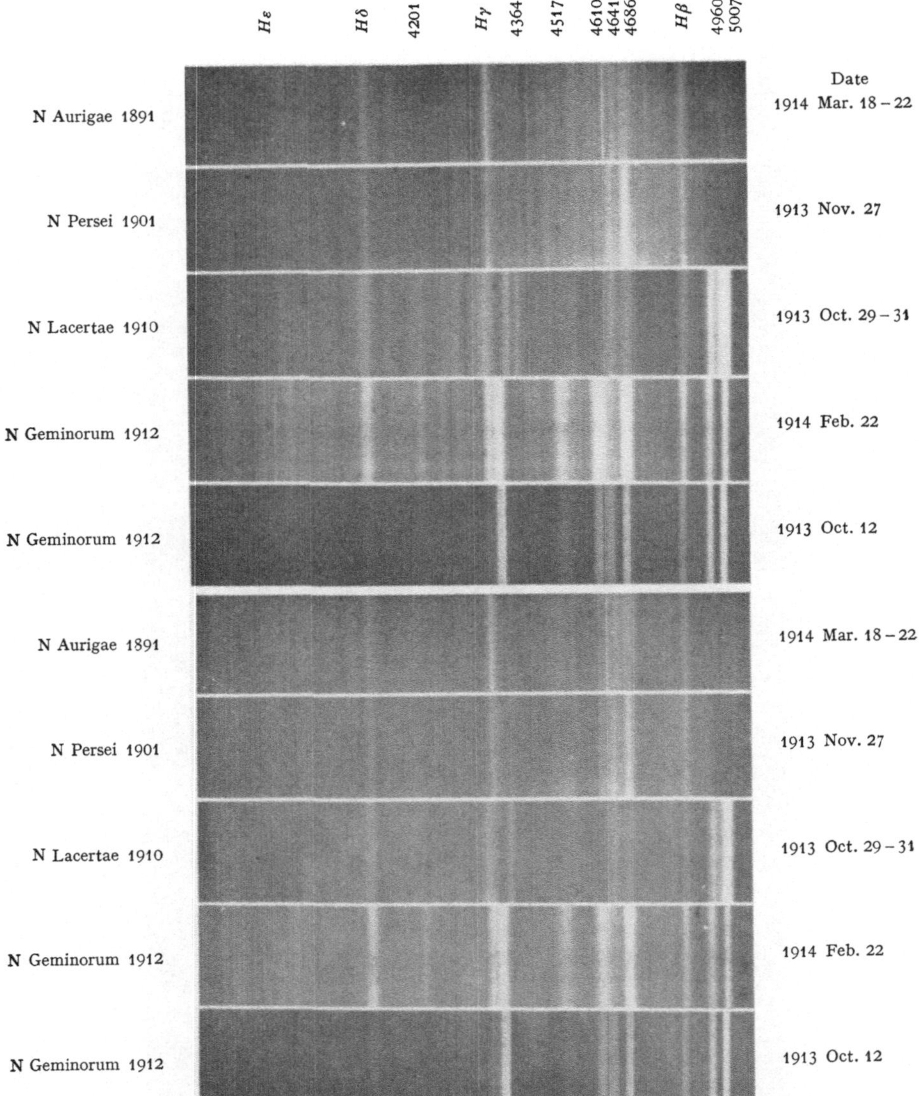

Fig. 11. Late stages of nova spectra (ADAMS and PEASE, Mt. Wilson); reproduced from the Astrophysical Journal. In the lower reproduction false lines introduced in widening the spectra have been touched out.

it would appear that novae in spiral nebulae have in general a different spectroscopic history to that of the ordinary galactic novae.

26. Exceptional Novae. In addition to the novae in the spiral nebulae there are one or two whose spectral history is different from the main sequence outlined in the above sections. We have already seen that N Coronae 1866 (T) has settled into a spectral type Maep—a most unusual spectrum for a nova. Two other novae, which like this star lie in unusually high galactic latitudes, have also an unusual spectroscopic history: these are N Carinae 1843 (η) and N Pictoris 1925 (RR).

N Carinae 1843 (η) has in the past had several bright maxima, the brightest being -1^m in 1843. Its spectrum is essentially a bright line spectrum, the chief lines being due to H and Fe^+; Cr^+ is also apparently present. There are in addition a number of bright lines as yet unidentified. There is some evidence for variation in the relative brightness of the lines from time to time. PERRINE with a long exposure has brought out the continuous background and finds that the bright hydrogen lines are bordered on the violet side by absorptions. Thus η Carinae seems to have settled down into the typical early stage of a nova and we have no evidence whether it ever passed through the normal sequence of the later stages.

N Pictoris 1925 (RR) had two or three maxima during the spring and summer of 1925 and in any case it had an unusually slow rise to maximum. It was discovered before maximum and showed at first a typical pre-maximum spectrum[1]— a continuous spectrum crossed by dark lines due to H, Fe^+, Ti^+, Sc^+, Cr^+, Sr^+, Mg^+, etc. The displacement of the lines to the violet was slight corresponding

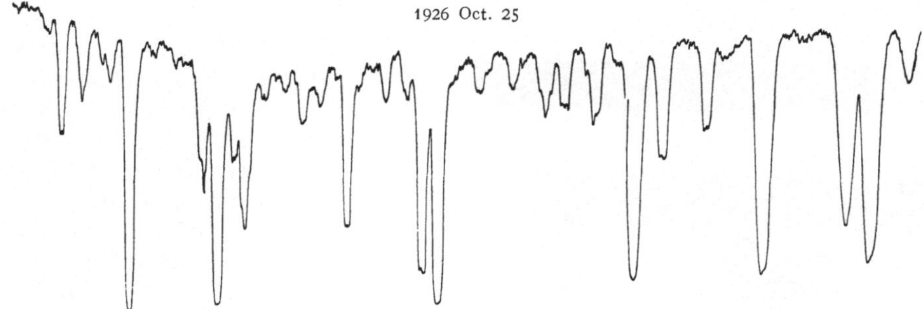

Fig. 12. Spectrophotometric tracings of Nova Pictoris 1925. Late stage. (DAVIDOVICH, Harvard.)

to a displacement factor of 2,4 unusually small for a nova, but this agreed with the slow rise to maximum. The lines were for the most part bordered on their red sides by narrow bright lines. At maximum June 8, 1925, the lines became diffuse and difficult to measure, the bright lines broadened both to the violet and to the red but in two days the spectrum settled down with the dark lines displaced with a factor 10, the original dark lines persisting also in many cases with a factor increased to 2,7. Many new lines also appeared due to Al^+, V^+, Mn^+, Y^+, Zr^+, and also a number of lower temperature Fe lines; and the dark lines became narrower and sharper. Two days later, on June 12, a third, more displaced, absorption line appeared for the hydrogen lines with a displacement factor of 40, diminishing in five days to 35. This greatly displaced absorption component appeared to accompany a few of the lines due to Fe^+ and Ti^+. The following month was occupied by a slow drop from 2^m to below 3^m and a return to a flat maximum at 2^m, a magnitude fainter than at the principal maximum. The chief changes lay in the continuous spectrum: as the star faded, it weakened and the bright bands became prominent only to merge into the continuous at the second brightening. The principal absorption spectrum with a displacement factor of 11 reappeared at this secondary maximum but with fewer lines than before and these more restricted to the ionized atoms. While this remained steady for the second month the hydrogen lines and a few other lines, notably Fe^+ 4233, underwent many striking changes in the width, multiplicity and displacement of the absorption components, a succession of

[1] LUNT, M N 86, p. 498 (1926).

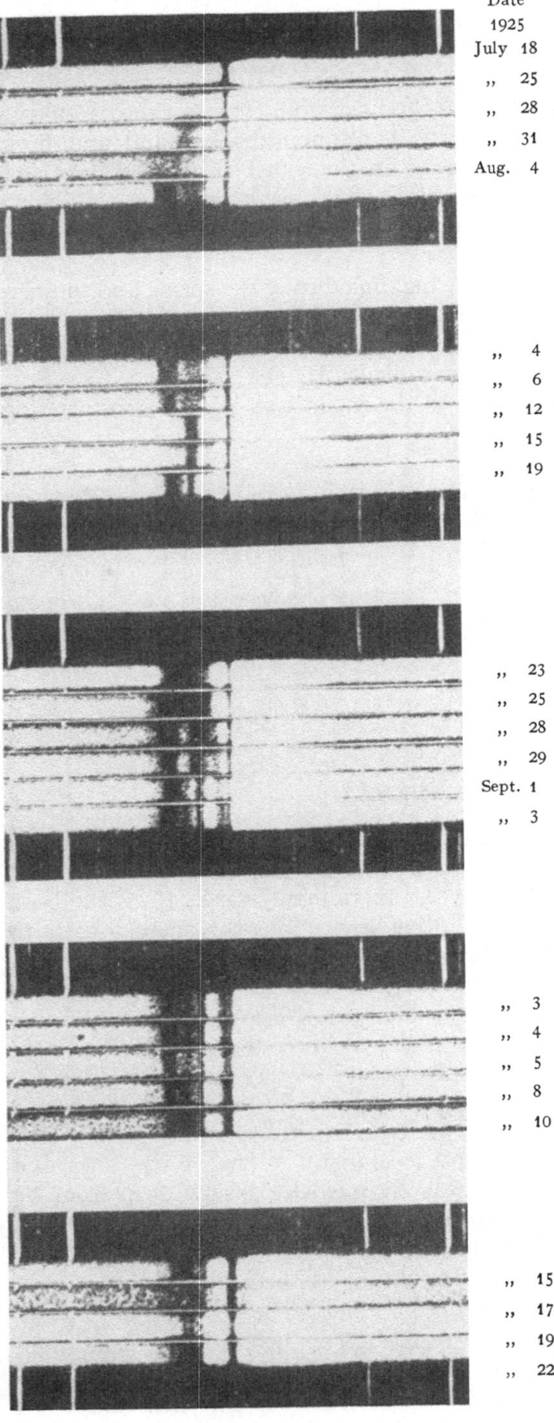

Fig. 13. Structure of $H\delta$ in Nova Pictoris 1925 (Lunt, Cape).

hydrogen outbursts being indicated (Fig. 13). Eight weeks after the principal outburst the star faded again and this time the principal absorption spectrum became blurred and settled down to a slightly more displaced position, factor 13. The hydrogen lines remained multiple and the continuous faded, all the absorption lines except the strongest vanishing. At the third month after maximum came the emergence as weak bright bands of the lines of unknown origin found in η Carinae, but these were shortly followed by a strong dark line He spectrum which agreed in displacement with the most displaced hydrogen absorptions. Save for the absence of nitrogen and for the presence of the η Carinae bands the sequence of this star had so far followed the normal line for novae.

From October to January the spectrum became more and more a bright band spectrum. The principal absorption spectrum still persisting weakly for a few lines gave a slightly increased displacement factor 14, while another hydrogen component and the helium lines gave a factor rising up to 51. Bright bands due to helium, hydrogen and the η Carinae lines at 4244, 4277, 4287, 4359, 4414, 4416, 4814 were all prominent. The bright bands due to ionized iron were fading. The first appearance of any nebular band was that of 4363 on January 27, 1926: 3869 appeared shortly afterwards and the usual nebular bright bands began to appear.

By March 13 the He^{+} band at 4686 had not been detected, but on March 15 this was the strongest band in the whole spectrum. He 4472 which had in January and February been present as a bright band but had faded came up again strongly after the middle of March. The nebular spectrum did not become strong until April 1926, ten months after the principal maximum. A lethargy in development may be noted as accompanying the unusual presence of the unknown η Carinae lines in its spectrum.

Another nova with an exceptional late spectral history is P Cygni which more than 300 years after its maximum remains a star visible to the naked eye and has a bright line B-type spectrum accompanied by absorption lines slightly displaced to the region of shorter wave-lengths (H, He, N, Si, O and S are all present)[1]. There is some evidence for variations from time to time in details of the spectrum of this star.

N Andromedae 1901 (Z) had several sudden outbursts of radiation accompanied by bright hydrogen and ionized helium (4686) bands. It appears to be a border line case as a nova.

27. Other Stars with Typical Nova-Spectra. As has been said already, practically every well marked stage in the ordinary evolution of a nova spectrum is apparently found as a stable or recurrent condition in one or more stars. Thus in BD + 11° 4673 we have dark and bright hydrogen lines—with the absorption lines on the side of shorter wave-lengths—and bright metallic lines. Again HUBBLE's variable nebula—and the nuclear star too—has a bright line metallic spectrum very reminiscent of the early bright line nova stage. J. S. PLASKETT has reported[2] the simultaneous presence in a single star's spectrum v Sagittarii of an A-type absorption spectrum due to ionized metals and a B-type helium spectrum, also absorption, accompanied by complex hydrogen lines; the hydrogen lines show varying emission and have several absorption components displaced to the violet with displacement factors 5, 6, 7, 10; the presence or absence of these components, but not their relative displacements are connected with the orbital phase of the star; the whole absorption spectrum shares in common periodic velocity changes. Another star similar in many respects to v Sagittarii is H D 50820. Here there is no change in the displacement of the lines but we have again a combined A 2 and B 3 absorption spectrum, the latter strengthening towards the ultraviolet. Again the hydrogen absorption is complex with displaced absorption components on both sides of the normal position. The small values of the displacement factors, + 5, − 5 and − 13, illustrate the point that where nova conditions seem to be stable, the displacements observed are less than is usually the case with a rapidly evolving nova.

In the case of γ Cassiopeiae we have the next stage illustrated where a bright α Cygni spectrum and a dark γ Orionis spectrum appear side by side[3]. Here too the hydrogen lines are complex, the displacements of the different absorption components having apparently simple numerical relations with each other. A still more striking star, from its analogy to novae, is H D 45910[4]. This star has a fundamental diffuse helium absorption spectrum: the hydrogen lines of the BALMER series are like those of novae—or rather like those of N Cygni 1600 (P) at the present time—bright with absorption companions on the violet side, with occasional bright centres to the absorption lines: there is a cyclical change in their displacement, with a period of 235 days: superposed upon this spectrum is an α Cygni spectrum, mainly emission, which waxes and wanes, but very occasionally and for a short time it may come out as an absorption spectrum—almost an exact replica of α Cygni. Where absorption

[1] FROST, Ap J 35, p. 286 (1912). [2] M N 87, p. 31 (1926).
[3] Publ. of the Observatory of Michigan, Vol. 2. [4] M N 87, p. 31 (1926).

and emission are both present the absorption lines are as usual on the side of shorter wave-length.

This star links up υ Sagittarii with the spectrum of P Cygni where we have a B-type spectrum with bright and dark components, the dark components being on the usual side. At least 21 other stars are known with the same type of spectrum[1]. These are all either in the larger Magellanic Cloud or else in low galactic latitudes, like P Cygni itself. They are mainly faint and their spectra lie between O6 and B8.

For the later stages of the novae we have parallels in B-type stars with nebular envelopes (or planetary nebulae with B-type nuclei), in planetary nebulae with or without O-type nuclei, in WOLF-RAYET or O-type stars and finally in faint stars about whose spectra very little is definitely known save that they are continuous with no marked feature.

Though it is legitimate to think that any of the stars of the above groups may have passed through the evolutionary phases of a nova and have settled down for a fairly long period—long only perhaps with regard to the normal rate of development of novae—in the stage in which they are now found, we only have evidence of this fact in the case of P Cygni and in the altogether exceptional case of η Carinae. In both these cases the star has had more than one well marked maximum—the outbursts being separated by an interval of some years.

g) Nebulosity and Novae.

28. Nebular Aureoles. The nebulous appearance of novae as compared with ordinary star images has been frequently noticed. Controversy raged over the question whether the fuzzy image reported by skilled observers using refracting telescopes were due to the unusual distribution of the light in the spectrum of a nova—several observers using reflectors finding sharp stellar images. BARNARD, NEWALL and later STEAVENSON all pointed out that a longer focus was wanted for most novae than for ordinary stars, but BARNARD did make the observation in 1903 that while N Coronae 1866 was looking at that date like an ordinary star, N Cygni 1876, N Aurigae 1891 and N Sagittarii 1898 still presented hazy images.

29. Nebulosity round N Persei 1901. The reality of these nebular aureoles round novae was left in dispute until August 19, 1901 when FLAMMARION and ANTONIADI photographed nebulosity of about 6′ in diameter surrounding N Persei 1901. This was quickly confirmed by MAX WOLF and by RITCHEY whose photographs taken a month later shewed the nebulosity extending out to 430″ from the nova. Later photographs secured by a number of observers shewed the nebulosity to be moving outwards with a velocity of 11′ a year (Fig. 14). The nebulosity was traced back by PERRINE on plates secured at the Lick Observatory to March 29, 1901, when two rings could be traced with radii 70″ and 140″. The enormous speed of growth of the nebulosity which by its structure suggested a number of spherical shells with the nova nearly central, indicated that unless the parallax of the nova was very large velocities approaching that of light were involved. The parallax subsequently determined by trigonometrical methods 0″,011 agreed very well with that obtained on the supposition that what was being observed was the spread of illumination from the central source of disturbance at the time of the nova's maximum brightness. This idea first put forward by KAPTEYN is now generally accepted and it gains support from the

[1] J Can R A S 20, p. 20 (1926).

fact that the spectrum of the nebulosity obtained by PERRINE in August 1902 from an exposure of $34^{h}\ 9^{m}$ resembled that of the nova at maximum brightness and not at all the later bright band stage or the ordinary nebular spectrum. PERRINE failed to get evidence of polarisation of the light from the nebulosity thus leaving the idea a possibility that the illumination of the nebulosity was not due to a simple reflection of light from the nova but due to some excitation set up by the intense radiation from the nova, in the way suggested more generally for diffuse nebulosity by HUBBLE. The nebulosity was traced out in 1902 to a distance of 20′ from the nova before it faded away.

b

a

Fig. 14. Nebulosity round Nova Persei 1901, as seen in 1901 (RITCHEY, Yerkes). (a) Sept. 30, 1901; (b) Nov. 13, 1901.

30. Nebular Discs. The extended nebulosity found round N Persei 1901, though it may play an important part in the ultimate explanation of novae is quite exceptional. It gave no clue as to the meaning of the nebulous images reported for earlier novae. BARNARD's skilful visual observations finally decided the point at issue when in December 1916 he found a faint nebulosity round N Persei 1901 much closer to the nucleus than the earlier nebulosity. This new nebulosity was moving outward at a rate of about 0″,4 a year and consisted of a faint ring with the nova at its centre which by August 1919 was about 10″ in diameter, with a brighter fan shaped wisp stretching in the direction preceding the nova to a distance of about 20″. This wisp resembled in shape and position angle a prominent portion of the earlier nebulosity but at a tenth of the distance from the nova (Fig. 15). In October 1918 BARNARD again made the discovery that N Aquilae 1918 showed a planetary disc of diameter 0″,65, growing to 1″,8 by December 1918. Subsequent examination by AITKEN and others showed a disc of 2″,4 diameter in August 1919, 3″,6 in June 1920 and 5″ in the late summer of 1921 and 16″ in June 1926. The curious structure of the bright bands has already been referred to and will be discussed further below; here we need only mention that, assuming the growth of the disc to be due to gases expanding with the velocity

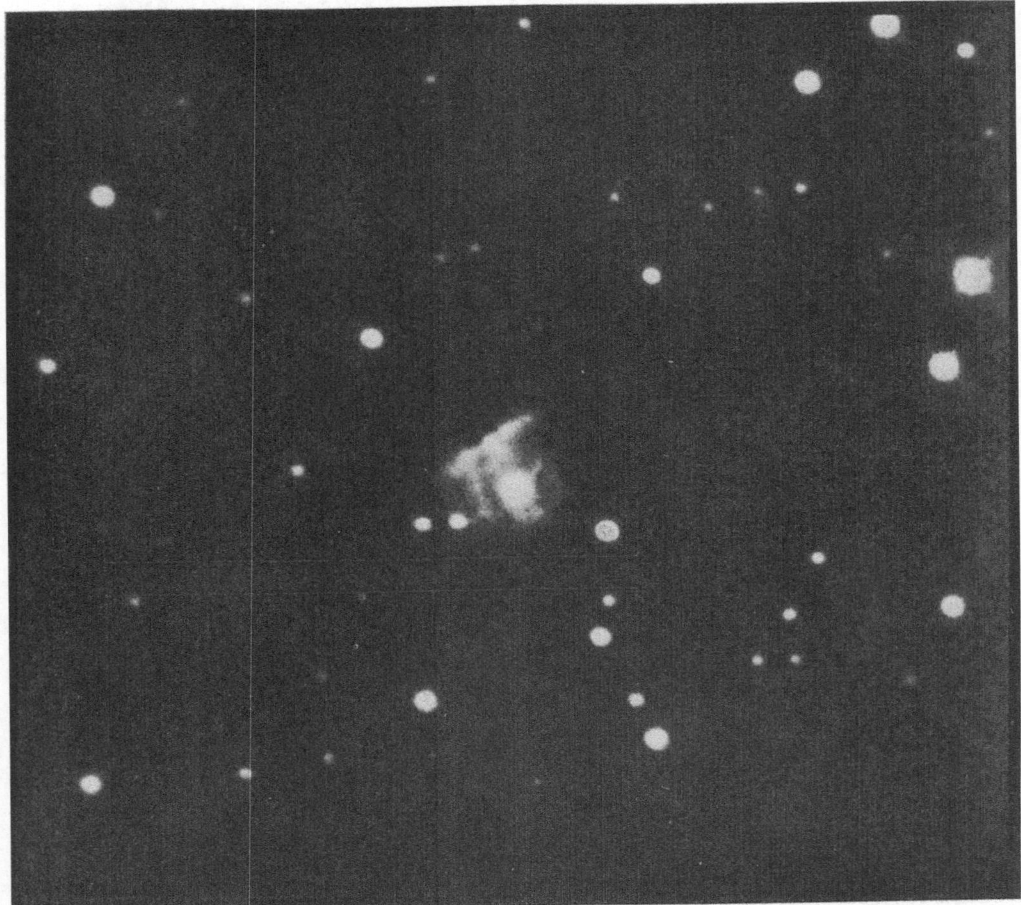

Fig. 15. Nebulosity round Nova Persei 1901, as seen in 1917 (RITCHEY, Mt. Wilson). (The stars on the right top corner of the plate can be recognised at the edge of the central glare in Fig. 14.)

indicated by the strong maxima of greatest displacement from the normal position in the bright bands of the nova, the rate of growth indicated quite a possible parallax, 0″,0025. Here it seems likely that we have witnessed the actual movement of matter outward from the centre of disturbance.

The two brightest novae since N Aquilae 1918, N Cygni 1920 and N Pictoris 1925, have both been observed to be surrounded by a growing nebulous disc while N Ophiuchi 1919 was also found by BARNARD in June 1920 to be embedded in nebulosity. The normal rate of growth of the diameters of these discs for the first year appears to be 2″ to 3″ a year.

31. Complex Structure. It is to be hoped that the remarkable phenomena observed in N Aquilae 1918 as the disc developed may throw light on the phenomena of these growing discs. To recapitulate some of the points of interest. The hydrogen disc as shown from slitless spectrograms only expanded at half the rate of the nebular discs, as shown from the bands of $H\beta$ and of the principal pair of nebular lines. At the same time the velocities of both gases in the line of sight as given by displaced maxima in the bright bands were equal. As time went by a slightly slower rate of expansion was coupled with a decrease in the displacement of the extreme maxima in the bands and at the same time the intensity of these extreme maxima increased greatly relative to the intensity of the less displaced maxima. What seems to be indicated, if the two velocities of expansion across the line of sight and approach are to be related to each other, is either that the hydrogen ceases to radiate as it moves further out or that it changes into some condition where the nebular spectrum is given out. By September 1920 the disc was sufficiently developed for some structure to be seen. Along a diameter in the position angle for the slit, which gave curiously displaced maxima as mentioned above, a bright bar was seen by PEASE crossing the disc, both in visual observations and in the slitless spectrograms. The latter gave for $H\alpha$ a dumb-bell shape, the bar being S-shaped. The reported length of the bar in the slitless spectrogram, 1',5 suggests that the figure there is much distorted by DOPPLER effects differing at different points of the disc. The latest observation of this nova, Sep. 21, 1926, at the Lick Observatory, shows a star of high intrinsic luminosity and of faint total brightness, possibly the result of an extended outer atmosphere which still surrounds the star and continues to expand. A slitless spectrogram secured in June 1926 at Mt. Wilson showed that He^+ 4686 had the brightest disc, that the BALMER lines were represented by discs apparently of the same size while of the nebular lines only faint traces were present[1].

h) Theories.

32. Early Views. It now remains to gather up the observed facts and to compare them with the many theories worked out by astronomers and physicists in explanation of novae. FORTUNIUS LICETUS in *De Novis Astris* (1622) propounded 20 theories including a special creation, the passage of clouds from in front of a star, the near approach and departure of a star, the conglomeration of matter (TYCHO's view), the conjunction of stars or planets, the brightening of existing stars, exhalations from the solar system or comets gathered together. RICCIOLUS in his *Almagest* added the idea of a rotating body, coupling with it the view that the outburst of brightness, when a particular part of the globe was turned to us, was a special warning act of God. NEWTON with a curious foretaste of certain modern ideas suggested that fixed stars, that have been gradually wasted by the light and vapours emitted from them for a long time, might be recruited by comets that fell upon them. LAPLACE on the other hand suggested that we had to do with a great surface conflagration of a star. The views of NEWTON and LAPLACE — collision or explosion from a simple star — have in varying forms reappeared in most speculations on the subject during the last sixty years, since spectroscopic results have been available to check theoretical views.

It must be borne in mind that there are two separate problems. First there is the question as to what underlying cause leads up to the outburst of a nova: secondly there is the question as to exactly what physical processes occur during

[1] Ap J 66, p. 59 (1927).

and after the highly disturbed stage of maximum and how the alleged cause leads up to them. As above mentioned the answers offered by many thinkers to the first question group themselves in support of the views of NEWTON or LAPLACE. We first mention the various forms taken by the collision theory.

33. Collision Theories. The evidence of the pre-existence of a faint star in the position of the nova seems now so general that NORMAN LOCKYER's view of the collision or near approach of two streams of meteorites must he abandoned. A star being one of the bodies concerned in the formation of the nova, for the other body or bodies various suggestions have been made. W. H. PICKERING has suggested a planetoid or cluster of planetoids. He claims that sufficient energy could be liberated and describes the phenomenon as being, on a very much larger scale, akin to the sudden outburst of two bright objects observed by CARRINGTON in 1859 over a sunspot. NÖLKE also suggested that a planet falling into the Sun produced a nova. SEE follows NEWTON in taking a huge comet as the second body while BICKERTON suggests a second star as required. The latter regards the nova as composed of a third star composed of matter struck off from the two colliding stars through their grazing impact. He regards the new star as an exploding sun which is surrounded by a series of expanding shells. ARRHENIUS explained the spectra in terms of jets of gas expelled from a rotating central mass formed by the collision of two stars. VOGEL, to account for the complex nature of the dark and bright line spectra, suggested that the nova outburst was caused by a star running into a group of several bodies, e. g. a collision of a dark star with a solar system of star and planets. In this way he suggested that the variations in the nature and the displacements of the spectra could be explained. But of all the collision theories the one that has been most fully worked out and which proves the most elastic in application to varying conditions is the hypothesis first worked out in detail by SEELIGER[1] of the entry of a star into a dust cloud or gaseous nebula.

34. SEELIGER's Theory. Opposing the explanation in terms of the collision or close approach between two stars, partly on account of the special conditions required which would have to be repeated for several novae, partly on account of the great stellar masses or initial velocities of approach that would be required, he shewed that results of the right order of magnitude for heating and brightening the star can be obtained by considering the motion of the particles of a nebula as a star enters it. For this solution there is a wide range of choice of reasonable values for nebular density and stellar mass without impossible or inconsistent results following. According to SEELIGER's views the star on entering the nebulosity would be heated up and gases would be evolved from its surface which would take up the velocity of the neighbouring streams of nebulosity. Secondary maxima are easily explained by varying density in the nebulosity while periodicity might be explained as due to rotation of the star. The range of velocities for particles in the nebula which pass the star at different distances accounts for the broad bright bands and the complex spectra in general.

HALM[2] considerably developed SEELIGER's theory adding an important corollary that the atmosphere round the star would expand under the influence of the heat generated in the bombardment. In addition to the streaming of matter past the star in hyperbolic orbits (and, for captured particles, the motions in more or less elliptical orbits round the star), there would be the expansion of shells round the central star. These shells would account for the ordinary dark line and bright band spectra. Absorption lines displaced to the violet

[1] A N 130, p. 393 (1892); 181, p. 81 (1909).
[2] Proc R S Edinburgh 25, p. 513 (1904).

could arise from the thin layer of gas proceeding out from the star towards the observer: broad bright bands would come from the gas moving out from the star across the line of sight and viewed through a much greater thickness. Modern views of atomic radiation make it less easy to accept the view that the same type of atom is giving absorption lines when moving towards the observer in front of the star and giving bright lines when seen clear of the brighter background, and a satisfactory explanation in terms of mere contrast effects can not easily be given. Considerations of intensity gradients and of the relative thicknesses of different layers give a possible solution which allows of the very dark absorption lines noted by WRIGHT.

HALM explained further how rotating rings would give maxima of brightness in the bands on opposite sides of the normal position of the corresponding spectral line and shewed that the observed displacements would represent reasonably possible velocities. He also pointed out how the bright belts should move slightly to the violet as the continuous spectrum faded, a result confirmed by CAMPBELL for N Aurigae 1891. Differences in the masses of various novae would chiefly affect the displacements of maxima in the bright bands; differences in the density of the cloud would affect the displacements of the dark lines. SEELIGER's hypothesis could be quite properly extended by making the star a double star. If the components were of different density then the corresponding pair of absorption lines would have different displacements. It is doubtful whether the introduction of the idea of a double star gives the most successful explanation of the double hydrogen spectrum. The different hydrogen absorptions appear first on different days and occasionally more than two dark lines appear. But the assumption of a double star gives a very natural explanation of some of the more complex data and in particular of periodicity.

E. W. BROWN[1] has with reference to HUBBLE's variable nebula, which is an appendage to R Monocerotis, worked out the simpler illumination effects which should be observed when a star enters a nebula. There is a strikingly close resemblance between the latest photographs of N Persei 1901 taken since 1916 and BROWN's picture of a spherical expanding shell with the star at its centre and a fan like appendage on one side of the star. Partly for its elasticity of treatment for varying conditions SEELIGER's theory is to be preferred amongst those based on celestial collisions. A fairly elementary theoretical discussion shows that it may lead to results of the right order of magnitude and its applicability to at least one case is suggested for N Persei 1901.

35. Explosion Theory. A bridge between the two views of NEWTON and LAPLACE is supplied by the tidal theory of KLINKERFUES. Although in this view the whole disturbance emanates from a single star its cause lies in the near approach of a second star with consequent tidal eruptions. The breakdown of the cooling crust of a dark star with consequent outrush of gas has been suggested by ZÖLLNER, HUGGINS, VOGEL, DESLANDRES and others. According to some writers (HARTMANN, KAYSER) radioactive forces, according to others (LOHSE, HUGGINS, VOGEL) chemical combination at low temperatures are the cause of the breakdown of equilibrium. In either case the star "blows up and bursts". The increasing brightness may be attributed to the expanding disc as much as to an increase of temperature. HARTMANN bases his view on the case of N Pictoris 1925 where a slow approach to maximum was followed by the sudden outburst of the typical nova spectrum at maximum. This he held to be inconsistent with any collision theory; his argument does not apply to

[1] Ap J 53, p. 169 (1921).

the passage of a not very massive star through the outer and more tenuous regions of a nebula. TURNER curiously inverts the explosion theory in a suggestion that central condensation of nebulosity surrounding a star is the effective cause of the outburst. SHAPLEY, noticing many likenesses to an ordinary variable star (e. g. the rapid rise in brightness followed by the slow decline and the simultaneous change of magnitude and spectrum), suggests that similar causes are at work in the case of a nova: the veil theory, pulsation, friction from a resisting medium, explosions, a close periastron passage might separately in some accentuated form contribute to the result.

36. Spectral Sequence. Whatever the cause may be of the disturbance the spectral changes that follow have to be explained as well as the unusual light curve. DESLANDRES suggested the presence of convection currents such as he had observed in the Sun; EBERT and JULIUS, anomalous dispersion; WILSING, VOGEL and DUFFIELD, the effect of a pressure shift coupled with rotation and expansion and the superposition of layers at different pressures; KAYSER, a canal-ray effect, basing his view on the alleged discovery of radium in the spectrum of N Geminorum 1912. It may be said that none of these explanations fit the detailed discoveries in the spectrum of N Aquilae 1918, and few of them are consistent with the generally accurate statement that the displacements of absorption lines or of maxima in bright bands vary directly as their wave-length. They may have their part in explaining the departures from this simple law but no details have been worked out as yet for any one star so as to show the presence of any one of these factors as definitely at work. WILSING, in applying his laboratory results on the pressure effect shewn by sparks in liquids, did indeed suggest that he might be only obtaining an explanation of deviations from an ordinary DOPPLER effect, but here again no detailed analysis studying the variations between the behaviour of different elements has yet been made in support of the view.

37. Outward Expansion. Whatever other factors may be concerned in the production of the spectrum of a nova it is almost certain that outward expansion of gases from some central body is taking place. This has been verified directly by a study of growing discs round a nova: it gives a simple, if incomplete, explanation of the dark lines displaced to the violet and the broad bright bands. It is consistent with the initial increase of light, which seems from a study of the continuous spectrum of a nova at the earliest stages to be not entirely due to increase in surface temperature. (DAVIDOVICH found that the pre-maximum displacements of the dark lines in N Pictoris 1925 were consistent with the change of linear dimensions in the star, obtained by a study of its effective temperature and its absolute magnitude — or spectroscopic parallax — during the stage before bright lines had become visible.)

38. MILNE's Theory. MILNE has given[1] an interesting explanation of the manner in which the immense outward velocities indicated by the displacements of the dark lines could be developed under the influence of an increase in the intrinsic brightness of a star's surface. A brightening of the photosphere of the star would result in atoms, previously balanced under gravity and radiation pressure, being driven outwards. These atoms moving outwards absorb radiation from the star in a wave-length shorter then usual, when referred to a frame of reference fixed relatively to the star. The atom moves into the wing of the absorption line, so far as the supporting radiation is concerned, it is exposed to stronger radiation and is accelerated further outwards. If enough atoms

[1] M N 86, p. 459 (1926).

move out together they will give an absorption line moving to the violet or suddenly appearing in a displaced position; this would happen as the atoms began to acquire their common limiting velocity in sufficient numbers. MILNE has calculated a value between 1000 and 2000 km/sec for the limiting value of the outward velocity of calcium atoms leaving a normal star and suggests that higher limiting velocities would be possible for a nova. A number of the complex changes in nova spectra can be accounted for in terms of MILNE's theory if we assume that the brightening of the photosphere is not uniform over the whole star and couple the outward expansion of the chromosphere with a rotation of the star. The fact that atoms of different elements move out with the same velocity would be due to the atoms mingling in a common jet or shell. Multiple absorptions corresponding to widely different velocities might be traced back to successive brightenings of the star, for which of course there is good evidence. If we accept SEELIGER's hypothesis we might even explain the higher velocities found for the later outbursts as due to the greater freedom from resistance from the nebula secured by the outward flow of the earlier shells which had met the approaching nebula. Such a resistance might explain the decrease of velocity found at first for a succession of shells in the case of N Geminorum 1912[1]. The term in λ^2 for the displacement of the absorption lines found by WRIGHT for this same star[2] still remains unexplained, unless the change in radiation pressure was some function of λ and atoms in a state to give absorption lines in a particular region of the spectrum tended to move together with a limiting velocity which was also a function of λ. The comparative crowding together of lines in the ultra-violet spectrum of a nova and the comparative freedom from lines in the visual spectrum would also tend to increase the velocity of such atoms as were in the state to give the lines of longer wave-length and might contribute towards the result found by WRIGHT.

The change of spectral type from A to B and then from B to P (with a nucleus of type O) need not be regarded as a problem in any way peculiar to novae; it should take its place as an instance of the manner in which the spectrum of a star or nebula is a function of the ionization and density of the gases involved in the external radiation, a somewhat complex illustration of SAHA's theory. The fact that in the case of N Aquilae 1918 hydrogen and nebulium lines showed equal velocities of expansion as judged by the DOPPLER effect while the hydrogen disc only grew at half the rate of the nebular disc might mean that whereas the hydrogen had become at a given distance from the central star too diffuse to radiate the BALMER series to a measurable extent the conditions of greater tenuity did not inhibit the presence of the nebular spectrum. An extension of a bright nebulous envelope beyond the limit of the bright hydrogen is not uncommon in planetary nebulae and it may be paralleled by hydrogen envelopes surrounding early type stars. It is probably the change in the physical conditions rather than the outflow of matter of changing chemical constitution which is operative in producing the sequence of spectra.

The long persistence of unchanged structure in the bright bands even with changing elements is not a serious difficulty in this theory though the presence of a number of well marked and widely separated maxima in these bands is a difficulty, which is only partly met by the fact that, as seen in the case of N Aquilae 1918, these maxima may be traced to different portions of the growing disc. The strengthening of the maxima shewing the largest velocities, at the expense of the others, is in accordance with MILNE's theory, and the existence of a group

[1] WRIGHT, Lick Obs Publ 14, p. 77.

[2] WRIGHT, Lick Obs Publ 14, p. 51.

of maxima corresponding to different limiting velocities may be easily explained by supposing that each of the different elements principally involved tends to reach its own limiting velocity while a stream or shell of atoms of different elements mixed together tends to settle down with the velocity corresponding to that particular element. The apparently complex system of rotation indicated by the later bands of N Aquilae 1918 may perhaps be interpreted in terms of the oblique entry of a binary star into a nebula, the plane of rotation of the star not being the same as the place of the principal motions set up in the nebula.

It is hardly to be expected that the many and various phenomena observed in the evolution of novae can be explained by a simple model. But, in the view of the writer, SEELIGER's theory that the initial disturbance is due to the entry of a stellar system into a nebular cloud, followed by the spread from one or more bodies, probably under increased radiation pressure, of outward moving shells or jets of gas, meets most of the facts so far known and gives the most consistent account so far offered, of what actually takes place.

Chapter 4.

Double and Multiple Stars.

By

F. C. HENROTEAU, Ottawa[1].

With 69 illustrations.

Introduction.

1. To interpret astronomical phenomena in the realm of double and multiple stars, is to disclose a series of startling discoveries and point the way to further revelation, all having a tremendous significance in general physics, mathematics and philosophy as well.

There is demand for endless research when one reflects that a star, assumed by the ancients to be a sacred fire forever burning on the hearth of the gods, may appear double through a telescope, formed of differently coloured points of light unequal in brightness and, at times, apparently so close to each other that long concentration is necessary to distinguish them; and when after years of measurement it is found that the chasm which separates them extends millions of miles, that they are perhaps larger than our Sun, itself so vast compared to the Earth, and swirl about a common centre-of-mass with the astonishing speed of miles per second, then one confronts the fathomless mystery of all ages where the most gifted mathematician and the wisest philosopher become as children who dream.

To-day, not only do we contemplate celestial bodies, revelling in colour, shape and lustre, but daring further we weigh them, ascertain temperatures, describe motions, pulsations and revolutions until at last the mechanism of the birth of satellites is no longer a wholly unsolved enigma.

Unremitting double-star work has resulted in knowledge relating to stellar bodies, revealing the extraordinary fact that the absolute luminosities of the stars are very nearly proportional to their masses irrespective of temperature or density, and it is a question whether energy of mass can be transformed into energy of radiation. In other words, we arrive at the revolutionary concept that light, hitherto considered imponderable, is a form of matter.

The most difficult propositions in dynamics are presented by multiple systems; even in double systems where the bodies are not spherical and tides or pulsations play an important rôle, there is opportunity for unlimited computations. The theoretical three-body problem remains a most intriguing puzzle.

The genesis of double stars has been looked upon as a most baffling question to which years of study have been devoted. The imagination can only conjecture the huge rotating spheroidal bodies which in time become unsymmetrical pseudo-ellipsoids, and after a series of cataclysmic changes in configuration, finally separate into unequal masses.

[1] To the Memory of Doctor OTTO KLOTZ, whose Life Work culminated in the Development of Astrophysics at the Dominion Observatory, Canada.

Planetary systems present an interesting philosophical aspect; although outside of our Sun no such system is positively known, there are indications that some super-giant stars have companions of comparatively small mass, in fact, according to EDDINGTON's theories, not large enough to be luminous.

Even to the remotest parts of the Universe, double and multiple star astronomy proves that the laws of gravitational attraction are true. These laws may be slightly modified by the theory of relativity or by a possible new relation between mass and radiation, but they remain universal.

It is comforting to know that the whole Universe, with all its complexities, is governed by a small number of simple rules. Nevertheless, the very existence of these bodies, the harmony of their motions and the intricacy of their relationships lure the astronomer from one inexplicable fact to another, until his ever deepening perception reaches beyond the measurements of time and space.

a) Double-Star Observers and their Work.

2. Knowledge Previous to the Telescope's Invention. Among the peoples of ancient civilization it is interesting to note that very early attention was called to double and multiple stars as subjects for observation and study, since even to the naked eye some stars appear to be in close proximity. The name Alcor, meaning eyesight, was used by the Arabs to designate the faint companion of ζ Ursae Majoris or Mizar; the ancient Arabs, and perhaps the Assyrians and Babylonians, when they could see Alcor considered their vision excellent.

PTOLEMY[1] called ν Sagittarii a double star, its two components being 14′ apart. The Italian astronomer RICCIOLI in 1651 claims that VAN DEN HOVE, otherwise called HORTENSIUS, indicated two double stars, one in Capricornus and the other in the Hyades, their components being approximately 5′ apart. The separation of such pairs of stars visible to the naked eye, is far too great to classify them as double.

In telescopes which magnify hundreds of times, double-star components are seen very near each other, often so close that it takes an expert eye to distinguish their division.

3. The First Double Stars Discovered. Optical double stars merely appear to be close to each other although they are perhaps at different distances from us, but physical doubles, also called binaries, are near enough to attract each other appreciably. GALILEO, possessing no high-power telescope, never specialized in the study of double stars but recognized the existence of optical pairs, and in his Giornata terza, mentions the usefulness of such stars in determining their parallaxes.

Among the first double stars discovered may be mentioned ζ Ursae Majoris (Mizar) found in 1650 by the Italian astronomer JEAN BAPTISTE RICCIOLI, and also found independently in 1700 by the German astronomer GOTTFRIED KIRCH.

In 1664, while observing a comet HOOKE discovered γ Arietis, and considered the circumstance so extraordinary that he wrote[2]: "I took notice that it consisted of two small stars very near together; a like instance to which I have not else met with in all the heavens." HUYGHENS in 1656 saw θ Orionis resolved into three stars, while J. D. CASSINI[3] in 1678 found the two double stars β Scorpii and α Geminorum.

---

[1] PTOLEMAEUS, Magna compositio. Venetiis 1515, lib. VIII, cap. 1.

[2] London R S Phil Trans, 1665, p. 150.

[3] Paris, Histoire de l'Académie des Sciences avant son renouvellement en 1699, 1, p. 216 (1733).

It is of interest to give the date of discovery of a few others until about the middle of the 18th century.

Star	Discoverer	Year of Discovery
α Crucis	Father FONTENAY[1]	1685
α Centauri	Father RICHAUD[2]	1689
γ Virginis	BRADLEY and POUND	1718
Castor	BRADLEY and POUND	1719
ζ Lyrae	BIANCHINI[3]	1737
ζ Crucis	LA CONDAMINE[4]	1749
61 Cygni	BRADLEY	1753

A first attempt to measure the relative positions of binary components was made by Father LOUIS FEUILLEE[5] in 1709, when he estimated the position angle and distance for α Centauri; however, toward the end of the 17th century FLAMSTEED, even at that early date, used a micrometer. Excluding those having components more than 32″ apart, by the middle of the 18th century scarcely twenty double stars were known.

4. LAMBERT and MICHELL. Two philosophers, J. H. LAMBERT[6] and the Rev. JOHN MICHELL, although not double-star observers, published the first definite ideas concerning the mutual attraction that exists between their components. LAMBERT, as well as KEPLER, thought that like our sun the distant suns must be surrounded by several dark bodies similar to planets or comets. As to stars which appear very close together, he was inclined to believe that they were satellites of a central dark body and suggested that these stars might also turn around their common centre of gravity in a relatively short interval of time.

MICHELL presented a different theory, computing the probability for stars to be double or multiple if their components have no relation whatever. On pages 243 and 249 of his interesting memoir in the Philosophical Transactions[7] he says: "We may conclude with the highest probability (the odds against the contrary opinion being many million millions to one) that stars form a kind of system by mutual gravitation. It is highly probable in particular and next to a certainty in general that such double stars as appear to consist of two or more stars placed near together, are under the influence of some general law, such perhaps as gravity ..."

5. CHRISTIAN MAYER (1709—1783). Father CHRISTIAN MAYER of Mannheim, Germany, was the first to do systematic work in double-star observation. The two memoirs entitled "Gründliche Vertheidigung neuer Beobachtungen von Fixsterntrabanten; 8°, Mannheim 1778" and "Dissert. de novis in Coelo sidereo Phaenomenis in miris Stellarum fixarum Comitibus, 1779" describe eighty double stars, sixty-seven being separated by less than 32″.

In his work Father MAYER used the excellent eight foot Bird mural quadrant with a power of 85. Some of the pairs discovered are very close, being difficult to distinguish even with powerful instruments, as for example ϱ and 71 Herculis, 5 Lyrae and ω Piscium. In order to obtain the differences in right ascension and declination of the components, only meridian instruments were used, and because of this when Father MAYER tried to compare his observations with previous ones to find changes in position, he could not always distinguish which displacements were due to proper motions. However, the same method was used by others long after Father MAYER's time.

[1] Idem 2, p. 19 (1733).
[2] Idem 7, p. 206 (1729).
[3] Observationes selectae astronomicae, Veronae, 1737, p. 208.
[4] London R S Phil Trans 1749, p. 139.
[5] Journal des Observations Physiques, Paris, 1, p. 425 (1714).
[6] Cosmologische Briefe über die Einrichtung des Weltbaues (1761).
[7] London R S Phil Trans 57, p. 231 (1767).

He was subjected to bitter criticism from the mathematician NICOLAS FUSS[1], who objected to designating the fainter components of double stars as satellites, and especially did he take exception to the statement that the two faint stars distant 2°30′ and 2°55′ from Arcturus were satellites of the latter. Could planetary bodies be seen reflecting the light of such distant sources? Father MAXIMILIAN HELL, director of the Imperial Observatory at Vienna, also criticized the astronomer, even doubting the authenticity of his observations; but the high value of MAYER's work was recognized by J. E. BODE of Berlin, who included his list of double stars or first catalogue ever published, in the Astronomisches Jahrbuch für 1784 (issued in 1781) under the title: "Verzeichnis aller bisher entdeckten Doppelsterne." Long after his death the priest's distinguished contribution to science was still mentioned by STRUVE and MÄDLER.

6. WILLIAM HERSCHEL. These meager but memorable beginnings were followed by Sir WILLIAM HERSCHEL's devoted work which crystallized all efforts in the field of double-star astronomy and built a strong foundation for further research. As an amateur, he revelled in observing Saturn's rings and the Orion nebula, but after his invention of the "revolving micrometer" in 1779, serious plans were formed in order to test GALILEO's idea relating to the determination of double-star parallaxes. On December 6th 1781, HERSCHEL read a paper before the Royal Society from which the following extract is taken: "As soon as I was fully satisfied that in the investigation of parallax, the method of double stars would have many advantages above any others, it became necessary to look out for proper stars. This introduced a new series of observations. I resolved to examine every star in the heavens with the utmost attention and a very high power, that I might collect such material for this research, as would enable me to fix my observations upon those that would best answer my ends." This was the beginning of his interesting work which is at the root of all that is known to-day about visual double stars. He published catalogues of double stars in 1782, 1783 and 1804, containing 846 stellar pairs practically all of them discovered and measured by himself[2]; also HERSCHEL's genius is reflected in all that relates to orbits, assumed periods of revolution, magnitudes and colours of components, and the classification of double stars according to their separation.

It is extraordinary that HERSCHEL's research, so prolific in discoveries, has been of little value in determining parallaxes. New and marvelous fields were opened and he was able to announce in the "Philosophical Transactions", 1803, Part I, p. 339, after years of investigation, that many of the stars are not only double in appearance but must be considered as a real combination of two stars intimately bound together by mutual attraction. This was an achievement of note. In the same memoir he also gave the approximate periods of revolution for five pairs which are:

Star	Revolution years
Castor	342
γ Leonis	1200
ε Boötis	1681
δ Serpentis	375
γ Virginis	708

Henceforth, NEWTON's law of gravitation was recognized as controlling, not only the solar system, but the stellar universe as well.

Although WILLIAM HERSCHEL had more powerful telescopes than were used before, they were unequipped with clock-work, their mountings of the altazimuth type, and his micrometers very crude compared with those built by modern mechanics. Consequently, his observations cannot compare with those of to-day.

[1] Acta Academiae scientiarum Petropolitanae, Petropoli 1780.

[2] London R S Phil Trans 1782, p. 40; 1783, p. 112; 1804, p. 87.

7. Wilhelm Struve. After William Herschel, Wilhelm Struve and Sir John Herschel devoted their no less admirable activities to further the knowledge of double-star astronomy. Struve began his work at Dorpat, in 1813, and published the results of his investigations in his "Catalogus 795 stellarum duplicium" in 1822; 500 of these stars were within the limiting distance of 32″.

The celebrated Fraunhofer refractor which Struve received in 1824, thirteen feet long with an aperture of nine Paris inches and possessing an excellent driving clock, marked a new era in double-star work. With this instrument, he examined some 120000 stars in 129 clear nights between November 1824 and February 1827. The "Catalogus Novus" appearing in 1827 was the result of all this labour. It contained 3112 double stars classified according to their distances apart, and briefly described. This book was followed by the publication in 1852 of "Positiones Mediae", giving the right ascensions and declinations of all the above stars. Previously, in "Mensurae Micrometricae", were given the relative positions of 2640 stars of the first catalogue. This last volume is huge, involving an enormous amount of work.

Struve adopted a new method of reckoning position angles in binaries by counting them from 0° to 360°, starting from the point north of the primary and turning from this origin toward the point east. They are thus measured counter-clock wise.

He analysed the colours of 600 double stars chosen among the brightest with the following results: for 375 stellar pairs both stars have similar colours with the same degree of intensity; for 101 pairs the colour is identical, but more pronounced in one of the components; many others differ completely in appearance. Perhaps half the stars investigated are white while those with two colours show yellow and blue as in ι Cancri, or orange and green as in γ Andromedae.

The Czar Nicholas, in 1837, called Struve to found the Russian Imperial Observatory at Pulkowa near St. Petersburg, and a refracting telescope of fifteen inches aperture was mounted, at that time the largest in the world. Wilhelm Struve remained at Pulkowa until 1858, his son Otto succeeding him as director of the observatory.

8. John Herschel and James South. Sir John Herschel, in 1816, decided to extend and elaborate his father's work, and Sir James South having independently arrived at the same decision, they perfected plans to cooperate during the years 1821 to 23, using South's five and seven foot refractors respectively of $3^3/_4''$ and 5″ aperture. The results were published in a catalogue comprising 380 stars[1].

Later both astronomers continued their researches separately; Herschel providing himself with a twenty foot reflector, the mirror of which he ground, also used the achromatic refractor which he purchased from South. Both observers measured practically all the double stars discovered by William Herschel, some of them many times, and found a large number of new pairs. The major proportion of these discoveries may be attributed to John Herschel; eight catalogues published in the Memoirs of the Royal Astronomical Society contain a mass of information concerning them.

During John Herschel's residence at Feldhausen near the Cape of Good Hope in the southern hemisphere, his work was epoch making in the history of astronomy. Besides examining about 4000 nebulae and star groups, he observed more than 2100 double stars[2], but the results were not published until 1847, his manuscript being prepared for the press entirely by himself.

[1] London R S Phil Trans 1824, part 3.

[2] J. Herschel, Results of Astronomical Observations made at the Cape of Good Hope, London 1847.

The Royal Astronomical Society presented gold medals to both Sir JOHN HERSCHEL and Sir JAMES SOUTH. As a matter of fact only a small proportion of double stars discovered by these astronomers have proven to be real physical systems, the majority of them being too wide to show orbital motion in the course of a century.

9. The Second Period of Visual Double-Star Astronomy. A few astronomers were engaged in studying double stars when JOHN HERSCHEL left England for the Cape of Good Hope, and these men may be said to have inaugurated the second period of research, succeeding systematic investigation begun by Father CHRISTIAN MAYER, and so capably carried on by the distinguished scientists previously mentioned.

This second period was a time of concentrated effort and exhaustive research, and while the results were significant they were not so vital to science as the work of the first pioneers. Following is a list of second period observers:

Astronomers	Place of Observation	Years of Activity	References
BESSEL	Königsberg	1814—34	1, 2
DUNLOP and BRISBANE .	Paramatta	1823—27	3
DAWES	Ormskirk	1830—44	4, 5
SMYTH	Bedford	1830—60	6
ENCKE and GALLE	Berlin	1838—48	7
KAISER	Leiden	1840—66	8
OTTO STRUVE	Pulkowa	1839—75	9
MÄDLER	Dorpat	1834—61	10
SIMMS, PHILPOTT, MORTON	Lord Wrottesley's Obs.	1843—59	11
JACOB	Madras	1848—52	12, 13
MITCHELL	Cincinnati	1846—48	14
DEMBOWSKI	Naples	1851—78	15
POWELL	Poonah	1853—62	16
WINNECKE	Berlin	1855—56	17
SECCHI	Rome	1856—66	18
ALVAN CLARK	Boston	1857—60	19
MAIN	Oxford	1862—77	20
ENGELMANN	Leipzig	1864—86	21
WILSON and SEABROKE .	England	1872—80	22

[1] Verzeichnis von 257 auf der Königsberger Sternwarte beobachteten Doppelsternen. A N 4, p. 301 (1825).

[2] Astronomische Untersuchungen, Königsberg, 1, p. 280 (1841).

[3] Mem R A S 3, p. 257 (1829).

[4] Mem R A S 8, p. 61 (1835); 19, p. 191 (1851).

[5] Catalogue of Micrometrical Measurements of Double Stars. Mem R A S 35, p. 137 (1867).

[6] A Cycle of Celestial Objects; 2 vol., London 1844.

[7] Astronomische Beobachtungen auf der Sternwarte zu Berlin 1, p. 141 (1840); 3, p. 234 (1848).

[8] A N 18, p. 1 (1840); 64, p. 97 (1865).

[9] Catalogue revu et corrigé des Etoiles Doubles et Multiples. Mémoires de l'Acad. Imp. des Sciences de St. Pétersbourg 7 (1853), see also 9 and 10 of the Pulkowa Observations.

[10] Dorpat Obs 9 (1842) to 15 (1859).

[11] Mem R A S 29, p. 85 (1861).

[12] JACOB, Astronomical Observations made at Madras 1848—1852, Madras.

[13] Mem R A S 17, p. 79 (1849).

[14] Published by O. STONE in Publ. of the Cincinnati Observatory No. 2 (1878).

[15] In numerous articles of the A N from 42 (1856) to 92 (1878). See also Misure Micrometriche di Stelle Doppie et Multiple fatte negli anni 1852—1878 dal Barone ERCOLE DEMBOWSKI, Roma, 1883.

[16] Mem R A S 25, p. 55 (1857); 32, p. 75 (1864).

[17] Astronomische Beobachtungen auf der Kgl. Sternwarte zu Berlin 5, S. 255 (1884).

[18] Atti dell'Accademia Pontificia dei Nuovi Lincei, Roma, 13, p. 8; 14, p. 1; 21, p. 159.

[19] M N 17, p. 257 (1857); 20, p. 55 (1860).

[20] Results of Astronomical and Meteorological Observations made at the Radcliffe Observatory, Oxford, 20 (1862) to 36 (1880).

[21] A N 64, p. 81 (1865); 70, p. 257 (1868).

[22] Mem R A S 42, p. 61 (1875); 43, p. 105 (1877).

The Rev. W. R. Dawes, Admiral W. H. Smyth, Otto Struve, Mädler, Baron Ercole Dembowski and Father Secchi are perhaps the most prominent.

The Rev. W. R. Dawes began his work using a three or an eight inch refractor, later having the use of larger telescopes. Gifted with a remarkable keenness of vision, he was given the name of the "eagle-eyed". In presenting the gold medal of the Royal Astronomical Society to the Rev. Dawes, Sir G. Airy said: "Distinguished as Mr. Dawes has been by an extraordinary acuteness of vision, and by a habitual and contemplative precision in the use of his instruments, his observations have commanded a degree of respect which has not often been obtained by the production of larger instruments."

The Bedford Catalogue and the Cycle of Celestial Objects of Admiral W. H. Smyth became very popular. The Cycle contains positions, micrometrical measures and other data, for 506 double and multiple stars, and Smyth received for it the medal of the Royal Astronomical Society.

The work of Otto Struve like that of his father, was invaluable; their double stars, being designated by Σ for Wilhelm Struve and by $O\Sigma$ for Otto, are still used extensively in many observing programmes.

Mädler was chiefly interested in problems relating to the construction of the Universe; he succeeded Wilhelm Struve at Dorpat, and maintained the high reputation of the observatory for double-star investigation. Numerous micrometric measures, discussions of binary-star motions and other valuable information, are to be found in "Dorpat Observations". There is also much of interest in his "Untersuchungen über die Fixsternsysteme".

Baron Ercole Dembowski was influenced in his choice of double-star research by Don Antonio Nobile, of the Capodimonte Observatory. In 1851 the Baron arranged a crowded programme which was to revise the brighter pairs of Wilhelm Struve's Mensurae Micrometricae. His telescope had an excellent object glass five inches in aperture but neither position circle nor driving clock, and only Dembowski realized the enormous amount of physical energy required to handle the instrument; however, with enthusiasm unabated he made in the course of eight years about 2000 sets of measures, until in 1859 he acquired a seven inch Merz refractor with circles, micrometer and a good driving clock. The following twenty years were prolific in results; 21000 sets of measures were made, including all Wilhelm Struve's stars and many of other astronomers. Baron Dembowski also was honoured by receiving the gold medal from the Royal Astronomical Society.

The work of Father Secchi in the measurement of double stars must also be acknowledged; he was distinguished in more than one field of research, particularly for his classification of stellar spectra.

10. Sherburne Wesley Burnham. For years it was thought that double-star discovery in the northern hemisphere was at an end, but a new impetus was given to this branch of astronomy by Burnham, an amateur observer in Chicago. He was an official reporter in the United States courts of that city, but rejoiced in a six inch Alvan Clark refractor mounted in his backyard, and possessed a library consisting of "Webb's Celestial Objects for Common Telescopes", a work then in its first edition. In 1873 the fruit of his labours was a list of 81 new double stars, the first of which, β 40, was discovered April 27th, 1870. Before the close of 1874 the Royal Astronomical Society received five lists including 300 new pairs, 252 being discovered with his six inch telescope. This small instrument became famous although later he used larger ones, the 9,4 inch refractor at Dartmouth College, the 16 inch of the Warner Observatory, the 26 inch Naval Observatory lens at Washington, and the huge refractors at Lick and Yerkes. In all, BURNHAM discovered practically 1340 new double stars which include many having a separation less than 0",2, and others with one component extremely faint, close to a bright primary; these types are almost never found in the catalogues of earlier astronomers. A high percentage of real binaries in rapid orbital motion are embraced in these discoveries.

BURNHAM's method was not the usual systematic survey of the sky, but having observed a certain double star, he proceeded to examine the field surrounding it.

Thousands of pairs were accurately measured, and the record is to be found in the Memoirs of the Royal Astronomical Society, the Publications of the Washburn, Lick and Yerkes Observatories, also in the Publications of the Carnegie Institution of Washington. The lists of BURNHAM's double stars discovered up to the year 1899, which had been published intermittently, were collected in the first volume of the Yerkes Observatory Publications and form his "General Catalogue of 1290 Double Stars".

Realizing the need of a catalogue containing all double stars, with characteristic thoroughness, BURNHAM proceeded to compile one by making pen copies of all available material. The manuscript catalogue was kept up to date by adding new discoveries and measurements. In 1906, this mass of information was published in two enormous volumes by the Carnegie Institution of Washington and they contain practically a complete history of every double star discovered north of $-31°$ declination. This work replaced CAMILLE FLAMMARION's older catalogue[1] of 1878.

11. The Third Period of Visual Double-Star Astronomy. It is generally conceded that BURNHAM's admirable work marks the beginning of this era, and the following list of astronomers associated with double star research is important:

Astronomer	Observatory	Reference	Astronomer	Observatory	Reference
ABETTI	Rome, Arcetri	2	HUSSEY . . .	Lick, La Plata	19
AITKEN	Lick	3	INNES	Johannesburg	20
BARNARD . . .	Yerkes	4	JACKSON . . .	Greenwich	
BIGOURDAN . .	Paris	5	JONCKHEERE .	Lille	21
BOWYER . . .	Greenwich	6	KRUMPHOLZ . .	Vienna	22
BRYANT . . .	Greenwich	6	KÜSTNER . . .	Hamburg	
BURNHAM . . .	Yerkes		LAU	Copenhagen	
COGSHALL . . .	Kirkwood	7	LEAVENWORTH	Minneapolis	23
COMSTOCK . . .	Washburn	8	LEWIS	Greenwich	24
DAWSON . . .	La Plata	9	LOHSE	Potsdam	25
DE JAEGHER .	Lille	10	LUPLAU-JANSSEN	Copenhagen	26
DEMBOWSKI . .	Naples		MAW	Kensington	27
DOBERCK . . .	Sutton	11	MILLER	Kirkwood	28
DOOLITTLE . .	Flower	12	OLIVIER . . .	McCormick	29
ESPIN	Tow Law	13	PERROTIN . . .	Nice	30
FRANKS	England	14	RABE	Breslau	31
FOX	Dearborn	15	ROE	Syracuse	32
FURNER . . .	Greenwich	16	SCHIAPARELLI .	Brera	
GIACOBINI . .	Paris		SCHOLL	Bad Tölz	33
GLASENAPP . .	St. Petersburg		SEE	Flagstaff	
HARGRAVE . .	Sydney		STRUVE, H. . .	Pulkowa	
HERTZSPRUNG .	Leiden		VAN BIESBROECK	Uccle, Yerkes	34
HOUGH	Dearborn	17	VOÛTE	Leiden	35
HOWARD . . .	Kirkwood	18	WIRTZ	Strassburg	36

[1] C. FLAMMARION, Astronomie sidérale, catalogue des étoiles doubles et multiples en mouvement relatif certain, comprenant toutes les observations faites sur chaque couple depuis sa decouverte, et les résultats conclus de l'étude des mouvements: Paris, 1878.

(Continuation next page!)

The most constructive work has been achieved by AITKEN, HUSSEY, INNES, JONCKHEERE, LEWIS and VAN BIESBROECK.

12. The Work of AITKEN. Dr. AITKEN, associate director of the Lick Observatory, has been connected with that institution since 1895, materially contributing to its high reputation. In July 1899, using the large telescope, he undertook with Prof. HUSSEY's collaboration to re-survey the stars as faint as magnitude 9,0, from the North Pole to $-22°$ declination; but in 1905 Prof. HUSSEY left the Lick Observatory, Dr. AITKEN continuing alone this colossal task, practically a consummation of BURNHAM's work.

To date he has discovered practically 3000 double stars, measured many known pairs and determined orbits, among them that of Sirius. Dr. AITKEN's work is most accurate and to his extraordinary vision is due the recent discovery of a companion to the variable star Mira Ceti. His book "The Binary Stars" is a valuable study and is the result of long research.

---

[2] Collegio Romano, Memorie ed Osservazioni (Serie III) 6, Part II (1916).

[3] Lick Bull 1, pp. 14, 66, 129, 190; 2, pp. 55, 139; 3, pp. 6, 61, 88, 90, 147, 169; 4, pp. 4, 75, 101, 107, 166, 170; 5, pp. 28, 43, 55, 115, 166; 6, pp. 1, 62, 70, 163; 7, pp. 3, 93, 186; 8, pp. 52, 93, 96, 99; 9, pp. 132, 184, 191; 11, p. 58; see also Lick Obs Publ 12; Publ A S P 33, p. 60; 34, p. 52.

[4] A J 28 (1915); A N 180, p. 159 (1909); M N 76 (1916); A J 30, p. 182 (1917) and 32, p. 56 (1919).

[5] Publications de l'Observatoire de Paris.

[6] Greenwich Results 1904 to 1909; M N 65 (1905) and succeeding.

[7] A J 22 (1902) and succeeding; A N 168, p. 213 (1905).

[8] Publ Washburn Obs 10, Part 4; 14, Part 1.

[9] Publ Obs Astr de la Plata 4, Part 1; A J 35, p. 147 (1923).

[10] J A 2.

[11] A N 168 (1905) to 212 (1916).

[12] Publ Flower Obs 4, Part 1 (1915); Part 2 (1923).

[13] M N 65 (1905) to 84 (1924).

[14] M N 74, p. 517 (1915); 79, p. 83 (1918); 80, p. 215 (1919); 81, p. 150 (1920).

[15] Annals of the Dearborn Obs. 1 (1915) and 2 (1925).

[16] Greenwich Results 1904 to 1909.

[17] Flower Obs Publ 3, Part 3 (1907).

[18] A J 25, p. 125 (1907).

[19] Lick, La Plata, Detroit Publications. Lick Obs Publ 5 (1901).

[20] Circulars of the Union Obs Johannesburg 1, pp. 31, 97, 185, 249, 285, 339 (1913 to 1918).

[21] Mem R A S 61 (1917); A J 31, p. 113 (1918); 32, p. 65 (1919) and 166 (1920); 35, p. 133 (1924).

[22] A N 222, p. 7 (1924).

[23] A J 29 pp. 17, 41 (1915).

[24] Greenwich Results 1895 to 1909; M N 65 (1905) and following; Mem R A S 56 (1906).

[25] Publ Astroph Obs Potsdam 20, No. 58 (1908).

[26] A N 210, pp. 225 (1919); 215, p. 1 (1921), 367 (1922); 222, p. 65 (1924).

[27] Mem R A S 50 (1892) and others.

[28] A J 23 (1903) to 25 (1908); A N 168, p. 213 (1905).

[29] Publ Leander McCormick Obs 3, Part 2; A J 30, p. 157 (1917); 31, p. 100 (1918); 33, p. 25 (1920); 35, p. 20 (1923).

[30] Publications de l'Observatoire de Nice.

[31] A N 217, p. 413 (1923).

[32] Pop Astr 18, p. 354 and 554 (1910); A N 181, p. 285 (1909); 183, p. 197 (1910); 187, p 39 (1911); 188, p. 373 (1911); 190, p. 137 (1911); 199, p. 169 (1914); 204, p. 21 (1917); A J 26, p. 93 (1910); 31, p. 149 (1918); 32, p. 6 and 25 (1919).

[33] A N 221, p. 297 (1924).

[34] Annales de l'Observatoire Royal de Belgique 14, 3e partie (1920); Pop Astr 29, p. 278 (1921).

[35] Annals of Leiden Obs 10, Part 2; M N 78, p. 683 (1918).

[36] Annalen Strassburg 4, Part 2 (1912); 5 (1923), (issued from Leipzig).

13. The Work of HUSSEY. Prof. HUSSEY, director of the Detroit Observatory, University of Michigan, and who also was director of the Argentine National Observatory, La Plata between 1914 and 1917, began his work at Lick Observatory in January 1896. He enriched scientific discovery by announcing 1329 new double stars.

Volume V of the Lick Observatory Publications, 1901, contains an account of Prof. HUSSEY's measurement of all the double stars discovered by OTTO STRUVE; not only are his own measures given but all those previously made; also there is a discussion of various orbital motions. To be noted is the account of δ Equulei which proved to be a visual binary of 5,7 years, the shortest known period.

In admiration for the work of HUSSEY, R. P. LAMONT of Chicago gave a large 27 inch refracting telescope to the Detroit Observatory; it is to be mounted in South Africa and dedicated to the work on southern double stars.

14. The Work of INNES. The southern hemisphere is Dr. INNES' particular field for astronomical research. Successively astronomer at Sydney Observatory, Australia, the Royal Observatory, Cape of Good Hope and Union Observatory, Johannesburg where he is Government Astronomer, the results of his work include more than one thousand new double stars, most of them close pairs.

In 1899, INNES published his "Reference Catalogue of Southern Double Stars", while his "New Reference Catalogue of Southern Double Stars" is ready for publication. This will be a valuable complement to Burnham's "General Catalogue" and to the "Extension to Burnham's General Catalogue", which will soon be published by Dr. AITKEN.

A 26 inch refractor is being installed at the Union Observatory, which is to be used by INNES. This, with the Lamont telescope, will adequately supplement the work done with huge northern instruments, such as the 36 inch Lick refractor used by AITKEN, and the 40 inch Yerkes refractor in the hands of VAN BIESBROECK.

15. The First Period in Stellar Spectroscopy. Although the last thirty or forty years were important in the development of visual double-star astronomy, they were particularly significant in having seen the birth and rapid growth of a new field of knowledge, that of double stars discovered by the spectroscope.

To trace the origin of astronomical spectroscopy it is necessary again to recognize the work of NEWTON, who not only endowed science with his famous law of gravitation, but was the first, in 1666, to produce a solar spectrum[1]. Many years later, in 1817, FRAUNHOFER[2] announced that each star has a spectrum of its own, while AMICI observed that various white stars have different spectral lines. These researches acquired greater significance through the work of BUNSEN and KIRCHHOFF who proved that these lines depend on the chemical constitution of their source[3]. A complete analysis of the solar spectrum, which served as a basis for wide research, was first published by KIRCHHOFF[4].

DONATI[5], HUGGINS and MILLER[6] produced the first systematic work on stellar spectra. The last two astronomers describe many spectra observed

---

[1] London Phil Trans 1671; see also NEWTON, Optics, Londini 1704, book 1, part 1, proposition 2.

[2] Bibliothèque universelle des sciences, Genève, 6, p. 21 (1817).

[3] Annalen der Physik und Chemie, begründet von J. C. POGGENDORFF, Berlin, 110, p. 160 (1860); 113, p. 337 (1861).

[4] Abhandlungen der Akademie der Wissenschaften zu Berlin, Berlin 1861, Phys. Kl. p. 63; 1862, Phys. Kl. p. 227.

[5] Annali del Museo di fisica e storia naturale di Firenze, Firenze 1, p. 1 (1866).

[6] London Phil Trans 1864, p. 413.

visually, with no attempt to correlate the different types. RUTHERFURD classified these[1], comparing all spectra with certain standards; this classification served as a starting point to the work of his successors. Among them, Father SECCHI[2] established four principal types in his memorable division of stellar spectra. This work was followed by that of VOGEL[3], and finally the DRAPER system of classification adopted by Harvard College Observatory came into general use.

16. Radial Velocities. It was recognized, in theory at least, that the spectroscope supplies a method of measuring radial velocity or component of stellar motion in the line of sight. In 1818, FRESNEL indirectly considered the influence of the earth's motion on optical phenomena[4], and made the statement that refraction of light from a star is the same whether the earth is advancing or receding. CHRISTIAN DOPPLER, a University of Prague scientist, was the first to dispel this theory and in his memoir[5] "Über das farbige Licht der Doppelsterne und einiger Gestirne des Himmels", published in 1842, presented the hypothesis that colour variation depends on the relative motions of earth and star. The change of colour does not occur precisely according to DOPPLER's conception; while there is a slight displacement of the spectrum's maximum intensity, the stretches of invisible light, infra-red and ultra-violet, will become red or violet or vice versa, and leave the limits of the visible spectrum unchanged.

FIZEAU was actually the first to interpret correctly DOPPLER's principle, asserting that motions of approach and recession caused the entire spectrum, including the dark lines of FRAUNHOFER, to shift. A lecture presented by him in 1848, in which he outlined methods of measuring radial velocity of celestial bodies, did not attract serious attention at the time; however, in 1870 it was published[6], finding even then antagonists, one of whom was VAN DER WILLIGEN who wrote in protest "Sur la fausseté de la proposition que la réfraction des rayons lumineux est modifiée par le mouvement de la source lumineuse et du prisme"[7].

By observation, Sir WILLIAM HUGGINS proved the DOPPLER-FIZEAU principle to be true, determining the radial velocity of certain stars and other celestial bodies. He published, in 1868, "Further observations of the spectra of some of the stars and nebulae, with an attempt to determine therefrom whether these bodies are moving towards or from the earth"[8], also in 1872, "On the spectrum of the great nebula in Orion, and on the motion of some stars toward or from the Earth"[9]. He prophesied the great possibilities of the spectrograph in astronomical research when he said: "From the beginning of our work upon the spectra of the stars, I saw in vision the application of the new knowledge to the creation of a great method of astronomical observation which could not fail in future to have a powerful influence on the progress of astronomy; indeed, in some respects greater than the more direct one of the investigation of the chemical nature and the relative physical conditions of the stars."

[1] The American Journal of Science and Arts, 2d series, New Haven, 35, p. 71, 407 (1863).

[2] Memorie di matematica e di fisica della società italiana delle scienze, serie III[a], Modena, 1, p. 67 (1876).

[3] A N 84, p. 113 (1874).

[4] Annales de Chimie et de Physique, par GAY-LUSSAC et ARAGO, Paris, 9, p 627 (1818).

[5] Abhandlungen der Böhmischen Gesellschaft der Wissenschaften, Vte Folge, Prag, 2, p. 465 (1842); 5, p. 293 (1847).

[6] Annales de Chimie et de Physique 19, p. 217—220 (1870); C R 70, p. 1062 (1870).

[7] Archives néerlandaises publiées par la Société hollandaise des sciences à Haarlem, La Haye, 9, p. 41 (1874).

[8] London Phil Trans 158, p. 529 (1868).

[9] London R S Proc 20, p. 379 (1872).

The observations of HUGGINS were confirmed by VOGEL, in Germany, who had published a memoir in 1872, entitled: "Versuche, die Bewegung der Sterne im Weltraume mit Hülfe des Spektroskops zu ermitteln"[1].

At the Greenwich Observatory, when AIRY was Astronomer Royal, measures of radial velocity were made by MAUNDER using a visual spectroscope, and various papers were published entitled "Spectroscopic results of the motions of stars in the line of sight obtained at the Royal Observatory, Greenwich"[2].

Contemporaneous work was being done by VOGEL who used improved methods of photography, determining the radial velocity of stars with an accuracy of about one mile per second. A few years later he published the great memoir "Untersuchung über die Eigenbewegung der Sterne im Visionsradius auf spektrographischem Wege"[3], which marked the beginning of modern research in radial velocity.

17. The Initial Work at Harvard by E. C. PICKERING. Before the publication of VOGEL's work, photographing stellar spectra was begun, in 1886, at Harvard College Observatory, objective prisms being used to obtain a large number of spectra on the same plate. It was Prof. PICKERING's purpose to ascertain the spectral characteristics of all stars down to the tenth magnitude. The Annals, Circulars and other publications of Harvard College Observatory, contain valuable data on the subject, but it is only of late, in 1924, that PICKERING's master work was completed in the well known "Henry Draper Catalogue"[4]. The formidable task of selecting different types of spectra on the numerous plates was accomplished by several women under Mrs. FLEMING's direction, and later under the supervision of Miss CANNON.

18. The First Spectroscopic Binaries Discovered. A short article by Prof. EDWARD C. PICKERING[5], on "A New Class of Binary Stars" dated 1890, is an introduction to the field of double stars discovered by spectroscopy. Two or three years after the study of photographic spectra had been introduced at Harvard, it was found that two plates taken in 1887 and 1889 showed the K-line in the spectrum of ζ Ursae Majoris to be double while on other plates it always appeared single. Immediately, Miss ANTONIA C. MAURY who was studying spectra of the brighter stars, undertook to examine sixty-two plates containing the spectra of ζ Ursae Majoris; there could be no doubt that this was a double star, its two components having similar dimensions and identical spectra. The spectral lines appear single when both components are moving at right angles to the line of sight, but when one star is approaching and the other receding their spectral lines are shifted in opposite directions and they appear double. The revolution period of the components was found to be 20,5 days.

After her work on ζ Ursae Majoris, Miss MAURY discovered β Aurigae which showed sometimes single and at other times double lines. Its period is only 3,96 days.

At Potsdam in November 1889, by using his spectrograph, Prof. H. C. VOGEL found that Algol (β Persei) is also a binary[6]. By referring its spectral lines to those of a comparison spectrum of a terrestrial source, he discovered the first spectroscopic binary not showing double lines, and at the same time proved that

[1] Beobachtungen, angestellt auf der Sternwarte des Kammerherrn VON BÜLOW zu Bothkamp, Leipzig, 1, p. 33 (1872).

[2] M N 36, p. 318 (1876); 37, p. 22 (1877); 38, p. 493 (1878); 41, p. 109 (1881).

[3] Publ Astroph Obs Potsdam 7, Teil I (1892).

[4] Harv Ann 91 (1918) to 99 (1924).

[5] M N 50, p. 296 (1890).

[6] A N 123, p. 289 (1889).

the periodic loss of light of this interesting variable star is due to partial eclipse by a relatively dark companion, a hypothesis previously presented by GOODRICKE in England when he investigated Algol's fluctuations of light more than a century before.

In 1890, Prof. VOGEL announced that Spica (α Virginis) is also a spectroscopic binary, having a period of 4,01 days. The spectrograms of this star yielded velocities varying through 200 kilometers per second[1].

Before 1898, thirteen spectroscopic binaries had been discovered by PICKERING, VOGEL, BÉLOPOLSKY, Miss MAURY, Mrs. FLEMING and BAILEY. In the years following, pioneer research workers have multiplied their discoveries and developed their studies to the point where new and startling problems are presented.

19. HERMANN CARL VOGEL. Just at the time when a great impetus was being given to spectroscopic astronomy by RUTHERFURD in America, SECCHI in Italy and HUGGINS in England, VOGEL was graduated at Leipzig in 1867 and at once realized the great possibilities suggested by the work of these observers. From 1870 to 1874 he was in charge of a private observatory founded by VON BÜLOW at Bothkamp, and many spectroscopic observations of the Sun, stars, star clusters, nebulae, comets and planets are the result. He used for this purpose, a visual spectroscope designed by himself, having great stability and capable of determining wave-lengths with reasonable accuracy.

When the Astrophysical Observatory at Potsdam was founded by the Prussian Government, VOGEL was immediately appointed observer, and in 1882 became its first director. With the assistance of Dr. MÜLLER he catalogued the spectra of all the stars[2] down to magnitude 7,5 in the zone $-1°$ to $+20°$, using the classification he proposed in 1874. This classification is an elaboration of SECCHI's system, recognizing ZÖLLNER's idea that the character of stellar spectra depends upon the temperature of the stars.

VOGEL's principal contribution to science was the introduction, in 1887, of the photographic method to obtain radial velocity with a slit spectrograph. Associated with Dr. SCHEINER, his astrophysical work became more definite in outline. Causes of error which might influence the minute shift of spectral lines were carefully studied and eliminated, and the results obtained became consistent and accurate to an unexpected degree. VOGEL's study of Algol and Spica has already been mentioned. He made a complete analysis of these two stars, also an exhaustive study of β Aurigae and ζ Ursae Majoris.

In 1899 a large telescope was erected with a view to continuing his work, and although VOGEL's health failed the year following, the marked success of radial velocity investigations at Potsdam, was due to his initiative.

20. HENRI DESLANDRES. Prof. DESLANDRES, director of the Astrophysical Observatory at Meudon, has for many years, as everyone knows, made a most brilliant record in the realm of solar physics. Inventor of the spectroheliograph and also the velocity recorder, he applied his genius to analysing the Sun at different levels of its atmosphere, and his discoveries in this field are too numerous to present here. His work in stellar spectroscopy is often lost sight of in comparison with the highly significant results of his solar studies. It is nevertheless most important, representing pioneer work with the spectrograph.

Admiral MOUCHEZ, director of the Paris Observatory, requested Dr. DESLANDRES in 1891 to organize spectroscopic work in that institution. DESLANDRES

[1] A N 125, p. 305 (1890); see also Sitzungsber. der Kgl. Akad. der Wissenschaften zu Berlin 1890, p. 401.

[2] Publ Astroph Obs Potsdam 3, No. 11 (1883).

developed solar work, undertook laboratory researches and proceeded to determine the radial velocity of stars; although using the photographic methods introduced by VOGEL, he replaced the comparison spectrum of hydrogen by that of iron, which is easier to produce. In this work, the first attempt was made to keep the spectrograph at a constant temperature, by mounting long strips of metal around and near the prisms and lenses. These were connected with a metallic thermometer in such a way that a slight fall in temperature would produce an electric contact sending a current over the strips. This current heating the spectrograph, in turn raised the temperature of the thermometer, breaking the contact at the proper point[1].

DESLANDRES found δ Orionis[2] and θ Aquilae[3] to be spectroscopic binaries and determined the orbit of the latter[4]. His extensive work on the spectra of planets and comets as well as researches in other fields cannot be detailed.

21. ARISTARCH BÉLOPOLSKY. Immediately after VOGEL in 1890, Prof. BÉLOPOLSKY had a spectrograph of the VOGEL type attached to the thirty inch Pulkowa refractor. Owing to the larger size of this telescope and the shorter exposures needed to secure spectrograms, the results were more accurate than at Potsdam.

Marked originality characterizes Prof. BÉLOPOLSKY's work. Concentrating on Cepheids, he found the radial velocity of δ Cephei[5], η Aquilae[6] and ζ Geminorum[7] to be variable, indicating that between the radial velocity curve and the light curve of the same star, there exists a peculiar relation. Between the years 1894 and 1908, a more exhaustive study of δ Cephei apparently showed a variation in the centre-of-mass velocity of the system[8]. Similar variation was suggested in the case of Algol[9].

Laboratory measurements of DOPPLER-FIZEAU effects were secured in 1900, confirming the correctness of the principle within the limits of unavoidable errors[10].

A method of determining the true dimensions and parallaxes of visual double stars by measuring their radial velocities was applied to γ Virginis and γ Leonis.

One of the most interesting researches by this astronomer, however, is that concerning the peculiar variation of α Canum Venaticorum[11]. The spectrum of this star possesses peculiar lines which vary in intensity and show radial velocity displacements, while the majority of other lines exhibit no variation whatever. According to BÉLOPOLSKY the variable lines probably belong to rare-earth elements, rare metals often of great atomic weight. These results have been confirmed and elaborated by KIESS[12].

22. WILLIAM WALLACE CAMPBELL. The visual measurement of stellar radial velocity was begun at Lick Observatory about 1890; a certain number of measures were made by Prof. KEELER and subsequently by Prof. CAMPBELL and Dr. CREW.

By using a grating spectroscope, KEELER completed his magnificent series of velocity measures on fourteen nebulae. In 1895 after the installation of the MILLS spectrograph designed by Dr. CAMPBELL, radial velocity measurements

---

[1] B A 15, p. 57 (1898).
[2] C R 130, p. 379 (1900).
[3] C R 136, p. 205 (1903).
[4] B A 20, p. 129 (1903).
[5] Ap J 1, p. 160 (1895).
[6] Ap J 6, p. 393 (1897).
[7] A N 149, p. 239 (1899).
[8] Pulkowa Mitteilungen 3, p. 63 (1909).
[9] Pulkowa Mitteilungen 4, p. 171 (1911).
[10] Ap J 13, p. 15 (1901); see also Bull. de l'Acad. des Sciences de St. Pétersbourg 13, p. 461 (1900).
[11] A N 196, p. 1 (1913).
[12] Publ. Obs Univers. of Michigan 3, p. 106 (1923).

by the photographic method formed part of a regular programme at Mount Hamilton.

It is to the generosity of D. O. MILLS, member of JAMES LICK's first board of trustees, that we are indebted for the construction of this spectrograph, as well as for later modifications and improvements. This instrument serving as a model for others, introduced modern accuracy in determining the radial velocity of celestial bodies. FROST and ADAMS said[1]: "The next great advance was made by CAMPBELL in his design of and work with the MILLS spectrograph of the Lick Observatory, described in the Ap J 8, p. 123–156 (1898). The use of iron as a comparison spectrum, previously tried by VOGEL and by DESLANDRES, but not regularly employed by them, together with the closest attention to the optical and mechanical construction of the instrument and great refinement in the measurement of the plates, enabled CAMPBELL to increase greatly the accuracy of the determinations, so that the natural unit became the kilometer per second, instead of the sevenfold greater German geographical mile employed by VOGEL."

Up to the present more than 10000 stellar spectrograms have been obtained at Lick Observatory, also, many spectrograms of nebulae analyzed by placing the slit of the spectrograph upon different sections or in different position angles.

Dr. CAMPBELL's research work ranks high and his discoveries are of utmost importance to modern science. His first and second catalogues of spectroscopic binaries[2] are invaluable and have been beautifully completed in Dr. MOORE's third catalogue published in 1924[3].

He found in the radial velocity curve of ζ Geminorum[4], a secondary oscillation, also a variation in the centre-of-mass velocity of Polaris[5]. Numerous spectroscopic binaries were discovered; among them CAPELLA[6] having a 104-day period, its spectroscopic orbit was determined by REESE[7] and its visual orbit from interferometer measures by MERRILL[8].

Dr. CAMPBELL's determination of the solar motion from radial velocities[9], of motions involving the brighter class B stars[10], his studies dealing with the relation between radial velocity and spectral type[11], and his collaboration with Dr. MOORE in the analysis of the spectrographic velocity of bright-line nebulae[12] testify to extraordinary ability.

23. Modern Research in the Field of Spectroscopic Binaries. Dr. CAMPBELL's work initiated wide interest in general astrophysics. Very few astrophysicists limited their research to spectroscopic binaries alone or made these studies the principal part of their work as in the case of visual double-star observers. Different problems relating to spectral characteristics, star streams, nebulae, statistical data, were also investigated.

Prof. EDWIN B. FROST, director of the Yerkes Observatory with Dr. W. S. ADAMS, initiated radial velocity work at WILLIAMS BAY. Prof. FROST discovered various spectroscopic binaries[13] with the BRUCE spectrograph[14]; Dr. ADAMS continued his work with the large telescopes of Mt. Wilson and is distinguished for his method of determining spectroscopic parallaxes of the stars.

[1] Publ. of the Yerkes Observatory 2, p. 145 (1904). [2] Lick Bull 6, p. 17 (1910).
[3] Lick Bull 11, p. 141 (1924). [4] Ap J 13, p. 93 (1901).
[5] Ap J 10, p. 180 (1899); see also Lick Bull 3, p. 86 (1905).
[6] Ap J 10, p. 177 (1899). [7] Lick Bull 1, p. 34 (1901).
[8] Ap J 56, p. 40 (1922). [9] Ap J 13, p. 80 (1901).
[10] Lick Bull 6, p. 101 (1911). [11] Lick Bull 6, p. 125 (1911).
[12] Lick Obs Publ 13, p. 75.
[13] See the many numbers of the Ap J of which Professor FROST is one of the editors.
[14] Yerkes Obs Publ 2, p. 145 (1903).

The stationary calcium lines in spectroscopic binaries were first detected in δ Orionis[1] by Dr. J. HARTMANN at Potsdam.

Dr. FRANK SCHLESINGER, at present director of the Yale Observatory, published three volumes treating of spectroscopic binary orbits[2]. Particular study was made of early class stars having wide and diffuse lines with enormous amplitudes of velocity variation.

Dr. R. H. CURTISS of Ann Arbor introduced methods of measuring spectra, studied the irregularities in W Sagittarii[3], the variable star β Lyrae[4] and the stars with bright emission lines[5].

The Lick Observers WRIGHT, MOORE and PADDOCK have accomplished notable work while Dr. H. D. CURTIS at present director of the Allegheny Observatory made a particularly interesting study of CASTOR[6].

Dr. J. S. PLASKETT with his collaborators contributed a large number of spectroscopic binary orbits[7], and among these one of the most massive couples in existence[8].

In Potsdam Dr. H. LUDENDORFF investigated variable stars and binaries, a most significant study being that of the complicated system ε Aurigae[9].

In Vienna there is HNATEK; at the Royal Observatory, Cape of Good Hope, LUNT carries out extensive work.

Among very recent observations may be mentioned the work of SANFORD at Mt. Wilson, OTTO STRUVE at Yerkes and researches at the Dominion Observatory in Ottawa[10].

An account of the distinguished work on photometric double stars by RUSSELL, SHAPLEY and others may be found treated adequately in books on variable stars.

b) Classification and Observation of Visual Double Stars.

24. Definition of the Term Double Star. During the first years of double-star astronomy many wide pairs, which at present would be rejected, were measured and catalogued. WILLIAM HERSCHEL divided his double stars into six classes, the first two comprising stars whose separation was smaller than 5″ and the others were between the limits 5″, 15″, 30″, 1′, 2′, or more. WILHELM STRUVE divided his double stars into eight classes between 0″, 1″, 2″, 4″, 8″, 12″, 16″, 24″ and 32″.

OTTO STRUVE placed his limit for retention at 16″, ESPIN at 10″, while other observers adopted different values.

Many, however, have come to recognize AITKEN's upper limits[11] which are used to determine whether or not a star is double. Thus:

1″ for combined magnitude of 11,1 or fainter,
3″ ,, ,, ,, of 11,0 or brighter,
5″ ,, ,, ,, of 9,0 or brighter.

This classification is not final because a faint binary system may be composed of dwarf stars having low luminosity, but large mass and parallax; consequently these components may be more than 1″ apart and still form a physical system governed by gravitational attraction. According to JONCKHEERE[12] the

---

[1] Ap J 19, p. 278 (1904). [2] Allegh Obs Publ 1, 2, 3.
[3] Lick Obs Bull 3, p. 31 (1904). [4] Allegh Obs Publ 2, p. 98 (1911).
[5] Publ. Obs Univ. of Michigan 2, 3. [6] Lick Bull 4, p. 58 and 64 (1906).
[7] Dom Obs Publ 1, 2, 3, 4 and Dom Ap Obs Publ 1, 2.
[8] M N 82, p. 447 (1922).
[9] Berliner Sitzungsber., Phys.-math. Kl. 9, p. 49 (1924).
[10] Dom Obs Publ 5, 8, 9. [11] A N 188, p. 381 (1911). [12] Obs 38, p. 426 (1915).

separation limit of 5″ could then, with a possibility of equal interest, stand for all magnitudes.

Double stars may be divided into two classes, optical systems and physical systems.

In optical systems the components are at a considerable distance from each other, appearing to be close together merely as a matter of perspective but having no connection whatever.

In physical systems the components are bound by their mutual gravitational attraction and during a certain number of years revolve about their common centre-of-mass. Double and multiple physical systems offer most intriguing problems: double systems are referred to as binaries or real double stars.

Sometimes the components of an optical system exhibit similar proper motion. BEATTIE[1] has suggested in this case that optical systems might be binary systems having open orbits, either parabolic or hyperbolic, in which the components long since passed periastron and are now well out from the common focus, their relative movements having become so slow as virtually to have ceased. An example of this is α Crucis, whose components are so identical in magnitude and general appearance that it is difficult to believe them not physically related. These components belong to the same star stream as other well known stars, but may be very distant from each other.

AITKEN made the assertion in 1912 that out of 12000 pairs of stars separated by less than 5″, only about twenty-five were not physical systems[2]. The adoption of 5″ as upper limit for the separation of components will give a fair proportion of interesting systems.

25. The Visual Method of Observing Double Stars. The relative position of the brighter component or primary, and the fainter or companion of a double-star is perfectly defined by the polar coordinates θ and ϱ. When the components are of equal brightness the discoverer usually selects the primary. θ, the angle at the primary between the lines connecting it with the North Pole and with the companion, is the position angle; it is counted from 0° to 360° from the north toward the east. ϱ, the angular separation between the two stars, is the distance.

In order to measure θ and ϱ, a parallel wire micrometer is used. Several important firms have constructed such micrometers; one of the oldest is that by TROUGHTON (1824) which fitted the HERSCHEL and SOUTH five foot equatorial telescope and was capable of measuring position angles to one minute of arc[3].

FRAUNHOFER designed a more accurate micrometer perfected later by MERZ; this type of instrument was used by the STRUVES, DEMBOWSKI, SECCHI and a large number of European astronomers. Reputable micrometers are those constructed by REPSOLD, ALVAN CLARK and WARNER and SWASEY.

On beginning observation the micrometer is examined to see if it is firmly attached to the telescope and the zero reading of the circle is determined. This reading is obtained by adjusting the wires in such a way that an equatorial star follows one of them and is constantly bisected by it while passing through the field of view in consequence of diurnal motion. It is well to check the zero at intervals through the night to guard against errors due to accidental changes.

Observers determine the position angle from four settings of the micrometer and then make three complete measures of double distance. More than three are made in the case of wide pairs because accidental errors are greater

[1] Obs 39, p. 177 (1916). [2] Publ A. S. P. 24, p. 128 (1912).
[3] Phil Trans 1824, part III, p. 10.

in the measurement of large angular distances. In all cases the stars must be in proximity to the optical axis of the instrument.

The determination of a position angle is made with the fixed wire, the movable one being placed far enough away to cause no inconvenience. In making a proper setting the wire is brought into position, parallel to the line joining the components, by rotating the micrometer back and forth while manipulating the wire by means of a screw which shifts the micrometer box longitudinally. These movements deliberately and concisely made are continued until the setting appears to be accurate, and the circle is then read. Before making the next adjustment the micrometer is rotated back and forth through a wide angle and the position of the wires in the field of view is changed. Thus the different settings are independent of each other.

In securing these measurements the imaginary line from eye to eye must be placed either perpendicularly or parallel to the line connecting the two stars.

To determine the double distance the fixed wire is first placed in the position angle indicated by the mean of previous settings; the micrometer being rotated by 90°, the fixed wire is placed on the primary, the movable wire on the companion, and the position of the latter read. Next, the system of wires is shifted to place the fixed wire on the companion; then, turning the micrometer screw which controls the movable wire, the latter is placed on the primary and its position again read. The difference between the two readings, multiplied by the value of a screw's revolution in seconds of arc on the celestial sphere, is the double distance required. In reading the micrometer screw its last motion must be made forward.

26. The Causes of Error in the Observations. All errors in observation may be classified as accidental or systematic.

To avoid accidental errors the micrometer must be properly tested and the scale in seconds of arc on the celestial sphere, accurately determined. Obviously it is most important that measurements be secured when observing conditions are favorable.

Systematic errors vary not only with the observer but with many other factors such as relative magnitudes of components, angular distance, the inclination of the line joining components with the horizontal, and in unequal pairs much depends upon the position of the companion with respect to the primary.

To eliminate these errors a number of astronomers have made measures on artificial stars whose position angles and distances were exactly known. The most elaborate investigation in this line is probably that made by OTTO STRUVE[1]. He measured what he called artificial double stars, formed by small ivory cylinders placed in holes in a black disk. Formulae were then derived by means of which he calculated corrections to be applied to all of his measures. Many observers, however, are doubtful whether these corrections improve the results. Others have made lists of carefully selected standard pairs to serve as a means of comparison. One list of fifty pairs has been suggested at the International Astronomical Union for general use by double-star observers.

SALET and BOSLER found that errors in position angle were eliminated by using an eyepiece with a reversing prism. [See B A 1908, p. 18 and also 1920, p. 136.]

An attempt was made to determine the personal error by means of the observations, assuming that, for a given epoch, the mean of the figures obtained by all the observers is correct, and comparing with this mean the measures of each observer.

[1] Observations de Pulkowa 9 et supplements.

This is the method followed by T. N. THIELE[1] who chose Castor. This binary, in fact, recommended itself on several scores: in particular, its relative motion is slow enough, and the number of observations as well as of observers considerable. But, on the other hand, it is a binary, which in the brilliancy of its components, differs too much from the average appearance of other double stars to permit generalization of the laws according to which its observations were analysed. Following on the determination of systematic errors, THIELE decided to bring to light certain variations. He was led to distinguish, for each observer, "seasons of observations", periods whose length varied from two or three days to several months, in accordance with which notable variations in the systematic errors were produced. It is true that AUWERS[2] called attention to a systematic difference between his observations in spring and fall, but the periods noticed by THIELE have no relation to season. Rather, the character of the variations is for them to be of short duration; it is known that there are days when for a while, sometimes for several hours, an observer obtains position angles and distances systematically too large or too small, and far more numerous observations than those made of Castor would be necessary to establish, with accuracy, periods of any length.

One might perhaps find the origin of the discrepancies which have led THIELE to admit these "seasons of observations" in the fact that, Castor having a very northerly declination, this binary has been measured at successively increasing hour angles whose extremities are considerably separated the one from the other.

27. The Separating Power of a Telescope. The value of observations depends to a large extent on the separating or resolving power of the telescope used. At the beginning of the nineteenth century the opinion was held that the separating power of a telescope was a function of its aperture and focal distance. Theoretically, it depends only on the aperture. Practically, it depends on the quality of the lens or mirror which is difficult to perfect, especially for small ratios of focal length to aperture.

The image formed by a circular lens, of any luminous point such as a star, is a diffraction pattern consisting of a series of concentric rings varying in intensity; measured along the radius this intensity exhibits decreasing maxima separated by zones of darkness. The diffraction through a theoretically perfect objective was worked out years ago by AIRY, who computed the exact distribution of light in the central disc and surrounding rings[3]. For unity brightness in the centre of the star disc, he found the maximum brightness of the first ring to be 0,017, of the second, 0,004 and the third, 0,0016. The resolving power of a telescope is determined by the scale of the diffraction system. The diffraction effect being caused by the edge of the objective, the size of disc and rings depends not on focal distance, but on the objective's diameter.

The matter of resolving power was thoroughly investigated by Rev. W. R. DAWES[4]; his study included years of observation with telescopes of different sizes and the final result established what has since been known as "Dawes' Limit".

If A is the aperture of the telescope in inches, the separating power or smaller distance inside of which two components cannot be separated, is $4'',56/A$ for stars of almost the same magnitude.

---

[1] CASTOR, Calcul du mouvement relatif et critique des observations de cette étoile double. Copenhague, 1879.

[2] A N 59, p. 9 (1862). [3] Cambridge Phil Trans 1834, p. 283.

[4] Mem R A S 35, p. 158 (1867).

"Dawes' Limit", however, is only approximate and subject to qualifying factors.

In an important paper, "On the Class of Double Stars which can be observed with Refractors of Various Apertures"[1], based on the analysis of work by observers using telescopes ranging in aperture from four to thirty-six inches, LEWIS has shown that the possibility of separating double stars is a function of physiological as well as instrumental factors. This possibility depends upon the total magnitude of a pair and upon the relative magnitudes of the components.

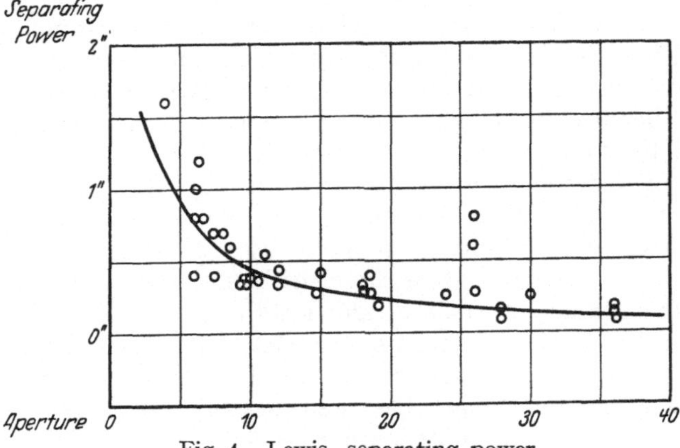

Fig. 1. Lewis, separating power.

Fig. 1 illustrates results obtained by LEWIS for components approximately of the fifth magnitude; the curve represents DAWES' empirical relation, and close agreement is evident between this curve and limits indicated. For fainter or unequal pairs, such agreement does not exist.

28. The Use of Diaphragms. The use of a circular diaphragm which decreases the aperture does not improve the quality of a double-star image, but in measuring stars like Sirius and Procyon where the primary is much brighter than the companion, experiments prove that a hexagonal diaphragm over the object-glass is of great value. In this case the diffraction pattern is entirely different, formed of a central disc from which six thin rays originate; the field between these rays being dark, a faint companion near the primary is detected readily. The field may be explored in any position angle as the rays are changed by rotating the diaphragm. BARNARD used a hexagonal diaphragm to advantage with the Yerkes telescope[2].

29. Discovery of Stars with Invisible Companions from Observation of Proper Motions. Variations of their proper motion indicate that some stars have invisible companions. BESSEL found by minute study of successive data relating to the position of fundamental stars, that the proper motions of Sirius and Procyon exhibited variations which could not be explained by errors of observation. "Über Veränderlichkeit der eigenen Bewegungen der Fixsterne"[3], a paper by BESSEL published in 1844, gives the following tables of variation in declination of Procyon and right ascension of Sirius; the coordinate for 1755 given in the Tabulae Regiomontanae has been taken as origin.

The observations of the variation in the right ascension of Sirius are plotted on Fig. 2. The circles indicate values taken from the table above; the crosses correspond to these circles displaced in time by the length of a period 49,3 years, as computed from modern observations.

This work is remarkably accurate for observations of that time.

BESSEL assumed that these variations in Procyon and Sirius were due to the presence of dark companions. However, his memoir was not accepted by all astronomers. STRUVE said in 1847: "If the irregularities in the motion of

[1] Obs 37, p. 372 (1914). [2] A N 182, p. 13 (1909).
[3] A N 22, p. 145, 169, 185 (1844).

Variation of the Declination of Procyon.

Fundam. Astr. . .	1755	0″,00
Maskelyne	1770	+1 ,54
Piazzi	1800	+1 ,99
Bessel	1820	0 ,00
Pond I	1822	−0 ,03
Pond II	1822	+0 ,16
Struve	1824	−0 ,15
Argelander	1830	+0 ,03
Airy	1830	+0 ,47
Pond	1832	+0 ,84
Henderson	1833	+0 ,89
Busch	1838	+1 ,59
Bessel	1844	+2 ,62

Variation in the Right Ascension of Sirius.

Lacaille	1750	+0ˢ,212
Fund. Astr.	1755	0 ,000
Maskelyne	1767	−0 ,079
Maskelyne	1790	+0 ,174
Piazzi	1800	+0 ,033
Maskelyne	1805	+0 ,032
Bessel	1815	−0 ,036
Pond	1819	−0 ,083
Bessel	1825	0 ,000
Struve	1825	−0 ,006
Argelander	1828	−0 ,003
Pond	1830	−0 ,085
Airy	1830	+0 ,049
Busch	1835	+0 ,188
Bessel	1843	+0 ,321

Sirius are real, this discovery would, without doubt, be one of the most important ever made in stellar astronomy, one of the most beautiful which science owes to the great astronomer of Königsberg. I have allowed myself yet to doubt of this reality until it is proved by more conclusive research." This research was undertaken by WILLIAM STRUVE himself, with the conclusion that no variation existed.

Today there is proof that BESSEL's theory was correct. By treating the observations anew, C. A. F. PETERS[1] in 1851, and later, A. AUWERS[2] in his "Untersuchungen über veränderliche Eigenbewegungen. Zweiter Teil: Bestimmung der Elemente der Siriusbahn" computed the orbit of Sirius by referring its motion to the centre of gravity of a binary system.

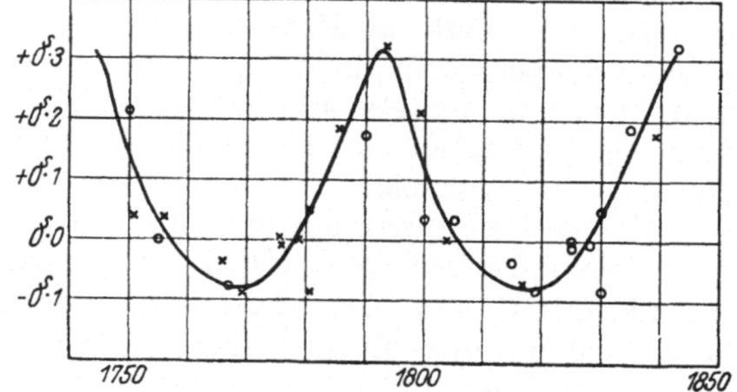

Fig. 2. Variation of the right ascension of Sirius deduced by Bessel.

The discovery of a faint companion to Sirius by ALVAN CLARK in 1861 and of a still fainter companion to Procyon by SCHAEBERLE in 1896 gave definite verification of BESSEL's hypothesis.

Periodic variations in the motion of components indicate the presence of an invisible third or fourth body in a few distinctive double-star systems. Among them may be mentioned:

ζ Cancri, consisting of three bright stars, two of which revolve about a common centre of gravity during a period of approximately sixty years, while the third one revolves with this pair in a much larger orbit. According to SEELIGER, irregularities noticed in the apparent motion of this third star may be attributed to the presence of an invisible fourth body revolving with the third around a common centre. ζ Cancri is treated with more detail in subdivision "k".

ξ Ursae Majoris, in which perturbations were discovered by NÖRLUND; this star is analyzed more extensively in subdivision "k".

ε Hydrae, whose primary was found to be a very close pair, the components of which complete a revolution in about fifteen years. SEELIGER[3] demonstrated that the orbital motion of this pair fully accounts for perturbations in the motion of the distant third body.

[1] A N 32, p. 1, 17, 33, 49 (1851). [2] Publ. d. Astron. Gesellschaft 7 (1868).
[3] A N 173, p. 325 (1906).

Perturbations due to the presence of invisible bodies are suspected in the systems ζ Aquarii, ε^2 Lyrae, ξ Scorpii and γ Leonis.

An interesting treatment of these celestial phenomena is developed by BURNHAM in his paper on "Invisible Double Stars"[1].

30. The Photographic Method. The first experiments in the use of photographic methods were begun in 1857, when BOND secured on a collodion plate the first measurable images of the double star ζ Ursae Majoris. Other experiments were made in America, by PICKERING and GOULD, in England at Greenwich observatory, and in France, by the HENRY brothers, who were the promoters of the "Carte photographique du Ciel".

Since that time improvements have been made in the quality of the plates, which are more sensitive and of finer grain, while marked advances characterize the construction of long-focus telescopes, accurately driven by clock-work or synchronized electric motor.

The fact that photographs can be measured any time and long series of them investigated by the same observer, incites to further endeavour.

The work of THIELE[2] and LAU[3] at Copenhagen and that of greater significance by HERTZSPRUNG at Potsdam, is followed in America by investigations of photographic parallax observers. OLIVIER at the Mc Cormick Observatory and PITMAN working at Sproul, have recently given results of great accuracy.

31. The Work of HERTZSPRUNG[4]. HERTZSPRUNG's photographic method cannot compete with the visual in the case of very close pairs, but for wider pairs it has advantages, and in the study of systems where there are deviations from single elliptic motion, such as ζ Cancri and ξ Ursae Majoris, it will be found highly valuable.

He used a 50 centimeter visual refractor and employed plates sensitive to yellow and green light (visual region), with or without a colour-filter (mean wave-length λ 5450). The mean scale value was 1 mm = 16″,39, corresponding to a focal distance of 12,58 metres. When ordinary photographic plates are used, on account of the great difference in effective wave-length, the effect of atmospheric dispersion is to displace red stars relatively to bluer ones. For instance at 45° zenith distance, there is a shift of 0″,18 for K-stars relatively to A-stars. With his special plates, HERTZSPRUNG obtained practically the same wave-length for stars of all colours, thus eliminating errors due to atmospheric dispersion.

In stellar photography, accidental shifts occur in the relative positions of images, if these have different intensities and are not perfectly symmetrical. HERTZSPRUNG reduces this magnitude error by using an objective grating, chosen from a number, formed by parallel bands of various widths and separations. When such an appliance is used, the image of a star is flanked by two symmetrically placed short spectra of stellar appearance. By proper selection of the grating used, the two secondary images of the primary will appear of nearly the same size and intensity as the central image of the companion. The two lateral images of the primary are then measured in Cartesian coordinates, and the mean of the two readings subtracted from similar measures of the companion.

Figs. 3 and 4 are images of the double stars ε Draconis and η Cassiopeiae as given in HERTZSPRUNG's memoir.

More than one hundred short exposures were sometimes secured on the same plate, the length of exposure varying from two seconds for a fourth magnitude star, to forty-five seconds for an eighth magnitude. Only the best images were

[1] M N 51, p. 388 (1891).
[2] A N 190, p. 78 (1911).
[3] A N 160, p. 353 (1902).
[4] Publ Astroph Obs Potsdam Nr. 75 (1920).

measured, because many were distorted through momentary unsteadiness or small errors in guiding. In all 16680 exposures on 408 plates were used for 126 double stars of W. STRUVE's Dorpat catalogue. The majority of these are wide pairs, the minimum distance recorded being about 1″,2. For all pairs

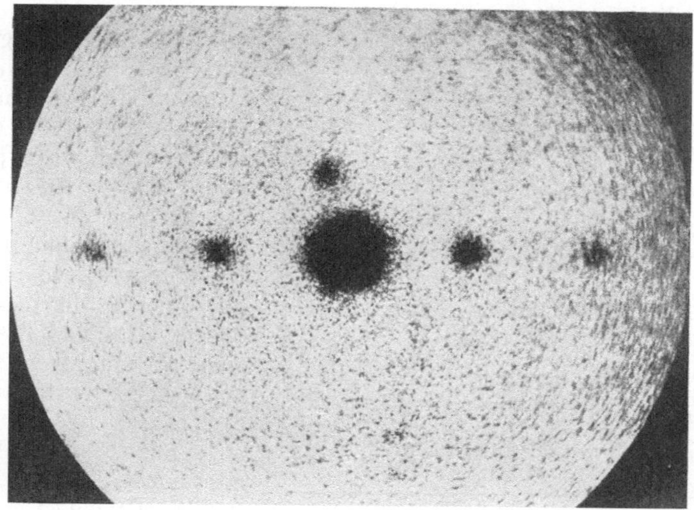

Fig. 3. Photographic image of ε Draconis.

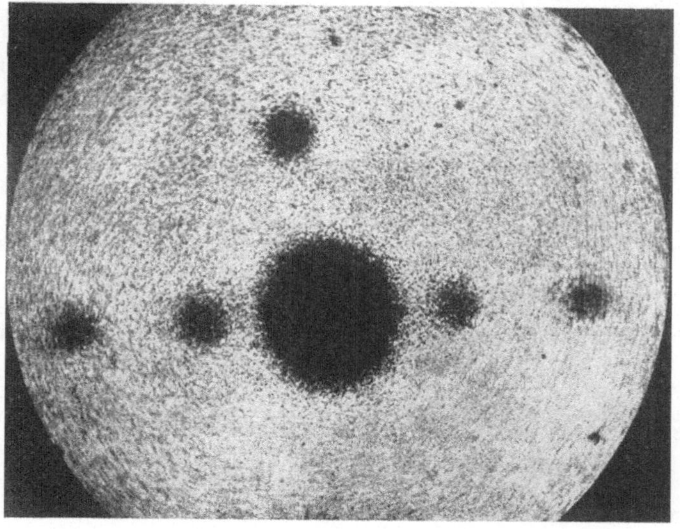

Fig. 4. Photographic image of η Cassiopeiae.

the inner agreement is quite satisfactory without, however, warranting the use of a second decimal place in the position angle, nor a third place in the distance.

Results for 70 Ophiuchi and ξ Boötis are given here as an example:

70 Ophiuchi (mags. 4,28, 5,98)

1915,195	140°,93 ± 0°,07	4″,554 ± 0″,005
1917,150	137 ,78 ± 0 ,07	4 ,377 ± 0 ,006
1919,376	134 ,56 ± 0 ,06	5 ,989 ± 0 ,006

ξ Boötis (mags. 4,80, 6,82)

1914,300	110°,12 ± 0°,38	(2″,084) ± 0″,014
1918,116	80 ,29 ± 0 ,24	2 ,290 ± 0 ,009
1919,385	71 ,83 ± 0 ,15	2 ,384 ± 0 ,006

For ξ Ursae Majoris the measures were especially numerous. 45 plates were secured during the interval 1914,3 to 1919,4.

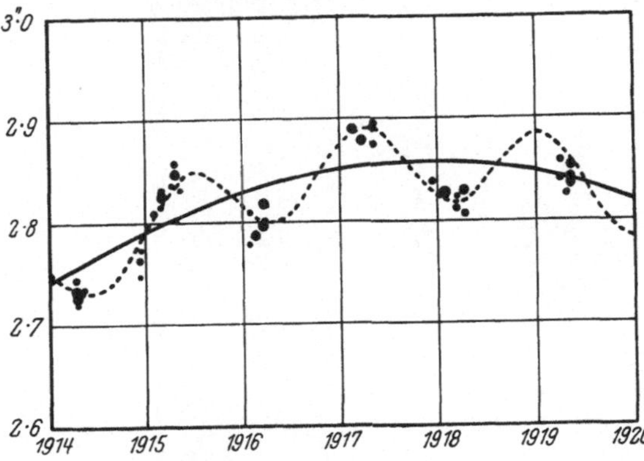

Fig. 5. ξ Ursae Majoris $\Delta\alpha \cos\delta$.

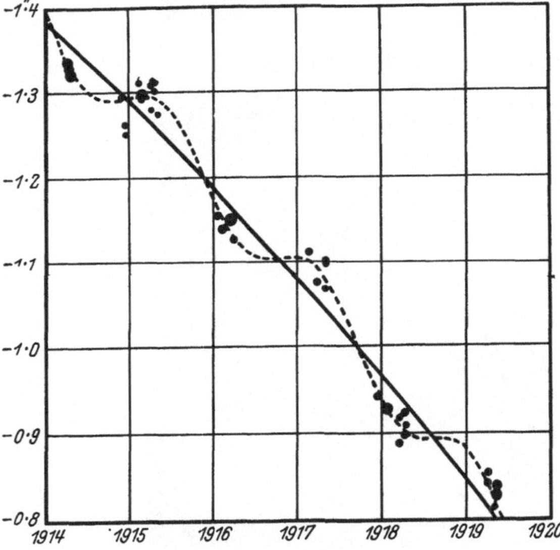

Fig. 6. ξ Ursae Majoris $\Delta\delta$.

Figs. 5 and 6 giving the variations of $\Delta\alpha \cos\delta$ and $\Delta\delta$ bring out in a striking manner the short period secondary motion of 1,8 years, previously suspected by NÖRLUND.

32. The Work of CHARLES P. OLIVIER[1]. Unlike the work of HERTZSPRUNG, whose ultimate purpose was the measuring of double stars and the development of accurate photographic methods, OLIVIER's work is, so to speak, a byproduct of the more important parallax determinations.

With this last end in view, the photographs obtained on yellow sensitive plates through a colour-filter, were found to exhibit a number of double stars. The scale is larger than that for HERTZSPRUNG's plates, 1 mm = 20″,8. OLIVIER's paper contains measures of ninety-five pairs and the estimates of four others, each of the latter appearing merely as one elongated star.

Altogether, thirty-two new pairs were discovered where the separation is less than 5″, and ten others with greater angular distance.

From these measures the conclusions are:

Without special care it is possible to discover and measure on the average McCORMICK plate any double of 3″ or wider, and as faint as the eleventh magnitude, if the components do not differ by more than two magnitudes.

With good seeing and with exposures properly timed for the purpose, OLIVIER calls a double of 2″ an "easy" object and states that if the components are nearly equal in magnitude, it will be possible to measure stars separated by as small a distance as 1″,5 or even 1″,0, under special circumstances.

No evidence of any systematic error is found in the measures of doubles on the McCORMICK plates.

[1] Publ Leander McCormick Obs 3, part 2.

OLIVIER's research offers proof that photographic measurement is comparable in accuracy with the best visual work.

33. The Research of JOHN H. PITMAN[1]. PITMAN used the large refractor of the Sproul Observatory, devoting his time to nearly equal pairs varying in magnitude from 2,8 to 9,7, and chiefly from 1″ 8 to 4″,0 in separation.

34. The Use of the Astrographic Plates. A few lists of the double stars found on astrographic plates have been published; among them are:

Astrographic Measures of Double Stars (Zone −40°) by R. W. WRIGLEY [M N 72, p. 34 (1912)] containing 238 pairs up to 10″ separation, and 51 over 10″.

PERTH, Western Australia, Zone −32°, by C. NOSSITER, containing 187 pairs up to 10″ separation, and 55 over 10″ (Perth Obs Bull No. 1).

Katalog von Doppelsternen der photographischen Himmelskarte aus der Zone von +31° bis +40° Deklination, von J. SCHEINER, containing 347 pairs up to 10″ separation, and 1217 over 10″ (Publ Astroph Obs Potsdam 20, No. 59).

According to ROBERT JONCKHEERE, who studied the matter carefully[2], it is not wise to rely implicitly on double-star information derived from ordinary astrographic rectangular coordinates. If objects were measured as double stars on several plates before their inclusion in a catalogue, the data would be more reliable. It is, however, manifest that the degree of accuracy obtainable from these plates is limited.

From BELLAMY's comparison[3] of determinations on Oxford astrographic plates with visual measures of 436 pairs, it appears that a systematic difference exists between photographic and visual values, in distance, for double stars separated by less than 6″. Plate measurements are the smaller.

35. The Discovery of Double Stars by Means of Their Spectra. Often the brighter component of a double star is red or yellow, while the fainter one is green or blue. This fact is due to the primary being of late spectral class G or K, with a companion of early class B or A, having strong hydrogen lines.

On photographs of star-fields taken with a prism-camera many double stars are signified by composite spectra, usually resembling the solar type with strong hydrogen lines superimposed. Prof. PICKERING published a paper in 1891, urging astronomers to search for such spectra[4], and in compliance with this suggestion three or four hundred of them were discovered at Harvard[5] by Miss MAURY and Miss CANNON. In a note on these stars[6] ERIC DOOLITTLE mentions that previous to obtaining their spectra many were known to be double, while a few others have been separated recently by AITKEN, with the large Lick refractor.

36. The Interferometer Method. The interferometer used for astronomical measurement consists of two apertures similar in shape and size, placed either before the objective or in the converging beam a short distance in front of the focal plane.

FIZEAU was the first[7] to conceive using the principle of interference in determining the diameter of a star. Applied later by STEPHAN[8] at the Marseilles Observatory, the instrument employed was not powerful enough; consequently the stars observed were merely shown to have diameters smaller than 0″,2.

In the Philosophical Magazine for July 1890, MICHELSON[9], carrying both theory and application further, describes his method for measuring the angular

[1] Pop Astr 32, p. 227 (1924). [2] Obs 40, p. 233 (1917).
[3] M N 77, p. 521 (1917). [4] A N 127, p. 155 (1891).
[5] Harv Ann 28, p. 93, 229; 56, p. 113, 160 and Harv Circ Nos. 178, 184, 196, 221
[6] A J 33, p. 8 (1920). [7] C R 66, p. 934 (1868).
[8] C R 78, p. 1008 (1874). [9] Phil Mag 30, p. 1 (1890).

magnitude of celestial objects when these are beyond the powers of the largest telescopes. The dimensions of planetoids and satellites, also the distance between double stars, could be determined at that time, but the exact measuring of fixed-star diameters is a recent achievement.

MICHELSON discussed astronomical applications of the interferometer in other papers[1]. In these, he again explained how diameters may be measured when the object is a uniformly luminous disc, or if not uniformly bright, how the amount of darkening toward the limb can be determined.

Further, if a body is not circular he indicates how its exact form may be derived.

A series of observations was taken on the satellites of Jupiter at the Lick Observatory in 1891, with results which amply confirmed the practicability and accuracy of MICHELSON's method[2].

It is of interest to mention here that SCHWARZSCHILD[3] gave a method of measuring the position angle and distance of a double star by making use of the interference fringes produced by a grating. A plane grating formed of wide parallel slits can be rotated in position angle before the telescope's lens and also its angle with the optical axis of the telescope can be varied, which is equivalent to modifying the widths and separations of the slits. A single star is then seen as a bright point flanked by symmetrical series of short spectra; for a double star, a position of the grating can be found so that one of the images of the secondary is in line with and completely fills the space between two images of the primary, and for this position, ϱ and θ can then be computed theoretically.

HAMY[4] considered theoretically the case of narrow slits, and also of two rectangular apertures where the width is appreciable compared to their separation, applying his deductions to the measurement of the diameter of Jupiter's satellites and a few bright asteroids.

The angular diameter of a body or the separation of a double star, can be ascertained by observing the interference fringes produced at the focus of a telescope, when only two portions of the objective located on the same diameter are used; it has been demonstrated that as the distance between the apertures increases, the visibility of the fringes varies and reaches a minimum when this distance is expressed by the formula $1{,}22\ \lambda/\alpha$ for a disc, or for a double star by $0{,}5\ \lambda/\alpha$, λ being the effective wave-length and α the desired angular value.

Originally, very good observing conditions were considered necessary for measuring diameters with a large telescope, but MICHELSON discovered by tests at Yerkes Observatory and at Mount Wilson, that even when these conditions were doubtful, clear and relatively steady fringes were produced.

Using the one hundred inch reflector, ANDERSON measured the distance between the components of Capella, obtaining an accuracy within less than one per cent. Separations as small as $0'',025$ can be measured with this same reflector without attachment, but to determine the diameter of a fixed star a distance of at least ten meters between apertures is required; therefore an interferometer with movable outer mirrors was built, in order to make tests with spacings up to twenty feet. A detailed description of this interferometer, attached to the one hundred inch telescope of Mount Wilson, may be found in the Astrophysical Journal[5]. An instrument of such great dimensions is of value when the diameters of stars similar to Betelgeuse and Antares are to be determined, but to investigate many double stars, an interferometer of moderate size is sufficient.

---

[1] Phil Mag 31, p. 338 (1891); 34, p. 280 (1892) and Americ. Journ. of Sc. 39, p. 115 (1890).
[2] MICHELSON, Light-Waves and their Uses, p. 142.
[3] A N 139, p. 353 (1896). [4] B A 16, p. 257 (1899). [5] Ap J 53, p. 249 (1921).

37. Measurement of Double Stars with the Interferometer. The interferometer consists of a plate A, having two parallel apertures placed in the converging beam of light coming from the objective. The interference fringes formed in the focal plane are viewed with a high-power eyepiece E. The entire instrument can be rotated about the axis of the telescope in order to modify the position angle of the slits. These openings themselves are so arranged that their distance apart can be altered, the actual separation being indicated on a scale at the eye end of the instrument where the circle is also located from which the position angle is read. The plate carrying the slits can be moved entirely out of the path of light to facilitate the accurate centering of the star under observation.

In measuring a double star, plate A is rotated until the fringes produced by the primary coincide exactly with those of the companion, the contrast between bright and dark fringes being as clearly marked as for a single star. When this takes place the fringes are parallel to the line joining the components. Then the position angle is changed 90°, bringing the fringes at right angles to the aforesaid line, the interference patterns usually not overlapping; and by modifying the space between the slits to an appropriate distance Do, the bright fringes of one star will merge with the dark bands of the other. For equal pairs, the fringes give place to a luminous band, and for components of different magnitudes their visibility is minimum.

For a single star, the angular distance between two bright fringes produced in the focal plane is

$$\delta = \frac{\lambda}{D}$$

where λ is the effective wave-length of the star-light, and D is the distance between the two slits expressed in a convenient unit.

For a double star, if the angular distance of the components is α at the minimum visibility of the interference pattern

$$\alpha = \frac{\delta}{2} = \frac{\lambda}{2\,Do}$$

since a bright fringe of one component is located exactly half way between two bright fringes of the other.

For a separation of the slits $D = 2\,Do$, the two systems of bright fringes will again coincide, and for successive values of D, multiple of Do, there will be alternate maxima and minima of visibility.

The position angle and the angular separation of a close double star is thus readily determined. Further detailed information concerning the method is given in an article by J. A. ANDERSON on "Application of MICHELSON's Interferometer Method to the Measurement of Close Double Stars"[1].

ANDERSON applied the method to CAPELLA, discovering the separation to be about 0″,05.

In a redetermination of CAPELLA's orbit, MERRILL[2] found that the average residuals for the measures are 1° for position angle, and 0″,0007 for separation. This is an extraordinary result for such a close pair, known before merely as a spectroscopic binary with a period of 104,022 days.

The study of this star as an extremely short-period visual binary and as a fairly long-period spectroscopic binary, bridges the gap between these two classes.

Similar investigations are recorded in "Measurement of the Spectroscopic Binary Star Mizar with the Interferometer" by F. G. PEASE[3].

---

[1] Ap J 51, p. 263 (1920). [2] Ap J 56, p. 40 (1922).
[3] Wash Nat Ac Proc 11, p. 356 (1925).

The theory of astronomical interferometer measurements has been elaborately developed by H. SPENCER JONES[1].

It is reported that the instrument is to be used for double-star observations at the Lick, Yerkes, Paris, Strasbourg and Catania observatories. At Strasbourg, DANJON conceived a new type of interferometer and the one used at Catania has been described by MAGGINI[2].

38. Catalogues and Lists of Visual Double Stars. The principal general catalogues of double stars arranged in chronological order are:

"A Catalogue of 10300 Multiple and Double Stars, Arranged in Order of Right Ascension" by Sir JOHN HERSCHEL, published in 1874 as one of the Memoirs of the Royal Astronomical Society (Vol. 40).

"Catalogue des Etoiles Doubles et Multiples en Mouvement relatif certain" by CAMILLE FLAMMARION. Published in 1878, this serviceable compendium contains 819 systems with a collection of published measures.

In the third section of the book "A Handbook of Double Stars" by CROSSLEY, GLEDHILL and WILSON, published in 1879, is found a catalogue of 1200 double stars with extensive lists of measures.

The most important catalogue is that of BURNHAM, "A General Catalogue of Double Stars within 121° of the North Pole" published in 1906 by the Carnegie Institution of Washington.

To this volume may be added that of JONCKHEERE "Catalogue and Measures of Double Stars Discovered visually from 1905 to 1916 within 105° of the North Pole and under 5″ Separation"[3].

For the southern hemisphere there is the "Reference Catalogue of Southern Double Stars" by INNES, published in 1899, and the "New Reference Catalogue of Southern Double Stars" by INNES, which is soon to be released.

At the Lick Observatory, a card catalogue of double and multiple stars is constantly revised to date, and a printer's copy of an Extension to BURNHAM's General Catalogue is being prepared by Dr. AITKEN.

A grant of 6000 francs was made by the International Astronomical Union, at the Cambridge meeting, 1925, to further the work on this catalogue.

Following are lists of double stars published recently by individual observers which testify of increasing activity in this field of science:

One Hundred New Double Stars, 24th List, R. G. AITKEN[4].

Cent nouvelles étoiles doubles, R. JONCKHEERE[5].

New Double Stars, E. D. ROE JR.[6].

Catalogue and Re-Measurement of 648 Double Stars Discovered by Prof. G. W. HOUGH, E. DOOLITTLE[7].

Verzeichnis von 66 neuen Doppelsternen, aufgefunden am Meridiankreise, F. KÜSTNER[8].

From time to time attention has been called to double stars of interest, showing rather rapid or peculiar relative motions. Notes concerning them are:

Some Interesting Double Stars, T. LEWIS[9].

Notes on Some Interesting Double Stars, R. G. AITKEN[10].

The Motion in Some A Double Stars, R. G. AITKEN[11].

---

[1] M N 82, p. 513 (1922). [2] CataniaObs Contributi No. 4 (1922). [3] Mem R A S 61 (1917).
[4] Lick Bull 9, p. 132 (1918). [5] A N 187, p. 393 (1911).
[6] A N 187, p. 39 (1910); 188, p. 374 (1911); 190, p. 137 (1911).
[7] Publ of the Univ of Pennsylvania, Astronomical Series 3, part III (1907).
[8] A N 160, p. 71 (1902).
[9] Obs 25, p. 269 (1902); 26, p. 97 (1903); 27, p. 236 (1904); 28, p. 249 (1905).
[10] Publ A S P 15, p. 217 (1903).
[11] Publ A S P 32, p. 56 (1920); 33, p. 60 (1921); 34, p. 52 (1922).

Einige Doppelsterne mit eben merklicher Bahnbewegung, C. LUPLAU-JANSSEN, S. FJELTOFTE[1].

39. Variable Double Stars. There is no reason why the component of a binary would not be variable, as well as any single star, but often, in the past, the suspected variability of a double star has been due simply to poor conditions of the atmosphere, or to an inferior telescope.

In 1888 Miss CLERKE published "An Historical and Descriptive List of Some Double Stars Suspected to Vary in Light"[2]. Among the double stars showing relative motions, she mentions γ Virginis, 44 Boötis and δ Cygni, as perhaps variable, but at present there is no positive evidence that this is true.

In 1906, attention was drawn to 95 Ceti[3], a star which BURNHAM pointed out as the most mysterious and intriguing body in the heavens. There appears to be a faint companion, the position of which is sometimes ascertained, but which becomes invisible intermittingly. Other double stars, exhibiting similar phenomena, are 80 Tauri, Draconis 205 and β 163. Since these stars have been observed by different astronomers, who at times failed to detect any companion, it is impossible to deny the authenticity of their disappearance. It is incredible that this absence can be due to poor seeing, and the angular separation is never small enough for the companion to be lost within the brighter light of the primary. Hence, light variability for the companion is the only logical hypothesis, although no positive observation proving this, is as yet on record.

There are very few evidences to-day, that components of visual double stars are variable. The most typical case is that of X Ophiuchi. Discovered by HUSSEY[4] in 1900, this pair, having a separation of 0″,22, consists of a long-period variable, class Me, magnitudes from 6,8 to about 12, and another star of class Ko, with magnitude 8,9. Spectroscopic observations of these components have been made by Dr. MERRILL at the Mt. Wilson Observatory[5].

Several long-period variables undoubtedly have companions, and for some of them there is proof.

BARNARD's research disclosed a companion to S Lyncis[6], but a most significant discovery, by AITKEN[7], which bids fair to play an important rôle in the theory of long-period variables, was that of a companion to the amazing variable Mira Ceti.

c) The Orbit of a Visual Binary Star.

40. Fundamental Assumptions. In this subdivision it will be assumed that two stars, which form a binary, attract each other according to the laws of gravitation, their relative motions not being sensibly affected by the disturbing action of other heavenly bodies. A further postulate is, that the dimensions of the components are small in comparison with the distance separating them. Thus, in analyzing their motions, the two stars are to be regarded as material points; that is, they must be treated as mathematical points so far as their position or movement is concerned, but endowed with attractive forces corresponding to their respective masses.

It is usually surmised that stars are nearly spherical, and at great distances from each other. Celestial mechanics proves that homogeneous spheres, and bodies composed of concentric homogeneous spherical shells, attract each other as if their masses were concentrated at their centres. Also it is shown that spheroidal bodies attract each other, sensibly in the same manner, when the distances between

[1] A N 214, p. 423 (1921). [2] Nature 39, p. 55 (1888).
[3] Publ A S P 18, p. 70 (1906). [4] A J 21, p. 35 (1900).
[5] Ap J 57, p. 251 (1923). [6] A N 149, p. 167 (1899). [7] Publ A S P 35, p. 323 (1923).

them are very great compared to their dimensions. It follows then, that the premises adopted are in accordance with the conditions which usually prevail in binary systems.

On account of rapid rotation, some heavenly bodies are spheroidal, and numbers of spectroscopic and photometric binary components revolve nearly in contact. When these conditions coexist, the shape of the bodies must be considered in the determination of their motions.

When a system consists of more than two bodies, each sufficiently massive to noticeably affect the motions of the others, these perturbations must be reckoned with to correctly determine the orbit of each individual body.

41. Position Angle and Distance. In nearly all investigations relating to the orbits of visual binaries, the origin is taken at the principal star, with the fundamental plane perpendicular to the line of sight. The axes of x and y are taken in this plane, the former directed toward the north and the latter toward the east. The axis of z is directed away from the earth.

The place of the companion is usually expressed in terms of the polar coordinates, position angle and distance. The position angle gives the direction of the companion from the principal star. It is the angle included between the hour circle passing through the principal star and the arc of the great circle connecting the two stars. It is reckoned from the north, toward the east, through 360°. The distance is the apparent angular separation of the two stars as seen from the earth, or it is the angular length of the arc of the great circle which joins them. It is usually expressed in seconds of arc.

42. Early Measures of Angle. The above manner of reckoning position angles was first proposed by Sir JOHN HERSCHEL in 1830, and since that time this system has been universally followed. Previously, observers found the position of the companion, by reckoning the angle, from the east or west, through 90° toward the north or south, and indicated the quadrant in which the companion was situated by designating it as nf, north following, sf, south following, np, north preceding, or sp, south preceding. Directions given in this way are easily converted into position angles. The accompanying diagram (Fig. 7) shows the relations of the two systems. The older system is somewhat cumbersome and liable to errors.

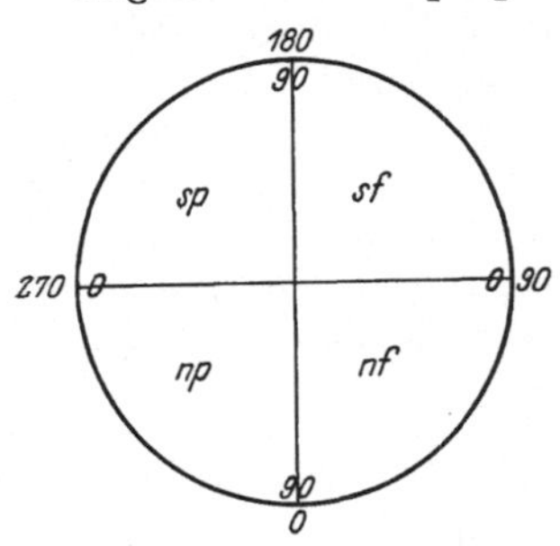

Fig. 7. Position angle.

43. Differences of Right Ascension and Declination. The relative positions of the components are sometimes expressed by differences of right ascension and declination. From such differences, the position angle and distance can be computed by the following equations:

$$\varrho \sin\theta = 1/15\,(\alpha' - \alpha)\cos\delta$$
$$\varrho \cos\theta = \delta' - \delta$$

where θ denotes the position angle, ϱ the distance, α and α' the right ascensions of the principal star and its companion, expressed in time, and δ, δ' their declinations.

44. Effect of Precession. Precession has no effect on the apparent distance, but it slowly shifts the hour circle from which the position angle is measured, and thereby produces a progressive change in its value which is independent of the companion's orbital motion. Consequently, when position angles corresponding to widely different dates are to be used in exact computations, they should be corrected for precession, in order that all may be referred to the same hour circle of some selected epoch.

$n \cos \alpha$ is the annual change in declination, due to precession, of a star having α for right ascension. The differential of this is $-n \sin \alpha \, d\alpha$. Hence $-n(\alpha'-\alpha)\sin\alpha$ can be taken as the difference of the annual precessions in declination of two stars, when they are so near each other that $\alpha'-\alpha$ can be written in place of $d\alpha$.

In the equation $\varrho \cos\theta = \delta' - \delta$, the quantities which are variable owing to precession, are θ, δ and δ'. Differentiating and taking finite differences instead of the differentials, we have

$$-\varrho \sin\theta \cdot \varDelta\theta = \varDelta\delta' - \varDelta\delta .$$

When a year is taken as the unit of time, $-n(\alpha'-\alpha)\sin\alpha$ and $\varDelta\delta' - \varDelta\delta$ may be regarded as identical. Therefore

$$\varDelta\theta = n \sin\alpha \sec\delta ,$$

which is the formula for computing the annual change in position angle due to precession. The value of the constant n, according to Newcomb, is

$$n = 20'',0468 - 0'',000086\,(t - 1900)$$

or

$$n = 0^\circ,00557$$

which is sufficiently correct for double-star investigations.

45. Effect of Proper Motion. The position angle of a binary is slightly affected by the proper motion of the system, without regard to changes which arise from any other cause.

In Fig. 8, let MN be the arc of the great circle along which the system is moving, A being the position of the principal star at the beginning, and B its position at the end of the year.

Let P be the pole of the equator, and AP, BP, the hour circles through A and B respectively. Draw BC, making the angles MAP and MBC equal. Then, PBC is the excess of MAP over MBP. Denote MBP by φ and PBC by $\varDelta\varphi$, thus $MAP = \varphi + \varDelta\varphi$.

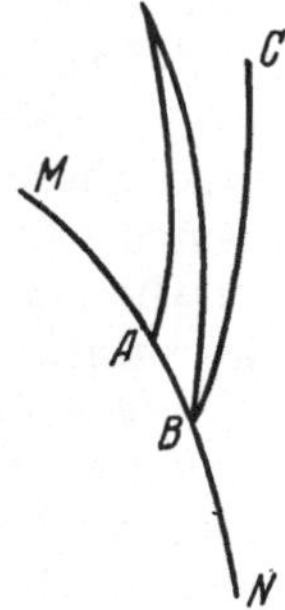

Fig. 8. Effect of proper motion.

Let μ and μ' be, respectively, the components of proper motion in right ascension and in declination, both expressed in seconds of arc. From the spherical triangle ABP (Napier's analogies),

$$\tan\tfrac{1}{2}(A+B) = \frac{\cot\frac{1}{2}P \cos\frac{1}{2}(a-b)}{\cos\frac{1}{2}(a+b)}$$

whence

$$\tan\tfrac{1}{2}\varDelta\varphi = \frac{\tan\frac{1}{2}\mu \sin(\delta + \frac{1}{2}\mu')}{\cos\frac{1}{2}\mu'} ,$$

by means of which the correction in position angle, due to proper motion, may be computed. Since $\varDelta\varphi$, μ and μ', are always small, the last equation can, without appreciable error, be written in the following form

$$\varDelta\varphi = \mu \sin\delta .$$

The correction for proper motion is always small and usually neglected; even in the interval of a hundred years, it seldom exceeds $0^\circ,1$, and no case is known where it amounts to $0^\circ,2$ per century.

46. The True and Apparent Orbits. The path which the companion describes in its revolution about the principal star is called the true orbit, and the projection of this path, upon a plane perpendicular to the line of sight, is called the apparent orbit. The stars being at enormous distances from us, the angular dimensions of all orbits are so small, that no sensible error is introduced by regarding the apparent orbit as the orthogonal projection of the true orbit upon the plane tangent to the celestial sphere at the point occupied by the principal star.

A true binary consists of two stars, one star revolving about the other in an ellipse. The true and apparent orbits are therefore ellipses. The quantities which determine the form, dimensions, and position of either orbit, are called geometrical elements. Since the apparent ellipse is the projection of the true ellipse, the two sets of geometrical elements are intimately related and can be expressed in terms of each other. Besides the geometrical elements, additional quantities are required to define the position of the companion in its orbit at some specified epoch.

47. The Apparent Ellipse. Practically all methods employed to compute elements of the true orbit use the apparent ellipse in some way; only the purely analytical methods dispense with it, and they are in disrepute, owing to the relatively large amount of labour involved, and the uncertain results obtained due to uneliminated errors in observation.

Graphical and semi-graphical methods are far more simple, but require an accurate drawing of the apparent ellipse, and sometimes its equation. The construction of the former is most important, because the accuracy of the final elements depends directly upon its truthfulness.

48. Testing the Accuracy of the Observations. Interpolating Curves. When corrected values of distance and position angle have been tabulated from observations, they will exhibit discrepancies due to accidental and systematic errors, and occasionally to actual mistakes. These values should be tested before being used in the construction of an apparent ellipse; a simple way is to plot upon coordinate paper, first the position angles, and then the distances, separately as ordinates, against the times of observation, expressed in years, as abscissae. Smooth free-hand interpolating curves are then drawn, to represent the general run of measures, more weight being attributed to the best observations, for instance those of reputed observers, or the average results based upon accordant measures. These interpolating curves clearly reveal incorrect observations and show whether the measures are sufficiently authentic to warrant the computation of an orbit.

Curves obtained for standard double stars, from a large number of measures by different observers, may be used to determine the systematic correction of measures by each individual; this correction is then applied to double stars of similar character.

In case of direct or photographic measures, the interpolating curve for position angle will satisfy the observations better than the curve for distance; with interferometer measures it may be the reverse.

The direct measures of distance between close components are often unreliable; corrected distances can then be computed using the interpolating curve for position angles in the following way:

The apparent orbit being the orthographic projection of the true orbit, its areal velocity is constant, and if ϱ and θ are the distance and the position angle at a certain time t, the following equation exists

$$\tfrac{1}{2}\varrho^2\, d\theta = C\, dt$$

or

$$\varrho^2 = 2C\,\frac{1}{\frac{d\theta}{dt}},$$

but $\frac{d\theta}{dt} = \tan\alpha$, where α is the angle which the tangent to the interpolating curve of position angle, for abscissa t, makes with the x-axis.

A table of values of $r = \frac{1}{\sqrt{\tan\alpha}}$ is made for each observed θ, $\tan\alpha$ being derived from the curve; the sum of the values of r is then divided by the sum of the observed values of ϱ and this gives the constant by which to multiply each value of r to obtain values of ϱ which are used in preference to the measured ones for determining the orbit.

49. Construction of the Apparent Ellipse. The different positions of the companion with respect to the primary are plotted on paper, using position angles and distances θ and ϱ as polar coordinates, the primary star being taken as origin. Using an ellipsograph or any other means, an ellipse is drawn through the plotted points and is adjusted by trial until it satisfies the law of areas. A speedy test for this law can be made by drawing radii from the origin to selected points on the ellipse and measuring the areas of the corresponding elliptic sectors with a planimeter. The comparison for well distributed sectors indicates what corrections the ellipse requires. A new ellipse is easily drawn and the areas again measured. The process is repeated until the ellipse is satisfactory.

Other ways of finding the apparent orbit are used, among them the method of Sir John Herschel[1].

The equation of an apparent ellipse is often derived by analytical procedure.

50. The True Orbit. When the apparent ellipse has been drawn, various methods are then available for finding the elements of the true orbit. Some of them are essentially graphical; others require a small amount of computation, but in principle all are simple.

Several important relations connecting the true and apparent ellipse follow at once from the principles of orthogonal projection. The following are so elementary that they need no demonstration here.

(a) The centre of the true ellipse coincides with that of the apparent ellipse.

(b) The position of the principal star in the apparent ellipse is the projection of one focus of the true ellipse and the projection of the other focus of the true ellipse is on the same diameter of the apparent ellipse at an equal distance from the centre.

(c) The diameter of the apparent ellipse which passes through the principal star is the projection of the major axis of the true ellipse and its extremities are the projections of periastron and apastron.

(d) The diameter of the apparent ellipse, conjugate to this, is the projection of the minor axis of the true ellipse. It is parallel to the projection of the latus rectum and to the tangents which may be drawn to the apparent ellipse at the projections of periastron and apastron.

(e) The chord of the apparent ellipse which is bisected by the principal star is the projection of the latus rectum of the true ellipse.

(f) Any chord of the true ellipse which is parallel to the line of nodes is projected into a parallel chord of the apparent ellipse without change of length, and consequently, that diameter of the true ellipse which is parallel to the line of nodes is projected into an equal and parallel diameter of the apparent ellipse.

(g) Any chord of the true ellipse which is perpendicular to the line of nodes, is projected into a chord of the apparent ellipse also perpendicular to the line of nodes, but its length is diminished in the ratio of the cosine of the angle included between the planes of the two ellipses.

[1] See for instance Aitken "The Binary Stars" (New York, 1918) p. 71.

51. Usual Notations. The elements of the true orbit are designated by the following letters; the first two are called the dynamical elements, while the others are the geometrical elements.

P = the period of revolution expressed in mean solar years.

T = the time of periastron passage.

e = the eccentricity.

a = the semi-axis major expressed in seconds of arc.

Ω = the position angle of that nodal point which lies between 0° and 180°; that is, the position angle of the line of intersection of the orbit plane with the plane perpendicular to the line of sight. Call this merely "the nodal point", disregarding the distinction between ascending and descending nodes.

ω = the angle in the plane of the true orbit between the line of nodes and major axis. It is to be measured from the nodal point to the point of periastron passage in the direction of the companion's motion, and may have any value from 0° to 360°. It should be stated whether the position angles increase or decrease with the time.

i = the inclination of the orbit plane, that is, the angle between the orbit plane and the plane at right angles to the line of sight. Its value lies between 0° and ± 90° and should always carry the double sign (±) until the indeterminateness has been removed by measures of the radial velocity. When these are available, i is to be regarded as positive (+) if the orbital motion at the nodal point is carrying the companion star away from the observer; negative (—), if it is carrying the companion star towards the observer.

The symbol μ denotes the mean annual motion of the companion expressed in degrees and decimals, measured always in the direction of motion.

52. Position Angle and Distance, from the Elements. When the elements are known, the apparent position angle θ and the angular distance ϱ for the time t are derived from the following equations:

$$\left.\begin{aligned} \mu &= \frac{360^\circ}{P} \\ M &= \mu\,(t - T) = E - e\sin E \\ r &= a\,(1 - e\cos E) \\ \tan\tfrac{1}{2}v &= \sqrt{\frac{1+e}{1-e}}\tan\tfrac{1}{2}E \end{aligned}\right\} \quad \text{(A)}$$

$$\left.\begin{aligned} \tan(\theta - \Omega) &= \pm\tan(v + \omega)\cos i \\ \varrho &= r\cos(v + \omega)\sec(\theta - \Omega) \end{aligned}\right\} \quad \text{(B)}$$

Equations (A) are the usual ones for elliptic motion, the symbols M, E and v representing respectively, the mean, eccentric and true anomaly, and r the radius vector.

Equations (B) convert the v and r of the companion in the true orbit into its position angle and distance in the projected or apparent orbit.

Another method of obtaining θ and ϱ is by using the Allegheny Tables of Anomalies[1]. Knowing M and e, the value of v can be found in the tables. Then

$$r = \frac{a\,(1 - e^2)}{1 + e\cos v}$$

and the equations (B) are used.

[1] Publ Allegheny Obs 2, p. 155 (1912).

53. Methods for Determining the Elements of the True Orbit. Among methods of historical interest may be mentioned those of SAVARY, J. HERSCHEL, VILLARCEAU, MÄDLER, KLINKERFUES, THIELE, GLASENAPP, SEELIGER, KOWALSKY and ZWIERS. Valuable information may be found in the following papers:

F. SAVARY, Sur la détermination des orbites que décrivent autour de leur centre de gravité, deux étoiles très rapprochées l'une de l'autre[1].

ENCKE, Über die Berechnung der Bahnen der Doppelsterne[2].

KLINKERFUES, Über die Berechnung der Bahnen der Doppelsterne[3].

Allgemeine Methode zur Berechnung von Doppelsternbahnen[4].

KOWALSKY, Sur la détermination des orbites des étoiles doubles[5].

S. GLASENAPP, On a Graphical Method for Determining the Orbit of a Binary Star[6].

THIELE, Über einen geometrischen Satz zur Berechnung von Doppelsternbahnen[7].

Neue Methode zur Berechnung von Doppelsternbahnen[8].

SCHWARZSCHILD, Methode zur Bahnbestimmung der Doppelsterne[9].

ZWIERS, Über eine neue Methode zur Bestimmung von Doppelsternbahnen[10].

H. C. PLUMMER, On the Analytical Method of Finding the Orbit of a Double Star[11].

H. N. RUSSELL, A New Graphical Method of Determining the Elements of a Double-Star Orbit[12].

G. C. COMSTOCK, On the Determination of Double-Star Orbits from Incomplete Data[13].

B. MEYERMAN, Eine neue graphische und halbgraphische Methode zur Bestimmung von Doppelsternbahnen[14].

G. SCHNAUDER, Über abzählende Methoden der Bahnbestimmung von Doppelsternen[15].

W. DOBERCK, Method of Calculating Double-Star Orbits[16].

R. JONCKHEERE, On the Corrections of Double-Star Orbits Determined graphically, and an Application to a Preliminary Orbit of Sirius[17].

C. RODRIGUEZ, Determinación de las orbitas de estrellas dobles[18].

R. G. AITKEN, The Binary Stars (New York 1918), Chapter IV.

Two extremely simple methods are those described in the following paragraphs.

54. The ZWIERS Method. The circle drawn in the plane of the true orbit, having the major axis of the true ellipse as a diameter, is the auxiliary circle of the true ellipse, and its projection upon the plane of the apparent orbit is the auxiliary ellipse. When a drawing of the apparent ellipse has been made, the auxiliary ellipse can be constructed in the following manner. The eccentricity of the true ellipse is first obtained; if C be the centre of the apparent ellipse, S the position of the principal star, and P the projection of periastron, or point

---

[1] Connaissance des Temps pour 1830.
[2] Berliner Astron. Jahrbuch für 1832, p. 253.
[3] A N 42, p. 81 (1855).
[4] A N 47, p. 253 (1858).
[5] Procès-Verbaux de l'Université Impériale de Kasan, 1873.
[6] M N 49, p. 276 (1889).
[7] A N 52, p. 39 (1860).
[8] A N 104, p. 245 (1883).
[9] A N 124, p. 215 (1890).
[10] A N 139, p. 369 (1896).
[11] Obs 33, p. 205 (1910).
[12] A J 19, p. 9 (1898).
[13] A J 33, p. 139 and 163 (1921).
[14] A N 215, p. 179 (1922).
[15] A N 216, p. 129 (1922).
[16] A N Jub.-Nr. p. 6 (1921).
[17] M N 78, p. 657 (1918).
[18] Memorias y Revista de la Sociedad Científica "Antonio Alzate", 31, p. 125.

where CS cuts the apparent ellipse, then $e = \frac{CS}{CP}$. The ratio of the semi-axes of the true ellipse is then found by the formula:

$$\frac{a}{b} = \frac{1}{\sqrt{1-e^2}}.$$

Chords of the apparent ellipse are drawn parallel to the projection of the minor axis of the true ellipse and lengthened in the ratio $\frac{a}{b}$. These lengthened chords will be chords of the auxiliary ellipse and their extremities will be points on it. In this way as many points of the auxiliary ellipse can be found as are needed for its construction.

The ZWIERS method makes use of the properties of the auxiliary ellipse without requiring its construction.

55. Derivation of Elements. The major axes of the true and auxiliary ellipses are equal, for each is equal to the diameter of the auxiliary circle.

The major axis of the auxiliary ellipse coincides with the line of nodes, for it is the projection of that diameter of the auxiliary circle which is parallel to the line of nodes and therefore not shortened by projection.

The minor axis of the auxiliary ellipse is perpendicular to the line of nodes. The ratio of its length to that of the diameter of the auxiliary circle is equal to $\cos i$.

Let a' and b' be respectively the projections of the major and minor semi-axes of the true ellipse, σ the acute angle included between the conjugate diameters $2a'$ and $2b'$, Ω the position angle of the ascending node, and λ the position angle of the projection of periastron. Further, let α and β be respectively the major and minor semi-axes of the auxiliary ellipse, and $b'' = \frac{a}{b} \cdot b'$.

Since the apparent and auxiliary ellipses are tangent at the projections of periastron and apastron, $2a'$ is a diameter common to the two ellipses and $2b'$ is the diameter of the auxiliary ellipse conjugate to $2a'$. Hence

$$\alpha^2 + \beta^2 = a'^2 + b''^2$$

and

$$\alpha\beta = a'b''\sin\sigma$$

therefore

$$\alpha + \beta = \sqrt{a'^2 + b''^2 + 2a'b''\sin\sigma}$$

$$\alpha - \beta = \sqrt{a'^2 + b''^2 - 2a'b''\sin\sigma},$$

by means of which α and β can be found. The elements a and i are then given by the equations:

$$a = \alpha$$

$$\cos i = \frac{\beta}{\alpha}.$$

If the major and minor axes of the auxiliary ellipse are taken as axes of coordinates and if it is remembered that ω is equal to the eccentric angle of P, where P is the projection of periastron, $a\cos\omega$ and $a\sin\omega$ are the coordinates of P. Since a' is the distance of P from the origin

$$a'^2 = \alpha^2\cos^2\omega + \beta^2\sin^2\omega$$

hence

$$\tan^2\omega = \frac{\alpha^2 - a'^2}{a'^2 - \beta^2}.$$

The position angle λ of the projection of periastron may be measured in the apparent ellipse, and then the direction of the line of nodes can be computed by the equation

$$\tan(\lambda - \Omega) = \cot\omega \cos i .$$

This completes the determination of the geometrical elements. The dynamical elements can be obtained easily, for instance by using the Allegheny tables of Anomalies[1], or the formulae

$$\tan(v + \omega) = \tan(\theta - \Omega) \sec i$$

$$\tan\tfrac{1}{2}E = \sqrt{\frac{1-e}{1+e}} \tan\tfrac{1}{2}v$$

$$M = E - e \sin E .$$

56. The HENROTEAU-STEWART Method[2]. When the position of the principal star in the apparent ellipse is known, the direction of the line of nodes is given by the bisector of the exterior angle formed by joining the position of the principal star to the two foci of the apparent ellipse. Then, take any line AB parallel to the line of nodes, cutting the projections of the major and minor axes of the true ellipse in A and B respectively, and erect upon AB a semi-circle on the side toward the centre of the ellipse. Draw through the centre C a line perpendicular to the line of nodes, cutting AB in D and the semi-circle in E. Find a point F in AB such that CF is equal to ED. Then the angle FCD is the inclination, and the angle BAE is the angle between the line of nodes and the major axis, while if A has been chosen coincident with the projection of the end of the major axis, AE will be equal to the semi-axis major.

The eccentricity is found as in the ZWIERS' method, from the ratio of the distance of the principal star from the centre to the length of the corresponding semi-diameter.

57. Formulae for the Improvement of Elements. Different methods for correcting orbital elements have been given, for instance by MARTH[3], DOBERCK[4] and COMSTOCK[5], but the usual procedure to improve the elements by least squares is given in the following lines:

The position angle θ is a function of the six elements Ω, i, ω, e $(= \sin\Phi)$, T and $\mu = \frac{360^\circ}{P}$, thus

$$d\theta = \frac{\partial F(\theta)}{\partial \Omega} d\Omega + \frac{\partial F(\theta)}{\partial i} di + \frac{\partial F(\theta)}{\partial \omega} d\omega + \frac{\partial F(\theta)}{\partial \Phi} d\Phi + \frac{\partial F(\theta)}{\partial T} dT + \frac{\partial F(\theta)}{\partial \mu} d\mu .$$

When the variations of the elements are finite, but small, this may be written

$$\Delta\theta = A\Delta\Omega + B\Delta i + C\Delta\omega + D\Delta\Phi + F\Delta T + G\Delta\mu .$$

The coefficients A, B, C, D, F, G are derived by partial differentiation from the equations

$$M = \mu(t - T) = E - e \sin E$$

$$\tan\tfrac{1}{2}v = \sqrt{\frac{1+e}{1-e}} \tan\tfrac{1}{2}E ,$$

$$\tan(\theta - \Omega) = \cos i \tan(v + \omega)$$

---

[1] Publ Allegheny Obs. 2, p. 155 (1912). [2] Pop Astr 33, p. 304 (1925).
[3] M N 47, p. 480 (1887). [4] A N 147, p. 343 (1898). [5] A J 31, p. 33 (1918).

and their values are

$$A = +1$$
$$B = -\sin i \tan(v+\omega)\cos^2(\theta-\Omega)$$
$$C = \cos^2(\theta-\Omega)\sec^2(v+\omega)\cos i$$
$$D = C\frac{2-e\cos E-e^2}{(1-e\cos E)^2}\sin E$$
$$F = C\frac{\cos\Phi}{(1-e\cos E)^2}$$
$$G = (t-T)\,F.$$

For different values of θ, there are thus different sets of values of A, B, C, D, F and G, as well as of $\Delta\theta = \theta_o - \theta_c$, or difference between the observed and the computed one for the time t.

For every time an observation has been secured there is an equation of condition, all of them forming the following system:

$$A_1\Delta\Omega + B_1\Delta i + C_1\Delta\omega + D_1\Delta\Phi + F_1\Delta T + G_1\Delta\mu - \Delta\theta_1 = 0$$
$$A_2\Delta\Omega + B_2\Delta i + C_2\Delta\omega + D_2\Delta\Phi + F_2\Delta T + G_2\Delta\mu - \Delta\theta_2 = 0$$
$$\cdots\cdots\cdots\cdots\cdots\cdots\cdots\cdots\cdots$$
$$A_n\Delta\Omega + B_n\Delta i + C_n\Delta\omega + D_n\Delta\Phi + F_n\Delta T + G_n\Delta\mu - \Delta\theta_n = 0.$$

According to the method of least squares, six normal equations are then formed from the above system of equations. They may be written

$$[AA]\,\Delta\Omega+[AB]\,\Delta i+[AC]\,\Delta\omega+[AD]\,\Delta\Phi+[AF]\,\Delta T+[AG]\,\Delta\mu-[A\,\Delta\theta]=0$$
$$[AB]\,\Delta\Omega+[BB]\,\Delta i+[BC]\,\Delta\omega+[BD]\,\Delta\Phi+[BF]\,\Delta T+[BG]\,\Delta\mu-[B\,\Delta\theta]=0$$
$$[AC]\,\Delta\Omega+[BC]\,\Delta i+[CC]\,\Delta\omega+[CD]\,\Delta\Phi+[CF]\,\Delta T+[CG]\,\Delta\mu-[C\,\Delta\theta]=0$$
$$[AD]\,\Delta\Omega+[BD]\,\Delta i+[CD]\,\Delta\omega+[DD]\,\Delta\Phi+[DF]\,\Delta T+[DG]\,\Delta\mu-[D\,\Delta\theta]=0$$
$$[AF]\,\Delta\Omega+[BF]\,\Delta i+[CF]\,\Delta\omega+[DF]\,\Delta\Phi+[FF]\,\Delta T+[FG]\,\Delta\mu-[F\,\Delta\theta]=0$$
$$[AG]\,\Delta\Omega+[BG]\,\Delta i+[CG]\,\Delta\omega+[DG]\,\Delta\Phi+[FG]\,\Delta T+[GG]\,\Delta\mu-[G\,\Delta\theta]=0,$$

where

$$[AA] = A_1^2 + A_2^2 + \cdots\cdots + A_n^2$$
$$[AB] = A_1B_1 + A_2B_2 + \cdots\cdots + A_nB_n$$
$$[AC] = A_1C_1 + A_2C_2 + \cdots\cdots + A_nC_n$$
$$\cdots\cdots\cdots\cdots\cdots\cdots$$
$$[A\,\Delta\theta] = A_1\Delta\theta_1 + A_2\Delta\theta_2 + \cdots\cdots + A_n\Delta\theta_n.$$

These equations are conveniently transformed by the method of substitution into the following ones:

$$\Delta\Omega+\frac{[AB]}{[AA]}\Delta i+\frac{[AC]}{[AA]}\Delta\omega+\frac{[AD]}{[AA]}\Delta\Phi+\frac{[AF]}{[AA]}\Delta T+\frac{[AG]}{[AA]}\Delta\mu-\frac{[A\Delta\theta]}{[AA]}=0$$
$$\Delta i+\frac{[BC\cdot1]}{[BB\cdot1]}\Delta\omega+\frac{[BD\cdot1]}{[BB\cdot1]}\Delta\Phi+\frac{[BF\cdot1]}{[BB\cdot1]}\Delta T+\frac{[BG\cdot1]}{[BB\cdot1]}\Delta\mu-\frac{[B\Delta\theta\cdot1]}{[BB\cdot1]}=0$$
$$\Delta\omega+\frac{[CD\cdot2]}{[CC\cdot2]}\Delta\Phi+\frac{[CF\cdot2]}{[CC\cdot2]}\Delta T+\frac{[CG\cdot2]}{[CC\cdot2]}\Delta\mu-\frac{[C\Delta\theta\cdot2]}{[CC\cdot2]}=0$$
$$\Delta\Phi+\frac{[DF\cdot3]}{[DD\cdot3]}\Delta T+\frac{[DG\cdot3]}{[DD\cdot3]}\Delta\mu-\frac{[D\Delta\theta\cdot3]}{[DD\cdot3]}=0$$
$$\Delta T+\frac{[FG\cdot4]}{[FF\cdot4]}\Delta\mu-\frac{[F\Delta\theta\cdot4]}{[FF\cdot4]}=0$$
$$\Delta\mu-\frac{[G\Delta\theta\cdot5]}{[GG\cdot5]}=0.$$

The form of the notation $[BB\cdot 1]$, $[BC\cdot 1]$, $[CD\cdot 2]$ .... may be symbolized thus

$$[\beta\gamma\cdot\mu] - \frac{[\alpha\beta\cdot\mu]}{[\alpha\alpha\cdot\mu]}[\alpha\gamma\cdot\mu] = [\beta\gamma\cdot(\mu+1)]$$

in which α, β, γ denote any three letters, and μ any numeral.

When $\Delta\Omega$, Δi, $\Delta\omega$, $\Delta\Phi$, ΔT and $\Delta\mu$ have been obtained, the correct values of the elements are

$$\left.\begin{array}{c} \Omega+\Delta\Omega \\ i+\Delta i \\ \omega+\Delta\omega \\ \Phi+\Delta\Phi \\ T+\Delta T \\ \mu+\Delta\mu \end{array}\right\}$$

The element a is usually regarded as constant.

With these new elements it may be necessary sometimes to carry on a second least squares solution.

58. Probable Errors of the Elements. When the least-square solution has been carried through the weights $p_{\Delta\Omega}$, $p_{\Delta i}$ ... of the elements are given by the following formulae:

$$p_{\Delta\mu} = [GG\cdot 5]$$

$$p_{\Delta T} = [FF\cdot 5] = \frac{[GG\cdot 5]}{[GG\cdot 4]}\cdot[FF\cdot 4]$$

$$p_{\Delta\Phi} = [DD\cdot 5] = \frac{[GG\cdot 5]}{[GG\cdot 4]_d}\cdot\frac{[FF\cdot 4]}{[FF\cdot 3]}\cdot[DD\cdot 3]$$

$$p_{\Delta\omega} = [CC\cdot 5] = \frac{[GG\cdot 5]}{[GG\cdot 4]_c}\cdot\frac{[FF\cdot 4]}{[FF\cdot 3]_c}\cdot\frac{[DD\cdot 3]}{[DD\cdot 2]}\cdot[CC\cdot 2]$$

$$p_{\Delta i} = [BB\cdot 5] = \frac{[GG\cdot 5]}{[GG\cdot 4]_b}\cdot\frac{[FF\cdot 4]}{[FF\cdot 3]_b}\cdot\frac{[DD\cdot 3]}{[DD\cdot 2]_b}\cdot\frac{[CC\cdot 2]}{[CC\cdot 1]}\cdot[BB\cdot 1]$$

$$p_{\Delta\Omega} = [AA\cdot 5] = \frac{[GG\cdot 5]}{[GG\cdot 4]_a}\cdot\frac{[FF\cdot 4]}{[FF\cdot 3]_a}\cdot\frac{[DD\cdot 3]}{[DD\cdot 2]_a}\cdot\frac{[CC\cdot 2]}{[CC\cdot 1]_a}\cdot\frac{[BB\cdot 1]}{[BB]}\cdot[AA].$$

The expressions for the new auxiliaries $[GG\cdot 4]_d$, $[GG\cdot 4]_c$, $[FF\cdot 3]_c$ ... are easily formed as follows:

$$[GG\cdot 4]_d = [GG\cdot 3] - \frac{[FG\cdot 3]}{[FF\cdot 3]}\cdot[FG\cdot 3]$$

$$[FG\cdot 3]_c = [FG\cdot 2] - \frac{[DF\cdot 2]}{[DD\cdot 2]}\cdot[DG\cdot 2]$$

$$[GG\cdot 3]_c = [GG\cdot 2] - \frac{[DG\cdot 2]}{[DD\cdot 2]}\cdot[DG\cdot 2]$$

$$[FF\cdot 3]_c = [FF\cdot 2] - \frac{[DF\cdot 2]}{[DD\cdot 2]}\cdot[DF\cdot 2]$$

$$[GG\cdot 4]_c = [GG\cdot 3]_c - \frac{[FG\cdot 3]_c}{[FF\cdot 3]_c}\cdot[FG\cdot 3]_c.$$

In like manner, expressions of the new auxiliaries introduced into the equations for $p_{\Delta i}$ and $p_{\Delta\Omega}$ may be derived. It will be expedient, however, in the actual application of the formulae, to eliminate first in the order $\Delta\Omega$, Δi, $\Delta\omega$, $\Delta\Phi$, ΔT, $\Delta\mu$ ... obtaining $p_{\Delta\mu}$, $p_{\Delta T}$ and $p_{\Delta\Phi}$ by the above formulae.

Then the elimination should be performed in the order $\varDelta\mu$, $\varDelta T$, $\varDelta\Phi$, $\varDelta\omega$, $\varDelta i$, $\varDelta\Omega$ obtaining $p_{\varDelta\Omega}$, $p_{\varDelta i}$ and $p_{\varDelta\omega}$ by formulae similar to the first three given above, or

$$p_{\varDelta\Omega} = [AA\cdot 5]$$

$$p_{\varDelta i} = \frac{[AA\cdot 5]}{[AA\cdot 4]}\cdot[BB\cdot 4]$$

$$p_{\varDelta\Phi} = \frac{[AA\cdot 5]}{[AA\cdot 4]_c}\cdot\frac{[BB\cdot 4]}{[BB\cdot 3]}\cdot[CC\cdot 3]$$

$$[AA\cdot 4]_c = [AA\cdot 3] - \frac{[AB\cdot 3]}{[BB\cdot 3]}\cdot[AB\cdot 3]\,.$$

If $v_1, v_2, \ldots v_n$ are the residuals $\theta_o - \theta'_c$, or differences between the observed θ and the θ computed using the corrected elements, and if $[vv]$ designates the sum of the squares of these residuals, the probable error of an observation of weight unity is

$$r = \pm 0{,}6745\sqrt{\frac{[vv]}{n-6}}\,,$$

where n is the number of observations or number of equations of condition, and 6 is the number of unknowns in these equations. The probable errors of the different elements are then

$$r_\Omega = \frac{r}{\sqrt{p_{\varDelta\Omega}}}$$

$$r_i = \frac{r}{\sqrt{p_{\varDelta i}}}$$

$$r_\omega = \frac{r}{\sqrt{p_{\varDelta\omega}}}$$

$$r_\Phi = \frac{r}{\sqrt{p_{\varDelta\Phi}}}$$

$$r_T = \frac{r}{\sqrt{p_{\varDelta T}}}$$

$$r_\mu = \frac{r}{\sqrt{p_{\varDelta\mu}}}\,.$$

59. Case when the Apparent Ellipse is Reduced to a Straight Line. The methods of ZWIERS, HENROTEAU-STEWART and all others based upon the construction of an apparent ellipse fail when the latter is reduced to a straight line, for then the observed motion is entirely in distance. Many systems are known where the inclination of the orbit plane is so near 90° that the apparent orbit is an extremely elongated ellipse; in the system 42 Comae Berenicis it actually is a straight line.

Dr. AITKEN in his book on the Binary Stars has called attention to this particular case. He finds that a and ω cannot be derived readily, and presents an elaborate method by MOULTON, for obtaining these elements[1]. However, the following solution which occured to me, is very simple and permits rapid computation.

In the present case, $i = 90°$, and Ω is read directly. An interpolating curve for distances is then drawn, that is, the observed distances are plotted as ordinates against the times as abscissae, and the most probable curve passed through the plotted points. The period P is easily obtained.

[1] AITKEN, Binary Stars, p. 96.

Let $PKBALD$, in Fig. 9, represent the true orbit, and $K'L'$ the apparent orbit, projection of the true. C', projection of the centre, must be the middle point of $K'L'$. S', projection of the focus is the apparent position of the principal star. Let $A''B''P''$ be the interpolating curve for distance, which projects also into $K'L'$; the points P'' and A'' on this curve, which correspond to periastron and apastron, must be separated in abscissae by exactly half the revolution period, and their abscissae measured from the line $C'C$ must be equal and of opposite sign. The point corresponding to periastron must lie on the same side of $C'C$ as S', and on the steeper branch of the curve. In practice, P'' and A'' are most readily found by tracing the curve on a piece of transparent paper, then turning the upper face of this paper down so that the curve seen on it appears like the symmetrical of the original curve with respect to $C'C$, then shifting the paper in the direction $C'C$ for a distance equal to half the revolution period of the binary. P'' will be one of the intersections of the original curve with the curve as seen on the transparent paper. P' is then the projection of P'' as indicated on the figure. Then,

$$e = \frac{C'S'}{C'P'}\,;$$

e being known, the angle PSB or true anomaly of B, extremity of the minor axis in the true ellipse, is easily found:

$$\sin PSB = \sqrt{1-e^2}\,.$$

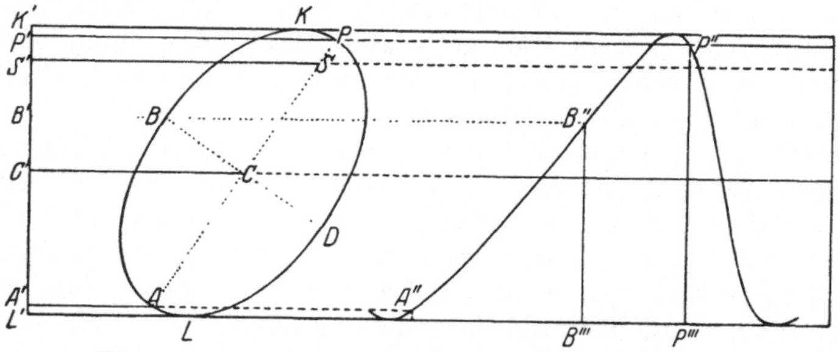

Fig. 9. Apparent ellipse reduced to a straight line.

The value of PSB between 90° and 180° must be taken.

The corresponding mean anomaly is then taken in the Allegheny Tables of Anomalies[1], or computed by the formulae (A) in paragraph 52. Then $P'''B'''$ is equal to this corresponding mean anomaly divided by μ, giving B''', B'', and B'. It is therefore easy to prove that

$$\cot\omega = \cot C'CB = \pm\frac{P'C'}{B'C'}\sqrt{1-e^2}$$

while

$$a = PC = P'C' \sec\omega\,.$$

60. The Second Method of HENROTEAU for the General Case. This method is based on the property that the projections of all points in the plane of the apparent ellipse on the line of nodes are identical with the projections of corresponding points in the plane of the true ellipse on the line of nodes.

The direction of the line of nodes on the apparent ellipse is given as in paragraph 56, by the bisector of the exterior angle formed by joining the position of the principal star to the two foci of the apparent ellipse.

All the necessary points, the centre, the position of the principal star, the extremities of the projections of the major and minor axes of the true ellipse, are then projected on the line of nodes. These projections are then considered like the apparent ellipse reduced to a straight line and the elements of the true orbit, e, ω and a are derived as in paragraph 59:

$$\cot\omega = \frac{P'C'}{B'C'}\sqrt{1-e^2}\,,\qquad a = P'C'\sec\omega\,.$$

[1] Publ Allegheny Obs 2, p. 155 (1912).

If ω' is the angle measured, in the plane of the apparent ellipse, between the direction of the line of nodes and the projection of the major axis on the true ellipse

$$\cos i = \frac{\tan \omega'}{\tan \omega}.$$

Let it be noted that when the apparent orbit is reduced to a straight line, there are two possible values of ω, one which we may call ω_1, and the other $\omega_1 + 90°$.

In the general case, the trigonometric quadrant in which to choose ω is the same as that of ω'.

d) The Visual Double Stars of Known Orbits. Some Interesting Systems.

61. Classes of Visual Binaries. The visual, unlike the spectroscopic binaries, usually belong to the dwarf branch of stars; none of them are known to be super-giant. This is due to the fact that a visual binary must be fairly near us in order to be seen separated. Some are stars of the same spectral type as the Sun or even the less advanced classes A, F and early G; others present interesting features, belonging to the very dwarf stars, of small mass, large density and very low temperature. The fact that none of them are super-giant places them at a disadvantage from the point of view of an astrophysicist.

Triple and multiple systems are not treated in this subdivision. Hence all the orbits to be considered are perfect ellipses; individual analyses of large numbers of them would be without interest, since finding more or less eccentric ellipses is not significant. For general statistical studies, however, knowledge of the accurate orbit elements of many visual binaries will be required, procuring results of extraordinary value.

According to present development of double-star research, visual binaries may be divided into the following classes, the division being more or less arbitrary:

(a) The 61 Cygni Class. Two fairly distant stars not vastly different in magnitude turn around their common centre of gravity in several centuries. Belonging to this group are 61 Cygni, η Cassiopeiae and σ Coronae Borealis.

(b) The α Centauri Class. Two stars having no great difference in magnitude turn around their common centre of gravity in a period ranging from a few years to approximately a century. Interesting systems are α Centauri, the brightest, and 42 Comae Berenicis.

(c) The Sirius Class. A fairly bright star having a companion much fainter than itself. So far this class has only two well known representatives, Sirius and Procyon. Sirius, composed of two stars, respective magnitudes $-1{,}6$ and 8,5, has a period of 50,2 years, while Procyon composed of two stars, respective magnitudes 0,5 and 13,5, has a period of 47,8 years.

(d) The Capella Class. These stars are really spectroscopic binaries which have been found, with the interferometer, to be also visual binaries. Of this type is Capella, having a period of 104,022 days. The importance of these stars rests in the fact that all their elements as well as their masses and parallaxes can be exactly determined. They constitute the link between what seemed before two separate kinds of celestial objects, the visual and the spectroscopic binaries.

The binaries mentioned as examples of the four classes, are discussed in the following paragraphs.

62. The Binary 61 Cygni. In 1755 BRADLEY discovered this star to be double; having a very large proper motion, it was the first star for which a parallax was found, by BESSEL, using the heliometer. The value for the parallax was $+0'',314 \pm 0'',014$, which agrees surprisingly well with that obtained by modern photographic methods $+ 0'',300 \pm 0'',003$. The spectroscopic parallaxes by Prof. ADAMS at Mt. Wilson are for each component, respectively $+ 0'',347$ and $+ 0'',302$. Thus, 61 Cygni is among the very nearest stars; the spectral classes of the components are K7 and K8, their magnitudes 5,57 and 6,28 and the spectroscopic absolute magnitudes 8,3 and 8,7, proving that they are dwarfs of almost identical physical character. This indicates that the two stars are intimately connected by the bonds of gravitational attraction.

Many observations of 61 Cygni as a binary are available; those preceding 1830 are uncertain or approximate and cannot be used in research. Later observations exhibit a very slow change evidenced by the following selected measures:

Year	θ	ϱ	Observer
1830,68	90°,6	15″,47	W. STRUVE
1837,06	94,8	16,01	W. STRUVE
1847,46	100,9	17,02	O. STRUVE
1860,80	108,8	18,23	O. STRUVE
1870,90	113,5	18,91	DUNÉR
1887,68	121,0	20,58	SCHIAPARELLI
1894,58	123,6	21,51	CHOFARDET
1905,41	127,2	22,50	BURNHAM
1909,50	128,6	22,85	BARNARD
1918,69	131,8	24,11	DE VOS VAN STEENWYK
1920,71	132,2	23,95	PHILLIPS

Some astronomers thought that the change in θ and ϱ indicated rectilinear motion, namely that the proper motions of the two components of 61 Cygni are not quite the same, there being slight differences in both amount and direction. FLAMMARION, in 1875, was of this opinion, while BESSEL maintained that orbital motion existed, having a period of about 400 years. From measures to 1883, PETERS found a period of 782,6 years[1].

By a thorough analysis of all observations and using the method of least squares[2], SCHLESINGER and ALTER, in 1910, definitely proved that orbital motion exists; the length of the period is several centuries, but so far cannot be determined with great accuracy.

63. The Binary η Cassiopeiae. At all times this binary is a very easy pair within reach of small telescopes; the great distance between its components, and the large parallax, $+ 0'',182 \pm 0'',005$, make it an object presenting great analogies with 61 Cygni. The magnitudes, spectral classes and absolute magnitudes of the two components are respectively 3,7, 7,4, F 8, K 5, 4,3, 8,3. The latter obtained by ADAMS give spectroscopic parallaxes 0″,132 and 0″,151.

This binary has been measured by many observers, and as the motion is very slow such large numbers of observations are superfluous. The following normal values from various measures are sufficient to show the relative change and total arc described. LOHSE[3] derived these values by graphical interpolation:

Year	θ	ϱ	Year	θ	ϱ
1770,0	46°,6	11″,25	1850,0	104°,5	8″,20
1780,0	52 ,8	11 ,23	1860,0	117 ,2	7 ,21
1790,0	59 ,0	11 ,15	1870,0	134 ,0	6 ,23
1800,0	65 ,3	11 ,00	1880,0	158 ,0	5 ,28
1810,0	71 ,7	10 ,77	1890,0	188 ,0	4 ,90
1820,0	78 ,3	10 ,42	1900,0	218 ,0	5 ,25
1830,0	85 ,7	9 ,93	1910,0	242 ,0	6 ,25
1840,0	94 ,2	9 ,20	1920,717	258 ,9	7 ,26 Observation by PHILLIPS

[1] A N 113, p. 321 (1885). [2] Publ Allegh Obs 2, p. 13 (1910).
[3] Publ Astroph Obs Potsdam 20, No 58, p. 80 (1909).

The curvature of the rather small arc described during the last century being uncertain, a wide range in values of the orbital elements has been found. Two of the best determinations are given:

	Elements by LOHSE[1]	Elements by DOBERCK[2]
P	345,59 years	507,60 years
T	1892,412	1890,03
e	0,376	0,522
a	10″,103	12″,21
Ω	92°,46	99°,2
i	39,05	31,6
ω	100,43	88,9
Angle	Increasing	Increasing

64. The Binary σ Coronae Borealis. The parallax of this binary is $+0'',048 \pm 0'',007$ showing it to be at a far greater distance than 61 Cygni and η Cassiopeiae; the spectroscopic absolute magnitudes of the components are 4,2 and 3,7 and their spectral classes F9 and F8. Consequently they are brighter than the Sun and perhaps more massive than in 61 Cygni.

At present, only a small arc of the apparent orbit is known. The following is a table of measured positions as given in BURNHAM's General Catalogue and completed from recent observations.

Year	θ	ϱ	Observer	Year	θ	ϱ	Observer
1827,02	89°,3	1″,31	W. STRUVE	1878,03	201°,0	3″,40	DEMBOWSKI
1830,11	104,9	1,22	,,	1878,50	202,0	3,51	,,
1832,99	118,8	1,30	,,	1882,01	202,8	3,70	A. HALL
1833,50	130,5	1,31	,,	1883,47	204,5	3,77	,,
1836,59	134,7	1,43	,,	1884,48	206,0	3,80	,,
1837,55	140,0	1,42	,,	1885,43	205,7	3,89	,,
1838,45	143,4	1,48	O. STRUVE	1886,48	206,9	3,96	,,
1840,82	150,2	1,54	,,	1887,44	205,5	3,99	,,
1847,02	168,7	1,74	,,	1888,44	206,6	3,91	,,
1849,92	170,7	1,97	,,	1890,33	207,8	4,08	BURNHAM
1852,13	173,4	2,06	,,	1892,62	209,9	4,06	COMSTOCK
1853,66	175,6	2,17	,,	1895,15	210,0	4,28	H. STRUVE
1855,29	179,1	2,27	,,	1897,77	209,9	4,31	AITKEN
1856,57	179,9	2,46	,,	1898,23	211,7	4,41	HUSSEY
1858,01	181,9	2,51	,,	1899,79	212,4	4,53	DOOLITTLE
1859,94	186,1	2,62	,,	1900,31	212,6	4,42	LOHSE
1861,58	187,4	2,69	,,	1901,29	213,2	4,32	BURNHAM
1863,32	188,5	2,77	,,	1903,39	214,4	4,51	DOOLITTLE
1865,36	191,9	2,94	,,	1905,53	214,9	4,59	BURNHAM
1866,63	193,0	3,00	,,	1906,46	214,8	4,76	LOHSE
1868,58	194,7	2,98	,,	1912,42	216,8	4,88	PHILLIPS
1872,57	195,3	3,26	,,	1916,82	218,3	4,82	,,
1873,56	197,6	3,14	,,	1917,42	217,2	5,00	FRANKS
1874,61	199,8	3,41	,,	1918,40	217,9	4,76	PHILLIPS

The interpolation curves on Fig. 10 give the following normal values:

Time	θ	ϱ	Time	θ	ϱ
1830,0	105°,9	1″,23	1880,0	202°,2	3″,62
1840,0	148,0	1,50	1890,0	208,0	4,04
1850,0	170,4	1,95	1900,0	212,7	4,45
1860,0	185,4	2,58	1910,0	216,1	4,75
1870,0	195,0	3,12	1920,0	218,3	4,95

[1] Publ Astroph Obs Potsdam 20, No 58, p. 80 (1909).
[2] A N 179, p. 383 (1909).

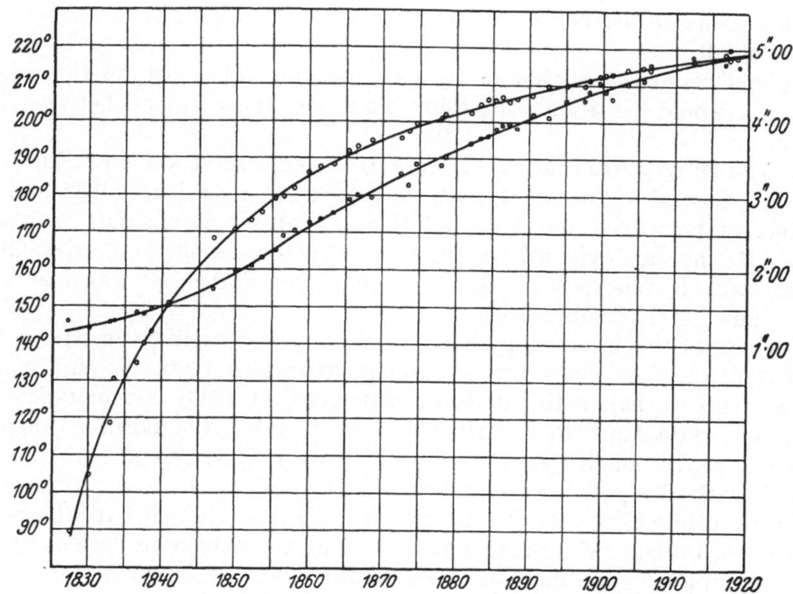

Fig. 10. σ Coronae Borealis. Interpolation curves.

These furnish the apparent orbit illustrated by Fig. 11. The orbital elements deduced by the writer, using the method evolved in paragraph 60 are computed here:

From measures of areas (one quarter of the apparent ellipse).

$$P = 7584 \text{ years}$$

$$e = \frac{CS}{CP} = \frac{791}{856} = 0{,}924,$$

$$\sqrt{1-e^2} = 0{,}382,$$

$$\cot\omega = \frac{36}{50}\cdot 0{,}382 = 0{,}2750,$$

$$\omega = 254°37' = 180° + 74°37',$$

$$P'C' = 7'',20,$$

$$a = P'C'\sec\omega = 27'',14,$$

$$\omega' = 180° + 65°24',$$

$$\cos i = \frac{\tan 65°24'}{\tan 74°37'} = \frac{2{,}1842}{3{,}6346} = 0{,}6009,$$

$$i = 53°4'.$$

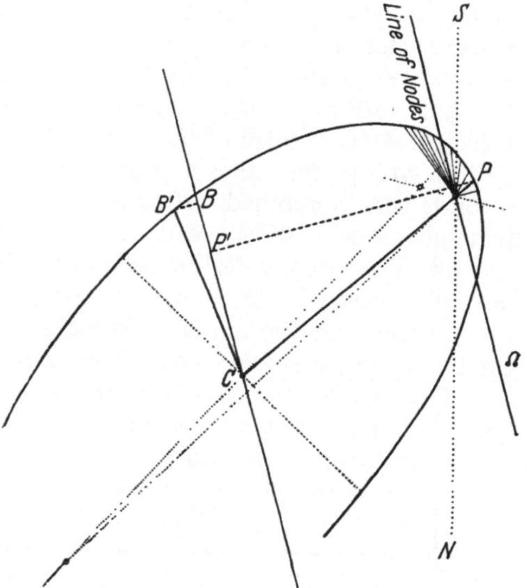

Fig. 11. Apparent orbit of σ Coronae Borealis.

Concluding are the elements of σ Coronae Borealis:

P	7584 years	Ω	15°,3
T	1835,0	i	53° 4′
e	0,924	ω	254° 37′
a	27″,14		

Angle Increasing.

These elements are no doubt uncertain, the arc covered by the observations being so small.

The net results show that orbits of stupendous size exist; in this case the major axis would be about one thousand times the distance between Earth and Sun.

65. The Binary α Centauri. α Centauri is our nearest known stellar neighbour. It is formed of two stars, respective apparent magnitudes 0,33 and 1,70; their spectra being G6 and K4, and their spectroscopic absolute magnitudes 4,8 and 6,2; they are evidently of the same structure as our Sun, especially the brighter star. It was discovered to be double by Father RICHAUD at Pondicherry, India, in December 1689.

Measures of this interesting binary have been made with the meridian circle, the micrometer, the heliometer, and on photographic plates; accurate micrometer measures began with Sir JOHN HERSCHEL in 1834, but meridian circle observations date back to LACAILLE's time in 1752; the position angle and distance derived from the latter's observation are 217°,84 and 20″,51, this agrees very well with the position computed from the orbit.

Several orbits of α Centauri have been published; the most reliable appears to be that of LOHSE[1], who made a survey of all available observations and corrected his values by the method of least squares; listed below are the results:

P	78,83 years	Ω	25°,05
T	1875,68	i	+79 ,04
e	0,512	ω	52 ,35
a	17″,65		

Angle Increasing.

The positive sign for the inclination has been determined from measures of radial velocity. WRIGHT finds, for the epoch 1904,7, a difference in radial velocity for the two components of 5,34 km[2]. Using ROBERTS' elements $P = 81{,}185$ years, $e = 0{,}529$ and $m_2/m_1 = 49/51$, he deduced $\pi = 0'',73$, $a = 3{,}64 \cdot 10^9$ km, $m_1 + m_2 = 2{,}2\,\odot$, $m_1 = 1{,}12\,\odot$, $m_2 = 1{,}08\,\odot$; the radial velocity of the centre of mass is $-22{,}18$ km. The accurate value of the proper motion of the system is 3″,688 corresponding to a cross motion of 23 km per second; this, combined with radial velocity, shows that α Centauri is travelling through space with a speed of 32 km per second.

66. The Binary 42 Comae Berenicis. Of all the STRUVE double stars, this one, discovered in 1827, has the shortest period.

In this case the apparent orbit is reduced to a straight line indicating that the binary possesses the remarkable property of an orbit plane in the line of sight.

The two components have the same magnitude 5,22 and the same spectral type F5. The spectroscopic parallax is 0″,055 and the trigonometric 0″,070.

Four orbits previously computed give these elements:

	STRUVE[3] in 1874	SEE[4] 1895	DOBERCK[5]	H. N. RUSSELL[6] 1917
P	25,71	25,556	25,335	25,87 years
T	1869,92	1885,69	1885,54	1911,74
e	0,480	0,461	0,496	0 ,522
a	0″,657	0″,642	0″,674	0″,665
Ω	11°,0	11°,0	11°,2	12°,6
i	90 ,0	90 ,0	90 ,0	89 ,87
ω	99 ,18	280 ,5	278 ,7	278 ,6

As an application of the method outlined in paragraph 59, and also since accurate observations by AITKEN are now available, it is of interest to compute a new set of elements.

[1] Publ Astroph Obs Potsdam 20, No 58, p. 105 (1909).
[2] Lick Bull 11, p. 154 (1924). [3] M N 35, p. 370 (1875). [4] M N 56, p. 512 (1896).
[5] A N 179, p. 55 (1908). [6] Pop Astr 25, p. 668 (1917).

The observations are:

Reduction with $P = 25{,}87$	Date	θ	ϱ	Observer
1907,01	1829,40	191°,6	0″,64	W. Struve
1914,02	1836,41	10,2	0,30	,,
1915,01	1837,40	11,0	0,39	,,
1916,02	1838,41	11,5	0,36	,,
1917,03	1839,42	12,2	0,59	Galle
1918,06	1840,45	15,7	0,55	O. Struve
1920,01	1841,40	14,7	0,32	Mädler
1921,02	1841,41	14,5	0,49	O. Struve
1922,01	1842,40	13,9	0,32	,,
1899,16	1847,42	195,5	0,20	,,
1900,16	1848,42	192,7	0,27	,,
1901,16	1849,42	188,6	0,42	,,
1902,13	1850,39	191,4	0,48	,,
1902,73	1850,99	193,3	0,40	Mädler
1903,01	1851,27	191,3	0,35	,,
1903,16	1851,42	187,0	0,49	O. Struve
1903,70	1851,96	194,5	0,45	Mädler
1904,16	1852,42	191,0	0,54	,,
1904,17	1852,43	190,9	0,56	O. Struve
1904,83	1853,09	194,2	0,62	Dawes
1905,09	1853,35	194,1	0,61	Mädler
1905,14	1853,40	190,8	0,57	O. Struve
1906,12	1854,38	194,1	0,60	,,
1906,13	1854,39	193,6	0,61	Mädler
1906,13	1854,39	192,8	0,55	Dawes
1907,12	1855,38	198,7	0,55	Mädler
1907,18	1855,44	189,1	0,62	O. Struve
1908,14	1856,40	192,7	0,52	Mädler
1908,16	1856,42	192,0	0,78	Winnecke
1908,70	1856,96	192,5	0,47	Secchi
1909,13	1857,39	198,3	0,50	Mädler
1909,23	1857,49	187,7	0,44	O. Struve
1910,18	1858,44	188,5	0,38	,,
1911,11	1859,37	single		,,
1913,16	1861,42	15,6	0,43	,,
1914,14	1862,40	11,6	0,54	,,
1915,18	1863,44	9,3	0,55	,,
1916,16	1864,42	12,5	0,51	,,
1916,17	1864,43	13,4	0,45	Dawes
1917,33	1865,59	13,7	0,54	Engelmann
1918,38	1866,64	8,5	0,40	O. Struve
1919,21	1867,47	13,0	0,36	,,
1920,18	1868,44	15,8	0,21	,,
1922,18	1870,44	single		,,
1897,30	1871,43	single		,,
1899,33	1873,46	189,0	0,20	,,
1900,28	1874,41	189,2	0,30	,,
1901,30	1875,43	190,4	0,51	Dembowski
1901,33	1875,46	189,7	0,39	O. Struve
1901,40	1875,53	191,5	0,32	Dunér
1902,25	1876,38	191,2	0,58	Dembowski
1902,27	1876,40	193,4	0,40	Hall
1902,29	1876,42	188,0	0,50	O. Struve
1903,28	1877,41	190,4	0,52	Schiaparelli
1903,32	1877,45	191,4	0,51	Dembowski
1903,33	1877,46	186,0	0,47	O. Struve
1904,24	1878,37	191,3	0,65	,,
1904,25	1878,38	189,6	0,51	Hall
1904,30	1878,43	190,8	0,57	Dembowski
1905,24	1879,37	192,1	0,68	Burnham

Reduction with $P = 25{,}87$	Date	θ	ϱ	Observer
1905,29	1879,42	193°,2	0″,51	HALL
1905,31	1879,44	190,9	0,65	O. STRUVE
1906,23	1880,36	191,7	0,52	HALL
1907,12	1881,25	190,9	0,60	DOBERCK
1907,24	1881,37	193,0	0,64	BURNHAM
1907,26	1881,39	192,6	0,53	HALL
1908,25	1882,38	191,9	0,54	,,
1908,33	1882,46	184,6	0,51	O. STRUVE
1908,80	1882,93	192,1	0,56	ENGELMANN
1909,29	1883,42	193,2	0,50	HALL
1910,27	1884,40	189,7	0,36	,,
1911,29	1885,42	single		SCHIAPARELLI
1912,29	1886,42	10,0	0,27	HALL
1912,38	1886,51	15,8	0,26	SCHIAPARELLI
1913,29	1887,42	13,1	0,38	,,
1913,31	1887,44	13,6	0,42	HALL
1914,14	1888,27	12,0	0,48	SCHIAPARELLI
1914,27	1888,40	13,8	0,45	HALL
1914,30	1888,43	8,7	0,42	O. STRUVE
1914,95	1889,08	10,5	0,56	LEAVENWORTH
1915,26	1889,39	11,8	0,61	O. STRUVE
1916,20	1890,33	9,3	0,70	BURNHAM
1917,31	1891,44	11,4	0,51	HALL
1917,31	1891,44	10,7	0,49	SCHIAPARELLI
1918,24	1892,37	11,7	0,47	LEAVENWORTH
1918,27	1892,40	10,7	0,42	SCHIAPARELLI
1918,31	1892,44	11,7	0,40	BIGOURDAN
1919,32	1893,45	10,2	0,32	SCHIAPARELLI
1920,20	1894,33	0,1	0,25	COMSTOCK
1920,33	1894,46	10,38	0,22	SCHIAPARELLI
1921,16	1895,29	13,9	0,14	SEE

AITKEN'S Values.

Date	θ	ϱ	Date	θ	ϱ
1898,309	3°,0	0″,22	1910,463	12°,3	0″,38
,312	13,0	0,17	,490	15,9	0,32
,315	10,1	0,18	,498	13,3	0,29
1899,254	4,9	0,22	1911,325	—	0,14
,363	9,4	0,21	,468	196,4	0,12
,380	10,8	0,25	1912,226	192,5	0,22
,456	13,8	0,22	,357	192,6	0,27
1901,165	10,9	0,27	1914,452	189,7	0,41
,176	10,7	0,33	,482	189,6	0,53
,449	8,5	0,30	1918,528	192,1	0,39
1903.410	11,3	0,48	1919,227	188,4	0,36
,424	9,4	0,38	,421	191,4	0,37
1904,959	13,4	0,65	1920,333	193,0	0,30
1905,146	13,0	0,88	,371	192,7	0,26
,222	11,4	0,65	1921,329	198,6	0,15
1906,265	11,2	0,70	,357	190,9	0,13
,285	13,0	0,75	,395	190,8	0,13
,342	10,2	0,58			

HUSSEY adopted for principal star the one taken as secondary in previous measures; this was no mistake, the two stars being identical in magnitude.

From these observations the interpolating curve for distances is illustrated in Fig. 12; the position P' of periastron was obtained by adopting $P = 25{,}87$ years. Then

$$e = \frac{0{,}035}{0{,}095} = 0{,}37\,, \qquad \sin PSB = \sqrt{1 - e^2} = 0{,}9285\,, \qquad PSB = 111^\circ{,}80\,.$$

From the Allegheny Tables of Anomalies, the mean anomaly corresponding to the true anomaly PSB for $e = 0{,}37$, is $69°{,}91$. This gives B'.

$$\cot\omega = \frac{P'C'}{B'C'}\sqrt{1-e^2} = \frac{0{,}095}{0{,}625}\cdot 0{,}9285 = 0{,}1411.$$

If the principal star is that of HUSSEY,

$$\omega = 98°2', \qquad \text{or } \omega = 261°58'$$

according to the adopted sense of motion.

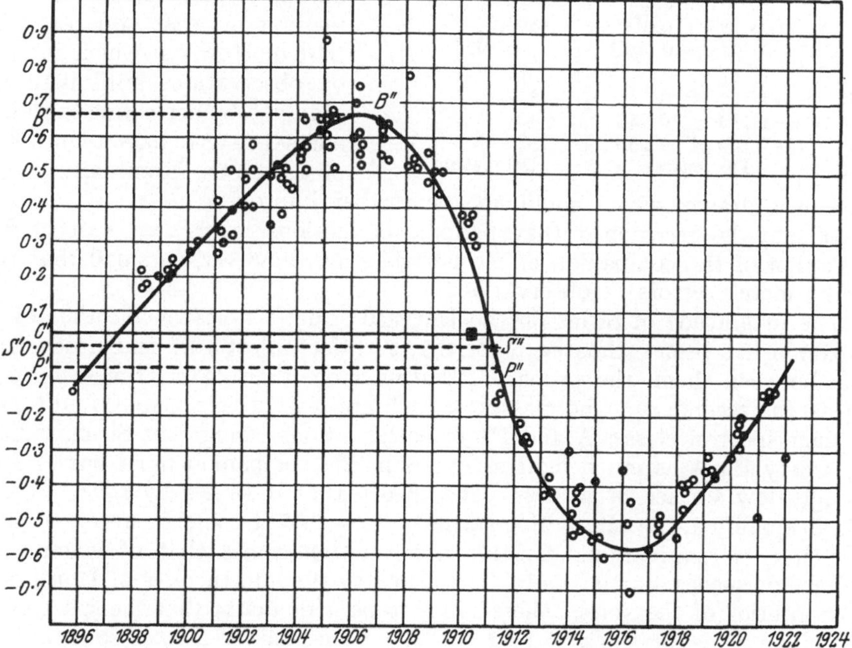

Fig. 12. 42 Comae Berenicis. Interpolating curve for distances.

If the principal star is the other one,

$$\omega = 81°58', \qquad \text{or } \omega = 278°2'.$$

$$a = P'C' \sec\omega = 0'',095 \cdot 7{,}1582 = 0'',68.$$

To sum up, the elements of 42 Comae Berenicis are

P	25,87 years
T	1911,50
e	0,37
a	0″,68
Ω	12°,6
i	90,0
ω	98°2′ or 261° 58′ for one star as primary 81°58′ or 278° 2′ for the other star as primary.

These elements might be corrected by the method of least squares but since the observations are of very different weights, it is a question whether the results would be improved.

Great uncertainty no doubt exists in the value of the eccentricity.

67. The Binary Sirius. That BESSEL discovered Sirius and Procyon to have invisible companions was mentioned in paragraph 29. The discovery of the companion to Sirius was actually achieved by ALVAN G. CLARK, on January 31st 1862, while pointing the new 18inch telescope of the Dearborn Observatory on that star.

Attention having been focused on the subject, micrometrical measurements accumulated and various computers made investigations of the orbit.

It will be sufficient to give the latest orbits determined, those of AITKEN[1] and HOWARD[2].

	AITKEN	HOWARD
P	50, 04 ± 0,09	50,17 years
T	1894,133 ± 0,011	1894,25
e	0,5945 ± 0,0023	0,5938
a	7″,570	7″,482
Ω	42°,71 ± 0°,33	44°,56
i	+43 ,31 ± 0 ,25	+42 ,01
ω	145 ,69 ± 0 ,38	138 ,38
Angle	Decreasing	Decreasing.

The orbit of AITKEN was obtained after correction by the method of least squares, while that of HOWARD which is based on observations until 1922, was obtained by correcting successive ellipses so as to have them fit the observations. Eventually, a 27th ellipse was drawn, which could not be further improved.

It may be mentioned that periodic oscillations have been suspected in the motion of the companion of Sirius[3]; they are, however, so small that they may be merely errors of observation.

The magnitude of Sirius on the Harvard scale is −1,58 while that of the companion has been estimated to be 8,5; a difference of 10,1 magnitudes thus exists between them, the primary radiating more than 11000 times as much light as the secondary. The masses of each are respectively 2,56 ⊙ and 0,74 ⊙ and their spectral classes A and F0 or earlier. Considering that Sirius radiates fully thirty times as much light as our Sun, the companion must have an exceedingly low surface brightness. The theory that it shines by reflected light being disqualified, the alternative suggests itself that it could be an exceedingly dwarf star of considerable density, about 60000 times that of water. The existence of such a star is possible according to the late theories of EDDINGTON on the interior of the stars. There seems to be little doubt that the spectrum of the companion is in some respects peculiar; the enhanced lines so prominent in the spectrum of Sirius are faint, λ 4481 of magnesium being especially noteworthy in this respect. This agrees with the results found for other white dwarf stars. The arc lines are also faint, while the hydrogen lines form the principal feature of the spectrum. The distribution of the light in the continuous spectrum is noticeably different from that of the scattered light of Sirius and resembles that of a class F star in being considerably more intense toward longer wave-lengths.

The companion of Sirius occupies thus a position almost unique as an object which might be expected to yield a very large gravitational displacement of its spectral lines according to the theory of generalized relativity. EDDINGTON has calculated a relativity shift of 20 km/sec on the basis of a spectral class F0 and an effective temperature of 8000° for the companion.

From measurements of spectra both on the spectrocomparator and with the photoelectric photometer, Prof. ADAMS, director of the Mount Wilson Observatory, found that the apparent relative radial velocity of Sirius and its companion for the mean date of the observations was +23 km. The real relative velocity may be computed readily from the elements of the visual orbit. For the mean

[1] Lick Bull 9, p. 186 (1918). [2] Pop Astr 30, p. 390 (1922).
[3] A N 215, p. 13 (1921).

epoch of the observations this is found to be 1,7 km, the companion showing a motion of recession from Sirius. Applying this correction to the observed value, there is thus a shift in the spectral lines of the companion of Sirius of +0,32 Ångströms corresponding to a radial velocity of +21 km. This value, interpreted as a relativity displacement, gives a radius for the star of about 18000 km. If the values derived by SEARES[1] for surface brightness are used, the following elements are found for the companion of Sirius and the assumed spectral classes:

	Fo	A 5
Surface brightness	−0,88	−1,45
Radius (km)	24000	18000
Density (water = 1)	30000	64000
Relativity displacement	+0,23	+0,32

The results obtained by ADAMS[2], compared with those predicted by EDDINGTON, may be considered as affording direct evidence for validity in the third test of the theory of general relativity and for the remarkable densities predicted by EDDINGTON for the dwarf stars of early type spectra.

Thus, Sirius may be considered not only as the brightest star in the heavens, but also as the most important as a subject for study in binary stars.

BESSEL, ALVAN G. CLARK and finally ADAMS, have successively startled the astronomical world by their announcements with respect to the problem. Their results are certainly well deserved rewards for the hours of patient observation and laborious collection of data.

68. The Binary Procyon. The history of Procyon is very similar to that of Sirius; its companion was discovered by BESSEL in the first part of the 19th century, but was not actually seen until 1896 when SCHAEBERLE beheld it through the 36 inch refractor of the Lick Observatory. It is estimated to be of magnitude 13,5, considerably fainter than that of Sirius. The following mean micrometrical measures have been given:

Date	θ	ϱ	Observer	Date	θ	ϱ	Observer
1896,93	320°,4	4″,63	SCHAEBERLE	1899,96	334°,5	4″,88	AITKEN
1897,00	321,1	4,83	AITKEN	1900,05	336,0	5,09	BARNARD
1897,16	319,8	4,65	HUSSEY	1900,23	338,3	4,83	LEWIS
1897,82	324,4	4,66	SCHAEBERLE	1900,29	332,4	4,60	SEE
1897,83	327,4	4,80	SEE	1901,20	338,5	5,13	AITKEN
1897,83	328,6	4,82	BOOTHROYD	1901,88	343,5	5,06	BARNARD
1897,88	323,8	4,70	AITKEN	1902,21	344,9	5,39	LEWIS
1898,21	325,3	4,82	AITKEN	1902,24	346,5	5,11	HUSSEY
1898,21	326,0	4,83	BARNARD	1903,15	351,0	5,16	BARNARD
1898,24	326,0	4,26	LEWIS	1905,14	6,2	5,14	AITKEN
1898,28	325,0	4,50	HUSSEY	1905,99	10,2	—	,,
1898,76	330,9	4,97	AITKEN	1906,77	28,2	—	,,
1899,07	330,6	4,91	BARNARD				

According to AITKEN, as a visible object Procyon's companion was on the decline from 1905 to 1906, and was entirely invisible under the best conditions from 1907 to 1921. JONCKHEERE, however, gives the following measurement:

1911,069 28°,6 4″,69.

This value for θ being nearly the same as that obtained by AITKEN in 1906, does not fit an interpolation curve for position angles obtained from the above

[1] Ap J 55, p. 165 (1922).

[2] Wash Nat Ac Proc 11, p. 382 (1925).

list of observations. This is easily seen from a consideration of the following values obtained by an interpolating curve:

Date	θ	ϱ
1897,0	320°,6	4″,68
1898,0	325,0	4,80
1899,0	329,3	4,92
1900,0	334,0	5,00
1901,1	339,1	5,08
1902,0	344,7	5,13
1903,0	350,8	5,14
1904,0	357,2	—
1905,0	5,0	—
1906,0	13,3	—
1907,0	28,0	—

There is no doubt from the curves that the values of ϱ are altogether unreliable. We know, however, that the values of θ are good, so values of ϱ can be computed from the interpolating curve of position angles by the method outlined in paragraph 48. It is known that $\varrho^2 = \frac{2C}{\tan\alpha}$ where α is the angle which the tangent to the interpolating curve of position angle makes with the x-axis. The following table is then obtained:

Date	θ	$r = \frac{1}{\sqrt{\tan\varphi}}$	$\varrho = \frac{5{,}0195}{\sqrt{\tan\alpha}}$
1897,0	320°,6	1,054	5″,29
1898,0	325,0	1,036	5,20
1899,0	329,3	1,027	5,16
1900,0	334,0	1,000	5,02
1901,0	339,1	0,966	4,85
1902,0	344,7	0,932	4,68
1903,0	350,8	0,908	4,56
1904,0	357,2	0,888	4,46
1905,0	5,0	0,821	4,12
1906,0	13,3	0,714	3,58
1907,0	28,0	0,636	3,19

These represent the apparent orbit of Fig. 13, taking the law of areas into consideration. The elements of the orbit computed by the method of paragraph 60 are then deduced in the following lines:

$$P = 47{,}819 \text{ years},$$

$$T = 1908{,}907,$$

$$e = \frac{CS}{CP} = 0{,}637,$$

$$\sqrt{1-e^2} = 0{,}770,$$

$$\cot\omega = 0{,}770 \cdot \frac{199}{1103} = 0{,}1389,$$

$$\omega = 262°6',$$

$$\omega' = 258°57',$$

$$\cos i = \frac{\tan\omega'}{\tan\omega} = \frac{5{,}1207}{7{,}2066} = 0{,}7105,$$

$$i = \pm 44°43',$$

$$a = 5'',48.$$

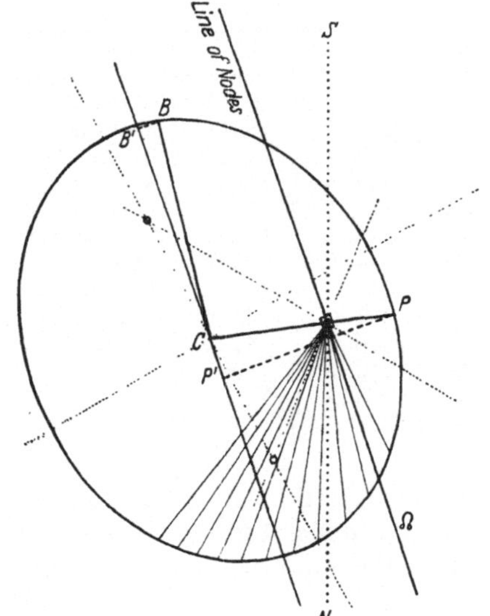

Fig. 13. Apparent orbit of Procyon.

To sum up, the elements of Procyon are:

P	47,819 years
T	1908,907
e	0,637
a	5″,48
Ω	19°
i	±44°43′
ω	262°6′,

Angle Increasing

These elements must be considered as approximations only, on account of the small arc of orbit covered by the observations.

From a study of the proper motion of the bright star SEE had found a period of 40,0 years[1], being the same figure found, some time previously, by AUWERS and published in the A N 58, p. 35 (1862).

An examination of the apparent orbit indicates why AITKEN failed to see the companion after 1907; as in the case of Sirius, it is probably because the separation was too small and the faint star was lost in the rays of the bright one.

The parallax of Procyon is $+0'',312$ and that of Sirius $+0'',371$, showing that in real dimensions the two orbits are very nearly the same.

69. The Binary Capella. Capella was discovered to be a spectroscopic binary independently by CAMPBELL[2] and NEWALL[3], in 1899. A little later, Prof. NEWALL published its radial-velocity curve[4], while the elements of the spectroscopic orbit were obtained by REESE[5] in 1901. Prof. NEWALL having called attention to the possibility of securing micrometric measures of Capella, DYSON and LEWIS examined it with the 28 inch Greenwich refractor on April 4th 1900. BOTH observers agreed that the star was elongated; various coloured shades and eye-pieces were used and the elongation appeared to be quite 0'',1. It was afterwards very diligently observed by several Greenwich astronomers until Dec. 10th of the same year. The distances were impossible to measure, but the change of position angle agreed on the whole with the calculated change, taking the period as 104 days[6]. The observations might have been of value in determining the inclination of the orbit, and hence its true dimensions as well as the masses of the components, the spectroscopic orbit being known from the double lines in the spectra. Considering the exceedingly small separation, however, these determinations would have been affected by a very large probable error.

A good visual orbit remained to be determined from interferometer measures and this is precisely what was done by P. W. MERRILL at the MT. WILSON Observatory in 1922[7], with the following resultant measurements:

Date	θ	ϱ	Observer
1919 Dec. 30,65	—	0'',0428	ANDERSON
1920 Feb. 13,63	5°,0	0,0469	,,
14,65	1,0	0,0462	,,
15,66	356,4	0,0454	,,
Mar. 15,63	242,0	0,0517	,,
Apr. 23,63	107,0	—	,,
Oct. 6,01	253,9	0,0474	MERRILL
Nov. 3,92	164,4	0,0439	,,
1921 Jan. 14,70	262,6	0,0461	,,
31,64	213,8	0,0544	,,
Feb. 2,66	208,9	0,0542	,,
Mar. 1,63	105,9	0,0409	,,
2,62	101,6	0,0421	,,
3,63	98,1	0,0420	,,
31,64	14,5	0,0514	,,
Apr. 1,65	11,5	0,0502	,,

By comparing recent spectroscopic observations at Mount Wilson with those made at the Lick Observatory in 1900, SANFORD[8] found that the original figures for the period, 104,022 days, are very nearly correct.

Accordingly this value has been used.

Preliminary trials showed that the eccentricity is probably smaller than 0,016 as given by REESE. Hence it did not seem necessary to vary both T and ω. The Julian date 2422596,79 was chosen for T by trial and retained thereafter. Corrections to a and to Ω were then found by direct algebraic processes. Finally a least-square solution was carried through for the elements ω, i and e, yielding only very small corrections to the preliminary values.

[1] A J 19, p. 58 (1898).
[2] Ap J 10, p. 177 (1899).
[3] M N 60, p. 2 (1899).
[4] M N 60, p. 418 (1900).
[5] Lick Bull 1, p. 34 (1901).
[6] M N 60, p. 595 (1900) and 61, p. 70 (1900).
[7] Ap J 56, p. 41 (1922).
[8] Publ A S P 34, p. 178 (1922).

The adopted elements follow:

P	104,022 days
T	J. D. 2422596,79
e	0,0086
a	0'',05360
Ω	38°,70
i	− 41 ,08
ω	114 ,30
Angle	Decreasing.

To these may be added the following quantities found by combining the foregoing elements with the spectroscopic results:

$$a_1 + a_2 = 126630000 \text{ km} \qquad m_1 = 4{,}2 \odot$$
$$\pi = 0'',0632 \qquad m_2 = 3{,}3 \odot .$$

Capella thus presents one of the very small number of cases in which a comparison of the visual and spectroscopic data leads to an accurate value of the parallax. The value found here, $+0'',063$, agrees well with the trigonometric parallax $+0'',075 \pm 0'',003$, and with the spectroscopic parallaxes found to be $0'',076$ by both ADAMS and RIMMER, but not so well with that obtained at Victoria, $+0'',120$.

70. Table of Orbits of Visual Double Stars. A table of all well determined orbits of double stars has been published by VAN DEN BOS[1]. Being the most complete list of orbital elements it was thought that its insertion here, omitting, however, his elements *A*, *F*, *B*, *G*, would be of great value.

The arrangement of the table is sufficiently explained by the headings. The number given in the first column is that of BURNHAM's General Catalogue, or INNES's Reference Catalogue. The year given in the last column is in most cases that of the last measures used, but in some cases the year of publication is given. (See Tables page 353 to 357.)

e) The Spectroscopic Binaries.

71. Stellar Radial Velocity. Detailed description of instruments and methods used for determining stellar radial velocity are outside the scope of this chapter, but every observatory engaged in these determinations includes in its work the investigation of spectroscopic binaries.

Stellar spectra are found in which the position of the lines varies from one epoch to another, indicative of a change in radial velocity often due to orbital motion in binaries. In this case, instead of finding periods which are counted in years as for visual binaries, reckoning has to be done in days, and even hours for stars like β Cephei and γ Ursae Minoris.

All radial velocity is obtained at present from photographs taken with a spectrograph attached to the telescope. Interesting articles on the construction of spectrographs, also, on the methods of measurement and reduction of spectrograms for radial velocity, are the following:

CAMPBELL, The Mills Spectrograph[2].

FROST, The Bruce Spectrograph[3].

HARTMANN, Remarks on the Construction and Adjustment of Spectrographs[4].

VOGEL, Description of the Spectrographs for the Great Refractor at Potsdam[5]. (Continuation page 357.)

[1] B A N 3, p. 149 (1926). [2] Ap J 8, p. 123 (1898). [3] Ap J 15, p. 1 (1902).
[4] Ap J 11, p. 400 (1900) and 12, p. 31 (1900). [5] Ap J 11, p. 393 (1900).

Table of Orbits of Visual Double Stars.

No α_{1900}	Star δ_{1900}	Mags Spec		P T	a n	e i	ω Ω	Authority
12755	Σ 3062	6,5	7,5	$105^y,55$	$1'',44$	0,4664	$98°,68$	Doberck 1905
$0^h1^m,0$	+ 57° 53′	G 5		1941,62	$3°,4107$	± 46°,08	$37°,42$	A N 173, p. 257
21	Σ 2	6,8	7,1	215	0,64	0,472	337,1	Russell 1917
0 3,8	+ 79 10	A 3		1890,8	1,674	± 109,1	166,5	Pop Astr 25, p. 668
104	OΣ 4	7,9	8,6	120	0,41	0,580	110,9	Russell 1917
0 11,5	+ 35 56	G 0		1907,0	3,00	± 153,8	133,6	Pop Astr 25, p. 668
270	A 111 AB	9,6	9,6	10,5	0,18	0,405	30,45	Aitken 1922
0 27,0	− 5 44	G 5		1919,75	34,3	± 142,15	125,5	Publ A S P 35, p. 259
314	Ho 212 AB	5,6	6,4	6,88	0,242	0,725	66,8	Aitken 1912
0 30,1	− 4 9	G 0		1925,91	52,33	± 53,45	38,7	Lick Obs Publ 12, p. 5
335	β 395	6,4	6,6	25,0	0,66	0,171	152,7	Aitken 1911
0 32,2	− 25 19	K 0		1924,50	14,40	± 76,0	112,8	Lick Obs Publ 12, p. 7
374	OΣ 18	7,8	9,8	182,75	0,96	0,50	202,9	Hussey 1899
0 37,2	+ 3 37	F 8		1949,50	1,9699	± 21,9	78,0	Lick Obs Publ 5, p. 37
426	η Cas	3,64	7,9	507,60	12,21	0,5220	88,92	Doberck 1908
0 43,0	+ 57 17	F 8		1890,03	0,70922	± 31,62	99,22	A N 179, p. 383, cf. § 63
482	Σ 73	6,1	6,7	114,8	1,01	0,75	71,3	Bowyer 1904
0 49,6	+ 23 5	K 0		1935,3	3,136	± 45,4	112,7	Mem R A S 56, p. 23
1^h 37	p Eri	6,00	6,03	218,9	8,025	0,721	301,40	Dawson 1919 (eq. 1880)
1 36,0	− 56 42	G 5	G 5	1806,14	1,6446	± 114,26	1,03	A J 32, p. 144
1015	Σ 186	6,9	6,9	136	1,15	0,67	226,8	Lewis 1903
1 50,7	+ 1 21	G 0		1894,0	2,65	± 73,9	42,6	Mem R A S 56, p. 48
1036	β 513	4,8	7,2	63,3	0,66	0,385	341,6	Bennot 1925
1 53,7	+ 70 25	A 3		1904,8	5,687	± 31,5	90,5	Pop Astr 33, p. 307
1070	γ And BC	5,4	6,6	55,0	0,346	0,82	201,2	Hussey 1900
1 57,8	+ 41 51	A 0		1892,0	6,545	± 103,4	113,5	Lick Obs Publ 5, p. 45
1144	Σ 228	6,4	7,3	167,4	0,974	0,313	303,7	Jackson 1912
2 7,6	+ 47 1	F 0		1894,50	2,150	± 61,3	99,7	Grw Cat, p. 206
1471	β 524 AB	5,6	6,7	33,33	0,16	0,60	325,0	Aitken 1912
2 47,4	+ 37 56	F 0		1928,33	10,801	± 146,5	127,1	Lick Obs Publ 12, p. 20
1761	Σ 412 AB	6,7	6,8	270,0	0,49	0,555	348,	Aitken 1919
3 28,5	+ 24 8	A 2		1917,3	1,3333	± 139,4	97,0	Lick Bull 11, p. 66
2093	OΣ 77 AB	8,1	8,2	51,6	0,44	0,846	287,6	van den Bos 1919
4 9,6	+ 31 27	F 8		1943,7	6,98	± 65,86	104,2	B A N 1, p. 82
2109	40 Eri BC	8,9	10,8	247,92	6,8945	0,4024	326,96	van den Bos 1925
4 10,7	− 7 49	B 9	Mdp	1848,93	1,4521	± 108,45	150,96	B A N 3, p. 128
2134	OΣ 79	7,2	9,0	88,9	0,57	0,625	129,8	Aitken 1912
4 14,2	+ 16 17	G 0		1897,8	4,049	± 56,2	66,0	Lick Obs Publ 12, p. 30
2154	OΣ 82	7,4	9,0	97,94	0,94	0,50	68,1	Hussey 1900
4 17,1	+ 14 49	G 0		1932,97	3,6755	± 120,2	39,8	Lick Obs Publ 5, p. 60
2159	β 774	6,6	6,6	100,6	0,74	0,48	278,0	Dawson 1920
4 17,4	− 25 58	F 0		1923,22	3,579	± 50,0	161,7	A J 34, p. 17
2187	β 1185	8,0	8,6	28,9	0,25	0,20	301,6	Aitken 1921
4 20,0	+ 18 38	G 0		1917,8	12,4566	± 104,35	24,3	Lick Bull 11, p. 70
2230	Σ 554	5,7	9,0	148,3	1,036	0,790	157,92	v. d. Bos 1921 (eq. 1925)
4 24,4	+ 15 25	F 0		1888,30	2,428	± 109,0	8,88	M N 81, p. 476
2381	β 883	7,7	7,7	16,61	0,19	0,445	190,0	Aitken 1912
4 45,7	+ 10 54	F 5		1907,03	21,67	± 9,35	34,2	Lick Obs Publ 12, p. 35
2383	β 552	6,8	10,0	86,0	0,56	0,51	309,6	Aitken 1917
4 46,2	+ 13 29	F 5		1886,35	4,186	± 39,35	145,8	Lick Bull 11, p. 71
2535	OΣ 98	5,9	6,7	190,48	1,22	0,2465	57,30	Gore 1887
5 2,4	+ 8 22	F 0 p		1959,05	1,8900	± 135,05	99,58	M N 47, p. 266
2597	Capella	0,8	0,8	104^d,022	0,05360	0,0086	114,30	Merrill 1921
5 9,3	+ 45 54	G 0		JD 2422596,79	3,46081	− 138,92	38,70	Ap J 56, p. 44, cf. § 69

No α_{1900}	Star δ_{1900}	Mags Spec		P T	a n	e i	ω Ω	Authority
3291	β 895 AB	7,9	7,9	$45^y,7$	0″,255	0,88	289°,9	van den Bos 1921
$6^h 13^m,6$	+ 28° 28′	A 3		1914,31	7°,87	± 60°,7	22°,7	B A N 2, p. 25
3596	Sirius	−1,6	8,4	50,04	7,570	0,5945	145,69	Aitken 1918
6 40,8	− 16 35	A 0	F	1894,133	7,194	+ 136,69	42,71	Lick Bull 9, p. 184, cf. § 67
3876	Σ 1037	7,2	7,2	120,4	0,870	0,932	254,1	van Biesbroeck 1925
7 6,7	+ 27 24	F 5		1920,57	2,991	± 141,0	31,4	M N 85, p. 480
4122	Castor	1,99	2,85	306,28	6,060	0,5593	278,031	Rabe 1920
7 28,2	+ 32 6	A 0		1954,728	1,1754	± 113,207	32,546	A N 216, p. 52
				477,5	6,573	0,2875	247,32	Doberck 1921
				1960,51	0,7539	± 116,10	42,12	A N Jub Nr, p. 7
4187	Procyon	0,48	13,5	39,0	4,05	0,324	36,8	Boss 1905
7 34,1	+ 5 29	F 5		1925,5	9,23	± 14,2	150,7	P G C 267, cf. § 68
4310	β 101	5,8	6,4	23,34	0,69	0,75	74,65	Aitken 1912
7 47,2	− 13 38	G 0		1892,60	15,424	± 79,8	99,7	Lick Obs Publ 12, p. 51
4355	OΣ 185	7,1	7,3	59,6	0,350	0,611	114,9	Jackson 1914
7 52,1	+ 1 24	F 8		1861,16	6,04	± 74,6	35,2	Grw Cat, p. 210
4414	β 581	8,7	8,7	44,0	0,38	0,39	292,3	Aitken 1924
7 58,8	+ 12 35	G 5		1909,75	8,182	± 47,7	116,1	Lick Bull 12, p. 47
4477	ζ Cnc AB[1]	6,02	6,26	57,891	0,874	0,3337	179,80	Schnauder 1914
8 6,5	+ 17 57	G 0		1928,139	6,2186	180,0	—	Dissertation
	C + D, C	5,56		16,92	0,162	0,039	94,5	
		G 0		1909,6	21,27	± 128,5	73,5	
4771	ε Hya AB	3,7	5,2	15,3	0,23	0,65	270,0	Aitken 1912
8 41,5	+ 6 47	F 8		1900,97	23,5	+ 49,95	104,4	Lick Obs Publ 12, p. 59
4923	σ_2 UMa	4,9	8,4	470	4,76	0,799	25,5	Russell 1919
9 1,6	+ 67 32	F 8		1920,2	0,766	± 127,0	121,5	Pop Astr 29, p. 97
5005	Σ 3121	7,8	8,1	34,00	0,6692	0,330	127,52	See 1895
9 12,0	+ 29 0	K 0		1912,30	10,588	± 75,00	28,25	Evol Stellar Systems, p. 96
5103	ω Leo	5,9	6,7	116,74	0,844	0,5601	122,10	Doberck 1903
9 23,1	+ 9 30	G 0		1957,56	3.0838	± 66,20	144,28	A N 173, p. 251
9^h 40	ψ Arg	3,7	5,7	34,90	0,914	0,37	219,2	Dawson 1924
9 26,8	− 40 2	F 5		1936,79	10,316	± 56,2	116,8	A J 36, p. 23
5223	φ UMa	5,1	5,5	112,663	0,34293	0,49745	9,237	Dick 1920 (eq. 1925)
9 45,3	+ 54 32	A 2		1883,576	3,19537	± 22,861	157,890	Diss. (Zwiers-Orbit)
5235	AC 5	5,8	6,1	72,76	0,41	0,60	133,1	Schoenberg 1906
9 47,6	− 7 38	A 2		1953,30	4,9478	± 142,86	17,95	A N 178, p. 189
5734	ξ UMa A+a, B	4,41	4,87	59,8096	2,5128	0,4108	129,213	Nørlund 1905
11 12,8	+ 32 6	G 0		1935,576	6,01912	+ 126,608	100,698	A N 170, p. 121[2]
11^h 22	Bris 3574	8,0	8,6	342,0	4,54	0,58	0,0	Dawson 1921
11 20,4	− 61 6	K 5		1918,48	1,0526	± 40,0	87,5	A J 34, p. 17
5805	OΣ 234	7,4	7,8	84,734	0,347	0,4225	218,370	Riechert 1922 (eq. 1925)
11 25,4	+ 41 50	F 5		1883,532	4,2485	± 54,075	151,628	A N 219, p. 229
5811	OΣ 235	5,7	7,0	71,9	0,78	0,40	135,0	Aitken 1912
11 26,7	+ 61 38	F 5		1909,0	5,007	± 43,6	78,5	Lick Obs Publ 12, p. 72
5951	β 794	7,0	8,3	63,1	0,34	0,41	126,9	Aitken 1921
11 48,3	+ 74 19	F 8		1911,0	5,705	± 34,5	149,7	Lick Bull 11, p. 77
6028	Σ 3123	7,8	8,0	103,3	0,32	0,49	79,1	See 1908
12 1,0	+ 69 15	F 5		1860,50	3,485	± 130,3	56,9	M N 68, p. 567
6158	Σ 1639	6,6	7,8	361	1,00	0,9258	300,9	Jackson 1920
12 19,4	+ 26 8	A 5		1888,10	0,997	± 136,4	78,4	Grw Cat, p. 212
				690	1,30	0,945	343,5	Russell 1917
				1891,95	0,5217	± 155,0	145,9	Pop Astr 25, p. 668

[1] An important perturbation term is to be added.
[2] There is a perturbation of 1,8 years period.

No α_{1900}	Star δ_{1900}	Mags Spec		P T	a n	e i	ω ☊	Authority
$^{\mathrm{h}}61$	γ Cen	3,1	3,1	$203^{\mathrm{y}},39$	1″,924	0,2958	285°,03	DAWSON 1919
$^{\mathrm{h}}36^{\mathrm{m}},0$	− 48° 25′	A0		1851,50	1°,7700	± 98°,22	3°,35	A J 32, p. 162
243	γ Vir	3,65	3,68	182,30	3,743	0,887	260,37	DOBERCK 1907
36,6	− 0 54	F0	F0	1836,42	1,9747	± 150,13	40,43	A N 177, p. 161
406	Σ 1728	5,22	5,22	25,87	0,665	0,522	278,6	RUSSELL 1917
5,1	+ 18 3	F5		1911,74	13,916	± 89,87	12,6	PopAstr 25, p. 668, cf. § 66
566	Σ 1768	5,2	7,5	220,4	1,205	0,8562	118,6	JACKSON 1919
33,0	+ 36 48	F0		1860,26	1,6335	± 132,6	52,8	Grw Cat, p. 213
578	β 612	6,3	6,3	23,05	0,225	0,52	357,95	AITKEN 1912
34,6	+ 11 15	F2		1930,27	15,618	± 50,4	33,85	Lick Obs Publ 12, p. 86
641	Σ 1785	7,8	8,2	193,55	2,549	0,4602	180,5	JACKSON 1919
44,5	+ 27 29	K2		1913,29	1,8600	± 39,4	156,3	Grw Cat, p. 214
711	β 1270	8,6	8,7	38,1	0,21	0,41	24,4	AITKEN 1920
58,8	+ 8 58	F5		1911,58	9,449	± 20,5	126,1	Lick Bull 11, p. 79
832	Σ 1834	7,9	8,1	295,6	0,93	0,823	169,2	VAN DEN BOS 1921
16,6	+ 48 58	F8		1901,73	1,2179	± 82,04	110,6	Proc Amst 30, p. 72
842	β 1111 BC	7,4	7,4	40,53	0,235	0,238	144,35	AITKEN 1919
18,5	+ 8 54	A0		1918,38	8,882	± 40,8	43,95	Publ A S P 31, p. 286
913	A 570	6,6	6,8	28,45	0,202	0,171	219,9	YOUNG 1923
27,9	+ 27 7	A2		1924,88	12,65	± 144,2	170,9	Publ A S P 35, p. 221
$^{\mathrm{h}}59$	α Cen	0,33	1,70	80,089	17,665	0,5208	52,132	FINSEN 1926
32,8	− 60 25	G0	K5	1875,759	4,4950	+ 79,233	25,445	Union Obs Circ, p. 68, cf. § 65
955	ζ Boo	4,6	4,6	130	0,62	0,96	180	HERTZSPRUNG 1916
36,4	+ 14 9	A2		1898,0	2,77	± 140,3	129	A N 203, p. 394
999	Σ 1879	7,6	8,6	177,9	0,789	0,623	148,3	JACKSON 1919
41,4	+ 10 5	F8		1868,14	2,023	± 128,8	70,6	Grw Cat, p. 215
001	OΣ 285	7,5	8,0	88,5	0,33	0,553	137,7	JACKSON 1917
41,7	+ 42 48	F5		1882,64	4,067	± 154,4	41,7	Grw Cat, p. 216
034	ξ Boo	4,7	6,7	151,425	4,874	0,5103	23,82	DOBERCK 1920
46,8	+ 19 31	G5		1909,36	2,37741	± 139,20	168,30	A N 214, p. 91
120	Σ 1909	5,2	6,0	204,74	3,578	0,4451	25,03	DOBERCK 1909
0,5	+ 48 3	G0		1998,22	1,7583	± 83,07	58,73	A N 182, p. 27
251	η CrB	5,58	6,08	41,56	0,89	0,2721	217,98	LOHSE 1906, Publ Astr. Obs Potsdam 20, p. 119
19,1	+ 30 39	G0		1933,829	8,662	± 58,48	25,26	
259	μ_2 Boo	7,2	7,8	224	1,30	0,53	339,0	COMSTOCK 1919
20,7	+ 37 42	K0		1864,6	1,607	± 138,0	177,2	A J 33, p. 144
$^{\mathrm{h}}55$	γ Lup	3,6	3,8	104,3	0,78	0,314	2,9	DAWSON 1919 (eq. 1910)
28,5	− 40 50	B3		1905,7	3,452	± 91,9	91,5	A J 32, p. 112
322	OΣ 298 AB	7,4	7,7	56,653	0,88349	0,58360	21,899	CELORIA 1887 (eq. 1888)
32,5	+ 40 8	K0		1939,510	6,3545	± 65,847	2,130	A N 119, p. 163
368	γ CrB	4,0	7,0	101	0,62	0,42	125	COMSTOCK 1919
38,6	+ 26 37	A0		1842,7	3,564	± 98	111	A J 33, p. 168
416	π_2 UMi	7,0	8,0	115	0,42	0,80	165,0	AITKEN 1912
45,1	+ 80 17	F2		1902,7	3,13	± 117,75	16,3	Lick Obs Publ 12, p. 100
487	ξ Sco AB	4,77	5,07	44,70	0,72	0,75	343,6	AITKEN 1912
58,9	− 11 6	F8		1905,39	8,054	± 29,1	27,2	Lick Obs Publ 12, p. 103
561	Σ 2026	9,0	9,5	215,0	1,53	0,695	7,2	COMSTOCK 1917
11,1	+ 7 37	K5		1908,07	1,677	± 135,9	178,7	A J 31, p. 35
642	Σ 2052	7,8	7,8	317,5	2,87	0,77	114,5	JACKSON 1920
24,5	+ 18 37	K0		1920,21	1,134	± 105,0	93,1	Grw Cat, p. 218
649	λ Oph	4,0	6,1	110,3	1,328	0,86	96,7	JACKSON 1919
25,9	+ 2 12	A0		1927,4	3,264	± 53,2	110,0	Grw Cat, p. 219
717	ξ Her	3,0	6,5	34,46	1,35	0,458	113,3	COMSTOCK 1916
37,5	+ 31 47	G0		1933,23	10,447	± 132,5	51,6	A J 30, p. 145

No α_{1900}	Star δ_{1900}	Mags Spec		P T	a n	e i	ω Ω	Authority
7748 16^h40^m,8	D 15 + 43° 40′	8,7 K5	9,0	126^y,1 1894,52	0″,935 2°,855	0,435 ± 120°,7	147°,0 147°,1	JACKSON 1919 Grw Cat, p. 220
7783 16 47,9	Σ 2107 + 28 50	6,8 F5	8,3	221,95 1896,64	0,853 1,6215	0,522 ± 23,35	123,5 179,6	RABE 1912 A N 198, p. 116
13364 17 4,5	Hu 1176 + 36 4	6,1 A5	6,1	15,5 1919,9	0,16 23,226	0,14 ± 124,0	308,4 90,8	AITKEN 1921 Lick Bull 11, p. 83
17^h31 17 11,4	Brisb. − 46 32	5,6 K0	8,4	100,9 1912,61	3,503 3,568	0,1675 ± 48,80	315,21 175,44	VAN DEN BOS 1923 B A N 2, p. 29[1]
7929 17 12,2	Melb. 4 AB − 34 53	6,0 K2	8,5	42,2 1891,48	1,83 8,53	0,551 ± 129,6	64,0 130,2	VOÛTE 1924 B A N 2, p. 181
8038 17 25,2	Σ 2173 − 0 59	5,9 G5	6,2	46,0 1915,2	1,06 7,83	0,18 ± 99,25	322,2 153,7	AITKEN 1912 Lick Obs Publ 12, p. 116
8099 17 34,0	β 962 + 61 57	5,3 F8	9,3	111 1893,3	1,56 3,24	0,23 ± 112,8	65,5 153,8	RUSSELL 1917 Pop Astr 25, p. 668
8162 17 42,6	μ Her BC + 27 47	10,0 Mb	10,5	43,23 1923,43	1,30 8,328	0,20 ± 63,15	182,0 60,8	AITKEN 1912 Lick Obs Publ 12, p. 119
8303 17 57,6	τ Oph − 8 11	5,34 F0	6,04	223,82 1814,79	1,307 1,6084	0,5338 ± 66,07	17,75 76,20	DOBERCK 1904 A N 170, p. 102
17^h129 17 59,6	h 5014 − 43 26	5,77 A3	5,77	153,96 1839,68	1,114 2,3383	0,480 ± 132,8	180,0 52,75	DAWSON 1924 Unpublished orbit I
				180,00 1841,94	1,146 2,0000	0,485 ± 138,2	202,5 60,41	Orbit II
				214,44 1843,83	1,208 1,6788	0,520 ± 144,8	219,0 65,82	Orbit III
8340 18 0,4	Σ 2272 + 2 31	4,1 K0	6,1	87,710 1895,965	4,495 4,10443	0,49873 − 121,257	193,352 122,184	PAVEL 1920 A N 212, p. 350
8353 18 1,6	OΣ 341 + 21 26	7,3 G0	8,2	19,75 1917,85	0,30 18,2275	0,96 ± 77,5	149,0 98,0	AITKEN 1920 Lick Bull 11, p. 85
8372 18 3,2	AC 15 + 30 33	5,2 F8	9,7	53,51 1941,35	1,11 6,728	0,763 ± 38,3	93,7 75,0	LOHSE 1906, Publ Astr. Obs Potsdam 20, p. 159
8380 18 4,6	Σ 2281 + 3 59	5,8 F2	7,3	423,5 1910,0	1,33 0,8501	0,70 ± 106,3	299,9 71,1	JACKSON 1919 Grw Cat, p. 223
				220,0 1913,5	1,02 1,6364	0,550 ± 102,8	324,9 74,4	RUSSELL 1917 Pop Astr 25, p. 668
8679 18 33,2	A 88 − 3 17	7,2 F8	7,2	12,12 1922,22	0,176 29,70	0,273 + 117,6	270,0 2,4	AITKEN 1912 Lick Obs Publ 12, p. 129
8933 18 53,3	β 648 + 32 46	5,2 G0	8,7	57,0 1911,2	1,24 6,32	0,20 ± 114,5	285,7 48,0	GUSHEE 1925 Pop Astr 33, p. 309
8966 18 55,8	Σ 2438 + 58 5	6,7 A2	7,3	233,0 1882,50	0,53 1,5451	0,916 ± 180,0	181,7 —	SEE 1908 M N 68, p. 568
8965 18 56,2	ζ Sgr − 30 1	3,4 A2	3,6	21,17 1921,54	0,565 17,005	0,185 ± 110,6	1,4 75,5	AITKEN 1900 Lick Obs Publ 12, p. 135
18^h113 18 59,7	γ CrA − 37 12	5,01 F8	5,01	124,65 1878,46	2,14 2,8881	0,3321 ± 148,10	169,55 53,47	DOBERCK 1909 A N 191, p. 125
9114 19 7,8	SE 2 BC + 38 37	8,9 G5	8,9	58 1894,0	0,40 6,2	0,50 ± 112	180 90	RUSSELL 1911 Lick Obs Publ 12, p. 138
9319 19 22,5	Σ 2525 + 27 7	8,4 F8	8,6	354,9 1887,31	1,205 1,014	0,933 ± 142,5	266,6 1,0	JACKSON 1920 Grw Cat, p. 224
9605 19 41,8	δ Cyg + 44 53	3,0 A0	7,9	321,0 1941,6	2,12 1,122	0,188 ± 132,2	201,0 87,9	JACKSON 1914 Grw Cat, p. 225
9643 19 44,5	ζ Sge + 18 53	5,2 A2	6,2	25,20 1914,11	0,32 14,286	0,85 ± 101,9	65,0 4,6	VAN BIESBROECK 1916 A J 29, p. 164

[1] The Campbell elements given in B A N 2, p. 29 are erroneous, but the apparent elements, formulae for ephemeris and ephemeris are correct. The sum of the masses with VOÛTE's parallax becomes 0,61 ⊙; a parallax of 0″,161 would give 1 ⊙.

No α_{1900}	Star δ_{1900}	Mags Spec		P T	a n	e i	ω ☊	Authority
9650	OΣ 387	7,0	8,0	128ʸ,0	0″,566	0,179	305°,0	JACKSON 1920
19ʰ45ᵐ,0	+ 35° 4′	F 5		1946,7	2°,813	± 128°,5	146°,4	Grw Cat, p. 226
9739	Ho 581	8,0	8,3	24,445	0,286	0,528	270,9	VAN BIESBROECK 1918
19 51,6	+ 41 36	K 0		1911,528	14,737	± 39,2	12,8	A J 31, p. 169
9979	OΣ 400	7,5	8,5	84,4	0,428	0,48	340,6	MEIER 1922 (eq. 1925)
20 6,9	+ 43 39	G 5		1885,1	4,263	± 117,5	143,9	A N 219, p. 232
10363	β Del	4,0	5,0	26,79	0,480	0,350	351,20	AITKEN 1912
20 32,9	+ 14 15	F 5		1936,62	13,438	± 62,25	178,55	Lick Obs Publ 12, p. 150
10559	Σ 2729	6,2	7,5	151,7	0,695	0,375	59,7	JACKSON 1918
20 46,1	− 6 0	F 2		1897,22	2,374	± 67,4	167,8	Grw Cat, p. 227
10643	ε Equ	5,8	6,3	97,4	0,61	0,72	9,0	RUSSELL 1916
20 54,1	+ 3 55	F 5		1922,2	3,696	± 94,5	106,8	A J 30. p. 125
10829	δ Equ	5,3	5,4	5,70	0,27	0,39	164,5	AITKEN 1912
21 9,6	+ 9 36	F 5		1924,15	63,2	± 99,0	21,0	Lick Obs Publ 12, p. 158
10846	τ Cyg	3,8	8,0	47,0	0,91	0,22	105,5	AITKEN 1912
21 10,8	+ 37 37	F 0		1936,60	7,66	± 137,3	149,8	Lick Obs Publ 12, p. 161
11125	24 Aqr	7,33	7,83	71,00	0,659	0,893	275,87	KUIPER 1923
21 34,4	− 0 30	F 8		1922,94	5,070	± 66,78	147,19	B A N 3, p. 147
11222	ϰ Peg AB	5,0	5,1	11,35	0,29	0,49	106,1	BURNHAM, LEWIS 1905
21 40,1	+ 25 11	F 5		1920,5	31,72	± 102,5	109,2	Mem R A S 56, p. 653
11761	KR 60 AB	9,3	10,8	44,27	2,46	0,38	171,0	AITKEN 1924
22 24,5	+ 57 12	Ma		1925,82	8,1319	± 154,0	113,8	Lick Bull 12, p. 45
11763	Σ 2912	5,6	7,0	136	0,72	0,534	180,0	VAN DEN BOS 1921
22 24,9	+ 3 55	F 5		1905,0	2,65	± 84,6	117,0	B A N 3, p. 127
12143	A 417	6,3	6,3	23,82	0,245	0,404	261,3	AITKEN 1918
22 59,9	− 8 14	F 0		1917,68	15,113	± 56,35	21,6	Lick Bull 9, p. 191
12290	β 80	8,8	8,9	95,2	0,72	0,77	98,0	AITKEN 1916
23 13,8	+ 4 52	K 0		1905,0	3,782	± 22,95	6,2	Publ A S P 28, p. 122
				85,7	0,79	0,773	288,9	JACKSON 1919
				1904,69	4,201	± 43,0	174,1	Grw Cat, p. 228
12404	β 1266	8,0	8,1	40	0,22	0,33	133	AITKEN 1923
23 25,5	+ 30 17	F 5		1909,8	9,0	± 132	46	Lick Bull 11, p. 97
12696	Hdn 60	9,1	9,5	40,76	0,50	0,35	245,8	JACKSON 1914
23 56,3	+ 39 3	G 5		1915,42	8,832	± 110,3	119,1	Grw Cat, p. 229
12701	β 733 AB	5,85	11,0	26,3	0,82	0,46	266,12	BOWYER, FURNER 1904
23 56,9	+ 26 33	G 0		1936,1	13,69	± 53,08	115,63	M N 66, p. 423

WRIGHT, Description of the Instruments and Methods of the D. O. MILLS Expedition[1].

CAMPBELL, The Reduction of Spectroscopic Observations of Motions in the Line of Sight[2].

FROST-SCHEINER, Astronomical Spectroscopy p. 338.

CURTISS, A Proposed Method for the Measurement and Reduction of Spectrograms for the Determination of the Radial Velocities of Celestial Objects[3].

HARTMANN, Über die Ausmessung und Reduktion der photographischen Aufnahmen von Sternspectren[4].

HARTMANN, The Spectrocomparator[5].

[1] Lick Obs Publ 9, part 3 p. 25 (1905).
[2] Astronomy and Astrophysics 11, p. 319 (1892).
[3] Lick Bull 3, p. 19 (1904).
[4] A N 155, p. 81 (1901).
[5] Ap J 24, p. 285 (1906); see also Publ Astroph Obs Potsdam 18, No 53 (1906).

J. S. PLASKETT, The Design of Spectrographs for Radial Velocity Determinations[1].

J. S. PLASKETT, Measurement and Reduction of Stellar Spectra[2].

SALET, Spectroscopie Astronomique.

AITKEN, The Binary Stars, Chapter V, by J. H. MOORE.

HENROTEAU, Les Etoiles Simples, Chapitre VIII, Vitesses radiales.

J. S. PLASKETT, Description of Building and Equipment of the Dominion Astrophysical Observatory[3].

Observatories interested in radial velocity work are those of Potsdam, Bonn and Babelsberg in Germany, Vienna in Austria, Pulkowa and Simeis in Russia, Paris and Meudon in France, Brera and Merate in Italy, Mount Wilson, Lick, Flagstaff, Detroit, Allegheny in the United States, Ottawa and Victoria in Canada and the Cape Royal Observatory in South Africa. The publications of these different observatories as well as a few standard periodicals such as the Astrophysical Journal, the Astronomische Nachrichten etc., contain pertinent articles on spectroscopic binaries.

72. Generalities on Spectroscopic Binaries. A spectroscopic binary is usually a star in which the radial velocity shows periodic oscillations indicated by shifts of spectral lines. Such shifts may not always be due to the motion of the star as a whole, or even to motion at all. To the astronomer familiar with radial velocity work, it is known that there are very different types of spectra even among those of the same spectral class. A binary may show lines narrow and sharp, giving radial velocities with a probable error as small as $\pm 0{,}5$ km, or its lines may be so wide and diffuse that their measurements have a very doubtful value. The periods of spectroscopic binaries range all the way from a few hours to many years, and the amplitudes of their velocity vary from three or four, to three or four hundred kilometers; their eccentricities may have any value, while the masses of their components are sometimes extremely large.

Many binaries show only one spectrum, that of the brighter component. Advantageous use has been made, however, of fine grained plates to bring out a secondary spectrum which would otherwise have escaped notice.

When visible, this secondary spectrum is usually a duplicate of the brighter, but cases are known where there is a difference.

73. Detached Calcium and Sodium Lines. In 1904 HARTMANN announced that in the spectrum of the Class B binary δ Orionis, the sharp K absorption line of calcium did not share the periodic variations of position of all other broad absorption lines[4] and, in his opinion, a similar result would be obtained for the H line. The observation of the latter is, however, more difficult being so near the $H\varepsilon$ line of hydrogen. He concluded that these H and K lines do not belong to the spectrum of the star itself, but are due to the absorption of enormous clouds of calcium situated between the star and the observer. These lines were first called stationary lines of calcium, but the term detached lines has now been adopted.

Following HARTMANN, a number of observers have discovered similar phenomena in other stars: V. M. SLIPHER was the first to acquire systematic information on the subject. Discovering that β Scorpii contained sharp detached calcium lines, he found that the same phenomenon was common to σ Scorpii, δ Scorpii, ζ Ophiuchi and some fainter stars between Antares and β Scorpii. After this when early class B spectroscopic binaries were investigated, detached

[1] Report of the Chief Astronomer (Dom. Obs. Ottawa) for the Year ending March 31, 1909, p. 152.

[2] Idem p. 175. [3] Publ Dom Astroph Obs 1, No. 1 (1920). [4] Ap J 19, p. 268 (1904).

H and K lines were often found. In some binaries these lines are quite stationary, although there are cases where they share the motion of the other lines but with a smaller amplitude. To this Miss HEGER[1] added the discovery that several stars with detached calcium lines also show detached D_1 and D_2 lines of sodium.

Summaries of the subject up to 1920 were published by R. K. YOUNG[2], and by the writer in "The Interstellar Clouds of Metallic Gases"[3].

From observations made in certain areas of the sky, J. S. PLASKETT[4] concluded that the detached calcium lines occur in most stars of spectral class earlier than B3 whether or not these stars are binaries. The radial velocities derived from these lines are nearly the same as the projections of the motion of the solar system on the directions of the stars investigated; λ Cephei, for instance, is approaching us with a velocity of 74 km, while its calcium lines show an approach of 14 km, the component of the solar motion in the direction of λ Cephei being 13 km.

The writer previously recognized the fact that calcium clouds may be considered almost at rest in space and that radial velocities, as measured from their lines, ought to give a fairly good value of the motion of the solar system. He has given for this motion[5] the direction $A = 271°$, $D = +42°$, and the velocity $V_\odot = 29{,}3$ km.

These results are provisional, being based on the radial velocities of six objects; later however STRÖMBERG[6] at the Mount Wilson Observatory made a second computation using 64 objects, his results being $A = 276°$, $D = +37°$, $V_\odot = 20{,}1$ km.

A comprehensive study of Calcium Clouds, as evidenced by detached lines, was made by OTTO STRUVE[7] at the Yerkes Observatory. He arrives at the following conclusions:

Any hypothesis that attempts to explain the phenomenon of detached Ca lines must comply with the following facts:

1. Detached Ca lines appear in the majority of stars of spectral class B3 and earlier.
2. They are very faint or entirely absent in many stars of high galactic latitude, but occasionally occur in such stars.
3. The radial velocities from the Ca lines clearly show the effect of the solar motion.
4. They have no K-effect.
5. Their individual velocities are considerably smaller than those of the B-type stars, but are well established in a large number of cases.
6. These individual velocities are not distributed at random, but indicate group motion for certain well defined regions of the sky.
7. These regions are associated with some of the more distinct Milky Way clouds or with regions of greatest star density.
8. The intensities of the detached lines are strong in certain regions of the sky and fainter in others.
9. The greatest intensity of these lines so far observed corresponds to a difference of about 1,8 stellar magnitudes between the brightness within the line and in the adjoining portion of the continuous spectrum.

---

[1] Lick Bull 10, p. 56 (1919). [2] Publ Dom Astroph Obs 1, p. 219 (1920).
[3] J Can R A S 15, p. 62 and 109 (1921).
[4] Publ Dom Astroph Obs 2, p. 335 (1924).
[5] J Can R A S 14, p. 234 (1920).
[6] Ap J 61, p. 372 (1925). [7] Pop Astr 33, p. 639 (1925) and 34, p. 1 (1926).

10. In an average of many stars they show a rapid increase in intensity from B 5 to B 2. Between B 2 and O the increase in intensity is much slower.

11. The stars which have the strongest detached K lines are appreciably redder than those in which these lines are faint.

12. The intensity curve of the ordinary stellar K line indicates that it should not be present in type B 3 or earlier. It begins to be appreciable in type B 5 and increases rapidly for later spectral subdivisions.

13. Certain spectroscopic binaries of spectral types B 0 to B 3 contain sharp Ca lines which oscillate with the same period as the other lines, but with a smaller amplitude and a different velocity of the system. This effect has not been noticed for types earlier than B 0. On the average the effect increases numerically with spectral type. However, there are stars like 16 Lacertae, belonging to spectral type B 3, in which no oscillation of the Ca lines occurs.

Not only is the problem of detached lines arising from the study of spectroscopic binaries important, since it concerns the constitution of the Universe, but because through it light may be shed on the hitherto unapproachable problem of the behaviour of tenuous particles surrounding binary systems.

A very noticeable peculiarity may here be mentioned, namely, the great number of close and rapid spectroscopic binaries found in important nebulae.

FROST was the first to call attention to such binaries in the trapezium of Orion[1], near which are found these five: 42 Orionis, θ^1 Orionis C, θ^1 Orionis A, θ^2 Orionis and ι Orionis, all belonging to class O or early B and showing tremendous ranges of radial velocity.

74. Anomalies Observed in Radial-Velocity Curves. In certain binaries the radial-velocity curves cannot be explained by simple elliptic motion. In some cases the observations are better satisfied by superimposing upon the elliptic motion a secondary oscillation, the period of which is commensurate or coincides with that of the orbit. In other cases the departures from elliptic motion seem rather erratic or non periodic; this occurrence is no doubt due to the presence of at least one additional body to the binary system.

If two stars are distinctly comparable in size with the dimensions of their orbits, that is, if their separation is small enough to mutually distort each other, their radial-velocity curves will be those of two revolving non-spherical or ellipsoidal masses of luminous matter. For such bodies these curves will vary appreciably from the mean velocity-curves; no doubt in many cases small secondary oscillations are not real, being due to a blend effect when double lines of very small separation are present, but in other cases non-elliptical orbits do exist.

A difficult problem in which radial-velocity curves cannot be explained by elliptic motion, is that of the Cepheid variables.

Dr. SCHLESINGER[2] has called attention to a number of orbits in which systematic departures from elliptic motion have been found. Here are cases he quotes:

α Andromedae (Allegh Obs Publ 1, p. 17). There seems to be some slight indication of a secondary curve having small amplitude and a period one-half that of the primary. The discrepancies here noted are such as would arise from the presence of the secondary spectrum, and their inconsiderable size suggests that this spectrum is relatively faint. A careful examination of our plates has failed to reveal its presence but the range in velocity is not sufficiently large for the diffuse lines of the two components to be separated at any time with our dispersion.

[1] Pop Astr 23, p. 361 (1915).
[2] Allegh Publ 1, p. 150 (1910).

ψ Orionis (Ap J 28, p. 266). In his investigation of this orbit PLASKETT used only plates of coarse grain and hence did not see the secondary spectrum. His measured velocities depart from the computed curve in conformity with the blend effect, and with this in view we have secured several spectrograms of this star on Seed 23 plates which show clearly the lines of the secondary. The discrepant points are thus readily explained.

ι Orionis (Ap J 28, p. 274). The secondary oscillation found by PLASKETT closely resembles the blend effect. Thus far we have not obtained a spectrogram at the maximum time of separation of the lines, and are unable to say whether or not the secondary spectrum is present.

o Leonis (Lick Bull 5, p. 21). The large displacements are all in the direction of the mean velocity, and this fact proves clearly that the disturbing cause is to be found in the influence of the second spectrum.

α Virginis (Publ Astroph Obs Potsdam 7, Part 1, p. 127; Allegh Obs Publ 1, p. 65). In VOGEL's orbit the effect of blending is present throughout, reducing K by one-fourth its true value. In the Allegheny orbit the influence of the secondary spectrum is conspicuous where the lines are blended.

ε Herculis (J Can R A S 3, p. 377). The assumption of a secondary disturbance of one-third the period of the main star greatly improved the agreement between theory and observation. The secondary spectrum is here present, and we believe that this is a case of blend effect.

θ Aquilae (Allegh Obs Publ 1, p. 45; J Can R A S 3, p. 87). The secondary oscillation noted by both writers can be explained as a blend effect remembering that this is proportionately greater near the V_0-line. The secondary spectrum is present on plates secured at the Allegheny Observatory.

75. Spectra in which the Radial-Velocity Curves Given by Some Lines Differ from the Curves Given by Other Lines. In the spectrum of a star lines are encountered which certainly are produced at different levels of the star's atmosphere.

Previously mentioned was the case of fine *H* and *K* lines of calcium giving radial-velocity curves of much smaller amplitude than other lines; perhaps this is due to a corona of wide extent entirely surrounding early class B stars.

When certain conditions exist in a hot star, matter, at high levels, is highly ionized and produces +lines, while neutral lines are found from other levels.

According to W. S. ADAMS and C. E. ST. JOHN[1], it may be assumed that convection currents are favoured by high temperature and low pressure, conditions which prevail especially at high levels; a rapid downward motion of the matter would therefore add a corrective term to radial velocity as obtained from lines of these levels.

If radial-velocity curves are not due to orbital motion but to some pulsation of the star, it may be that gases at different levels will have different speeds of expansion and contraction and consequently their corresponding lines will give curves of different amplitudes. This has been suggested by R. H. CURTISS in the case of Cepheids.

A remarkable star that demands particular mention is α Canum Venaticorum. Its spectrum has been studied by Miss MAURY[2], LOCKYER and BAXANDALL[3], LUDENDORFF[4], and BÉLOPOLSKY[5].

---

[1] Pop Astr 32, p. 623 (1924). [2] Harv Ann 28, p. 96 (1897).

[3] London R S Proc (A) 77, p. 550 (1906).

[4] A N 173, p. 1 (1906).

[5] Bull. de l'Acad. Imp. des Sciences de St-Pétersbourg (Sér. VI) 7, p. 689 (1913); also A N 196, p. 1 (1913).

After extensive investigation by means of the photo-electric cell, GUTHNICK and PRAGER announced the star to be variable, the period agreeing with that found by BÉLOPOLSKY and the amplitude of variation being 0,051 magnitude.

Many faint lines in the star belonging exclusively to rare earth elements such as europium, exhibit periodic variations of intensity, and a few indicate a variation of radial velocity in the same period, as obtained by BÉLOPOLSKY, 5,50 days. All the other lines in the spectrum, produced by normal elements such as hydrogen, iron, silicon, calcium and magnesium, give a constant velocity and always have the same intensity.

In an important paper on α Canum Venaticorum KIESS[1] arrived at the following conclusions:

1. BÉLOPOLSKY's discovery of two groups of variable lines in the spectrum of α Canum Venaticorum and the period of variation have been confirmed.

2. His discovery that certain lines belonging to the two groups have variable velocities is also verified.

3. The identity of the variable lines with lines in the spectra of europium and terbium is indicated.

4. The identity of many of the faint lines with lines of the spectra of other rare earth elements such as gadolinium, dysprosium, yttrium, lanthanum etc., is suggested.

5. The symmetry axis of the velocity-curve of the variable lines λ 3930,65 and λ 4205,20 is identical with the constant velocity of the star.

It is apparently a rule that whenever various spectral lines give different radial-velocity curves some of them, if not all, exhibit remarkable changes in width and intensity.

76. Photometric Tests of Spectroscopic Binaries. Slight variations of brightness in spectroscopic binaries must be expected and may occur from several causes. For instance, eclipsing pairs are not unusual and naturally exhibit some degree of variation; again, if a secondary is faint it may reflect the light of the primary in quantities varying according to its phases in the orbit, while some fluctuations of light are still unexplained.

In most binaries, however, changes in magnitude are small and cannot be detected easily by visual or photographic methods. Fortunately the photo-electric cell is considerably more sensitive than either, and by this means most enlightening researches have been made especially by GUTHNICK and PRAGER[2] at the Berlin-Babelsberg Observatory and by STEBBINS at the University of Illinois[3].

An interesting article "La mesure de l'énergie lumineuse des étoiles par la cellule photoélectrique" has been published by G. ROUGIER[4].

77. Classification of Spectroscopic Binaries. The following classification of spectroscopic binaries may be proposed:

1. Binaries having spectra of advanced type from M dwarf to G dwarf. They usually have long periods amounting to several months or even years; their radial velocity curves are well determined, the spectra yielding very accurate measurements on account of their many faint well defined lines. Some of them like λ Andromedae and σ Geminorum of class K0 have periods as short as twenty days; in these cases the presence of a third companion is suspected and may be responsible for the shortness of the period.

---

[1] Publ Obs Univ Michigan 3, p. 106 (1923).

[2] Veröff. der K. Sternwarte zu Berlin-Babelsberg 1, Heft 1 (1914) and 2, Heft 3 (1918).

[3] Ap J 39, p. 459 (1914).

[4] Bull Soc Astr de France 39, p. 58 (1925).

Lal. 46867 has a still shorter period $6^d,72$; evidently a very dwarf star, it exhibits an extraordinary feature, namely bright H and K lines superimposed upon the usual broad absorption bands and yielding the same radial velocities as the other lines. Lal. 29330 has a period $4^d,28$ [1].

2. Binaries having spectra of early type, from F dwarf to O, with fairly long periods, for example: π Andromedae, class B3, period $143^d,67$, and μ Sagittarii, class B8p, period $180^d,2$.

3. Binaries having spectra of early type, from F dwarf to O, and having short periods of a few days or less than a day. These often have enormous ranges of radial velocity.

4. Cepheid variables and stars of the β Canis Majoris type. There is a question as to the authenticity of these stars as binaries; they are treated in LUDENDORFF's chapter on variable stars and in Subdivision i.

5. Super-giant stars. A number of stars from all spectral classes exhibit numerous ionized lines in their spectra and like the Cepheids, perhaps, have very small densities and large volumes. Some of them present slight periodic or irregular variations of radial velocity; W. S. ADAMS called these stars Pseudo-Cepheids; typical examples are α Cygni, σ Cygni and ε Aurigae. These types are never encountered among visual binaries and therefore open an entirely new field in the realm of astrophysics.

f) The Orbit of a Spectroscopic Binary Star.

78. Usual Notation. The orbital elements of a spectroscopic binary star are defined as follows:

P = apparent period of revolution in mean solar days, unless specified in mean solar years. The true period of revolution, P_0, may be found from the apparent period by means of the equation

$$P_0 = \frac{P}{1 + \frac{V_0}{300000}},$$

where V_0 = the radial velocity of the centre of mass of the system.

μ = mean daily motion $= \frac{2\pi}{P}$.

T = Greenwich mean time of periastron passage expressed in Julian Days; for variable stars T is the mean solar time interval after maximum or minimum brightness, as specified in each case.

ω = angular distance of periastron from the ascending node measured in the direction of orbital motion. The radial velocity of the observed body has its maximum value at the ascending node. Spectrographic observations enable the computer to distinguish between the ascending and descending nodes, but observations of visual double stars secured with a micrometer do not distinguish between the two nodes, leaving the value of ω uncertain by 180°.

e = eccentricity of the orbit.

K_1 = semi-amplitude of radial-velocity curve of the primary member of the system.

K_2 = semi-amplitude of radial-velocity curve of the secondary member of the system.

$K_1 + K_2$ = semi-amplitude of the radial-velocity curve of one member of the system with reference to the other member.

[1] Ap J 53, p. 210 (1921).

a_1 = semi-axis major of the orbit of the primary with reference to the centre of mass.

a_2 = semi-axis major of the orbit of the secondary with reference to the centre of mass.

$a_1 + a_2$ = semi-axis major of the orbit of one member of the system with reference to the other member.

i = inclination of the orbit plane, conveniently defined as the angle between the line of sight and the normal to the orbit plane. Spectrographic observations leave the value of i undetermined; micrometer observations of a visual binary leave the quadrant of i undetermined; spectrographic and micrometer observations combined determine i completely.

m_1 = mass of primary.

m_2 = mass of secondary.

$m_1 + m_2$ = mass of system.

$\odot$ = mass of Sun.

V_0 = radial velocity of the centre of mass of the system.

r_1 = radius of primary.

r_2 = radius of secondary.

r_s = radius of Sun.

d_1 = mean density of primary.

d_2 = mean density of secondary.

D_0 = mean density of Sun.

To these notations may be added formulae in the three following cases:

1. One stellar spectrum, with comparison spectrum observed

$$a_1 \sin i = 13751\,(1 - e^2)^{\frac{1}{2}}\, K_1 P\,,$$

$$\frac{m_2^3 \sin^3 i}{(m_1 + m_2)^2} = 0{,}00000010385\,(1 - e^2)^{\frac{3}{2}}\, K_1^3 P \odot\,.$$

This second formula, or function of masses, is derived from the third law of KEPLER, which says that for any double star system

$$\frac{a^3}{P^2\,(m_1 + m_2)} = \text{const}\,.$$

If R is the semi-axis major of the Earth's orbit, 1 the mass of the Sun, m' the mass of the Earth and T the duration of the sidereal year,

$$\frac{a^3}{P^2\,(m_1 + m_2)} = \frac{R^3}{T^2\,(1 + m')}\,,$$

since m' is very small and if a is expressed in astronomical units and P in years

$$m_1 + m_2 = \frac{a^3}{P^2}\,, \qquad \text{or } m_1 + m_2 = \frac{(a_1 + a_2)^3}{P^2}\,,$$

multiplying by $\sin^3 i$ and since

$$\frac{a_1}{a_2} = \frac{m_2}{m_1}\,, \qquad \text{or } a_1 + a_2 = a_1 \cdot \frac{m_1 + m_2}{m_2}\,,$$

$$(m_1 + m_2) \cdot \sin^3 i = \frac{1}{P^2} \left(\frac{m_1 + m_2}{m_2}\right)^3 a_1^3 \sin^3 i\,,$$

replacing $a_1 \sin i$ by its value given above and taking account that P is expressed in days and a_1 in kilometers, the function of masses is easily derived.

2. Two stellar spectra observed, without comparison spectrum, as by means of objective prism,

$$(a_1 + a_2) \sin i = 13751\,(1 - e^2)^{\frac{1}{2}}\,(K_1 + K_2)\, P\,,$$

$$(m_1 + m_2) \sin^3 i = 0{,}00000010385\,(1 - e^2)^{\frac{3}{2}}\,(K_1 + K_2)^3\, P \odot\,.$$

3. Two stellar spectra, with comparison spectrum observed

$$m_1 \sin^3 i = 0{,}00000010385\,(1 - e^2)^{\frac{3}{2}}\,(K_1 + K_2)^2\,K_2 P\odot,$$
$$m_2 \sin^3 i = 0{,}00000010385\,(1 - e^2)^{\frac{3}{2}}\,(K_1 + K_2)^2\,K_1 P\odot,$$
$$\frac{m_1}{m_2} = \frac{K_2}{K_1}.$$

79. Fundamental Equations. If ϱ_1 is the radius vector of any point S in the primary's orbit with reference to the centre of mass of the system, and if z is the projection of ϱ_1 on the line of sight, and v the true anomaly of S,

$$z = \varrho_1 \sin i \sin(v + \omega)\,. \tag{1}$$

Thus the observed radial velocity of point S is

$$V = V_0 + \frac{dz}{dt}, \tag{2}$$

t being the time at which the primary is at S.

Differentiating equation (1), i and ω being constants,

$$\frac{dz}{dt} = \sin i \sin(v + \omega)\frac{d\varrho_1}{dt} + \varrho_1 \sin i \cos(v + \omega)\frac{dv}{dt}\,.$$

The known laws of elliptic motion give

$$\varrho_1 \frac{dv}{dt} = \frac{\mu a_1 (1 + e\cos v)}{\sqrt{1 - e^2}},$$

$$\frac{d\varrho_1}{dt} = \frac{\mu a e \sin v}{\sqrt{1 - e^2}},$$

therefore

$$\frac{dz}{dt} = \frac{\mu a_1 \sin i}{\sqrt{1 - e^2}}\,[e\cos\omega + \cos(v + \omega)]\,,$$

and equation (2) becomes

$$V = V_0 + \frac{\mu a_1 \sin i}{\sqrt{1 - e^2}}\,[e\cos\omega + \cos(v + \omega)]\,. \tag{3}$$

Six constants enter the right-hand member of equation (3), they are:

V_0, $a_1 \sin i$, e, μ, ω and v (from which T is derived).

The quantity $a_1 \sin i$ is usually considered as an element, the inclination of the orbit not being determinable.

Theoretically six different values of V would suffice to determine the six elements; but V is sometimes far from accurate, and the equations being transcendant the system would be very difficult to solve.

Usually the elements of the orbit are not computed until numbers of radial velocity measurements are available, each separated by short intervals throughout the entire period of revolution. A radial-velocity curve is then obtained; great care must be taken in drawing this curve to satisfy closely all the observations as well as the laws of elliptic motion. Weights should be assigned to the observations according to their value.

If A_1 and B_1 are absolute values of the maximum and the minimum radial velocity as counted from the V_0-axis on the curve, and K_1 is the semi-amplitude of this curve, then

$$K_1 = \frac{\mu a_1 \sin i}{\sqrt{1 - e^2}},$$

$$A_1 = K_1 (1 + e\cos\omega)\,,$$

$$B_1 = K_1 (1 - e\cos\omega)\,,$$

therefore

$$\left.\begin{aligned} \frac{A_1 + B_1}{2} &= K_1\,, \\ \frac{A_1 - B_1}{2} &= K_1 e \cos\omega\,, \\ \frac{A_1 - B_1}{A_1 + B_1} &= e \cos\omega\,, \end{aligned}\right\} \tag{4}$$

and equation (3) becomes

$$V = V_0 + \frac{A_1 - B_1}{2} + \frac{A_1 + B_1}{2} \cos(v + \omega)\,. \tag{5}$$

If the mean axis of the radial-velocity curve is taken as origin of ordinates, the equation of the curve will take the simple form

$$V = K_1 \cos(v + \omega), \tag{6}$$

where the time is the abscissa and V the ordinate.

Several methods of determining orbital elements are based on the fundamental equations given above.

80. The Different Methods of Obtaining Orbital Elements. Many methods have been proposed since 1891, a year distinguished by RAMBAUT's elaborate publication[1]. There followed that of WILSING[2] in 1893, but a far more simple method by LEHMANN-FILHÉS[3] is now commonly used.

Other methods have been devised by SCHWARZSCHILD[4], RUSSELL[5], NIJLAND[6], LAVES[7], ZURHELLEN[8], W. F. KING[9], PLUMMER[10], PADDOCK[11], CURTISS[12], HENROTEAU[13], E. S. KING[14], WOLTJER[15] and HALM[16].

Those given in the next paragraphs will suffice for general purposes.

81. LEHMANN-FILHÉS' Method. The velocity-curve having been drawn, the V_0-axis is determined. Using a planimeter, its position is obtained when the two areas it determines with the radial-velocity curve are equal.

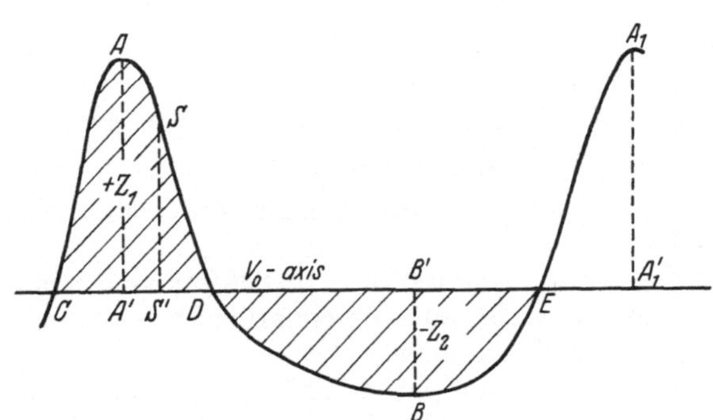

Fig. 14. Velocity-curve.

From equation (1) in paragraph 79 it is known that

$$z = \int_0^t \frac{dz}{dt} \cdot dt\,.$$

The orbit being a closed curve it follows that

$$\int_0^t \frac{dz}{dt}\, dt = \int_0^{t+P} \frac{dz}{dt} \cdot dt\,,$$

or

$$\int_t^{t+P} \frac{dz}{dt} \cdot dt = 0\,,$$

or the total area of the velocity-curve must be zero, that is the positive area = the negative area.

[1] M N 51, p. 316 (1891). [2] A N 134, p. 90 (1893). [3] A N 136, p. 17 (1894).
[4] A N 152, p. 65 (1900). [5] Ap J 15, p. 252 (1902) and 40, p. 282 (1914).
[6] A N 161, p. 105 (1903). [7] Ap J 26, p. 164 (1907).
[8] A N 173, p. 353 (1907); 175, p. 245 (1907); 177, p. 321 (1908); 187, p. 433 (1911).
[9] Ap J 27, p. 125 (1908). [10] Ap J 28, p. 212 (1908).
[11] Lick Bull 8, p. 153 (1915). [12] Publ Obs Univ Michigan 2, p. 178 (1916).
[13] Publ A S P 29, p. 195 (1917). [14] Harv Ann 81, p. 231.
[15] B A N 1, p. 95 (1922). [16] M N 87 p. 628 (1927).

It can be demonstrated that the radial velocity is maximum at the ascending node and minimum at the descending node, A and B thus correspond to these two nodes. Since the nodes are both in the orbit plane and in the plane perpendicular to the line of sight passing through the centre of mass, $z = 0$ when the star passes through the nodes; thus

$$\int_0^{A'} \frac{dz}{dt} \cdot dt = 0,$$

$$\int_0^{B'} \frac{dz}{dt} \cdot dt = 0,$$

and the difference

$$\int_{A'}^{B'} \frac{dz}{dt} \cdot dt = 0,$$

or

$$\text{area } AA'D = \text{area } DB'B = Z_1.$$

Similarly

$$\text{area } CA'A = \text{area } BB'E = Z_2.$$

According to the shape of the velocity-curve, it is often simpler to determine one of the two points A or B. The easiest is then determined, and the other is checked, using the planimeter, to make certain that one of the last two relations is satisfied.

By a series of transformations LEHMANN-FILHÉS shows that

$$e \sin\omega = \frac{2\sqrt{A_1 B_1}}{A_1 + B_1} \cdot \frac{Z_2 - Z_1}{Z_2 + Z_1}. \qquad (7)$$

Since the areas enter as a ratio, the unit of area used is entirely immaterial.

K_1 is then measured on the curve, while the last of equations (4) and equation (7) furnish e and ω.

At the time of periastron passage $v = 0°$, hence its ordinate measured from the mean axis is $K_1 \cos\omega$, as derived from equation (6). Two points of the curve will have the same ordinate, but since $v + \omega$ equals $0°$ and $180°$ for the points A and B respectively, the proper periastron point may be easily distinguished. The abscissa of this point gives T.

It has been seen in paragraph 79 that

$$K_1 = \frac{\mu a_1 \sin i}{\sqrt{1 - e^2}},$$

or

$$a_1 \sin i = 86400 \frac{K_1}{\mu} \sqrt{1 - e^2}, \qquad (8)$$

where the factor 86400 had to be introduced, since the unit of time for K_1 is the second, while for μ it is the day.

This equation may be written

$$a_1 \sin i = 13751\,(1 - e^2)^{\frac{1}{2}}\, K_1 P \qquad (9)$$

as in paragraph 78.

This completes the determination of the elements.

82. SCHWARZSCHILD's Method of Determining the Periastron Point. SCHWARZSCHILD gave for this the following process:

Lay a piece of semi-transparent paper over the velocity-curve and trace the curve and the mean axis, also marking the points 0, $P/2$, P, and $3\,P/2$. Shifting this copy along the mean axis for a distance $P/2$, or half the period, rotate the copy 180° about the mean axis; in other words turn the tracing face downward on the original curve, keeping the mean axis in coincidence and bringing the point 0 or P of the copy over the point $P/2$ of the curve. As a rule the curves will then cut one another in four points, of which two will be the points of periastron and apastron. There will be no difficulty in selecting these two points, for periastron and apastron must be separated in time by one half a revolution and must, moreover, lie on different branches of the velocity curve. To determine which is periastron we have this criterion that the curve is for a shorter time on that side of the mean axis where the point of periastron lies.

These constructions are demonstrated by using equation (6)

$$\text{At periastron, } V = K_1 \cos\omega,$$
$$\text{at apastron, } V = K_1 \cos(\omega + 180°) = -K_1 \cos\omega.$$

The respective mean anomalies of these two points are 0° and 180°; consequently they are determined from the two conditions: that their abscissae differ by $P/2$ and that their ordinates are equal with opposite signs.

83. Henroteau's First Method. The computation of orbital elements by this method is based on the fact that, if v_1 is the value of v for point C, the value of v for points A, D, B, E, are respectively $v_1 + 90°$, $v_1 + 180°$, $v_1 + 270°$, $v_1 + 360°$. (See Fig. 14, in which CE is considered as the mean axis.)

If S is the periastron point, the mean anomalies of points A, D, B, E, A_1, are respectively

$$-S'A' \cdot \frac{360°}{P}, \quad S'D \cdot \frac{360°}{P}, \quad S'B' \cdot \frac{360°}{P}, \quad S'E \cdot \frac{360°}{P}, \quad S'A_1' \cdot \frac{360°}{P}.$$

Since points C, D and E will usually be more accurately determined than points A, B and A_1, it will be of advantage to use the two points which are directly to the left and to the right of S'.

Choosing the points C and D on Fig. 14, let

$$M_x = S'C \cdot \frac{360°}{P}, \quad M_y = S'D \cdot \frac{360°}{P}.$$

Then, referring to the Allegheny Tables of Anomalies[1], the eccentricity of the orbit is found at the top of the column in which the corresponding true anomalies, v_x and v_y have a sum equal to 180°.

The value of ω will always be the true anomaly corresponding to $A'S' \cdot \frac{360°}{P}$, in the present case $v_y - 90°$.

If V' is the radial velocity of the mean axis, then

$$V_0 = V' - K_1 e \cos\omega = V' - e \cdot \overline{SS'},$$

which gives the radial velocity of the centre of mass, and as formerly,

$$a_1 \sin i = 13751\,(1 - e^2)^{\frac{1}{2}}\, K_1 P,$$

completing the determination of the orbital elements.

If S has been well determined by SCHWARZSCHILD's method,

$$\cos\omega = \frac{SS'}{K_1}.$$

As a test it is a wise practice to see if this equation is satisfied.

The points A' and B' can be used, or even the points A' and D instead of C and D as above.

[1] Allegh Obs Publ 2, p. 155 (1912).

If

$$M_x = S'A'\frac{360^\circ}{P} \text{ and } M_y = S'B'\frac{360^\circ}{P}$$

the corresponding v_x and v_y must have a sum equal to $180°$.

If

$$M_x = S'A'\frac{360^\circ}{P} \text{ and } M_y = S'D\frac{360^\circ}{P}$$

the corresponding v_x and v_y must have a sum equal to $90°$. Recently this first method has been much simplified, making use of a short table especially computed[1].

84. Henroteau's Second Method. To simplify computation, let it be assumed that the undetermined value of i be equal to $90°$. In such a case the line of sight is in the plane of the orbit; the only thing to be remembered is that the value obtained for a_1 is really the value of $a_1 \sin i$.

On the radial-velocity curve the V_0-axis will be determined by the use of a planimeter as in the case of the Lehmann-Filhés method.

A certain number of well distributed points will be chosen on the curve, 1, 2, 3, ... and the areas 011′, 022′, 033′, ... measured with the planimeter counting the areas below the V_0-axis as negative. Using the same abscissae as in the velocity-curve and the corresponding measures of areas as ordinates, a new curve, which may be called the integrated curve, will be traced. This curve, having (1) for equation, shows the variation with time of the projection z of ϱ_1 on the line of sight. The ordinates corresponding to the abscissae of A' and B' will be equal and will mark the value of z at ascending and descending nodes.

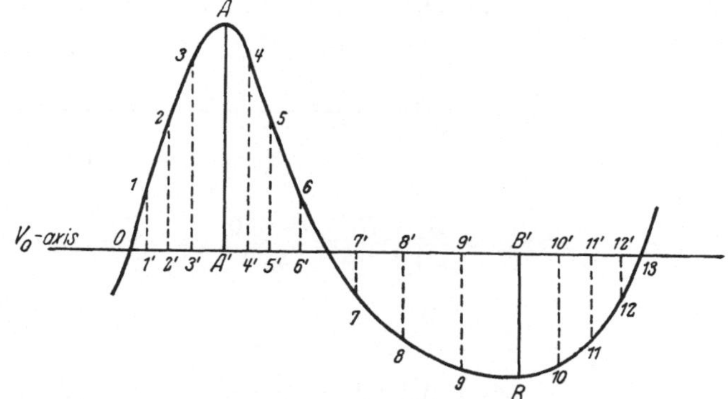

Fig. 15. Theoretical velocity-curve, treated according to the second method.

The integrated curve, projected on the axis of ordinates, represents the apparent orbit as seen from a point at infinity in the orbit plane, its direction being at right angles to the line of sight. In this case the integrated curve is similar to that of Fig. 9, and the elements of the orbit may be determined exactly as in paragraph 60, except that the value of ω given there is really the value of $\omega + 90°$ for the spectroscopic orbit, and that the value of a_1 will be found in kilometers instead of in seconds.

This second method is of interest in the determination of the true dimensions of an orbit when spectroscopic as well as visual data are available.

85. Practical Application of these Three Methods. As an illustration, the radial-velocity curve of the spectroscopic binary H. R. 6385 = HD 155375 (Fig. 16), obtained by W. E. Harper at the Dominion Astrophysical Observatory[2], will be considered, and the elements of the orbit determined.

[1] J Can R A S 21, p. 265 (1927).

[2] Publ Dom Astroph Obs 1, p. 201 (1920).

1. Method of LEHMANN-FILHÉS.

Having determined the V_0-axis it follows that

$$A_1 = 20{,}1 \text{ km}, \quad B_1 = 35{,}1 \text{ km}, \quad K_1 = 27{,}6 \text{ km}, \quad Z_1 = 133, \quad Z_2 = 243.$$

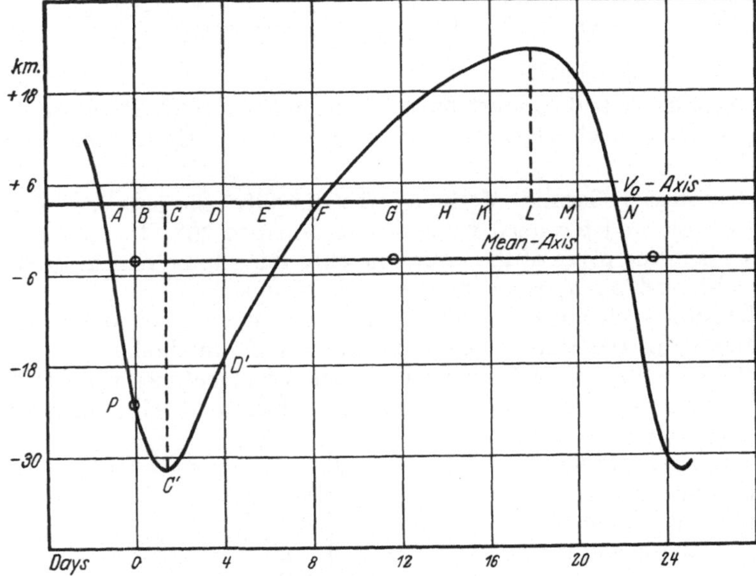

Fig. 16. Velocity-curve of H. R. 6385.

$$e\cos\omega = \frac{A_1 - B_1}{A_1 + B_1} = \frac{-15{,}0}{55{,}2} = -0{,}27174,$$

$$e\sin\omega = \frac{2\sqrt{A_1 B_1}}{A_1 + B_1} \cdot \frac{Z_2 - Z_1}{Z_2 + Z_1} = + \frac{2\sqrt{35{,}1 \cdot 20{,}1}}{55{,}2} = 0{,}29771,$$

$$e^2 = [0{,}272]^2 + [0{,}298]^2 = 0{,}162788,$$

$$e = 0{,}407,$$

$$\tan\omega = -\frac{29771}{27174} = -1{,}0956,$$

$$\omega = 132^\circ 24',$$

$$v_0 = +3{,}51 \text{ km},$$

$$a_1 \sin i = 13751\,(1 - e^2)^{\frac{1}{2}} K_1 P = 13751 \cdot (0{,}837212)^{\frac{1}{2}} \cdot 27{,}6 \cdot 23{,}245 = 8072239 \text{ km},$$

$$K_1 \cos\omega = -27{,}6 \cdot 0{,}6743 = -18{,}61.$$

This ordinate gives the position of the periastron point with abscissa zero.

2. HENROTEAU's First Method.

The position of the periastron point is obtained by SCHWARZSCHILD's method and is found to coincide exactly with the position found by the LEHMANN-FILHÉS method.

$$M_x = 6{,}5 \cdot \frac{360^\circ}{23{,}245} = 6{,}5 \cdot 15^\circ{,}487 = 100^\circ{,}7,$$

$$M_y = 1{,}1 \cdot \frac{360^\circ}{23{,}245} = 1{,}1 \cdot 15^\circ{,}487 = 17^\circ{,}0.$$

From the Allegheny Tables it is found that for $e = 0{,}40$,

$$v_x = 41^\circ{,}2,$$
$$v_y = 138^\circ{,}8,$$
$$v_x + v_y = 180^\circ.$$

Thus $e = 0{,}40$,

$$\omega = v_x + 90^\circ = 131^\circ{,}2.$$

The values of other elements are easily obtained.

With careful drawing and proper interpolation, this method is more rapid than that of LEHMANN-FILHÉS and quite as accurate.

3. HENROTEAU's Second Method.

Areas are measured comprised between the V_0-axis, the velocity-curve beginning at point A and the parallels to BP passing respectively through $B, C, D, E, F, G, H, K, L, M$ and N. Care must be taken to include the points where the V_0-axis cuts the curve, and those indicating the minimum and maximum of this curve.

The measurement is accomplished very quickly with the planimeter by moving its point along $A, B, P, A, C, C', A, D, D', A, E, E', A$, and reading the instrument every time the point is at A. The differences between successive readings give the areas according to the planimeter's scale. Their values are negative below and positive above the V_0-axis.

The following negative areas are obtained

A	0	G	339
B	43	H	276
C	135	K	203
D	297	L	138
E	361	M	43
F	380	N	0

From these, the integrated curve of Fig. 17 is obtained.

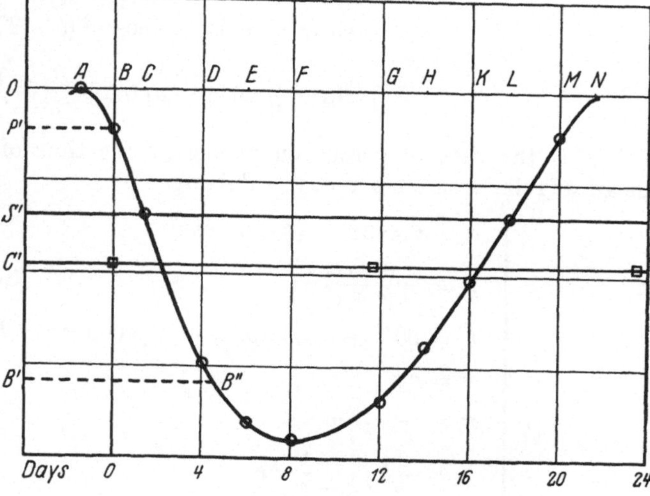

Fig. 17. Integrated curve of H. R. 6385.

Then,

$$e = \frac{C'S'}{C'P'} = \frac{55}{147} = 0{,}37,$$

$$\sin PSB = \sqrt{1 - e^2}$$
$$= \sqrt{0{,}8631} = 0{,}9290,$$
$$PSB = 111^\circ 43'.$$

The corresponding mean anomaly for $e = 0{,}37$, is $69^\circ{,}5$.

But

$$\mu = 15^\circ{,}487,$$

therefore

$$P'''B''' = \frac{69^\circ{,}5}{15^\circ{,}487} = 4^d{,}48,$$

and $C'B' = 127$,

$$\tan\omega = \frac{P'C'}{B'C'}\sqrt{1 - e^2} = \frac{147}{127} \cdot 0{,}93 = 1{,}0765,$$

$$\omega = 132^\circ\, 54'.$$

The scale of the ordinates in Fig. 17 is to be determined. Measured with the planimeter, the area of one square is 100; on the scale of Fig. 16, it represents the distance travelled by a body moving with a speed of 12 km/sec in 4 days, or

$$12 \cdot 4 \cdot 86400 = 4147200 \text{ km.}$$

$$\text{One unit area} = 41472 \text{ km.}$$

$$a_1 = P'C' \operatorname{cosec} \omega \cdot 41472 = \frac{147 \cdot 41472}{0{,}7325} = \frac{6096384}{0{,}7325} = 8322708 \text{ km}.$$

The three methods give satisfactory preliminary elements. In exact determinations, these always have to be corrected by the method of least squares.

86. Corrections of the Elements by the Method of Least Squares[1]. The method of obtaining these corrections by least squares has been presented in a simple way by Dr. SCHLESINGER[2]. The difference δV is derived by computing V for each time of observation, from preliminary elements (normal places are usually taken) and subtracting it from the observed V.

The equations of condition for the least-square solution take the form

$$\begin{aligned} \delta V = {} & \delta V_0 + [\cos(v+\omega) + e\cos\omega]\,\delta K_1 \\ & + \left[\cos\omega - \frac{\sin(v+\omega)\sin v}{1-e^2}(2 + e\cos v)\right] K_1 \delta e \\ & - [\sin(v+\omega) + e\sin\omega]\,K_1\delta\omega \\ & - \sin(v+\omega)\,[1 + e\cos v]^2\,(t - T)\,\frac{K_1\delta\mu}{(1-e^2)^{\frac{3}{2}}} \\ & + \sin(v+\omega)\,[1 + e\cos v]^2\,\frac{K_1\mu\,\delta T}{(1-e^2)^{\frac{3}{2}}}. \end{aligned}$$

T is the time of periastron passage, t the time of observation and μ the mean angular motion per day. Putting

$$\text{(A)} \left\{ \begin{aligned} \alpha &= 0{,}452 \sin v\,(2 + e\cos v) && \text{(SCHLESINGER's table II)} \\ \beta &= \frac{(1 + e\cos v)^2}{(1+e)^2} && \text{(SCHLESINGER's table III)} \\ \Gamma &= \delta V_0 + e\cos\omega\,\delta K_1 + K_1\cos\omega\,\delta e - K_1 e\sin\omega\,\delta\omega \\ \varkappa &= \delta K_1 \\ \pi &= -K_1\delta\omega \\ \varepsilon &= -K_1\frac{2{,}21}{1-e^2}\,\delta e \\ \tau &= K_1\mu\sqrt{\frac{1+e}{1-e}}\cdot\frac{1}{1-e}\,\delta T \\ m &= -K_1\sqrt{\frac{1+e}{1-e}}\cdot\frac{1}{1-e}\,\delta\mu, \end{aligned} \right.$$

the equations of condition assume this simple form:

$$\begin{aligned} \delta V = {} & \Gamma + \cos(v+\omega)\cdot\varkappa + \sin(v+\omega)\cdot\pi + \alpha\sin(v+\omega)\cdot\varepsilon + \beta\sin(v+\omega)\cdot\tau \\ & + \beta\sin(v+\omega)\cdot(t - T)\cdot m. \end{aligned}$$

There are as many equation of conditions as there are observations.

---

[1] See also § 57. [2] Allegh Obs Publ 1, p. 33 (1908).

Providing $v + \omega = u$, if p_1, p_2 etc. be the weights of the observations, the following six normal equations are obtained:

$$\begin{array}{llllll}
\varkappa & \pi & \varepsilon & \tau & m & \\
+[p\cos u] & +[p\sin u] & +[p\,\alpha\sin u] & +[p\,\beta\sin u] & +[p\,\beta\sin u\,(t-T)] & =[p\,\delta V] \\
p]-[p\sin^2 u] & +\tfrac{1}{2}[p\sin 2u] & +\tfrac{1}{2}[p\,\alpha\sin 2u] & +\tfrac{1}{2}[p\,\beta\sin 2u] & +\tfrac{1}{2}[p\,\beta\sin 2u\,(t-T)] & =[p\cos u\,\delta V] \\
\times & +[p\sin^2 u] & +[p\,\alpha\sin^2 u] & +[p\,\beta\sin^2 u] & +[p\,\beta\sin^2 u\,(t-T)] & =[p\sin u\,\delta V] \\
\times & \times & +[p\,\alpha^2\sin^2 u] & +[p\,\alpha\,\beta\sin^2 u] & +[p\,\beta^2\sin^2 u\,(t-T)] & =[p\alpha\sin u\,\delta V] \\
\times & \times & \times & +[p\,\beta^2\sin^2 u] & +[p\,\beta^2\sin^2 u\,(t-T)] & =[p\,\beta\sin u\,\delta V] \\
\times & \times & \times & \times & +[p\,\beta^2\sin^2 u\,(t-T)^2] & =[p\,\beta\sin u\,(t-T)\,\delta V].
\end{array}$$

The coefficients of Γ in the second, third etc. . . . equations are the same as the coefficients of $\varkappa$, π etc. . . . in the first equation. The coefficients of $\varkappa$ in the third, fourth etc. equations are the same as the coefficients of π, ε etc. . . . in the second equation. In other words the coefficients which are replaced by crosses in the above equations are the same as their symmetricals with respect to the diagonal $[p]$, $[p] - [p\sin^2 u]$, etc. . . .

The tables for α and β have been arranged by SCHLESINGER so as to render the normal equations "homogeneous" and to enable all the multiplications to be made with CRELLE's tables without interpolation.

It is to be remembered that:

$$[p] = p_1 + p_2 + \cdots + p_n\,,$$

$$[p\cos u] = p_1\cos u_1 + p_2\cos u_2 + \cdots + p_n\cos u_n\,,$$

. . . .

$$[p\,\beta\sin u\,(t-T)] = p_1\beta_1\sin u_1\,(t_1-T) + p_2\beta_2\sin u_2\,(t_2-T) + \cdots + p_n\beta_n\sin u_n(t_n-T)\,,$$

$$[p\,\delta V] = p_1\,\delta V_1 + p_2\,\delta V_2 + \cdots + p_n\,\delta V_n\,.$$

. . . .

Unity should be the maximum weight assigned to the observations, and if convenient to do so, the unit and the zero of time should be so chosen that all the values of $t - T$ come between -1 and $+1$.

After the unknowns and the residuals (Δ_1, Δ_2, etc.) have been evaluated, a complete check upon the numerical work is furnished by the equations:

$$[p\cdot\Delta] = 0\,,\qquad [p\cdot\sin u\cdot\Delta] = 0\,,\qquad [p\cdot\beta\cdot\sin u\cdot\Delta] = 0\,,$$

$$[p\cdot\cos u\cdot\Delta] = 0\,,\qquad [p\cdot\alpha\cdot\sin u\cdot\Delta] = 0\,,\qquad [p\cdot\beta\cdot\sin u\cdot\Delta\,(t-T)] = 0\,.$$

By means of equations (A) the corrections to the elements and their final values are easily obtained. To compute their probable errors, it is only necessary to multiply the errors of ε, π, etc. by the factors contained in equations (A). This method is essentially the same as that outlined in paragraph 58.

If the preliminary elements are not well enough determined, it may happen that the corrected values found by a least-square solution diverge from the true values. Also if they converge, since the formulae used in least squares do not contain terms of the second order, the convergence may not be appreciable. In that case a second least-square solution is necessary using the new elements as preliminaries. Generally one such solution is enough, but there are cases where a succession of solutions are needed to obtain satisfactory elements.

87. SCHLESINGER's Criterion for Spectroscopic Binaries[1]. Sometimes after having measured a large number of radial velocities, it is impossible to find a period of variation or even to decide whether the star is single or multiple. This is especially true when the measurements differ by quantities comparable to the accidental errors of observation.

[1] Ap J 41, p. 162 (1915).

A frequency-curve is then constructed by dividing the total range exhibited by the measured velocities into equal intervals, for instance, of two kilometers each, and counting the number of velocities which fall respectively within each interval. Taking the kilometers as abscissae and the corresponding numbers of velocities as ordinates, a smooth curve is drawn by joining the ends. The shape of this frequency-curve is then compared with that of the well known error-curve having for equation $y = \frac{1}{\sqrt{\pi}} e^{-x^2}$, adopting the proper units for x and y.

If the two are reasonably the same, the star is not a spectroscopic binary within the limits that the measurements are capable of defining. If the frequency-curve is different, the object is most likely a binary, and SCHLESINGER shows that by the shape of this curve an idea can be formed of the orbit's nature both as to eccentricity and the approximate position of the periastron point. However, when these properties can be fairly well determined from the frequency-curve it is usually easy to determine the period and the radial-velocity curve.

88. Corrections to the Measured Velocity of a Spectroscopic Binary Depending upon the Length of Exposure of the Spectrograms. When a spectroscopic binary is faint and its period short, the length of time needed to secure a spectrogram may consume a large fraction of the period.

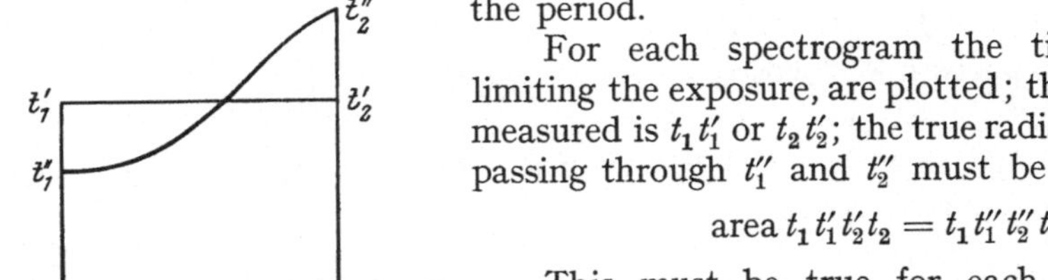

Fig. 18. Correction to velocity when period is short.

For each spectrogram the times t_1 and t_2, limiting the exposure, are plotted; the radial velocity measured is $t_1 t_1'$ or $t_2 t_2'$; the true radial-velocity curve passing through t_1'' and t_2'' must be such that

$$\text{area } t_1 t_1' t_2' t_2 = t_1 t_1'' t_2'' t_2 .$$

This must be true for each spectrogram so that, if a fairly large number of them have been secured it may be possible to determine the radial-velocity curve by the rather difficult method of trial and error.

Short-period spectroscopic binaries have usually circular or nearly circular orbits, and SCHLESINGER treated this simple case[1]. Calling the phase angles at the beginning and end of the exposure u_1 and u_2 respectively, he finds that the measured velocity needs the correction

$$K_1 \cos \frac{u_2 + u_1}{2} \left(1 - \frac{2}{u_2 - u_1} \sin \frac{u_2 - u_1}{2}\right).$$

In order to find the value of K_1 which is still unknown, the observations are plotted in the usual way and from this is derived a first approximation K_1' to the semi-amplitude.

Then
$$K_1 = K_1' \left(1 - \frac{2}{u_2 - u_1} \sin \frac{u_2 - u_1}{2}\right)^{-1},$$

when u_2 and u_1 are such that $\frac{u_2 + u_1}{2} = 0$.

According to the writer, when long exposures are necessary, the best way to obtain a velocity-curve is to secure consecutive observations limited by the successive times $t_1, t_2, t_3, t_4, \ldots$

A curve is then traced having for abscissae $t_1, t_2, t_3, \ldots$ and for ordinates the values of the areas 0; $t_1 t_1' t_2' t_2$; $t_1 t_1' t_2' t_2 + t_2 t_2' t_3' t_3$; $t_1 t_1' t_2' t_2 + t_2 t_2' t_3' t_3 + t_3 t_3' t_4' t_4$; ... areas measured below $t_1 t_2$ being negative.

[1] Ap J 43, p. 167 (1916).

If $y = f(x)$ is the equation of this curve, the equation of the radial-velocity curve will be

$$y = \frac{dy}{dx}.$$

Graphical methods for deriving the second curve from the first are not lacking.

This can be used also in case the brightness of variable stars is under consideration and the method may be developed to include series of observations covering different cycles.

89. True Dimensions of the Orbit. Several cases when the true dimensions of the orbit can be computed are given here:

1. The visual orbit and two radial velocities of the primary are known. The true dimensions of the orbit of the primary around the centre of mass can be determined.

By formula (3) of paragraph 78,

$$V_1 = V_0 + \frac{\mu a_1 \sin i}{\sqrt{1-e^2}} [e \cos\omega + \cos(v_1 + \omega)],$$

$$V_2 = V_0 + \frac{\mu a_1 \sin i}{\sqrt{1-e^2}} [e \cos\omega + \cos(v_2 + \omega)].$$

Subtracting,

$$V_1 - V_2 = \frac{\mu a_1 \sin i}{\sqrt{1-e^2}} [\cos(v_1 + \omega) - \cos(v_2 + \omega)],$$

$$a_1 = \frac{(V_1 - V_2)\sqrt{1-e^2}}{\mu \sin i \,[\cos(v_1+\omega) - \cos(v_2+\omega)]};$$

all the quantities in the second member are known from the visual orbit and from the radial velocities; μ is the mean daily motion.

If two radial velocities of the secondary are known, a similar formula would give the length of the semi-axis major of its orbit.

2. The visual orbit and the relative radial velocity of the primary with reference to the secondary are known. The true dimensions of the relative orbit can then be determined.

Let V' be the relative radial velocity, then

$$V' = \frac{\mu (a_1 + a_2) \sin i}{\sqrt{1-e^2}} [e \cos\omega + \cos(v + \omega)],$$

$$a_1 + a_2 = \frac{V'\sqrt{1-e^2}}{\mu \sin i} \cdot \frac{1}{e\cos\omega + \cos(v+\omega)};$$

this expresses the value of $a_1 + a_2$ in kilometers; from the visual orbit a is the value of $a_1 + a_2$ in seconds of arc.

Thus the parallax of the binary is given by the formula

$$\pi = \frac{a \text{ (in seconds)} \cdot 150000000}{a_1 + a_2 \text{ (in kilometers)}}.$$

150000000 kilometers is the average distance from the Sun to the Earth.

3. The spectroscopic orbit, two values of the angular distance, and two of the position angle are known.

Referring to paragraph 52 we find that

$$\tan(\theta - \Omega) = \pm \tan(v + \omega) \cos i,$$

$$\varrho = \frac{a(1-e^2)}{1 + e\cos v} \cos(v + \omega) \sec(\theta - \Omega),$$

and knowing ϱ_1, θ_1 and θ_2 these equations follow:

$$\tan(\theta_1 - \Omega) = \pm\tan(v_1 + \omega)\cos i\,,$$

$$\tan(\theta_2 - \Omega) = \pm\tan(v_2 + \omega)\cos i\,,$$

$$\varrho_1 = \frac{a(1-e^2)}{1 + e\cos v_1}\cos(v_1 + \omega)\sec(\theta_1 - \Omega)\,,$$

$$\varrho_2 = \frac{a(1-e^2)}{1 + e\cos v_2}\cos(v_2 + \omega)\sec(\theta_2 - \Omega)\,,$$

containing the three unknowns i, Ω and a (in seconds of arc), with ambiguity of sign in the first two equations.

If $(a_1 + a_2)\sin i$ is known from the spectroscopic orbit, the value of the parallax is easily determined.

It may also be possible from measures of proper motion to obtain ϱ_1', the angular distance between the primary and its centre of mass; then a', the angular semi-axis of the orbit of the primary can be found, and the parallax

$$\pi = \frac{a'\,(\text{in seconds}) \cdot 150000000}{a_1\,(\text{in kilometers})}\,.$$

g) The Spectroscopic Binaries of Known Orbits; Typical Systems.

90. Introduction. This subdivision comprises different types of spectroscopic binaries, excluding complex systems such as those with at least three components as well as the Algol, β Lyrae and β Canis Majoris types which are reserved for other subdivisions.

The following systems, classified according to the method of paragraph 77 are treated here:

First class: α Phoenicis, β Capricorni, BD $+66°878$, α Aurigae, ζ Aurigae, 1 Geminorum, Lal. 46867.

Second class: π Andromedae, χ Aurigae, φ Persei.

Third class: θ^2 Orionis, α Virginis, β Aurigae, Boss 5026, BD$+6°1309$.

Fifth class: σ Cygni.

91. The Spectroscopic Binary α Phoenicis. α Phoenicis, $0^h 21^m,3$, $-42°51'$, visual magnitude 2,44 and spectral class K0 was announced to be a spectroscopic binary by WRIGHT[1] in 1905. Its orbit determined by LUNT at the Cape Observatory[2] makes it the longest period spectroscopic binary so far known. The radial-velocity curve is based on fifty-four observations taken between 1903 Sept. 15 and 1914 Dec. 28; these have been grouped into thirteen normal places from which the accompanying curve (Fig. 19) has been constructed.

Elements as obtained from this curve and corrected by the method of least squares are

P	$3848^d,83 \pm 53^d,37$ or $10^y,538 \pm 0^y,146$
T	J. D. $2416201,85 \pm 57^d,62$
ω	$10°,822 \pm 6°,21$
e	$0,335 \pm 0,04$
K_1	$5,76$ km $\pm 0,43$ km
V_0	$+75,21$ km $\pm 0,19$ km
$a_1 \sin i$	290000000 km
$\dfrac{m_2^3 \sin^3 i}{(m_1 + m_2)^2}$	$0,0639\ \odot$.

[1] Lick Bull 3, p. 110 (1905).

[2] Annals Cape Obs 10, Part VII, p. 24G (1924).

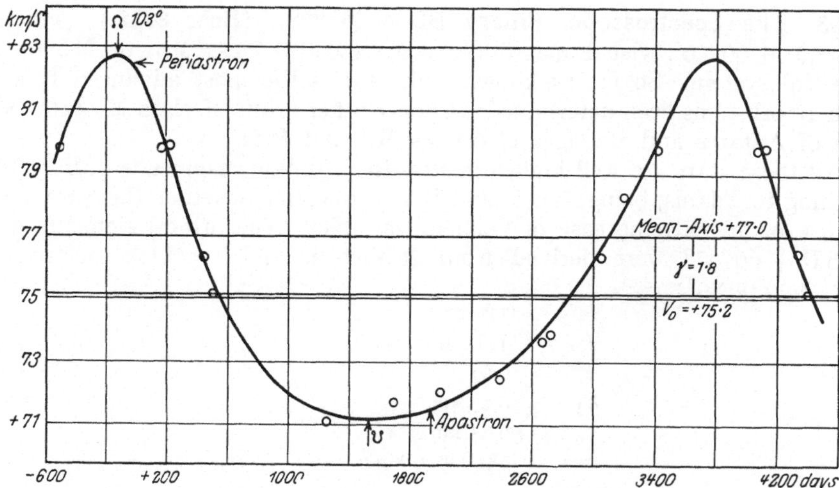

Fig. 19. Velocity-curve of α Phoenicis.

92. The Spectroscopic Binary β Capricorni. The position of this star is $\alpha = 20^h 15^m,4$, $\delta = -15°6'$, visual magnitude 3,25 and spectral class G0. Campbell[1] sponsored this binary in 1899, forty-five three-prism spectrograms being secured between the years 1898 and 1909. The spectrum is composite[2], a secondary spectrum of earlier class having been detected by the Harvard observers; the Mills spectrograms however record only the principal component.

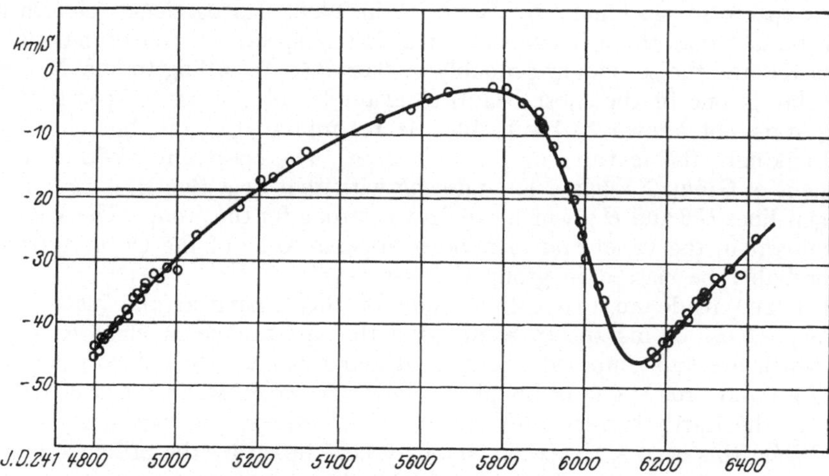

Fig. 20. Velocity-curve of β Capricorni.

The radial-velocity curve in Fig. 20 has been obtained from the Lick radial velocities.

From it, Merrill[3] obtained the following orbital elements:

P	$1375^d,3$	K_1	22,2 km
T	J. D. 2416035	V_0	−18,8 km
ω	124°,0	$a_1 \sin i$	377000000 km
e	0,44	$\dfrac{m_2^3 \sin^3 i}{(m_1 + m_1)^2}$	1,13 $\odot$.

[1] Ap J 10, p. 241 (1899).
[2] Harv Ann 28, p. 93 (1908).
[3] Lick Bull 6, p. 5 (1910).

93. The Spectroscopic Binary BD + 66°878. (Boss 3827.) Position, 14^h56^m,0, +66°20′, visual magnitude 4,86, spectral class Mb, or M6 on the numerical system. So far as known, this star is the most advanced type for which an orbit has been determined; the only other orbits of class M stars being those of Antares and Betelgeuse, classes M2 and M1.

All these stars are well known giants, the absolute magnitude of Boss 3827, according to Adams, being −0,6, and it is a question whether they are simple binaries, or an extreme case of Cepheid variables. The orbital elements given for BD + 66°878 were derived from 26 Victoria and five Lick spectrograms by R. K. YOUNG[1]:

P	750 days
T	J. D. 2422065,0
e	0,0
K_1	6,67 km
V_0	+6,85 km
$a_1 \sin i$	68800000 km
$\frac{m_2^3 \sin^3 i}{(m_1 + m_2)^2}$	0,023 ⊙.

94. The Spectroscopic Binary α Aurigae (Capella). It is only necessary to mention this unusual spectroscopic binary as one of the few bright stars of late spectral class exhibiting double lines. For a full discussion read paragraph 69.

95. The Spectroscopic Binary ζ Aurigae. Position 4^h55^m,5, +40°56′, visual magnitude 3,94, spectral class K0 (composite). This star was discovered to be a spectroscopic binary by WRIGHT[2] in 1908. Its particular significance lies in the fact that there is evidence of a double spectrum, the primary being of class K0 and the secondary probably of class B5. According to Miss MAURY[3]: "This star is one of the most beautiful examples of composite spectra found among stars not known to be double. It resembles 31 Cygni, but shows still more strikingly the features of the first type. The spectrum predominant in the blue is of Group XV (class K), agreeing with that of α Boötis, except in the hydrogen lines $H\beta$ and $H\gamma$ which are far too strong for the group. The spectrum predominant in the violet and ultraviolet appears to be of the Orion type and may probably be classed in group V (class B5)".

HARPER[4] made an exhaustive study of this binary at the Victoria Observatory. Most of his spectrograms give the appearance of class K spectra upon which are superimposed nearly continuous bands; their absorption lines are diffuse and broad; one of his plates, however, taken when the primary was closest to the Earth is very different from the others, showing sharp intense lines. When this occurs, the secondary may be eclipsed by the primary, leaving a single spectrum of class K; however, the chances of an eclipse in an orbit of such dimensions are few.

BOTTLINGER has called attention to the possibility of investigating this with the photo-electric cell[5].

The spectroscopic absolute magnitude obtained from this last mentioned spectrogram is −0,5, corresponding to a parallax of 0″,013.

In addition to spectrograms taken at Lick, Bonn, Mt. Wilson and Ottawa, twenty-eight plates are listed at Victoria. From the latter, after

[1] Pop Astr 32, p. 628 (1924).
[2] Lick Bull 5, p. 62 (1908).
[3] Harv Ann 28, p. 99 (1897).
[4] Publ Dom Astroph Obs 3, p. 151 (1924).
[5] A N 226, p. 239 (1926).

correction by a least-square solution, HARPER found the following orbital elements:

P	973 days
T	J. D. 2415122,471 $\pm$ 4,454 days
ω	$330°,13 \pm 2°,04$
e	$0,411 \pm 0,011$
K_1	23,78 km $\pm$ 0,32 km
V_0	+10,73 km $\pm$ 0,32 km
$a_1 \sin i$	294300000 km
$\frac{m_2^3 \sin^3 i}{(m_1 + m_2)^2}$	1,030 $\odot$.

The graph (Fig. 21) shows the radial-velocity curve with the Victoria observations grouped into normal places. The probable error of a velocity is $\pm$ 1,0 km.

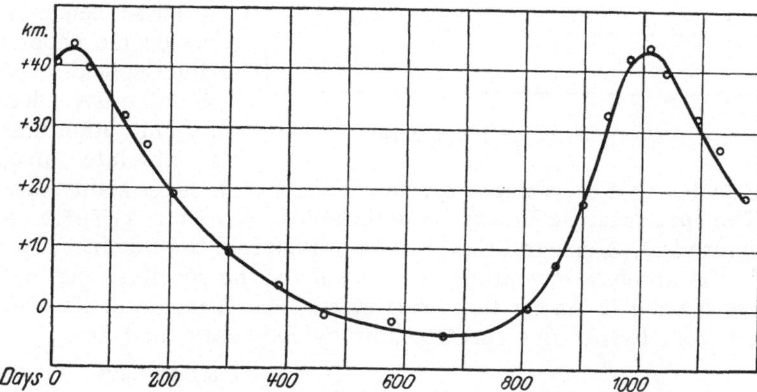

Fig. 21. Velocity-curve of ζ Aurigae.

96. The Spectroscopic Binary 1 Geminorum. The position of this star is $\alpha = 5^h58^m,0$, $\delta = +23°16'$, its visual magnitude 4,30, and spectral class G5. Discovered by MOORE to be a binary[1], the study of its orbit has been published by R. K. YOUNG[2]. Its period being less than ten days, it belongs to a class of short-period late type stars of which very few representatives are known, such as λ Andromedae, σ Geminorum, Lal. 46867 and Lal. 29330. I have shown definitely that the centre-of-mass velocities of λ Andromedae and σ Geminorum vary; such may be the case also for 1 Geminorum, since YOUNG says:

"The determination of the period has proved to be difficult. Fifty-one observations have been secured here (Victoria) and these are in harmony with a period of 9,590 days. This same period will satisfy the Ottawa observations in 1917 very well and the Bonn observations in 1909 and 1912. The residuals are somewhat larger than one would be led to expect from the fine quality of the spectrum but on the whole they indicate that the period 9,590 days is satisfactory. The Lick observations taken from 1903 to 1906 could not be made to agree, and it would seem that there are peculiarities in the orbit not yet explained. The double amplitude of the curve is twenty-three kilometers only, and the irregularities in the curve are of the order of five kilometers or a little over, so that a complete investigation with one-prism dispersion would prove a difficult task. Three-prism dispersion would be much better, but until such are available the present elements serve as a good approximation to give the general form of the curve."

[1] Lick Bull 4, p. 96 (1906). [2] Publ Dom Astroph Obs 1, p. 119 (1919).

The orbital elements corrected by a least-square solution are:

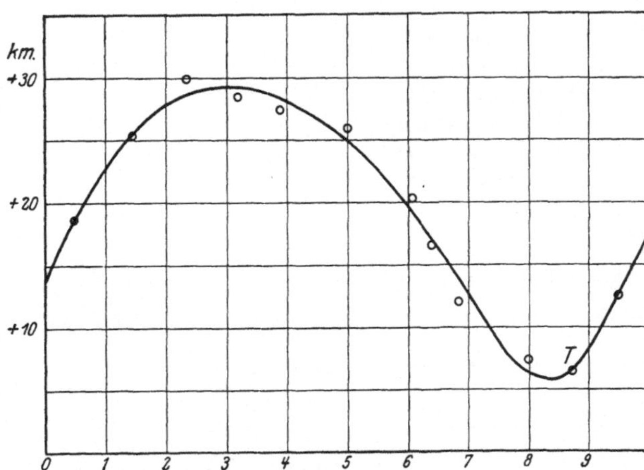

Fig. 22. Velocity-curve of 1 Geminorum.

P	9,590 days
T	J. D. 2421898,741 $\pm$ 0,196 day
ω	203°,28 $\pm$ 6°,80
e	0,2065 $\pm$ 0,0344
K_1	11,74 km $\pm$ 0,33 km
V_0	+19,71 km $\pm$ 0,22 km
$a_1 \sin i$	1 510000 km
$\frac{m_2^3 \sin^3 i}{(m_1 + m_2)^2}$	0,0015 $\odot$.

The radial-velocity curve is given in Fig. 22. These elements are to some degree characteristic of the Cepheids. However, 1 Geminorum does not vary in brightness, and its absolute magnitude +2,82 indicates that it is not a typical Cepheid or a super-giant star.

97. The Spectroscopic Binary Lalande 46867. This star, $23^h 50^m,0$, +28°5′, visual magnitude 7,30, spectral class K2, was discovered to be a binary by ADAMS and JOY[1]. Its absolute magnitude, as derived by the spectroscopic method is +6,0, thus placing it among the dwarf stars. The elements of its orbit were determined from twenty-five spectrograms by SANFORD[2], and are:

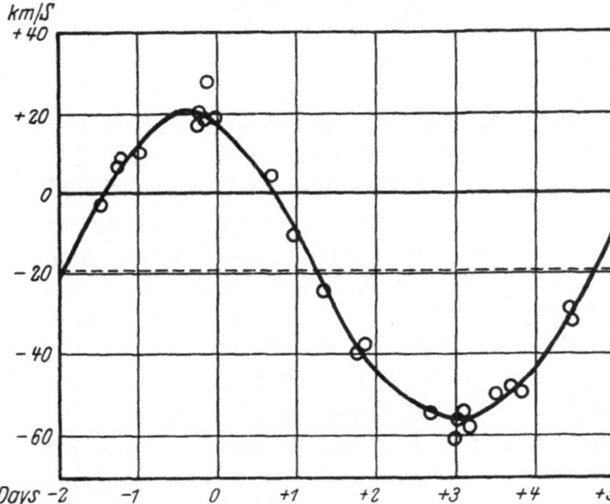

Fig. 23. Velocity-curve of Lalande 46867.

P	6,7217 days
T	J.D. 2422220,7403 $\pm$ 0d,3005
ω	18°,6 $\pm$ 16°,3
e	0,059 $\pm$ 0,019
K_1	38,50 km $\pm$ 0,76 km
V_0	−19,8 km
$a_1 \sin i$	3 552 300 km
$\frac{m_2^3 \sin^3 i}{(m_1 + m_2)^2}$	0,0125 $\odot$.

The radial-velocity curve is given in Fig. 23.

A remarkable feature about Lal. 46867 is that it shows sharp H and K calcium emission lines superimposed on the usual broad absorption bands and yielding the same velocities as the lines of the absorption spectrum. The latter seem to be less sharp than usual for spectral class K2; also this is the case for σ Geminorum, in which SCHWARZSCHILD[3] observed H and K emission lines on objective-prism spectrograms. On a spectrogram taken by DUNCAN at Mt. Wilson the emission lines were found, but weaker than in Lal. 46867; the velocity recorded from these was in close agreement with that obtained from the absorption lines. The spectroscopic absolute magnitude of σ Geminorum is + 1,1, while for Lal. 46867 it is +6,0, and it is a phenomenon worthy of

[1] Publ A S P 31, p. 42 (1919). [2] Ap J 53, p. 221 (1921).
[3] Ap J 38, p. 292 (1913).

further investigation in other stars, that the emission lines H and K are stronger in the absolutely fainter star.

98. The Spectroscopic Binary π Andromedae. Position $0^h 31^m,5$, $+33°10'$, visual magnitude 4,44, spectral class B3. The discovery of the variable radial velocity of this star is due to FROST and ADAMS[1]. At the Allegheny Observatory the orbital elements were computed by Dr. JORDAN[2] from one hundred and eleven plates secured with the MELLON spectrograph.

According to JORDAN: "The spectrum has all lines, except the hydrogen, sharp and well measurable. A very decided shading which is present on the violet side of the hydrogen lines at the epoch of high positive velocity may indicate the presence of a secondary spectrum, but no distinct lines can be detected, nor do any of the other lines show this effect. The nucleus of the hydrogen lines is well defined, so this shading has probably but slight effect upon the measures, though it is somewhat noticeable in the case of the $H\delta$ line,

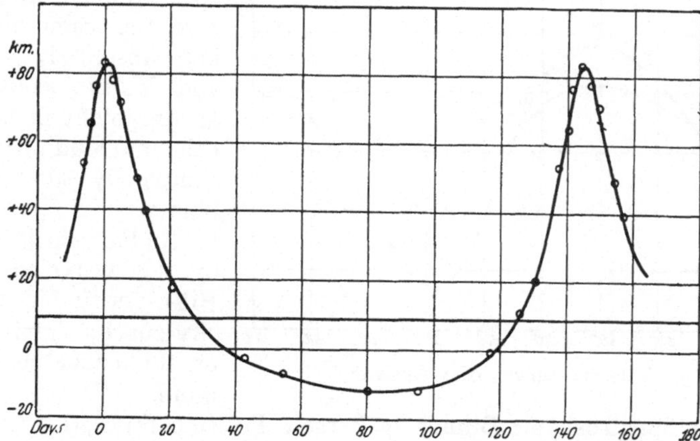

Fig. 24. Velocity-curve of π Andromedae.

which at these epochs gives on the average a little smaller velocity than that shown by the other lines. An examination of the measures to see what effect the omission of the hydrogen lines would have, showed that the orbit would remain sensibly the same." The elements of the orbit after correction by a least-square solution are:

P	143,67 days $\pm$ 0,02 day
T	J. D. 2418564,144 $\pm$ 0,171 day
ω	350°,53 $\pm$ 0°,90
e	0,573 $\pm$ 0,006
K_1	47,66 km $\pm$ 0,36 km
V_0	+8,83 km
$a_1 \sin i$	77200000 km
$\frac{m_2^3 \sin^3 i}{(m_1+m_2)^2}$	0,8894 $\odot$.

The radial-velocity curve is given in Fig. 24.

99. The Spectroscopic Binary χ Aurigae. Position $5^h 26^m,2$, $+32°7'$, visual magnitude 4,88, spectral class B1. The distinguishing characteristic of this star as a binary was discovered by FROST and ADAMS[3] in 1903. An orbit was

[1] Ap J 18, p. 384 (1903). [2] Allegh Obs Publ 2, p. 45 (1910).
[3] Ap J 18, p. 383 (1903).

determined by R. K. YOUNG from 88 single-prism spectrograms taken at the Ottawa Observatory during 1913, 14, 15 and 16; the elements found are:

P	655,16 days ± 5,26 days
T	J. D. 2420629,78 ± 9,56 days
ω	135°,52 ± 5°,2
e	0,171 ± 0,026
K_1	20,53 km ± 0,57 km
V_0	−0,15 km ± 0,35 km
$a_1 \sin i$	182300000 km
$\dfrac{m_2^3 \sin^3 i}{(m_1 + m_2)^2}$	0,56 ⊙.

The velocity-curve is given in Fig. 25. The curve of smaller amplitude is derived from the H and K lines of calcium alone. Its amplitude is about one-half that for the other spectral lines; in the present case the calcium lines are not absolutely stationary, and YOUNG considers this an argument in favour of a calcium cloud of vast extent completely surrounding the binary.

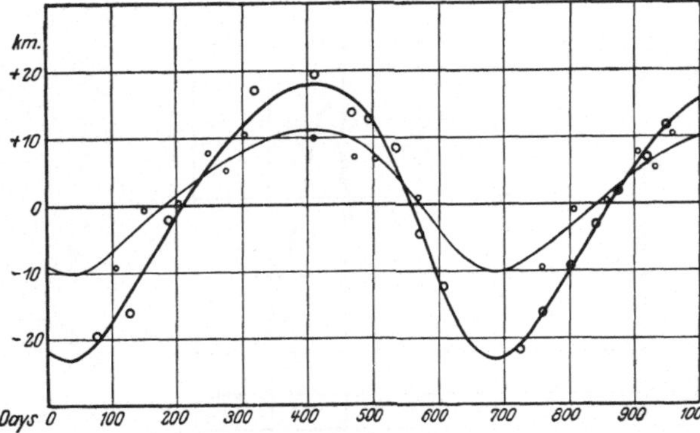

Fig. 25. Velocity-curves of χ Aurigae.

Radial velocities obtained by the writer in December 1920 verify YOUNG's velocity-curve, giving points on this curve near its minimum.

100. The Spectroscopic Binary φ Persei. Position $1^h37^m,4$, $+5°11'$, visual magnitude 4,19, spectral class B0p. This star offers an intricate problem to the astrophysicist. Important studies have been made of it by CANNON[1], LUDENDORFF[2], JORDAN[3], GUTHNICK[4], LOCKYER[5] and FREDETTE[6]. The Dominion Observatory is still at work securing numerous slit spectrograms; the Lockyer Observatory is taking prism-camera spectrograms, and Miss GÜSSOW is continuing its study with the photo-electric photometer at the Berlin-Babelsberg Observatory. From a survey of the Dominion Observatory spectrograms, the character of the spectrum is found to be essentially as follows:

1. A certain number of wide diffuse absorption lines, at times very difficult to distinguish from the continuous spectrum, and some narrow diffuse lines. The hydrogen lines are especially wide, gradually decreasing in intensity toward their edges.

2. Strong wide emission lines of hydrogen superimposed on much wider absorption bands. The emission $H\beta$ is very strong, $H\gamma$ weaker, and other emission lines of the BALMER series are much weaker or non-existent. Periodically these emission lines vary in intensity. Occasionally there are wide faint emission lines of other elements superimposed on broader absorption bands, but these are difficult to see.

---

[1] Report of the Chief Astr Dom Obs 1, p. 150 (1910); see also J Can R A S 4, p. 195 (1910).
[2] A N 186, p. 17 (1910), 192 p. 173 (1912), 197 p. 219 (1914).
[3] Allegh Obs Publ 3, p. 31 (1913).
[4] Veröff. der K. Sternwarte Berlin-Babelsberg 2, Heft 3, p. 87 (1918).
[5] M N 85, p. 580 (1925). [6] J Can R A S 19, p. 185 (1925).

3. Sharp, narrow absorption lines of hydrogen divide the emission in two; sometimes the parts are symmetrical, but frequently unsymmetrical.

The above statements are confirmed by JORDAN's description of his slit spectrograms. Using 112 of these, he measured the radial velocities from the sharp absorption lines of hydrogen as well as from the emission lines. His results are shown graphically in Fig. 26.

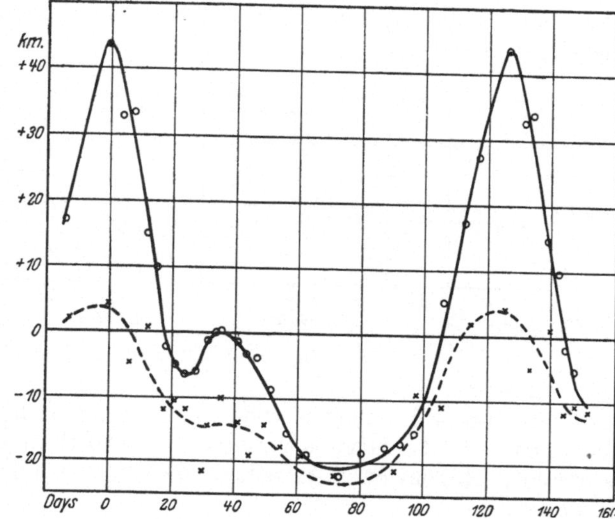

Fig. 26. Curves of absorption and emission lines of φ Persei.

The spectrograms taken by LOCKYER are superior for an analysis of characteristic lines; using a 9 inch camera of 120 inches focal length and a prism of 45° angle, he obtains a dispersion of 46,0 mm between $H\beta$ and $H\delta$. Besides the hydrogen emission lines, he finds numerous other bright lines due chiefly to ionized iron. These appear double, the relative intensities of the components of all lines being alike. Absorption lines besides hydrogen and λ 4481,3 of ionized magnesium are due to the singlet and doublet systems of helium; at different phases of the velocity variation the helium lines become alternately nebulous and sharp.

Measuring the intensities of the central absorption lines of hydrogen, LOCKYER pointed out the fact that $H\zeta$ apparently did not change in intensity, while the other lines varied; Fig. 27 shows the curves of intensity differences $H\zeta - H\varepsilon$ and $H\zeta - H\delta$. While the individual measures, represented by small circles, are connected by short lines, an approximate smooth mean curve has been drawn to indicate the larger variation. There is, no doubt, a short-period change with a rather large amplitude superimposed on the primary variation. According to LOCKYER the only regular secondary variation resulting from the measures is one having a period of 21 days, or equal in length to a sixth of the primary period. Such a short-period change is indicated in the lower curve of Fig. 27.

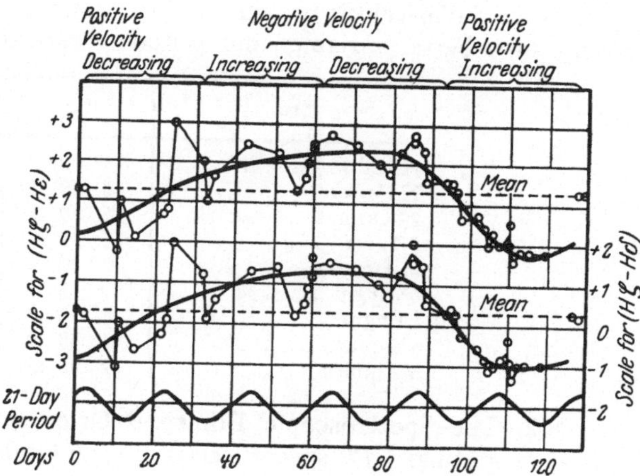

Fig. 27. φ Persei. Variation of hydrogen absorption.

LOCKYER's results are paralleled by the data of GUTHNICK and PRAGER; the light-curve obtained with the photo-electric cell is formed of a series of equidistant waves, each having an approximate period one seventh that of the spectroscopic period. Sometimes one of these waves collapses into a straight line, the light of the star being constant; the length of time separating two collapsed waves being equal to the longer period. GUTHNICK and PRAGER studied φ Persei for years and from observations between September 1916 and January 1917, they have published a light-curve (Fig. 28) which represents the peculiar variation mentioned above.

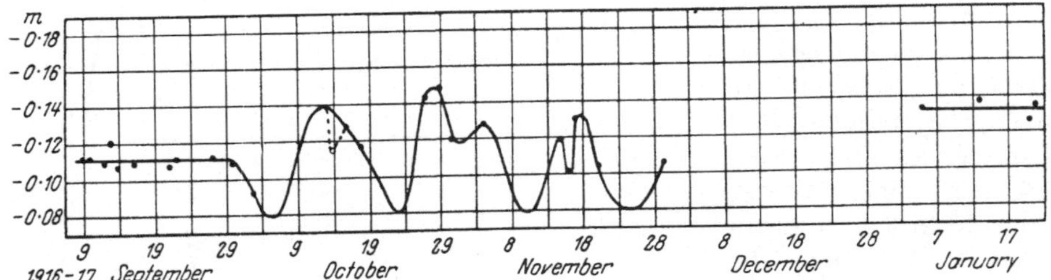

Fig. 28. Light-curve of φ Persei.

This star still offers a wide field for research; LUDENDORFF found that the radial-velocity curve is variable and also that the spectrum of a secondary is present. This latter observation, however, has not yet been verified by other observers.

101. Stars of the φ Persei Type. There are stars which resemble φ Persei in behaviour. They belong to a class called Be, the symbol e meaning emission, in which hydrogen and sometimes other elements as well, produce emission lines. These stars, of which more than ninety examples exist, have been studied particularly by MERRILL at the Mt. Wilson Observatory[1].

CURTISS[2] has found about eleven φ Persei type stars characterized by a difference in the strength and especially in the relative strengths of the two components of hydrogen emission lines. Most of them seem to have much longer periods of radial-velocity variation than φ Persei, amounting to several years. A list of φ Persei variables and periods found at Ann Arbor, follows:

Star	R. A.	Period
H. D. 20336	$3^h\ 11^m,2$	1690 days
X Persei	3 49 ,1	Several years
25 Orionis	5 19 ,5	1875 days
ζ Tauri	5 31 ,7	15 ± years
$\varkappa_2$ Draconis	12 29 ,2	4000 days
b^2 Cygni	20 5 ,7	1373 days
v Cygni	21 13 ,8	1800 days
o Aquarii	21 58 ,1	1900 ± days
π Aquarii	22 20 ,2	2000 ± days

102. The Spectroscopic Binary θ^2 Orionis. Position $5^h30^m,5$, $-5°29'$, visual magnitude 5,17, spectral class B1; θ^2 Orionis is one of the brighter stars located in the Orion nebula. Its variable velocity was announced by FROST and ADAMS[3].

A peculiar property noticed by the writer is the appearance in the spectrum, at indefinite intervals, of bright hydrogen lines[4].

[1] Ap J 61, p. 389 (1925). [2] Pop Astr 33, p. 537 (1925). [3] Ap J 19, p. 153 (1904).
[4] Publ Dom Obs 5, p. 19 (1920); see also C R 176, p. 1210 (1923).

The orbital elements were derived by OTTO STRUVE from the measurements of thirty-nine plates obtained with the BRUCE spectrograph of the Yerkes Observatory[1]. They are:

P	21,029 days
T	J. D. 2423741,362 ± 0,314 day
ω	154°,7 ± 3°,2
e	0,27 ± 0,02
K_1	93,7 km ± 3,6 km
V_0	+36,8 km
$a_1 \sin i$	27000000 km
$\frac{m_2^3 \sin^3 i}{(m_1+m_2)^2}$	1,795 ⊙.

The radial-velocity curve is given in Fig. 29.

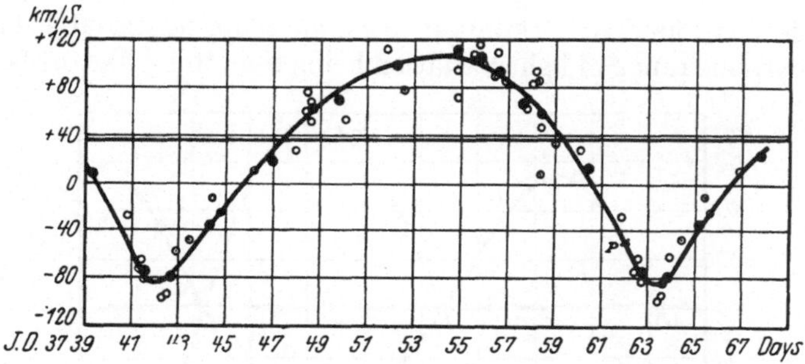

Fig. 29. Velocity-curve of 43 θ^2 Orionis.

During a twenty year interval there has been no appreciable change in the period; therefore it is evident that should the star be plunged in the Orion nebula, the resistance of the nebulous gases is extremely slight.

The calcium lines as well as the emission lines of hydrogen do not oscillate; their mean velocity is the same as that of the nebula, justifying the theory that they are produced by its gases. This mean velocity is about 20 km less than the centre-of-mass velocity of the binary. The gases producing the hydrogen emission and the detached calcium lines in θ^2 Orionis are nearly at rest with respect to the system of stars from which the motion of the solar system has been derived.

Since the bright lines do not participate in the large periodic oscillation of the absorption lines, they are at times superimposed upon the continuous spectrum and are rendered invisible on account of their comparative dimness.

103. The Spectroscopic Binary α Virginis. Position $13^h19^m,9$, $-10°38'$, visual magnitude 1,21, spectral class B2. The bright star, Spica, was one of the first to be known as a spectroscopic binary. In 1890 it was discovered by VOGEL[2] who, assuming the orbit to be circular, determined its dimensions from 29 spectrograms taken at Potsdam between April 1889 and May 1891; they are:

P	4,0134 days
V_0	−15 km
K_1	91 km.

At the Allegheny Observatory there are eighty-three spectrograms of α Virginis on fine grain plates. They show the presence of two components in the

[1] Ap J 60, p. 159 (1924). [2] A N 125, p. 305 (1890).

spectra. The following orbital elements[1] and radial-velocity curves (Fig. 30) are BAKER's work:

P	4,01416 days $\pm$ 0,00001 day
T	J. D. 2417955,846 $\pm$ 0,054 day
ω_1	328° $\pm$ 5°,3
ω_2	148° $\pm$ 5°,3
e	0,10 $\pm$ 0,014
K_1	126,1 km $\pm$ 1,4 km
K_2	207,8 km $\pm$ 1,6 km
V_0	+1,6 km
$a_1 \sin i$	6930000 km
$a_2 \sin i$	11400000 km
$m_1 \sin^3 i$	9,6 $\odot$
$m_2 \sin^3 i$	5,8 $\odot$.

α Virginis was found by STEBBINS[2] to be an eclipsing variable having two minima, the extreme range of light-variation being something like 0,10 magnitude.

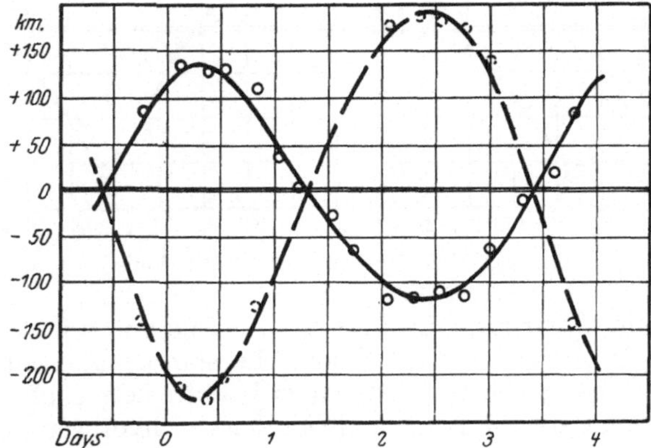

Fig. 30. Velocity-curves of α Virginis.

104. The Spectroscopic Binary β Aurigae. Position $5^h 52^m,2$, $+44°56'$, visual magnitude 2,07, spectral class A0p.

To Miss MAURY[3] is due the credit for discovering this binary, one of the first known.

A complete analysis of this interesting star was made by BAKER[4] in 1910, and this elaborate study includes many details necessary to the reader. The orbital elements derived from the two components visible on his spectrograms are:

P	$3^d,960027 + 0^d,000010\,t$ (variable period) $\pm 0^d,000004$
Time of passage through ascending node	J. D. 2417100,7324 $\pm$ 0,0024 day
e	0,00 $\pm$ 0,0057
K_1	108,96 km $\pm$ 0,50 km
K_2	111,04 km $\pm$ 0,55 km
V_0	−18,1 km $\pm$ 1,0 km
$a_1 \sin i$	5934000 km
$a_2 \sin i$	6047000 km
$m_1 \sin^3 i$	2,21 $\odot$
$m_2 \sin^3 i$	2,17 $\odot$.

[1] Allegh Obs Publ 1, p. 65 (1909). [2] Ap J 39, p. 475 (1914).
[3] Fourth Report of the Draper Memorial, 1890. [4] Allegh Obs Publ 1, p. 163 (1910).

The velocity-curves of the two components are given in Fig. 31.

Considering the tremendous range of velocity variation and the very short period, β Aurigae is a remarkable star in the fact that the spectral lines are narrow and sharp; as seen from the above elements the probable errors of K_1 and K_2 are very small. β Aurigae is thus of first importance for testing light dispersion in space; its parallax being small, less than 0″,02, if the red rays travel faster through space than the blue rays, lines near the red end of the spectrum would announce the variation in velocity more readily than rays near the violet end.

Previous to BAKER's work two studies on β Aurigae, with positive results indicating dispersion in space, were published. They were:

G. A. TIKHOV[1], Deux méthodes de recherche de la dispersion des espaces célestes (1908).

A. BÉLOPOLSKY[2], Untersuchungen über die radialen Geschwindigkeiten von β Aurigae in Beziehung zur Frage über die Dispersion im Weltenraum (1909).

BAKER's research in 1910, however, seemed to question the reality of such a dispersion or difference in speed through space of different light-waves.

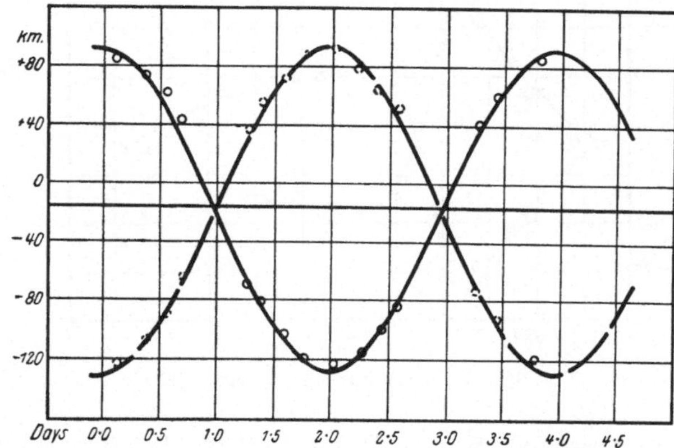

Fig. 31. Velocity-curves of β Aurigae.

A very thorough investigation was made in 1924 by TIKHOV who afterwards published[3] "Les orbites spectrales de l'étoile β Aurigae calculées séparément d'après les raies différentes", in which he analyzed 222 spectrograms obtained by BÉLOPOLSKY at Pulkowa between 1902 and 1909 and also the 46 spectrograms secured at the Allegheny Observatory (1908—1909), formerly used by BAKER.

The spectral lines λ 4550, 4481, 4352, 4078, 4046, and 3934 were measured, and orbits determined separately for each of them. Except in the case of 3934, K line of calcium which shows individual discrepancy, the other lines indicate a shift in time of one velocity-curve with respect to another, amounting to an approximate retardation of 15 minutes for 50 $\mu\mu$ toward the violet.

105. The Spectroscopic Binary Boss 5026. Position $19^h 36^m,4$, $+54°44'$, visual magnitude 5,86, spectral class F5.

From the presence of double lines on one plate HARPER deduced this star to be a binary. He determined its orbit from 25 spectrograms[4], proving it to

[1] Mitt. der Nikolai-Hauptsternwarte zu Pulkowo 2, p. 141 (1908).
[2] Mitt. der Nikolai-Hauptsternwarte zu Pulkowo 3, p. 91 (1909).
[3] Bull. de l'Obs. central de Russie à Poulkovo 10, p. 205 (1924).
[4] Publ Dom Astroph Obs 1, p. 161 (1919).

be an extraordinary system on account of its short period and unusually large eccentricity. Stars of similar periods commonly have small eccentricities. The orbital elements obtained are:

P	7,6383 days ± 0,0019 day
T	J. D. 2422201,398 ± 0,007 day
ω_1	46°,74 ± 0°,81
ω_2	226°,74 ± 0°,81
e	0,527 ± 0,006
K_1	89,91 km ± 0,69 km
K_2	91,12 km ± 0,71 km
V_0	−15,59 km ± 0,36 km
$a_1 \sin i$	8017000 km
$a_2 \sin i$	8134000 km
$m_1 \sin^3 i$	1,854 ⊙
$m_2 \sin^3 i$	1,827 ⊙.

The velocity-curves are given in Fig. 32.

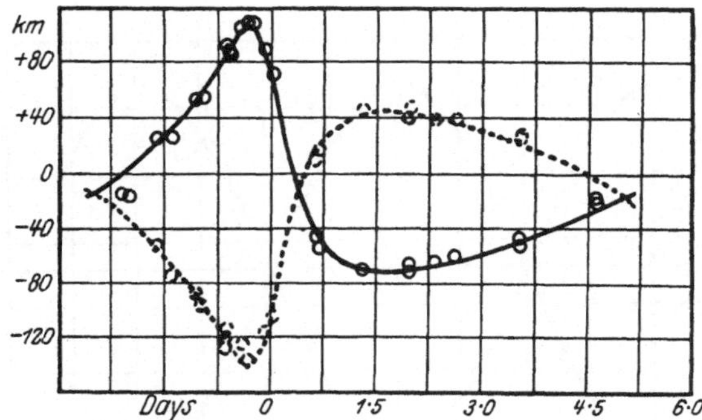

Fig. 32. Velocity-curves of Boss 5026.

106. The Spectroscopic Binary BD + 6°1309. Position $6^h32^m,0$, $+6°13'$, visual magnitude 6,06, spectral class midway between Oe and Oe5.

Investigated by J. S. PLASKETT[1], this binary system proved to be a very massive one.

The orbit was satisfactorily determined from thirty spectrograms, the binary character of the star being discovered from the first of these plates on which the spectra of both components appear.

The orbital elements obtained are:

P	14,414 days ± 0,0157 day
T	J. D. 2423031,870
ω_1	181°,95 ± 2°,54
ω_2	1°,95 ± 2°,54
e	0,0349 ± 0,0114
K_1	206,38 km ± 1,95 km
K_2	246,7 km
V_0	+23,94 km ± 1,4 km
$a_1 \sin i$	40880000 km
$a_2 \sin i$	48870000 km
$m_1 \sin^3 i$	75,6 ⊙
$m_2 \sin^3 i$	63,3 ⊙.

[1] Publ Dom Astroph Obs 2, p. 147 (1922); see also M N 82, p. 447 (1922).

Velocity of Ca lines $+16,0$ km $\pm$ 0,25 km.

The radial-velocity curve is seen in Fig. 33.

$\sin^3 i$ being smaller than unity, it follows that the mass of each component is probably more than one hundred times the mass of the Sun. This feature is especially significant when one remembers that bodies having masses ten times that of the Sun were formerly considered the largest among the stars.

If one of the components of BD $+ 6°1309$ has a mass 75,6 times and a density 0,01 times that of the Sun, a surface brightness $-4,0$, its absolute magnitude would be $-5,65$; this, with a visual magnitude 6,71 corresponds to a parallax 0″,00035, indicating a distance of about 10000 light-years.

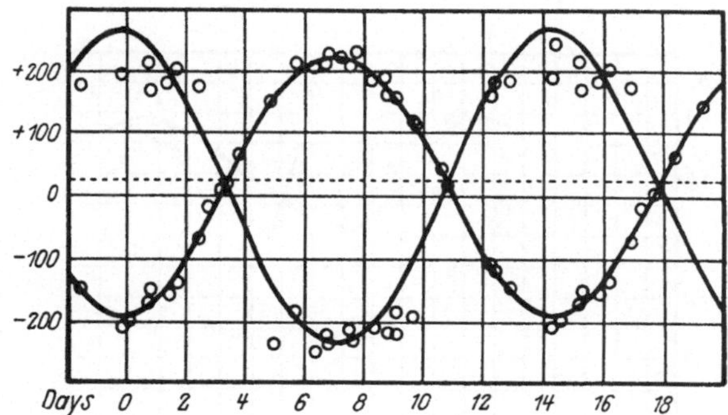

Fig. 33. Velocity-curves of BD 6° 1309.

107. The Spectroscopic Binary σ Cygni. Position $21^h13^m,5$, $+38°59'$, visual magnitude 4,28, spectral class B 8 p. Evidently this body belongs to the class of super-giant stars such as the Cepheids and the Pseudo-Cepheids. Absorption lines of ionized iron and other elements, signifying a very small density, are found in its spectrum.

Discovered by CAMPBELL to be a spectroscopic binary[1], the orbital elements were computed, by the writer[2], from one hundred spectrograms and are:

P	11,043 days $\pm$ 0,042 day
T	J. D. 2421069,27 $\pm$ 0,34 day
ω	119° 6′ $\pm$ 14° 46′
e	0,40 $\pm$ 0,08
K_1	1,98 km $\pm$ 0,186 km
V_0	$-3,80$ km
$a_1 \sin i$	275700 km
$\dfrac{m_2^3 \sin^3 i}{(m_1+m_2)^2}$	0,000007 ⊙.

Velocities derived from the hydrogen line $H\varepsilon$ do not agree with those from other lines. The curve deduced from the $H\varepsilon$ line resembles roughly the general velocity-curve, supposing the latter to be shifted about 5,5 days along the time axis and displaced about ten kilometers upward along the velocity axis.

The unprecedented behaviour of the $H\varepsilon$ line may be explained by anomalous dispersion through an interstellar calcium cloud unless it is due merely to the photographic effect of a faint variable emission H line of calcium close to $H\varepsilon$. The exact cause of this phenomenon demands continued research.

[1] Lick Bull 5, p. 176 (1909). [2] Publ Obs Univ Michigan 3, p. 39 (1917).

In 1920, Prof. GUTHNICK, who had no knowledge of the existence of the star's orbit, found with his photo-electric cell that σ Cygni was slightly variable[1]. This is Prof. GUTHNICK's report: "σ Cygni, a spectroscopic binary with still unknown period, has been found to be variable. The amplitude of variation of magnitude is $0^m,05$. The measures made to date indicate Algol type. The star was at or very near minimum brightness on October 11, 12, 13 and again on October 30, 31; it was of normal brightness on October 20, 21, 22, 28, 29."

Considering σ Cygni as an Algol variable, the duration of the eclipse being two or three days and assuming the mass of the secondary to be between plausible limits, the writer[2] computed the possible dimensions of the system. For all chosen values of m_2 the resulting density of the primary changes very

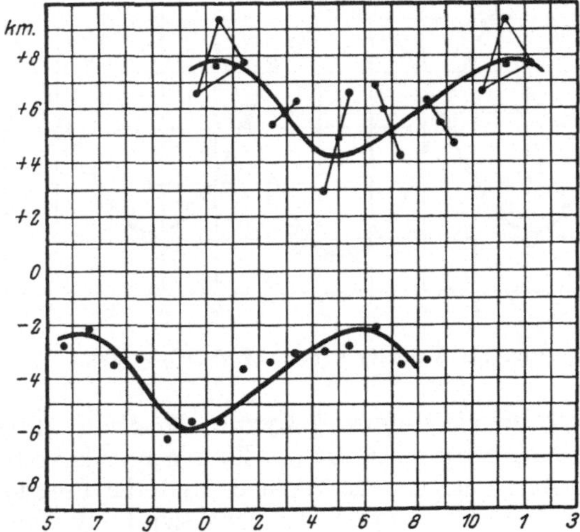

Fig. 34. Velocity-curves of σ Cygni.

little and may be stated as 0,00033 that of the Sun. The variation in the diameter of the primary is between 10,3 and 46,6 times that of the Sun, and knowing this diameter, the parallax of the system may be found by the formula BOTTLINGER[3] used, which is:

$$\log 1000\pi = 2{,}25 - 0{,}2\,m' + 2\log\left(\frac{c_2}{T}\right) - \log\varrho\,,$$

where π is the parallax, m' the bolometric magnitude, $\frac{c_2}{T}$ a value given in BOTTLINGER's catalogue, and ϱ the diameter of the star, that of the Sun being unity.

For the assumed masses of the secondary the value of the parallax lies between 0'',000664 and 0'',000008. The trigonometric parallax as determined at the Allegheny Observatory is

$$-0'',001 \pm 0'',009\,,$$

proving that it is extremely small although still practically unknown.

[1] A N 212, p. 167 (1920).
[2] J Can R A S 18, p. 339 (1924).
[3] Veröff. der Univ.-Sternwarte zu Berlin-Babelsberg 3, Heft 4 (1923).

108. The Variation of Radial Velocity in Super-Giant Stars. σ Cygni is a super-giant star, by this meaning that it has an exceedingly small density as shown by the narrow sharp lines and the large number of them belonging to ionized elements. Such is the case also for α Cygni, ε Aurigae, γ Cygni and other stars called by W. S. ADAMS Pseudo-Cepheids because their spectra are similar to those of the Cepheids.

The majority of these Pseudo-Cepheids do not exhibit variations of radial velocity, and when they do show some change, as in the case of σ Cygni, it amounts to a few kilometers per second only. Similar conditions are recorded by LEE[1] for α Cygni. Many spectrograms of this star have been secured at Potsdam and at Ottawa but, so far, the curve of variation seems to be irregular and no period has been found.

Apparently a triple system, ε Aurigae presents to the astronomer unparalleled features of interest. A complete description is contained in Subdivision k.

As a rule, stars highly ionized and of very low density are Cepheids when the variation of radial velocity is extreme.

On the contrary, for small amplitudes, the secondary component is likely to have a rather diminutive mass; for instance, the writer computed a mass ratio of about 1/100 for σ Cygni, the secondary bearing the same relation to the primary as a huge planet to the Sun.

The suggestion follows that in cases similar to α Cygni, where the velocity-curve is irregular, there are perhaps two or more companions, a series of planet-like bodies having much smaller masses than the primary itself.

109. Dr. MOORE's Table of Orbital Elements. A table of all determined orbital elements of spectroscopic binaries has been published by Dr. MOORE at the Lick Observatory[2]. Considering its considerable importance it is reproduced here with some abbreviations and corrections. We omit the Cepheids which are treated in the chapter on variable stars, and some erroneous or uncertain orbits (Boss 1275, α Ori, γ Gem, m Vel, τ Lup, α Sco). Out of several determinations of the orbit of the same star we give, as a rule, only one (in most cases the last one).

Under the heading f the quantity $\frac{m_2^3 \sin^3 i}{(m_1 + m_2)^2}$ is given; but if the spectra of both components of the system are measured, we give under f the quantities $m_1 \sin^3 i$ and $m_2 \sin^3 i$ (put into brackets).

Under the heading "Reference" the following abbreviations not used elsewhere in this work are introduced:

A O = Publications of the Allegheny Observatory.
C A = Annals of the Cape Observatory.
D A O = Publications of the Dominion Astrophysical Observatory Victoria.
D O = Publications of the Dominion Observatory Ottawa.
Det O = Publications of the Astronomical Observatory of the University of Michigan (Detroit Observatory).
L O B = Lick Observatory Bulletins.

Since the appearance of Dr. MOORE's catalogue in 1924, the orbits of many other spectroscopic binaries have been determined; a list of them will appear in Vierteljahrsschrift der Astronomischen Gesellschaft 1927, Heft 4.

[1] Ap J 31, p. 177 (1910).
[2] Lick Bull 11, p. 166 (1924).

Spectroscopic

Boss	Star	R. A. 1900	Decl.	Mag.	Class	P d	T Jul. Day 2410000 +	ω	e	K_1 km
10	α And	0h3m,2	28°32′	2,1	A0p	96,67	7882,40	76°,2	0,525	30,7
46	....	12,4	50 53	6,1	B3	3,5225	11231,86	323,0	0,094	217,4
....	TV Cas	13,9	58 35	var	B9	1,812635	−0,0045*	......	0	87,9
										150
78	α Phe	21,3	−42 51	2,4	K0	3848,83	6201,85	19,8	0,335	5,8
82	....	22,8	43 50	5,1	A2	3,95583	8841,59	233,2	0,152	41,7
116	13 Cet	30,1	− 4 9	5,2	G0	2,0818	7484,49	222,7	0,09	36,4
123	π And	31,5	33 10	4,4	B3	143,67	8564,14	350,5	0,573	47,7
145	π Cas	37,9	46 29	5,0	A5	1,96408	9970,03	45,1	0,010	117,3
								225,1	.....	119
146	ρ Tuc	38,2	−66 1	5,5	F5	4,820223	9299,1	269,0	0,02	25,7
159	23 Cas	41,1	74 18	5,4	B8	33,75	10577,41	269,7	0,405	16,3
164	ζ And	42,0	23 43	4,3	K0	17,7673	10024,88	182,2	0,037	25,7
175	ν And	44,3	40 32	4,4	B3	4,28284	8155,66*	......	0,0	75,6
										104
329	γ Phe	1 24,0	−43 50	3,4	K5	193,79	7945,0	267	0,005	15,8
373	Σ 145 btr.	35,7	25 14	6,3	F5	4,43474	11940,99	295,6	0,108	81,5
								115,6	.....	88,6
384	φ Per	37,4	50 11	4,2	B0p	126,5	8290,42	347,3	0,428	26,9
						63,25*	8324,37	257,1	0,107	7,0
421	α Tri	47,4	29 6	3,6	F5	1,73652	10793,82	135,6	0,121	12,1
425	ω Cas	48,2	68 12	5,0	B8	69,92	10426,02	50,0	0,30	29,6
428	β Ari	49,1	20 19	2,7	A5	107,0	7632	19,7	0,88	32,6
497	ι Tri btr.	2 6,6	29 50	5,2	G0	14,732	1243,1	5,4	0,043	56,5
								185,4	.....	57
497	ι Tri ftr.	6,6	29 50	6,4	A5	2,2365	1246,70	3,7	0,010	95,4
								183,7	.....	101,0
514	δ Tri	10,9	33 46	5,1	G0	9,92912	13309,43	21,5	0,059	8,8
530	ο Cet	14,3	− 3 26	var	Me	330		265,2	0,20	5,9
593	Σ 5 btr.	31,2	24 13	6,6	F5	9,851	11586,90	129,6	0,145	22,5
....	RZ Cas	39,9	69 13	var	A0	1,19525	9449,73	154,7	0,052	69,3
641	π Ari	43,7	17 3	5,3	B5	3,854	10370,26	78,3	0,042	24,8
708	β Per	3 1,7	40 34	var	B8	2,86730	1,51*	277,5	0,038	44,1
						**1y,885	1902,79	0	0,13	10,0
804	f Tau	25,4	12 36	4,3	K0	960	4889,56	326,3	0,397	8,2
844	ο Per	38,0	31 58	3,9	B1	4,41916	8217,92	......	0,00	111,9
										160,0
913	ξ Per	52,5	35 30	4,0	Oe5	6,951	8248,3	99,2	0,034	7,9

Orbital Elements

f	$a_1 \sin i$ 1000 km	V_0 km	Computer	Reference	Remarks
0,18	34790	−11,5	BAKER	A O 1, p. 22 (1908)	
3,71	10480	−44,9	ADAMS, STRÖMBERG	Ap J 47, p. 329 (1918)	Two spectra. H and K give small variation.
[1,60] [0,94]	2141 3739	+ 0,5	PLASKETT	D A O 2, p. 141 (1922)	Eclipsing variable. *Spec.-photm. phase. a (sep) $= 8{,}86\, r_s$; $r_1 = 2{,}50\, r_s$ $r_2 = 2{,}83\, r_s$; $m_1 = 1{,}83\, \odot$; $m_2 = 1{,}01\, \odot$; $d_1 = 0{,}118\, D_0$; $d_2 = 0{,}044\, D_0$.
0,064	290000	+75,2	LUNT	C A 10, p. 27G (1924)	
0,029	2240	+ 2,0	MISS UDICK	A O 2, p. 194 (1912)	
0,010	1037	+10,3	PARASKÉVOPOULOS	Ap J 52, p. 110 (1920)	cf. § 135.
0,89	77200	+ 8,8	JORDAN	A O 2, p. 49 (1910)	
[1,35] [1,34]	3168 3213	+12,4	HARPER	D O 4, p. 149 (1917)	
0,0085	1700	+13,5	LUNT	C A 10, p. 41G (1924)	
0,012	7020	− 4,1	YOUNG	D O 2 p. 194 (1915)	
0,031	6272	−29,8	CANNON	D O 2, p. 157 (1915)	
[1,50] [1,10]	4454 6125	−23,9	JORDAN	A O 1, p. 196 (1910)	*Time of greatest positive velocity.
0,079	42100	+25,8	WILSON	L O B 9, p. 116 (1918)	
[1,16] [1,06]	4942 5372	+ 4,6	SANFORD	Ap J 53, p. 201 (1921)	
0,19	44803	+ 3,2	CANNON	J Can R A S 4, p. 195 (1910)	*Elliptical elements of secondary oscillation. T is here the time when up curve crosses zero line. LUDENDORFF and JORDAN find that different revolutions of the system give different velocity-curves. cf. § 100.
0,00031	287	−12,6	HARPER	D O 3, p. 133 (1915)	
0,16	27190	−24,8	YOUNG	D O 2, p. 102 (1915)	
0,042	22880	− 0,6	LUDENDORFF	Ap J 25 p. 325 (1907)	
[1,12] [1,12]	11441 11532	−19,1	HARPER	D A O 2, p. 129 (1921)	
[0,91] [0,86]	2934 3107	−19,8	HARPER	D A O 2, p. 129 (1921)	
0,00070	1200	− 5,8	PEARCE	L O B 11, p. 131 (1923)	
0,007	28200	+58,2	JOY	Ap J 63, p. 281 (1926)	Positive maximum velocity coincident with maximum brightness.
0,011	3013	+14,4	ADAMS, JOY	Ap J 49, p. 186 (1919)	
0,041	1137	−38,3	JORDAN	A O 3, p. 137 (1914)	Algol variable. From spectrographic and photometric elements DUGAN, Ap J 44, p. 115 (1916), deduces absolute elements of system.
0,0061	1312	+ 7,8	YOUNG	D O 4, p. 69 (1917)	
0,025	1736 93000	+16,9 + 5,7	MCLAUGHLIN	Ap J 60, p. 22 (1924)	Most recent orbit, epoch 1923, 66. *after light-minimum. Velocity of system variable. **Orbit of centre of mass. cf. § 114.
0,043	99950	+14,2	HARPER	D A O 3, p. 145 (1924)	
[5,42] [3,79]	6801	+18,5	JORDAN	A O 2, p. 68 (1910)	H and K give constant velocity and same as that of system.
0,00035	752	+15,4	CANNON	D O 1, p. 370 (1912)	Orbit from H and K lines.

Spectroscopic

Boss	Star	R. A. 1900	Decl.	Mag.	Class	P d	T Jul. Day 2410000 +	ω	e	K_1 km
920	λ Tau	3h55m,1	12°12′	var	B3	3,952917	7945,12	77°,5	0,061	56,2
						†34,60	7831,30	......	0,0	10,4
967	μ Per	4 7,6	48 9	4,3	G0	284	10061,97	302,0	0,061	20,5
986	b¹ Per	10,7	50 3	4,6	A2	1,52732	8956,17	151,7	0,22	41,9
								331,7	.....	152,5
....	HD 27130	11,9	16 42	8,3	G5	5,6100	13102,19	88,6	0,04	58,6
1001	υ Eri	14,1	−34 3	3,6	B9	5,0105	7562,27	124,3	0,014	63,8
								304,3	.....	64,8
....	O Σ 82 btr.	17,1	14 49	7,0	F8	4,00000	12274,81	12,7	0,060	36,1
1018	63 Tau	17,7	16 33	5,7	A2	8,425	9819,0	190,7	0,16	36,5
1046	θ² Tau	22,9	15 39	3,6	F0	140,70	8054,72	54,2	0,717	27,1
1076	d Tau	30,2	9 57	4,4	A3	3,57122	*9734,99	......	0,0	72,7
1082	3 Cam	32,0	52 53	5,3	K0	121	11137,55	285	0,019	28,2
1107	94 Tau	36,2	22 46	4,3	B5	1,5047	7892,50	242,9	0,087	44,3
1139	9 Cam	44,1	66 10	4,4	B0	7,9957	6480,35	90 or 270	0,30	9,0
1147	π_4 Ori	45,9	5 26	3,8'	B3	9,5191	8279,64	152,3	0,027	25,9
1159	π_5 Ori	49,0	2 17	3,4	B3	3,70045	*7921,64	......	0,00	57,9
1161	7 Cam	49,3	53 35	4,4	A2	3,8846	8281,18	217,1	0,013	35,1
1190	ζ Aur	55,5	40 56	3,9	K0	973	5122,47	330,1	0,411	23,8
1213	66 Eri	5 1,8	− 4 47	5,2	B9	5,52242	13087,58	335,9	0,074	97,0
								155,9	.....	111,0
1244	14 Aur	8,9	32 34	5,1	A2	3,789	10802,71	19,7	0,033	21,6
1246	α Aur	9,3	45 54	0,2	G0	104,022	4899,5	117,3	0,016	25,8
								297,3	.....	32,4
1250	β Ori	9,7	− 8 19	0,3	B8p	21,90	7968,80	254,8	0,296	3,8
1301	η Ori	19,4	− 2 29	3,4	B1	7,9896	5720,82	42,3	0,016	144,7
								†222,3	.....	152,4
						††7,99303	*6,62	41,2	0,006	143,2

Orbital Elements

f	$a_1 \sin i$ 1000 km	V_0 km	Computer	Reference	Remarks
0,073	3050	+12,9	SCHLESINGER	A O 3, p. 173 (1914)	Algol variable. †Orbit of centre of mass. SCHLESINGER considers system composed of bright star, eclipsing satellite and distant satellite of masses respectively 2,5, 1,0 and 0,4 times Sun; and at distances from centroid of bright and eclipsing bodies of 3 200 000, 8 000 000 and 50 000 000 km, respectively.
0,0040	4950	− 0,9	SCHLESINGER	A O 3, p. 169 (1914)	
0,25	80000	+ 7,8	CANNON	D O 2, p. 365 (1915)	
[0,85]	858	+23,1	CANNON	D O 1, p. 298 (1911)	
[0,23]	3124				
0,12	4517	+38,2	SANFORD	Ap J 59, p. 358(1924)	
[0,56]	4393	+17,8	PADDOCK	L O B 8, p. 175(1915)	
[0,55]	4468				
0,019	1980	+37,4	SANFORD	Ap J 53, p.201 (1921)	
0,041	4170	+36,4	JANTZEN	A N 196, p. 120(1913)	
0,099	37471	+42,6	PLASKETT	D O 2, p. 80 (1915)	Some evidence of departure from elliptic motion.
0,14	3570	+29,2	DANIEL	A O 3, p. 96 (1914)	Earlier orbit HARPER D O 1, p. 125 (1913), from spectrum of secondary gives $m_2/m_1=0{,}47$; $m_1 \sin^3 i = 2{,}7\odot$. *Time when secondary crosses line of sight nearest to solar system. Eclipsing binary, SHAPLEY, AN 196, p. 383 (1913); HERTZSPRUNG A N 199, p. 142 (1914).
0,28	46900	−40,5	CANNON	D O 4, p. 175 (1918)	
0,013	914	+13,5	PARKER	Rep. Chief Astronomer Ottawa 1, p. 166 (1910)	
0,00083	944	− 2,2	LEE	Ap J 37, p. 1 (1913)	Orbit is from sharp H and K lines of calcium.
0,017	3393	+23,3	BAKER	A O 1 p. 110 (1909)	
0,075	2945	+24,2	LEE	Ap J 38, p. 180 (1913)	*Epoch of maximum positive velocity.
0,018	1877	− 8,9	HARPER	J Can R A S 5, p. 115 (1911)	
1,03	294300	+10,7	HARPER	D A O 3, p. 151(1924)	Two spectra.
[2,5]	7300	+30,9	FROST, STRUVE	Ap J 60, p. 313 (1925)	A faint third component is visible between other two lines but does not partake of their velocities. LICK, Publ A S P 30, p. 351 (1918).
[2,2]	8400				
0,0039	1123	−10,7	HARPER	D O 3, p. 232 (1916)	
[1,19]	36848	+30,2	REESE	L O B 1, p. 34 (1901)	Visual orbit from interferometer measures cf § 69.
[0,94]	46430				
0,00011	1109	+22,6	PLASKETT	Ap J 30, p.26 (1909)	
[11,2]	15901	+35,5	ADAMS	Ap J 17, p. 71 (1903)	Eclipsing binary. BEAL, Amer. Astr. Soc. 3, p. 117 (1918) finds velocity of system variable in a period of nine to ten years. †Orbit of secondary, based on BAKER's ratio of masses, A O 1, p. 136 (1910).
[10,6]	16750				
2,44	15740	+37,0	HNATEK	A N 217, p. 53 (1922)	††BASED on ADAMS' observations and those of 1920, Vienna. *T time of zero velocity when changing from positive to negative. H and K sharp and give constant velocity, SLIPHER.

Spectroscopic

Boss	Star	R.A. 1900	Decl.	Mag.	Class	P d	T Jul. Day 2410000 +	ω	e	K_1 km
1314	ψ Ori	5h21m,6	3° 0′	4,7	B2	2,52588	7916,36	184°,7	0,065	144,1
								4,7	.....	190,0
1333	χ Aur	26,2	32 7	4,9	B1	655,16	10629,78	135,5	0,171	20,5
						*655,16	10629,78	135,5	0,171	10,5
1339	δ Ori	26,9	− 0 22	2,5	B0	5,732585	12391,58	337,9	0,107	101,5
1349	VV Ori	28,5	− 1 14	5,4	B2	1,48540	9836,026*	......	0,00	132,4
						**120	9819	40	0,30	13,0
1365	θ^2 Ori	30,5	− 5 29	5,2	B1	21,029	13741,36	154,7	0,27	93,7
1366	ι Ori	30,5	− 5 59	2,9	Oe5	29,136	7587,99	113,3	0,754	109,9
1375	ζ Tau	31,7	21 5	3,0	B3p	138	5769,9	9,8	0,180	14,9
1381	28 G Dor	32,4	−64 18	5,3	G5	180,8757	13108,42	333,0	0,51	22,4
1384	β Dor	32,8	−62 33	3,8	F5p	9,84072	12702,48	5,3	0,25	15,0
1388	125 Tau	33,5	25 50	5,0	B3	27,864	10471,61	335	0,55	25,5
1399		35,8	− 1 11	5,0	B3	27,160	7961,46	87,0	0,765	93,0
						*27,160	7973,79	......	0,00	10,1
1452	31 Cam	46,0	59 52	5,3	A0	2,9332	11938,37	359,2	0,030	76,0
1457	136 Tau	47,0	27 35	4,5	A0	5,969	9362,52	191,4	0,022	48,9
								11,4	.....	71
1478	β Aur	52,2	44 56	2,1	A0p	3,960027	7100,73*	......	0,00	109,0
							....	......	.....	111,0
1492	2 Mon	54,3	− 9 34	5,1	A5	9,3553	9673,81	35,4	0,208	57,1
1501	μ Ori	56,9	9 39	4,2	A2	4,44746	13863,17	......	0	30,8
						*18 yrs.	(1911y,7)	98	0,6	4,0
1508	1 Gem	58,0	23 16	4,3	G5	9,590	11898,74	203,3	0,206	11,7
1515	40 Aur	59,7	38 29	5,3	A3	28,28	10468,20	178,4	0,556	51,4
								1,6	.....	62,5
1525	ν Ori	6 1,9	14 47	4,4	B2	131,26	7975,16	1,6	0,599	34,1
1593	45 Aur	13,6	53 30	5,4	F5	6,5013	13634,17	331	0,019	31,7
	RR Lyn	18,0	56 20	5,5	A3	9,944	9341,78	152,9	0,081	67,2
1609	β CMa	18,3	−17 54	2,0	B1	0,25714	8367,70	225	0,04	9,1
						†42	13070	90	0,27	4,7
1610	δ Col	18,4	−33 23	4,0	G5	870	17305	115	0,67	11,0
1635	ν Gem	23,0	20 17	4,1	B5	9y,6	(1909y,75)	285	0,20	30,0
1646	WW Aur	25,9	32 32	6,0	A0	2,5248	11623,36*	......	0,00	115,6
										135,1

Orbital Elements

f	$a_1 \sin i$ 1000 km	V_0 km	Computer	Reference	Remarks
[5,33] [4,19]	4995 6570	+12,0	PLASKETT	Ap J 28, p. 272 (1908)	BAKER, A O 1, p. 100 (1909): $m_2/m_1 = 0,76$.
0,56	182300	− 0,1 + 1,5	YOUNG	D O 4, p. 1 (1916)	*Orbit from H and K lines. cf. § 99.
0,61	7958	+21,2	HNATEK	A N 213, p. 24 (1920)	Algol variable.
0,36	2704 20460	+20,8 + 3,0	DANIEL DANIEL	A O 3, p. 184 (1915) A O 3, p. 185 (1915)	Algol variable. *Time when radial component of orbital velocity is zero and changing from negative to positive. **Provisional orbit for centre of mass.
1,79	27000	+36,8	STRUVE	Ap J 60, p. 159 (1924)	
1,14	28907	+21,3	PLASKETT	Ap J 28, p. 275 (1908)	Secondary disturbance. Ap J 30, p. 339 (1909).
0,045	27900	+16,4	ADAMS	Ap J 22, p. 118 (1905)	
0,13	47900	+ 9,8	LUNT	C A 10, part 7, p. 22G (1924)	
0,0031	1965	+ 5,3	LUNT	C A 10, part 7, p. 32G (1924)	The elements are probably variable.
0,028	8160	+14,8	CANNON	D O 3, p. 424 (1916)	
0,61	22380 3791	+26,1	PLASKETT, HARPER	Ap J 30, p. 376 (1909)	Orbit when superimposed oscillation of same period is introduced. *Orbit of secondary disturbance.
0,13	3065	− 3,9	HARPER	D A O 3, p. 159 (1924)	Two spectra. LICK, Publ A S P 34, p. 168 (1922).
[0,6] [0,4]	4011	−17,1	CANNON	D O 2, p. 119 (1915)	
[2,21] [2,17]	5934 6047	−18,1	BAKER	A O 1, p. 190 (1910)	*T is the time of passage through ascending node. cf. § 104.
0,17	7200	+22,2	ELVEY	Ap J 60, p. 320 (1924)	Secondary spectrum is faintly visible at times of max. and min. vel. $m_2/m_1 = 0,86$.
0,013	2000 300000	+42,7 +40,8	FROST, STRUVE	Ap J 60, p. 192 (1924)	Epoch of orbit 1921,7. Velocity of system variable. *Orbit of centre of mass of short period system. Close visual double, AITKEN, L O B 8, p. 93 (1914).
0,0015	1510	+19,7	YOUNG	D A O 1, p. 119 (1919)	
[1,35] [1,11]	16550 20140	+16,9	YOUNG	D O 4, p. 95 (1917)	
0,28	49270	+22,1	HARPER	J Can R A S 5, p. 24 (1911)	
0,022	2837	− 1,5	HARPER	D A O 3, p. 189 (1925)	
0,31	9127	−13,7	HARPER	D O 2, p. 177 (1915)	Eclipsing binary.
0,00002 0,00041	31 2633	+33,1 +29,3	HENROTEAU HENROTEAU	L O B 9, p. 155 (1918) D O 8, p. 41 (1922)	†Approximate elements for a possible orbit of centre of mass. cf. § 118.
0,049	97600	− 5,9	WILSON, HUFFER	L O B 10, p. 16 (1918)	
9,24	1417000	+38,4	HARPER	D O 4, p. 279 (1919)	Two Spectra. There probably is a short period oscillation. H and K lines give velocities differing from other lines.
[2,2] [1,9]	4010 4690	− 5,8	JOY	Publ A S P 30, p. 254 (1918)	Algol variable. *Primary minimum.

Spectroscopic

Boss	Star	R.A. 1900	Decl.	Mag.	Class	P d	T Jul. Day 2410000 +	ω	e	K_1 km
....	BD 6°1309	$6^h32^m,0$	6°13′	6,1	B0p	14,414	13031,87	181°,9	0,035	206,4
										246,7
1773	A Car	47,7	−53 30	4,4	G5	195,32	11344,0	......	0,0	24,8
....	HD 54371	7 3,5	25 54	7,0	G0	32,8092	13071,94	82,1	0,080	27,5
1899	29 CMa	14,5	−24 23	4,9	Oe	4,3934	7240,25	37,6	0,156	218,4
1906	19 Lyn	14,7	55 28	5,6	B8	2,25960	9031,63	126,1	0,076	104,4
1909	R CMa	14,9	−16 12	var	F0	1,1359514	7966,58	195,9	0,138	28,6
1972	σ Pup	26,1	−43 6	3,3	K5	257,8	10418,6	349,3	0,17	18,5
1979	α_1 Gem	28,2	32 6	2,8	A0	2,928285	6828,06	102,5	0,01	31,8
1979	α_2 Gem	28,2	32 6	2,0	A0	9,218826	6746,38	265,3	0,503	13,6
2023	σ Gem	37,1	29 8	4,3	K0	19,605	5824,02	330,2	0,022	34,2
2119	V Pup	55,4	−48 58	var	B1p	1,454475	2778,33	72	0,08	......
2179	ε Vol	8 7,6	−68 19	4,5	B5	14,16833	9453,56*	......	0,00	66,7
2227	1 Hya	19,6	− 3 26	5,7	F5	1,562975	12650,08	123,9	0,051	30,3
2285		30,5	6 58	6,0	F5	14,296	11599,47	220,8	0,276	22,7
2354	ε Hya	41,5	6 47	3,5	F8	5588	5375	90,0	0,65	8,4
2421	w Vel	56,3	−40 52	4,4	F8	74,1469	12728,63	90,0	0,05	17,8
2445	ϰ Cnc	9 2,3	11 4	5,1	B8	6,393	6486,90	162,3	0,149	67,8
2447	75 Cnc	2,9	27 3	6,0	G1	19,4589	12426,63	252,5	0,206	20,2
2463	c UMa	6,4	61 50	5,2	F8	16,2382	13049,62	169,2	0,09	34,8
2473	a Car	8,4	−58 33	3,6	B3	6,744	6533,81	115,8	0,18	21,5
2484	....	10,8	47 14	5,7	A0	15,986	9408,03	355,2	0,504	63,3
								175,2	.....	73,6
2490	23 Hya	11,7	− 5 56	5,4	K0	914,0	11263	57,0	0,36	11,7
2526	ϰ Vel	19,0	−54 35	2,6	B3	116,65	6459,00	96,2	0,19	46,5
....	S Ant	27,9	−28 11	var	F0	0,64833872	....	......	0	81
										148
2602	o Leo	35,8	10 21	3,8	F5	14,4980	4656,48*	......	<0,02	54,0
										63,1
....	W UMa	36,7	56 25	var	G0	0,3336392	....	......	0,00	134
										188
2665	19 L Mi	51,6	41 32	5,2	F4	9,283	13498,60	351,1	0,048	15,2
2754	30 H UMa	10 16,9	66 4	4,9	A0	11,5832	8468,21	171,9	0,381	34,1
2830	p Vel	33,1	−47 42	4,1	F2	10,210955	10259,38	184,6	0,541	42,3
								4,6	.....	52

Orbital Elements

f	$a_1 \sin i$ 1000 km	V_0 km	Computer	Reference	Remarks
[75,6] [63,3]	40880	+23,9	Plaskett	D A O 2, p.147 (1922)	cf. § 106.
0,31	66880	+25,4	Wilson, Huffer	L O B 10, p.17 (1918)	
0,070	12383	+19,7	Sanford	Ap J 56, p.446 (1922)	
4,58	13035	−12,1	Harper	D O 4, p.123 (1917)	
0,26	3235	+ 7,4	Harper	D O 4, p.235 (1918)	Two Spectra
0,0027	443	−39,7	Jordan	A O 3, p. 52 (1913)	Eclipsing variable.
0,16	64800	+87,3	Wilson	L O B 9, p.119 (1918)	
0,0097	1279	− 1,0	Curtis	L O B 4, p. 58 (1906)	cf. § 136.
0,0015	1485	+ 6,2	Curtis	L O B 4, p. 64 (1906)	
0,081	9220	+45,8	Harper	D O 1, p.279 (1911)	
....	....	......	Miss Maury	Pop Astr 29, p. 22. (1922)	Relative orbit. $K_1+K_2=604$ km; $(a_1+a_2)\sin i = 12041000$ km; $(m_1+m_2)\sin^3 i = 33\odot$. Algol variable. Two spectra.
0,44	12999	+ 9,7	Sanford	L O B 8, p.130 (1914)	*T epoch of perihelion passage of primary.
0,0045	650	+71,3	Sanford	Ap J 55, p. 30 (1922)	
0,015	4300	+24,3	Joy, Abetti	Ap J 50, p.391 (1919)	
0,15	493000	+36,8	Aitken	Publ A S P 24, p.218 (1912)	Visual orbit, Aitken, Publ A S P 24, p. 216 (1912): $P=15^y,3$; $e=0,65$; $T=1900,97$; $\omega=270°,0$; $i=49°,95$; $\Omega=104°,4$; $a=0'',23$; distance between components = 1359000000 km; $\pi=0'',025$; $m_1=1,75\odot$; $m_2=1,58\odot$.
0,043	18130	− 7,4	Lunt	C A 10, p. 37G (1924)	
0,20	5890	+26,3	Ichinohe	Ap J 25, p.318 (1907)	Two spectra. The elements refer to brighter component.
0,016	5296	+12,3	Sanford	Ap J 55, p. 35 (1922)	
0,07	7730	−15,0	Young	D A O 2, p.205 (1923)	
0,0066	1960	+23,3	Curtis	L O B 4, p.154 (1907)	
[1,48] [1,27]	12026 13981	−13,1	Harper	D O 3, p.405 (1916)	
0,12	137190	− 5,0	Lunt	Unpublished	
1,15	73200	+21,9	Curtis	L O B 4, p.156 (1907)	
[0,52] [0,29]	722 1320	− 5,0	Joy	Ap J 64, p.287 (1916)	Algol variable. From photometric and spectroscopic elements $m_1 = 0,75\odot$; $m_2 = 0,42\odot$; $d_1 = 0,31 D_0$; $d_2=0,38 D_0$; π (surface brightness = 4 × Sun) = $0'',016$.
[1,30] [1,12]	10775 12571	+27,1	Plummer	L O B 5, p. 21 (1908)	*Time when observed velocity of each component is equal to V_0, with principal component between the centre of mass and the solar system.
[0,67] [0,48]	610 860	− 5	Adams	Ap J 49, p.189 (1919)	Algol variable. Using Shapley's photometric orbit (uniform solution) and spectroscopic elements ($i=77°,6$); $m_1=0,69\odot$; $m_2 = 0,49\odot$; $d_1 = 2,8 D_0$; $d_2 = 1,9 D_0$.
0,0034	1943	−10,8	Harper	D A O 3, p.194 (1925)	
0,038	5020	− 0,1	Schlesinger	A O 2, p.149 (1912)	K line gives orbit with simple elliptic motion.
[0,28] [0,23]	4981 6107	+19,2	Sanford	L O B 9, p.182 (1918)	

Spectroscopic

Boss	Star	R.A. 1900	Decl.	Mag.	Class	P d	T Jul. Day 2410000 +	ω	e	K_1 km
2900	ω UMa	$10^h48^m,2$	43°43′	4,8	A0	15,8401	7991,10	11°,9	0,264	20,6
....	HD 96511	11 2,2	82 17	7,1	G0	18,8922	13154,07	333,0	0,282	40,1
2984	ξ UMa	12,9	32 6	4,4	G0	665	5252,7	322,6	0,41	7,0
2987	55 UMa	13,7	38 44	4,8	A2	2,5	11412,76	173,4	0,11	38,5
								353,4	.....	54,5
3098	93 Leo	42,8	20 46	4,5	F8	71,70	8088,40	270,8	0,008	26,5
3138	31 Crt	55,7	−19 6	5,3	B3	1,50307	10917,57	185,1	0,078	118,2
3146	θ^1 Cru	57,9	−62 45	4,5	A5	24,483	12807,26	354,8	0,629	48,2
3182	4 H Dra	12 7,5	78 10	5,1	A5	1,27100	10685,26*	......	0,00	63,2
3210	η Vir	14,8	− 0 7	4,0	A0	71,9	6206,93	180,0	0,254	26,8
						71,9	7643,50	185,0	0,40	(27,6)*
....	HD 107760	17,9	73 48	8,2	G5	5,41454	12853,12*	......	0,00	64,9
										74,0
3281	$\varkappa$ Dra	29,2	70 20	3,9	B5p	8,986	10609,02	326	0,23	15,7
3323	32 Vir	40,7	8 13	5,2	A5	38,3	10573,45	223,3	0,072	41,0
								43,3	.....	80
3363	ε UMa	49,6	56 30	1,7	A0p	$4^y,15$	$(1907^y,40)$	55,8	0,31	3,5
3371	α^2 CVn	51,4	38 51	2,9	A0p	5,50	3,84	110	0,3	21,5
....	RS CVn	13 6,0	36 28	var	{F3	4,797851	....	......	0	86
					{K0					86
3474	ζ^1 UMa btr	19,9	55 27	2,4	A2p	20,53644	9477,74	104,0	0,535	69,2
										68,8
3476	α Vir	19,9	−10 38	1,2	B2	4,01416	7955,85	328	0,10	126,1
								148	.....	207,8
3511	....	30,3	37 42	5,0	F0	1,61100	7018,02	199,0	0,054	10,1
3564	ν Cen	43,5	−41 11	3,5	B2	2,62516	10301,39	......	0,00	20,6
3586	h Cen	47,5	−31 26	4,8	B5	6,927	8733,25	147,2	0,23	21,4
3593	ζ Cen	49,3	−46 48	2,6	B2	8,024	2266,81	287	0,5	.....
3596	η Boo	49,9	18 54	2,8	G0	497,1	8240,60	315,2	0,236	8,7
3626	α Dra	14 1,7	64 51	3,6	A0p	51,38	7403,28	19,0	0,384	46,2
3635	d Boo	5,8	25 34	4,8	F5	9,6045	7679,52	273	0,169	68,4
								93	.....	72,0
3644	....	8,5	− 0 22	5,8	F5	2,6960	12744,10*	......	0,00	24,3
3673	A Boo	13,8	35 58	6,8	K0	211,95	10561,18	223,4	0,54	18,0
3793	39 Boo ftr	46,3	49 8	4,8	F5	12,822	12379,49	97,0	0,394	58,3
								277,0	.....	72,2

Orbital Elements.

f	$a_1 \sin i$ 1000 km	V_0 km	Computer	Reference	Remarks
0,013	4336	−18,4	PARKER	J Can R A S 5, p. 382 (1911)	Two spectra. From 3 measures of secondary, PARKER derives $K_2 = 120$ km; $m_2/m_1 = 0{,}17$.
0,11	9975	−46,7	SANFORD	Ap J 59, p.360 (1924)	
0,018	58300	−16,4	ABETTI	Soc. Degli Spet. Ital. 8, Ser. 2a, p. 113 (1919)	cf. § 132.
[0,12]	1315	− 3	HENROTEAU	Pop Astr 27, p. 29 (1919)	
[0,08]	1862				
0,14	26170	+ 0,2	CANNON	J Can R A S 4, p. 458 (1910)	
0,25	2435	+ 1,7	CANNON	D O 4, p. 125 (1917)	
0,13	12600	− 2,0	LUNT	C A 10, p. 17G (1924)	
0,033	1104	+ 0,3	LEE	Ap J 43, p.320 (1916)	*Time of maximum velocity of recession.
0,13	25290	− 0,4	ICHINOHE	Ap J 26, p.282 (1907)	Two spectra; faint companion with non-elliptic curve suggesting third component, gives $V_0 = +30$ km. *Estimated from published curve. $m_2/m_1 = 0{,}7$, SCHLESINGER and BAKER, A O 1, p. 145 (1910).
0,13	25750	+ 2,2	HARPER	Ap J 27, p.160 (1908)	
[0,80]	4833	−97,4	SANFORD	Ap J 56, p.452 (1922)	*T is time of smallest negative velocity of primary.
[0,70]	5510				
0,0032	1890	− 7,2	BAKER	Det O 3, p. 33 (1920)	Elements from Hβ emission and broad He 4471 absorption. Narrow H absorption lines and K do not share in short period oscillation but slowly change position.
[4,6]	21530	− 8,9	CANNON	D O 2, p.383 (1915)	
[2,4]					
0,0058	69360	−12,9	LUDENDORFF	A N 195, p.370 (1913)	
0,0049	1500	+ 1	BÉLOPOLSKY	A N 196, p. 1 (1913)	cf. § 75.
[1,3]	5700	− 9	JOY	Publ A S P 34, p.221 (1922)	Algol variable.
[1,3]	5700				
[1,62]	16400	− 9,6	HADLEY	Det O 2, p.101 (1915)	
[1,70]	16400				
[9,6]	6930	+ 1,6	BAKER	A O 1, p. 72 (1909)	cf. § 103.
[5,8]	11400				
0,0002	222	+ 6,6	HARPER	D O 4, p. 232 (1918)	
0,0024	745	+ 9,0	WILSON	L O B 8, p. 131 (1914)	
0,0065	1984	+ 5,2	PADDOCK	L O B 9, p. 45 (1916)	
....	....	......	MISS MAURY	Harv Circ 233 (1922)	Relative orbit. $K_1 + K_2 = 312$ km; $(a_1 + a_2) \sin i = 29800000$ km; $(m_1 + m_2) \sin^3 i = 16{,}4 \odot$.
0,031	57735	− 0,2	HARPER	J Can R A S 4, p. 194 (1910)	
0,42	30173	−17,0	HARPER	J Can R A S 4, p. 92 (1910)	
[1,36]	8904	+ 9,8	HARPER	D O 1, p. 329 (1911)	
[1,29]	9380				
0,0040	902	+17,6	DUNCAN	Ap J 54, p.226 (1921)	*Epoch of maximum positive velocity.
0,076	44000	−25,6	YOUNG	D O 3, p. 107 (1915)	
[1,27]	9450	−28,2	HARPER	D A O 2, p.167 (1922)	
[1,03]	11700				

Spectroscopic

Boss	Star	R.A. 1900	Decl.	Mag.	Class	P d	T Jul. Day 2410000+	ω	e	K_1 km
3825	δ Lib	$14^h 55^m,6$	− 8° 7′	var	A0	2,32735†	1,89*	29°,2	0,054	76,5
3827	2 H UMi	56,0	66 20	4,9	Mb	750	12065,0	......	0	6,7
....	U CrB	15 14,1	32 1	var	B8	3,4522269	....	......	0	69,5
										181,9
3928	γ UMi	20,9	72 11	3,1	A2	0,108449	13204,73	160,0	0,09	14
3940	β CrB	23,7	29 27	3,7	F0p	40,9	8739,29	240	0,4	3,1
						†490,8	8804,4*	......	0,0	2,4
3961	α CrB	30,5	27 3	2,3	A0	17,36	7742,55	312,2	0,387	34,9
....	TW Dra	32,4	64 14	var	A5	2,80654	0,02†	90	0,054	65,8
4008	A Ser	40,9	− 1 29	5,4	B8	38,95	9528,60	208,5	0,773	50,5
4086	β Sco	59,6	−19 32	2,9	B1	6,8283	9163,92	20,1	0,270	125,7
								200,1	.....	197
						6,8284	8126,00	20	0,27	126
								200	.....	166
4090	θ Dra	16 0	58 50	4,1	F8	3,0708	5368,96	126,1	0,014	23,5
....	HD 144426	0,8	8 22	6,1	A2	8,855	1846,70	265,4	0,376	31,6
....	HD 144515	1,2	10 57	8,5	G5	4,28503	12418,05	82,5	0,089	38,1
4158	σ Sco	15,1	−25 21	3,1	B1	0,246834	11687,97	15	0,11	41,2
						†34,08	11715,35	270	0,33	33
4166	ζ TrA	17,7	−69 52	4,9	G0	13,7155	13629,68	270,0	0,00	7,9
4174	ι TrA	18,7	−63 50	5,3	F0	39,888	13275,7	82,0	0,33	39,2
4204	β Her	25,9	21 42	2,8	K0	410,575	5500,37	24,6	0,550	12,8
....	HD 149632	30,9	17 15	6,3	A0	10,56	12422,24	4,1	0,430	62,4
								184,1	.....	101,4
4277	μ^1 Sco	45,1	−37 53	3,1	B3p	1,44627	2375,57	190	0,05	......
4322	h Dra	55,5	65 17	4,8	F5	51,710	4813,75	329,3	0,128	18,0
4327	ε UMi	56,2	82 12	4,4	G5	39,482	8005,75	359,5	0,011	40,0
4328	ε Her	56,5	31 4	3,9	A0	4,0235	8086,25	180	0,023	70,4
								0	.....	112,1
....	HD 155375	17 6,1	12 35	6,5	A0	23,245	11780,29	129,8	0,427	27,7
4374	α Her ftr.	10,1	14 30	5,4	F9	51,590	12468,58	27,9	0,028	29,6

Orbital Elements.

f	$a_1 \sin i$ 1000 km	V_0 km	Computer	Reference	Remarks
0,11	2450	−45,0	SCHLESINGER	A O 1, p. 126 (1910)	Algol variable. *Interval after light minimum. †Apparent period. True period = 2,32770 days. From photometric and spectroscopic data SCHLESINGER finds ($i = 81°,5$), $m_1 + m_2 = 1{,}5 \odot$; assuming density of two bodies equal, $d = 0{,}026\, D_0$; $m_1 = 0{,}87 \odot$; $m_2 = 0{,}63 \odot$.
0,023	68800	+ 6,8	YOUNG	J Can R A S 21, p. 35 (1927)	
[4,1] [1,6]	3295	− 7,5	PLASKETT	D A O 1, p. 189 (1920)	Eclipsing variable. From SHAPLEY's photometric orbit ($i = 81°,4$) $m_1 = 4{,}27 \odot$; $m_2 = 1{,}63 \odot$; $d_1 = 0{,}175\, D_0$; $d_2 = 0{,}015\, D_0$.
0,00003	35	− 2,0	STRUVE	Pop Astr 31, p. 90 (1923)	Range and shape of velocity-curve varies. Light variable.
0,0001 0,70	1600	−21,3 + 0,8	CANNON	D O 1, p.405 (1912)	Probably triple system. †Approximate elements of long period oscillation, assumed to be a sine curve. *Time when "up-curve" crosses zero line.
0,060	7671	+ 0,4	JORDAN	A O 1, p. 89 (1909)	Eclipsing variable. STEBBINS, Ap J 39, p. 478 (1914).
0,083	2535	− 0,3	PLASKETT	D A O 1, p.147 (1919)	Eclipsing variable. †From primary minimum.
0,13	17170	−11,6	JORDAN	A O 3, p. 157 (1915)	
[13,0] [8,3]	11360 17800	−11,0	DANIEL, SCHLESINGER	A O 2, p. 135 (1912)	Brightest of visual triple. SLIPHER and DUNCAN find calcium lines constant, $V = -16{,}6$ km. DANIEL and SCHLESINGER find K line may show oscillation of small amplitude in same period as brighter component, but if constant $V = -8{,}5$ km. May be eclipsing variable, STEBBINS, Ap J 39, p. 481 (1914).
[9,0] [6,8]	10990 14450	− 8,0	DUNCAN	Lowell Obs Bull 2, p. 21 (1912)	
0,0041	990	− 8,4	CURTIS	L O B 4, p. 157 (1907)	
0,023	3568	−21,5	J. W. CAMPBELL	D A O 1, p.315 (1921)	
0,024	2238	−60,6	SANFORD	Ap J 53, p.201 (1921)	
0,0018	138	var	HENROTEAU	L O B 9, p.176 (1918)	†Elements of long period variation. cf. § 139.
0,11	14600	− 3,2	HENROTEAU	L O B 9, p.177 (1918)	
0,00070	1490	+ 7,4	LUNT	Unpublished	
0,21	19835	− 5,2	LUNT	Unpublished	
0,052	60280	−25,5	PLUMMER	L O B 5, p. 25 (1908)	
[2,19] [1,35]	8180 13280	− 9,9	YOUNG	D A O 1, p.233 (1920)	
....	....	......	MISS MAURY	Pop Astr 29, p. 22 (1921)	Eclipsing variable. Relative orbit. $K_1 + K_2 = 480$ km; $(a_1 + a_2) \sin i = 9\,534\,000$ km; $(m_1 + m_2) \sin^3 i = 16{,}5 \odot$.
0,03	12625	−22,6	HARPER	D O 4, p. 243 (1918)	
0,26	17346	−11,4	PLASKETT	J Can R A S 4, p.464 (1910)	
[1,6] [1,0]	3890 6200	−24,0	BAKER	A O 2, p. 21 (1910)	
0,038	7997	+ 3,5	HARPER	D A O 1, p.200 (1920)	
0,14	2103	−37,2	SANFORD	Ap J 53, p.214 (1921)	

Spectroscopic

Boss	Star	R.A. 1900	Decl.	Mag.	Class	P d	T Jul. Day 2410000+	ω	e	K_1 km
....	U Oph	$17^h 11^m$,4	1°19′	var	B8	1,6773476	....	......	0,00	179,8
										204,6
4388	u Her	13,6	33 12	var	B3	2,05102	8125,80	66°,1	0,053	99,5
								246,1	.....	253
....	TX Her	15,4	42 0	var	A5	2,059786	0,04*	......	0,0	121,0
										140,2
4423		21,3	−5 0	4,6	F0	26,2742	8411,52	14,5	0,491	47,5
								194,5	.....	50,7
....	HD 159082	27,6	12 0	6,2	B9	6,7984	12878,15	116,3	0,069	50,2
4462	ξ Ser	31,9	−15 20	3,6	A5	2,292285	9209,62*	......	0	19,3
4483	ω Dra	37,5	68 48	4,9	F5	5,27968	7415,49	333,8	0,011	36,3
4507	....	44,4	47 39	6,3	A0	2,82424	12106,71	30	0,017	60,1
....	Z Her	53,6	15 9	var	F5p	3,992775	....	......	0,00	88,2
										101,8
4602	40 Dra	18 7,5	79 59	6,2	F5	10,5217	11764,65	256,8	0,314	46,2
								76,8	.....	51,5
4604	μ Sgr	7,8	−21 5	4,0	B8p	180,2	7495,64	79,1	0,447	66,8
4622	....	12,9	56 33	6,4	F0	2,0476	12147,63	195,1	0,039	105,1
								15,1	.....	108,1
4643	108 Her	17,1	29 48	5,5	A2	5,51460	9551,74*	......	0,00	70,1
										101,7
4669	....	22,1	29 46	5,7	A2	9,6120	12048,71	326,4	0,468	28,5
4672	χ Dra	22,9	72 41	3,7	F8	281,8	4864,3	119,0	0,423	17,9
....	RX Her	26,0	12 30	var	A0	1,7785740	9658,59*	......	0,00	104
										111
4752	ζ¹ Lyr	41,3	73 30	4,3	A3	4,29991	8109,72*	......	0,00	51,2
4756	β Sct	41,9	−4 51	4,5	G6	834	12480,9	33,9	0,35	16,6
....	HD 174343	45,0	49 19	7,2	{F2	3,76468	12159,77*	......	0,00	97,7
					{A					98,3
4776	β Lyr	46,4	33 15	var	B8p	12,922	8,54*	325,0	0,014	183,7

Orbital Elements.

t	$a_1 \sin i$ 1000 km	V_0 km	Computer	Reference	Remarks
[5,3]	4147	−11,5	PLASKETT	D A O 1, p.137 (1919)	Eclipsing variable. SHAPELY's photometric data (uniform solution) gives ($i=85°,7$) $m_1=5,31$ ⊙; $m_2=4,66$; $d_1=0,20D_0$; $d_2=0,18D_0$. For darkened solution ($i=83°,97$) $m_1=5,36$ ⊙; $m_2=4,71$ ⊙; $d_1=0,18D_0$; $d_2=0,16D_0$.
[4,6]	4718				
[6,8]	2800	−21,2	BAKER	A O 1, p. 82 (1909)	Eclipsing variable. From photometric and spectroscopic data, A O 2, p. 51 (1910), SCHLESINGER and BAKER find ($i=75°$), $m_1=7,5$ ⊙; $m_2=2,9$ ⊙; $d_1=0,37D_0$; $d_2=0,14D_0$; distance between centres of stars 10200000 km.
[2,6]	7120				
[2,03]	3430	− 6,4	PLASKETT	D A O 1, p.209 (1920)	Eclipsing variable. *After principal minimum. From SHAPLEY's photometric orbit (darkened) $i=86°,27$, then $m_1=2,04$ ⊙; $m_2=1,77$⊙; $d_1=0,87D_0$; $d_2=0,75D_0$. Separations of two stars = 7413000 km; radius of each star 927000 km.
[1,75]					
[0,9]	14950	+ 0,4	PARKER	D O 2, p. 348 (1915)	
[0,8]	....				
0,089	4680	−12,8	J. W. CAMPBELL	D A O 2, p.162 (1922)	
0,0017	610	−42,8	YOUNG	L O B 6, p.161 (1911)	*Epoch of perihelion.
0,026	2632	−13,7	TURNER	L O B 4, p.164 (1907)	
0,06	2336	−27,3	HARPER	D A O 1, p.128 (1919)	
[1,5]	4800	−46,5	ADAMS, JOY	Ap J 49, p.194 (1919)	Eclipsing variable. From SHAPLEY's photometric orbit (darkened) $i=82°$; $m_1=1,6$ ⊙; $m_2=1,3$ ⊙; $d_1=0,3D_0$; $d_2=0,04D_0$; radius of bright star = 1230000 km; radius of secondary = 2290000 km.
[1,3]	5600				
[0,46]	6341	+ 2,9	BOOTHROYD	D A O 1, p.245 (1920)	
[0,41]	7074				
4,0	148110	− 8,2	KOHL	A N 219, p.213 (1923)	
[1,04]	2957	− 8,5	HARPER	D A O 1, p.312 (1921)	
[1,01]	3042				
[1,72]	5320	−20,2	DANIEL, MISS JENKINS	A O 3, p. 152 (1914)	*Time when radial velocity of principal star is decreasing and equal to that of system.
[1,18]	7710				
0,016	3330	+ 7,5	YOUNG	D A O 1, p.134 (1919)	
0,13	63000	+32,4	WRIGHT	Ap J 11, p.133 (1900)	
[0,94]	2540	−19,5	BAKER, MISS CUMMINGS	Laws Obs Bull 2, p. 154 (1916)	Algol variable. *Primary minimum. From photometric orbit (darkened) and spectroscopic data they find $m_1=0,96$ ⊙; $m_2=0,91$ ⊙; $d_1=0,24D_0$; $d_2=0,22D_0$.
[0,89]	2710				
0,060	3030	−26,0	JORDAN	A O 1, p.118 (1909)	*T is time of maximum positive velocity.
0,33	178000	−21,9	YOUNG	J Can R A S 21, p. 37 (1927)	
[1,48]	5058	−18,8	SANFORD	Ap J 53, p.217 (1921)	*Epoch of maximum positive velocity for star 1.
[1,47]	5089				
8,32	32639	−19,0	ROSSITER	Ap J 60, p. 15 (1924)	β Lyrae variable. *After principal minimum. cf. § 115.

Spectroscopic

Boss	Star	R.A. 1900	Decl.	Mag.	Class	P d	T Jul. Day 2410000+	ω	e	K_1 km
4788	50 Dra	18h49m,6	75°19′	5,4	A0	4,1175	10293,52	107°,6	0,012	79,1
								287,6	.....	83,9
4790	ο Dra	49,7	59 16	4,8	K0	138,420	9258,16	274,3	0,114	23,5
4794	δ^1 Lyr	50,2	36 51	5,5	B3	88,112	9220,73	204,5	0,28	33,7
4797	113 Her	50,5	22 31	4,6	G0	245,3	9805,0	169,5	0,12	16,0
4864	18 Aql	19 2,3	10 55	5,1	B8	1,30226	8157,50*	......	0,00	27,6
4870	....	3,0	41 16	6,1	B3	1,03088	11735,65	20,0	0,015	12,1
....	RS Vul	13,4	22 16	var	B8	4,477325	1,90*	236,3	0,053	55,0
										175,9
....	U Sge	14,4	19 26	var	B9	3,380603	8428,18	44,1	0,035	66,4
4934	υ Sgr	16,0	−16 9	4,6	B8p	137,937	9648,72	28,6	0,087	48,1
....	Z Vul	17,5	25 23	var	B3	2,45492	−0,01*	......	0,0	96,3
										213,7
4947	2 Sge	19,8	16 45	6,0	A0	7,390	10943,23	332,6	0,05	52,9
										73,8
5018	σ Aql	34,3	5 10	5,2	B3	1,95022	10054,33	......	0,0	163,5
										199
5026	....	36,4	54 44	5,9	F5	7,6383	12201,40	46,7	0,527	89,8
								226,7	.....	91,1
5070	....	47,2	40 21	5,6	B2	12,427	9636,23	120,1	0,199	94,4
5099	φ Aql	51,5	11 9	5,3	A2	3,3204	13324,04	56,0	0,055	38,2
5108	θ^1 Sgr	53,2	−35 33	4,4	B3	2,10514	1140,64	......	0,0	15,9
5171	θ Aql	20 6,1	− 1 7	3,4	A0	17,1245	8261,91	14,9	0,681	46,0
								194,9	.....	63,0
5173	18 Vul	6,4	26 36	5,5	A2	9,316	10304,69	103,1	0,012	78,5
								283,1	.....	86,3
5200	θ^2 Cyg	12,4	47 24	4,2	K0 +A3	1170	10700,39	281,0	0,182	16,6
						*390	10515,82	......	.....	5,9
5216	β Cap	15,4	−15 6	3,2	G0 +A0	1375,3	6035	124,0	0,44	22,2
5223	α Pav	17,7	−57 3	2,1	B3	11,753	6379,90	224,8	0,01	7,2
5292	ι Del	33,0	11 2	5,4	A2	10,9960	12095,82	60,8	0,252	26,7
....	Y Cyg	48,1	34 17	var	B2	2,996332	....	......	0,0	223,9
										242,5
5375	57 Cyg	49,7	44 0	4,7	B3	2,8546	8554,77	45	0,137	110,4
								225	.....	118,8

Orbital Elements.

f	$a_1 \sin i$ 1000 km	V_0 km	Computer	Reference	Remarks
[0,95]	4480	− 8,8	HARPER	J Can R A S 13,	
[0,90]	4751			p. 236 (1919)	
0,18	28030	−19,5	YOUNG	D A O 1, p.265 (1920)	
0,31	39220	−25,8	JORDAN	A O 3, p.119 (1914)	
0,10	53580	−23,2	WILSON	L O B 7, p.107 (1913)	
0,0028	494	−18,6	JORDAN	A O 3, p. 82 (1914)	*Time of maximum positive velocity.
0,00019	172	−21,2	BOOTHROYD	D A O 2, p.173 (1922)	
[4,35]	3388	−22,0	PLASKETT	D A O 1, p.143 (1919)	Eclipsing variable. *From minimum. From STEWART's photometric orbit (darkened) and spectrographic elements ($i=68°,41$), $r_1=2{,}05$ Sun's radius; $r_2=10{,}25$ Sun's radius; $m_1=5{,}40\odot$; $m_2=1{,}69\odot$; $d_1=0{,}63D_0$; $d_2=0{,}0016D_0$.
[1,35]	10842				
0,10	3090	−19,1	MISS FOWLER	A O 3, p. 14 (1912)	Algol variable.
1,58	91010	+12,1	WILSON	L O B 8, p.133 (1914)	
[5,24]	3253	−15,1	PLASKETT	D A O 1, p.253 (1920)	Eclipsing variable. *From photometric phase. From SHAPLEY's photometric orbit (darkened) and spectrographic elements $r_1=4{,}23$ Sun's radius; $r_2=4{,}46$ Sun's radius; $m_1=5{,}24\odot$; $m_2=2{,}36\odot$; $d_1=0{,}085\,D_0$; $d_2=0{,}033D_0$; a (separation of two stars) $=15{,}05$ Sun's radius.
[2,36]	7215				
[0,91]	5370	+11,0	YOUNG	D O 4, p. 66 (1917)	
[0,65]	7490				
[5,3]	4380	− 5,0	JORDAN	A O 3, p. 192 (1915)	Eclipsing variable. K line has constant velocity.
[4,4]	5340				
[1,46]	8017	−15,6	HARPER	D A O 1, p.161 (1919)	
[1,44]	8134				
1,02	15808	− 6,2	HARPER	D A O 1, p.261 (1920)	
0,019	1745	−27,6	HARPER	D A O 2, p.182 (1922)	
0,00088	460	+ 0,9	WILSON, HUFFER	Pop Astr 29, p. 86 (1921)	
[0,52]	7930	−30,5	BAKER	A O 2, p. 43 (1910)	
[0,38]	10860				
[2,27]	10054	−13,0	HARPER	D O 4, p. 219 (1918)	
[2,06]	11055				
0,53	263250	−14,3	CANNON	D O 4, p. 172 (1918)	*Residuals from elliptic motion represented by sine curve of 1/3 the period of principal curve.
1,13	377000	−18,8	MERRILL	L O B 6, p. 6 (1910)	
0,00046	1170	+ 2,0	CURTIS	L O B 4, p.154 (1907)	
0,020	3910	− 5,5	HARPER	D A O 1, p.156 (1919)	
[16,44]	9230	−49,1	PLASKETT	D A O 1, p.215 (1920)	Eclipsing variable. H and K stationary. From SHAPLEY's (darkened) photometric orbit and spectroscopic data, $m_1=16{,}6\odot$; $m_2=15{,}3\odot$; $d_1=0{,}170\,D_0$; $d_2=0{,}158\,D_0$; $r_1=r_2=3216000$ km; $a=19294000$ km.
[15,11]					
[1,79]	4200	−16,2	BAKER	A O 2, p. 38 (1910)	
[1,67]	4620				

Spectroscopic

Boss	Star	R.A. 1900	Decl.	Mag.	Class	P d	T Jul. Day 2410000+	ω	e	K_1 km
....	HD 199579	20^h53^m,1	44°33′	6,0	Oe 5	48,608	12892,23	66°,7	0,099	42,2
5442		21 4,4	29 48	5,6	A 0	3,3137	12521,23*	......	0,0	26
5460	τ Cyg	10,8	37 37	3,8	F 0	0,1425	12522,64	263,0	0,306	8,0
5469	σ Cyg	13,5	38 59	4,3	A 0p	11,043	11069,27	119,1	0,40	2,0
....	HR 8170	17,2	39 55	6,5	F 8	3,24343	11801,55	357,5	0,022	62,2
5532	β Cep	27,4	70 7	3,3	B 1	0,1904795	9638,81	2,6	0,040	15,8
5565		35,8	57 2	5,6	Oe 5	1,36372	13300,55	289,9	0,334	77,0
5566	42 Cap	36,1	−14 30	5,3	G 5	13,25	11529,16	175	0,20	22,7
5591		39,9	28 19	6,9	A 5	3,74860	12175,16	262,7	0,189	92,1
								82,7	.....	93,2
5592	ϰ Peg	40,1	25 11	4,3	F 5	5,9715	5239,25	4,7	0,04	39,8
5600	δ Cap	41,5	−16 35	3,0	A 5	1,02275	11451,86	149,1	0,019	65,7
....	H.R. 8427	22 2,0	47 45	6,2	B 3	2,1721	12138,17*	......	0,0	127,7
5688	ι Peg	2,4	24 51	4,0	F 5	10,21312	4820,97	251,8	0,008	48,0
5764	2 Lac	16,9	46 2	4,7	B 5	2,6164	8193,30	180	0,015	80,3
								0	.....	98,8
5856	12 Lac	37,0	39 42	5,2	B 2	0,193089	*10761,15	......	0,0	16,9
5865	η Peg	38,3	29 42	3,1	G 0	818,0	5288,7	5,6	0,155	14,2
5900		48,1	16 19	5,7	K 0	24,65	12240,99	......	0	33,2
....	H.R. 8800	23 2,7	45 33	6,6	B 5	3,3372	12151,27	125,7	0,233	87,7
....	H.R. 8803	3,0	59 13	6,3	B 3	7,25050	11825,04	71,6	0,376	90,5
5965	ι Gru	4,7	−45 47	4,1	K 0	409,444	16116,14	251	0,75	15,4
5996	9 And	13,6	41 14	5,9	A 3	3,2195	11059,91	40,6	0,036	73,6
6046	1 H Cas	25,4	58 0	4,9	B 3	6,067	8223,76	3,3	0,224	59,1
6071	λ And	32,7	45 55	4,0	K 0	20,546	6683,46	301,0	0,086	7,1
....	Lal 46867	50,0	28 5	7,3	K 0	6,72127	12220,74	18,6	0,059	38,5
6142		50,5	56 51	6,0	B 0	13,435	10800,63	339,6	0,105	115,5
										167
6148		52,1	55 9	5,7	F 5	12,155	12162,60	213,4	0,278	71,5
								33,4	.....	73,0

Orbital Elements.

f	$a_1 \sin i$ 1000 km	V_0 km	Computer	Reference	Remarks
0,37	28076	− 5,8	PLASKETT	D A O 2, p.186 (1922)	Velocity Ca lines = − 11,3 km.
0,0060	1185	−26,8	YOUNG	D A O 1, p.321 (1921)	*Perihelion.
0,000006	15	−22,0	PARASKÉ-VOPOULOS	Ap J 53, p.146 (1921)	Brighter component of visual pair.
0,0000007	276	− 3,8	HENROTEAU	Det O 3, p. 49 (1917)	cf. § 107.
0,081	2773	+ 0,2	PLASKETT	D A O 1, p.115 (1919)	
0,00007	405	−14,1	CRUMP	Det O 2, p.152 (1915)	Variable. Residuals from elliptic curve represented by secondary curve of 1/3 the major period and semi-amplitude 1,25 km. There is some evidence of a variation in the velocity of centre of mass and in the amplitude of the velocity-curve.
0,054	1354	− 1,0	PLASKETT	D A O 2, p.273 (1923)	
0,016	4061	− 4,4	LUNT	C A 10, p. 3F (1921)	cf. § 137.
[1,19]	4662	+ 4,2	SANFORD	Ap J 53, p.219 (1921)	
[1,17]	4728				
0,044	3266	− 2,8	HENROTEAU	L O B 9, p.122 (1918)	cf. § 134.
0,030	926	− 5,8	CRUMP	Ap J 54, p.131 (1921)	
0,47	3810	−17,8	YOUNG	D A O 1, p.195 (1920)	*Perihelion passage. H and K lines of calcium yield a constant velocity of + 1,5 km.
0,12	6740	− 4,1	CURTIS	L O B 2, p.172 (1904)	
[0,87]	2890	− 9,0	BAKER	A O 1, p. 98 (1909)	
[0,71]	3550				
0,0001	45	−13,7	YOUNG	D O 3, p. 88 (1915)	*Perihelion. YOUNG D A O 1, p. 105 (1918) finds amplitude of velocity-curve varies from night to night with a range from 0 to 70 km. The shape of the curve is not constant and spectral lines vary in width. Velocity of system subject to long period variation. Amplitude from calcium lines show same variation as that from other lines. See also Pop Astr 30, p. 20 (1922).
0,23	157800	+ 4,3	CRAWFORD	L O B 1, p. 29 (1901)	
0,094	11254	−12,8	HARPER	D A O 1, p.205 (1920)	
0,21	3910	−15,1	YOUNG	D A O 1, p.243 (1920)	Velocity Ca lines = − 9,1 km.
0,35	7751	− 7,4	BOOTHROYD	D A O 1, p.284 (1921)	K line stationary. On a few plates two components of K line are visible, one on each side of stationary line.
0,045	57500	− 3,6	LUNT	C A 10, p. 12G (1924)	
0,13	3240	− 4,9	YOUNG	D O 4, p. 92 (1917)	
0,12	4920	−14,8	BAKER	A O 2, p. 28 (1910)	Eclipsing variable.
0,0009	1990	+ 7,4	BURNS	L O B 4, p. 89 (1906)	
0,040	3552	−19,8	SANFORD	Ap J 53, p.221 (1921)	cf. § 97.
[18,5]	21200	−26,7	YOUNG	D O 3, p. 387 (1916)	H and K lines stationary.
[12,7]	30700				
[1,70]	11483	+12,0	HARPER	D A O 2, p.267 (1923)	
[1,67]	11716				

h) The Algol Variables and β Lyrae Stars.

110. Light-Curve of an Eclipsing Binary. If the inclination i of the orbit plane of a binary system is sufficiently near 90° to cause a partial or total eclipse of one component by the other, each time such an eclipse occurs the total light of the system will appear to fluctuate.

When the companion is completely dark and incapable of reflecting the light of the primary, if the eclipse is only partial, the light-curve will consist of a straight line with periodic depressions as indicated in Fig. 35.

If the eclipse is annular and the disc of the primary uniformly illuminated, the bottom of the depression will also be a straight line as in Fig. 36.

Fig. 35. Partial eclipse, dark companion.

Fig. 36. Annular eclipse, dark companion.

When the eclipse of the primary by the secondary is total, the lowest extremity of the depression should be at infinity, but such cases are practically impossible.

If the system consists of a bright primary and a faintly luminous companion there will be two eclipses per period and the light-curve will resemble Fig. 37, the greater depression representing the eclipse of the primary with consequent great reduction of light, and the minor depression representing the eclipse of the secondary.

When the two stars are spherical, and the orbit circular, the position of the secondary minimum will be half-way between two primary minima. Otherwise, there may be a shift depending on the values of e, ω, i, and the shape of the components.

Fig. 37. Double eclipse, luminous companion.

Fig. 38. Phase effect.

If the secondary reflects the light of the primary, there will be a phase effect, due to the fact that slightly before or after the small eclipse, the second body presents a wider illuminated area than at any other time. The light-curve will then be similar to that of Fig. 38.

During an eclipse, account must be taken of the darkening toward the limb of the star's discs, also of the ellipsoidal shape of these discs, since these factors influence the theoretical shape of the light-curve.

Any star exhibiting variations of light that can be represented by curves similar to those above is called an Algol variable, the brightest and best known representative being β Persei or Algol, which is described in the present subdivision.

If the two components of a binary system are very large compared with their orbits, the effect of the reflection on the secondary may become exaggerated, and combined with a real physical action of one component upon the other, form a curve similar to that of Fig. 39.

Fig. 39. β Lyrae type.

This is a variable of the β Lyrae type, so named after the well known bright variable, which is considered in the present subdivision.

In case the two stars are identical in size, shape and brightness, the primary and secondary minima would be exactly the same. Then the light variation represented by Fig. 40 resembles a sine curve.

Fig. 40. W Ursae Majoris type.

Typical of this case is W Ursae Majoris. Other stars, however, such as ζ Geminorum, present light-curves of a similar type, but they are not eclipsing variables; for instance, they may be Cepheids, or ellipsoidal variables, or in the case of variations of small amplitude as detected with photo-electric cells, merely a phase effect as described for Fig. 38, but without the eclipses.

The "Katalog und Ephemeriden veränderlicher Sterne für 1927" published by the Astronomische Gesellschaft, gives the elements of 229 Algol variables and β Lyrae stars[1]. They are given in the Tables I and II, p. 412 to 416 (D = duration of eclipse, d = duration of minimum. Photographic magnitudes are italicised).

111. Determination of the Orbital Elements, Dimensions and Luminosity of the Components in an Eclipsing Binary. In his important work on the determination of orbits, BAUSCHINGER said: "Der Zusammenhang zwischen den Größen-, Formen- und Helligkeitsverhältnissen der Körper und den Elementen der elliptischen Bahn einerseits und der Lichtkurve anderseits ist aber so kompliziert, daß man eine allgemeine Theorie wohl kaum aufstellen kann, sondern die Lösung von Fall zu Fall den vorliegenden Verhältnissen anpassen muß."

Few astronomers have attached the problem of determining orbital elements, dimensions and luminosity of the components, and then usually for restricted cases containing one or more assumptions such as circular orbits, spherical stars, completely dark companions, etc.... A bibliography of the subject is:

E. C. PICKERING[2], Dimensions of the Fixed Stars, with Special Reference to Binaries and Variables of the Algol Type.

J. HARTING[3], Untersuchungen über den Lichtwechsel des Sterns β Persei.

A. W. ROBERTS[4], On the Relation Existing between the Light Changes and the Orbital Elements of a Close Binary System, with Special Reference to the Figure and Density of the Variable Star RR Centauri.

A. W. ROBERTS[5], On a Method of Determining the Absolute Dimensions of an Algol Variable Star.

J. H. JEANS[6], On the Density of Algol Variables.

S. BLAŽKO[7], Sur les étoiles variables du type Algol.

H. N. RUSSELL[8], On the Determination of the Orbital Elements of Eclipsing Variable Stars.

H. N. RUSSELL and H. SHAPLEY[9], On Darkening at the Limb in Eclipsing Variables.

J. WOLTJER[10], On a Special Case of Orbit Determination in the Theory of Eclipsing Variables.

J. FETLAAR[11], A Contribution to the Theory of Eclipsing Binaries.

B. W. SITTERLY[12], A Graphical Method for Obtaining the Elements of Eclipsing Variables.

---

[1] Kleinere Veröff. der Univ.-Sternwarte zu Berlin-Babelsberg No. 1 (1926).
[2] Proc Amer Acad of Arts and Sc 16, p. 1 (1881).
[3] Diss. München (1898).
[4] M N 63, p. 527 (1903).
[5] M N 66, p. 123 (1906).
[6] Ap J 22, p. 93 (1905).
[7] Ann. de l'Obs. de Moscou (2e série) 5, p. 76 (1911).
[8] Ap J 35, p. 315 (1911); 36, p. 54 (1912).
[9] Ap J 36, p. 239 and 385 (1912).
[10] B A N 1, p. 93 (1922).
[11] Recherches Astron. de l'Obs. d'Utrecht 9, Part 1 (1923).
[12] Pop Astr 32, p. 231 (1924).

Table I. Algol Stars.

Star		Epoch Jul. Day G. M. T.	Period	Brightness		*D*	*d*
				M	*m*		
RT	And	2424051,315	0^{d},6289316	9^{m},1	10^{m},3	2^{h},9	0^{h}
SY	,,	2418948,330	34,90796	10,7	12,2	41,8	21,1
TT	,,	9629,474	2,765054	11,3	12,6	8,6	1,0
TW	,,	2420051,618	4,122745	8,6	11,5	8,8	2,4
UU	,,	2607,380	1,486310	10,5	[12		
VV	,,	2419417,403	0,95974	9,7	10,2		
WW	,,	2422719,3	23,285	10,3	11,4		>2
WX	,,	2547,531	3,00104	10,7	14	10?	?
WZ	,,	4018,488	0,695649[1]	10,6	12,3	5,0	0,4
XZ	,, *	4152,2840	1,35730	9,6	11,8		
Y	Ant	2410002,96	3,0519	*9,1*	*9,9*		
RY	Aqr	9618,4493	1,966598	8,8	10,4	6	
SU	,,	2423980,5065	1,044562[2]	9,3	9,9		
XZ	Aql	4439,275	2,1387	*9,8*	*11,1*	6	
YZ	,,	4466,27	4,672142	11,5	14,2		
R	Ara	2415025,316	4,42509	6,8	7,9	9,2	0
RW	,,	2421794,34	4,3677	8,5	11,5	12	2,8
RS	Ari	4020,366	8,8031	9,0	10,3	>11,5	5,5
RY	Aur	2418661,292	2,72541	10,7	[12,3	7,5	
RZ	,,	8028,362	3,01058	11,6	13,3	11	
WW	,,	2420959,2734	2,525022[3]	6,0	6,5	4,5	0
β	,,	2418966,907	3,96008	2,1	2,2	6	
ε	,,	5850	9900	3,3	4,1	700^{d}	340^{d}
SS	Boo	2420707,420	7,60546	10,0	10,5	19^{h}	0^{h}
SU	,,	1071,397	1,56112	10,5	11,0	9	0
Y	Cam	2416306,388	3,305568[4]	9,7	11,8	10,1	1,4?
SS	,,	2420842,594	4,82438	9,0	9,8		
S	Cnc	2419865,470	9,484549	8,0	10,1	20,4	3,8
RU	,,	2420168,512	10,072878[5]	10,0	11,5	16	7
RY	,,	2784,4498	1,49293	12,5	15,3	3?	7
RS	CVn	3579,3474	4,79799	7,8	8,9	11	0
R	CMa	4151,441	1,13595	5,8	6,4	6	
RW	Cap	2418924,594	3,3923923	9,2	10,5	7,2	2
X	Car	5021,114	0,54132	7,9	8,7		
SS	,,	2423828,36	3,300759	*12,3*	*13,0*	13?	
ST	,,	3901,675	0,90165	*9,3*	*10,2*	4	
SW	,,	2423941,923	8,16608	*10,0*	*11,9*	18	0
YY	,,	3901,932	2,64264	*10,0*	*10,8*	7,5	0
AS	,,	3921,655	2,76593	*11*	*12,2*	8,5	0
AW	,,	3997,328	6,51070	*12,2*	*16,5*	14	3
BL	,,	3900,874	3,35531	*14,5*	*15,2*	10	0
BP	,,	3879,931	9,6449	*11,5*	*12,5*	9	0
CD	,,	3900,198	2,96756	*12*	*13*	6	0
CI	,,	3901,246	2,81849	*12*	*13*	7,5	0
CO	,,	3877,983	8,3121	*9,3*	*10,6*	20	0
CP	,,	3898,707	2,3942	*13,5*	*14,2*	7	0
CU	,,	3928,944	4,10112	*13,5*	*14,5*	5	0
CV	,,	3900,209	14,4157	*10,3*	*12*	35	17
CW	,,	3940,986	1,97398	*13,5*	*14,3*	6	0
CX	,,	3909,914	3,34717	*11,0*	*12,2*	10	0
CZ	,,	3900,498	2,2850	*13*	*14*	7	0
DD	,,	3904,270	1,44272	*13,5*	*14,2*	5	0
DE	,,	3891,343	3,71306	*12*	*15*	10	0
DF	,,	3906,697	1,8663	*13*	*13,5*	4,5	0
DK	,,	3918,458	11,335	*12*	*13*	27	0
DL	,,	3901,355	4,8215	*13*	*14,5*	10	0
DM	,,	3911,648	5,3163	*13*	*15*	18	3
DN	,,	3871,050	1,4525	*13*	*14,3*	5	
DO	,,	3868,292	3,8543	*8,9*	*9,1*	9,5	
DP	,,	3877,795	7,5637	*13*	*13,8*	18	

Table I (Continued).

Star		Epoch Jul. Day G. M. T.	Period	Brightness M	m	D	d
DQ	Car	2423872,929	$0^d,86691$	*$11^m,1$*	*$11^m,5$*	4^h	
DR	,,	3849,5	3,9945	*12*	*12,4*		
DS	,,	3843,447	1,0988	*12,5*	*12,8*	2,5	
DT	,,	3869,293	4,2866	*14,5*	*14,9*	11	
DU	,,	3856,352	4,9695	*13,5*	*14,5*	12	
DV	,,	3840,321	0,8405	*10,0*	*10,2*	5	
DW	,,	3820,770	0,66382	*9,5*	*10,0*	5,5	
DX	,,	3810,13	10,466	*10,6*	*10,8*	24	
DZ	,,	3899,235	2,3921	*13*	*13,6*	8,5	
EF	,,	3899,42	6,578	*13*	*13,5*	13	
EG	,,	3874,264	6,908	*13,5*	*14*	16,5	
EH	,,	3916,48	13,367	*12,5*	—	26	
EI	,,	3870,557	0,90145	*14*	*15,2*	3,5	
EK	,,	3864,54	1,75106	*14*	*14,4*	7	
EL	,,	3912,93	3,46824	*13,5*	*14*	10	
EM	,,	3906,125	1,7071	*8,9*	*9,2*	10	
EN	,,	3879,894	1,53498	*10,4*	*10,7*	5,5	
EO	,,	3870,129	2,15225	*13,5*	*14,3*	9,5	
EU	,,	3918,540	2,52589	*14*	*16*	5,5	
EX	,,	3997,641	1,396366	*10,0*	*12,2*	4,8	
EZ	,,	3995,432	0,594312	*9,7*	*10,0*	5,8	
FG	,,	3903,885	2,262515	*13*	*14,5*	7,7	
FL	,,	3940,386	0,92576	*12,5*	*12,8*	5,5	
FP	,,	1725,224	176,027	*10,1*	*11,9*	5^d	$1^d,8$?
RZ	Cas	2417355,4233	1,1952506	6,4	7,7	$5^h,7$	$0^h,4$
TV	,,	2420117,7464	1,8126096	7,3	8,2	6,2	0
TW	,,	2419819,3771	2,856635	8,3	9,0	6	
XX	,,	2423369,4211	3,067317	9,3	10,5	11,4	
ZZ	,,	4245,365	1,24358	*11,3*	*12,0*	2,8	
SS	Cen	2410000,350	2,47871	*8,7*	*10,2*	14,5?	
SU	,,	0004,595	5,35442	*8,7*	*9,5*	19?	
SW	,,	0002,90	5,21943	*9,1*	*11,6*	14,5?	
SY	,,	0001,275	6,6313	*9,9*	*10,7*	2,4	
AR	,,	2423893,2	8,971	*12,5*	*14*		
AV	,,	3899,932	1,5804	*14*	*15*	8,5	
U	Cep	3947,0282	2,4929409	6,8	9,2	11,1	2,0
RS	,,	2417140,469	12,4204	10,3	12,0	9,6	
TV	,,	2421366,6	3,8571	*12,7*	[*14*		
U	Col	2411080,371	1,24617[6]	9,6	10,2		
TY	CrA	2422527,75	2,8888	8,6	10,5		
U	CrB	4152,946	3,4522008	7,8	9,0	10,6	1,2
RW	,,	0401,3132	0,7264171	9,3	9,8	4,3	0
Y	Cyg	3317,354	2,99634	7,1	7,9	4	0
SW	,,	2418971,206	4,572820	9,9	12,5	13,4	3,4
SY	,,	2420001,571	6,005716	11,1	13,2	16,3	5,7
UW	,,	2419624,501	3,450797	10,4	13,0	10,5	1,3
UZ	,,	2420095,221	31,30586	10,1	12,1	60	22
VV	,,	0270,451	1,477039	12,9	13,6	4,8	
VW	,,	0327,733	8,430274	9,9	12,6	18,2	5,8
WW	,,	2418882,372	3,317676	9,4	13,4	12,0	0,7
ZZ	,,	2421137,438	0,6286185	10,4	11,5	4,0	0
AE	,,	2415704,451	0,96919	10,8	11,4	5,5	0,0
BO	,,	2422191,473	3,51244	11,4	12,3		
BR	,,	3279,3477	1,3325547	9,1	10,0	4,5	0,5
W	Del	2418183,243	4,806044	9,4	12,1	12,3	2,1
RR	,,	7976,449	4,59952	10,5	11,8	17,7	0,0
Z	Dra	7837,519	1,3574184	10,5	12,7	4,7	0,3
RR	,,	7026,680	2,83110[7]	10,0	12,9	9,9	1,0
RX	,,	7502,411	1,89318	9,9	10,7	4,3	
SX	,,	8652,425	5,169258	9,5	11,8	17,5	4,2

Table I (Continued).

Star		Epoch Jul. Day G. M. T.	Period	Brightness M	Brightness m	D	d
TW	Dra	2419661,414	2^d,806566	7^m,6	10^m,1	9^h,0	1^h,8
UZ	,,	9429,306	1,630632[2]	9,0	9,8		
WW	,,	9890,777	3,501	*9,8*	*10,7*		
S	Equ	2424151,943	3,436038	8,1	10,3	8	0,7
RU	Eri	2419685,508	0,6322	8,9	9,5	5,5	0
RY	,,	2420174,240	4,99	10,3	11,5	9,6	
RZ	,,	3834,32	19,64	*7,8*	*9,5*		
RU	Gem	0533,35	8,6707	12,0	13,0		
RW	,,	2418222,419	2,865507	9,8	11,9	12,0	1,6
RX	,,	7970,00	12,20868	8,6	10,7	9?	
RY	,,	8015,324	9,3009	8,6	10,4		
SV	,,	8662,46	4,00604	10,0	11,3	15?	
SX	,,	9031,27	1,36692	10,8	11,5	4,5	
TX	,,	9848,412	2,80007	10,0	11,9	9	
YY	,,	2424595,4105	0,81430	9,0	9,6	2,2	0
W	Gru	2410001,60	1,47603	*9,5*	*10,0*		
Z	Her	3086,365	3,992795	7,4	8,0	9,6	2,2
RX	,,	2423298,4521	1,77857	7,2	7,9	4,6	0
SZ	,,	2418495,4036	0,81809741[8]	9,5	11,2	3,3	0,0
TT	,,	8132,39	20,755	8,9	9,5	48	
TU	,,	2424152,445	2,26715	9,5	11,7	8	0,8
TX	,,	2419999,3651	2,059797	7,8	8,8	5	0
UX	,,	9876,489	1,54885	8,8	10,5	5	
AD	,,	2423257,420	9,766617	9,3	10,5	26,2	6
RX	Hya	0239,306	4,56326	9,3	11,4	9	2,5
SX	,,	2410001,235	2,89570	*8,6*	*12,7*		
TT	,,	2424615,40	6,96	*7,6*	*10,1*	40	5
RW	Lac	2418652,19	5,18453	10,2	11,2		
SS	,,	2420845,395	1,201499[9]	9,2	9,6	7	0
TW	,,	4387,334	3,03754	10,2	12,7		
Y	Leo	2418054,4223	1,686097	9,4	12,3	4,8	0
RT	,,	2420224,36	7,448	9,9	10,7		
RW	,,	1342,365	1,68252	11,7	13,2	4,0	0,3
T	LMi	2053,450	3,019875	10,0	12,4	9,3	0
SS	Lib	0251,472	1,43792	9,7	10,7	2?	
δ	,,	3601,7643	2,32735	5,0	5,9	10	
RR	Lyn	3835,33	9,945	5,8	6,2	8	
RV	Lyr	2431,576	3,599001	11,6	13,5	9,1	1,9
TT	,,	2419680,649	5,243754	9,2	11,3	20,6	0
TZ	,,	2420669,480	0,5288227	10,5	10,9	2,2	0
UZ	,,	2511,5040	1,8912500[10]	9,8	11,5	6,8	0
RU	Mon	2417262,351	0,89615	9,8	10,3	5	0
RW	,,	7943,701	1,906095	8,9	12,0	6,9	1,1
U	Oph	8152,507	1,6773476	6,0	6,8	7,7	0
RV	,,	2424152,243	3,68713	9,5	11,0		
RZ	,,	2418630,6	261,97	10,0	10,9	18^d	6^d,5
SW	,,	0002,320	2,44578	*9,2*	*10,0*		
SX	,,	0001,704	2,06330	10,5	11,2	4^h,5	0^h
SZ	,,	2420664,455	7,41699	10,0	11,5	12,0	0
WZ	,,	2897,4557	2,091748	8,9	10,1	4,8	0,0
Z	Ori	2419857,338	5,203	9,6	10,7	15?	0
VV	,,	9835,278	1,485382	*5,1*	*5,4*		
BM	,,	2422717,24	6,47075	8,1	8,8	?	?
BN	,,	4386,0	27	9,5	10,3	4^d,5	
TY	Peg	3676,464	3,092253	9,9	12,1	$<8^h,7$	
UX	,,	2183,496	3,0892	10,7	11,2		
VW	,,	0777,128	2,64276	9,9	10,6	3,2	0
AQ	,,	4749,323	5,548	9,7	12,5	>12	4
Z	Per	2418173,663	3,056422	9,9	12,4	9,5	2,1
RT	,,	7861,6304	0,8494185[11]	10,7	12,0	4,1	0

Table I (Continued).

Star		Epoch Jul. Day G. M. T.	Period	Brightness		D	d
				M	m		
RV	Per	2418629,516	1^d,973491	10^m,8	13^m,0	7^h	0^h
RW	,,	8711,41	13,1989	8,8	11,0	24	
RY	,,	8216,701	6,863571	8,2	10,6	22,8	3,6
ST	,,	8548,499	2,648382	9,7	11,8	9,1	1,0
WY	,,	2421827,483	3,32711	11,0	13,7	8,0	0
AB	,, *	2987,250	7,15965	9,1	9,8		
β	,,	3603,2674	2,86731	2,3	3,5	9,3	0
Y	Psc	2410002,844	3,76582	9,0	12,1	>8,3	0
RV	,,	2424388,403	0,276998	11,3	12,2	2,9	0,0
W	PsA	2410001,8	8,18	*10,0*	*[11,5*		
RR	Pup	5021,859	6,42984	9,5	10,7	14,1	7,9
U	Sge	7130,4126	3,3806234	6,6	9,4	12,5	1,8
RS	Sgr	5023,085	2,41570	6,6	7,6	12,5	8,0
SX	,,	0001,980	2,07698	*8,7*	*10,0*		
WX	,,	9317,2590	2,129095	9,3	11,1	5,3	0
WY	,,	9313,272	4,6708	9,5	10,6	14?	0
XY	,,	9979,375	2,022919	10,8	12,3	8,2	0,0
XZ	,,	2420311,519	3,275537	8,5	11,3	7	2,2
YY	,,	2419305,299	1,0610893	9,4	10,1	4	0
AI	,,	2421495,33	8,7695	12,3	[13,5	22±	?
BQ	,,	2412592,7905	8,01964	*9,7*	*12,5*	15	3,7
EG	,,	5929,40	2,48620	*9,8*	*12,0*		
W	Sct	2420665,47	10,26980	9,5	10,2	30	0?
RS	,,	4334,630	1,32880	9,3	10,3	4,8	0
RZ	,,	2419640,90	15,1895	7,3	8,5	77	
AC	,,	2424350,558	4,79745	*10,0*	*12,5*	7,7	4,0
AD	,,	4669,50	1,075	*12,5*	*13,2*		
BN	,,	4244,58	7,3008	*12,0*	*15,0*	30	
RW	Tau	2417198,419	2,768874[12]	7,1	11,0	7,7	1,2
SV	,,	2424202,446	2,16690	9,4	10,6	10	
TY	,,	4383,68	1,165	11,6	12,2	4	
λ	,,	2399607,538	3,95295[13]	3,8	4,2	10,5	
X	Tri	2423130,314	0,971537	8,6	10,6	5,0	0,4
V	Tuc	2416662,840	0,87092	10,0	11,8	4,3	
RW	UMa	8987,409	7,32820	10,4	11,4	14,6	4,8
W	UMi	2421219,543	1,701213	*8,4*	*9,3*		
S	Vel	2415021,155	5,93358	7,8	9,3	15,0	6,3
RR	,,	8867,226	1,854210	10,0	10,9	4,5	0
TT	,,	0001,57	2,10838	*10,2*	*[12,0*		
XY	,,	2423901,375	2,51037	*11,5*	*12,8*	6,5	0
YY	,,	3871,249	4,1636	*10,6*	*11,1*	12	
YZ	,,	3915,6	5,4895	*13*	—		
ZZ	,,	3700,420	2,87615	*9,7*	*9,9*	8	
AA	,,	3852,212	1,1679	*9,4*	*9,5*	10	
AC	,,	3934,011	2,28083	*9,1*	*9,5*	15,4	
UY	Vir	4302,399	1,99955	*8,2*	*8,9*		0,8
Z	Vul	0364,602	2,45492	7,1	8,8	11,0	0,0
RR	,,	0661,475	5,05026[2]	9,6	10,9	10,8	3,3
RS	,,	0606,623	4,477666	7,4	8,1	11,0	4,8

Remarks concerning Table I.

* XZ And and AB Per are probably β Lyrae-Stars.

1 $+ 0^d,017 \sin (1°,104 \cdot E + 168°,96)$.

2 Variable period.

3 $+ 0^d,0120 \sin 0°,360 (E - 12,8)$.

4 $- 0^d,60 \cdot 10^{-7} E^2$.

5 $+ 0^d,325 \cdot 10^{-6} E^2$.

6 $+ 0^d,0163 \sin (0°,0965\, E + 203°,21)$.

7 $- 0^d,30 \cdot 10^{-7} E^2$.

8 $+ 0^d,0060 \sin (0°,1230\, E + 30°,25)$.

9 $+ 0^d,120 \sin (1°,248\, E + 83°,0)$.

10 $+ 0^d,0045 \sin (0°,44\, E + 325°,7)$.

11 $- 0^d,238 \cdot 10^{-8} E^2$.

12 $+ 0^d,200 \sin 0°,108 (E - 335)$.

13 $- 0^d,0255 \sin (0°,2637\, E + 59°,1)$.

Table II. β Lyrae Stars.

Star	Epoch Jul. Day G. M. T.	Period	Brightness M	m_1	m_2
S Ant	2410741,525	0^d,64833872	6^m,4	6^m,8	6^m,8
AL Aql	2424382,40	2,6627	*10,6*	*13,0*	
SX Aur**	0446,654	3,06468	*7,8*	*8,6*	*8,6*
TT ,,	2605,642	1,332728	8,4	9,1	8,8
TU Boo	4609,332	0,32440	*11,8*	*12,5*	*12,5*
TY ,,	4610,365	0,31730	*11,8*	*12,3*	*12,2*
TZ ,,	4609,542	0,29712[1]	*10,7*	*11,2*	*11,2*
RZ Cnc*	2100,40	21,643	8,6	10,1	9,2
BS Car	3866,298	0,292506	*14,7*	*15,3*	*15,3*
RX Cas	2416250,9	32,315	8,5	9,1	
SX ,,	7983,45	36,5668	9,0	9,7	9,1
TX ,,	9708,339	2,92694	9,2	9,8	
RR Cen	5021,050	0,605680	7,4	7,8	7,8
RZ ,,	0000,155	1,87592	*8,5*	*9,0*	
SV ,, *	2421627,946	1,6605382[2]	*8,8*	*9,7*	*9,1*
SZ ,,	2410001,789	4,10796	*8,2*	*8,8*	
VZ ,,	0001,7	2,4645	*8,3*	*8,7*	
AS ,,	2423819,259	0,305218	*13*	*13,3*	*13,3*
RW Com	2419127,234	0,2373492	*11,5*	*12,1*	*12,1*
RU CrB*	9488,02	7,610	8,8	9,5	9,1
W Cru	0158	198,5	*8,9*	*9,5*	*9,2*
WZ Cyg	4936,549	0,58446	9,9	10,8	10,2
CG ,,	2422967,4283	0,631143	9,9	10,4	10,2
CP ,,	3404,555	0,4984	6,2	6,4	6,4
RW Dor	4172,537	0,285458	*10,8*	*11,4*	*11,2*
RZ Dra	2417673,2507	0,5508772	9,9	10,4	10,1
W Eqa	2424765,29	2,114	10,5	11,4	
AK Her**	2935,948	0,421519	8,2	8,9	8,9
u ,,	2411431,384	2,051028	4,8	5,3	4,9
RT Lac	2421913,483	5,073921	9,1	10,1	9,7
SW ,,	2419144,1635	0,32071382	*9,2*	*9,9*	*9,9*
β Lyr	2398590,57	12,908006[3]	3,5	4,1	3,8
U Peg	2423735,3396	0,374784005	9,3	9,9	9,7
V Pup	2415021,212	1,45448	4,1	4,9	4,7
RT Scl	1736,114	0,51156935	*9,6*	*10,5*	*9,8*
U Sct	6366,300	0,95499	9,7	10,7	9,9
V Ser	0002,562	3,45356	9,5	10,5	9,8
RS ,,	2420656,1902	0,598142	10,8	11,5	11,4
T Sex	4525,590	0,4809	10,0	10,6	
RZ Tau	4177,4172	0,3440033	*10,5*	*11,1*	*11,1*
V Tri	0830,238	0,58520	10,6	11,5	10,9
W UMa	0416,4587	0,3336381566[4]	7,9	8,5	8,5

* RZ Cnc, SV Cen and RU CrB are probably Algol stars.
** Character of the light-curve still uncertain.
[1] The period is probably 0^d,25865. [2] $- 0^d,2682 \cdot 10^{-6} E^2$.
[3] $+ 0^d,3914 \cdot 10^{-5} E^2$. [4] $- 0^d,676 \cdot 10^{-10} E^2$.

112. Russell's Method. A résumé of Russell's method for determining the elements of eclipsing variables is vital and nothing better can be done than to reproduce here the summary he has published in Ap J 36, pp. 404—408 (1912).

I. General Notation.

The unit of length is the radius of the relative orbit; the unit of light is the combined light of the two stars; the unit of mass is their combined mass.

The subscript 1 refers always to the larger star. The principal quantities used are defined as follows:

Spherical Stars.

r_1, radius of large star.
$r_2 = k r_1$, radius of small star.
δ, apparent distance of centers.
i, inclination of the orbit.
t, time from principal minimum in days.
P, period of revolution in days.
$\theta = \frac{2\pi t}{P}$, true longitude in orbit.
$\frac{J_1}{J_2}$, ratio of surface brightness.
ϱ_1, ϱ_2, densities of the stars.

l, light at any time.
L_1, light of large star.
$L_2 = 1 - L_1$, light of small star.
λ_1, minimum light at eclipse when large star is in front.
λ_2, minimum light, small star in front.
α, loss of light at any time in terms of loss for complete eclipse.
α_0, greatest loss at minimum.
n, fraction of greatest loss.
y, mass of larger star.

Elliptical Stars.

a_1, a_2, semi-major axes.
b_1, b_2, semi-minor axes.

ε, eccentricity of meridian section.
$z = \varepsilon^2 \sin^2 i$.

Darkened Stars.

x, coefficient of darkening (Ap J 36, p. 240). Z, defined by $z = \frac{5}{4} Z - \frac{5}{28} Z^2$.
α'', loss of light at any time during an eclipse where the small star is in front, in terms of the loss at the moment of internal tangency.

II. Functional Notation and Index of Tables.

Function	Use	Where found: Uniform	Darkened: Big Star in Front	Darkened: Small Star in Front
$p(k, \alpha)$ Ap J 35, p. 332	Relation between loss of light and distance of centres	Table I	Table Ix	Table Iy
$\Psi(k, \alpha_1)$ Ap J 35, p. 321	To find light-curve when L_1 and L_2 are known	Table II	Table IIx	Table IIy
$\Phi(k, \alpha)$ Ap J 36, p. 254	To find light-curve for central total eclipse	Table IV	Table IVx	
$\varphi_1(k)$, $\varphi_2(k)$ Ap J 35, p. 336	To find the elements when L_1 and L_2 are known	Table IIa	Table IIa x	
$\chi(k, \alpha_0, \frac{1}{4})$ etc. Ap J 35, p. 326	To find elements in case of partial eclipse	Table III	Table IIIx	Table IIIy
$\omega_1(n)$, $\omega_2(n)$ Ap J 35, p. 329	To find light-curve in case of partial eclipse (highly approximate)	Table IIIa		
$Q(k, \alpha_0)$ Ap J 36, p. 389	To aid in solution of partial eclipse of completely darkened stars		Table V	
$X(k, \alpha)$, $Y(k, \alpha)$ Ap J 36, p. 402	To aid in discussion of intermediate degrees of darkening		Table VI	

Table A is for converting stellar magnitudes into light intensities.

Table B is for converting θ into $\sin\theta$.

Table C is for converting elements derived on the "uniform" hypothesis into "darkened" elements (highly approximate).

III. Fundamental Equations for Spherical Stars.

The references to the original discussion precede the equations.

θ_1 corresponds to $\alpha = \alpha_1$, θ_2 to $\alpha = 0{,}6$, θ_3 to $\alpha = 0{,}9$, $\theta(n)$ to $\alpha = n\,\alpha_0$, θ' to $\alpha = 0$, and θ'' to $\alpha = 1$.

Ap J 35, p. 321 $\delta^2 = \cos^2 i + \sin^2 i \sin^2\theta,$ (a)

Ap J 35, p. 321 $\delta = r_1\{1 + k p(k, \alpha)\},$ (b)

Ap J 35, p. 321 $A = \sin^2\theta_2,\ B = \sin^2\theta_2 - \sin^2\theta_3,\ \sin^2\theta_1 = A + B\,\psi(k, \alpha_1),$ (c)

Ap J 35, p. 323 $\cot^2 i = \frac{B}{\varphi_2(k)} - A,\quad r_1^2 \operatorname{cosec}^2 i = \frac{B}{\varphi_1(k)},$ (d)

Ap J 36, p. 254 $\sin^2\theta_1 = A\,\Phi(k, \alpha_1),$ (e)

Ap J 35, p. 326 $\frac{\sin^2\theta(n)}{\sin^2\theta(\frac{1}{2})} = \chi(k, \alpha_0, n),$ (f)

Ap J 35, p. 329 $\sin^2\theta(n) = C\,\omega_2(n) + D\,\omega_1(n),\quad \chi(k, \alpha_0, \tfrac{1}{4}) = \frac{C}{D},$ (g)

Ap J 35, p. 331 $\left\{\begin{aligned}\cos^2 i \cos^2\theta' + \sin^2\theta' &= r_1^2(1+k)^2\\ \cos^2 i &= r_1^2\{1 + k p(k, \alpha_0)\}^2\end{aligned}\right\}.$ (h)

For Uniform Disks

$$\alpha_0 = 1 - \lambda_1 + \frac{1-\lambda_2}{k^2}, \qquad \text{(j)[1]}$$

$$l_1 = 1 - \alpha L_2,\ 1 - \lambda_1 = \alpha_0 L_2, \qquad \text{(k)[2]}$$

$$l_2 = 1 - k^2(1 - l_1)\frac{L_1}{L_2}, \qquad \text{(m)}$$

For Darkened Disks

$$Q(k, \alpha_0) = \frac{(1-\lambda_2)}{\alpha_0 - (1-\lambda_1)}, \qquad \text{(n)}$$

$$l_1 = 1 - \alpha L_2,\ 1 - \lambda_1 = \alpha_0 L_2;$$

α_0 from Table Ix. (p)

$$l_2 = 1 - Q(k, \alpha)\frac{L_1}{L_2}(1 - l_1) = 1 - \alpha_0'' L_1 Q(k, 1), \qquad \text{(q)}$$

$$\alpha_0'' Q(k, 1) = \alpha_0 Q(k, \alpha_0), \qquad \text{(r)}$$

$$\varrho_1 = 0{,}01344\frac{y}{P^2 r_1^3},\qquad \varrho_2 = 0{,}01344\frac{1-y}{P^2 r_2^3}, \qquad \text{(s)[3]}$$

$$\frac{J_1}{J_2} = \frac{k^2 L_1}{L_2}. \qquad \text{(t)}$$

IV. Directions for the Computer.

Preliminary: Reduce the observed curve to intensity-curve, taking maximum intensity as unity. Determine the epoch of minimum, and draw the best symmetrical curve to represent it.

A. The light is constant between minima, therefore the stars are spherical.

1. There is a constant phase at principal minimum of brightness λ_1. Then $L_1 = \lambda_1$, and $L_2 = 1 - \lambda_1$.

---

[1] Ap J 35, p. 320 (1912). [2] Ap J 35, p. 318 (1912).
[3] Ap J 36, p. 73 (1912).

Solution on uniform hypothesis: For tabular values of α_1 compute l_1 by (k), read θ from the light-curve, take $\sin\theta$ from Table B, find $\psi(k, \alpha_1)$ from (c), hence k from Table II, and other elements from (d) and Table IIa. If $\cot^2 i$ comes out negative, use (e) and Table IV. Compute the light-curve for principal minimum from (c), (k), and Table II, and for the secondary minimum from (m).

For completely darkened stars substitute Tables IIx, IIax, and IVx, for II, IIa, and IV; and compute secondary minimum by (q) and Table V.

2. There is no constant phase at principal minimum, therefore the eclipses are partial (or perhaps the principal eclipse is annular if the stars are darkened). Knowledge of the secondary minimum is essential in this case.

For uniform stars use (k) and the light-curve to find $\theta(n)$, then find C and D from (g), using Table IIIa. For assumed values of α_0 compute k from (j) and $\chi(k, \alpha_0, \frac{1}{4})$ from Table III, and find what value of α_0 gives $\chi = \frac{C}{D}$. Then compute θ' from (g) (setting $n = 0$), the elements from (h), using Table I, and L_2 from (k).

For darkened stars find $\theta(n)$ as above, then $\chi(k, \alpha_0, \frac{3}{4})$, $\chi(k, \alpha_0, \frac{1}{4})$, and $\chi(k, \alpha_0, 0)$ from (f). For assumed values of α_0 find k from (n) and Table V, and α_0'' from (r). Then take the χ-functions from Table IIIx, with arguments k, α_0, if the principal minimum is assumed to correspond to λ_1, or from Table IIIy, with arguments k, α_0'', if to λ_2, and find what values of α_0 and k give the best agreement. Compute the other elements by (h), using Table Ix, and find L_2 from (p). For details regarding annular eclipses see Ap J 36, pp. 393ff.

B. The light is not constant between minima, therefore the stars are elliptical. Rectify the intensity-curve: (1) for ellipticity of the stars, Ap J 36, pp. 64, 65 (illustration, Ap J 36, pp. 139ff.); (2) for radiation-effect (if necessary), Ap J 36, pp. 67—69. Then discuss as under (A) except that for equations (c), (d), (f), (g), (h) must be substituted, respectively,

$$\psi(k, \alpha_1) = \frac{(\sin^2\theta_1 - \sin^2\theta_2)(1 - z\cos^2\theta_3)}{(\sin^2\theta_2 - \sin^2\theta_3)(1 - z\cos^2\theta_1)}, \qquad (c')^1$$

$$\left.\begin{aligned} \cos^2 i + \sin^2 i \sin^2\theta' &= \alpha_1^2 (1 - z\cos^2\theta')(1 + k)^2, \\ \cos^2 i + \sin^2 i \sin^2\theta'' &= \alpha_1^2 (1 - z\cos^2\theta'')(1 - k)^2, \\ \cos^2 i &= \alpha_1^2 (1 - z)\{1 + kp(k, \alpha_0)\}^2. \end{aligned}\right\} \qquad (d' \text{ and } h')^2$$

$$\frac{\sin^2\theta(n)\{(1 - z\cos^2\theta(\frac{1}{2}))\}}{\sin^2\theta(\frac{1}{2})\{(1 - z\cos^2\theta(n))\}} = \chi(k, \alpha_0, n). \qquad (f')^3$$

$$\frac{\sin^2\theta(n)}{1 - z\cos^2\theta(n)} = C\,\omega_2(n) + D\,\omega_1(n), \quad \frac{C}{D} = \chi(k, \alpha_0, \tfrac{1}{4}). \qquad (g')$$

C. For the discussion of eccentric orbits refer to Ap J 36, pp. 54—60, 396 to 397, and the illustration, Ap J 36, pp. 146ff. For the discussion of intermediate degrees of darkening refer to pp. 401—403, and Table VI.

113. Tables Used in RUSSELL's Method. By permission of Professor RUSSELL the tables used in his method are presented here.

[1] Ap J 36, p. 65 (1912). [2] Ap J 36, p. 66 (1912).
[3] Ap J 36, p. 66 (1912).

Table I, from Astrophysical Journal, vol. 35, p. 333.
Relation between the Eclipsed Area α and the Distance of Centres δ.

α	k=1,0	0,9	0,8	0,7	0,6	0,5	0,4	0,3	0,2	0,1	0,0
0,00	+ 1,000	+ 1,000	+ 1,000	+ 1,000	+ 1,000	+ 1,000	+ 1,000	+ 1,000	+ 1,000	+ 1,000	+ 1,000
,01	0,919	0,921	0,922	0,924	0,925	0,927	0,929	0,930	0,932	0,934	0,935
,02	,868	,871	,873	,876	,879	,881	,884	,887	,890	,892	,895
,05	,755	,759	,764	,769	,774	,779	,785	,790	,795	,800	,805
,10	,610	,618	,624	,631	,638	,645	,653	,661	,670	,678	,687
0,15	+ 0,488	+ 0,496	+ 0,504	+ 0,513	+ 0,523	+ 0,533	+ 0,544	+ ,554	+ 0,565	+ 0,576	+ 0,585
,20	,374	,388	,398	,408	,419	,430	,443	,456	,469	,481	,492
,25	,267	,284	,297	,310	,322	,335	,348	,363	,378	,391	,405
,30	,168	,186	,200	,216	,230	,244	,258	,272	,288	,303	,321
,35	+ ,075	,084	,110	,127	,143	,160	,175	,190	,207	,222	,239
0,40	− 0,015	+ 0,005	+ 0,024	+ 0,041	+ 0,059	+ 0,077	+ 0,094	+ 0,109	+ 0,126	+ 0,143	+ 0,159
,45	− ,106	− ,081	− ,061	− ,042	− ,023	− ,004	+ ,013	+ ,028	+ ,045	+ ,062	+ ,079
,50	− ,194	− ,166	− ,145	− ,124	− ,103	− ,084	− ,067	− ,051	− ,034	− ,017	− ,000
,55	− ,280	− ,250	− ,226	− ,204	− ,184	− ,165	− ,148	− ,131	− ,113	− ,096	− ,079
,60	− ,364	− ,332	− ,306	− ,284	− ,263	− ,244	− ,226	− ,209	− ,192	− ,175	− ,159
0,65	− 0,447	− 0,413	− 0,386	− 0,363	− 0,343	− 0,323	− 0,305	− 0,288	− 0,271	− 0,255	− 0,239
,70	− ,528	− ,492	− ,465	− ,441	− ,420	− ,401	− ,383	− ,367	− ,350	− ,336	− ,321
,75	− ,607	− ,571	− ,544	− ,520	− ,498	− ,481	− ,463	− ,448	− ,432	− ,419	− ,405
,80	− ,686	− ,649	− ,622	− ,600	− ,580	− ,563	− ,546	− ,532	− ,517	− ,504	− ,492
,85	− ,765	− ,728	− ,701	− ,680	− ,663	− ,648	− ,633	− ,620	− ,607	− ,596	− ,585
0,90	− 0,843	− 0,807	− 0,783	− 0,764	− 0,749	− 0,736	− 0,725	− 0,715	− 0,705	− 0,696	− 0,687
,95	− ,922	− ,890	− ,872	− ,858	− ,847	− ,838	− ,830	− ,823	− ,817	− ,811	− ,805
,98	− ,967	− ,945	− ,935	− ,928	− ,922	− ,915	− ,910	− ,905	− ,900	− ,896	− ,892
,99	− ,983	− ,967	− ,960	− ,955	− ,951	− ,948	− ,945	− ,942	− ,939	− ,937	− ,934
1,00	− 1,000	− 1,000	− 1,000	− 1,000	− 1,000	− 1,000	− 1,000	− 1,000	− 1,000	− 1,000	− 1,000

Table II, from Astrophysical Journal, vol. 35, p. 335.
For Use in Case of Total Eclipse. Values of $\psi(k, \alpha_1)$.

α_1	k=1,00	0,90	0,80	0,70	0,60	0,50	0,40	0,30	0,20	0,10	0,00
0,00	+ 9,464	+ 7,478	+ 6,200	+ 5,279	+ 4,556	+ 3,984	+ 3,503	+ 3,104	+ 2,755	+ 2,454	+ 2,199
,02	8,095	6,457	5,373	4,606	4,000	3,504	3,106	2,768	2,478	2,216	2,000
,05	7,042	5,616	4,047	4,407	3,534	3,118	2,777	2,488	2,241	2,017	1,829
0,10	+ 5,759	+ 4,625	+ 3,895	+ 3,364	+ 2,960	+ 2,627	+ 2,358	+ 2,131	+ 1,934	+ 1,754	+ 1,603
,15	4,755	3,839	3,248	2,826	2,504	2,240	2,024	1,841	1,682	1,537	1,412
,20	3,906	3,184	2,712	2,374	2,110	1,898	1,726	1,581	1,453	1,336	1,235
0,25	+ 3,158	+ 2,600	+ 2,232	+ 1,969	+ 1,760	+ 1,591	+ 1,453	+ 1,344	+ 1,242	+ 1,146	+ 1,070
,30	2,522	2,088	1,803	1,603	1,443	1,314	1,205	1,115	1,039	0,968	0,911
,35	1,979	1,641	1,425	1,276	1,157	1,061	0,982	0,911	0,854	0,797	0,756
0,40	+ 1,490	+ 1,245	+ 1,087	+ 0,978	+ 0,894	+ 0,825	+ 0,770	+ 0,721	+ 0,675	+ 0,633	+ 0,604
,45	1,040	0,881	0,777	,705	,649	,603	,566	,530	,501	,473	,453
,50	0,648	,555	,491	,451	,418	,392	,370	,348	,331	,314	,302
0,55	+ 0,300	+ 0,258	+ 0,233	+ 0,217	+ 0,202	+ 0,191	+ 0,181	+ 0,171	+ 0,164	+ 0,156	+ 0,151
,60	0,000	0,000	0,000	0,000	0,000	0,000	0,000	0,000	0,000	0,000	0,000
,65	− ,258	− 0,231	− 0,214	− 0,202	− ,191	− 0,181	− 0,174	− 0,167	− 0,160	− 0,156	− 0,152
0,70	− 0,480	− 0,435	− 0,408	− 0,387	− 0,369	− 0,354	− 0,344	− 0,331	− 0,320	− 0,314	− 0,306
,75	− ,660	− ,613	− ,584	− ,558	− ,539	− ,522	− ,508	− ,494	− ,483	− ,475	− ,465
,80	− ,805	− ,765	− ,738	− ,717	− ,700	− ,684	− ,670	− ,659	− ,647	− ,639	− ,632
0,85	− 0,922	− 0,893	− 0,877	− 0,863	− 0,854	− 0,843	− 0,833	− 0,825	− 0,818	− 0,812	− 0,808
,90	− 1,000	− 1,000	− 1,000	− 1,000	− 1,000	− 1,000	− 1,000	− 1,000	− 1,000	− 1,000	− 1,000
,95	− 1,045	− 1,085	− 1,112	− 1,134	− 1,152	− 1,166	− 1,179	− 1,190	− 1,203	− 1,214	− 1,226
0,98	− 1,062	− 1,126	− 1,176	− 1,220	− 1,256	− 1,284	− 1,308	− 1,329	− 1,350	− 1,369	− 1,391
,99	1,0643	− 1,139	− 1,199	− 1,250	− 1,293	− 1,328	− 1,362	− 1,390	− 1,419	− 1,444	− 1,471
1,00	1,0650	− 1,155	− 1,231	− 1,297	− 1,354	− 1,402	− 1,445	− 1,484	− 1,525	− 1,556	− 1,596

Table III, from Astrophysical Journal, vol. 35, p. 337.
For Use in Case of Partial Eclipse. Values of $\chi(k, \alpha_0, \frac{1}{4})$.

α_0	$k=1{,}00$	0,95	0,90	0,80	0,70	0,60	0,50	0,40	0,30	0,20	0,10	0,00
1,00	2,462	2,316	2,194	2,010	1,870	1,756	1,667	1,596	1,541	1,491	1,441	1,407
0,98	2,402	2,270	2,168	2,005	1,872	1,770	1,690	1,628	1,572	1,522	1,479	1,437
,95	2,328	2,208	2,121	1,985	1,872	1,779	1,703	1,638	1,586	1,540	1,499	1,460
,90	2,218	2,135	2,062	1,950	1,860	1,778	1,708	1,650	1,605	1,560	1,522	1,485
,85	2,127	2,067	2,010	1,919	1,840	1,771	1,710	1,660	1,618	1,577	1,540	1,501
0,80	2,054	2,001	1,968	1,890	1,823	1,763	1,711	1,665	1,626	1,589	1,522	1,510
,70	1,956	1,926	1,896	1,842	1,792	1,749	1,711	1,675	1,640	1,606	1,570	1,532
,60	1,887	1,860	1,838	1,799	1,764	1,734	1,707	1,677	1,648	1,620	1,586	1,551
,50	1,828	1,805	1,789	1,762	1,740	1,719	1,699	1,675	1,650	1,626	1,597	1,568
,40	1,772	1,761	1,751	1,732	1,717	1,700	1,685	1,669	1,650	1,629	1,607	1,584
0,30	1,727	1,722	1,717	1,709	1,700	1,688	1,674	1,661	1,646	1,635	1,615	1,598
,20	1,693	1,692	1,690	1,687	1,682	1,675	1,665	1,655	1,642	1,635	1,623	1,608
,10	1,673	1,672	1,670	1,666	1,660	1,655	1,650	1,646	1,639	1,633	1,626	1,616
,00	1,630	1,630	1,630	1,630	1,630	1,630	1,630	1,630	1,630	1,630	1,630	1,630

Table IV, from Astrophysical Journal, vol. 36, p. 252.
Values of $\Phi(k, \alpha_1)$ for Uniformly Illuminated Disks.
For Computing the Values of k and the Light-Curve in the Case when the Eclipse is Central.
$\sin^2\theta = \Phi(k, \alpha_1)\sin^2\theta_6 = A\,\Phi(k, \alpha_1)$.

α	$k=1{,}0$	0,9	0,8	0,7	0,6	0,5	0,4	0,3	0,2
0,00	9,887	7,345	5,683	4,500	3,608	2,920	2,368	1,925	1,556
,10	6,407	4,922	3,940	3,233	2,695	2,266	1,919	1,627	1,391
,20	4,668	3,700	3,047	2,572	2,209	1,914	1,673	1,472	1,294
,30	3,368	2,761	2,363	2,062	1,826	1,635	1,470	1,332	1,210
,40	2,399	2,056	1,822	1,648	1,522	1,398	1,300	1,215	1,136
,50	1,609	1,471	1,371	1,299	1,239	1,189	1,144	1,104	1,067
,60	1,000	1,000	1,000	1,000	1,000	1,000	1,000	1,000	1,000
,70	0,549	0,631	0,692	0,744	0,789	0,829	0,870	0,901	0,935
,80	0,244	0,351	0,443	0,525	0,599	0,670	0,738	0,804	0,869
,90	0,061	0,152	0,245	0,337	0,428	0,518	0,610	0,702	0,798
,95	0,019	0,080	0,151	0,248	0,341	0,438	0,540	0,645	0,751
,98	0,002	0,045	0,112	0,191	0,281	0,381	0,490	0,604	0,727
1,00	0,000	0,020	0,070	0,140	0,225	0,324	0,436	0,558	0,692

Table V, from Astrophysical Journal, vol. 36, p. 394.
Values of $Q(k, \alpha_0)$ for Partial Eclipses of Disks Completely Darkened toward the Edge.

α_0	$k=1{,}0$	0,9	0,8	0,7	0,6	0,5	0,4
0,00	1,000	0,769	0,573	0,410	0,279	0,177	0,102
0,10	1,000	0,777	0,585	0,426	0,292	0,188	0,108
0,20	1,000	0,786	0,596	0,438	0,302	0,195	0,113
0,30	1,000	0,796	0,605	0,446	0,310	0,200	0,116
0,40	1,000	0,805	0,614	0,453	0,316	0,205	0,119
0,50	1,000	0,814	0,624	0,461	0,321	0,210	0,123
0,60	1,000	0,823	0,634	0,470	0,328	0,215	0,126
0,70	1,000	0,832	0,646	0,483	0,339	0,222	0,131
0,80	1,000	0,844	0,661	0,499	0,352	0,232	0,137
0,90	1,000	0,861	0,685	0,522	0,371	0,247	0,147
0,95	1,000	0,872	0,704	0,538	0,384	0,256	0,153
0,98	1,000	0,880	0,717	0,553	0,398	0,267	0,160
0,99	1,000	0,890	0,728	0,563	0,405	0,273	0,164
1,00	1,000	0,904	0,750	0,587	0,427	0,289	0,175
$1+x$	1,000	0,917	0,784	0,636	0,488	0,351	0,230

Table VI, from Astrophysical Journal, vol. 36, p. 403.

α	k=1,0	0,9	0,8	0,7	0,6	0,5	0,4
				Values of $X(k, \alpha)$.			
0,10	— 0,256	— 0,261	— 0,266	— 0,270	— 0,276	— 0,282	— 0,286
,20	0,117	0,125	0,131	0,137	0,142	0,146	0,151
,30	— 0,020	— 0,033	— 0,042	— 0,049	— 0,256	0,062	0,068
,40	+ 0,035	+ 0,022	+ 0,013	+ 0,005	— 0,002	— 0,008	— 0,015
,50	0,072	0,060	0,052	0,044	0,038	+ 0,032	+ 0,027
,60	0,097	0,083	0,073	0,067	0,062	0,058	0,055
,70	0,103	0,090	0,082	0,074	0,069	0,066	0,064
,80	0,088	0,079	0,073	0,067	0,064	0,062	0,060
,90	0,058	0,051	0,047	0,045	0,044	0,043	0,042
,95	0,035	0,030	0,028	0,027	0,026	0,026	0,025
,98	+ 0,016	+ 0,015	+ 0,014	+ 0,014	+ 0,014	+ 0,013	+ 0,013
1,00	0,000	0,000	0,000	0,000	0,000	0,000	0,000
				Values of $Y(k, \alpha)$.			
0,10	— 0,256	— 0,297	— 0,341	— 0,386	— 0,435	— 0,490	— 0,552
,20	0,117	0,154	0,191	0,228	0,277	0,332	0,399
,30	— 0,020	— 0,051	0,093	0,138	0,190	0,251	0,322
,40	+ 0,035	+ 0,019	— 0,020	0,068	0,126	0,194	0,268
,50	0,072	0,067	+ 0,033	— 0,011	0,071	0,139	0,209
,60	0,097	0,105	0,075	+ 0,031	— 0,024	0,086	0,160
,70	0,103	0,127	0,108	0,075	+ 0,025	— 0,036	0,111
,80	0,088	0,137	0,134	0,111	0,066	+ 0,010	0,064
,90	0,058	0,130	0,150	0,140	0,107	0,057	— 0,010
,95	0,035	0,122	0,156	0,160	0,133	0,087	+ 0,022
,98	+ 0,016	0,117	0,161	0,176	0,156	0,114	0,050
1,00	0,000	+ 0,116	+ 0,172	+ 0,198	+ 0,186	+ 0,152	+ 0,093
Central annular	0,000	+ 0,145	+ 0,225	+ 0,298	+ 0,356	+ 0,404	+ 0,437

Table II a, from Astrophysical Journal, vol. 35, p. 337.
For Computing the Elements in the Case of Total Eclipse.

k	$\varphi_1(k)$	$\varphi_2(k)$
1,00	0,380	0,939
0,95	,401	,894
,90	,417	,848
0,85	0,427	0,802
,80	,431	,755
,75	,431	,709
0,70	0,427	0,663
,65	,419	,617
,60	,406	,572
0,55	0,390	0,527
,50	,371	,482
,45	,349	,436
0,40	0,323	0,390
,35	,294	,345
,30	,262	,298
0,25	0,226	0,250
,20	,187	,202
,15	,145	,153
0,10	0,100	0,103
,05	,052	,052
,00	,000	,000

Table II a x, from Astrophysical Journal, vol. 36, p. 246.
For Computing the Inclination and the Radius of Larger Star.

k	$\varphi_1^1(k)$	$\varphi_2^1(k)$
1,00	0,444	0,865
0,95	,451	,824
,90	,454	,782
,85	,454	,740
,80	,451	,698
0,75	0,444	0,656
,70	,433	,614
,65	,420	,572
,60	,403	,530
0,55	0,383	0,488
,50	,361	,445
,45	,336	,402
,40	,308	,359
0,35	0,278	0,317
,30	,246	,274
,25	,212	,230
0,20	0,176	0,186

Table III a, from Astrophysical Journal, vol. 35, p. 338.
For Computing the Form of the Light-Curve in Case of Partial Eclipse.

n	$\omega_1(n)$	$\omega_2(n)$
0,00	— 3,94	+ 4,10
,10	— 1,45	+ 2,21
,20	— 0,399	+ 1,330
,25	,000	+ 1,000
0,30	+ 0,316	+ 0,720
,35	+ ,567	+ ,488
,40	+ ,758	+ ,295
,45	+ ,899	+ ,133
0,50	+ 1,000	0,000
,55	+ 1,065	— 0,107
,60	+ 1,095	— ,190
,65	+ 1,090	— ,249
0,70	+ 1,046	— 0,285
,75	+ 0,967	— ,297
,80	+ ,846	— ,285
,85	+ ,693	— ,250
0,90	+ 0,503	— 0,191
,95	+ ,273	— ,108
,98	+ ,114	— .047
,99	+ ,058	— ,024
1,00	0,000	0,000

Table Ix, from Astrophysical Journal, vol. 36, p. 243.
Relation between k, α^1 and $p(k, \alpha^1)$ for Total Eclipses of Disks Completely Darkened toward the Edge.

α^1	$k=1{,}0$	0,9	0,8	0,7	0,6	0,5	0,4	0,3	0,2
0,00	+1,000	+1,000	+1,000	+1,000	+1,000	+1,000	+1,000	+1,000	+1,000
,02	0,796	0,800	0,804	0,808	0,812	0,816	0,820	0,823	0,827
,05	,685	,689	,694	,698	,703	,708	,712	,717	,721
,10	,543	,550	,558	,564	,569	,576	,583	,591	,599
0,15	+0,428	+0,437	+0,447	+0,455	+0,462	+0,471	+0,480	+0,492	+0,503
,20	,328	,338	,349	,359	,369	,379	,389	,402	,415
,25	,241	,251	,263	,273	,283	,295	,306	,320	,334
,30	,160	,169	,181	,192	,203	,216	,229	,244	,259
0,35	+0,081	+0,091	+0,104	+0,116	+0,128	+0,142	+0,156	+0,172	+0,187
,40	,004	,017	,032	,045	,057	0,72	,085	,101	,117
,45	−0,070	−0,055	−0,039	−0,025	−0,011	,003	,018	,033	0,49
,50	,143	,126	,108	,093	,078	,064	−0,049	−0,034	−0,018
0,55	−0,214	−0,196	−0,176	−0,161	−0,145	−0,130	−0,115	−0,100	−0,084
,60	,284	,265	,245	,229	,213	,197	,182	,167	,151
,65	,353	,334	,314	,298	,282	,265	,249	,235	,219
,70	,424	,403	,383	,367	,351	−0,334	,318	,304	,290
0,75	−0,495	−0,475	−0,454	−0,438	−0,423	−0,406	−0,390	−0,377	−0,365
,80	,570	,549	,528	,513	,498	,482	,467	,454	,443
,85	,650	,628	,607	,592	,578	,563	,549	,538	,528
,90	7,37	,716	,697	,682	,668	,656	,644	,635	,626
0,95	−0,837	−0,818	−0,800	−0,784	−0,770	−0,761	−0,753	−0,747	−0,741
,98	,910	,889	,876	,865	,855	,850	,846	,840	,836
,99	,945	,930	,917	,907	,898	,894	,892	,888	,885
1,00	−1,000	−1,000	−1,000	−1,000	−1,000	−1,000	−1,000	−1,000	−1,000

Table IIx, from Astrophysical Journal, vol. 36, p. 245.
Relation between k, α_1^1 and $\psi(k, \alpha_1^1)$ for Total Eclipses of Disks Completely Darkened toward the Edge.

α_1^1	$k=1{,}0$	0,9	0,8	0,7	0,6	0,5	0,4	0,3	0,2
0,00	+7,865	+6,676	+5,752	+5,050	+4,476	+3,979	+3,571	+3,189	+2,853
,02	6,115	5,238	4,558	4,042	3,618	3,240	2,928	2,642	2,383
,05	5,248	4,509	3,936	3,483	3,135	2,828	2,566	2,322	2,104
,10	4,215	3,649	3,206	2,875	2,583	2,344	2,142	1,958	1,789
0,15	+3,440	+3,000	+2,651	+2,394	+2,170	+1,983	+1,820	+1,674	+1,546
,20	2,825	2,473	2,196	1,993	1,821	1,669	1,542	1,430	1,328
,25	2,314	2,034	1,810	1,649	1,513	1,395	1,298	1,210	1,130
,30	1,869	1,649	1,473	1,347	1,238	1,149	1,076	1,011	0,946
0,35	+1,475	+1,306	+1,173	+1,078	+0,994	+0,927	+0,871	+0,822	+0,772
,40	1,115	0,994	0,901	0,831	0,770	0,721	0,679	0,644	0,609
,45	0,794	0,711	0,651	0,603	0,563	0,528	0,500	0,475	0,452
,50	0,504	0,453	0,418	0,391	0,368	0,346	0,329	0,314	0,300
0,55	+0,240	+0,217	+0,201	+0,190	+0,181	+0,170	+0,163	+0,156	+0,149
,60	0,000	0,000	0,000	0,000	0,000	0,000	0,000	0,000	0,000
,65	−0,217	−0,199	−0,188	−0,180	−0,173	−0,165	−0,159	−0,154	−0,150
,70	0,411	0,385	0,365	0,352	0,340	0,328	0,318	0,309	0,302
0,75	−0,585	−0,556	−0,534	−0,518	−0,505	−0,490	−0,478	−0,468	−0,460
,80	0,740	0,714	0,695	0,680	0,667	0,654	0,643	0,634	0,628
,85	0,879	0,861	0,850	0,840	0,832	0,823	0,815	0,810	0,807
,90	1,000	1,000	1,000	1,000	1,000	1,000	1,000	1,000	1,000
0,95	−1,097	−1,125	−1,147	−1,161	−1,173	−1,188	−1,203	−1,216	−1,227
,98	1,138	1,191	1,237	1,272	1,305	1,336	1,369	1,388	1,419
,99	1,149	1,220	1,276	1,323	1,362	1,403	1,477	1,474	1,501
1,00	−1,156	−1,257	−1,347	−1,424	−1,495	−1,559	−1,617	−1,669	−1,715

Table IIIx, from Astrophysical Journal, vol. 36, p. 392.

Values of the χ-Functions for Total Eclipses of Disks Completely Darkened toward the Edge.

k	$\alpha_0=1{,}0$	0,98	0,95	0,9	0,8	0,6	0,4	0,2	0,0
					$\chi(k, \alpha_0, \frac{3}{4})$.				
1,00	0,344	0,358	0,370	0,381	0,399	0,427	0,438	0,448	0,457
0,90	0,410	0,406	0,408	0,410	0,418	0,434	0,442	0,450	0,457
0,80	0,460	0,443	0,439	0,435	0,435	0,441	0,446	0,452	0,457
0,70	0,499	0,470	0,461	0,455	0,450	0,448	0,450	0,454	0,457
0,60	0,531	0,495	0,483	0,474	0,464	0,455	0,453	0,455	0,457
0,50	0,560	0,522	0,504	0,492	0,476	0,460	0,456	0,457	0,457
0,40	0,585	0,544	0,524	0,508	0,486	0,465	0,459	0,458	0,457
					$\chi(k, \alpha_0, \frac{1}{4})$.				
1,00	2,086	2,065	2,026	1,977	1,922	1,854	1,814	1,768	1,707
0,90	1,923	1,926	1,912	1,891	1,866	1,824	1,794	1,759	1,707
0,80	1,789	1,818	1,822	1,819	1,814	1,800	1,778	1,750	1,707
0,70	1,694	1,736	1,749	1,759	1,767	1,777	1,759	1,739	1,707
0,60	1,615	1,665	1,689	1,705	1,732	1,753	1,740	1,727	1,707
0,50	1,550	1,609	1,632	1,652	1,690	1,726	1,720	1,715	1,707
0,40	1,498	1,558	1,578	1,610	1,653	1,691	1,699	1,703	1,707
					$\chi(k, \alpha_0, 0)$.				
1,00	5,44	5,30	5,16	4,95	4,64	4,24	3,94	3,64	3,41
0,90	4,64	4,61	4,58	4,48	4,32	4,06	3,84	3,60	3,41
0,80	4,02	4,12	4,12	4,09	4,04	3,91	3,75	3,55	3,41
0,70	3,58	3,72	3,76	3,77	3,80	3,75	3,63	3,50	3,41
0,60	3,20	3,38	3,46	3,50	3,58	3,60	3,52	3,46	3,41
0,50	2,90	3,10	3,19	3,26	3,37	3,46	3,44	3,42	3,41
0,40	2,66	2,82	2,94	3,05	3,19	3,32	3,35	3,38	3,41

Table IVx, from Astrophysical Journal, vol. 36, p. 253.

Values of $\Phi(k, \alpha_1)$ for Disks Completely Darkened toward the Edge.

For Computing the Values of k and the Light-Curve in the Case of Central Eclipse

α	$k=1{,}0$	0,9	0,8	0,7	0,6	0,5	0,4	0,3	0,2
0,00	7,805	6,220	5,015	4,100	3,373	2,770	2,281	1,874	1,530
,10	4,648	3,856	3,238	2,763	2,369	2,044	1,769	1,536	1,333
,20	3,446	2,917	2,532	2,223	1,966	1,742	1,554	1,392	1,247
,30	2,617	2,290	2,028	1,826	1,656	1,512	1,386	1,277	1,176
,40	1,965	1,778	1,629	1,510	1,408	1,321	1,244	1,177	1,113
,50	1,435	1,354	1,292	1,240	1,195	1,154	1,118	1,086	1,056
,60	1,000	1,000	1,000	1,000	1,000	1,000	1,000	1,000	1,000
,70	0,644	0,698	0,746	0,784	0,820	0,854	0,886	0,915	0,944
,80	0,360	0,442	0,515	0,583	0,646	0,709	0,769	0,826	0,883
,90	0,135	0,218	0,302	0,386	0,470	0,555	0,641	0,726	0,814
,95	0,051	0,120	0,200	0,287	0,378	0,471	0,568	0,667	0,772
,98	0,016	0,069	0,137	0,219	0,309	0,406	0,509	0,620	0,736
1,00	0,000	0,017	0,061	0,127	0,209	0,307	0,419	0,544	0,681

Table Iy, from Astrophysical Journal, vol. 36, p. 390.
Values of $p(k, \alpha'')$ for Annular Eclipses of Disks Completely Darkened toward the Edge.

α''	$k=1{,}0$	0,9	0,8	0,7	0,6	0,5	0,4
0,00	+ 1,000	+ 1,000	+ 1,000	+ 1,000	+ 1,000	+ 1,000	+ 1,000
0,02	0,796	0,786	0,778	0,772	0,768	0,765	0,762
0,05	0,685	0,660	0,644	0,635	0,728	0,624	0,620
0,10	0,543	0,511	0,495	0,485	0,479	0,473	0,468
0,15	0,428	0,390	0,370	0,357	0,350	0,344	0,338
0,20	0,328	0,287	0,263	0,250	0,240	0,232	0,226
0,25	0,240	0,195	0,168	0,150	0,142	0,134	0,127
0,30	0,158	0,109	+ 0,079	+ 0,060	+ 0,049	+ 0,042	+ 0,036
0,35	0,080	+ 0,029	− 0,003	− 0,024	− 0,036	− 0,043	− 0,048
0,40	+ 0,004	− 0,048	0,083	0,104	0,117	0,124	0,129
0,50	− 0,143	0,200	0,237	0,257	0,272	0,280	0,285
0,60	0,284	0,341	0,379	0,401	0,413	0,421	0,427
0,70	0,424	0,482	0,521	0,542	0,553	0,561	0,567
0,80	0,570	0,627	0,663	0,682	0,693	0,700	0,706
0,85	0,650	0,704	0,736	0,754	0,763	0,770	0,776
0,90	0,737	0,787	0,814	0,828	0,835	0,842	0,847
0,95	0,837	0,880	0,899	0,907	0,913	0,918	0,922
0,98	0,910	0,943	0,955	0,959	0,962	0,964	0,967
1,00	− 1,000	− 1,000	− 1,000	− 1,000	− 1,000	− 1,000	− 1,000
$1+0{,}2x$	− 1,000	− 1,014	− 1,030	− 1,047	− 1,070	− 1,100	− 1,145
$1+0{,}4x$	1,000	1,030	1,062	1,099	1,147	1,216	1,312
$1+0{,}6x$	1,000	1,048	1,100	1,163	1,241	1,355	1,516
$1+0{,}8x$	1,000	1,069	1,146	1,246	1,361	1,533	1,780
$1+x$	− 1,000	− 1,111	− 1,250	− 1,429	− 1,667	− 2,000	− 2,500
x	0,000	0,015	0,047	0,084	0,143	0,220	0,322

Table IIy, from Astrophysical Journal, vol. 36, p. 391.
Values of $\psi(k, \alpha_1'')$ for Annular Eclipses of Disks Completely Darkened toward the Edge.

α_1''	$k=1{,}0$	0,9	0,8	0,7	0,6	0,5	0,4
0,00	+ 7,865	+ 7,920	+ 7,570	+ 6,961	+ 6,333	+ 5,648	+ 5,082
0,02	6,115	6,161	5,900	5,445	4,979	4,472	4,027
0,05	5,248	5,213	4,977	4,605	4,224	3,813	3,474
0,10	4,215	4,181	4,021	3,747	3,472	3,152	2,876
0,15	3,440	3,403	3,278	3,073	2,860	2,606	2,410
0,20	2,825	2,790	2,692	2,530	2,360	2,160	2,003
0,25	2,307	2,280	2,194	2,065	1,945	1,791	1,665
0,30	1,858	1,835	1,772	1,668	1,574	1,455	1,369
0,35	1,474	1,448	1,400	1,320	1,247	1,160	1,098
0,40	1,115	1,099	1,061	1,007	0,953	0,891	0,849
0,45	0,794	0,777	0,750	0,714	0,676	0,636	0,608
0,50	0,504	0,484	0,470	0,454	0,432	0,404	0,388
0,55	+ 0,240	+ 0,231	+ 0,227	+ 0,217	+ 0,209	+ 0,196	+ 0,188
0,60	0,000	0,000	0,000	0,000	0,000	0,000	0,000
0,65	− 0,217	− 0,211	− 0,206	− 0,200	− 0,194	− 0,187	− 0,182
0,70	0,411	0,406	0,397	0,388	0,376	0,367	0,358
0,75	0,585	0,580	0,572	0,556	0,546	0,535	0,526
0,80	0,740	0,735	0,723	0,715	0,706	0,697	0,689
0,85	0,879	0,876	0,872	0,863	0,856	0,850	0,847
0,90	1,000	1,000	1,000	1,000	1,000	1,000	1,000
0,95	1,097	1,106	1,118	1,128	1,141	1,148	1,154
0,98	1,138	1,158	1,181	1,202	1,223	1,232	1,234
1,00	− 1,156	− 1,190	− 1,224	− 1,255	− 1,284	− 1,296	− 1,308
$1+0{,}2x$		− 1,196	− 1,249	− 1,309	− 1,383	− 1,460	− 1,570
$1+0{,}4x$		1,201	1,273	1,362	1,483	1,630	1,845
$1+0{,}6x$		1,207	1,295	1,416	1,583	1,804	2,126
$1+0{,}8x$		1,211	1,316	1,470	1,683	1,976	2,414
$1+x$		− 1,215	− 1,335	− 1,519	− 1,790	− 2,164	− 2,746

Table IIIy, from Astrophysical Journal, vol. 36, p. 393.

Values of the χ-Functions for Annular Eclipses of Disks Completely Darkened toward the Edge

k	$\alpha_0''=1+x$	1,0	0,95	0,9	0,8	0,6	0,4	0,2	0,0
					$\chi(k, \alpha_0'', \frac{3}{4})$.				
1,00	0,344	0,344	0,370	0,381	0,399	0,427	0,438	0,484	0,457
0,90	0,361	0,365	0,378	0,386	0,401	0,422	0,434	0,446	0,457
0,80	0,388	0,385	0,390	0,396	0,405	0,423	0,435	0,446	0,457
0,70	0,430	0,409	0,407	0,411	0,415	0,425	0,437	0,447	0,457
0,60	0,477	0,430	0,428	0,426	0,427	0,430	0,441	0,448	0,457
0,50	0,526	0,448	0,445	0,442	0,440	0,435	0,444	0,450	0,457
0,40	0,579	0,462	0,457	0,454	0,448	0,444	0,446	0,452	0,457
					$\chi(k, \alpha_0'', \frac{1}{4})$.				
1,00	2,086	2,086	2,026	1,977	1,922	1,854	1,814	1,768	1,707
0,90	2,086	2,073	2,026	1,983	1,925	1,855	1,814	1,768	1,707
0,80	2,038	2,021	1,991	1,964	1,915	1,850	1,808	1,765	1,707
0,70	1,923	1,941	1,936	1,924	1,883	1,840	1,800	1,760	1,707
0,60	1,812	1,877	1,883	1,874	1,843	1,815	1,787	1,751	1,707
0,50	1,685	1,814	1,820	1,814	1,801	1,788	1,770	1,743	1,707
0,40	1,566	1,760	1,767	1,762	1,760	1,756	1,748	1,732	1,707
					$\chi(k, \alpha_0'', 0)$.				
1,00	5,44	5,44	5,16	4,95	4,64	4,24	3,94	3,64	3,41
0,90	5,50	5,44	5,22	5,02	4,71	4,31	4,03	3,70	3,41
0,80	5,29	5,19	5,04	4,90	4,65	4,27	3,99	3,69	3,41
0,70	4,77	4,82	4,74	4,64	4,46	4,17	3,92	3,66	3,41
0,60	4,25	4,44	4,41	4,35	4,24	4,01	3,83	3,60	3,41
0,50	3,68	4,08	4,09	4,06	4,00	3,88	3,74	3,55	3,41
0,40	3,11	3,77	3,79	3,78	3,75	3,71	3,65	3,51	3,41

Table A, from Astrophysical Journal, vol. 35, p. 339.

Loss of Light Corresponding to an Increase Δm in Stellar Magnitude.

Δm	0	1	2	3	4	5	6	7	8	9
0,0	0,0000	0,0092	0,0183	0,0273	0,0362	0,0450	0,0538	0,0624	0,0710	0,0795
,1	,0880	,0964	,1046	,1128	,1210	,1290	,1370	,1449	,1528	,1605
,2	,1682	,1759	,1834	,1909	,1983	,2057	,2130	,2202	,2273	,2344
,3	,2414	,2484	,2553	,2621	,2689	,2756	,2822	,2888	,2953	,3018
,4	,3082	,3145	,3208	,3270	,3332	,3393	,3454	,3514	,3573	,3632
0,5	0,3690	0,3748	0,3806	0,3862	0,3919	0,3974	0,4030	0,4084	0,4139	0,4192
,6	4246	,4298	,4351	,4402	,4454	,4505	,4555	,4605	,4654	,4703
,7	,4752	,4848	,4848	,4895	,4942	,4988	,5034	,5080	,5125	,5169
,8	,5214	,5258	,5301	,5344	,5387	,5429	,5471	,5513	,5554	,5594
,9	,5635	,5675	,5715	,5754	,5793	,5831	,5870	,5907	,5945	,5982
1,0	0,6019	0,6055	0,6092	0,6127	0,6163	0,6198	0,6233	0,6267	0,6302	0,6336
1,1	,6369	,6403	,6435	,6468	,6501	,6533	,6564	,6596	,6627	,6658
1,2	,6689	,6719	,6749	,6779	,6808	,6838	,6867	,6895	,6924	,6952
1,3	,6980	,7008	,7035	,7062	,7089	,7116	,7142	,7169	,7195	,7220
1,4	,7246	,7271	,7296	,7321	,7345	,7370	,7394	,7418	,7441	,7465
1,5	0,7488	0,7511	0,7534	0,7557	0,7579	0,7601	0,7623	0,7645	0,7667	0,7688
1,6	,7709	,7730	,7751	,7772	,7792	,7812	,7832	,7852	,7872	,7891
1,7	,7911	,7930	,7949	,7968	,7986	,8005	,8023	,8041	,8059	,8077
1,8	,8095	,8112	,8129	,8146	,8163	,8180	,8197	,8214	,8230	,8246
1,9	,8262	,8278	,8294	,8310	,8325	,8340	,8356	,8371	,8386	,8400
2,0	0,8415	0,8430	0,8444	0,8458	0,8472	0,8486	0,8500	0,8515	0,8528	0,8541
2,1	,8555	,8568	,8581	,8594	,8607	,8620	,8632	,8645	,8657	,8670
2,2	,8682	,8694	,8706	,8718	,8729	,8741	,8753	,8764	,8775	,8787
2,3	,8798	,8809	,8820	,8831	,8841	,8852	,8862	,8873	,8883	,8893
2,4	,8904	,8914	,8924	,8933	,8943	,8953	,8962	,8972	,8981	,8991
2,5	0,9000	0,9009	0,9018	0,9027	0,9036	0,9045	0,9054	0,9062	0,9071	0,9080

For values of Δm greater than 2,5, the loss of light is 0,9000 plus 1/10 of the loss of light corresponding to $\Delta m - 2,5$

Table B, from Astrophysical Journal, vol. 35, p. 339.

Values of $\theta - \sin\theta$.

	0,0	0,1	0,2	0,3	0,4	0,5	0,6	0,7	0,8	0,9
0,00	0,0000	0,0002	0,0013	0,0045	0,0105	0,0206	0,0354	0,0558	0,0826	0,1167
,01	,0000	,0002	,0015	,0049	,0114	,0218	,0372	,0582	,0857	,1205
,02	,0000	,0003	,0018	,0055	,0122	,0231	,0390	,0607	,0888	,1243
,03	,0000	,0004	,0020	,0060	,0131	,0244	,0409	,0632	,0920	,1283
,04	,0000	,0005	,0023	,0066	,0141	,0258	,0428	,0658	,0953	,1324
0,05	0,0000	0,0006	0,0026	0,0071	0,0151	0,0273	0,0448	0,0684	0,0987	0,1365
,06	,0000	,0007	,0029	,0078	,0160	,0288	,0469	,0711	,1022	,1407
,07	,0001	,0008	,0033	,0084	,0171	,0304	,0490	,0739	,1057	,1450
,08	,0001	,0010	,0037	,0091	,0183	,0320	,0512	,0767	,1093	,1494
,09	,0001	,0011	,0041	,0098	,0194	,0337	,0535	,0796	,1130	,1539

Table C, from Astrophysical Journal, vol. 36, p. 250.

For the Transformation of "Uniform" into "Darkened" Elements, or Vice Versa.

k_d	1,000	0,900	0,800	0,700	0,600	0,500	0,400
k_u	0,850	0,735	0,630	0,525	0,420	0,310	0,200
α	1,019	1,027	1,040	1,065	1,107	1,168	1,283
β	0,865	0,840	0,821	0,798	0,775	0,724	0,642
γ'	0,180	0,254	0,314	0,366	0,431	0,528	0,650
δ	− 0,020	− 0,010	0,000	0,000	0,000	+ 0,010	+ 0,020
α_u''	0,722	0,541	0,397	0,276	0,176	0,096	0,040
α_d''	1,000	0,917	0,784	0,636	0,488	0,350	0,229
α_d'''	1,000	0,904	0,750	0,581	0,427	0,289	0,175

114. The Eclipsing Variable Algol. β Persei (Algol) has been carefully observed for many decades and as an object of long research is treated in the valuable work "Geschichte und Literatur des Lichtwechsels der veränderlichen Sterne" of MÜLLER and HARTWIG 1, pp. 82—86.

GOODRICKE studied this star in England in 1783 and his assumption that its light variation was an eclipse phenomenon has been verified and generally adopted.

ARGELANDER has given a history[1] of Algol up to the year 1840 and has shown the presence of slight variations in the lapse of time between successive light-minima. CHANDLER[2] using observations of 684 minima from 1783 to 1887 considered that these variations were periodic. The present theory is that they are due to the perturbation of a third body.

STEBBINS' use of a selenium photometer[3] in 1909—1910 brought out evidence of a slight secondary minimum about half-way between two successive primary minima. Observations during 1919—1920 made also by STEBBINS[4] with a photoelectric photometer, create a new light-curve for β Persei (Fig. 41) which is more accurate than the former.

Besides confirming the secondary depression and the continuous variation between minima, the last results show an effect due to the ellipsoidal shape of the components. The phase of the secondary minimum indicates that there can be no rapid motion of the line of apsides of an elliptical orbit; in fact all the photometric evidence points to a circular orbit.

[1] Bonner Beobachtungen 7, p. 343 (1869).
[2] A J 7, p. 165, 169, 177 (1888).
[3] Ap J 32, p. 185 (1910).
[4] Ap J 53, p. 105 (1921).

VOGEL in 1889 found Algol to be a spectroscopic binary[1]. Several orbits have been given by BÉLOPOLSKY[2], also by SCHLESINGER and CURTISS[3]. In 1906 BÉLOPOLSKY announced a long-period oscillation in the radial velocity of Algol; this was confirmed by CURTISS[4] in 1911, and more recently by the complete investigation of the system by MCLAUGHLIN[5].

Fig. 41. Light-curve of Algol. Heavy line 1920; broken line 1910.

Mentioned here are remarkable results of this last investigation based on 156 spectrograms taken at the Detroit Observatory:

From plates taken during one month in 1923 the following elements have been obtained for the short period:

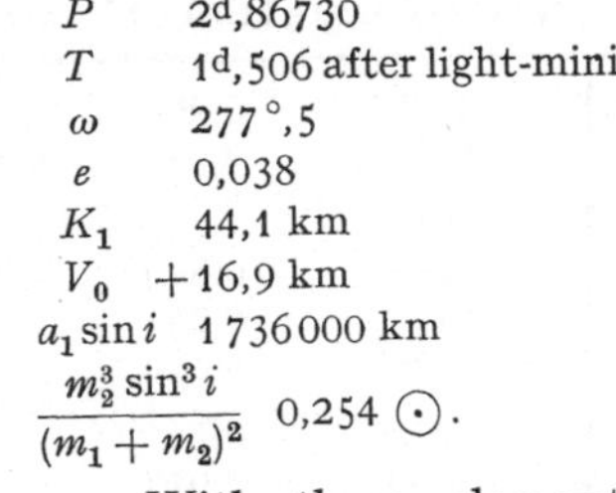

P	$2^d,86730$
T	$1^d,506$ after light-minimum
ω	$277°,5$
e	0,038
K_1	44,1 km
V_0	+16,9 km
$a_1 \sin i$	1 736 000 km
$\dfrac{m_2^3 \sin^3 i}{(m_1+m_2)^2}$	0,254 $\odot$.

With these elements as standard the velocity of the centre of mass is derived for other epochs by means of simple residuals. Supposing a third body to be present, the following are preliminary elements of the orbit of the centre of mass of the close binary system around the centre of mass of the triple system:

E (epoch of minimum velocity)	1901,85
P	1,885 years
T (after minimum velocity)	0,943 years
ω	0°
e	0,13
K_1	10,0 km
V_0	+5,7 km
$a_1 \sin i$	93 000 000 km.

During the first half of the eclipse, it was found that spectrograms of Algol gave radial velocities with large positive residuals from the mean curve, while during the second half the residuals are negative. The interpretation of this is no doubt a rotation of the partially eclipsed star.

Such a result was formerly forecast by FORBES[6] who was unaware that SCHLESINGER had discovered this particular fact in the case of the eclipsing binary δ Librae[7].

Determined by MCLAUGHLIN from forty plates taken during eclipse, the curve of residuals for rotational effect in Algol is given in Fig. 42.

[1] A N 123, p. 289 (1889).
[2] Mitt. der Nikolai-Hauptsternwarte zu Pulkowa 3, p. 72 (1908).
[3] Publ Allegh Obs 1, p. 30 (1908).
[4] Ap J 28, p. 150 (1908).
[5] Ap J 60, p. 22 (1924).
[6] M N 71, p. 578 (1911).
[7] Publ Allegh Obs 1, p. 134 (1909); see also M N 71, p. 719 (1911).

Theoretical deductions lead to the hypothesis that the rotational period of Algol and the period of revolution in its orbit are equal. The tidal action in the Algol system might compel this equality and an argument in favour of this is STEBBINS' detection of an ellipsoidal shape of the primary as shown by slight curvature of the light-curve during maximum (see Fig. 41).

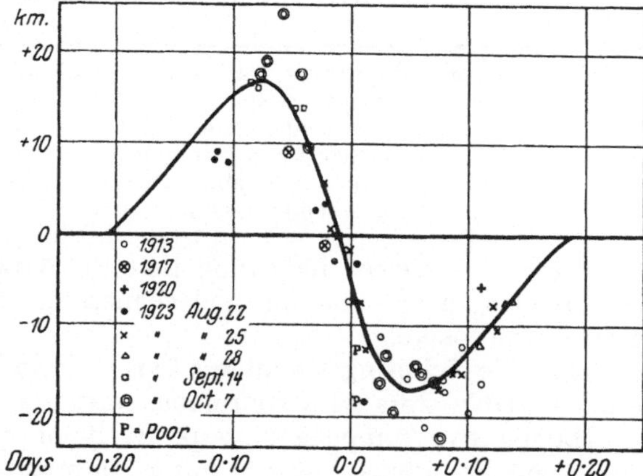

Fig. 42. Curve of the rotational effect in Algol.

The period of rotation being known, and the speed of rotation inferred from Fig. 42, the dimensions of the eclipsing system can then be found. The light-curve gives the relative dimensions and, on some assumption of masses, the means of computing the range of the rotational effect. The observed range of the effect gives us the one unknown value needed or the absolute scale of the system. The true dimensions of the Algol system as found by McLAUGHLIN are:

Radius of bright body	3,12 ⊙
Radius of faint body	3,68
Mass of bright body	4,72
Mass of faint body	0,95
Density of bright body	0,16
Density of faint body	0,02
Distance between centres	10522000 km.

IDA BARNEY's[1] research on the spectrum of Algol is based on the study of about 250 spectrograms secured at the Allegheny Observatory during the years 1907—1912.

When measuring these plates for the calculation of orbital elements, SCHLESINGER used the eight prominent lines which are the only ones constantly visible in the Algol spectrum. He noticed, however, that some of the spectrograms exhibited many other lines and the largest number, sometimes from eighty to ninety, were on those taken near the primary minimum.

There are three possible sources from which these additional lines may come; since they appear in greatest numbers when the primary body is partially eclipsed they may originate from the edge of the primary, from the eclipsing body, or from the third body since at the principal minimum, the light from that body may form an appreciable part of the total light of the system.

According to Miss BARNEY these occasional lines belong to the spectrum of the primary in the Algol system as well as to that of the secondary, and furthermore they are relatively stronger in the former. These lines are, at the same time, just on the border of visibility in the spectrum of the primary and do not appear unless the light from this body is reinforced by that from the eclipsing companion. Since so many of them are iron lines their presence suggests a spectral class later than B8, not only for the secondary but for the brighter body itself.

[1] A J 35, p. 95 (1923).

An investigation of the light-curves of Algol for monochromatic radiations of wave lengths λ 6450 and λ 4120 was made by MENTORE MAGGINI[1]; the principal facts obtained are given in the following table:

	λ 6450	λ 4120
Duration of principal variation of brightness	$9^h 54^m$	$9^h 15^m$
Magnitude during constant brightness . . .	$2^m,250$	$2^m,340$
,, at principal minimum	3,423	3,655
,, at secondary minimum	2,308	2,380
Amplitude of principal minimum	1,173	1,315
,, of secondary minimum	0,058	0,040

There is a marked indication that the darkening at the limb varies with the wave-length of radiation, and chances are that the secondary has an extensive atmosphere.

115. The Eclipsing Variable β Lyrae. Beta Lyrae is the prototype of short-period variable stars with continuous variation and two unequal minima.

Having two components respectively of spectral classes cB8 and B5e, it is accepted as a binary system formed of prolate ellipsoidal stars of low density and unequal masses which revolve about their centre of mass in a period of 12,92 days. The longest axes of these stars are in the same straight line and their adjacent vertices are practically in contact.

On the spectrograms of the variable each star is represented by a dark-line spectrum. However, a third spectrum of bright hydrogen and helium lines is present but there is a difference of opinion as to its interpretation.

According to Miss MAURY[2] the two hundred Harvard spectrograms of β Lyrae, extending over thirty-six years, show complex changes due to revolution and double eclipse, repeated with great regularity in agreement with the light-curve. The following extract from her note is of interest:

"The tidal fringe is seen on the violet side of the dark lines when, following primary minimum, the approaching limb of the primary emerges from eclipse. The bright bands of the B5e companion, although it is then in front, fade gradually during the eclipse owing to the intense light of the primary cB8 star shining through from behind, and flash out on the red side of the dark lines as the receding limb passes beyond the limb of the primary. At the opposite eclipse they flash out on the blue side on coming out from occultation.

The permanent displacement toward the blue in the dark reversals of the hydrogen and helium lines of the B5e star is verified and shown to increase with wave-length for hydrogen as well as helium, varying also with the line series.

The unexplained companions of magnesium 4481 and of helium 3964 and 3889 appear, while $H\gamma$ and $H\delta$ and 4026 are suspected of being triple in the opposed phases of maximum velocity."

The best light-curves of β Lyrae have been determined with the photo-electric cell photometer both by STEBBINS[3] and by GUTHNICK[4]. Fig. 43 is STEBBINS' light-curve.

The main features of this curve agree with the former visual curves such as that of LUIZET[5], but during different cycles of variation the star does not

[1] B A 35, p. 131 (1918). [2] Pop Astr 32, p. 559 (1924).

[3] Lick Bull 8, p. 186 (1916).

[4] Veröff. der K. Sternwarte zu Berlin-Babelsberg 2, Heft 3, p. 102 (1918); see also Sirius 50, p. 158 (1917) and Berlin Sitzungsber. 1917, p. 222.

[5] B S A F 21, p. 38 (1907).

exhibit exactly the same changes, discordances existing up to 0,10 magnitude. These discrepancies are apparently real and not due to instrumental errors or to variations of the comparison stars.

The curve is unsymmetrical, the decrease of light being more rapid than the increase. Such lack of symmetry is found also in other stars of similar type. Prof. GUTHNICK calculated the following elements of β Lyrae as derived from the photo-electric light-curve:

Ellipticity of the components	0,59	
Ratio of the axes of an equatorial section	1:0,8	
Inclination of the orbit	90°	
Semi-axis major of B5 component	0,5	(1 = semi-axis major of orbit)
Semi-axis major of B8 component	0,3	
Maximum brightness of B5 component	0,55	1 = total light
Maximun brightness of B8 component	0,45	
Ratio of surface intensities	B5 : B8 = 0,35	

Densities $\left\{ \begin{array}{l} D_{B5} = 0{,}0022 \\ D_{B8} = 0{,}0060 \end{array} \right\}$ Density of the Sun = 1.

At principal minimum the B8 component is hidden, one eclipse being total and the other annular.

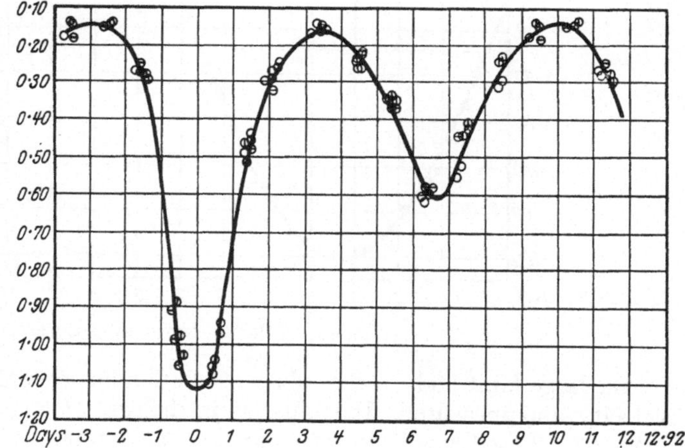

Fig. 43. The light-curve of β Lyrae.

An extensive study of β Lyrae, including a comprehensive history of the system, has been made by R. H. CURTISS[1]. Previous important spectrographic observations were made by VOGEL, BÉLOPOLSKY[2] and others.

MYERS[3] computed an orbit based on BÉLOPOLSKY's observations and, combining spectrographic and photometric data, estimated within certain assumptions the complete elements of the system. He recognized the possibility that the line of apsides of the orbit did rotate between 1855 and 1892, and that the eccentricity increased.

Recently the star has been studied by ROSSITER[4]. Elements from numerous spectrograms prove that the velocity of the system's centre of mass is constant and that elliptical motion is very well satisfied. One series of plates was taken at the Allegheny Observatory in 1907, and from the year 1911 to 1921, seven series at Ann Arbor.

Obviously, no third body exists since perturbations in the two-body system cannot be detected.

On the velocity-curve (Fig. 44), the dots are the normal places representing radial-velocity observations of the B8 component.

[1] Allegh Obs Publ 2, p. 73 (1911). [2] Ap J 6, p. 328 (1897).
[3] Ap J 7, p. 1 (1898). [4] Ap J 60, p. 15 (1924).

Rossiter deduced the following orbital elements for the B8 component:

P	$12^d,922$
T	$8^d,541$ after principal minimum light
ω	$325^\circ,05$
e	0,014
K_1	183,71 km
V_0	−19,02 km
$a_1 \sin i$	32639000 km
$\frac{m_2^3 \sin^3 i}{(m_1+m_2)^2}$	$8{,}32\odot$.

Confined to the region of the velocity curve extending 1,6 days on each side of the principal minimum light, that is, the eclipse of the small bright body

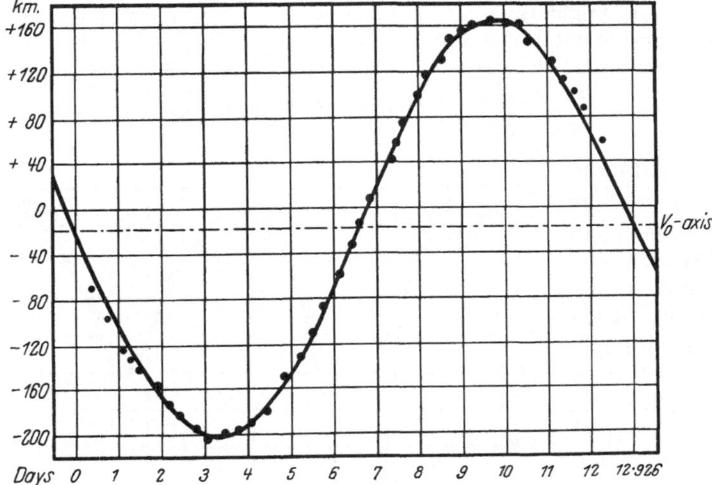

Fig. 44. Velocity-curve of β Lyrae.

by the large faint body, a secondary oscillation has been isolated from the orbital velocity and measured. It is due, as in the case of Algol, to the rotation of the partially eclipsed smaller component, and designated by Rossiter as rotational effect. This effect in β Lyrae, or deviation of the measured velocity from the computed velocity should there be no eclipse, is shown distinctly in Fig. 45; it has a total range of 26 kilometers.

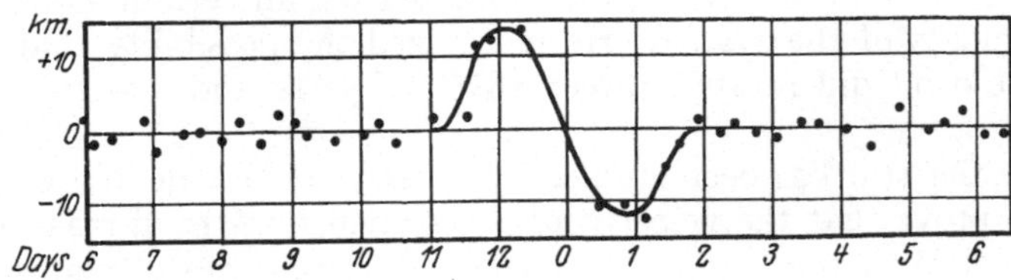

Fig. 45. β Lyrae. Rotational effect.

116. Other Eclipsing Binary Systems. Centres of observation for the study of eclipsing variables are the Harvard College Observatory where material has been collected for nearly forty-five years, the Princeton University Observatory where the theoretical work of Russell and Shapley as well as the accurate observations of Dugan suggest new fields of research; extensive work may be credited to Nijland in the Netherlands, Hellerich, Graff and Hoffmeister in Germany, Maggini in Italy and Gadomski in Poland.

Under the auspices of the International Astronomical Union, Prof. BANACHIEWICZ of Cracow, Poland, publishes annual ephemerides of eclipsing variables[1]; these give in fractions of a day, for each day of the year, the times at which minima occur (with three decimals).

On the determination of spectroscopic orbits of bright eclipsing binaries, the admirable work of PLASKETT should be mentioned.

There is a distinct variation in the relative sizes of the two members of an eclipsing pair. The following table gives the principal dimensions for some of them when the spectroscopic orbit is known, using results from the darkened solution when available.

The linear dimensions are in terms of the Sun's radius and other units are the Sun's mass and density. Bright and faint component are designated by *b* and *f*.

Star	Greatest Radii		Masses		Densities		Distance of Centres
	r_b	r_f	m_b	m_f	ϱ_b	ϱ_f	
β Aurigae[2]	2,81	2,81	2,40	2,36	0,11	0,11	17,7
u Herculis[2]	4,56	5,35	7,5	2,9	0,095	0,022	14,8
V Puppis[2]	8,45	7,70	19,4	19,4	0,042	0,055	12,7
RX Herculis[3]	1,54	1,38	0,96	0,91	0,25	0,34	7,5
W Ursae Maj.[4]	0,78	0,78	0,69	0,49	2,8	1,9	2,2
Z Herculis[5]	1,77	3,29	1,6	1,3	0,3	0,04	15,1
U Ophiuchi[6]	3,23	3,23	5,36	4,71	0,18	0,16	12,8
RS Vulpeculae[6]	2,05	10,25	5,40	1,69	0,63	0,0016	22,0
U Coronae Bor.[7]	2,90	4,74	4,27	1,63	0,175	0,015	17,36
TX Herculis[8]	1,33	1,33	2,04	1,77	0,87	0,75	10,66
Y Cygni[9]	4,6	4,6	16,6	15,3	0,170	0,158	27,7
Z Vulpeculae[10]	4,23	4,46	5,24	2,36	0,085	0,033	15,05
TV Cassiopeiae[11]	2,50	2,83	1,83	1,01	0,118	0,044	8,86

It is impossible to include here a complete description of the many eclipsing systems, each one unique and of individual interest. The reader may, however, refer to the following papers:

A. W. ROBERTS[12], On the Orbits of the Algol Variables RR Puppis and V Puppis.

F. SCHLESINGER[13], The Algol Variable δ Librae.

E. S. HAYNES and H. SHAPLEY[14], The Algol Variable RZ Draconis.

E. S. HAYNES[15], The Algol Variable RX Draconis.

R. H. BAKER and E. E. CUMMINGS[16], The Eclipsing Binary RX Herculis.

H. N. RUSSELL and H. SHAPLEY[17], Elements of the Eclipsing Variable Stars Z Draconis and RT Persei.

H. SHAPLEY[18], The Orbits of Eighty-seven Eclipsing Binaries. A Summary.

H. SHAPLEY[19], New Light Elements and Revised Orbit of UZ Cygni.

H. SHAPLEY[20], The Orbits of RZ Ophiuchi and ε Aurigae Treated as Eclipsing Binaries.

---

[1] Rocznik Astronomiczny Obserwatorjum Krakowskiego No. 1, 2, 3, 4 (1923/24/25/26).
[2] Ap J 38, p. 169 (1913). [3] Ap J 40, p. 399 (1914).
[4] Ap J 49, p. 189 (1919).. [5] Ap J 49, p. 192 (1919).
[6] Publ Dom Astroph Obs 1, p. 137 (1919). [7] Publ Dom Astroph Obs 1, p. 189 (1920).
[8] Publ Dom Astroph Obs 1, p. 209 (1920). [9] Publ Dom Astroph Obs 1, p. 217 (1920).
[10] Publ Dom Astroph Obs 1, p. 254 (1920). [11] Publ Dom Astroph Obs 2, p. 144 (1922).
[12] Ap J 13, p. 177 (1901). [13] Allegh Obs Publ 1, p. 123 (1909)..
[14] Laws Bull No. 19 (1911). [15] Laws Bull No. 18 (1911). [16] Laws Bull No. 25 (1916).
[17] Ap J 39, p. 405 (1914). [18] Ap J 38, p. 158 (1913).
[19] Publ A S P 28, p. 31 (1916). [20] A N 194, p. 225 (1913).

H. SHAPLEY[1], The Visual and Photographic Ranges and the Provisional Orbits of Y Piscium and RR Draconis.

R. J. MCDIARMID[2], The Eclipsing Variable Star SS Camelopardalis.

R. S. DUGAN[3], A Photometric Study of the Eclipsing Variable RV Ophiuchi.

R. S. DUGAN[4], Photometric Studies of SZ Herculis and RS Vulpeculae.

R. S. DUGAN[5], The Eclipsing Variables RV Ophiuchi and RZ Cassiopeiae.

R. S. DUGAN[6], Inequalities in the Period of the Eclipsing Variable RT Persei.

R. S. DUGAN[7], Inequalities in the Period of the Eclipsing Variable Z Draconis.

R. S. DUGAN[8], An Inequality in the Period of the Eclipsing Variable RZ Cassiopeiae.

J. STEBBINS[9], The Eclipsing Variable Star δ Orionis.

J. STEBBINS[10], The Eclipsing Variable Star λ Tauri.

K. GRAFF[11], Untersuchung des Lichtwechsels einiger veränderlicher Sterne vom Algoltypus.

ERICH KRON[12], Der Algol-Variable δ Librae.

W. S. ADAMS and A. H. JOY[13], Spectroscopic Observations of W Ursae Majoris.

P. GUTHNICK[14], Der Lichtwechsel von Boss 46.

F. C. JORDAN[15], The Algol Variable X Trianguli.

CH'ING-SUNG YÜ[16], Light-Curve and Orbit of CG Cygni.

J. GADOMSKI[17], TV Cassiopeiae.

J. GADOMSKI[18], RZ Cassiopeiae.

Dr. SCHLESINGER finds in Algol, δ Librae, and u Herculis that the phase of mean primary eclipse comes, perhaps, an hour or so later than the time when $u = 90°$ in the spectroscopic orbit, while in theory the two should coincide.

PLASKETT finds the reverse for RS Vulpeculae and TW Draconis, the spectroscopic phase for the first star being later than the photometric by about three hours, and twenty-seven minutes later for the second.

HELLERICH discusses similar anomalies in his paper "Untersuchungen über Bedeckungsveränderliche"[19]. According to him they are of two kinds: (1) ascending branch of light-curve during eclipse steeper than descending branch; (2) brightness at the end of an eclipse greater than at the beginning, the minimum being symmetrically placed.

This would indicate a second cause of light variation superimposed on that due to eclipse; it is probably a type of variation like that found in Cepheids due perhaps to interaction between the two components or between each component and the orbital motion. In the first case tides are suggested, and in the second, motion in a surrounding medium. HELLERICH thinks the latter theory which resembles DUNCAN's hypothesis for Cepheids, is the more logical.

i) Stars of the β Canis Majoris Type.

117. The β Canis Majoris Type. The Cepheids being treated adequately in the chapter on variable stars by Dr. H. LUDENDORFF, their discussion will be avoided here. A few paragraphs on the stars of the β Canis Majoris type are, however, necessary.

[1] Ap J 37, p. 154 (1913). [2] Ap J 45, p. 50 (1917). [3] Ap J 43, p. 130 (1916).
[4] Ap J 58, p. 164 (1923). [5] Princeton Contr 4 (1916). [6] M N 75, p. 692 (1915).
[7] M N 75, p. 702 (1915). [8] M N 76, p. 729 (1916). [9] Ap J 42, p. 133 (1915).
[10] Ap J 51, p. 193 (1920). [11] Hamb. Mitt. 11 (1907). [12] Inaug.-Diss. Berlin 1907.
[13] Pop Astr 26, p. 634 (1918). [14] A N 211, p. 391 (1920).
[15] Pop Astr 32, p. 224 (1924). [16] Ap J 58, p. 75 (1923).
[17] Bull. de l'Académie Polonaise des Sciences et des Lettres, Série A, 1923.
[18] Bull. de l'Académie Polonaise des Sciences et des Lettres, Série A, 1924.
[19] A N 223, p. 369 (1925).

Stars of the β Canis Majoris type are distinguished by the following principal characteristics:

1. Very short-period variation of radial velocity often accompanied by a slight parallel variation of magnitude in stars of spectral classes B or A.

2. In the majority of cases there is a variation of amplitude of the short-period velocity-curve, variation which is also found in the light-curve; for example it is marked in the case of 12 Lacertae (Fig. 46). Changes of the same character, at first considered erratic, appear in δ Ceti[1].

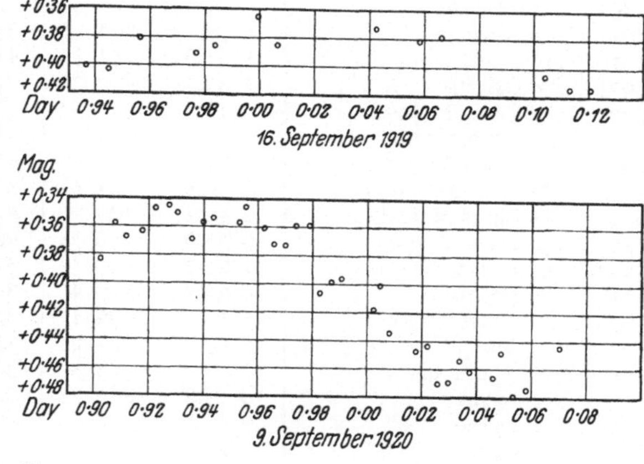

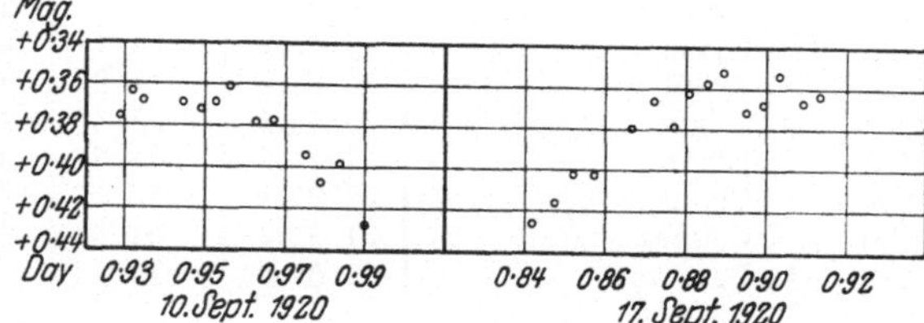

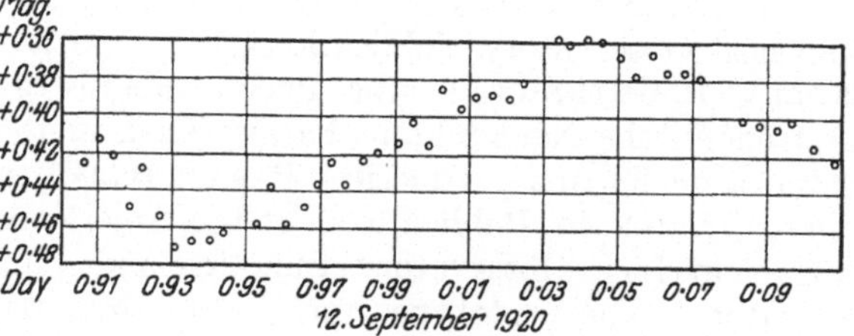

Fig. 46. Light-curves of 12 Lacertae.

3. In many cases a long-period variation of velocity indicates that the star is moving in an orbit of that period. The variation of amplitude referred to is, no doubt, a function of the star's position in this orbit. Occasionally the short-period variation is not constant—which may well be a consequence, real or apparent, of the motion in the orbit; but this variation itself cannot be ascribed to orbital motion.

4. Width and intensity of spectral lines vary in a constant period which is almost equal to the short-period of radial-velocity oscillation.

All these variations are similar to those which occur in RR Lyrae, according to SHAPLEY and HERTZSPRUNG.

Obviously, research devoted to the β Canis Majoris type stars as a particular case of Cepheid variation is of the utmost significance. The majority of these stars are comparatively bright and hence can be investigated with instruments of precision, the spectrograph and the photo-electric cell; certain ones are distinguished by comparatively large parallaxes, which is not the case among stars of the δ Cephei type. The shortness of their periods permits variation to be disclosed which would be difficult to detect otherwise.

Frequently among the more advanced classes the spectra are exceptionally poor, and this renders it difficult to find the variation with any degree of certainty when only spectrographic means are used; however, this handicap might be eliminated in case of photo-electric investigations.

---

[1] Dom Obs Publ 9, No. 1, Chapter 2 (1925).

Following is a list of stars which are known or suspected to be of the β Canis Majoris type:

Stars of the β Canis Majoris Type (known or suspected).

H. R.	Star	α 1900	δ 1900	Visual Mag.	Spect.	Discoverer
779	δ Ceti	$2^h34^m,4$	$-0°\ 6'$	4,04	B2	
1149	20 Tauri	3 39,9	+24 4	4,02	B5	
1320	μ Tauri	4 10,1	+ 8 39	4,32	B5	
1463	ν Eridani	4 31,3	− 3 33	4,12	B2	
1641	η Aurigae	4 59,5	+41 6	3,28	B3	
1810	114 Tauri	5 21,6	+21 51	4,83	B3	
1931	σ Orionis	5 33,7	− 2 39	3,78	B	
2294	β Canis Majoris	6 18,3	−17 54	1,99	B1	ALBRECHT
2344	10 Monocerotis	6 23,0	− 4 42	4,98	B3	
2387	4 Canis Majoris	6 27,6	−23 21	4,35	B1	
2490	42 Camelopardalis	6 40,5	+67 41	5,04	B3	
2571	15 Canis Majoris	6 49,2	−20 6	4,66	B1	
4295	β Ursae Majoris	10 55,8	+56 55	2,44	A	GUTHNICK
4422	57 Ursae Majoris	11 23,7	+39 54	5,26	A2	OTTO STRUVE
5435	γ Boötis	14 28,1	+38 45	3,00	F	GUTHNICK
5735	γ Ursae Minoris	15 20,9	+72 11	3,14	A2	OTTO STRUVE
6084	σ Scorpii	16 15,1	−25 21	3,08	B1	SELGA
6453	θ Ophiuchi	17 15,9	−24 54	3,37	B3	
7178	γ Lyrae	18 55,2	+32 33	3,30	A	OTTO STRUVE
7298	η Lyrae	19 10,4	+38 58	4,46	B3	
7372	2 Cygni	19 20,2	+29 26	4,86	B2	
7377	δ Aquilae	19 20,5	+ 2 55	3,44	F	
7426	8 Cygni	19 28,1	+34 14	4,85	B3	
7447	ι Aquilae	19 31,6	− 1 31	4,28	B5	
7977	55 Cygni	20 45,5	+45 35	4,89	B2	
8130	τ Cygni	21 10,8	+37 37	3,82	F	PARASKÉVOPOULOS
8238	β Cephei	21 27,4	+70 7	3,32	B1	FROST
8279	9 Cephei	21 35,2	+61 38	4,87	B2	
8640	12 Lacertae	22 37,0	+39 43	5,18	B2	YOUNG

Individual studies have been made by R. K. YOUNG in Dominion Astrophysical Observatory Publications, C. C. CRUMP in Detroit Observatory Publications, P. GUTHNICK in the Berlin-Babelsberg Observatory Publications and Astronomische Nachrichten, Father Selga in the Revista de la Sociedad Astronómica de España y América, PARASKÉVOPOULOS in the Astrophysical Journal, OTTO STRUVE in Publications of the American Astronomical Society, and the writer in Lick Observatory Bulletins and Publications of the Dominion Observatory. The prototype star, β Canis Majoris, is comprehensively treated.

118. β Canis Majoris ($\alpha = 6^h18^m,3$, $\delta = -17°55'$, vis. mag. = 2,0, Class B1). At the Lick Observatory in 1908[1] Dr. ALBRECHT discovered a short period variation of radial velocity for this star. Between 1904 and 1910 there were secured a number of three-prism spectrograms of which only one series was taken in succession through a continuous period (March 1st, 1909). Many other spectra were photographed by the writer[2] at Lick Observatory during 1917/18, and an effort was made to procure continuous series of plates covering the same cycles of variation.

Previous to 1909 most radial-velocity measures were due to ALBRECHT and Miss HOBE; subsequently the writer continuing the same work, using a HARTMANN spectro-comparator, derived velocities that were quite in harmony with other values. His velocities for March 1st, 1909 are plotted on Fig. 47 as

[1] Lick Bull 5, p. 62 (1908). [2] Lick Bull 9, p. 155 (1918).

circles, and crosses represent former measures; the comparison of these values gives a clear idea of the precision reached in these measurements.

After careful survey of the plates, the aspect of the spectrum was seen to change. Sometimes the lines are narrower, at other times wider and more diffuse than the general average condition. The maximum as well as the minimum widths are attained approximately once in each period of velocity-variation, and the maximum is separated from the minimum by about half a period. The maximum width of the lines in different periods of velocity-variation occurs at different places on the corresponding velocity-curves, disclosing a slightly different period for the variation of widths than for the variation in velocity. Also there is clearly a variation in the intensity of the lines. It is improbable that the variation in width results from the combination of two spectra but it is likely due to physical changes in one body only. No trace of a second component has ever been found on the spectrograms.

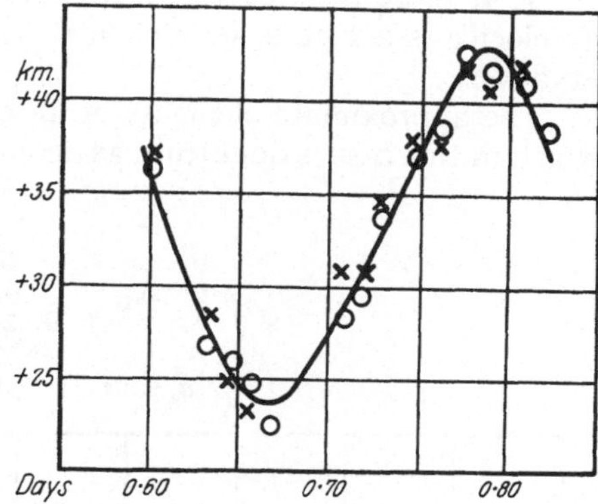

Fig. 47. Velocity-curve of β Canis Majoris.

The writer's preliminary work on the velocity-curves brings out a difference in range so appreciable that it is impossible to combine all the observations

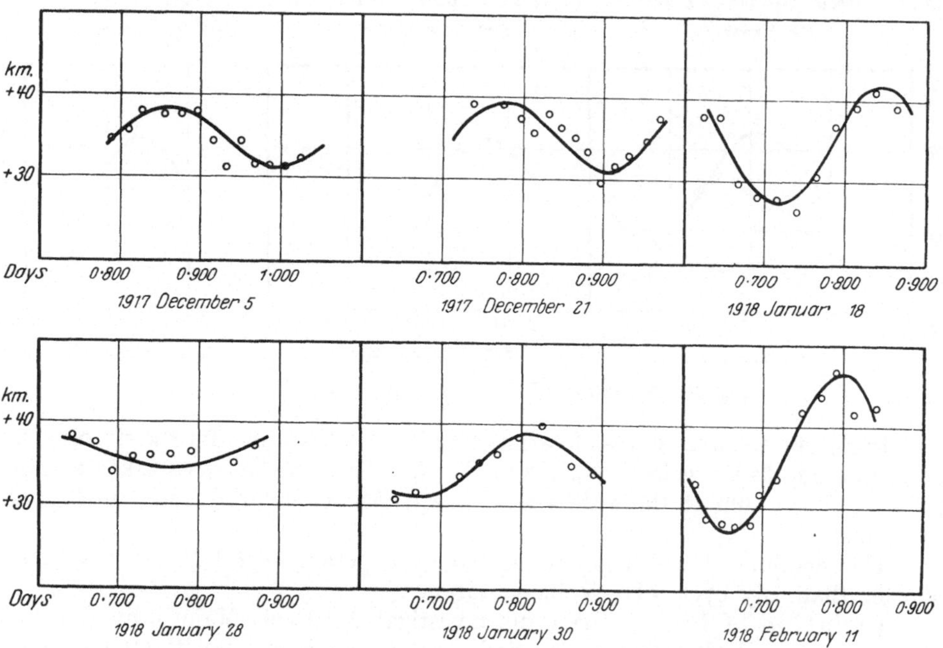

Fig. 48. Velocity-curves of β Canis Majoris.

into a single curve and treat them by a least-squares adjustment. The different velocity-curves for six dates in 1917 and 1918 are given in Fig. 48, while the curve of March 1, 1909 is in Fig. 47.

The considerable variation of range is of real import and now well established for stars of this type. On 1918 January 28 the velocity-curve was almost a straight line, having a variation of about 3 km; it seems to change very rapidly in a few days.

That there is no period which connects and represents the different minima of velocity is a fact to be reckoned with, confirmed by a great number of observations.

The approximate elements of an orbit, supposing the motion to be elliptic (which in this case is doubtful), as deduced from the curve for March 1, 1909, are:

P	$0^d,25$
e	0,04
ω	225°
K_1	9,10 km
T	J. D. 2418367,700
V_0	+33,06 km
$a_1 \sin i$	31 300 km.

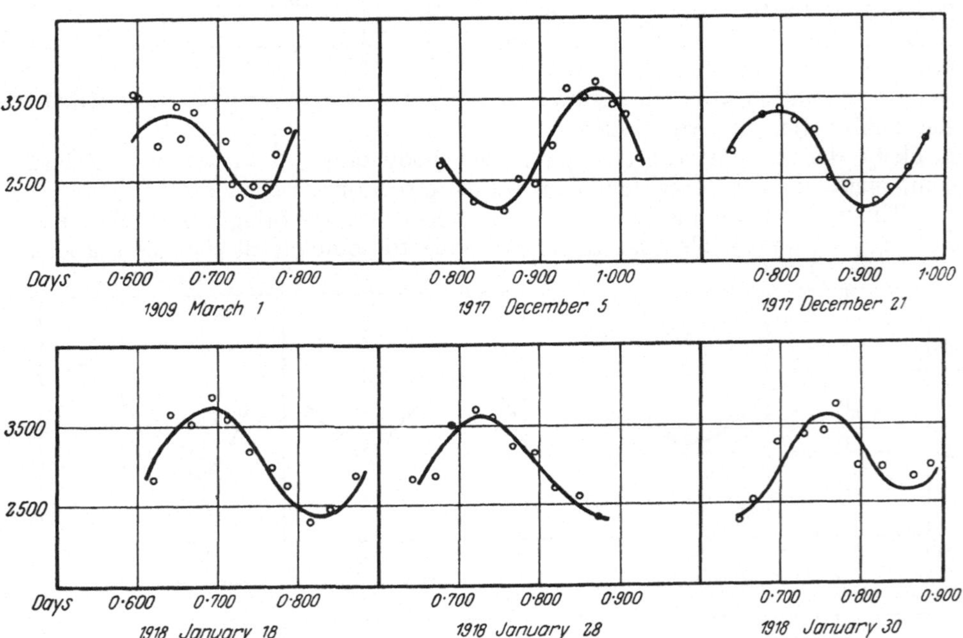

Fig. 49. Variation of line-width in β Canis Majoris.

In Fig. 49 we have curves of variation in line-width, drawn for the six periods corresponding to the velocity-curves in Fig. 48. The time has been taken as abscissa and the sum of the widths as ordinate. All these curves show analogous variations.

The maxima of width are connected by a period, $0^d,25130$. It is a little shorter than any cycle of velocity-variation.

Variation of Light. STEBBINS investigated β Canis Majoris by the use of an electric cell photometer. He finds that the star does not fluctuate more than one tenth of a magnitude and suspects that it varies on the basis of the period $0^d,25130$ which is the period of change in width. However, he infers that irregular changes are taking place because his best observations are discordant by nearly a tenth of a magnitude. If these discrepancies are real they may correspond to variations of amplitude in the velocity-curve.

Research on β Canis Majoris was continued by the writer in OTTAWA[1]. The principal results of the former study were:

1. A radial-velocity oscillation of very short duration (about six hours) mentioned by Dr. ALBRECHT in 1908.
2. A considerable variation in amplitude of the velocity-curve.
3. The certainty that the maxima of different velocity-curves cannot be connected by a period.
4. A short-periodic variation in width of the spectral lines, $0^d,25130$.

From recent work it appears that the variation of amplitude is periodical, while the radial-velocity variation evidently follows a rather complicated law.

Altogether the velocity curves furnish the following values:

Results obtained for β Canis Majoris.

Julian Day	Maximum Velocity km	Amplitude of Velocity Curve km	Estimated Mean Velocity km	Time of Maximum (decimal fraction of a day)
243045	+37 *			,770*
051	+38,0	15,0	+30,5	,655
053	+31,0			,752
056	+38,0	14,7	+30,4	,640
057	+40,0	17,5	+31,2	,687
061	+42,0	20,0*	+32,0*	,700*
067	+41,0	19,0	+31,5	,675
077	+32,5	11,5	+26,5	,699
078	+30,5	14,0	+23,7	,690
079	+29,0*	10,0*		
080	+32,5*	15,0*	+25,0*	
081	+31,5	12,0	+25,5	,766
084	+32,0	10,0	+27,0	,788
085	+31,6			,560
093	+36,5	13,0	+30,0	,540
102	+40,4	15,0	+33,0	,672

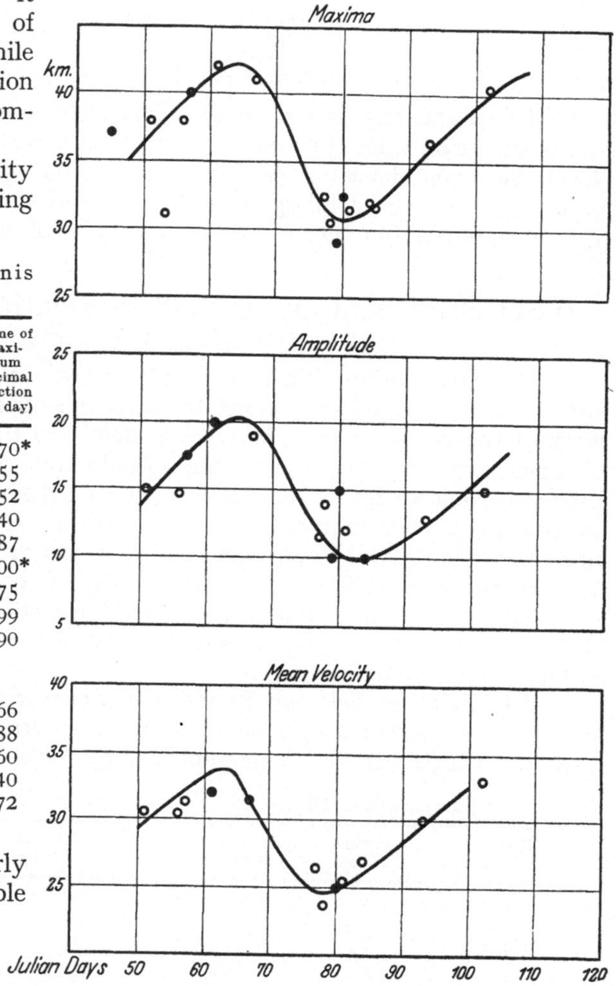

Fig. 50. β Canis Majoris. Maximum, amplitude and mean velocity.

Some curves are too poorly determined to furnish reliable data and doubtful values in the table are marked with an asterisk. Columns two, three and four, combined with column one, yield the curves of Fig. 50 where doubtful points are indicated by darkened circles.

The three curves show unmistakably analogous tendencies,—probably passing from one maximum to the next in about 42 days—as if the centre of mass of the short-period system, if such it is, moved in a rather eccentric orbit. Providing this orbit exists, the amplitude of the short-period velocity-variation is a function of the position of its centre of mass in the orbit.

[1] Dom Obs Publ 8, No. 3, p. 31 (1922).

Granting the curve of variation of mean velocity to represent also the variation of the centre-of-mass velocity of the short-period system, the following approximate elements for a possible orbit of that centre of mass are derived:

K_1	4,75 km
e	0,27
P	42^d
ω	90°
T	J. D. 2423070
V_0	+29,3 km
$a_1 \sin i$	2633000 km
$\frac{m_2^3 \sin^3 i}{(m_1+m_2)^2}$	0,00041 $\odot$.

The eccentricity is not small. This adds weight to the hypothesis that the short-period velocity-curve is due to tidal action, the variation of amplitude depending upon change of position in the 42-day orbit. This last effect would be more pronounced for large eccentricities and small orbits, tidal action being approximately a function of the inverse cube of the distance.

j) Statistical and Other Studies Relating to Characteristics of Double Stars.

119. Introduction. Binary star orbits present peculiar features of absorbing interest when studied as a group. If one reflects upon the endless and unremitting work of the astronomers, it becomes evident that the question is not merely one of dimensions or physical properties of individual systems, but concerns a comprehensive knowledge of the relation between these dimensions and consequent bearing upon the evolution of celestial bodies. The reason for their existence must be made clear, the necessary conditions to warrant their stability, analyzed. Indeed, the most definite information is often best acquired from the study of physical doubles, and data en masse are imperative to check up mathematical theories and verify the hypotheses involved in cosmogonical problems of extraordinary significance.

120. Relation between Period and Eccentricity. There is a general tendency for the eccentricity in binary systems, spectroscopic and visual, to increase with the period. In the table published by AITKEN[1] this increase is apparent.

Spectroscopic Binaries.

Number of Systems	Average Period	Average Eccentricity
46	$2^d,75$	0,047
19	7,80	0,147
12	15,17	0,202
13	30,24	0,437
15	106,4	0,371
14	1035	0,328

Visual Binaries.

Number of Systems	Average Period	Average Eccentricity
30	$31^y,3$	0,423
20	74,4	0,514
6	124,5	0,558
12	243	0,529

A study on "The Period-Eccentricity Relation in Binary Systems"[2], by R. E. WILSON, discusses the influence of the subdivision of binary systems into giant and dwarf stars. He collected the data for 237 orbits of which 151 are spectroscopic systems. These do not include Cepheid variables. He obtained the following table:

[1] The Binary Stars, p. 196 (1918).
[2] A J 33, p. 147 (1921).

Mean Period-Eccentricity Relation.

Group	Period	All		Spectral Type I		Spectral Type II		Giant		Dwarf	
		e	number	e	number	e	number	e	number	e	number
1	0—4 days	0,046	44	0,044	37	0,057	7	0,047	38	0,010	1
2	4—8	115	24	109	20	142	4	114	19	020	3
3	8—25	231	30	268	18	176	12	250	22	132	4
4	25—100	403	19	473	14	208	5	468	15	010	1
5	100—5 years	311	28	395	13	239	15	394	16	420	1
6	5—50	416	35	405	6	418	29	433	11	414	24
7	50—100	530	26	555	6	522	20	612	8	494	18
8	100—200	580	18	574	5	483	13	626	7	552	11
9	200	605	13	553	3	620	10	632	4	592	9
	Total		237		122		115		140		72

All spectroscopic binaries having periods greater than five years have been combined with the visual binaries in their respective groups.

WILSON arrives at the following conclusions: "Though the data is too meagre perhaps to warrant any definite conclusions, the indications are then: (1) that the relatively rapid increase in eccentricity in the first four period-groups is due almost wholly to the predominance of giant systems in these groups and, consequently, that an increase in the number of dwarf systems would tend to reduce the apparent maximum in groups 3 and 4; (2) that the minimum in groups 5 and 6 is due partially to the systematically low eccentricities of the dwarf stars in these groups and partially to the greater probability of the discovery of the less eccentric orbits among the long-period spectroscopic binaries. As more spectroscopic binaries with long periods are discovered, or as orbits are determined for many which are now recognized as long-period systems, it is probable that most of the discordances from a uniform mean period-eccentricity curve will be removed, especially when taking into account a probable difference in the period-eccentricity relation for giant and dwarf systems."

PERRINE[1] has pointed out that a dependence of orbital eccentricity upon the relative masses of the components must also be taken into account; in general, the systems with small secondaries have the larger orbital eccentricities.

121. Relation between Period and Semi-Amplitude of Velocity-Variation in Spectroscopic Binaries. OTTO STRUVE at the Yerkes observatory made two investigations of this relation. The first[2], based on the observational data for 144 spectroscopic binaries, shows that spectroscopic binaries of periods longer than 2,4 days arranged in order of increasing period have a continuously decreasing orbital velocity which agrees excellently with KEPLER's third law written in the form

$$K = C P^{-\frac{1}{3}},$$

where P is the period, K the semi-amplitude of the velocity-curve, and C a constant.

For P smaller than 2,4 days the last equation is not satisfied except for a few stars like W Ursae Majoris which usually are eclipsing variables and certainly binary systems.

The Cepheid variables do not satisfy the equation, K having a constant average value of $+16$ km for all periods.

The chance is that stars of shortest periods, having an average value of K almost identical with that of the Cepheid variables, are in some way related to the latter; indeed, most of them belong to the β Canis Majoris type.

[1] A J 33, p. 180 (1921). [2] Ap J 60, p. 167 (1924).

Beyond question, the stars for which the equation is satisfied are actual binaries.

STRUVE's second paper[1] is based on the material contained in the Third Catalogue of Spectroscopic Binary Stars, by J. H. MOORE[2]. It verifies the conclusions of the first paper except that C is different for various spectral types. For these types have been determined the values of the shortest periods possible for double stars, which correspond to actual contact of the components. Beyond this, obviously, double stars cannot exist.

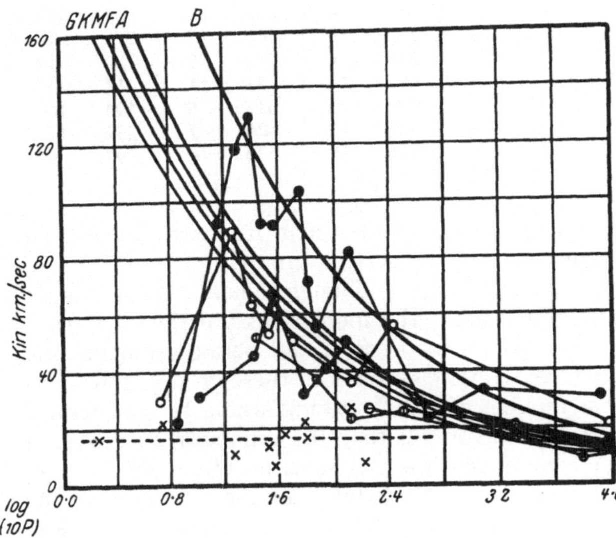

Fig. 51. Relation between period and semi-amplitude of velocity-variation.

● = Type B. ○ = Type A. o = Type F. ⦶ = Type G. ⊖ = Type K, M. × = Cepheid variables.

Fig. 51 shows theoretical curves computed from the above equation by substitution of the various values of C for each spectral type.

Attention may be directed to the agreement of the distribution of K with the computed curves as being satisfactory except for the shortest periods of classes B, A and F, where there is a drop in the values of K joining the theoretical curves to the straight line corresponding to Cepheid variables.

122. The Colours and Spectral Classes of the Components in Visual Double Stars. Relative colours of double stars have always presented a peculiar fascination to the observer. Since the beginning of telescopic work colours were recorded, especially by W. HERSCHEL and W. STRUVE, but only comparatively recent study recognizes the bearing of this phenomenon upon the theory of stellar evolution.

The reader may consult these papers:

TH. LEWIS[3], Colours and Magnitudes of Double Stars.

H. E. LAU[4], Über die Farbe der Doppelsterne.

C. WIRTZ und W. SCHREYER[5], Die Farben der Doppelsterne nach den Beobachtungen von ERCOLE DEMBOWSKI.

P. DOIG[6], Double Star Colours and their Bearing on the Theory of Stellar Development.

G. ABETTI[7], Determinazioni degli indici di colore di stelle doppie.

G. RAYMOND[8], Catalogue des étoiles doubles ou multiples dont les composantes ont des couleurs variables.

Since the direct detection of colour is rather difficult and often questionable, it is rational to rely upon the more accurate determination of spectral class, as this is an index of colour.

[1] M N 86, p. 63 (1925). [2] Lick Bull 11, p. 141 (1924). [3] Obs 29, p. 314 (1906).
[4] A N 208, p. 179 (1920). [5] A N 210, p. 297 (1920). [6] J B A A 33, p. 66 (1922).
[7] Osserv. e. Mem. del R. Oss. Astrof. di Arcetri, Fasc. No. 40 (1923).
[8] J O 2, p. 167 (1918); see also B S A F 22, p. 163 (1908).

Miss CLERKE's account[1] of early work along these lines is comprehensive, and the following papers contain valuable data.

W. HUGGINS[2], Sur les spectres des composantes colorées des étoiles doubles.

E. C. PICKERING[3], The Discovery of Double Stars by Means of their Spectra.

Miss CANNON[4], The Spectra of 745 Double Stars.

R. G. AITKEN[5], The Spectral Classification of 3919 Visual Binary Stars.

P. DOIG[6], The Spectra of Physically Connected Pairs and the Giant and Dwarf Theory.

A work of consequence is LEONARD's, which includes a résumé of previous research as well as conclusions reached through his own extensive investigations[7], namely:

"1. The spectrum of the secondary component of a dwarf star is generally of later class than that of the primary, whereas the spectrum of the fainter component of a giant star is usually of earlier class than that of the primary.

2. In the cases of both giant and dwarf systems, the absolute difference in spectral class between the components seems ordinarily to be related to their disparity in magnitude, being practically zero when they are equal in luminosity, or nearly so, with some exceptions, notably among giant stars (pairs of the γ Circini type), and increasing on the whole as the difference in magnitude increases; but, for a given difference in magnitude, the absolute inequality in spectral class is greater for giants with late-type primaries and less for those whose components are of early type, than for any other kind of stars.

3. The spectrum of each component of a double star appears to be a function mainly of its absolute magnitude; or, in other words, the spectra of the components of double stars are so related to each other that, with but few exceptions, these systems conform to the HERTZSPRUNG-RUSSELL arrangement for individual stars plotted according to spectral class and absolute magnitude. In this configuration the fainter component normally precedes the brighter one, regardless of whether the latter be a giant or a dwarf, in the order prescribed by the LOCKYER-RUSSELL theory of stellar evolution. In view of this conclusion, all of those previously enumerated obviously represent but special phases or necessary consequences of the foregoing generalization.

4. The difference in spectral class between the components of double stars is apparently independent of their projected linear separation.

5. From a consideration of thirteen binary systems, all dwarf stars, there is some indication that, on the average, as the sum of the masses of the components increases their disparity in spectral class approaches zero; and from seven or eight of these systems for which the value of the mass-ratio is known, there is evidence that the difference in spectral class exhibits a tendency to decrease as the ratio of the masses approaches unity.

6. Of any two stars whose masses are unequal but whose remaining physical properties are identical, the less massive star will in general pass through its life-history in advance of the more massive one."

These decisions are in harmony with facts obtained by PERRINE[8]. The latter asserts that stars for which the fainter component is of earlier type are frequently closer together than bodies where the secondary is of later type.

In a survey of the orbits of fifty-nine visual double stars DOBERCK[9] finds it relevant that pairs belonging to classes A and F which are very hot, describe

---

[1] Problems in Astrophysics (1903), Part II, Chapt. XIV.
[2] C R 125, p. 512 (1897). [3] A N 127, p. 155 (1891). [4] Harv Ann 56, p. 227 (1912).
[5] Pop Astr 26, p. 635 (1918). [6] M N 82, p. 372 (1922). [7] Lick Bull 10, p. 169 (1921).
[8] Wash Nat Ac Proc 4, p. 71 (1918). [9] A N 197, p. 397 (1914).

orbits of greater average eccentricity than those of class K which appear to have cooled down in course of time, no doubt because the orbits, once of great eccentricity, have become more circular.

Dr. CAMPBELL and Dr. AITKEN approve the evidence in favour of longer periods for binary stars, spectroscopic or visual, of more advanced spectral classes, and this tendency must be accepted; but as we have seen from the classification of spectroscopic binaries, this relation is not infallible since there are long-period binaries of class B and short-period of class K.

123. The Masses of the Binary Stars. The total mass of a visual binary can be obtained by the formula

$$m_1 + m_2 = \frac{a^3}{\pi^3 P^2},$$

in which π is the parallax of the system, P the period, and a the semi-axis major of the orbit, and the units of mass, length, and time are respectively, the Sun's mass, the astronomical unit and the year. If the proper motion of one of the components is known, it is easy to compute m_1 and m_2.

The formulae for the masses of components of spectroscopic binaries are in paragraph 78.

When spectroscopic as well as visual observations are available, an accurate parallax of the system may be derived by formulae in paragraph 89, and hence the values of m_1 and m_2.

For eclipsing binaries, the value of $\sin i$ is computed from the light-curve, therefore if spectral lines of both components are present on the spectrograms, m_1 and m_2 can be found.

The mass of the Sun being unity we have for visual binaries:

Star	Mass Brighter Component	Mass Fainter Component	Spectral Class Brighter	Spectral Class Fainter
Capella	4,18	3,32	G0	F0
Sirius	2,45	—	A0	—
α Centauri	1,14	0,97	G5	K5
Krueger 60	0,248	0,182	Ma	Ma
ε Hydrae	3,64	2,29	F9	—
β Aurigae	2,38	2,34	A0p	A0p
δ Equulei	1,01	1,00	F5	F5
η Cassiopeiae	0,72	0,41	F8	K5
ζ Herculis	1,09	0,51	G0	—
70 Ophiuchi	1,05	0,77	K0	K4
ξ Boötis	0,62	0,47	G6	K4
85 Pegasi	0,62	0,31	G0	—
μ Herculis	0,46	0,42	Mb	Mb
o^2 Eridani	0,21	0,20	B9	Md

A N 216, pp. 301—302 contains MEYERMANN's list of masses of the two components in fifty-nine pairs.

Masses for eclipsing binaries whose spectroscopic orbits have been determined are:

Star	Spectral Class	Mass of Brighter Component	Mass of Fainter Component	Star	Spectral Class	Mass of Brighter Component	Mass of Fainter Component
V Puppis	B1	19,4	19,4	U Coronae	B3	4,27	1,63
Y Cygni	B2	16,6	15,3	β Aurigae	A0p	2,40	2,36
u Herculis	B3	7,5	2,9	TX Herculis	A2	2,04	1,77
RS Vulpeculae	A0	5,40	1,69	TV Cassiopeiae	A0	1,83	1,01
U Ophiuchi	B9	5,36	4,71	Z Herculis	F2	1,6	1,3
Z Vulpeculae	B3	5,24	2,36	RX Herculis	A0	0,96	0,91
Algol	B8	4,72	0,95	W Ursae Majoris	F8p	0,69	0,49

This includes practically all the positive data relating to the masses of the stars.

To procure further information it is necessary to establish a relation between absolute magnitude and mass. EDDINGTON's theory of this relation[1] is based on his mathematical calculations as to the interior of a star, and seems to be verified by the measured absolute magnitudes and masses of the binaries referred to and of a few Cepheids, providing certain preliminary assumptions are recognized.

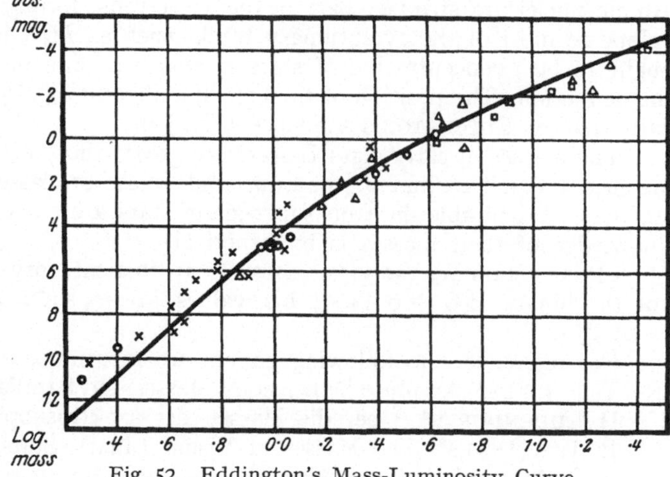

Fig. 52. Eddington's Mass-Luminosity Curve.
○ First Class. × Second Class. □ Cepheids. △ Eclipsing Variables.

The curve in Fig. 52 represents this relation and its verification by observations. When first published it created more or less sensation, first, on account of the influence it might have on the determination of stellar masses, and again because it apparently antagonized Prof. RUSSELL's theory of stellar evolution. A very massive giant star could not, even through the most protracted process of evolution, become a dwarf star. At present, however, this difficulty is obviated by the hypothesis that stars are losing mass, which is being transformed into energy of radiation.

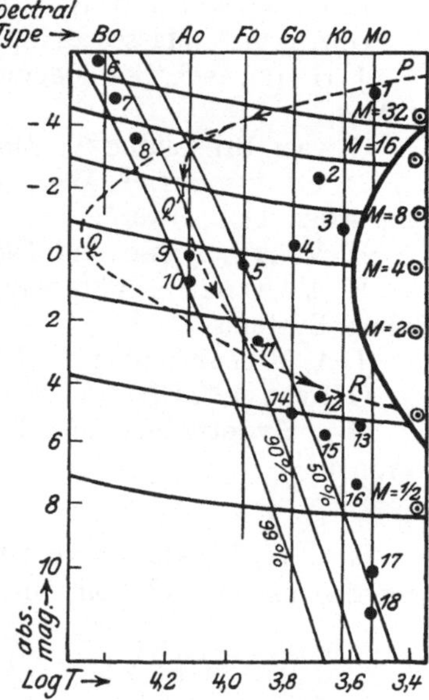

Fig. 53. Jeans' Mass-Rate-of-Emission Relation.

JEANS does not entirely agree with EDDINGTON's idea. His mathematical work on conditions of stellar equilibrium[2] shows that the mass M, rate of emission E (where $-2^1/_2 \log E$ is the star's absolute magnitude) and surface temperature T of a spherical star in equilibrium, are connected approximately by an equation of the form

$$2\tfrac{1}{2}\log E = \Phi(M) + c\log T + \text{a constant},$$

where $\Phi(M)$ is a function of the mass, which is roughly proportional to $\log M$.

This equation is solved graphically in Fig. 53.

If c were strictly constant and positive the curves along which M has constant values would be a series of parallel straight lines slanting upwards to the left. When the slight variations in c are taken into account these lines run in the way indicated by the heavy lines in the diagram, the unstable parts for which c is negative being omitted. The accentuated curve on the right

[1] MN 84, p. 308, p. 372 (1924). [2] MN 85, p. 196, p. 394, p. 792 (1925).

passes through the points at which c changes from positive to negative and so marks the limit between stable and unstable configurations. Data from observed binaries are plotted on the figure as points 1 to 18.

The belt of configurations in which from 80 to 99 per cent of the star's atomic nuclei are stripped bare of their electrons, forms a sort of common envelope to most of the evolutionary tracks such as PQR in the diagram. There ought to be a concentration of stars in this belt, and observations corroborate the deduction. The position of the belt described in Fig. 53 coincides very closely with that of EDDINGTON's sequence (Fig. 52).

The masses of early class B stars are great; those of classes F and G dwarf are nearly equal to that of the Sun, while those of classes K and M dwarf are still less. Procurable data on super-giant stars are extremely limited, and our knowledge of their masses is hypothetical.

Class O stars are the most massive. In this category PLASKETT has pointed out the binary B D + 6°1309; his values are $m_1 \sin^3 i = 75{,}6\,\odot$ and $m_2 \sin^3 i = 63{,}3\,\odot$.

For reference, the following papers are mentioned:

T. J. J. SEE[1], Absolute Dimensions, Masses and Parallaxes of Visual Binaries.

H. LUDENDORFF[2], Über die Massen der spektroskopischen Doppelsterne.

R. T. A. INNES[3], The Masses of Visual Binary Stars.

R. G. AITKEN[4], An Observing List for the Determination of Relative Masses in Visual Binary Systems.

R. G. AITKEN[5], Note on the Masses of Visual Binary Stars.

W. DOBERCK[6], On the Masses of Double Stars.

M. FOUCHÉ[7], Sur les masses des étoiles doubles dont on connaît l'orbite.

E. B. WILSON and W. J. LUYTEN[8], On the Mass Ratios of Binary Stars.

G. ABETTI[9] (Masses of Spectroscopic Binaries).

J. H. JEANS[10], Cosmogonic Problems Associated With a Secular Decrease of Mass.

G. VAN BIESBROECK[11], Mass-Ratios in Visual Binary Stars.

H. LUDENDORFF[12], Weitere Untersuchungen über die Massen der spektroskopischen Doppelsterne.

H. VOGT[13], Masse und Dichteverhältnis bei Doppelsternveränderlichen.

H. L. ALDEN[14], Photographic Determinations of Mass Ratios in Visual Binary Systems.

J. A. MILLER and J. H. PITMAN[15], The Masses of Visual Binary Stars.

O. STRUVE[16], Über das Massenverhältnis der spektroskopischen Doppelsterne.

124. Hypothetical and Dynamical Parallaxes of Double Stars. If in the formula

$$m_1 + m_2 = \frac{a^3}{\pi^3 P^2}$$

$m_1 + m_2$ is assumed to be $= 2$ (mass of the Sun), an assumption generally accepted as not far from the truth, then

$$\pi = \sqrt[3]{\frac{a^3}{P^2}} = \frac{a}{P^{\frac{2}{3}}}\,.$$

[1] A N 139, p. 17 (1895). [2] A N 189, p. 145 (1911).
[3] The South African Journal of Science 12, p. 453.
[4] Lick Bull 7, p. 3 (1912). [5] Pop Astr 18, p. 483 (1910).
[6] A N 191, p. 425 (1912). [7] B S A F 30, p. 90 (1916). [8] Pop Astr 32, p. 626 (1924).
[9] Reale Accad. Naz. dei Lincei, 1922, June, Sept.
[10] M N 85, p. 2 (1924). [11] A J 29, p. 173 (1916). [12] A N 211, p. 105 (1920).
[13] A N 211, p. 123 (1920). [14] Pop Astr 33, p. 164 (1925).
[15] A J 34, p. 127(1922). [16] A N 227, p. 113 (1926).

This value of the parallax is the hypothetical parallax; JACKSON[1], AITKEN[2] and others have devoted much time to this work and at Princeton University Observatory, where Dr. RUSSELL is director, a complete card catalogue of more than sixteen hundred pairs has been prepared.

RUSSELL, ADAMS and JOY[3] acquired an extensive knowledge of the masses of double stars by comparing the hypothetical parallaxes of three hundred and twenty-seven visual pairs, including giant and dwarf stars of all spectral types from O8 to M6, with the spectroscopic parallaxes derived from Mt. Wilson spectrograms. For class B stars, however, KAPTEYN's moving-cluster parallaxes were used instead of the spectroscopic parallaxes, his values being more authentic. Calling s/h_1 the ratio of the spectroscopic to the hypothetical parallax, and M_1 the absolute magnitude corresponding to the combined light of the pair based on the hypothetical parallax, the following relation is found

$$\frac{s}{h_1} = 0{,}62 + 0{,}045\, M_1 .$$

Fig. 54 shows in what close proximity to the theoretical line represented by the above equation are the mean individual observations for the average giant or dwarf star of a certain spectral class.

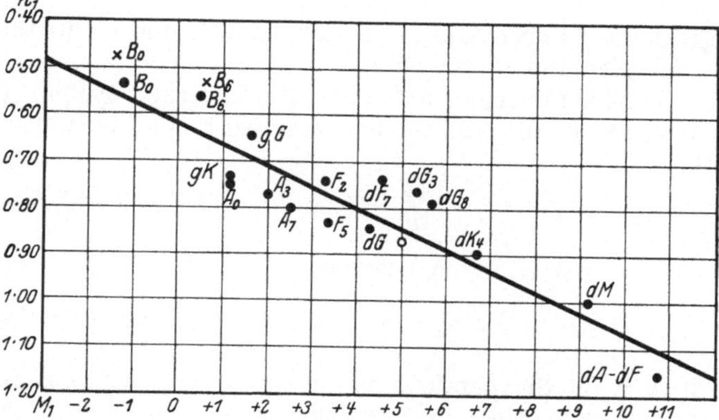

Fig. 54. Relation between s/h_1 and absolute magnitude.

In other words, the results obtained by RUSSELL, ADAMS and JOY indicate a quasi-linear relation between absolute magnitude and mass similar to EDDINGTON's plotted on Fig. 52.

If $a/P^{\frac{2}{3}}$ represents the hypothetical parallax, then

$$M_1 = m + 5 + 5 \log \frac{a}{P^{\frac{2}{3}}},$$

where m is the total apparent magnitude of the pair, and what RUSSELL calls the dynamical parallax of the double star, is

$$\pi_s = \frac{a}{P^{\frac{2}{3}}} \left[0{,}62 + 0{,}045 \left(m + 5 + 5 \log \frac{a}{P^{\frac{2}{3}}}\right)\right].$$

Reliable dynamical parallaxes are determined without knowledge of the spectral type and as a rule these calculations give values that approximate true parallaxes; however, the last formula is not final since the trigonometrical parallaxes have been omitted in the calibration.

125. Densities of Binary Stars. Most of our information relating to stellar density is acquired from two main sources:

First, from eclipsing binaries when the spectra of two components are visible on the plate. Paragraph 116 contains a tabulation of density in such binaries.

Second, from research directed to ascertain the density of visual binaries feasible only if PLANCK's theory of quanta is adopted.

[1] M N 81, p. 2 (1920); 83, p. 441 (1923).
[2] Pop Astr 31, p. 575 (1923). [3] Publ A S P 35, p. 189 (1923).

Developed by ÖPIK[1] in "The Densities of Visual Binary Stars", and extended by BERNEWITZ's[2] publication of "Über die Dichten der Doppelsterne", the substance of this research is the following:

Having derived the radius of a component from its colour-index and parallax, the volume becomes known, and if the mass has been calculated from the absolute orbit, a simple division yields the density.

Let h_* and $h_\odot$ be respectively the bolometric brightness of the star and the Sun, ϱ_* the radius of the star, $\varrho_\odot = 1$, and r_* the distance of the star in astronomical units (Sun's distance), then

$$\frac{h_*}{h_\odot} = \frac{\varrho_*^2}{r_*^2}\left[\frac{\frac{c_2}{T_\odot}}{\frac{c_2}{T_*}}\right].$$

c_2 is the second constant of PLANCK while T_* and $T_\odot$ are the absolute temperatures of the star and of the Sun.

If K' is the colour-index of a star in KING's system (Harvard), BOTTLINGER[3] arrives at an empirical relation between this colour-index and the quantity c_2/T, which is:

$$\frac{c_2}{T} = 1{,}20 + 2{,}071\,K' - 0{,}088\,K'^2\,.$$

The first equation becomes

$$\log\varrho_* = 0{,}2m'_\odot - 0{,}2m'_* - \log\pi_* + 5{,}3144 + \log\frac{c_2}{T_*} - \log\frac{c_2}{T_\odot}\,,$$

where π_* is the parallax of the star and m' the bolometric magnitude. Using EDDINGTON's table[4] for the purpose of transforming the visual magnitude m into the bolometric m', we have:

$\frac{c_2}{T}$	$m'-m$	$\frac{c_2}{T}$	$m'-m$	$\frac{c_2}{T}$	$m'-m$
0,5	−3,30	1,4	−0,25	3,0	−0,20
0,6	−2,80	1,5	−0,20	3,5	−0,50
0,7	−2,30	1,6	−0,15	4,0	−0,90
0,8	−1,90	1,8	−0,05	4,5	−1,30
0,9	−1,50	2,0	0,00	5,0	−1,80
1,0	−1,15	2,2	0,00	5,5	−2,30
1,1	−0,85	2,4	0,00	6,0	−2,80
1,2	−0,60	2,6	−0,05	6,5	−3,40
1,3	−0,40	2,8	−0,10	7,0	−4,00

If the magnitude of the Sun is taken as $-26^{\mathrm{m}},9$, and $c_2/T_\odot = 2{,}20$ then

$$\log\varrho_* = 2{,}25 - 0{,}2\,m'_* - \log 1000\,\pi + 2\log\frac{c_2}{T_*}\,.$$

If ϱ_1 is the radius of the primary and ϱ_2 the radius of the secondary, while m_1 and m_2 are the respective masses, the densities become:

$$\frac{m_1}{\frac{4}{3}\pi\varrho_1^3} \quad \text{and} \quad \frac{m_2}{\frac{4}{3}\pi\varrho_2^3}\,,$$

the unit of mass being the mass of the Sun, and the unit of length for the radii being the radius of the Sun.

---

1 Ap J 44, p. 292 (1916). 2 A N 213, p. 1 (1921).
3 Veröff. der Univ.-Sternwarte zu Berlin-Babelsberg 3, Heft 4, p. 18 (1923).
4 M N 77, p. 605 (1917).

In BERNEWITZ' article mentioned a moment ago the densities of sixty-three double star systems are listed, the majority being of the same order as that of the Sun; but from the tabulation there must be others extraordinarily large. A few of these values follow:

Star		Density
η Cassiopeiae	A	0,59
	B	1,18
o^2 Eridani	B	5600
Sirius	A	0,22
	B	88000
ξ Ursae Maj.	A	0,71
	B	1,15
α Centauri	A	0,49
	B	0,05
Krueger 60	A	3,12
85 Pegasi	A	0,44

Attention was drawn in paragraph 67 to the unusual density of the companion of Sirius. If this unparalleled feature is actually true, an illuminating field for study will open before the physicist.

o^2 Eridani is another remarkable body; its density belies the spectral class A to which it really belongs, because stars in this category are supposed to be less dense than the Sun, while o^2 Eridani discloses considerable density. Being a type of celestial body called white dwarf in which the atoms are practically stripped of their attending electrons, its compact character is due to the consequent crowding of these atoms into an extremely small volume.

126. The Orientation of Orbit Planes. The problem of the parallelism of orbit planes in visual binaries is an old one. MÄDLER[1], in 1838, after examining fifty-one physical pairs, inferred from the positions and figures of apparent orbits that there is a celestial equator to which the orbit planes are nearly parallel; the position of the pole corresponding to this equator is 73° right ascension and +52° declination.

There is no record of further serious work for a long interval of time. More than half a century later, studies on the parallelism of orbit planes were published by Miss EVERETT[2], SEE[3], DOBERCK[4], LEWIS and TURNER[5], BOHLIN[6], and POOR[7]. The first three find no authority for preferential orientation. POOR is opposed to this opinion. The thesis upon which he based his problem is: "Were the orbit-planes of binary stars parallel, then because the apparent orbits of those situated on the great circle parallel to their orbit-planes would be straight lines, while at the poles of this great circle the apparent orbits would be ellipses, the parallelism would show itself in a statistical study as a variation in correlation between position angle and distance of doubles in different parts of the sky." The solution depends upon the data given in BURNHAM's General Catalogue and later lists of double stars up to 1913, and proves that a preferential pole of the orbit planes exists near the vertex of preferential motions of the stars.

GUTHNICK and PRAGER[8] did excellent work on the orbits of double stars belonging to the Ursae Majoris stream.

A quantity of unreliable data is responsible for wide discrepancy in conclusions, therefore it is not wise to consider the present opinions relating to the parallelism of orbit planes as incontrovertible. The question, however, is extremely important because of its bearing upon cosmogonical theories.

127. Other Statistical Investigations. Much further statistical work has been published relating to number, separation, galactic distribution and other characteristics of particular classes and binary stars in general, but most of

---

[1] C R 6, p. 920 (1838). [2] M N 56, p. 462 (1896).
[3] T. J. J. SEE, Evolution of the Stellar Systems 1, p. 247 (1896).
[4] A N 147, p. 251 (1898) and 179, p. 199 (1908).
[5] M N 67, p. 498 (1907). [6] A N 176, p. 197 (1907).
[7] A J 28, p. 145 (1914). [8] Festschrift ELSTER und GEITEL (1915).

this data is incomplete and of minor importance, or needs revision on account of more recent accumulation of material.

The reader will find additional interesting statistics in these papers:

F. SCHLESINGER and H. BAKER[1], A Comparative Study of Spectroscopic Binaries.

H. LUDENDORFF[2], Zur Statistik der spektroskopischen Doppelsterne.

H. N. RUSSELL and H. SHAPLEY[3], On the Distribution of Eclipsing Variables in Space.

C. D. PERRINE[4], The Distribution and Some Possible Characteristics of the Spectroscopic Binaries of class M.

P. STROOBANT[5], Distribution des étoiles doubles spectroscopiques.

R. G. AITKEN[6], The Relation between the Separation and the Number of Visual Double Stars.

R. G. AITKEN[7], A Statistical Study of the Visual Double Stars in the Northern Sky.

G. ABETTI[8], Relazioni fra le caratteristiche fisiche delle stelle doppie.

E. HERTZSPRUNG[9], On the Median Period of Binary Systems near our Sun.

S. WICKSELL[10], Contributions to the Statistics of Spectroscopic Binary Stars.

128. Origin of Binary Stars. Statistical research is of immeasurable advantage when considering the origin of binary stars. This problem, which at a superficial glance appears so simple, becomes when thoroughly analyzed one of the cosmogonical questions most beset with difficulties. Some of its ramifications lead to such intricate propositions of celestial mechanics as to tax the genius of men like DARWIN, POINCARÉ and JEANS.

Astronomers have assumed three possible origins for binary stars:

A. Fission. A single rotating star divides into two components through centrifugal action reinforced by internal radiation pressure and perhaps other disruptive forces.

B. Independent nuclei. The material in a primal luminous or dark nebula condenses about two nuclei sufficiently close to each other.

C. Capture. Two stars, hitherto independent, approach each other closely, and following the laws of gravitation are forced to revolve forever about a common centre of gravity.

Dr. JEANS thinks that capture does not account for any appreciable proportion of observed orbits. He says: "Apart from observation, I do not think this hypothesis can explain any large number of binary systems, for it can be shown by mathematical calculation that capture would be an excessively rare event. In a universe of 1 000 000 000 stars moving with about the present velocities of the stars and spaced at about their present distances, the number of captures in a period of 1 000 000 000 years would at most be of the order of 10 000, a proportion of 0,001 per cent."

The theories (A) and (B) are thus the only two plausible ones to hold the field, and it is admitted that both are valid. (A) being the cause of most spectroscopic binaries and close visual doubles, while (B) applies to wide visual doubles.

One of the original explanations of the formation of a double star by fission is due to Sir GEORGE DARWIN[11]; we quote from him: "Originally the star must

---

[1] Allegh Obs Publ 1, p. 135 (1910). [2] A N 184, p. 373 (1910).
[3] Ap J 40, p. 417 (1914). [4] Ap J 42, p. 370 (1915). [5] C R 156, p. 37 (1913).
[6] Lick Bull 6, p. 1 (1910). [7] Wash Nat Ac Proc 1, p. 530 (1915).
[8] Arcetri Publ 39, p. 25 (1922). [9] B A N 1, p. 149 (1922).
[10] Lund Medd, Série I, No. 63 (1913).
[11] Darwin and Modern Science p. 563 (Cambridge 1909).

have been single, it must have been widely diffused, and must have been endowed with a slow rotation. In this condition the strata of equal density must have been of the planetary form. As it cooled and contracted the symmetry round the axis of rotation must have become unstable through the effects of gravitation assisted perhaps by the increasing speed of rotation. The strata of equal density must then become somewhat pear-shaped, and afterwards like an hour-glass, with the constrictions more pronounced in the internal than in the external strata. The constrictions of the successive strata then begin to rupture from the inside progressively outwards, and when at length all are ruptured we have the twin stars portrayed by ROBERTS and by others."

RUSSELL in a valuable paper on "The Origin of Binary Stars"[1], maintains that in the absence of a well developed theory of formation an appeal to the facts of observation would be advisable. His attention was particularly drawn toward triple and multiple systems, because they present certain general characteristics supporting the fission theory. Naturally it would be impossible for such systems, if originating from independent nuclei, to betray any relation of mass or of relative distance, but this would not be true for triple or multiple stars formed by successive fissions.

In its simplest form, the fission hypothesis considers merely the separation of a body into two parts through increasing angular velocity induced by a continuous shrinkage of volume. When condensation reaches a certain point the consequent rupture of the original mass is an obvious means to check excessive rotation. The tendency to separation being a function of the square of the angular velocity divided by the density, each constituent of the fissured body will at the beginning be immune from further division, but as concentration continues these components may in turn break up.

RUSSELL arrives at the conclusions:

1. The distance of centres at the time of separation is greater, and the density less, the more unequal the parts are.

2. The ratio in which the initial distance can be increased by tidal action, increases as the masses become more unequal.

3. The smaller mass has the greater density just after separation.

4. The ratio of contraction necessary to bring about a second fission (other things being equal, and tidal friction absent) is less for the greater mass.

5. The ratio of the dimensions of the separating masses at the time of the second fission to that of the first, involves a factor c^2 (function of the internal constitution of the mass and of its form) and is always small.

6. The same is true of the final orbits resulting from the successive fissions.

7. The increase of density between the fissions is very great.

The results of RUSSELL's computation are quite definite. Multiple systems formed by fissions must be pairs, one or both components being themselves double, and separated by a distance less than about one-fifth that of the wide pair, usually much less. Some components of these close pairs may be still closer pairs after the same fashion.

Various double and multiple star catalogues contain abundant data with which to test this conclusion. The majority of multiple stars belong to the type under consideration in RUSSELL's work.

Exceptions, however, become marked as we approach systems where the separation of the wide pair is of the order of one thousand years proper motion, and for still wider pairs the law fails completely. Therefore it is evident in

[1] Ap J 31, p. 185 (1910).

the last case that we must search for some origin other than the disruption of a single mass, and the most plausible answer to our quest would be the theory of independent nuclei.

The theory of fission has been developed in a masterly way by Dr. JEANS; his wonderful book: "Problems of Cosmogony and Stellar Dynamics" contains the most complete mathematical treatment as yet available of this intricate and complex problem.

Dr. JEANS' hypothesis is in substance the following:

In the process of contraction a stellar nucleus is formed in a nebular arm, such as may be seen on modern photographs of spiral nebulae supposed to be other universes. In this early period of its existence its mean density is very low, perhaps 10^{-17} grammes per cubic centimetre, and its surrounding atmosphere is contiguous with that of neighbouring stars. During this time it shares in the rotation of the parent nebula, the period being of the order of 160000 years.

Later, the nebular arms continue to expand while individual stars contract, becoming more distinct one from the other until finally they can be regarded as entirely separate bodies, each describing an independent orbit under the gravitational attraction of the other stars.

When the increasing density has attained the value 10^{-6} grammes per cubic centimetre, the star should begin to disintegrate. A sharp edge forms, and projections of matter are thrown off from the equator.

There is more than one reasonable argument to show that this material does not condense into filaments but forms a surrounding atmosphere which increases at the expense of the nucleus as dissolution progresses. Dynamically this phenomenon is exceedingly difficult to follow even in its main outlines.

That the atmosphere may in time condense into nuclei and ultimately form planets is a possibility concerning which we can merely conjecture.

The equatorial ejection of matter will continue until a further critical density ϱ_0 is reached, at which the pseudo-spheroidal figure for the nucleus becomes unstable and gives place to a pseudo-ellipsoidal form. Material is thrown off from the extremities of this new figure which ultimately becomes transformed into a pear-shaped body and divides into two detached masses.

Accordingly, in the last analysis the result of disintegration will be a binary star, the two components revolving about one another in a more or less chaotic atmosphere through which they plough their way; the time arrives, however, when the atmosphere condenses around the two bodies leaving an ordinary binary star.

Incidentally the former condition recalls DUNCAN's tentative explanation of Cepheid variables.

Moreover JEANS' conclusions are:

1. No binary star which has formed by fission can have a density of less than about $^1/_4$.

2. No giant binary star can have been formed by fission.

3. The temperature of a binary star which has formed by fission must decrease as its evolution progresses.

These conclusions are partially verified by observation.

No super-giant star, not even when it is a Cepheid, seems to be a binary formed of components having nearly equal masses.

However, the Cepheids may be single bodies in the process of division.

The writer has previously pointed out the correlation existing between the light-curve and the probable constitution of the system[1], for example:

---

[1] L'Astronomie 1924, p. 339.

Smooth curve such as VX Cygni: symmetrical body. Curve with a secondary hump such as VY Cygni: pear-shaped or unsymmetrical body. Curve with two unequal minima such as β Lyrae: two bodies very close to each other.

No doubt it is to the super-giant stars and particularly to the Cepheid variables that astronomers will appeal in their search for authentic data to unveil this very puzzling secret of nature, the origin of double and multiple stars.

k) Triple and Other Multiple Systems.

129. Visual Triple and Other Multiple Systems. One of the most amazing discoveries of modern research is that of triple and multiple stars; it has been found that a physical relation may exist between three or more bodies close together, just as we have globular clusters and other systems composed of enormous groups of stars.

A very attractive problem in dynamics is that of a close binary, with rather short revolution period, accompanied by a third body at an appreciable distance but physically related to the first pair as evidenced by common proper motion. It is characteristic that every triple system, visual or spectroscopic, is without exception a close double widely separated from a third body.

Among the few real physical and visual triple or multiple systems the following may be mentioned:

No. in Burnham's General Cat.	Star	
1471	20 Persei	triple
2109	*o* Eridani	triple
4414	β 581	triple
4477	ζ Cancri	quadruple
4771	ε Hydrae	quadruple
7259	μ_2 Boötis	triple
7487	ξ Scorpii	triple
8162	μ Herculis	triple
9114	Secchi 2	triple

In these systems the orbit of the close pair is usually well determined, but the orbital motion of the third component around the centre of gravity of the close pair is very slow, if detected at all, and covers only a small arc. Perhaps the finest typical illustration of a visual triple system is ξ Scorpii.

130. The Visual Triple System ξ Scorpii. Sir William Herschel discovered this strange system, and his measurements constitute our first practical knowledge regarding it. The separation of the close pair is about 1″, while the third star is distant about 7″. Uninterrupted observations were made from 1825 to the present time. An exceptionally good orbit for the close pair has been computed by Dr. Aitken[1] who gives the elements:

P	44,70 years
T	1905,39
e	0,75
a	0″,72
Ω	27°,2
i	±29°,1
ω	343°,6
Angle	Increasing.

The following positions of the third component C with respect to AB are:

Date	θ	ϱ	Observer	Date	θ	ϱ	Observer
1828,47	77°,6	6″,73	Struve	1894,89	65°,9	7″,22	Comstock
1836,34	75 ,1	7 ,13	,,	1896,47	63 ,2	7 ,45	Hussey
1852,22	74 ,6	6 ,99	,,	1903,50	63 ,2	7 ,34	Jouffray
1864,38	70 ,6	7 ,15	Dembowski	1912,43	62 ,0	7 ,44	Aitken
1868,89	69 ,9	7 ,07	,,	,56	61 ,5	7 ,53	,,
1874,95	69 ,0	7 ,11	,,	1921,51	56 ,2	7 ,96	,,
1883,34	65 ,9	7 ,28	Asaph Hall				

[1] Lick Obs Publ 12, p. 103 (1914).

Orbital motion is obvious, but many decades may pass before an accurate orbit will be established. After long and patient research one may be rewarded by a rational analysis of the dynamical behaviour of this difficult system and perhaps find evidences of perturbation in the close pair. On the whole, however, micrometric visual observations are too crude to disclose any information of great value to problems in celestial mechanics, but using the spectrograph and the interferometer we anticipate more reliable data in the near future.

131. The System of ζ Cancri. This system, as the subject of much discussion, is no doubt quite familiar to the average reader of astronomical lore. Composed of a close pair AB, magnitudes 5,5 and 5,7, separation about $1''$, there is a third body C of about the sixth magnitude in this group, discovered by TOBIAS MAYER in 1756, the distance between A and C being nearly $5''$. This division was the preliminary to HERSCHEL's further analysis in 1781, and was the earliest example of the decomposition of a double into a triple star. The next distinct view of these close objects was obtained by JAMES SOUTH in 1825, but STRUVE's nine inch FRAUNHOFER telescope showed them easily and they have never since been lost. DOBERCK's orbital elements[1] of the close pair are:

P	60,083 years
T	1870,65
e	0,339
a	$0'',856$
Ω	Indeterminable
i	$0°,0$
ω	$183°\ 33'$
Angle	Decreasing.

But the orbital movements of AB are only part of a more elaborate scheme of displacements. The star C apparently retrogrades round AB at an average rate of half a degree a year, roughly indicating a revolution period of 600 to 700 years. But this average rate is subject to very remarkable irregularities. The path traced out on the sky, far from being a smooth curve, is looped into a series of epicycles, in traversing which the star alternately quickens and slackens, or even altogether desists from its advance while increasing or diminishing, by proportionate amounts, its distance from the centre of motion. This anomalous behaviour, detected by FLAMMARION (see his Catalogue) in 1873, was both detected and interpreted by OTTO STRUVE[2], and later discussed by SEELIGER[3] in several important papers. The latter attributes these irregularities to the existence of a dark physical companion gravitating about C, the two forming a binary of period 18,0 years.

It is interesting to give here the method of SEELIGER to compute the elements of C's orbit.

All the measures of C having been made with respect to the middle point G_1 of (AB), this last point may be considered as fixed. Let the plane passing through G_1 at right angles to the line of sight be the plane of reference, and let G_1 be the origin of coordinates; the orbit of C around the centre of mass G_2 of its partial system being an ellipse having G_2 as a focus, its apparent orbit in the plane of reference will be an ellipse surrounding the projection of G_2, therefore,

[1] A N 173, p. 241 (1906).

[2] C R 79, p. 1463 (1874).

[3] Denkschr. der Wiener Akad. 44 (1881); Abh. der k. bayer. Akad. der Wiss., Klasse II, 17, 1. Abt. (1888). Sitzber. der math.-phys. Klasse der k. bayer. Akad. der Wiss. 24, Heft 3 (1894).

if x and y are the coordinates of C, while ξ and η are the coordinates of G_2's projection, this ellipse is represented by the equation

$$A x'^2 + B y'^2 + 2 C x' y' + 2 D x' + 2 E y' - 1 = 0, \qquad (1)$$

where

$$x' = x - \xi, \qquad y' = y - \eta.$$

x and y are obtained directly from the observations; as to ξ and η they are not exactly known, but it may be assumed that G_2's projection moves uniformly in a circle around G_1 during the short interval of 18 years which is the period of C's perturbations or C's unknown orbit; hence ϱ_0 being the mean distance between G_2 and G_1, and μ and ν being constants determined from the observations, it may be written

$$\xi = \varrho_0 \cos(\mu t + \nu), \qquad \eta = \varrho_0 \sin(\mu t + \nu),$$

equations which determine ξ and η for any value of the time; each observation of C will then give an equation similar to (1) and all the equations so obtained will give a provisional system of values for the coefficients $A, B, \ldots E$.

These values are indeed only approximate, such as those of ξ and η; but if φ represents the first member of equation (1) considered as a function of the variables $A, B, \ldots, E, \xi, \eta$, and φ_1 its numerical value after having replaced these variables by their approximate values, if $dA, dB, \ldots d\xi$ and $d\eta$ are the necessary corrections to these values, we have evidently

$$\varphi_1 + \Delta\varphi = 0$$

or

$$x'^2 dA + y'^2 dB + x'y' dC + x' dD + y' dE + R\, d\xi + Q\, d\eta + \varphi_1 = 0, \qquad (2)$$

where

$$R = -2(A x' + C y' + D), \qquad Q = -2(B y' + C x' + E),$$

and

$$d\xi = \frac{d\varrho_0}{\varrho_0}\xi - (d\nu + t\, d\mu)\eta, \qquad d\eta = \frac{d\varrho_0}{\varrho_0}\eta + (d\nu + t\, d\mu)\xi$$

and, consequently

$$R\, d\xi + Q\, d\eta = \frac{d\varrho_0}{\varrho_0}(R\xi + Q\eta) + (Q\xi - R\eta)(d\nu + t\, d\mu),$$

and replacing in (2):

$$\left.\begin{aligned} x'^2 dA + y'^2 dB + x'y' dC + x' dD - y' dE + \frac{d\varrho_0}{\varrho_0}(R\xi + Q\eta) \\ + (Q\xi - R\eta)(d\nu + t\, d\mu) + \varphi_1 = 0. \end{aligned}\right\} \qquad (3)$$

Each observation of C will then give an equation of this form; all of them treated by the method of least squares will give the most probable values for the corrections and, consequently for the unknowns

$$A, B, C, D, E, \varrho_0, \mu \text{ and } \nu.$$

The determination of the orbit of C around the centre of mass G_2 becomes then that of an ordinary visual double star orbit. SEELIGER obtained for the elements of this orbit

P	18 years
a	0″,217
e	0,1106
Ω	71°,96
ω	109°,68
i	17°,35

BURNHAM, doubtful regarding the values of the observations of ζ Cancri[1], strongly opposed SEELIGER's theory[2]; according to him the plotted positions of component C indicate rectilinear motion with irregularities probably due to errors of observation.

In a more recent article, "Über das mehrfache Sternsystem ζ Cancri"[3], SEELIGER defended his original ideas.

It is difficult to say, at present, whether or not these views should be generally accepted. Perhaps radial velocities of component C will elucidate the question; two determinations by ADAMS[4] at the Yerkes Observatory are:

1902 December	17	−12 km
1904 March	19	−12 km.

Other values are not available except one by Victoria observers[5], — 14,7 km.

More accurate measures of relative distances made with an interferometer would go far to solve the problem of ζ Cancri.

"Über das mehrfache System ζ Cancri", published not long ago by G. SCHNAUDER[6], advocates the supposition that component C really possesses a dark companion. Elements of its orbit were computed obtaining a period of 16,92 years.

132. The System of ξ Ursae Majoris. The well known double star ξ Ursae Majoris has its two components of nearly the same brightness; they revolve around each other in 60,8 years. In his investigation of their motion NÖRLUND[7] found a perturbation with an apparent period of 1,8 years. The brighter component had been found by WRIGHT[8], in 1900, to be a spectroscopic binary of fairly long period, and the elements of its orbit were determined in 1919 by ABETTI[9] and found to be

P	665^d
T	J. D. 2415252,7
ω	322°, 65
e	0,41
K_1	7,0 km
V_0	−16,4 km
$a_1 \sin i$	58300000 km
$\frac{m_2^3 \sin^3 i}{(m_1 + m_2)^2}$	0,018 ⊙.

The fainter component is also a spectroscopic binary[10] with a period of about ten days.

There is thus no doubt that the perturbations in the visual motion discovered by NÖRLUND are due to the fact that the bright component is a wide spectroscopic binary.

As mentioned in paragraph 31, HERTZSPRUNG confirmed the existence of these visual perturbations (see Fig. 5 and 6). Neglecting the eccentricity, HERTZSPRUNG found the radius of the 1,8 year visual orbit to be nearly 0″,05 and its inclination is so near 90° that there is a possibility of eclipses, but these have so far not been detected. It is interesting to consider that on account of the large proper motion of ξ Ursae Majoris, this inclination may be modified rapidly and be such that within two or three thousand years the close couple may change from eclipsing binary to ordinary binary or vice versa. According

[1] M N 51, p. 388 (1891).
[2] General Catal. of Double Stars, Part II, p. 513 (1906).
[3] A N 199, p. 273 (1914).
[4] Ap J 19, p. 355 (1904).
[5] Publ Dom Astroph Obs 3, p. 76 (1924).
[6] A N 215, p. 441 (1921).
[7] A N 170, p. 123 (1905).
[8] Ap J 12, p. 254 (1900).
[9] Mem. Spet. It. Ser. 2a, 8, p. 113 (1919).
[10] Publ A S P 30, p. 353 (1918).

to HERTZSPRUNG, the exact status of ξ Ursae majoris as an eclipsing binary remains to be determined from more accurate observations.

From visual observations alone, and from the parallax 0″,155, HERTZSPRUNG found the masses of the three components to be 0,43 ⊙, 0,60 ⊙ and 0,16 ⊙.

133. The System of 44 i Boötis. 44 i Boötis is the well known visual binary Σ 1909. The two components show a well-marked relative motion. The companion reached apastron about 1870 at a distance of 5″ from the brighter star. SCHILT[1] discovered that the fainter component is an eclipsing binary of the W Ursae Majoris type, and its light-curve was determined photographically using the 60 inch telescope of the Mt. Wilson Observatory.

Thus this system constitutes a new and remarkable kind of triple body.

134. The Triple System $\varkappa$ Pegasi. Of greater interest than the strictly visual, are triple systems discovered both from visual observation and with the spectrograph. The observational material, obtained when the spectra are good, facilitates the study in a double-star orbit, of perturbations due to a third body.

Purely spectrographic triples have been discovered during the last few years, but investigation in this particular direction has not produced sufficient observational data upon which to base the theoretical developments of celestial mechanics. There is hope that some day mere observation will bring definite knowledge to bear on the three body problem, a conundrum baffling to the ablest mathematicians and of which so little is relatively known.

In 1880, BURNHAM found that $\varkappa$ Pegasi was a visual binary having two yellowish components, magnitudes 4,8 and 5,3; later a very short period was ascertained, 11,37 years. By discovering one of the components to be a spectroscopic binary, Dr. CAMPBELL[2] contributed another triple system of great possibilities.

Numerous spectrograms of $\varkappa$ Pegasi were obtained at the Lick Observatory especially in 1900, 1912, and later a complete study of the system was made by the writer[3] in 1917, obtaining for the three epochs indicated the following elements:

Element	1900,7	1912,7	1917,6	Probable error for 1917
P	$5^d,9715$	$5^d,9715$	$5^d,9715$	—
T	J. D. 2415239,25	2419593,17	2421440,69	$\pm 0^d,23$
ω	4°,7	47°,5	189°,4	±13°,9
e	0,04	0,03	0,034	±0,007
K_1	39,81 km	40,35 km	41,53 km	±0,34 km
V_0	−2,80 km	−3,79 km	−14,62 km	—
$a_1 \sin i$	3266000 km	3312000 km	3408000 km	±29000 km

The change of V_0 is at once evident, but this must be expected since V_0 represents the centre-of-mass velocity of the spectroscopic system which revolves around the centre of mass of the visual system.

A very important fact is the variation of ω which indicates a rotation of the line of apsides. It seems highly probable that the line of apsides rotates 360° during the time that the spectroscopic body revolves around the centre of mass of the visual system; that is, in 11,35 years. This rotation, however, is not likely to be uniform.

The three velocity-curves for 1900, 1912 and 1917 are drawn in Fig. 55, 56 and 57.

[1] Ap J 54, p. 215 (1926). [2] Ap J 12, p. 257 (1900).
[3] Lick Bull 9, p. 120 (1917).

If the three velocity-curves are superimposed so that their centre-of-mass lines coincide (Fig. 58), a large change due to the variation of ω is immediately apparent. This rotation of the line of apsides reflects the perturbations occuring in the spectroscopic binary orbit under the influence of the third body.

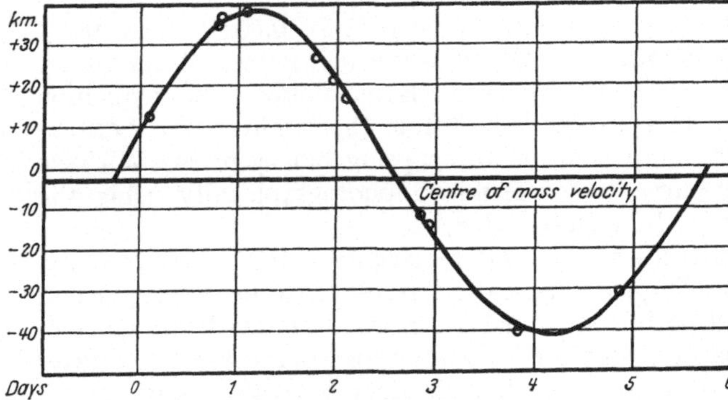

Fig. 55. $\varkappa$ Pegasi. Velocity-curve for 1900.

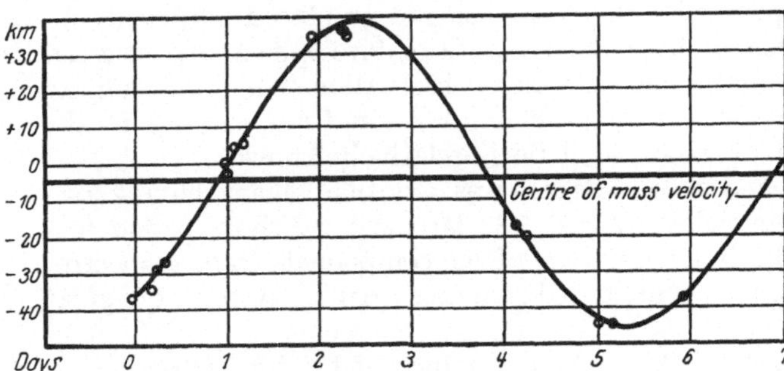

Fig. 56. $\varkappa$ Pegasi. Velocity-curve of 1912.

There is but slight change in eccentricity, if any; however, a small variation of $a \sin i$ is possible, caused by an inconsequential fluctuation of the major axis or a deviation of the orbit plane.

The visual orbital elements by LEWIS are most reliable:

P	11,35 years
T	1897,8
e	0,49
a	0″,29
Ω	109°,2
i	77°,5
ω	106°,1
Angle	Decreasing.

From the measurements on three plates of well isolated lines belonging to the spectrum of the second visual component the following radial velocities were computed:

1917 Aug. 5	+3,20 km
10	+2,33
28	+3,87

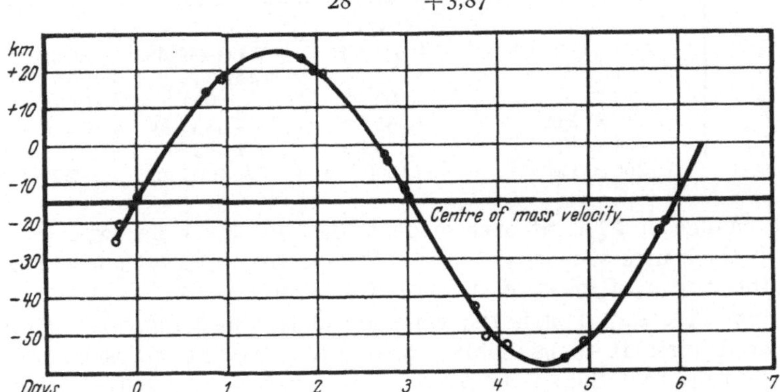

Fig. 57. $\varkappa$ Pegasi. Velocity-curve for 1917.

All the above values furnish the elements:

Value of a_1 (visual orbit) 511 100 000 km.

Value of $a = a_1 + a_2$ (visual orbit) 1 826 000 000 km.

Parallax 0″,025 ± 0″,002.

Mass M_1 (visual) $= m_1 + m_2$ (spectroscopic) $= 10{,}33\odot$.
M_2 (visual) $= 4{,}00\odot$.

Assuming that the planes of the two orbits (visual and spectroscopic) coincide:

$$m_1 = 6{,}62\odot,$$

$$m_2 = 3{,}71\odot.$$

For comparison we mention the values of the trigonometric parallax of $\varkappa$ Pegasi:

$+0'',028 \pm 0'',043$ (FLINT),
$+0'',026 \pm 0'',005$ (ALLEGHENY),

and of the spectroscopic parallax:

$+ 0'',030$ (ADAMS),
$+ 0'',032$ (RIMMER),
$+ 0'',033$ (VICTORIA).

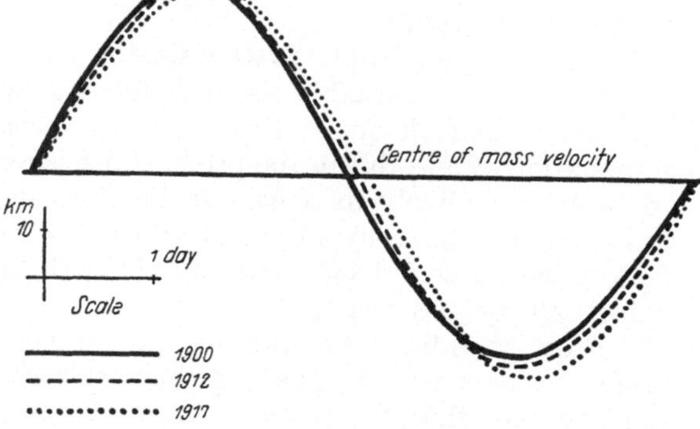

Fig. 58. $\varkappa$ Pegasi. Velocity-curves superimposed.

135. The Triple System 13 Ceti. PARASKÉVOPOULOS' survey[1] of plates taken at the Yerkes Observatory proved that this system is analogous to $\varkappa$ Pegasi except that lines of the second visual component were not measured in the spectra, and consequently no value of the parallax could be derived.

The visual binary has a period 6,88 years and the spectroscopic about two days.

The visual orbital elements by AITKEN[2] are:

P	6,88 years
T	1905,27
e	0,725
a	$0'',242$
Ω	$38°,7$
i	$53°,45$
ω	$66°,8$
Angle	Increasing.

For the epochs given, the elements of the spectroscopic orbit by PARASKÉVOPOULOS are:

Element	1906—07	1908	1912—13	Probable Error for 1912—13
P	$2^d,0818$	$2^d,0810$	$2^d,0866$	$\pm 0^d,0006$
T	J. D. 2417484,494	2418265,260	2419716,571	$\pm 0^d,098$
ω	$222°,7$	$230°,6$	$285°,0$	$\pm 17°,2$
e	0,09	0,07	0,10	$\pm 0,04$
K_1	36,4 km	36,3 km	34,2 km	$\pm 1,2$ km
V_0	$+10,3$ km	$+9,0$ km	$+17,2$ km	
$a_1 \sin i$	1037262 km	1036214 km	975000 km	± 34000 km

From a comparison of these elements the perturbations due to the fainter visual component are clearly distinguished:

The period of the spectroscopic binary is shorter near the apastron (1908) than near the periastron (1912—13) of the visual orbit. The same perturbation observed in the motion of the moon around the earth is well known.

The eccentricity shows no appreciable variation.

[1] Ap J 52, p. 110 (1920). [2] Lick Obs Publ 12, p. 5 (1914).

The value of $a_1 \sin i$ exhibits a slight variation, but it is impossible to say whether a_1, or i, or both vary.

The motion of the line of apsides is distinctly indicated, its rotation, however, is slower than in $\varkappa$ Pegasi.

Another system similar to $\varkappa$ Pegasi and 13 Ceti is μ Orionis, studied by FROST and STRUVE[1].

136. The Quadruple System Castor. α_1 and α_2 Geminorum together constitute a really remarkable visual double star with unmistakable orbital motion. The period and all other elements of the orbit are wholly unknown. The determination of an orbit in the present case is extremely uncertain and according to BURNHAM[2], likely to remain so for a century or more. DOBERCK[3] however, has computed three hypothetical orbits, one of which he considers fairly satisfactory with a period 347 years. In his work LOHSE[4] adopted for this period the values 249 or 298 years.

From meridian observations CROMMELIN asserts[5] that the path of the brighter component α_2 has a greater curvature than that of the fainter α_1, which proves that the latter is more massive.

Both α_1 and α_2 Geminorum are spectroscopic binaries; we owe the first to BÉLOPOLSKY[6], and the second to H. D. CURTIS[7], who thoroughly studied the entire system[8]. His spectroscopic elements are:

Element	α_1 Geminorum	α_2 Geminorum
P	$2^d,928285$	$9^d,218826$
T	J. D. 2416828,057	2416746,385
ω	102°,52	265°,35
e	0,01	0,503
K_1	31,76 km	13,56 km
V_0	−0,98 km	+6,20 km
$a_1 \sin i$	1279000 km	1485000 km
$\frac{m_2^3 \sin^3 i}{(m_1+m_2)^2}$	0,0097	0,0015

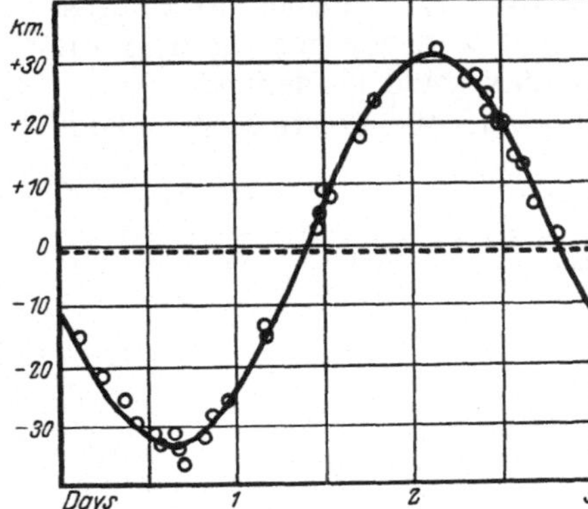

Fig. 59. Velocity-curve of α_1 Geminorum.

The corresponding radial-velocity curves are found in Fig. 59 and 60.

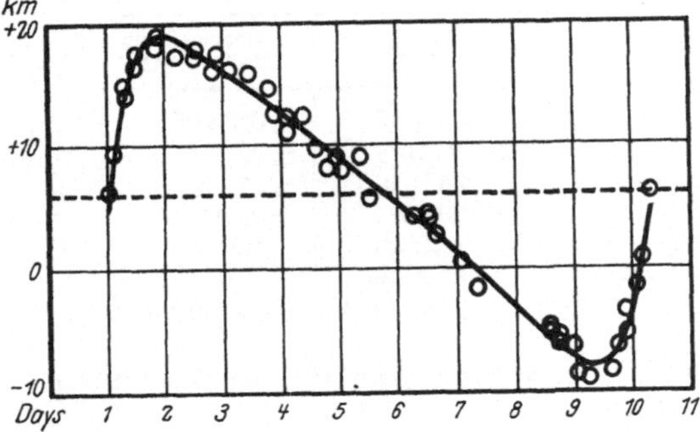

Fig. 60. Velocity-curve of α_2 Geminorum.

Castor is thus composed of two spectroscopic systems, the orbital dimensions of which are nearly of the same order of magnitude. The great difference in eccentricity, however, is striking. The fainter component which is doubtless the most massive has an almost circular orbit, while the brighter and less massive has an extremely eccentric orbit. Both components are of spectral class A0. Accordingly this system must be regarded as another profound enigma.

[1] Ap J 60, p. 192 (1924).
[2] Pop Astr 4, p. 286 (1896).
[3] A N 166, p. 145 (1904).
[4] Potsdam Publ 20, No. 58, p. 92 (1908).
[5] M N 67, p. 140 (1906).
[6] Ap J 5, p. 1 (1897).
[7] Lick Bull 3, p. 84 (1905).
[8] Lick Bull 4, p. 55 (1906).

There is also a dwarf companion to Castor which, although it may not belong to the system of Castor proper, is of extraordinary interest. The spectral type of Castor C is dM1e. $H\alpha$, $H\beta$, $H\gamma$, $H\delta$, and H and K of calcium are present as emission lines. In 1916 the star was found to be a spectroscopic binary, both bright and dark lines being double and both spectra being nearly alike except that one was somewhat stronger than the other. The period was found to be 0,814266 day with no indication of eccentricity. Studied by JOY and SANFORD[1] its spectroscopic elements are:

P	0,814266 day
K_1	114,0 km
K_2	126,7 km
T	J. D. 2423746,524
V_0	+4,3 km
$(a_1+a_2)\sin i$	2695200 km
$m_1\sin^3 i$	0,63 ⊙
$m_2\sin^3 i$	0,57 ⊙.

The intensities of the two absorption spectra are in the ratio of 5 : 4. The emission lines seem to be stronger when on the side of greater wave-length. They give the same velocities as the absorption lines. The spectroscopic absolute magnitude of the brighter component is estimated from plates showing single lines to be 9,2, which corresponds to a parallax of 0",083.

With the aid of VAN GENT's light-curve[2], elements and absolute dimensions have been computed: $a_0 = 0{,}88$; $k = 0{,}89$; $i = 86°{,}4$; $a_1 + a_2 = 2\,700\,000$ km; $r_1 = 0{,}76$⊙; $r_2 = 0{,}68$⊙; $m_1 = 0{,}63$⊙; $m_2 = 0{,}57$⊙; $\varrho_1 = 1{,}4$⊙; $\varrho_2 = 1{,}8$⊙. The surface brightness of both stars is 3,6 magnitudes fainter than that of the Sun.

137. The Spectroscopic Triple 42 Capricorni. 42 Capricorni, considered as a purely spectroscopic triple system, was investigated by J. LUNT[3] at the Cape Observatory. The spectroscopic orbital elements for 1917 and 1920 are:

Element	I 1917	II 1920
P	$13^d{,}25$	$13^d{,}16582$
T	J. D. 2421529,16	J. D. 2421529,16
ω	175°	157°
e	0,20	0,24
K_1	22,75 km	24,5 km
V_0	−4,4 km	−1,5 km
$a_1 \sin i$	4061000 km	4306000 km

The velocity-curves are given in Fig. 61.

The 1917 curve (I) is shown as a broken line, and the 1920 curve (II) as a continuous line.

Curve I is advanced a third of a day along the time line in order to bring the ascending branches of the two curves into coincidence, therefore each curve has a separate time scale.

Comparing the two sets of elements a definite change is seen in ω and V_0, similar to those found for $\varkappa$ Pegasi and 13 Ceti. P has also changed slightly, and again we evidently have an instance of perturbations due to the existence of an invisible third body.

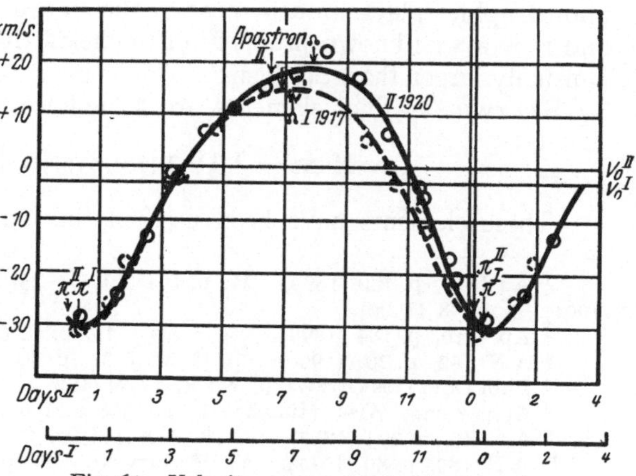

Fig. 61. Velocity-curves of 42 Capricorni.

[1] Ap J 54, p. 253 (1926).
[2] B A N No. 3, p. 121 (1926).
[3] Cape Annals 10, Part 6 (1921).

Other cases analogous to 42 Capricorni, where V_0 varies and the line of apsides rotates, are λ Andromedae and σ Geminorum.

A complete investigation of these two systems, by the writer, is being conducted at the Dominion Observatory, and a long series of spectrograms have been secured during the last few years with convincing results.

138. α Ursae Minoris. The star α Ursae Minoris (Polaris) studied spectroscopically by CAMPBELL[1], FROST[2], HARTMANN[3], BÉLOPOLSKY[4], KÜSTNER[5], and the writer[6], was found to act as a binary with a period $3^d{,}9681$; moreover Dr. CAMPBELL discovered a variable centre-of-mass velocity, so that at first it was looked upon as a triple system. At present it is universally agreed that Polaris is a Cepheid, and the short-period variation of radial velocity is perhaps not that of a binary but of some kind of pulsation or other physical phenomenon. If this last instance is true Polaris can no longer be considered as a triple system. Miss HOBE, at the Lick Observatory, gave a period of 11,9 years for the variation of the centre-of-mass velocity[7], but the Ottawa observations do not verify this period.

Previous to our investigation a series of observations during a short interval of time had not been available, and I thought that although the amplitude is small, a long succession of clear one-prism spectrograms would yield results practically as authentic as those obtained from three-prism plates.

The question relating to variation in light had been discussed by several observers, among them SEIDEL[8], SCHMIDT[9] and PANNEKOEK[10].

HERTZSPRUNG, employing his method of placing a grating in front of a camera's objective to produce side images, used the period derived from radial-velocity observations and succeeded in describing a photographic light-curve[11]. The amplitude of this curve is only $0^m{,}171$ and the elements of variation are:

$$\text{Max.} = \text{J. D. } 2418985{,}856 + 3^d{,}9681 \text{ E.}$$

HERTZSPRUNG, not only from this light-curve but also from the character of the spectrum, established without any doubt that Polaris is a Cepheid variable. This was verified again from the Harvard photographic observations by KING[12], and from the Harvard visual observations by E. C. PICKERING[13].

A very good light-curve of Polaris was determined by Prof. STEBBINS[14] with a selenium photometer. This curve has an amplitude $0^m{,}078$, considerably smaller than that computed by HERTZSPRUNG, but the light affecting the selenium cell has a longer wave-length than that affecting the ordinary photographic plate (much nearer the mean wave-length of visual light), and it is a well-known property of Cepheids that their photographic amplitude is usually larger than the visual. The elements given by STEBBINS and adopted by HARTWIG as final elements are as follows:

$$\text{Max.} = \text{J. D. } 2418985{,}936 + 3^d{,}9681 \text{ E.}$$

These elements have been used in the writer's programme.

---

[1] Ap J 10, p. 180 (1899); 21, p. 191 (1905); 25, p. 59 (1907), and Lick Bull 1, p. 23 (1901); 4, p. 98 (1906).

[2] Ap J 10, p. 184 (1899). [3] Ap J 14, p. 52 (1902).

[4] A N 152, p. 201 (1900). [5] Ap J 27, p. 304 (1908).

[6] Publ Dom Obs Ottawa 9, No. 1, p. 49 (1925). [7] Lick Bull 6, p. 18 (1910).

[8] Abh. Akad. Wiss. München 6, pp. 568 and 603 (1852); 9, p. 578 (1863).

[9] A N 46, p. 293 (1857). [10] A N 194, p. 359 (1913).

[11] A N 189, p. 89 (1911). [12] Harv Ann 59, p. 249 (1912).

[13] Harv Circ No. 174 (1912). [14] A N 192, p. 189 (1912).

GRAMATZKI in Germany[1], using a visual photometer of his own invention, redetermined the visual light-curve; he finds elements slightly different but a difference not affecting the results to any great extent. His elements are:

$$\text{Max.} = \text{J.D. } 2422954{,}2147 \text{ helioc. G.M.T.} + 3^{d}{,}96835 \text{ E.}$$

All the Ottawa radial velocities, obtained between April 29 and June 29, 1923, are plotted on the graph shown in Fig. 62, the open circles indicating the observations from April 29 to May 13 inclusive, the squares those from May 14 to June 18, and the darkened circles those from June 24 to June 29.

The amplitude of the general mean curve is about 4 kilometres and is perhaps slightly different for the various groups of observations represented in Fig. 62, while the mean amplitude given by Campbell is 6,08 kilometres.

It cannot be said with authority that there is a variation of amplitude resembling that found for the β Canis Majoris type stars, but measurements of exceptionally good high-dispersion spectrograms might confirm this possibility.

The radial-velocity curve is in agreement with the theory of Cepheid variation, the maximum velocity of approach coinciding with the maximum light.

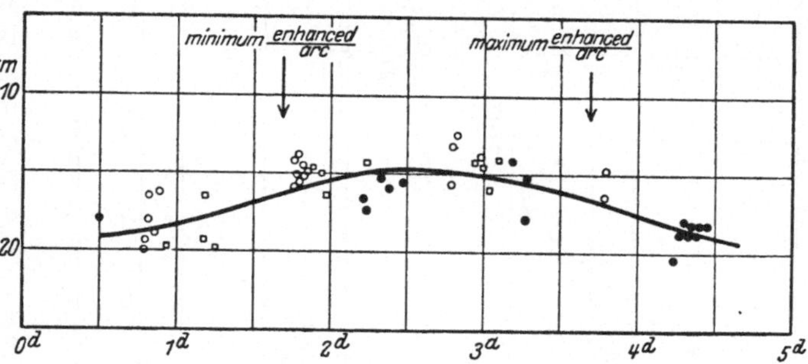

Fig. 62. Velocity-curve of α Ursae Minoris.

The centre-of-mass velocity wholly disagrees with the provisional orbital elements of the centre of mass given by Miss HOBE. This discrepancy might be due to a systematic difference between the Ottawa and Lick radial velocities, but it is doubtful since Ottawa standard velocities are quite in harmony with the Mt. Hamilton values.

After an examination of all our plates of α Ursae Minoris definitely showing the existence of a variation in spectral type, three separate investigations were made establishing that the minimum value of the ratio between the intensities of enhanced and arc lines occurs at about phase $1^{d}{,}2$, and the maximum at about phase $3^{d}{,}2$; the maximum is more vaguely determined, however, and might without difficulty be interpreted to occur at phase $0^{d}{,}0$, coinciding with the maximum light.

139. The Spectroscopic System σ Scorpii. σ Scorpii, discovered by V. M. SLIPHER[2] to be a spectroscopic binary, may be classified among the β Canis Majoris type stars, since Father SELGA's investigation discloses a very short period.

Study of the system was first conducted by the writer[3] at the Lick Observatory in 1918; the centre-of-mass velocity-curve was found to be periodical, signifying that this centre of mass presumably moves in an elliptical orbit of rather high eccentricity having a thirty-three day period.

Three additional researches are to be found in the Dominion Observatory Publications[4] which point with certainty to σ Scorpii as a real triple system

[1] A N 217, p. 454 (1923). [2] Lowell Obs Bull 1, p. 57 (1904) and 2, p. 1 (1909).
[3] Lick Bull 9, p. 173 (1918).
[4] Publ Dom Obs Ottawa 5, p. 303 (1921); 8, p. 45 (1922); 9, p. 91 (1924). See also Report British Ass Adv Sc Toronto 1924, p. 374.

composed of a close pair having a revolution period $33^d,0$ and a centre of mass moving in a larger orbit of several years' period, controlled by the third body.

If we except cases like α Ursae Minoris, where the shorter period of radial-velocity oscillation may be ascribed to some kind of pulsation (Cepheid), and like $\varkappa$ Pegasi, where the longer period of revolution was discovered by visual observation, σ Scorpii would be one of the rare triple systems discovered entirely from spectrographic measures.

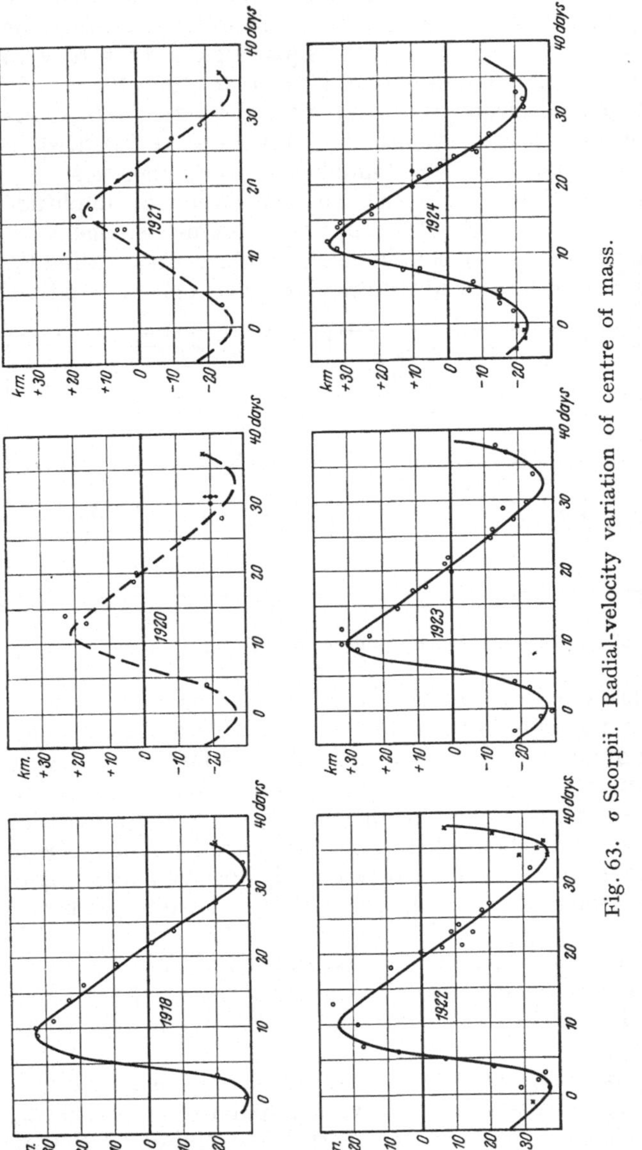

Fig. 63. σ Scorpii. Radial-velocity variation of centre of mass.

Besides being of the β Canis Majoris type σ Scorpii also presents two relevant features pointed out by the writer, namely, H and K stationary lines and the possibility of computing long-period orbital elements from the equation of light affecting the six hour variation; the system is therefore unique in stellar physics.

So far, centre-of-mass velocity-curves have been determined every year from 1918 to 1927 (excepting 1919), and in years to come will be an important subject for continued research.

The velocities of 1922 to 1926 furnish excellent radial-velocity curves for the centre of mass, but the velocities of 1920 and 1921 are too few for good determinations. Fig. 63 shows the different curves for 1918, 1920, 1921, 1922, 1923 and 1924.

An extraordinary change in the amplitude of these curves is doubtful, but the fact remains that there is a slight degree of variation in shape and amplitude which is necessarily caused by the gravitational effect of a third body. Indeed, it is a conspicuous fact that the centre-of-mass velocity computed from these curves is not constant, which strongly indicates an orbital revolution with a period of several years.

Fluctuations appear in the observed value of the very short period of velocity-variation; this may be explained by assuming that the true period is constant but seems to be variable by reason of the star's displacement in a large orbit; consequently, the observed times of maxima and minima require correction for the equation of light.

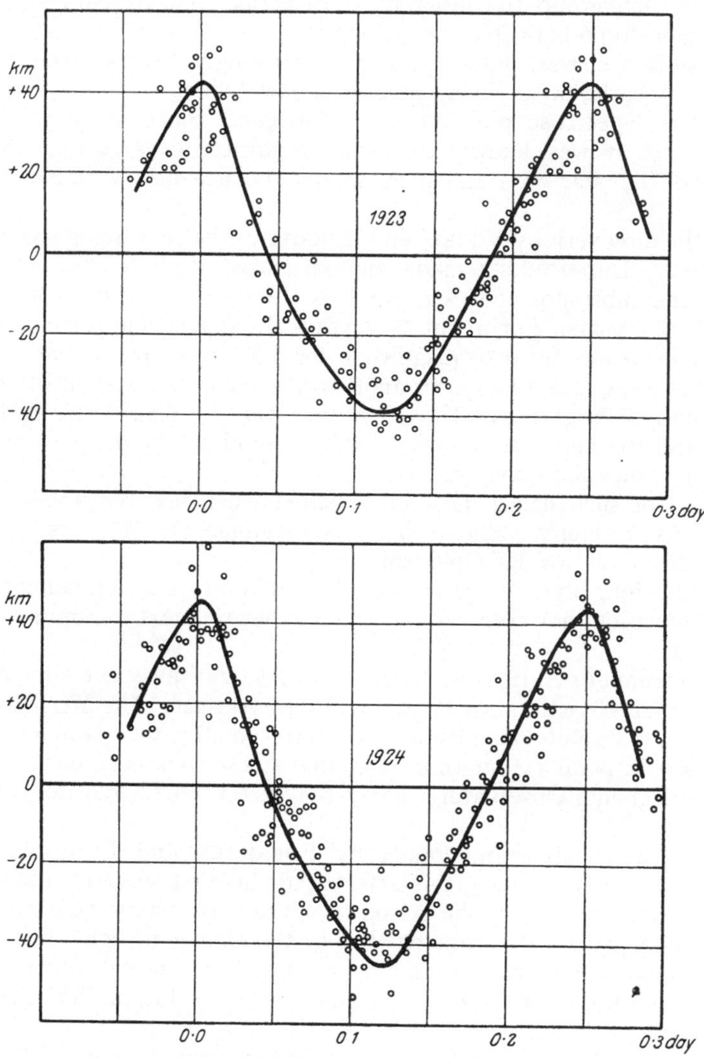

Fig. 64. σ Scorpii. Short-period velocity-curve.

A circular orbit of twelve year's period with a semi-amplitude of radial-velocity variation of ten kilometers would give for $a_1 \sin i$ nearly the same value as an equation of light $0^d{,}05$. We have, approximately, from the spectroscopic orbit

$$a_1 \sin i = 602\,293\,800 \text{ km.}$$

and from the equation of light

$$a_1 \sin i = 648\,000\,000 \text{ km.}$$

The parallax of σ Scorpii is no doubt very small; Prof. KAPTEYN's value, obtained from star-streaming, is 0″,0086 (Boss 4158)[1]. The apparent radius of a visual twelve-year period orbit would be practically 0″,04, a quantity impossible to detect by direct measurement unless by interferential methods.

I have taken the two groups of observations of 1923 and 1924 and superimposed for each group the different short-period velocity-curves so that all have the same hypothetical mean velocity.

The resultant curves, defined in Fig. 64, strikingly illustrate the remarkable radial-velocity variation occuring in only a few hours.

140. The Spectroscopic System ε Aurigae. Since early in the nineteenth century, it was known that the brightness of ε Aurigae fluctuated slightly, and the star was classified among the irregular variables for many years.

After the discovery by VOGEL[2] and EBERHARD that ε Aurigae has a variable radial velocity, LUDENDORFF made an exhaustive study of the variation of light[3], and the substance of this research is the following: The light-curve has a period of 27,1 years. For about 25,2 years the magnitude remains constant, then slowly decreases by 0,74 magnitude for 180 days, remains at a constant minimum 340 days, and, finally, for another 180 days increases until once more it reaches normal brightness. Thus ε Aurigae may be placed among the Algol type stars and the light-curve explained by assuming a primary to be eclipsed by a darker companion every 27 years.

A period of such length is unprecedented when one remembers that the longest periods of Algol variables, other than ε Aurigae, are 759 days for δ Cassiopeiae[4] and 262 days for RZ Ophiuchi.

After only four observed returns of the minimum, a period cannot be computed with precision, however, we can predict that the next minimum will occur in May 1929.

The spectrum of ε Aurigae is classified as cF5, and since all c-stars are supergiants, this characteristic must be adopted for ε Aurigae. The Mt. Wilson value confirms this, an absolute magnitude $-2{,}0$ corresponding to a parallax $+0'',008$; the spectroscopic parallax determined at the Victoria Observatory is $+0'',006$, while the Allegheny Observatory gives the direct trigonometric parallax as $+0'',002 \pm 0'',007$.

Based upon spectrograms taken by EBERHARD and LUDENDORFF from November 1901 to March 1905, the latter[5] made his first spectrographic analysis of ε Aurigae, not only proving the existence of a radial-velocity variation throughout a very long period, the same as that of the change in light, but moreover disclosing pronounced minima or maxima separated by intervals of a few months. Therefore LUDENDORFF arrives at the reasonable conclusion that ε Aurigae is a triple system.

With the collaboration of EBERHARD, and afterwards alone, LUDENDORFF continued to secure spectrograms of the star until April 1913. His research forms an elaborate study published in the "Sitzungsberichte der Preußischen Akademie der Wissenschaften", 1924, p. 49.

From a graphical representation of 194 radial velocities it is clearly seen that the short oscillations between two consecutive maxima are of widely

[1] Mt Wilson Contr 4, p. 417 (1914).
[2] Sitzber. d. Kgl. Preuß. Akad. d. Wiss. 1902, p. 1068.
[3] A N 164, p. 81 (1903) and 192, p. 389 (1912).
[4] Veröff. der K. Sternwarte Berlin-Babelsberg 2, Heft 3, p. 112 (1918).
[5] A N 171, p. 49 (1906).

differing duration, in which event the mean values for the epochs indicated are:

Epoch	Duration of Oscillation Days
1903—06	143
1906—08	150
1908—09	170
1909—11	150
1911—13	125

Adopting orbital motion as the cause of individual oscillations, LUDENDORFF derived velocity-curves from the combined observations of these three periods:

I 1908 Nov. — 1909 April
II 1909 Oct. — 1910 April
III 1910 Oct. — 1911 April.

Fig. 65, 66 and 67 illustrate the curves from which the following orbital elements were calculated:

Element	I	II	III
P	161^d	152^d	143^d
T	J. D. 2418266	2418615	2419005
ω	$120°,8$	$122°,4$	$4°,0$
e	0,34	0,58	0,26
K_1	7,8 km	4,8 km	4,8 km
V_0	—11,1 km	—10,7 km	—13,1 km
$a_1 \sin i$	16240000 km	8171000 km	9018000 km
$\frac{m_2^3 \sin^3 i}{(m_1 + m_2)^2}$	0,0066 ⊙	0,00094 ⊙	0,00143 ⊙

The three systems of elements are decidedly dissimilar and if orbital motion really exists there are indications of strong perturbations inevitably due to a massive third body.

Estimated values of V_0, for the assumed short-duration orbit, were then obtained by LUDENDORFF. In the tabulation the quantities between brackets are very doubtful.

Year	V_0	Year	V_0
1897,0	(+ 9) km	1910,0	—11 km
1900,0	(+ 4)	11,0	—13
01,8	(+ 4)	12,0	—13
03,0	— 8	13,0	—13
04,0	— 7	17,5	— 4
05,0	(— 8)	18,5	(— 3)
07,0	—10	20,5	+13
09,0	—11		

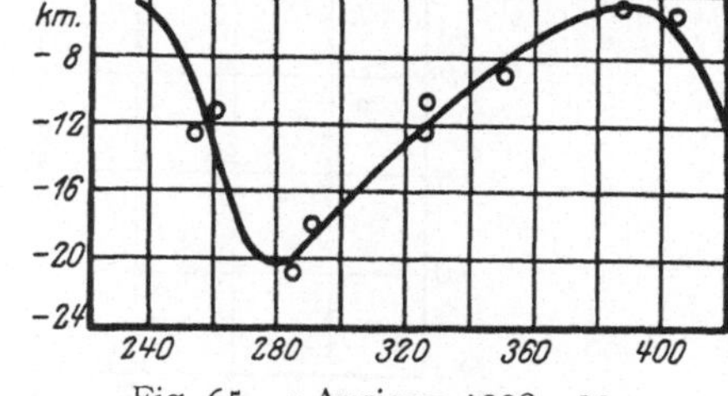

Fig. 65. ε Aurigae 1908—09.

The last three are mean velocities derived at the Yerkes Observatory from observations made respectively on fourteen, four and ten days of each year.

From these values the following approximate elements of the long-period orbit were computed:

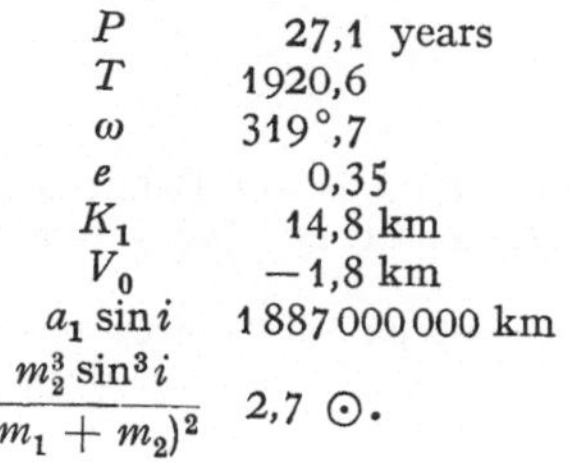

P	27,1 years
T	1920,6
ω	$319°,7$
e	0,35
K_1	14,8 km
V_0	—1,8 km
$a_1 \sin i$	1887000000 km
$\frac{m_2^3 \sin^3 i}{(m_1 + m_2)^2}$	2,7 ⊙.

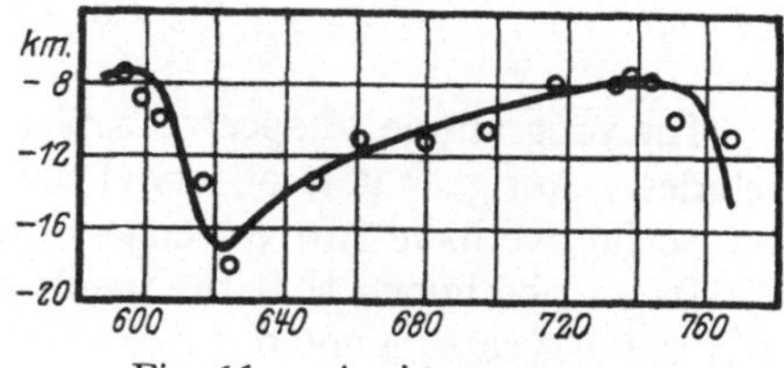

Fig. 66. ε Aurigae 1909—10.

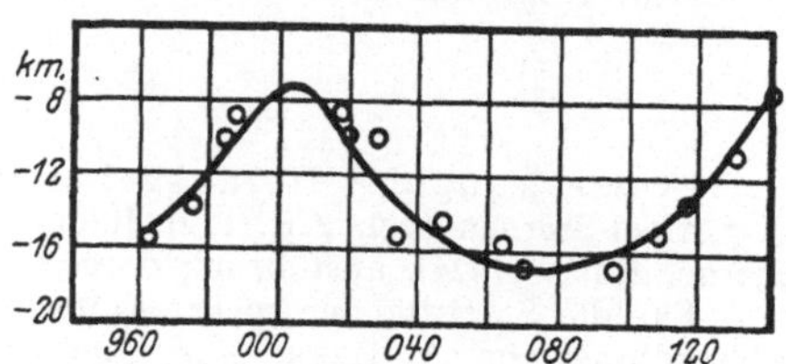

Fig. 67. ε Aurigae 1910—11.

Although these figures are most uncertain, if additional assumptions are recognized, they lead to roughly estimated densities and masses of the three components, and to original theories developed by LUDENDORFF in his interesting paper.

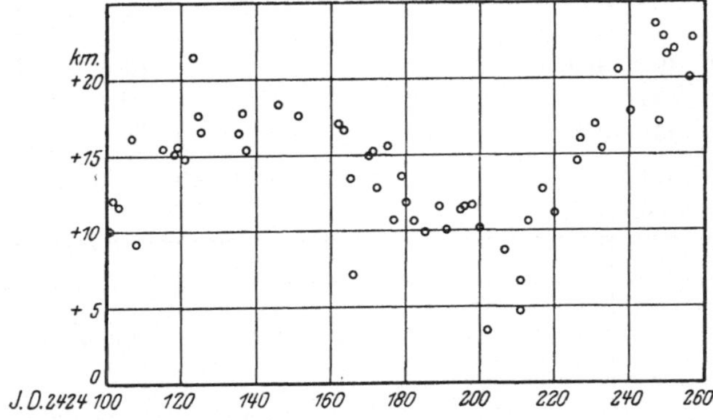

Fig. 68. ε Aurigae 1924—25.

Classified by ADAMS among the Pseudo-Cepheids, there seems to be no further question that ε Aurigae is a super-giant star of exceedingly low density.

LUDENDORFF's analysis of this unique system reveals the expediency of thorough investigation.

Scientists will eagerly anticipate the eclipse of 1928-29, and many observers will await that event with equipment representing the last word in modern invention.

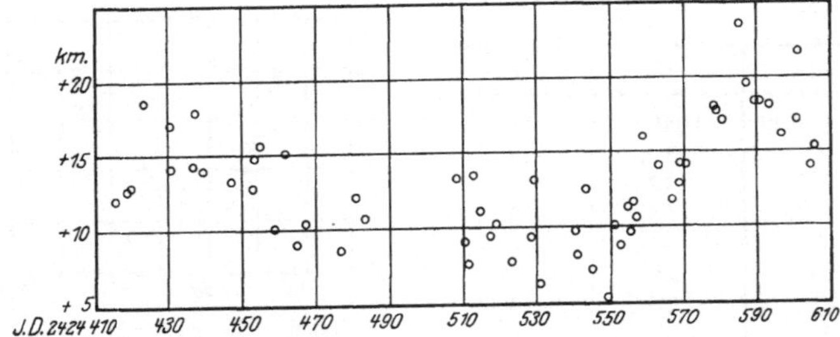

Fig. 69. ε Aurigae 1925—26.

The programme of spectroscopic observations at the Dominion Observatory includes ε Aurigae; it is observed once every clear night when in good position and, so far, we have two velocity-curves for 1924-25 and 1925-26 (Fig. 68 and 69).

It is to be hoped that the result of our investigations may in some measure justify the years devoted to research, contribute to a reasonable explanation of celestial phenomena and the solving of new problems in astronomy.

[Anmerkung der Herausgeber: Die RUSSELLsche Methode der Bahnbestimmung von Algol-Sternen (vgl. Ziff. 112) findet sich auch in HAGEN-STEIN, Die veränderlichen Sterne, Bd. 2 (1924), ausführlich dargelegt.

Für die Statistik der spektroskopischen Doppelsterne ist auf die soeben erschienene Abhandlung von A. BEER, Zur Charakterisierung der spektroskopischen Doppelsterne (Veröff. der Sternwarte Berlin-Babelsberg 5, Heft 6), zu verweisen.]

Sternverzeichnis.

(Die in Tabellen vorkommenden Sterne sind nicht in dieses Verzeichnis aufgenommen.)

Sachverzeichnis.

Berichtigung.

Seite 97, Zeile 8 von oben lies $2S_y - 1$ statt $2S_y - 12$.